T0325106

Springer Series in Statistics

Perspectives in Statistics

Advisors
J. Berger, S. Fienberg, J. Gani,
K. Krickeberg, I. Olkin, B. Singer

Springer Series in Statistics

(continued after index)

Samuel Kotz   Norman L. Johnson
Editors

Breakthroughs in Statistics
Volume I
Foundations and Basic Theory

Springer Science+Business Media, LLC

Samuel Kotz
College of Business
  and Management
University of Maryland
  at College Park
College Park, MD 20742
USA

Norman L. Johnson
Department of Statistics
Phillips Hall
The University of North Carolina
  at Chapel Hill
Chapel Hill, NC 27599
USA

Library of Congress Cataloging-in-Publication Data
Breakthroughs in statistics / Samuel Kotz, Norman L. Johnson, editors.
    p.   cm. — (Springer series in statistics. Perspectives in
statistics)
    Includes bibliographical references and index.
    Contents: v. 1. Foundations and basic theory — v. 2. Methodology
    and distribution.
    ISBN 978-0-387-97566-5   ISBN 978-1-4612-0919-5 (eBook)
    DOI 10.1007/978-1-4612-0919-5
    1. Mathematical statistics. I. Kotz, Samuel. II. Johnson,
Norman Lloyd. III. Series.
QA276.B68465   1993
519.5—dc20   93-3854

Printed on acid-free paper.

© 1992 Springer Science+Business Media New York
Originally published by Springer-Verlag Berlin Heidelberg New York in 1992

All rights reserved. This work may not be translated or copied in whole or in part without the
written permission of the Springer-Science+Business Media, LLC, except for brief excerpts in
connection with reviews or scholarly analysis. Use in connection with any form of information
storage and retrieval, electronic adaptation, computer software, or by similar or dissimilar
methodology now known or hereafter developed is forbidden.
The use of general descriptive names, trade names, trademarks, etc., in this publication, even if
the former are not especially identified, is not to be taken as a sign that such names, as understood
by the Trade Marks and Merchandise Marks Act, may accordingly be used freely by anyone.

Typeset by Asco Trade Typesetting Ltd., Hong Kong.

9 8 7 6 5 4 3 2

To the memory of Guta S. Kotz
1901–1989

Preface

McCrimmon, having gotten Grierson's attention, continued: "A breakthrough, you say? If it's in economics, at least it can't be dangerous. Nothing like gene engineering, laser beams, sex hormones or international relations. That's where we don't want any breakthroughs." (Galbraith, J.K. (1990) *A Tenured Professor*, Houghton Mifflin; Boston.)

To judge [*astronomy*] in this way [*a narrow utilitarian point of view*] demonstrates not only how poor we are, but also how small, narrow, and indolent our minds are; it shows a disposition always to calculate the payoff before the work, a cold heart and a lack of feeling for everything that is great and honors man. One can unfortunately not deny that such a mode of thinking is not uncommon in our age, and I am convinced that this is closely connected with the catastrophes which have befallen many countries in recent times; do not mistake me, I do not talk of the general lack of concern for science, but of the source from which all this has come, of the tendency to everywhere look out for one's advantage and to relate everything to one's physical well-being, of the indifference towards great ideas, of the aversion to any effort which derives from pure enthusiasm: I believe that such attitudes, if they prevail, can be decisive in catastrophes of the kind we have experienced. [Gauss, K.F.: *Astronomische Antrittsvorlesung* (cited from Buhler, W.K. (1981) *Gauss: A Biographical Study*, Springer: New York)].

This collection of papers (reproduced in whole or in part) is an indirect outcome of our activities, during the decade 1979–88, in the course of compiling and editing the *Encyclopedia of Statistical Sciences* (nine volumes and a Supplementary volume published by John Wiley and Sons, New York). It is also, and more directly, motivated by a more recent project, a systematic rereading and assessment of Presidential Addresses delivered to the Royal

Statistical Society, the International Statistical Institute, and the American Statistical Association during the last 50 years.

Our studies revealed a growing, and already embarrassingly noticeable, diversification among the statistical sciences that borders on fragmentation. Although our belief in the unified nature of statistics remains unshaken, we must recognize certain dangers in this steadily increasing diversity accompanying the unprecedented penetration of statistical methodology into many branches of the social, life, and natural sciences, and engineering and other applied fields.

The initial character of statistics as the "science of state" and the attitudes summed up in the Royal Statistical Society's original motto (now abandoned) of *aliis exterendum* ("let others thresh")—reflecting the view that statisticians are concerned solely with the collection of data—have changed dramatically over the last 100 years and at an accelerated rate during the last 25 years.

To trace this remarkably vigorous development, it seemed logical (to us) to search for "growth points" or "breakthrough" publications that have initiated fundamental changes in the development of statistical methodology. It also seemed reasonable to hope that the consequences of such a search might result in our obtaining a clearer picture of likely future developments. The present collection of papers is an outcome of these thoughts.

In the selection of papers for inclusion, we have endeavored to identify papers that have had lasting effects, rather than search to establish priorities. However, there are Introductions to each paper that do include references to important precursors, and also to successor papers elaborating on or extending the influence of the chosen papers.

We were fortunate to have available S.M. Stigler's brilliant analysis of the history of statistics up to the beginning of the 20th century in his book, *The History of Statistics: The Measurement of Uncertainty* (Belknap Press, Cambridge, Mass., 1986), which, together with Claire L. Parkinson's *Breakthroughs: A Chronology of Great Achievements in Science and Mathematics 1200–1930* (G.K. Hall, Boston, Mass., 1985), allowed us to pinpoint eleven major breakthroughs up to and including F. Galton's *Natural Inheritance*. These are, in chronological order, the following:

C. Huyghens (1657). *De Ratiociniis in Aleae Ludo* (Calculations in Games of Dice), in *Exercitationum Mathematicarum* (F. van Schooten, ed.). Elsevier, Leiden, pp. 517–534.
(The concept of *mathematical expectation* is introduced, as well as many examples of combinatorial calculations.)
J. Graunt (1662). *Natural and Political Observations Mentioned in a Following Index and Made upon the Bills of Mortality.* Martyn and Allestry, London.
(Introduced the idea that vital statistics are capable of scientific analysis.)
E. Halley (1693). An estimate of the degrees of mortality of mankind, drawn from the curious "*Tables of the Births and Funerals* at the City of Breslaw;

with an attempt to ascertain the price of *annuities* upon *lives*," *Philos. Trans. Roy. Soc., Lon.*, **17**, 596–610, 654–656.
[Systematized the ideas in Graunt (1662).]
J. Arbuthnot (1711). An argument for Divine Providence, taken from the constant regularity observed in the births of both sexes, *Philos. Trans. Roy. Soc., Lon.*, **27**, 186–190.
(This is regarded as the first use of a test of significance, although not described as such explicitly.)
J. Bernoulli (1713). *Ars Conjectandi* (The Art of Conjecture). Thurnisorium, Basel.
(Development of combinatorial methods and concepts of statistical inference.)
A. De Moivre (1733, 1738, 1756). *The Doctrine of Chances*, 1st-3rd eds. Woodfall, London.
(In these three books, the normal curve is obtained as a limit and as an approximation to the binomial.)
T. Bayes (1763). Essay towards solving a problem in the doctrine of chances, *Philos. Trans. Roy. Soc., Lon.*, **53**, 370–418.
(This paper has been the source of much work on inverse probability. Its influence has been very widespread and persistant, even among workers who insist on severe restrictions on its applicability.)
P.S. Laplace (1812). *Théorie Analytique des Probabilités*. Courcier, Paris.
(The originating inspiration for much work in probability theory and its applications during the 19th century. Elaboration of De Moivre's work on normal distributions.)
K.F. Gauss (1823). *Theoria Combinationis Observationum Erroribus Minimis Obnoxiae*. Dieterich, Gottingen.
(The method of least squares and associated analysis have developed from this book, which systematized the technique introduced by A.M. Legendre in 1805. Also, the use of "optimal principles" in choosing estimators.)
L.A.J. Quetelet (1846). *Lettres à S.A.R. le Duc Regnant de Saxe-Cobourg et Gotha, sur la Théorie des Probabilités, appliquée aux Sciences Morales et Politiques*. Hayez, Brussels. [English Translation, Layton: London 1849.]
(Observations on the stability of certain demographic indices provided empirical evidence for applications of probability theory.)
F. Galton (1889). *Natural Inheritance*. Macmillan, London.
[This book introduces the concepts of correlation and regression; also mixtures of normal distributions and the bivariate normal distribution. Its importance derives largely from the influence of Karl Pearson. In regard to correlation, an interesting precursor, by the same author, is 'Co-relations and their measurement, chiefly from anthropometric data,' *Proc. Roy. Soc., Lon.*, **45**, 135–145 (1886).]

In our efforts to establish subsequent breakthroughs in our period of study (1890–1989), we approached some 50 eminent (in our subjective evaluation) statisticians, in various parts of the world, asking them if they would supply us with "at least five (a few extra beyond five is very acceptable) possibly suitable references...". We also suggested that some "explanations of reasons for choice" would be helpful.

The response was very gratifying. The requests were sent out in June–July 1989; during July–August, we received over 30 replies, with up to 10 references each, the modal group being 8. There was remarkable near-unanimity recommending the selection of the earlier work of K. Pearson, "Student," R.A. Fisher, and J. Neyman and E.S. Pearson up to 1936. For the years following 1940, opinions became more diverse, although some contributions, such as A. Wald (1945), were cited by quite large numbers of respondents. After 1960, opinions became sharply divergent. The latest work cited by a substantial number of experts was B. Efron (1979). A number of replies cautioned us against crossing into the 1980s, since some time needs to elapse before it is feasible to make a sound assessment of the long-term influence of a paper. We have accepted this viewpoint as valid.

Originally, we had planned to include only 12 papers (in whole or in part). It soon became apparent, especially given the diversity of opinions regarding the last 50 years, that the field of statistical sciences is now far too rich and heterogeneous to be adequately represented by 12 papers over the last 90 years. In order to cover the field satisfactorily, it was decided that at least 30 references should be included. After some discussion, the publisher generously offered to undertake two volumes, which has made it possible to include 39 references! Assignment to the two volumes is on the basis of broad classification into "Foundations and Basic Theory" (Vol. I) and "Methodology and Distribution" (Vol. II). Inevitably, there were some papers that could reasonably have appeared in either volume. When there was doubt, we resolved it in such a way as to equalize the size of the two volumes, so far as possible. There are 19 introductions in the first volume and 20 in the second. In addition, we have included Gertrude Cox's 1956 Presidential Address "Frontiers of Statistics" to the American Statistical Association in Vol. I, together with comments from a number of eminent statisticians indicating some lines on which statistical thinking and practice have developed in the succeeding years.

Even with the extension to two volumes, in order to keep the size of the books within reasonable limits, we found it necessary to reproduce only those parts of the papers that were relevant to our central theme of recording "breakthroughs," points from which subsequent growth can be traced. The necessary cutting caused us much "soul-searching," as did also the selection of papers for inclusion. We also restricted rather severely the lengths of the introductions to individual items. We regret that practical requirements made it necessary to enforce these restrictions. We also regret another consequence of the need to reduce size—namely, our inability to follow much of the advice of our distinguished correspondents, even though it was most cogently advocated. In certain instances the choice was indeed difficult, and a decision was reached only after long discussions. At this point, we must admit that we have included two or three choices of our own that appeared only sparsely among the experts' suggestions.

Although the division between the two volumes is necessarily somewhat arbitrary, papers on fundamental concepts such as probability and mathematical foundation of statistical inference are clearly more Vol. I than Vol. II material (concepts however can influence application).

There have been laudable and commendable efforts to put the foundations of statistical inference, and more especially probability theory on a sound footing, according to the viewpoint of mathematical self-consistency. Insofar as these may be regarded as attempts to reconcile abstract mathematical logic with phenomena observed in the real world—via interpretation (subjective or objective) of data—we feel that the aim may be too ambitious and even doomed to failure. We are in general agreement with the following remarks of the physicist H.R. Pagels:

"Centuries ago, when some people suspended their search for absolute truth and began instead to ask how things worked, modern science was born. Curiously, it was by abandoning the search for absolute truth that science began to make progress, opening the material universe to human exploration. It was only by being provisional and open to change, even radical change, that scientific knowledge began to evolve. And ironically, its vulnerability to change is the source of its strength." (From *Perfect Symmetry: The Search for the Beginning of Time*, Simon and Schuster, New York 1985, p. 370).

To continue along these lines, in his famous essay "Felix Klein's Stellung in der Mathematischen Gegenwart" (Felix Klein's Position in Mathematical Modernity) which appeared originally in *Naturwissenschaffen 18*, pp. 4–11, 1930, Herman Weyl notes Klein's restrained attitude towards foundations of mathematics: "Klein liked to emphasize that cognition starts from the *middle* and leads to an unexplored in both directions—when the movement is upward as well as when a downward trend occurs. Our task is to gradually diffuse the darkness on both sides, while an absolute foundation—the huge elephant which carries on his mighty back the tower of truth-exists perhaps only in fairy tales."

It is evident that this work represents the fruits of collaboration among many more individuals than the editors. Our special thanks go to the many distinguished statisticians who replied to our inquiries, in many cases responding to further "follow-up" letters requiring additional effort in providing more details that we felt were desirable; we also would like to thank those who have provided introductions to the chosen papers. The latter are acknowledged at the appropriate places where their contributions occur.

We take this opportunity to express our gratitude to Dean R.P. Lamone of the College of Business and Management and Professor B.L. Golden,

Chairman of the Department of Management Science and Statistics at the
University of Maryland at College Park, and to Professor S. Cambanis,
Chairman of the Department of Statistics at the University of North Carolina
at Chapel Hill, for their encouragement and the facilities they provided in
support of our work on this project.

We are also grateful to the various persons and organizations who have
given us reprint permission. They are acknowledged, together with source
references, in the section "Sources and Acknowledgments."

We welcome constructive criticism from our readers. If it happens that our
first sample proves insufficiently representative, we may be able to consider
taking another sample (perhaps of similar size, *without replacement*).

Samuel Kotz
College Park, Maryland
November 1990

Norman L. Johnson
Chapel Hill, North Carolina
November 1990

Contents

Contents

Volume II: Methodology and Distribution

Contributors

GEISSER, S. School of Statistics, 270 Vincent Hall, University of Minnesota, 206 Church St., S.E., Minneapolis, MN 55455 USA.

ANDERSON, T.W. Department of Statistics, Sequoia Hall, Stanford University, Stanford, CA 94305 USA.

LEHMANN, E.L. Department of Statistics, University of California, Berkeley, CA 94720 USA.

FRASER, D.A.S. Department of Statistics, University of Toronto, Toronto, Canada M5S 1A1.

BARLOW, R.E. Department of Operations Research, 3115 Etcheverry Hall, University of California, Berkeley, CA 94720 USA.

LEADBETTER, M.R. Department of Statistics, University of North Carolina, Chapel Hill, NC 27599-3260 USA.

SMITH, R.L. Department of Statistics, University of North Carolina, Chapel Hill, NC 27599-3260 USA.

PATHAK, P.K. Department of Mathematics & Statistics, University of New Mexico, Albuquerque, NM 87131 USA.

GHOSH, B.K. Department of Mathematics, Christmas-Saucom Hall 14, Lehigh University, Bethlehem, PA 18015 USA.

SEN, P.K. Department of Statistics, University of North Carolina, Chapel Hill, NC 27599-3260 USA.

WEISS, L. College of Engineering, School of Operations Research and Industrial Engineering, Upson Hall, Cornell University, Ithaca, NY 14853-7501 USA.

LINDLEY, D.V. 2 Periton Lane, Minehead, Somerset TA 24 8AQ, United Kingdom.

GOOD, I.J. Department of Statistics, Virginia Polytechnic and State Univ., Blacksburg, VA 24061-0439 USA.

WYNN, H.P. School of Mathematics, Actuarial Science and Statistics, City University, Northampton Square, London EC1V OHB, United Kingdom.

EFRON, B. Department of Statistics, Sequoia Hall, Stanford University, Stanford, CA 94305 USA.

BJØRNSTAD, J.F. Department of Mathematics and Statistics, College of Arts and Sciences, University of Trondheim, N-7055 Dragvoll, Norway.

du MOUCHEL, W.H. BBN Software Corporation, 10 Fawcett Street, Cambridge, MA 02138 USA.

REID, N. Department of Statistics, University of Toronto, Toronto Canada M5S 1A1.

de LEEUW, J. Social Statistics Program, Depts. of Psychology and Mathematics, University of California, 405 Hilgard Avenue, Los Angeles, CA 90024-1484 USA.

Sources and Acknowledgments

Cox Gertrude M. (1957) Statistical frontiers. *J. Amer. Statist. Assoc.*, **52**, 1–10. Reproduced by the kind permission of the American Statistical Society.

Fisher R.A. (1922) On the mathematical foundations of theoretical statistics. *Philos. Trans. R. Soc. London, Ser. A.*, **222A**, 309–368. Reproduced by the kind permission of the Royal Society.

Hotelling H. (1931) The generalization of Student's ratio. *Ann. Math. Statist.*, **2**, 368–378. Reproduced by the kind permission of the Institute of Mathematical Statistics.

Neyman J. and Pearson E.S. (1933) On the problem of the most efficient test of statistical hypotheses. *Philos. Trans. R. Soc. London, Ser. A.*, **231**, 289–337. Reproduced by the kind permission of the Royal Society.

Bartlett M.S. (1937) Properties of sufficiency and statistical tests. *Proc. R. Soc. London, Ser. A*, **168**, 268–282. Reproduced by the kind permission of the Royal Society.

de Finetti B. (1937) Foresight: Its logical laws, its subjective sources. *Ann. Inst. H. Poincaré*, **7**, 1–68, (english translation by Henry E. Kyberg, Jr.). Reproduced by the kind permission of Robert E. Krieger Publishing Company.

Cramér H. (1942) On harmonic analysis in certain functional spaces. *Ark. Mat. Astr. Fys.*, **28B**(12). Reproduced by the kind permission of the Royal Swedish Academy of Sciences.

Gnedenko B.V. (1943) On the limiting distribution of the maximum term in a random series. *Ann. Math.*, **44**, 423–453 (English translation).

Rao C.R. (1945) Information and the accuracy attainable in the estimation of statistical parameters. *Bull. Calcutta Math. Soc.*, **37**, 81–91. Reproduced by the kind permission of the Calcutta Mathematical Society.

Wald A. (1945) Sequential tests of statistical hypotheses. *Ann. Math. Statist.*, **16**, 117–196. Reproduced by the kind permission of the Institute of Mathematical Statistics.

Hoeffding W. (1948) A class of statistics with asymptotically normal distributions. *Ann. Math. Statist.*, **19**, 293–325. Reproduced by the kind permission of the Institute of Mathematical Statistics.

Wald A. (1949) Statistical decision functions. *Ann. Math. Statist.*, **29**, 165–205. Reproduced by the kind permission of the Institute of Mathematical Statistics.

Good I.J. (1952) Rational decisions. *J. R. Statist. Soc. Ser. B.*, **14**, 107–114. Reproduced by the kind permission of the Royal Statistical Society and Basil Blackwell, Publishers.

Robbins H.E. (1955) An empirical Bayes approach to statistics. *Proc. 3rd Berkeley Symp. Math. Statist. Prob.* **1**, 157–163. Reproduced by the kind permission of the Regents of the University of California and the University of California Press.

Kiefer J.C. (1959) Optimum experimental designs. *J. R. Statist. Soc. Ser. B.*, **21**, 272–304. Reproduced by the kind permission of the Royal Statistical Society and Basil Blackwell, Publishers.

James W. and Stein C.M. (1961) Estimation with quadratic loss. *Proc. 4th Berkeley Symp. Math. Statist. Prob.*, **1**, 311–319. Reproduced by the kind permission of the Regents of the University of California and the University of California Press.

Birnbaum A.W. (1962) On the foundations of statistical inference. *J. Amer. Statist. Assoc.*, **57**, 269–306. Reproduced by the kind permission of the American Statistical Association.

Edwards W., Lindman H., and Savage L.J. (1963) Bayesian statistical inference for psychological research. *Psychol. Rev.*, **70**, 193–242. Reproduced by the kind permission of the American Psychological Association.

Fraser D.A.S. (1966) Structural probability and a generalization. *Biometrika*, **53**, 1–9. Reproduced by the kind permission of the Biometrika Trustees.

Akaike H. (1973) Information theory and an extension of the maximum likelihood principle. *2nd Intern. Symp. Inf. Theory*, (B.N. Petrov and F. Csàki, eds.) Akad. Kiàdo, Budapest, 267–281.

Editorial Note

To illustrate the enormous strides in Statistical Sciences during the last three and a half decades and to exhibit the direction of these developments the Editors decided to reproduce the well-known American Statistical Association Presidential Address by Gertrude Cox *Statistical Frontiers* delivered on September 9, 1956, at the 116th Annual Meeting of the ASA in Detroit and printed in the March 1957 issue of the *Journal of American Statistical Association*.

Gertrude Cox (1900–1978), an illustrious representative of the classical school of modern statistics in the K. Pearson - R.A. Fisher tradition, delivered her address on the state of statistical sciences just before the major impact and eventual dominating position of computer technology in statistical methodology and practice, and the expansion of appreciation of statistical methodology to various new branches of medical engineering and behavioral sciences. Although the comparison between the state of statistical sciences in the fall of 1956 and in the fall of 1990 (when these lines are being written) is self-evident for readers of these volumes, we thought that it would be expedient to solicit comments on this subject. Each person was requested to provide a 200–400 word commentary on *Statistical Universe in* 1990 *versus* 1956. Respondents' comments are printed, with minor editorial alterations, following G. Cox's address.

Statistical Frontiers*

Gertrude M. Cox
Institute of Statistics,
University of North Carolina

1. Introduction

I am going to ask you to look forward as we try to discern, as best we can, what the future holds for statisticians. If ten years ago we had predicted some of the things we are doing today, we would have been ridiculed. Now, my concern is that we may become too conservative in our thinking.

Civilization is not threatened by atomic or hydrogen bombs; it is threatened by ourselves. We are surrounded with ever widening horizons of thought, which demand that we find better ways of analytical thinking. We must recognize that the observer is part of what he observes and that the thinker is part of what he thinks. We cannot passively observe the statistical universe as outsiders, for we are all in it.

The statistical horizon looks bright. Exciting experiences lie ahead for those young statisticians whose minds are equipped with knowledge and who have the capacity to think constructively, imaginatively, and accurately.

Will you, with me, look upon the statistical universe as containing three major continents: (1) descriptive methods, (2) design of experiments and investigations, and (3) analysis and theory. As we tour these continents, we shall visit briefly a few selected well developed countries, where statisticians have spent considerable time. As tourists, we shall have to stop sometimes to comment on the scenery, culture, politics, or the difficulties encountered in securing a visa. With our scientific backgrounds, we should spend most of our time, seeking out the new, the underdeveloped, the unexplored or even the dangerous areas.

* Presidential address, at the 116th Annual Meeting of the American Statistical Association, Detroit, Michigan, September 9, 1956.

It is one of the challenges of the statistical universe that, as new regions are discovered and developed, the horizon moves further away. We cannot visit all the frontiers for they are too numerous. I believe that we should try to visualize the challenges of the future by looking at typical types of unsolved problems. I hope you will find the trip so interesting that you will revisit some of these statistical frontiers not as tourists but as explorers.

You know how many folders and guide books one can accumulate while traveling. I am not going even to list the ones used. This will leave you guessing whether I am quoting or using original ideas. Many people in this audience will recognize their statements used with no indication that they are quotations.

2. Descriptive Methods Continent

In planning our tour, I decided to take you first to the descriptive methods continent, for it is the oldest and has the densest settlement. The lay conception of descriptive methods ordinarily includes these countries: (1) collection of data; (2) summarization of data including such states as tabulation, measures of central tendency and dispersion, index numbers and the description of time series; and (3) the presentation of data in textual, tabular, and graphic form.

The collection of data is the largest country on this descriptive methods continent. This country is of common interest and concern to the whole statistical universe and is by far the oldest country. Official statistics existed in the classic and medieval world. In fact, in 1500 B.C. in Judea the population is given as 100,000 souls. Practical necessity forced the earliest rulers to have some count of the number of people in their kingdom.

The collection of official statistics has increased in importance over the years as evidenced by the large units of our Federal Government such as Census, Agriculture, and Labor, organized to collect all kinds of useful data.

Before going into the frontier area to collect more data, one should check carefully the sources of data in the settled areas to be sure that he is not about to perform needless duplication. The decision will have to be made whether to take a census, or to take a sample from the population. Here, as we stand on a ridge, we look over into the sampling country which we shall visit later.

Between the collection and the summarization of data countries, there is this border area, where the police (editors) check our schedule to make sure the blanks are filled and that no absurd or highly improbable entries have been made. As we continue our tour, our papers and passports will be checked frequently.

Our first stop in the summarization country is at the state called tabulation. Here the data on all items from the individual schedules are tabulated and cross-tabulated. A visit here is prerequisite to all further study of the data by statistical methods.

I shall have to ask you to pass up a visit to the well-known array, ranking, and frequency tables states. There still exists disputed area around the frequency table, such as the choice of the beginning and extent of class intervals. These historic frontiers and the political devices such as ratios, proportions and percentages are visited by many tourists.

Let us proceed to two other states, where the calculations of measures of central tendency and dispersion are made. The central tendency state has several clans. In one, the arithmetic mean is the rule. A second group has a median rule, and a third group prefers the mode rule.

Near the mainland, there are islands between it and the analysis and theory continent. Even on these islands mathematical definitions are required for the rules used for measuring central tendencies such as the geometric and harmonic means.

As we go on into the dispersion state you will note that the topography is becoming less familiar. Yet variation of individuals in a measurable characteristic is a basic condition for statistical analysis and theory. If uniformity prevailed, there would be no need for statistical methods, though descriptive methods might be desired.

This variation state also has several clans. One advocates the range as the simplest measure to describe the dispersion of a distribution. Another prefers the use of the mean deviation, while the most densely populated clan advocates the standard deviation. Nearby is a frontier area where dwell less familiar and relatively uninteresting groups such as the quartile deviation and the 10–90 percentile range.

In this descriptive methods continent, placed in the summarization of data country are other states settled by special purpose groups. Let us now visit two, the index number and the description of time series states, to look at some of their unsettled and disputed frontier problems.

The index number state, consisting of one or a set of measures for one or a group of units, evaluates indirectly the incidence of a characteristic that is not directly measurable. We do not have time to visit the single factor index area, but proceed directly to the wide open frontiers of multi-factor indexes. For example, the price and level-of-living indexes are well known and of vital interest. On this frontier: (1) Which multi-factor index formula is the best? (2) What items should be included? (3) What is the proper weighting of items? (4) Is the fixed base or chain method best? (5) How frequently should the base be changed? (6) When and how can you remove obsolete commodities and add new ones into the index? and (7) If the index number has no counterpart in reality, should it be discarded? To settle these frontiers, developments are needed on the borders with the theory continent.

In the description of time series state, we find measures recorded on some characteristic of a unit (or a group of units) for different periods or points of time. There are several method groups governing this state such as inspection, semi-averages, moving averages and least squares. Of course, there are disputes about which method is best. One of the frontier problems is how to handle nonlinear trends. One group of statisticians exploring in this state deals with time series accounting for secular trend, cyclical, periodic, and irregular movements.

Note that most of the folks in this area are economists. The public health and industrial scientists are beginning to explore here. They have such problems as fatigue testing, incubation period of a disease, and the life time of radioactive substances.

This is rather an exhausting tour, so much to be seen in so short a time. However, before you leave the descriptive methods continent, I want you to visit the presentation of results country. The availability and usefulness of whatever contribution to scientific knowledge the project has yielded are dependent upon the successful performance in this country.

As we enter the presentation of results country, you will be asked to swear allegiance to logical organization, preciseness, and ease of comprehension. In this country, certain conventions in structure and style of the form of presentation have developed and are generally accepted.

The methods of presentation of results divide into several states: textual, tabular, and graphic. The textual state gives only statements of findings and interpretation of results. The tabular state has two types of tables, the general and the special purpose tables, according to their functions. In the graphic state, presentation of quantitative data is represented by geometric designs. It is obvious that the tourist naive in mathematics will enjoy this state. Some of the named community settlements are: the bar diagram, area diagram, coordinate chart, statistical map, and pictorial statistics.

3. Design of Experiments and Investigations Continent

Later in discovery and development was the analytical statistics hemisphere where the tools and techniques for research workers are provided and used. The northern continent, called Design of Experiments and Investigations, is divided into two major sections, the design of experiments and the design of investigations.

My own random walks have taken me into the design of experiment section of this continent more frequently and extensively than into any other area we shall visit.

This section is divided into four major countries: (1) completely randomized, (2) randomized block, (3) latin square, and (4) incomplete block designs. The first three countries are the oldest and are well developed. However, in the latin square country, let us visit a newly explored state, where the latin square is adjusted so as to measure residual effects which may be present when the treatments are applied in sequence.

We might inquire about the uprisings in the latin square country when nonrandom treatments are assigned to the rows and columns. This takes you over into the incomplete block design country. It is hoped that this area will be placed in the incomplete block design country without further trouble.

The selection of the treatment combinations to go into these countries takes us into another dimension of this statistical universe. We have single factor and multiple factor treatment combinations. Small factorial groups fit nicely into our design countries. If several factors are involved, we may need to introduce confounding. This requires settlement in the incomplete block design country, where there are more blocks than replications. Some confounded areas are settled, such as those where confounding on a main effect, the split-plot design country. Here you find political parties with platforms ranging from randomization to stripping the second factor. This latter complicates its trade relations with the analysis countries.

Let us continue in the incomplete block design country and cross the state where confounding on high-order interactions is practiced. Right near, and often overlapping, is a new state using confounding on degree effects. These two states are being settled, with good roads already constructed, but the border has not been defined or peacefully occupied.

A rather new and progressive group of settlers are the fractional replication folks. Their chief platform is that five or more factors can be included simultaneously in an experiment of a practicable size so that the investigator can discover quickly which factors have an important effect on their product. In this area the hazard of misinterpretation is especially dangerous when one is not sure of the aliases. The penalties may be trivial. However, it seems wise not to join this group unless you know enough about the nature of the factor interactions.

The balanced and partially balanced incomplete block states are being settled very rapidly. So far as experimental operations are concerned, the incomplete block design country is no more difficult to settle than the complete block design country. It will take some extra planning and analysis to live in the incomplete block country and you will have to have adjusted means. The weights to use to adjust the means are still in a frontier status.

There are numerous frontier areas in this incomplete block country where roads and communications have been established. There are 376 partially balanced incomplete block design lots with $k > 2$ and 92 lots with $k = 2$ from which to choose. These lots have two associate classes.

We should look at some of the newer settlements as (1) the chain block and the generalized chain block design states; (2) the doubly-balanced incomplete block design state where account can be taken of the correlation between experimental units; and (3) the paired comparison design areas for testing concordance between judges, together with the appropriate agreements with the analysis continent. Beyond the latin square country dikes have been built to provide new land. There are latin squares with a row and column added or omitted, or with a column added and a row omitted. Further work covering more general situations will give this design continent more areas for expansion.

Let us go now to another large new country which, after negotiations, has been established by taking sections of the design and analysis continents. The process has raised some political issues and questions of international control. The development came about because, in the design continent, there is a two-party system with data measured (1) on a continuous scale (quantitative variable) or (2) on a discontinuous scale (qualitative variable). These party members have settled side by side in the design continent for single-factor groups.

If we have factorial groups, we have to consider both whether the measures are continuous or discontinuous and whether the factors are independent or not. To handle these problems, some of the continuous scale statisticians have established a response surface country. To prepare for the peaceful settlement of this response surface country a portion of the regression analysis state has been transferred. Whether this separation of portions of countries to make up a new country will hold, only time will tell.

Here in this rather new response surface country, observe that major interest lies in quantitative variables, measured on a continuous scale. In this situation, it is often natural to think of response as related to the levels of the factors by some mathematical function. The new methods are applicable when the function can be approximated, within the limits of the experimental region, by a polynomial.

In this tropical and exciting response surface country, the central composite and non-central composite states have been settled for some time. Some of the other borders are not firmly fixed, as would be expected in a new country. New states identified as first, second, third, and higher-order designs are seeking admittance to this country. They overlap with some of the older countries. We can stand over here on this mountain top and see many frontiers as the very special central composite rotatable design area, which has been named and partially settled with some roads constructed. Over there is the evaluation frontier where the relative efficiency of these designs and methods needs to be determined.

Progress has been made on strategies to be used for determining the optimum combination of factor levels. In addition to locating the maximum of y, it is often desirable to know something about how y varies when the factor levels are changed from their optimum values. The efficient location of an optimum combination of factor levels often requires a planned sequential series of experiments.

Most experimentation is sequential, since the treatments are applied to the experimental units in some definite time sequence. To explore in this area, the process of measurement must be rapid so that the response on any unit is known before the experimenter treats the next unit. A method of sequential analysis gives rules that determine, after any number of observations, whether to stop or continue the experiment.

the use of both continuous and discontinuous variables and even the fitting of regression models for the fixed continuous variable. This latter region is being urged to establish an alliance with the response surface country.

We have time to observe only a few frontier problems: (1) What is the power of analysis of variance to detect the winner? (2) How do you analyze data which involve both a quantal and a graded response? (3) How do you attach confidence limits to proportions? (4) What about nonhomogeneity of variance when making tests of significance? and (5) Should we enter these countries with nonnormal data? I may just mention a subversive area, at least it is considered so by some, that is, the region where effects suggested by the data are tested.

Are you ready now to visit the correlation country? Bivariate populations are often interesting because of the relationship between measurements. First, let us visit the well developed product moment correlation section, where the cultural level is high due to theoretical verifications. Around here are several unincorporated areas, quite heavily populated by special groups, but not too well supported by theory. You should be careful if you visit the method of rank difference, ρ (rho), the non-linear, η (eta), the biserial or the tetrachoric coefficients of correlation districts.

While we travel across to the regression country, I might mention that its constitution has several articles like the constitution of the correlation country. The two are confused by some users of statistics and even by statisticians.

We had better check to see if you have your visa before we enter the regression country. Some of the acceptable reasons for granting visas are: (1) to see if Y depends on X and if so, how much, (2) to predict Y from X, (3) to determine the shape of the regression line, (4) to find the error involved in experiments after effect of related factor is discounted or (5) to seek cause and effect.

Some near frontier areas are being settled, such as those where there are errors in both the X and the Y variables. Other frontiers include the test of the heterogeneity of two or more regressions. How do we average similar ones? What about the nonlinear regression lines?

As we leave the bivariate countries of the analysis and theory continent and enter the multivariate countries, we find that life becomes more complicated. All kinds of mechanical, electrical and electronic statistical tools have come into use. These countries have been developed from, but are not independent of, the univariate and bivariate areas by a process of successive generalizations. For example, people were taken from the t-test country and by generalization they developed the statistics T country. This T group does all the things done by the t group for any number of variates simultaneously, be they mutually correlated or independent.

In this multivariate area, new territory related to the analysis of variance has been explored and is called the multivariate analysis of variance. Here are theoretical frontiers to be explored. Some are (1) What are the values of the roots of a determinantal equation and what particular combination of them should be used for a particular purpose? (2) What are the limitations and usefulness of the multivariate analysis of variance country? and (3) What are the confidence bounds on parametric functions connected with multivariate normal populations?

The next time you come this way, I wish you would stop to explore the areas where the discriminant function and factor analysis methods are used. There may be some danger that the latter will not be able to withstand the attacks being made by those who advocate replacing factor analysis by other statistical methods. I personally believe the factor analysis area will resist its attackers and will remain in the statistical universe as a powerful country.

The simple correlation ideas were generalized into two new countries, the multiple correlation country and the less well known canonical correlation country, which has two sets of variates.

Crossing the multiple regression country, we look at the frontiers. There are situations where it is desirable to combine scored, ranked, and continuous data into a multiple regression or factor analysis. How can this be done legally? What about the normal distribution assumptions?

I cannot resist having you visit the analysis of covariance country for it accomplishes some of the same purposes as do the design countries. Covariance helps to increase accuracy of estimates of means and variances. However, dangerous mountains exist in this country. The explorers may need to develop added theory to enable the applied statistician to reach the top of such cliffs as the one where the X variable is affected by the treatments. If the treatments do affect X, a covariance analysis may add information about the way in which the treatments produce their effects. The interpretation of the results when covariance is used requires care, since an extrapolation danger may be involved. Now that I have acknowledged that we are in a dangerous area, I might state that the dangers of extrapolation exist in all regression and related areas, and especially back in the response surface country.

We are ready to enter the variance component country, where separate sources of variability are identified. Estimates of these variance components are desired. These estimates are used to plan future experiments, to make tests of significance, and to set confidence limits.

This country is relatively new, so that adequate statistical theory has not been developed, thus leaving rugged frontiers: (1) The assumption of additivity needs to be explored in detail, (2) A clear statement is needed of how to decide whether the interaction in a two-way classification is nonrandom or random, (3) More exact methods of assigning confidence limits for the variance components need to be developed, (4) How does one handle the mixed model? (5) How can one detect correlated errors? (6) What can be done to simplify the analysis of data with unequal variances? (7) What are the effects of various types of nonnormality on the consistency and efficiency of estimates? and (8) Some study needs to be made of the proper allocation of samples in a nested sampling problem when resources are limited and good estimates of all components are desired.

Another section of the variance component country is called components of error. The problem of choosing the correct error term in the analysis of two or more factors depends upon whether the factors are random or nonrandom or upon the question you ask. Do you want the mean difference between treatments averaged over these particular areas with narrow confidence limits, or do you want mean differences averaged over a population of areas of which these areas are a sample with broad confidence limits?

So far, we have visited almost exclusively the parametric inference countries. Let us take a glimpse at the frontier in the nonparametric inference territory. When the experimenter does not know the form of his population distribution, or knows that it is not normal, then he may either transform his data or use methods of analysis called distribution free or nonparametric methods. This territory is being settled. The area dealing with the efficiency of certain tests for two by two tables has been partially settled and some general theorems on the asymptotic efficiency of tests have been proved.

Some of the frontiers are: (1) What is the general theory of power functions for distribution free tests? (2) What is the efficiency of nonparametric tests? (3) Can sequential methods be applied to nonparametric problems, and (4) How can two nonnormal populations be compared?

There are three more general frontiers I wish to mention. (1) How far are we justified in using statistical methods based on probability theory for the analysis of nonexperimental data? Much of the data used in the descriptive methods continent are observational or nonexperimental records. (2) What are the effects of nonnormality, heterogeneity, nonrandomness and nonindependence of observations to which standard statistical methods are applied? And (3) How can we deal with truncated populations in relationship problems?

As we complete our tour of the three continents, I wish to emphasize the fact that there are many important problems of design and statistical inference which remain unexplored.

5. Training Frontier

Our travels took us to only a part of the statistical universe, but we managed to observe many frontier areas. I hope one thing impressed you: that is, the extent of the need for statisticians to explore these areas. In recent years, there have been advances in statistical theory and technology, but the prompt application of these to our biological, social, physical, industrial, and national defense needs has created an unprecedented demand for intelligent and highly trained statisticians. Research workers in many fields are requesting the statistician to help both in planning experiments or surveys and in drawing conclusions from the data. Administrators are facing the quantitative aspects of problems, such as optimum inventories, production schedules, sales efforts, pricing policies and business expansion, which call for new mathematical methods for solving problems concerned with decision making.

Comments on
Cox (1957) Statistical Frontiers

G.A. Barnard

Perhaps the most obvious omission from her survey is any mention of computers, which might be thought of as large mechanised tractors which are in course of ploughing all the land she traversed, and bringing about in every area *new* and luxuriant growth. When the University of London took delivery of its Mercury Computer in the mid fifties I recall saying that at last we could really draw likelihood curves and contour plots. Yet it was not until the late seventies that a set of likelihood plots appeared in the "Annals", in a paper by Morris De Groot. Of course it would not have been possible in the mid fifties to foresee all the graphics we now can have on desk top computers, nor the computer-intensive possibilities of projection pursuit and the bootstrap. The statistical world has yet to adjust to the full possibilities opened up by computers.

A feature that has developed since Gertrude wrote—and which is to a large extent a consequence of initiatives which she promoted—is the direct involvement of statisticians in applications of their ideas. The Kettering award to David Cox is an instance of what I mean—the award normally goes to a piece of medical research. Related to this sort of development are the powerful pressures exerted both in the US and in the UK, as well as in other countries, for guaranteeing public access to accurate statistics as an important bulwark of democracy.

I.J. Good

In 1977, Gertrude Cox presented a colloquium at Virginia Tech with the title "A Consulting Statistician: Facing a Real World." She encouraged students to become consultants and see the world, and she received a standing ovation. Here's one anecdote from this colloquium. Gertrude had been invited to do some statistical consulting for a mining company and she insisted that she should be allowed to go down the mine. She was one of the rare women at that time (Eleanor Roosevelt was another) to do so. This anecdote reveals her determination and her hands-on down-into-the-earth approach to consulting. Her love of traveling, which was clear from the colloquium, would help to explain the "geographical" structure of her presidential address in 1957. Perhaps she had specific continents and countries in mind and used them to organize her address.

Her address contained about fifty suggestions for research projects, many of which are still topical. For example, she mentioned the problem of combining discrete and continuous data, an area of considerable current interest for medical diagnosis. She said she'd let us guess which ideas were original to her, but I think her main aim in this address was to be useful rather than original.

Ideas in one continent often affect others, and can even affect another world. For example, one of the earliest ideas in Gertrude's continent, Yates's adding-and-subtracting algorithm for obtaining the interactions in a 2^n factorial experiment (Yates, 1933, pp. 15 and 29; Cochran & Cox, 1957, §5.24a) led to an influential Fast Fourier Transform. It was probably anticipated by Gauss.

Gertrude's address had little to say about Computerica (two sentences on page 7), nothing on multivariate categorical data, and, apart from the two words "decision making" on page 10, she didn't mention Bayesiana. Fisher had pushed that continent away but by 1957 it was already drifting back fast.

The prediction on page 11 that "statisticians are destined for a larger role" was correct and probably influential. It was anticipated by Wilks (1950) who acknowledged the prophet H.G. Wells but without a citation. In fact Wells (1932, pp. 372 and 391) said "The science of statistics is still in its infancy—a vigorous infancy", and on page 391 "... the movement of the last hundred years is all in favour of the statistician."

References

Cochran, W.G. and Cox, G.M. (1957). *Experimental Designs*. 2nd ed. New York: Wiley.

Wells, H.G. (1932). *The Work, Wealth and Happiness of Mankind*. London: Heinemann. American printing (1931), two volumes, Doubleday, Doran & Co., Garden City.

Wilks, S.S. (1950). Undergraduate statistical education, *Journal of the American Statistical Association* **46**, 1–18.

Yates, F. (1937). *The Design and Analysis of Factorial Experiments*. Harpenden, England: Imperial Bureau of Soil Science.

D.V. Lindley

Guide books do not ordinarily concern themselves with the politics or philosophy of the countries they are describing and tourists, save in exceptional cases, ignore the manner of government. In this respect, Dr. Cox really is a tourist, not mentioning the philosophy of statistics. In 1956, this was reasonable, since Savage, the revolutionary text, had only just appeared. Jeffreys lay unread and de Finetti was still only available in Italian. The statistical world, at least in the United States, looked to be soundly governed by the Wald-Neyman-Pearson school. Few had doubts that power, confidence intervals and unbiased estimates were not completely sound. Basu had not produced his counter-examples.

Today, the travellers would surely look at the philosophy of statistics and its implication for practice. They would not be quite so confident that their methods were sound. Fisher still flourishes like an awkward government that is slightly suspect. The upstart Bayesian movement is being contained, largely by being ignored, but represents a continual threat to the establishment. Even the arithmetic mean has fallen from its pedestal and we argue about whether or not to shrink our census returns.

To leave the travel-guide analogy, there are three features that would be present in a contemporary survey yet are omitted by Cox. First, there would be a discussion about computers; about their ability to handle large data sets, to perform more complicated and larger analyses than hitherto, to simulate in procedures like the bootstrap and Gibbs sampling. Second, the topic of probability would loom larger. The ideas of influence diagrams, expert systems and artificial intelligence have led to an appreciation of probability manipulations, and especially of independence, that are important. Third, there would be some consideration of decision-making. Cox's view of a statistician's role was passive; we observe and report. There is increasing awareness today, for example in Ron Howard's recent address, to the more active statistician who contemplates risk and utility, and is prepared to advise not just about beliefs but about the actions that might spring from those beliefs.

F. Mosteller

Gertrude Cox loved travel so much that we are not surprised that she chose this analogy for her paper. Although statisticians have made progress on many of the issues that she mentions in 1956, her list leaves plenty of room for a decade more of thoughtful doctoral dissertations in the last decade of her century.

One omission I note is that in dealing with descriptive statistics, both graphical and tabular, she does not invoke the need for behavioral science to help us decide what methods of presentation come through to the viewers as especially helpful. We have had little progress in this area, though Cleveland's group has made some contributions. I look forward to big progress as computer technology offers us plenty of options for flexible and attractive presentations and for easy assessment.

An example of the kind of research needed is given in Ibrekk and Morgan (1987) where these authors explore for nontechnical users the communication merits of nine pictorial displays related to the uncertainity of a statistic.

In learning how to use graphics to improve analysis, statistics alone may well be adequate, but in improving presentation, we have to find out what methods are better at communicating, and for this nothing can replace the findings for actual users.

Reference

H. Ibrekk and M. G. Morgan, Graphical communication of uncertain quantities to nontechnical people, *Risk Analysis*, 1987, 7: 519–529.

P.K. Sen

Looking back at this remarkable article written almost thirty-five years ago, I have nothing but deep appreciation for the utmost care with which (the late) Gertrude M. Cox depicted the *statistical universe* (in 1956) as well as for her enormous foresight. In fact, to appreciate fully this (ASA) presidential address delivered to a very wide audience (from all walks in statistical methodology and applications), it would be very appropriate to bear the remembrance of her prime accomplishments in creating such a universe in the Research Triangle Park in the heart of North Carolina, and even after 35 years, we are proudly following her footsteps.

The three major *fortresses* in her *statistical frontiers* are (i) descriptive methods, (ii) design of experiments and investigations, and (iii) analysis and theory. Collection of data, their summarization and presentation in textual/ tabular/graphical forms constitute the first aspect. The advent of modern computer and statistical packages has made this job somewhat easier and mechanical, albeit the abuses of such statistical packages have been increasing at an alarming rate. The main burden lies with a good *planning* (synonymous to design) of experiments/investigations, so that the collected data convey meaningful information, can be put to valid statistical analysis, and suitable statistical packages can be incorporated in that context. In spite of the fact that most of us have our bread and butter from statistical theory and analysis, we often digress from applicable methodology onto the wilderness of abstractions. Gertrude was absolutely right in pointing out that there is a compelling need to ensure that statistical methodology is theoretically sound and at the same time adoptable in diverse practical situations. The scenario has not changed much in the past three decades, although introduction of new disciplines has called for some shifts in emphasis and broadening of the avenues emerging from the Cox fortresses.

The genesis of statistical sciences lies in a variety of disciplines ranging from agricultural science, anthropometry, biometry, genetics, sociology, economics, physical and engineering sciences, and bio-sciences to modern medicine, public health and nascent bio-technology. While Cox's thoughtful observations pertain to a greater part of this broad spectrum, there may be some need to examine minutely some of the frontiers which were mostly annexed to the Cox universe later on. In this respect, I would like to place the utmost emphasis on Energy, Ecology and Environmetrics. Our planet is endangered with the thinning of the ozone layer, extinction of several species, massive atmospheric pollution, nuclear radiation, genotoxicity, ecological imbalance and numerous other threatening factors. The thrust for energy-sufficiency and economic stability has led to global tensions, and the mankind is indeed in a perilous state. Statisticians have a basic role to play in conjunction with the scientists in other disciplines in combating this extinction. The design of such investigations may differ drastically from that of a controlled experiment. The collection of data may need careful scrutiny in order that valid statistical analysis can be done, and more noticably, novel statistical methodology has to be developed to carry out such valid and efficient statistical analysis. Lack of a control, development of proper scientific instruments to improve the measurement system, proper dimension reduction of data for efficient analysis and above all good modelling are essential factors requiring close attention from the statisticians. To a lesser extent, similar problems cropped up in the area of epidemiological investigations including clinical trials and retrospective studies, and the past two decades have witnessed phenomenal growth of the literature of statistical methodology to cope with these problems. Nonstationarity of concomitant variates (over time or space), measurement errors, doubts about the appropriateness of linear, log-linear or logistic models, and above all, the relevance of 'random sampling' schemes (particularly, equal probability sampling with/without replacement) all call for non-standard statistical analysis, for which novel methodology need to be developed. As statisticians, we have the obligation to bridge the gap between the classical theory and applicable methodology, so that valid statistical conclusions can be made in a much broader spectrum of research interest. Last year, at the Indian Science Congress Association Meeting in Madurai, I have tried to summarize this concern, and as such, I would not go into the details. Rather, I would like to conclude this discussion with the remark that most of the problems relating to multivariate analysis, nonparametric methods and sequential analysis referred to in this Cox address has been satisfactorily resolved in the past three decades, and we need to march forward beyond these traditional quarters onto the rough territories which are as yet deprived of the statistical facilities, and towards this venture, we need to accommodate a plausible shift in our statistical attitude too. Nevertheless, the Cox milestone remains a good exploration point.

Reference

Sen, P.K. (1989). Beyond the traditional frontiers of statistical sciences: A challenge for the next decade. Platinum Jubilee Lecture in Statistics, Indian Science Congress Association Meeting, Madurai. *Inst. Statist., Univ. N. Carolina Mimeo. Rep.* 1861.

Introduction to Fisher (1922) On the Mathematical Foundations of Theoretical Statistics

Seymour Geisser
University of Minnesota

1. General Remarks

This rather long and extraordinary paper is the first full account of Fisher's ideas on the foundations of theoretical statistics, with the focus being on estimation. The paper begins with a sideswipe at Karl Pearson for a purported general proof of Bayes' postulate. Fisher then clearly makes a distinction between parameters, the objects of estimation, and the statistics that one arrives at to estimate the parameters. There was much confusion between the two since the same names were given to both parameters and statistics, e.g., mean, standard deviation, correlation coefficient, etc., without an indication of whether it was the population or sample value that was the subject of discussion. This formulation of the parameter value was certainly a critical step for theoretical statistics [see, e.g., Geisser (1975), footnote on p. 320 and Stigler (1976)]. In fact, Fisher attributed the neglect of theoretical statistics not only to this failure in distinguishing between parameter and statistic but also to a philosophical reason, namely, that the study of results subject to greater or lesser error implies that the precision of concepts is either impossible or not a practical necessity. He sets out to remedy the situation, and remedy it he did. Indeed, he did this so convincingly that for the next 50 years or so almost all theoretical statisticians were completely parameter bound, paying little or no heed to inference about observables.

Fisher states that the purpose of statistical methods is to reduce a large quantity of data to a few that are capable of containing as much as possible of the relevant information in the original data. Because the data will generally supply a large number of "facts," many more than are sought, much information in the data is irrelevant. This brings to the fore the Fisherian dictum that statistical analysis via the reduction of data is the process of extracting the relevant information and excluding the irrelevant information. A way of accomplishing this is by modeling a hypothetical population specified by relatively few parameters.

Hence, the critical problems of theoretical statistics in 1920, according to Fisher, were (1) specification, choice of the hypothetical parametric distribution; (2) estimation, choice of the statistics for estimating the unknown parameters of the distribution; (3) sampling distributions, the exact or approximate distributions of the statistics used to estimate the parameters. For a majority of statisticians, these have been and still are the principal areas of statistical endeavor, 70 years later. The two most important additions to this view are that the parametric models were, at best, merely approximations of the underlying process generating the observations, and in view of this, much greater emphasis should be placed on observable inference rather than on parametric inference.

2. Foundational Developments

In this paper, Fisher develops a number of concepts relevant to the estimation of parameters. Some were previously introduced but not generally developed, and others appear for the first time. Here, also, the richness of Fisher's *lingua statistica* emerges, yielding poignant appelatives for his concepts, vague though some of them are. This activity will continue throughout all his future contributions. First he defines consistency: A statistic is consistent if, when calculated from the whole population, it is equal to the parameter describing the probability law. This is in contradistinction to the usual definition which entails a sequence of estimates, one for each sample size, that converges in probability to the appropriate parameter. While Fisher consistency is restricted to repeated samples from the same distribution, it does not suffer from the serious defect of the usual definition. That flaw was formally pointed out later by Fisher (1956): Suppose one uses an arbitrary value A for an estimator for $n < n_1$, where n is as large as one pleases, and for $n > n_1$ uses an asymptotically consistent estimator T_n. The entire sequence, now corrupted by A for $n < n_1$ and then immaculately transformed to T_n thereafter, remains a useless, but perfectly well-defined, consistent estimator for any n. Fisher is not to be trifled with!

Indicating that many statistics for the same parameter can be Fisher-consistent, in particular, the sample standard deviation and sample mean deviation for the standard deviation of a normal population, he goes on to suggest a criterion for efficiency. It is a large sample definition. Among all estimators for a parameter that are Fisher-consistent and whose distributions are asymptotically normal, the one with the smallest variance is efficient. Later, he shows that when the asymptotic distribution of the method of moments estimator is normal for the location of a uniform distribution while that of the "optimum" estimator is double exponential, he realizes that the variance does not necessarily provide a satisfactory basis for comparison, especially for small samples. Thus, he also recognizes that his large sample definition of intrinsic accuracy (a measure of relative efficiency) should not be based on variances and a definition appropriate for small samples is required. In later papers, e.g., Fisher (1925), vague concepts of intrinsic accuracy will be replaced by the more precise amount of information per observation. At any rate, the large sample criterion is incomplete and needs to be supplemented by a sufficiency criterion. The "remarkable" property of this concept was previously pointed out when introduced for a special case without giving it a name [Fisher (1920)]. A statistic, then, is sufficient if it contains all the information in the sample regarding the parameter to be estimated; that is, given a sufficient statistic, the distribution of any other statistic does not involve the parameter. This compelling concept of his, including the factorization result, is still in vogue. Assuming a sufficient statistic and any other statistic whose joint distribution is asymptotically bivariate normal with both means being the parameter estimated, he then "demonstrates" that the sufficient statistic has an asymptotic variance smaller than that of the other statistic by a clever conditioning argument that exploits the correlation between the statistics. Hence, he claims that a sufficient* statistic satisfies the criterion of (large sample) efficiency. This "proof" of course could only apply to those statistics whose asymptotic bivariate distribution with the sufficient statistic was normal.

He comments further on the method of moments estimation procedure. While ascribing great practical utility to it, he also exposes some of its shortcomings. In particular, in estimating the center of a one-parameter Cauchy distribution, he points out that the first sample moment, the sample mean, which is the method of moments estimator is not consistent but the median is. He also cautions against the statistical rejection of outliers unless there are other substantive reasons. Rather than outright rejection, he proposes that it seriously be considered that the error distribution is not normal. Fisher effectively argues that the specification of the underlying probability law will generally require the full set of observations. A sufficient reduction is only meaningful once the probability law has been adequately established.

3. Maximum Likelihood

Fisher begins this part of his discourse acknowledging, first, that properties such as sufficiency, efficiency, and consistency per se were inadequate in directly obtaining an estimator. In solving any particular problem, we would

---

* In the author's note, Fisher (1950), there is a handwritten correction to the definition of intrinsic accuracy replacing sufficiency by efficiency, possibly based on his later recognition that maximum likelihood estimators were not always sufficient.

require a method that would lead automatically to the statistic which satisfied these criteria. He proposes such a method to be that of maximum likelihood, while admitting dissatisfaction with regard to the mathematical rigor of any proof that he can devise toward that result. Publication would have been withheld until a rigorous proof was found, but the number and variety of new results emanating from this method pressed him to publish. With some uncharacteristic humility, he says, "I am not insensible of the advantage which accrues to Applied Mathematics from the cooperation of the Pure Mathematician and this cooperation is not infrequently called forth by the very imperfections of writers on Applied Mathematics." This totally disarming statement would preclude any harsh commentary on the evident lack of rigor in many of his "proofs" here. Such evident modesty and good feelings toward mathematicians would never again flow from his pen.

Fisher (1912) had earlier argued for a form of maximum likelihood estimation. He had taken what superficially appeared to be a Bayesian approach because the maximizing procedure resembled the calculation of the mode of a posterior probability. In the present paper, he is very concerned to differentiate it from the Bayesian approach. He also argues against the "customary" Bayesian use of flat priors on the grounds that different results are obtained when different scales for the parameters are considered.

To illustrate Fisher's argument, suppose x denotes the number of successes out of n independent trials with probability of success; then the likelihood function is

$$L(p) = \frac{n!}{x!(n-x)!} p^x (1-p)^{n-x} \qquad (0 < p < 1),$$

which is maximized when p is chosen to be x/n. Now, if a uniform distribution on $(0, 1)$ is taken to be the prior distribution of p, then Bayesian analysis would yield

$$\pi(p) \propto p^x (1-p)^{n-x}$$

as the posterior density of p. But if we parameterize this Bernoulli process in a different way, say, in terms of θ with $\sin\theta = 2p - 1$, then the likelihood function of θ is

$$L(\theta) = \frac{n!}{x!(n-x)!} \frac{(1+\sin\theta)^x}{2^x} \frac{(1-\sin\theta)^{n-x}}{2^{n-x}} \qquad \left(-\frac{\pi}{2} < \theta < \frac{\pi}{2} \right),$$

which, when maximized with respect to θ, gives $\sin\hat{\theta} = (2x - n)/n = 2\hat{p} - 1$. Thus, the maximum likelihood estimate is invariant under a 1-1 transformation. For the Bayes approach, he questions the assignment of a prior assigned to θ. The uniformity of θ on $(-\pi/2, \pi/2)$ leads to the posterior density of p as

$$\pi(p) \propto p^{x-1/2}(1-p)^{n-x-1/2},$$

which is different from the previous result above. Due to this inconsistency and other reasons, Fisher derides the arbitrariness of the Bayes prior and chooses not to adopt the Bayesian approach. He apparently was strongly influenced in this regard by the prior criticisms of inverse probability by Boole (1854) and Venn (1866). Venn's criticism led to the elimination of the material on inverse probability from the very prominent textbook on algebra at this time by Chrystal (1886).

Fisher's argument regarding invariance was convincing to many and undoubtedly was a setback for the Bayesian approach until Jeffreys (1946) proposed a transformation-invariant procedure for calculating a prior density. There is a bit of irony here in that Fisher's expected information quantity, used in this paper but precisely defined later by Fisher (1925), was the effectuating ingredient for Jeffreys. If $\tau = g(\theta)$ satisfies certain conditions, then the expected amount of information is

$$I(\theta) = E\left[\frac{\partial \log L(\theta)}{\partial \theta}\right]^2 = \left[\frac{d\tau}{d\theta}\right]^2 I(\tau),$$

and therefore

$$I^{1/2}(\theta)\, d\theta = I^{1/2}(\tau)\, d\tau.$$

Hence, using the positive square root of the expected amount of information as a prior on θ will transform invariantly on a prior for τ and vice-versa.

Fisher also suggests using $L(\theta)$ as a relative measure of the plausibility of various values of θ and introduces the term "likelihood" to distinguish the concept from probability, confessing that earlier he had confused the two. He says that "likelihood is not here used loosely as a synonym for probability, but simply to express the relative frequencies with which such values of the hypothetical quantity θ would in fact yield the observed sample." The likelihood is then an alternative measure of the degree of rational belief when inverse probability was not applicable, which he believed was true most of the time. In more recent years this has led to a school of inference using the likelihood in various ways [Barnard et al. (1963), Edwards (1973)].

Assuming that the distribution of the maximum likelihood estimator tends to normality, Fisher demonstrates that the variance is the reciprocal of the Fisher information. That is, if T is an "optimal" statistic satisfying

$$\frac{\partial \log L(\theta)}{\partial \theta} = 0$$

at $\theta = T$, then the variance† of the large-sample normal distribution of T is the inverse of

$$-\frac{\partial^2 \log L(\theta)}{\partial \theta^2}.$$

The general consistency of the maximum likelihood estimator is not used but essentially assumed, and the "proof" relies heavily on the initial assumption.

---

† In the ultimate formula on page 31 of the following abridged version, x should be replaced by n.

Fisher consistency of the maximum likelihood estimator is also assumed without proof.

Fisher derives the explicit form of the limiting normal distribution for the maximum likelihood estimator, after having "demonstrated" that a sufficient estimate has the smallest-variance normal distribution in large samples. Now the theory would be complete if the maximum likelihood estimator were found to be sufficient, since then the reciprocal of Fisher-expected information would be a lower bound against which efficiency could be measured. Fisher claims to show that it is sufficient, although his wording is ambiguous in some passages. Fisher's "demonstration" begins with the joint distribution of the maximum likelihood estimator $\hat{\theta}$ and an arbitrary statistic T, whose joint density f satisfies

$$\frac{\partial \log f(\hat{\theta}, T | \theta)}{\partial \theta} = 0$$

at $\theta = \hat{\theta}$. The factorization

$$f(\hat{\theta}, t; \theta) = g(\hat{\theta}; \theta) h(\hat{\theta}, T)$$

is then deduced, and the sufficiency of $\hat{\theta}$ follows.

Using the inverse of the expected information,

$$\frac{1}{-E\left[\dfrac{\partial^2 \log L(\theta)}{\partial \theta^2}\right]}$$

as a lower bound on variance, Fisher illustrates the calculations on the Pearson-type III error curve with density function

$$f(x | m, a, p) \propto \left[\frac{x - m}{a}\right]^p \exp\left\{\frac{-(x - m)}{a}\right\},$$

where only m is unknown and to be estimated. The method of moments estimator is $m = X - a(p + 1)$ with variance $a^2(p + 1)/n$, whereas the asymptotic variance of the maximum likelihood estimator is

$$\frac{a^2(p - 1)}{n} = -\frac{1}{E\left[\dfrac{\partial^2 \log L(m)}{\partial m^2}\right]}.$$

Thus, the method of moments is not efficient for any n and approaches zero efficiency as $p \to 1$. In addition, he points out that for estimating the location parameter of a Cauchy distribution, the method of moments is useless. No moment of the Cauchy exists and the distribution of the sample mean is independent of the sample size! Savage (1976) later provides a curious, if not pathological, example in which a Fisher-consistent estimator is derived that is sufficient and hence loses no information in the Fisher sense, while the maximum likelihood estimator θ is not sufficient and hence does lose information. But $\hat{\theta}$ has a smaller mean squared error for a sufficiently large sample size.

The method of maximum likelihood appears to have been anticipated by Edgeworth (1908–9) according to Pratt (1976). Although there is less than universal consensus for this view, there is ample evidence that Edgeworth derived the method in the translation case directly and also using inverse probability. It appears he also conjectured the asymptotic efficiency of the method without giving it a name.

4. Other Topics

The remainder of the paper contains mainly applications of maximum likelihood techniques and various relative efficiency calculations. There is a long discussion of the Pearson system of frequency curves. This section serves mainly to display Fisher's analytic virtuosity in handling the Pearson system, also displaying graphs that serve to characterize the system in a more useful form than previously. This enables him to calculate for the various Pearson frequency curves,‡ regions for varying percent efficiencies of the method of moments estimators of location and scale. He also determines the conditions that make them fully efficient. In the latter case, he shows that if the log of a density is quartic, under certain conditions it will be approximately normal and fit the Pearson system. In dealing with the Pearson-type III curve, he now demonstrates that the asymptotic variance of the maximum likelihood estimators of scale a and shape p is smaller than that of their method of moments counterparts. However, he fails to remark or perhaps notice the anomaly of the nonregular case. Here the asymptotic variance of the maximum likelihood estimator of a is larger when m and p are given than when only p is given. Similarly, the maximum likelihood estimator of p has smaller asymptotic variance when a and m are unknown than when a and m are known.

Interest in the Pearsonian system has declined considerably over the years, being supplanted by so-called nonparametric and robust procedures, and revival appears unlikely unless Bayesians find use for them. The final part of the paper looks at discrete distributions, where the method of minimum chi-square is related to maximum likelihood, and the problem of Sheppard's correction for grouped normal data is addressed in detail. This and the material on the Pearson system actually make up the bulk of the paper. No doubt of considerable interest 70 years ago, it is of far less interest than the preceding work on the foundations. Fisher implies as much in his author's note. However, there is a final example that deals with the distribution of observations in a dilution series that is worthy of careful examination.

‡ The density on the bottom of page 342 as well as the one on page 343 of the original paper contain misprints. The section involving this material has been omitted in the abridged version of Fisher's paper which follows.

After earlier displaying the potential lack of efficiency inherent in an un-critical application of the method of moments, Fisher in an ingenious *volte-face* produces an estimation procedure for a dilution series example, which, though inefficient, is preferable to a fully efficient one essentially for economic and practical reasons. To be sure, in later years Fisher fulminated against the wholesale introduction of utility or decision theory into scientific work, but rarely again were such principles so elegantly and unobtrusively applied to such a significant practical problem. The analysis here represents a peerless blend of theory and application.

An important monitoring procedure, of ongoing interest and wide applica-bility, used in this instance for estimating the density of protozoa in soils, was brought to Fisher's attention. A series of dilutions of a soil sample solution were made such that each is reduced by a factor a. At each dilution, a uniform amount of the solution is deposited on s different plates containing a nutrient. After a proper incubation period, the number of protozoa on each plate is to be counted. A reasonable model for such a situation is that the chance of z protozoa on a plate is Poisson-distributed with expected value θ/a^x, where θ is the density or number of protozoa per unit volume of the original solution, and x the dilution level. A large number of such series were made daily for a variety of organisms. It proved either physically impossible or economically prohibitive to count the number of such organisms on every plate for many such series in order to estimate θ. First, Fisher suggests that only those plates containing no organisms be counted; the chance of such an occurrence at level x is $p_x = \exp(-\theta/a^x)$. By this device, an experimentally feasible situation is attained that produces a joint likelihood for Y_x, the number of sterile plates at level x, as

$$L = \prod_{x=0}^{k} \binom{s}{y_x} p_x^{y_x}(1 - p_x)^{s-y_x}$$

for dilution levels $x = 0, 1, \ldots, k$. He then calculates the contribution of a plate at level x to the information about $\log \theta$ to be

$$w_x = p_x(1 - p_x)^{-1}(\log p_x)^2.$$

This is informative as to the number of dilution levels necessary in such exper-iments. Further, the total expected information is approximately given as

$$s \sum_x w_x \approx \frac{s\pi^2}{(6 \log a)}.$$

The maximum likelihood solution to the problem, however, required a heavy investment of time and effort given the computational facilities of 1922. (Of course, it can easily be done today.)

At this point, Fisher employs a second wrinkle that makes the problem tractable. He suggests that the expected total number of sterile plates be equated to the observed total number in order to obtain an estimate of θ. This "rough" procedure has expected information with respect to $\log \theta$ of approximate value

$$\frac{s}{\log 2 \log a}.$$

This results in a very quick and easy procedure possessing an efficiency, independent of the dilution factor, of about 88%.

This may very well be one of the earliest statistical applications of a decision like approach to the analysis of data.

5. Summary

Clearly Fisher's paper was a landmark event in theoretical statistics. While it suffered from a lack of mathematical rigor, long analytic excursions into areas of lesser interest, and some confusion in parts, the novelty and number of ideas expressed here, both those developed from previous work and newly introduced, are still compelling for most statisticians. Although this paper is infrequently cited, its influence completely pervades the subsequent paper [Fisher (1925)],§ which presents a clearer exposition of his views. However, he poured into the 1922 paper, pell-mell, all his creative thinking and work on the foundations of statistics, the major exception being the fiducial argument. This work, filtered through the 1925 paper, has had a profound impact on statistical thinking unto this day. One has only to scan any serious work on the foundations to see that these ideas still have relevance in statistical theory, although citation is almost always to the 1925 paper.

References

Barnard, G., Jenkins, G.M. and Winsten, C.B. (1963). Likelihood inference and time series, *Jo. Roy. Statist. Soc., Ser. A*, **125**, 321–372.

Boole, G. (1854). *The Laws of Thought.* Dover, New York.

Chrystal, G. (1886). *Algebra.* Adam and Charles Black, London.

Edgeworth, F.Y. (1908–9). On the probable errors of frequency-constants, *J. Roy. Statist. Soc.*, 71 381–397, 499–512, 651–678. Addendum ibid. **72**, 81–90.

Edwards, A.W.F. (1972). *Likelihood.* Cambridge University Press, New York.

Fisher, R.A. (1912). On an absolute criterion for fitting frequency curves, *Messenger of Math.*, **41**, 155–160.

Fisher, R.A. (1920). A mathematical examination of the methods of determining the accuracy of an observation by the mean error, and by the mean square error, *Monthly Notices Roy. Astro. Soc.*, **80**, 758–770.

§ This paper is cited a number of times in Fisher (1956), his final work on the foundations, while the seminal 1922 paper is not mentioned. In fact, the dozen or so times that Fisher subsequently cites the 1922 paper, he misdates it about half the time as 1921. This Fisherian slip, making him a year younger at its publication, accords with the author's note attributing certain deficiencies in the paper to youth.

Fisher, R.A. (1925). Theory of statistical estimation, *Proc. Cambridge Philos. Soc.*, **22**, 700–725.

Fisher, R. A. (1950). *Contributions to Mathematical Statistics*. Wiley, New York.

Fisher, R.A. (1956). *Statistical Methods and Scientific Inference*. Oliver and Boyd, Edinburgh.

Geisser, S. (1975). The predictive sample reuse method with applications, *Jo. Amer. Statist. Assoc.*, **70**, 320–328.

Jeffreys, H. (1946). An invariant form for the prior probability in estimation problems, *Proc. Roy. Soc. London, Ser. A*, **186**, 453–454.

Pratt, J.W. (1976). F.V. Edgeworth and R.A. Fisher on the efficiency of maximum likelihood estimation, *Ann. Statist.*, 501–514.

Savage, L.J. (1976). On Rereading R.A. Fisher (with discussion), *Ann. Statist.*, **4**, 441–500.

Stigler, S.M. (1976). Discussion of Savage (1976), *Ann. Statist.*, **4**, 498–500.

Venn, J. (1866). *The Logic of Chance*. Macmillan, London.

On the Mathematical Foundations of Theoretical Statistics

R.A. Fisher
Fellow of Gonville and Caius College,
Chief Statistician, Rothamsted Experimental Station

Definitions

Centre of Location. That abscissa of a frequency curve for which the sampling errors of optimum location are uncorrelated with those of optimum scaling. (9.)

Consistency. A statistic satisfies the criterion of consistency, if, when it is calculated from the whole population, it is equal to the required parameter. (4.)

Distribution. Problems of distribution are those in which it is required to calculate the distribution of one, or-the simultaneous distribution of a number, of functions of quantities distributed in a known manner. (3.)

Efficiency. The efficiency of a statistic is the ratio (usually expressed as a percentage) which its intrinsic accuracy bears to that of the most efficient statistic possible. It expresses the proportion of the total available relevant information of which that statistic makes use. (4 and 10.)

Efficiency (Criterion). The criterion of efficiency is satisfied by those statistics which, when derived from large samples, tend to a normal distribution with the least possible standard deviation. (4.)

Estimation. Problems of estimation are those in which it is required to estimate the value of one or more of the population parameters from a random sample of the population. (3.)

Intrinsic Accuracy. The intrinsic accuracy of an error curve is the weight in large samples, divided by the number in the sample, of that statistic of location which satisfies the criterion of efficiency. (9.)

Isostatistical Regions. If each sample be represented in a generalized space of which the observations are the co-ordinates, then any region throughout which any set of statistics have identical values is termed an isostatistical region.

Likelihood. The likelihood that any parameter (or set of parameters) should have any assigned value (or set of values) is proportional to the probability that if this were so, the totality of observations should be that observed.

Location. The location of a frequency distribution of known form and scale is the process of estimation of its position with respect to each of the several variates. (8.)

Optimum. The optimum value of any parameter (or set of parameters) is that value (or set of values) of which the likelihood is greatest. (6.)

Scaling. The scaling of a frequency distribution of known form is the process of estimation of the magnitudes of the deviations of each of the several variates. (8.)

Specification. Problems of specification are those in which it is required to specify the mathematical form of the distribution of the hypothetical population from which a sample is to be regarded as drawn. (3.)

Sufficiency. A statistic satisfies the criterion of sufficiency when no other statistic which can be calculated from the same sample provides any additional information as to the value of the parameter to be estimated. (1.)

Validity. The region of validity of a statistic is the region comprised within its contour of zero efficiency. (10.)

1. The Neglect of Theoretical Statistics

Several reasons have contributed to the prolonged neglect into which the study of statistics, in its theoretical aspects, has fallen. In spite of the immense amount of fruitful labour which has been expended in its practical applications, the basic principles of this organ of science are still in a state of obscurity, and it cannot be denied that, during the recent rapid development of practical methods, fundamental problems have been ignored and fundamental paradoxes left unresolved. This anomalous state of statistical science is strikingly exemplified by a recent paper (1) entitled "The Fundamental Problem of Practical Statistics," in which one of the most eminent of modern statisticians presents what purports to be a general proof of Bayes' postulate, a proof which, in the opinion of a second statistician of equal eminence, "seems to rest upon a very peculiar—not to say hardly supposable—relation." (2.)

Leaving aside the specific question here cited, to which we shall recur, the obscurity which envelops the theoretical bases of statistical methods may perhaps be ascribed to two considerations. In the first place, it appears to be widely thought, or rather felt, that in a subject in which all results are liable to greater or smaller errors, precise definition of ideas or concepts is, if not impossible, at least not a practical necessity. In the second place, it has happened that in statistics a purely verbal confusion has hindered the distinct formulation of statistical problems; for it is customary to apply the same name, *mean, standard deviation, correlation coefficient*, etc., both to the true value which we should like to know, but can only estimate, and to the particular value at which we happen to arrive by our methods of estimation; so also in applying the term probable error, writers sometimes would appear to suggest that the former quantity, and not merely the latter, is subject to error.

It is this last confusion, in the writer's opinion, more than any other, which has led to the survival to the present day of the fundamental paradox of inverse probability, which like an impenetrable jungle arrests progress towards precision of statistical concepts. The criticisms of Boole, Venn, and Chrystal have done something towards banishing the method, at least from the elementary text-books of Algebra; but though we may agree wholly with Chrystal that inverse probability is a mistake (perhaps the only mistake to which the mathematical world has so deeply committed itself), there yet remains the feeling that such a mistake would not have captivated the minds of Laplace and Poisson if there had been nothing in it but error.

2. The Purpose of Statistical Methods

In order to arrive at a distinct formulation of statistical problems, it is necessary to define the task which the statistician sets himself: briefly, and in its most concrete form, the object of statistical methods is the reduction of data. A quantity of data, which usually by its mere bulk is incapable of entering the mind, is to be replaced by relatively few quantities which shall adequately represent the whole, or which, in other words, shall contain as much as possible, ideally the whole, of the relevant information contained in the original data.

This object is accomplished by constructing a hypothetical infinite population, of which the actual data are regarded as constituting a random sample. The law of distribution of this hypothetical population is specified by relatively few parameters, which are sufficient to describe it exhaustively in respect of all qualities under discussion. Any information given by the sample, which is of use in estimating the values of these parameters, is relevant information. Since the number of independent facts supplied in the data is usually far greater than the number of facts sought, much of the information supplied by any actual sample is irrelevant. It is the object of the statistical processes employed in the reduction of data to exclude this irrelevant information, and to isolate the whole of the relevant information contained in the data.

When we speak of the *probability* of a certain object fulfilling a certain condition, we imagine all such objects to be divided into two classes, according as they do or do not fulfil the condition. This is the only characteristic in them of which we take cognisance. For this reason probability is the most elementary of statistical concepts. It is a parameter which specifies a simple dichotomy in an infinite hypothetical population, and it represents neither more nor less than the frequency ratio which we imagine such a population to exhibit. For example, when we say that the probability of throwing a five with a die is one-sixth, we must not be taken to mean that of any six throws with that die one and one only will necessarily be a five; ar that of any six million throws, exactly one million will be fives; but that of a hypothetical population of an infinite number of throws, with the die in its original condition, exactly one-sixth will be fives. Our statement will not then contain any false assumption about the actual die, as that it will not wear out with continued use, or any notion of approximation, as in estimating the probability from a finite sample, although this notion may be logically developed once the meaning of probability is apprehended.

The concept of a *discontinuous frequency distribution* is merely an extension of that of a simple dichotomy, for though the number of classes into which the population is divided may be infinite, yet the frequency in each class bears a finite ratio to that of the whole population. In *frequency curves*, however, a second infinity is introduced. No finite sample has a frequency curve: a finite sample may be represented by a histogram, or by a frequency polygon, which to the eye more and more resembles a curve, as the size of the sample is increased. To reach a true curve, not only would an infinite number of individuals have to be placed in each class, but the number of classes (arrays) into which the population is divided must be made infinite. Consequently, it should be clear that the concept of a frequency curve includes that of a hypothetical infinite population, distributed according to a mathematical law, represented by the curve. This law is specified by assigning to each element of the abscissa the corresponding element of probability. Thus, in the case of the normal distribution, the probability of an observation falling in the range dx, is

$$\frac{1}{\sigma\sqrt{2\pi}} e^{-(x-m)^2/2\sigma^2} \, dx,$$

in which expression x is the value of the variate, while m, the mean, and σ, the standard deviation, are the two parameters by which the hypothetical population is specified. If a sample of n be taken from such a population, the data comprise n independent facts. The statistical process of the reduction of these data is designed to extract from them all relevant information respecting the values of m and σ, and to reject all other information as irrelevant.

It should be noted that there is no falsehood in interpreting any set of independent measurements as a random sample from an infinite population; for any such set of numbers are a random sample from the totality of numbers produced by the same matrix of causal conditions: the hypothetical population which we are studying is an aspect of the totality of the effects of these conditions, of whatever nature they may be. The postulate of randomness thus resolves itself into the question, "Of what population is this a random sample?" which must frequently be asked by every practical statistician.

It will be seen from the above examples that the process of the reduction of data is, even in the simplest cases, performed by interpreting the available observations as a sample from a hypothetical infinite population; this is *a fortiori* the case when we have more than one variate, as when we are seeking the values of coefficients of correlation. There is one point, however, which may be briefly mentioned here in advance, as it has been the cause of some confusion. In the example of the frequency curve mentioned above, we took it for granted that the values of both the mean and the standard deviation of the population were relevant to the inquiry. This is often the case, but it sometimes happens that only one of these quantities, for example the standard deviation, is required for discussion. In the same way an infinite normal population of two correlated variates will usually require five parameters for its specification, the two means, the two standard deviations, and the correlation; of these often only the correlation is required, or if not alone of interest, it is discussed without reference to the other four quantities. In such cases an alteration has been made in what is, and what is not, relevant, and it is not surprising that certain small corrections should appear, or not, according as the other parameters of the hypothetical surface are or are not deemed relevant. Even more clearly is this discrepancy shown when, as in the treatment of such fourfold tables as exhibit the recovery from smallpox of vaccinated and unvaccinated patients, the method of one school of statisticians treats the proportion of vaccinated as relevant, while others dismiss it as irrelevant to the inquiry. (3.)

3. The Problems of Statistics

The problems which arise in reduction of data may be conveniently divided into three types:

(1) Problems of Specification. These arise in the choice of the mathematical form of the population.
(2) Problems of Estimation. These involve the choice of methods of calculating from a sample statistical derivates, or as we shall call them statistics, which are designed to estimate the values of the parameters of the hypothetical population.
(3) Problems of Distribution. These include discussions of the distribution of statistics derived from samples, or in general any functions of quantities whose distribution is known.

It will be clear that when we know (1) what parameters are required to specify the population from which the sample is drawn, (2) how best to calculate from the sample estimates of these parameters, and (3) the exact form of the distribution, in different samples, of our derived statistics, then the theoretical aspect of the treatment of any particular body of data has been completely elucidated.

As regards problems of specification, these are entirely a matter for the practical statistician, for those cases where the qualitative nature of the hypothetical population is known do not involve any problems of this type. In other cases we may know by experience what forms are likely to be suitable, and the adequacy of our choice may be tested *a posteriori*. We must confine ourselves to those forms which we know how to handle, or for which any tables which may be necessary have been constructed. More or less elaborate forms will be suitable according to the volume of the data. Evidently these are considerations the nature of which may change greatly during the work of a single generation. We may instance the development by Pearson of a very extensive system of skew curves, the elaboration of a method of calculating their parameters, and the preparation of the necessary tables, a body of work which has enormously extended the power of modern statistical practice, and which has been, by pertinacity and inspiration alike, practically the work of a single man. Nor is the introduction of the Pearsonian system of frequency curves the only contribution which their author has made to the solution of problems of specification: of even greater importance is the introduction of an objective criterion of goodness of fit. For empirical as the specification of the hypothetical population may be, this empiricism is cleared of its dangers if we can apply a rigorous and objective test of the adequacy with which the proposed population represents the whole of the available facts. Once a statistic, suitable for applying such a test, has been chosen, the exact form of its distribution in random samples must be investigated, in order that we may evaluate the probability that a worse fit should be obtained from a random sample of a population of the type considered. The possibility of developing complete and self-contained tests of goodness of fit deserves very careful consideration, since therein lies our justification for the free use which is made of empirical frequency formulae. Problems of distribution of great mathematical difficulty have to be faced in this direction.

Although problems of estimation and of distribution may be studied separately, they are intimately related in the development of statistical methods. Logically problems of distribution should have prior consideration, for the study of the random distribution of different suggested statistics, derived from samples of a given size, must guide us in the choice of which statistic it is most profitable to calculate. The fact is, however, that very little progress has been made in the study of the distribution of statistics derived from samples. In 1900 Pearson (15) gave the exact form of the distribution of χ^2, the Pearsonian test of goodness of fit, and in 1915 the same author published (18) a similar result of more general scope, valid when the observations are regarded as subject to linear constraints. By an easy adaptation (17) the tables of probability derived from this formula may be made available for the more numerous cases in which linear constraints are imposed upon the hypothetical population by the means which we employ in its reconstruction. The distribution of the mean of samples of n from a normal population has long been known, but in 1908 "Student" (4) broke new ground by calculating the distribution of the ratio which the deviation of the mean from its population value bears to the standard deviation calculated from the sample. At the same time he gave the exact form of the distribution in samples of the standard deviation. In 1915 Fisher (5) published the curve of distribution of the correlation coefficient for the standard method of calculation, and in 1921 (6) he published the corresponding series of curves for intraclass correlations. The brevity of this list is emphasised by the absence of investigation of other important statistics, such as the regression coefficients, multiple correlations, and the correlation ratio. A formula for the probable error of any statistic is, of course, a practical necessity, if that statistic is to be of service: and in the majority of cases such formulae have been found, chiefly by the labours of Pearson and his school, by a first approximation, which describes the distribution with sufficient accuracy if the sample is sufficiently large. Problems of distribution, other than the distribution of statistics, used to be not uncommon as examination problems in probability, and the physical importance of problems of this type may be exemplified by the chemical laws of mass action, by the statistical mechanics of Gibbs, developed by Jeans in its application to the theory of gases, by the electron theory of Lorentz, and by Planck's development of the theory of quanta, although in all these applications the methods employed have been, from the statistical point of view, relatively simple.

The discussions of theoretical statistics may be regarded as alternating between problems of estimation and problems of distribution. In the first place a method of calculating one of the population parameters is devised from common-sense considerations: we next require to know its probable error, and therefore an approximate solution of the distribution, in samples, of the statistic calculated. It may then become apparent that other statistics may be used as estimates of the same parameter. When the probable errors of these statistics are compared, it is usually found that, in large samples, one particular method of calculation gives a result less subject to random errors than those given by other methods of calculation. Attacking the problem more thoroughly, and calculating the surface of distribution of any two statistics, we may find that the whole of the relevant information contained in one is contained in the other: or, in other words, that when once we know the other, knowledge of the first gives us no further information as to the value of the parameter. Finally it may be possible to prove, as in the case of the Mean Square Error, derived from a sample of normal population (7), that a particular statistic summarises the whole of the information relevant to the corresponding parameter, which the sample contains. In such a case the problem of estimation is completely solved.

4. Criteria of Estimation

The common-sense criterion employed in problems of estimation may be stated thus:—That when applied to the whole population the derived statistic should be equal to the parameter. This may be called the *Criterion of Consistency*. It is often the only test applied: thus, in estimating the standard deviation of a normally distributed population, from an ungrouped sample, either of the two statistics—

$$\sigma_1 = \frac{1}{n} \sqrt{\frac{\pi}{2}} S(|x - \bar{x}|) \qquad \text{(Mean error)}$$

and

$$\sigma_2 = \sqrt{\frac{1}{n} S(x - \bar{x})^2} \qquad \text{(Mean square error)}$$

will lead to the correct value, σ, when calculated from the whole population. They both thus satisfy the criterion of consistency, and this has led many computers to use the first formula, although the result of the second has 14 per cent. greater weight (7), and the labour of increasing the number of observations by 14 per cent. can seldom be less than that of applying the more accurate formula.

Consideration of the above example will suggest a second criterion, namely: —That in large samples, when the distributions of the statistics tend to normality, that statistic is to be chosen which has the least probable error.

This may be called the *Criterion of Efficiency*. It is evident that if for large samples one statistic has a probable error double that of a second, while both are proportional to $n^{-1/2}$, then the first method applied to a sample of $4n$ values will be no more accurate than the second applied to a sample of any n values. If the second method makes use of the whole of the information available, the first makes use of only one-quarter of it, and its efficiency may therefore be said to be 25 per cent. To calculate the efficiency of any given method, we must therefore know the probable error of the statistic calculated by that method, and that of the most efficient statistic which could be used. The square of the ratio of these two quantities then measures the efficiency.

The criterion of efficiency is still to some extent incomplete, for different methods of calculation may tend to agreement for large samples, and yet differ for all finite samples. The complete criterion suggested by our work on the mean square error (7) is:

That the statistic chosen should summarise the whole of the relevant information supplied by the sample.

This may be called the *Criterion of Sufficiency*.

In mathematical language we may interpret this statement by saying that if θ be the parameter to be estimated, θ_1 a statistic which contains the whole of the information as to the value of θ, which the sample supplies, and θ_2 any other statistic, then the surface of distribution of pairs of values of θ_1 and θ_2, for a given value of θ, is such that for a given value of θ_1, the distribution of θ_2 does not involve θ. In other words, when θ_1 is known, knowledge of the value of θ_2 throws no further light upon the value of θ.

It may be shown that a statistic which fulfils the criterion of sufficiency will also fulfil the criterion of efficiency, when the latter is applicable. For, if this be so, the distribution of the statistics will in large samples be normal, the standard deviations being proportional to $n^{-1/2}$. Let this distribution be

$$df = \frac{1}{2\pi\sigma_1\sigma_2\sqrt{1-r^2}} e^{-1/(1-r^2)\{(\overline{\theta_1-\theta^2})/2\sigma_1^2 - (2r\overline{\theta_1-\theta}\,\overline{\theta_2-\theta})/2\sigma_1\sigma_2 + (\overline{\theta_2-\theta^2})/2\sigma_2^2\}} \, d\theta_1 \, d\theta_2,$$

then the distribution of θ_1 is

$$df = \frac{1}{\sigma_1\sqrt{2\pi}} e^{-(\overline{\theta_1-\theta^2})/2\sigma_1^2} \, d\theta_1,$$

so that for a given value of θ_1 the distribution of θ_2 is

$$df = \frac{1}{\sigma_2\sqrt{2\pi}\sqrt{1-r^2}} e^{-1/(2\overline{1-r^2})\{(r\overline{\theta_1-\theta})/\sigma_1 - (\overline{\theta_2-\theta})/\sigma_2\}^2} \, d\theta_2;$$

and if this does not involve θ, we must have

$$r\sigma_2 = \sigma_1;$$

showing that σ_1 is necessarily less than σ_2, and that the efficiency of θ_2 is measured by r^2, when r is its correlation in large samples with θ_1.

Besides this case we shall see that the criterion of sufficiency is also applicable to finite samples, and to those cases when the weight of a statistic is not proportional to the number of the sample from which it is calculated.

5. Examples of the Use of the Criterion of Consistency

In certain cases the criterion of consistency is sufficient for the solution of problems of estimation. An example of this occurs when a fourfold table is interpreted as representing the double dichotomy of a normal surface. In this case the dichotomic ratios of the two variates, together with the correlation, completely specify the four fractions into which the population is divided. If these are equated to the four fractions into which the sample is divided, the correlation is determined uniquely.

In other cases where a small correction has to be made, the amount of the correction is not of sufficient importance to justify any great refinement in estimation, and it is sufficient to calculate the discrepancy which appears when the uncorrected method is applied to the whole population. Of this nature is Sheppard's correction for grouping, and it will illustrate this use of the criterion of consistency if we derive formulae for this correction without approximation.

Let ξ be the value of the variate at the mid point of any group, a the interval of grouping, and x the true value of the variate at any point, then the kth moment of an infinite grouped sample is

$$\sum_{p=-\infty}^{p=\infty} \int_{\xi-(1/2)a}^{\xi+(1/2)a} \xi^k f(x)\, dx,$$

in which $f(x)\, dx$ is the frequency, in any element dx, of the ungrouped population, and

$$\xi = \left(p + \frac{\theta}{2\pi}\right)a,$$

p being any integer.

Evidently the kth moment is periodic in θ, we will therefore equate it to

$$A_0 + A_1 \sin\theta + A_2 \sin 2\theta \ldots$$
$$+ B_1 \cos\theta + B_2 \cos 2\theta \ldots.$$

Then

$$A_0 = \frac{1}{2\pi} \sum_{p=-\infty}^{p=\infty} \int_0^{2\pi} d\theta \int_{\xi-(1/2)a}^{\xi+(1/2)a} \xi^k f(x)\, dx$$

$$A_s = \frac{1}{\pi} \sum_{p=-\infty}^{p=\infty} \int_0^{2\pi} \sin s\theta\, d\theta \int_{\xi-(1/2)a}^{\xi+(1/2)a} \xi^k f(x)\, dx,$$

$$B_s = \frac{1}{\pi} \sum_{p=-\infty}^{p=\infty} \int_0^{2\pi} \cos s\theta\, d\theta \int_{\xi-(1/2)a}^{\xi+(1/2)a} \xi^k f(x)\, dx.$$

But

$$\theta = \frac{2\pi}{a}\xi - 2\pi p,$$

therefore

$$d\theta = \frac{2\pi}{a} d\xi,$$

$$\sin s\theta = \sin \frac{2\pi}{a} s\xi,$$

$$\cos s\theta = \cos \frac{2\pi}{a} s\xi,$$

hence

$$A_0 = \frac{1}{a}\int_{-\infty}^{\infty} d\xi \int_{\xi-(1/2)a}^{\xi+(1/2)a} \xi^k f(x)\, dx = \frac{1}{a}\int_{-\infty}^{\infty} f(x)\, dx \int_{\xi-(1/2)a}^{\xi+(1/2)a} \xi^k\, d\xi.$$

Inserting the values 1, 2, 3 and 4 for k, we obtain for the aperiodic terms of the four moments of the grouped population

$$_1A_0 = \int_{-\infty}^{\infty} xf(x)\, dx,$$

$$_2A_0 = \int_{-\infty}^{\infty} \left(x^2 + \frac{a^2}{12}\right) f(x)\, dx,$$

$$_3A_0 = \int_{-\infty}^{\infty} \left(x^3 + \frac{a^2 x}{4}\right) f(x)\, dx,$$

$$_4A_0 = \int_{-\infty}^{\infty} \left(x^4 + \frac{a^2 x^2}{2} + \frac{a^4}{80}\right) f(x)\, dx.$$

If we ignore the periodic terms, these equations lead to the ordinary Sheppard corrections for the second and fourth moment. The nature of the approximation involved is brought out by the periodic terms. In the absence of high contact at the ends of the curve, the contribution of these will, of course, include the terms given in a recent paper by Pearson (8); but even with high contact it is of interest to see for what degree of coarseness of grouping the periodic terms become sensible.

Now

$$A_s = \frac{1}{\pi} \sum_{p=-\infty}^{p=\infty} \int_0^{2\pi} \sin s\theta \, d\theta \int_{\xi-(1/2)a}^{\xi+(1/2)a} \xi^k f(x)\, dx,$$

$$= \frac{2}{a} \int_{-\infty}^{\infty} \sin \frac{2\pi s\xi}{a} \, d\xi \int_{\xi-(1/2)a}^{\xi+(1/2)a} \xi^k f(x)\, dx,$$

$$= \frac{2}{a} \int_{-\infty}^{\infty} f(x)\, dx \int_{\xi-(1/2)a}^{\xi+(1/2)a} \xi^k \sin \frac{2\pi s\xi}{a} \, d\xi.$$

But

$$\frac{2}{a} \int_{x-(1/2)a}^{x+(1/2)a} \xi \sin \frac{2\pi s\xi}{a} \, d\xi = -\frac{a}{\pi s} \cos \frac{2\pi sx}{a} \cos \pi s,$$

therefore

$$_1A_s = (-)^{s+1} \frac{a}{\pi s} \int_{-\infty}^{\infty} \cos \frac{2\pi sx}{a} f(x)\, dx;$$

similarly the other terms of the different moments may be calculated.

For a normal curve referred to the true mean

$$_1A_s = (-)^{s+1} \frac{2\varepsilon}{s} e^{-(s^2\sigma^2/2\varepsilon^2)},$$

$$_1B_s = 0,$$

in which

$$a = 2\pi\varepsilon.$$

The error of the mean is therefore

$$-2\varepsilon(e^{-(\sigma^2/2\varepsilon^2)} \sin \theta - \tfrac{1}{2}e^{-(4\sigma^2/2\varepsilon^2)} \sin 2\theta + \tfrac{1}{3}e^{-(9\sigma^2/2\varepsilon^2)} \sin 3\theta - \cdots).$$

To illustrate a coarse grouping, take the group interval equal to the standard deviation: then

$$\varepsilon = \frac{\sigma}{2\pi},$$

and the error is

$$-\frac{\sigma}{\pi}e^{-2\pi^2} \sin \theta$$

with sufficient accuracy. The standard error of the mean being $\dfrac{\sigma}{\sqrt{n}}$, we may calculate the size of the sample for which the error due to the periodic terms becomes equal to one-tenth of the standard error, by putting

$$\frac{\sigma}{10\sqrt{n}} = \frac{\sigma}{\pi}e^{-2\pi^2},$$

whence

$$n = \frac{\pi^2}{100}e^{4\pi^2} = 13{,}790 \times 10^{12}.$$

For the second moment

$$B_s = (-)^s 4\left(\sigma^2 + \frac{\varepsilon^2}{s^2}\right)e^{-(s^2\sigma^2/2\varepsilon^2)},$$

and, if we put

$$\frac{\sqrt{2\sigma^2}}{10\sqrt{n}} = 4\sigma^2 e^{-2\pi^2},$$

there results

$$n = \tfrac{1}{800}e^{4\pi^2} = 175 \times 10^{12}.$$

The error, while still very minute, is thus more important for the second than for the first moment.

For the third moment

$$A_s = (-)^s \frac{6\sigma^4 s}{\varepsilon}\left\{1 + \frac{\varepsilon^2}{s^2\sigma^2} - \frac{\varepsilon^4}{3s^4\sigma^4}(\pi^2 s^2 - 6)\right\}e^{-(s^2\sigma^2/2\varepsilon^2)};$$

putting

$$\frac{\sqrt{15\sigma^3}}{10\sqrt{n}} = 12\pi\sigma^3 e^{-2\pi^2},$$

$$n = \frac{1}{960\pi^2}e^{4\pi^2} = 147 \times 10^{12}.$$

While for the fourth moment

$$B_s = (-)^{s+1}\frac{8\sigma^6 s^2}{\varepsilon^2}\left\{1 - (\pi^2 s^2 - 3)\frac{\varepsilon^4}{s^4\sigma^4} - (\pi^2 s^2 - 6)\frac{\varepsilon^6}{s^6\sigma^6}\right\}e^{-(s^2\sigma^2/2\varepsilon^2)},$$

so that, if we put,

$$\frac{\sqrt{96\sigma^4}}{10\sqrt{n}} = 32\pi^2\sigma^4 e^{-2\pi^2},$$

$$n = \frac{3}{3200\pi^4}e^{4\pi^2} = 1.34 \times 10^{12}.$$

In a similar manner the exact form of Sheppard's correction may be found for other curves; for the normal curve we may say that the periodic terms are exceedingly minute so long as a is less than σ, though they increase very rapidly if a is increased beyond this point. They are of increasing importance as higher moments are used, not only absolutely, but relatively to the increasing probable errors of the higher moments. The principle upon which the correction is based is merely to find the error when the moments are calculated from an infinite grouped sample; the corrected moment therefore fulfils the criterion of consistency, and so long as the correction is small no greater refinement is required.

Perhaps the most extended use of the criterion of consistency has been developed by Pearson in the "Method of Moments." In this method, which is without question of great practical utility, different forms of frequency curves are fitted by calculating as many moments of the sample as there are parameters to be evaluated. The parameters chosen are those of an infinite population of the specified type having the same moments as those calculated from the sample.

The system of curves developed by Pearson has four variable parameters, and may be fitted by means of the first four moments. For this purpose it is necessary to confine attention to curves of which the first four moments are finite; further, if the accuracy of the fourth moment should increase with the size of the sample, that is, if its probable error should not be infinitely great, the first eight moments must be finite. This restriction requires that the class of distribution in which this condition is not fulfilled should be set aside as "heterotypic," and that the fourth moment should become practically valueless as this class is approached. It should be made clear, however, that there is nothing anomalous about these so-called "heterotypic" distributions except the fact that the method of moments cannot be applied to them. Moreover, for that class of distribution to which the method can be applied, it has not been shown, except in the case of the normal curve, that the best values will be obtained by the method of moments. The method will, in these cases, certainly be serviceable in yielding an approximation, but to discover whether this approximation is a good or a bad one, and to improve it, if necessary, a more adequate criterion is required.

A single example will be sufficient to illustrate the practical difficulty al-

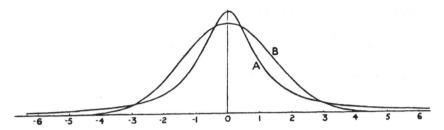

Figure 1. Symmetrical error curves of equal intrinsic accuracy:

$$A \ldots \ldots df = \frac{1}{\pi} \frac{dx}{1 + x^2}.$$

$$B \ldots \ldots df = \frac{1}{2\sqrt{\pi}} e^{-x^2/4}$$

luded to above. If a point P lie at known (unit) distance from a straight line AB, and lines be drawn at random through P, then the distribution of the points of intersection with AB will be distributed so that the frequency in any range dx is

$$df = \frac{1}{\pi} \cdot \frac{dx}{1 + (x - m)^2},$$

in which x is the distance of the infinitesimal range dx from a fixed point 0 on the line, and m is the distance, from this point, of the foot of the perpendicular PM. The distribution will be a symmetrical one (Type VII.) having its centre at $x = m$ (fig. 1). It is therefore a perfectly definite problem to estimate the value of m (to find the best value of m) from a random sample of values of x. We have stated the problem in its simplest possible form: only one parameter is required, the middle point of the distribution. By the method of moments, this should be given by the first moment, that is by the mean of the observations: such would seem to be at least a good estimate. It is, however, entirely valueless. The distribution of the mean of such samples is in fact the same, identically, as that of a single observation. In taking the mean of 100 values of x, we are no nearer obtaining the value of m than if we had chosen any value of x out of the 100. The problem, however, is not in the least an impracticable one: clearly from a large sample we ought to be able to estimate the centre of the distribution with some precision; the mean, however, is an entirely useless statistic for the purpose. By taking the median of a large sample, a fair approximation is obtained, for the standard error of the median of a large sample of n is $\frac{\pi}{2\sqrt{n}}$, which, alone, is enough to show that by adopting adequate statistical methods it must be possible to estimate the value for m, with increasing accuracy, as the size of the sample is increased.

This example serves also to illustrate the practical difficulty which observers often find, that a few extreme observations appear to dominate the value of the mean. In these cases the rejection of extreme values is often advocated, and it may often happen that gross errors are thus rejected. As a statistical measure, however, the rejection of observations is too crude to be defended: and unless there are other reasons for rejection than mere divergence from the majority, it would be more philosophical to accept these extreme values, not as gross errors, but as indications that the distribution of errors is not normal. As we shall show, the only Pearsonian curve for which the mean is the best statistic for locating the curve, is the normal or gaussian curve of errors. If the curve is not of this form the mean is not necessarily, as we have seen, of any value whatever. The determination of the true curves of variation for different types of work is therefore of great practical importance, and this can only be done by different workers recording their data in full without rejections, however they may please to treat the data so recorded. Assuredly an observer need be exposed to no criticism, if after recording data which are not probably normal in distribution, he prefers to adopt some value other than the arithmetic mean.

6. Formal Solution of Problems of Estimation

The form in which the criterion of sufficiency has been presented is not of direct assistance in the solution of problems of estimation. For it is necessary first to know the statistic concerned and its surface of distribution, with an infinite number of other statistics, before its sufficiency can be tested. For the solution of problems of estimation we require a method which for each particular problem will lead us automatically to the statistic by which the criterion of sufficiency is satisfied. Such a method is, I believe, provided by the Method of Maximum Likelihood, although I am not satisfied as to the mathematical rigour of any proof which I can put forward to that effect. Readers of the ensuing pages are invited to form their own opinion as to the possibility of the method of the maximum likelihood leading in any case to an insufficient statistic. For my own part I should gladly have withheld publication until a rigorously complete proof could have been formulated; but the number and variety of the new results which the method discloses press for publication, and at the same time I am not insensible of the advantage which accrues to Applied Mathematics from the co-operation of the Pure Mathematician, and this co-operation is not infrequently called forth by the very imperfections of writers on Applied Mathematics.

If in any distribution involving unknown parameters θ_1, θ_2, θ_3, ..., the chance of an observation falling in the range dx be represented by

$$f(x, \theta_1, \theta_2, \ldots)\, dx,$$

then the chance that in a sample of n, n_1 fall in the range dx_1, n_2 in the range dx_2, and so on, will be

$$\frac{n!}{\prod(n_p!)}\prod\{f(x_p, \theta_1, \theta_2, \ldots)\, dx_p\}^{n_p}.$$

The method of maximum likelihood consists simply in choosing that set of values for the parameters which makes this quantity a maximum, and since in this expression the parameters are only involved in the function f, we have to make

$$S(\log f)$$

a maximum for variations of $\theta_1, \theta_2, \theta_3$, &c. In this form the method is applicable to the fitting of populations involving any number of variates, and equally to discontinuous as to continuous distributions.

In order to make clear the distinction between this method and that of Bayes, we will apply it to the same type of problem as that which Bayes discussed, in the hope of making clear exactly of what kind is the information which a sample is capable of supplying. This question naturally first arose, not with respect to populations distributed in frequency curves and surfaces, but with respect to a population regarded as divided into two classes only, in fact in problems of *probability*. A certain proportion, p, of an infinite population is supposed to be of a certain kind, *e.g.*, "successes," the remainder are then "failures." A sample of n is taken and found to contain x successes and y failures. The chance of obtaining such a sample is evidently

$$\frac{n!}{x!\, y!} p^x (1-p)^y.$$

Applying the method of maximum likelihood, we have

$$S(\log f) = x \log \hat{p} + y \log(1 - \hat{p})$$

whence, differentiating with respect to p, in order to make this quantity a maximum,

$$\frac{x}{p} = \frac{y}{1 - \hat{p}}, \quad \text{or} \quad \hat{p} = \frac{x}{n}.$$

The question then arises as to the accuracy of this determination. This question was first discussed by Bayes (10), in a form which we may state thus. After observing this sample, when we know $\hat{p}$, what is the *probability* that p lies in any range dp? In other words, what is the frequency distribution of the values of p in populations which are selected by the restriction that a sample of n taken from each of them yields x successes. Without further data, as Bayes perceived, this problem is insoluble. To render it capable of mathematical treatment, Bayes introduced the *datum*, that among the populations upon which the experiment was tried, those in which p lay in the range dp were equally frequent for all equal ranges dp. The probability that the value of p lay in any range dp was therefore assumed to be simply dp, before the sample was taken. After the selection effected by observing the sample, the probability is clearly proportional to

$$p^x(1 - p)^y\, dp.$$

After giving this solution, based upon the particular datum stated, Bayes adds a *scholium* the purport of which would seem to be that in the absence of all knowledge save that supplied by the sample, it is reasonable to assume this particular *a priori* distribution of p. The *result*, the *datum*, and the *postulate* implied by the *scholium*, have all been somewhat loosely spoken of as Bayes' Theorem.

The postulate would, if true, be of great importance in bringing an immense variety of questions within the domain of probability. It is, however, evidently extremely arbitrary. Apart from evolving a vitally important piece of knowledge, that of the exact form of the distribution of values of p, out of an assumption of complete ignorance, it is not even a unique solution. For we might never have happened to direct our attention to the particular quantity p: we might equally have measured probability upon an entirely different scale. If, for instance,

$$\sin \theta = 2p - 1,$$

the quantity, θ, measures the degree of probability, just as well as p, and is even, for some purposes, the more suitable variable. The chance of obtaining a sample of x successes and y failures is now

$$\frac{n!}{2^n x! y!}(1 + \sin \theta)^x (1 - \sin \theta)^y;$$

applying the method of maximum likelihood,

$$S(\log f) = x \log(1 + \sin \theta) + y \log(1 - \sin \theta) - n \log 2,$$

and differentiating with respect to θ,

$$\frac{x \cos \theta}{1 + \sin \theta} = \frac{y \cos \theta}{1 - \sin \theta}, \quad \text{whence} \quad \sin \theta = \frac{x - y}{n},$$

an exactly equivalent solution to that obtained using the variable p. But what *a priori* assumption are we to make as to the distribution of θ? Are we to assume that θ is equally likely to lie in all equal ranges $d\theta$? In this case the *a priori* probability will be $d\theta/\pi$, and that after making the observations will be proportional to

$$(1 + \sin \theta)^x (1 - \sin \theta)^y\, d\theta.$$

But if we interpret this in terms of p, we obtain

$$p^x(1 - p)^y \frac{dp}{\sqrt{p(1 - p)}} = p^{x-1/2}(1 - p)^{y-1/2} \, dp,$$

a result inconsistent with that obtained previously. In fact, the distribution previously assumed for p was equivalent to assuming the special distribution for θ,

$$df = \frac{\cos \theta}{2} \, d\theta,$$

the arbitrariness of which is fully apparent when we use any variable other than p.

In a less obtrusive form the same species of arbitrary assumption underlies the method known as that of inverse probability. Thus, if the same observed result A might be the consequence of one or other of two hypothetical conditions X and Y, it is assumed that the probabilities of X and Y are in the same ratio as the probabilities of A occurring on the two assumptions, X is true, Y is true. This amounts to assuming that before A was observed, it was known that our universe had been selected at random from an infinite population in which X was true in one half, and Y true in the other half. Clearly such an assumption is entirely arbitrary, nor has any method been put forward by which such assumptions can be made even with consistent uniqueness. There is nothing to prevent an irrelevant distinction being drawn among the hypothetical conditions represented by X, so that we have to consider two hypothetical possibilities X_1 and X_2, on both of which A will occur with equal frequency. Such a distinction should make no difference whatever to our conclusions; but on the principle of inverse probability it does so, for if previously the relative probabilities were reckoned to be in the ratio x to y, they must now be reckoned $2x$ to y. Nor has any criterion been suggested by which it is possible to separate such irrelevant distinctions from those which are relevant.

There would be no need to emphasise the baseless character of the assumptions made under the titles of inverse probability and Bayes' Theorem in view of the decisive criticism to which they have been exposed at the hands of Boole, Venn, and Chrystal, were it not for the fact that the older writers, such as Laplace and Poisson, who accepted these assumptions, also laid the foundations of the modern theory of statistics, and have introduced into their discussions of this subject ideas of a similar character. I must indeed plead guilty in my original statement of the Method of the Maximum Likelihood (9) to having based my argument upon the principle of inverse probability; in the same paper, it is true, I emphasised the fact that such inverse probabilities were relative only. That is to say, that while we might speak of one value of p as having an inverse probability three times that of another value of p, we might on no account introduce the differential element dp, so as to be able to say that it was three times as probable that p should lie in one rather than the other of two equal elements. Upon consideration, therefore, I perceive that the word probability is wrongly used in such a connection: probability is a ratio of frequencies, and about the frequencies of such values we can know nothing whatever. We must return to the actual fact that one value of p, of the frequency of which we know nothing, would yield the observed result three times as frequently as would another value of p. If we need a word to characterise this relative property of different values of p, I suggest that we may speak without confusion of the *likelihood* of one value of p being thrice the likelihood of another, bearing always in mind that likelihood is not here used loosely as a synonym of probability, but simply to express the relative frequencies with which such values of the hypothetical quantity p would in fact yield the observed sample.

The solution of the problems of calculating from a sample the parameters of the hypothetical population, which we have put forward in the method of maximum likelihood, consists, then, simply of choosing such values of these parameters as have the maximum likelihood. Formally, therefore, it resembles the calculation of the mode of an inverse frequency distribution. This resemblance is quite superficial: if the scale of measurement of the hypothetical quantity be altered, the mode must change its position, and can be brought to have any value, by an appropriate change of scale; but the optimum, as the position of maximum likelihood may be called, is entirely unchanged by any such transformation. Likelihood also differs from probability* in that it is not a differential element, and is incapable of being integrated: it is assigned to a particular point of the range of variation, not to a particular element of it. There is therefore an absolute measure of probability in that the unit is chosen so as to make all the elementary probabilities add up to unity. There is no such absolute measure of likelihood. It may be convenient to assign the value unity to the maximum value, and to measure other likelihoods by comparison, but there will then be an infinite number of values whose likelihood is greater than one-half. The sum of the likelihoods of admissible values will always be infinite.

Our interpretation of Bayes' problem, then, is that the likelihood of any value of p is proportional to

$$p^x(1 - p)^y,$$

and is therefore a maximum when

$$p = \frac{x}{n},$$

* It should be remarked that likelihood, as above defined, is not only fundamentally distinct from mathematical probability, but also from the logical "probability" by which Mr. Keynes (21) has recently attempted to develop a method of treatment of uncertain inference, applicable to those cases where we lack the statistical information necessary for the application of mathematical probability. Although, in an important class of cases, the likelihood may be held to measure the degree of our rational belief in a conclusion, in the same sense as Mr. Keynes' "probability," yet since the latter quantity is constrained, somewhat arbitrarily, to obey the addition theorem of mathematical probability, the likelihood is a quantity which falls definitely outside its scope.

which is the best value obtainable from the sample; we shall term this the *optimum* value of *p*. Other values of *p* for which the likelihood is not much less cannot, however, be deemed unlikely values for the true value of *p*. We do not, and cannot, know, from the information supplied by a sample, anything about the probability that *p* should lie between any named values.

The reliance to be placed on such a result must depend upon the frequency distribution of *x*, in different samples from the same population. This is a perfectly objective statistical problem, of the kind we have called problems of distribution; it is, however, capable of an approximate solution, directly from the mathematical form of the likelihood.

When for large samples the distribution of any statistic, θ_1, tends to normality, we may write down the chance for a given value of the parameter θ, that θ_1 should lie in the range $d\theta_1$ in the form

$$\Phi = \frac{1}{\sigma\sqrt{2\pi}} e^{-(\theta_1-\theta)^2/2\sigma^2} \, d\theta_1.$$

The mean value of θ_1 will be the true value θ, and the standard deviation is σ, the sample being assumed sufficiently large for us to disregard the dependence of σ upon θ.

The likelihood of any value, θ, is proportional to

$$e^{-(\theta_1-\theta)^2/2\sigma^2},$$

this quantity having its maximum value, unity, when

$$\theta = \theta_1;$$

for

$$\frac{\partial}{\partial\theta} \log \Phi = \frac{\theta_1 - \theta}{\sigma^2}.$$

Differentiating now a second time

$$\frac{\partial^2}{\partial\theta^2} \log \Phi = -\frac{1}{\sigma^2}.$$

Now Φ stands for the total frequency of all samples for which the chosen statistic has the value θ_1, consequently $\Phi = S'(\phi)$, the summation being taken over all such samples, where ϕ stands for the probability of occurrence of a certain specified sample. For which we know that

$$\log \phi = C + S(\log f),$$

the summation being taken over the individual members of the sample.

If now we expand log *f* in the form

$$\log f(\theta) = \log f(\theta_1) + \overline{\theta - \theta_1} \frac{\partial}{\partial\theta} \log f(\theta_1) + \frac{\overline{\theta - \theta_1}^2}{\underline{|2}} \frac{\partial^2}{\partial\theta^2} \log f(\theta_1) + \cdots,$$

or

$$\log f = \log f_1 + a\overline{\theta - \theta_1} + \frac{b}{2}\overline{\theta - \theta_1}^2 + \cdots,$$

we have

$$\log \phi = C + \overline{\theta - \theta_1}S(a) + \tfrac{1}{2}\overline{\theta - \theta_1}^2 S(b) + \cdots;$$

now for optimum statistics

$$S(a) = 0,$$

and for sufficiently large samples $S(b)$ differs from $n\bar{b}$ only by a quantity of order $\sqrt{n\sigma_b}$; moreover, $\theta - \theta_1$ being of order $n^{-1/2}$, the only terms in $\log \phi$ which are not reduced without limit, as n is increased, are

$$\log \phi = C + \tfrac{1}{2}n\bar{b}\,\overline{\theta - \theta_1}^2;$$

hence

$$\phi \propto e^{(1/2)n\bar{b}\,\overline{\theta - \theta_1}^2}.$$

Now this factor is constant for all samples which have the same value of θ_1, hence the variation of Φ with respect to θ is represented by the same factor, and consequently

$$\log \Phi = C' + \tfrac{1}{2}n\bar{b}\,\overline{\theta - \theta_1}^2;$$

whence

$$-\frac{1}{\sigma_{\theta_1}^2} = \frac{\partial^2}{\partial \theta^2} \log \Phi = n\bar{b},$$

where

$$b = \frac{\partial^2}{\partial \theta^2} \log f(\theta_1),$$

θ_1 being the optimum value of θ.

The formula

$$-\frac{1}{\sigma_{\hat\theta}^2} = x\frac{\overline{\partial^2}}{\partial \theta^2} \log f$$

supplies the most direct way known to me of finding the probable errors of statistics It may be seen that the above proof applies only to statistics obtained by the method of maximum likelihood.*

* A similar method of obtaining the standard deviations and correlations of statistics derived from large samples was developed by Pearson and Filon in 1898 (16). It is unfortunate that in this memoir no sufficient distinction is drawn between the *population* and the *sample*, in consequence of which the formulae obtained indicate that the likelihood is always a maximum (for continuous distributions) when the *mean* of each variate in the sample is equated to the corre-

For example, to find the standard deviation of

$$\hat{p} = \frac{x}{n}$$

in samples from an infinite population of which the true value is p,

$$\log f = x \log p + y \log(1 - p),$$

$$\frac{\partial}{\partial p} \log f = \frac{x}{p} - \frac{y}{1 - p},$$

$$\frac{\partial^2}{\partial p^2} \log f = -\frac{x}{p^2} - \frac{y}{1 - p^2}.$$

Now the mean value of x is pn, and of y is $(1 - p)n$, hence the mean value of $\frac{\partial^2}{\partial p^2} \log f$ is

$$-\left(\frac{1}{p} + \frac{1}{1 - p}\right)n;$$

therefore

$$\sigma_{\hat{p}}^2 = \frac{p(1 - p)}{n},$$

the well-known formula for the standard error of p.

sponding mean in the population (16, p. 232, "$A_r = 0$"). If this were so the mean would always be a sufficient statistic for location; but as we have already seen, and will see later in more detail, this is far from being the case. The same argument, indeed, is applied to all statistics, as to which nothing but their *consistency* can be truly affirmed.

The probable errors obtained in this way are those appropriate to the method of maximum likelihood, but not in other cases to statistics obtained by the method of moments, by which method the examples given were fitted. In the 'Tables for Statisticians and Biometricians' (1914), the probable errors of the constants of the Pearsonian curves are those proper to the method of moments; no mention is there made of this change of practice, nor is the publication of 1898 referred to.

It would appear that shortly before 1898 the process which leads to the correct value, of the probable errors of *optimum* statistics, was hit upon and found to agree with the probable errors of statistics found by the method of moments for *normal* curves and surfaces; without further enquiry it would appear to have been assumed that this process was valid in all cases, its directness and simplicity being peculiarly attractive. The mistake was at that time, perhaps, a natural one; but that it should have been discovered and corrected without revealing the inefficiency of the method of moments is a very remarkable circumstance.

In 1903 the correct formulae for the probable errors of statistics found by the method of moments are given in 'Biometrika' (19); references are there given to Sheppard (20), whose method is employed, as well as to Pearson and Filon (16), although both the method and the results differ from those of the latter.

7. Satisfaction of the Criterion of Sufficiency

That the criterion of sufficiency is generally satisfied by the solution obtained by the method of maximum likelihood appears from the following considerations.

If the individual values of any sample of data are regarded as co-ordinates in hyperspace, then any sample may be represented by a single point, and the frequency distribution of an infinite number of random samples is represented by a density distribution in hyperspace. If any set of statistics be chosen to be calculated from the samples, certain regions will provide identical sets of statistics; these may be called *isostatistical* regions. For any particular space element, corresponding to an actual sample, there will be a particular set of parameters for which the frequency in that element is a maximum; this will be the optimum set of parameters for that element. If now the set of statistics chosen are those which give the optimum values of the parameters, then all the elements of any part of the same isostatistical region will contain the greatest possible frequency for the same set of values of the parameters, and therefore any region which lies wholly within an isostatistical region will contain its maximum frequency for that set of values.

Now let θ be the value of any parameter, $\hat{\theta}$ the statistic calculated by the method of maximum likelihood, and θ_1 any other statistic designed to estimate the value of θ, then for a sample of given size, we may take

$$f(\theta, \hat{\theta}, \theta_1) \, d\hat{\theta} \, d\theta_1$$

to represent the frequency with which $\hat{\theta}$ and θ_1 lie in the assigned ranges $d\hat{\theta}$ and $d\theta_1$. The region $d\hat{\theta} \, d\theta_1$ evidently lies wholly in the isostatistical region $d\hat{\theta}$. Hence the equation

$$\frac{\partial}{\partial\theta} \log f(\theta, \hat{\theta}, \theta_1) = 0$$

is satisfied, irrespective of θ_1, by the value $\theta = \hat{\theta}$. This condition is satisfied if

$$f(\theta, \hat{\theta}, \theta_1) = \phi(\theta, \hat{\theta}) \cdot \phi'(\hat{\theta}, \theta_1);$$

for then

$$\frac{\partial}{\partial\theta} \log f = \frac{\partial}{\partial\theta} \log \phi,$$

and the equation for the optimum degenerates into

$$\frac{\partial}{\partial\theta} \log \phi(\theta, \hat{\theta}) = 0,$$

which does not involve θ_1.

But the factorisation of f into factors involving $(\theta, \hat{\theta})$ and $(\hat{\theta}, \theta_1)$ respectively is merely a mathematical expression of the condition of sufficiency; and it appears that any statistic which fulfils the condition of sufficiency must be a solution obtained by the method of the optimum

It may be expected, therefore, that we shall be led to a sufficient solution of problems of estimation in general by the following procedure. Write down the formula for the probability of an observation falling in the range dx in the form

$$f(\theta, x)\, dx,$$

where θ is an unknown parameter. Then if

$$L = S(\log f),$$

the summation being extended over the observed sample, L differs by a constant only from the logarithm of the likelihood of any value of θ. The most likely value, $\hat{\theta}$, is found by the equation

$$\frac{\partial L}{\partial \theta} = 0,$$

and the standard deviation of $\hat{\theta}$, by a second differentiation, from the formula

$$\frac{\partial^2 L}{\partial \theta^2} = -\frac{1}{\sigma_{\hat{\theta}}^2};$$

this latter formula being applicable only where $\hat{\theta}$ is normally distributed, as is often the case with considerable accuracy in large samples. The value $\sigma_{\hat{\theta}}$ so found is in these cases the least possible value for the standard deviation of a statistic designed to estimate the same parameter; it may therefore be applied to calculate the efficiency of any other such statistic.

When several parameters are determined simultaneously, we must equate the second differentials of L, with respect to the parameters, to the coefficients of the quadratic terms in the index of the normal expression which represents the distribution of the corresponding statistics. Thus with two parameters,

$$\frac{\partial^2 L}{\partial \theta_1^2} = -\frac{1}{1 - r_{\hat{\theta}_1 \hat{\theta}_2}^2} \cdot \frac{1}{\sigma_{\hat{\theta}_1}^2}, \qquad \frac{\partial^2 L}{\partial \theta_2^2} = -\frac{1}{1 - r_{\hat{\theta}_1 \hat{\theta}_2}^2} \cdot \frac{1}{\sigma_{\hat{\theta}_2}^2},$$

$$\frac{\partial^2 L}{\partial \theta_1 \partial \theta_2} = +\frac{1}{1 - r_{\hat{\theta}_1 \hat{\theta}_2}^2} \cdot \frac{r}{\sigma_{\hat{\theta}_1} \sigma_{\hat{\theta}_2}},$$

or, in effect, $\sigma_{\hat{\theta}}^2$ is found by dividing the Hessian determinant of L, with respect to the parameters, into the corresponding minor.

The application of these methods to such a series of parameters as occur in the specification of frequency curves may best be made clear by an example. ...

12. Discontinuous Distributions

The applications hitherto made of the optimum statistics have been problems in which the data are ungrouped, or at least in which the grouping intervals are so small as not to disturb the values of the derived statistics. By grouping, these continuous distributions are reduced to discontinuous distributions, and in an exact discussion must be treated as such.

If p_s be the probability of an observation falling in the cell (s), p_s being a function of the required parameters $\theta_1, \theta_2 \ldots$; and in a sample of N, if n_s are found to fall into that cell, then

$$S(\log f) = S(n_s \log p_s).$$

If now we write $\bar{n}_s = p_s N$, we may conveniently put

$$L = S\left(n_s \log \frac{n_s}{\bar{n}_s}\right),$$

where L differs by a constant only from the logarithm of the likelihood, with sign reversed, and therefore the method of the optimum will consist in finding the *minimum* value of L. The equations so found are of the form

$$\frac{\partial L}{\partial \theta} = -S\left(\frac{n_s}{\bar{n}_s} \frac{\partial \bar{n}_s}{\partial \theta}\right) = 0. \tag{6}$$

It is of interest to compare these formulae with those obtained by making the Pearsonian χ^2 a minimum.

For

$$\chi^2 = S\frac{(n_s - \bar{n}_s)^2}{\bar{n}_s},$$

and therefore

$$1 + \chi^2 = S\left(\frac{n_s^2}{\bar{n}_s}\right),$$

so that on differentiating by $d\theta$, the condition that χ^2 should be a minimum for variations of θ is

$$-S\left(\frac{n_s^2}{\bar{n}_s^2} \frac{\partial \bar{n}_s}{\partial \theta}\right) = 0. \tag{7}$$

Equation (7) has actually been used (12) to "improve" the values obtained by the method of moments, even in cases of normal distribution, and the Poisson series, where the method of moments gives a strictly sufficient solution. The discrepancy between these two methods arises from the fact that χ^2 is itself an approximation, applicable only when $\bar{n}_s$ and n_s are large, and the difference between them of a lower order of magnitude. In such cases

$$L = S\left(n_s \log \frac{n_s}{\bar{n}_s}\right) = S\left(\overline{m + x} \log \frac{m + x}{m}\right) = S\left\{x + \frac{x^2}{2m} - \frac{x^3}{6m^2} \cdots\right\},$$

and since

$$S(x) = 0,$$

we have, when x is in all cases small compared to m,

$$L = \frac{1}{2}S\left(\frac{x^2}{m}\right) = \frac{1}{2}\chi^2$$

as a first approximation. In those cases, therefore, when χ^2 is a valid measure of the departure of the sample from expectation, it is equal to 2L; in other cases the approximation fails and L itself must be used.

The failure of equation (7) in the general problem of finding the best values for the parameters may also be seen by considering cases of fine grouping, in which the majority of observations are separated into units. For the formula in equation (6) is equivalent to

$$S\left(\frac{1}{\bar{n}_s} \frac{\partial \bar{n}_s}{\partial \theta}\right)$$

where the summation is taken over all the observations, while the formula of equation (7), since it involves n_s^2, changes its value discontinuously, when one observation is gradually increased, at the point where it happens to coincide with a second observation.

Logically it would seem to be a necessity that that population which is chosen in fitting a hypothetical population to data should also appear the best when tested for its goodness of fit. The method of the optimum secures this agreement, and at the same time provides an extension of the process of testing goodness of fit, to those cases for which the χ^2 test is invalid.

The practical value of χ^2 lies in the fact that when the conditions are satisfied in order that it shall closely approximate to 2L, it is possible to give a general formula for its distribution, so that it is possible to calculate the probability, P, that in a random sample from the population considered, a worse fit should be obtained; in such cases χ^2 is distributed in a curve of the Pearsonian Type III.,

$$df \propto \left(\frac{\chi^2}{2}\right)^{(n'-3)/2} e^{-\chi^2/2} \, d\left(\frac{\chi^2}{2}\right)$$

or

$$df \propto L^{(n'-3)/2} e^{-L} \, dL,$$

where n' is one more than the number of degrees of freedom in which the sample may differ from expectation (17).

In other cases we are at present faced with the difficulty that the distribution L requires a special investigation. This distribution will in general be discontinuous (as is that of χ^2), but it is not impossible that mathematical research will reveal the existence of effective graduations for the most important groups of cases to which χ^2 cannot be applied.

We shall conclude with a few illustrations of important types of discontinuous distribution.

1. The Poisson Series

$$e^{-m}\left(1, m, \frac{m^2}{2!}, \ldots, \frac{m^x}{x!}, \ldots\right)$$

involves only the single parameter, and is of great importance in modern statistics. For the optimum value of m,

$$S\left\{\frac{\partial}{\partial m}(-m + x \log m)\right\} = 0,$$

whence

$$S\left(\frac{x}{\hat{m}} - 1\right) = 0,$$

or

$$\hat{m} = \bar{x}.$$

The most likely value of m is therefore found by taking the first moment of the series.

Differentiating a second time,

$$-\frac{1}{\sigma_{\hat{m}}^2} = S\left(-\frac{x}{m^2}\right) = -\frac{n}{m},$$

so that

$$\sigma_{\hat{m}}^2 = \frac{m}{n},$$

as is well known.

2. Grouped Normal Data

In the case of the normal curve of distribution it is evident that the second moment is a sufficient statistic for estimating the standard deviation; in investigating a sufficient solution for grouped normal data, we are therefore in reality finding the optimum correction for grouping; the Sheppard correction having been proved only to satisfy the criterion of consistency.

For grouped normal data we have

$$p_s = \frac{1}{\sigma\sqrt{2\pi}} \int_{x_s}^{x_{s+1}} e^{-(\bar{x}-m^2)/2\sigma^2}\, dx,$$

and the optimum values of m and σ are obtained from the equations,

$$\frac{\partial L}{\partial m} = S\left(\frac{n_s}{p_s}\frac{\partial p_s}{\partial m}\right) = 0,$$

$$\frac{\partial L}{\partial \sigma} = S\left(\frac{n_s}{p_s}\frac{\partial p_s}{\partial \sigma}\right) = 0;$$

or, if we write,

$$z = \frac{1}{\sigma\sqrt{2\pi}} e^{-(\bar{x}-m^2)/2\sigma^2},$$

we have the two conditions,

$$S\left(\frac{n_s}{p_s}\overline{z_s - z_{s+1}}\right) = 0$$

and

$$S\left\{\frac{n_s}{p_s}\left(\frac{x_s}{\sigma}z_s - \frac{x_{s+1}}{\sigma}z_{s+1}\right)\right\} = 0.$$

As a simple example we shall take the case chosen by K. Smith in her investigation of the variation of χ^2 in the neighbourhood of the moment solution (12).

Three hundred errors in right ascension are grouped in nine classes, positive and negative errors being thrown together as shown in the following table:

$0'' \cdot 1$ arc	0–1	1–2	2–3	3–4	4–5	5–6	6–7	7–8	8–9
Frequency . .	114	84	53	24	14	6	3	1	1

The second moment, without correction, yields the value

$$\sigma_v = 2.282542.$$

Using Sheppard's correction, we have

$$\sigma_\mu = 2.264214,$$

while the value obtained by making χ^2 a minimum is

$$\sigma_{\chi^2} = 2.355860.$$

If the latter value were accepted we should have to conclude that Sheppard's correction, even when it is small, and applied to normal data, might be altogether of the wrong magnitude, and even in the wrong direction In order to obtain the optimum value of σ, we tabulate the values of $\dfrac{\partial L}{\partial \sigma}$ in the region under consideration; this may be done without great labour if values of σ be chosen suitable for the direct application of the table of the probability integral (13, Table II.). We then have the following values:

$\dfrac{1}{\sigma}$	0.43	0.44	0.45	0.46
$\dfrac{\partial L}{\partial \sigma}$	+ 15.135	+ 2.149	− 11.098	− 24.605
$\Delta^2 \dfrac{\partial L}{\partial \sigma}$		− 0.261	− 0.260	

By interpolation,

$$\frac{1}{\hat{\sigma}} = 0.441624$$

$$\hat{\sigma} = 2.26437.$$

We may therefore summarise these results as follows:—

Uncorrected estimate of σ 2.28254
Sheppard's correction . . . . −0.01833
Correction for maximum likelihood . . . . −0.01817
"Correction" for minimum χ^2 . . . . +0.07332

Far from shaking our faith, therefore, in the adequacy of Sheppard's correction, when small, for normal data, this example provides a striking instance of its effectiveness, while the approximate nature of the χ^2 test renders it unsuitable for improving a method which is already very accurate.

It will be useful before leaving the subject of grouped normal data to calculate the actual loss of efficiency caused by grouping, and the additional loss due to the small discrepancy between moments with Sheppard's correction and the optimum solution.

To calculate the loss of efficiency involved in the process of grouping normal data, let

$$v = \frac{1}{a} \int_{\xi - (1/2)a}^{\xi + (1/2)a} f(\xi)\, d\xi,$$

when a_σ is the group interval, then

$$v = f(\xi) + \frac{a^2}{24}f''(\xi) + \frac{a^4}{1920}f^{iv}(\xi) + \frac{a^6}{322,560}f^{vi}(\xi) + \cdots$$

$$= f(\xi)\left\{1 + \frac{a^2}{24}(\xi^2 - 1) + \frac{a^4}{1920}(\xi^4 - 6\xi^2 + 3)\right.$$

$$\left. + \frac{a^6}{322,560}(\xi^6 - 15\xi^4 + 45\xi^2 - 15) + \cdots\right\},$$

whence

$$\log v = \log f + \frac{a^2}{24}(\xi^2 - 1) - \frac{a^4}{2880}(\xi^4 + 4\xi^2 - 2)$$

$$+ \frac{a^6}{181,440}(\xi^6 + 6\xi^4 + 3\xi^2 - 1) - \cdots,$$

and

$$\frac{\partial^2}{\partial m^2}\log v = -\frac{1}{\sigma^2} + \frac{1}{\sigma^2}\left\{\frac{a^2}{12} - \frac{a^4}{720}(3\xi^2 + 2) + \frac{a^6}{30,240}(5\xi^4 + 12\xi^2 + 1) - \cdots\right\},$$

of which the mean value is

$$-\frac{1}{\sigma^2}\left\{1 - \frac{a^2}{12} + \frac{a^4}{144} - \frac{a^6}{1778} + \frac{31a^8}{25.12^4} - \frac{313a^{10}}{175.12^5}\right\}$$

neglecting the periodic terms; and consequently

$$\sigma_{\hat{m}}^2 = \frac{\sigma^2}{n}\left(1 + \frac{a^2}{12} - \frac{a^8}{86,400}\cdots\right).$$

Now for the mean of ungrouped data

$$\sigma_{\hat{m}}^2 = \frac{\sigma^2}{n},$$

so that the loss of efficiency due to grouping is nearly $\dfrac{a^2}{12}$.

The further loss caused by using the mean of the grouped data is very small, for

$$\sigma_{v_1}^2 = \frac{v_2}{n} = \frac{\sigma^2}{n}\left(1 + \frac{a^2}{12}\right),$$

neglecting the periodic terms; the loss of efficiency by using v_1 therefore is only

$$\frac{a^8}{86,400}.$$

Similarly for the efficiency for scaling,

$$\frac{\partial^2}{\partial \sigma^2} \log v$$

$$= \frac{1}{\sigma^2} - \frac{3\xi^2}{\sigma^2} + \frac{1}{\sigma^2} \left\{ \frac{a^2}{12}(10\xi^2 - 3) - \frac{a^4}{360}(9\xi^4 + 21\xi^2 - 5) \right.$$

$$+ \frac{a^6}{30,240}(26\xi^6 + 110\xi^4 + 36\xi^2 - 7)$$

$$\left. - \frac{a^8}{1,814,400}(51\xi^8 + 315\xi^6 + 351\xi^4 - 55\xi^2 + 9) + \cdots \right\},$$

of which the mean value is

$$-\frac{2}{\sigma^2} \left\{ 1 - \frac{a^2}{6} + \frac{a^4}{40} - \frac{a^6}{270} + \frac{83a^8}{129,600} \cdots \right\},$$

neglecting the periodic terms; and consequently

$$\sigma_{\hat{\sigma}}^2 = \frac{\sigma^2}{2n} \left\{ 1 + \frac{a^2}{6} + \frac{a^4}{360} - \frac{a^8}{10,800} \cdots \right\}.$$

For ungrouped data

$$\sigma_{\hat{\sigma}}^2 = \frac{\sigma^2}{2n},$$

so that the loss of efficiency in scaling due to grouping is nearly $\frac{a^2}{6}$. This may be made as low as 1 per cent by keeping a less than $\frac{1}{4}$.

The further loss of efficiency produced by using the grouped second moment with Sheppard's correction is again very small, for

$$\sigma_{v_2}^2 = \frac{v_4 - v_2^2}{n} = \frac{2\sigma^4}{n} \left(1 + \frac{a^2}{6} + \frac{a^4}{360} \right)$$

neglecting the periodic terms.

Whence it appears that the further loss of efficiency is only

$$\frac{a^8}{10,800}.$$

We may conclude, therefore, that the high agreement between the optimum value of σ and that obtained by Sheppard's correction in the above example is characteristic of grouped normal data. The method of moments with Sheppard's correction is highly efficient in treating such material, the gain in efficiency obtainable by increasing the likelihood to its maximum value is trifling, and far less than can usually be gained by using finer groups. The loss of efficiency involved in grouping may be kept below 1 per cent. by making the group interval less than one-quarter of the standard deviation.

Although for the normal curve the loss of efficiency due to moderate grouping is very small, such is not the case with curves making a finite angle with the axis, or having at an extreme a finite or infinitely great ordinate. In such cases even moderate grouping may result in throwing away the greater part of the information which the sample provides....

13. Summary

During the rapid development of practical statistics in the past few decades, the theoretical foundations of the subject have been involved in great obscurity. Adequate distinction has seldom been drawn between the sample recorded and the hypothetical population from which it is regarded as drawn. This obscurity is centred in the so-called "inverse" methods.

On the bases that the purpose of the statistical reduction of data is to obtain statistics which shall contain as much as possible, ideally the whole, of the relevant information contained in the sample, and that the function of Theoretical Statistics is to show how such adequate statistics may be calculated, and how much and of what kind is the information contained in them, an attempt is made to formulate distinctly the types of problems which arise in statistical practice.

Of these, problems of Specification are found to be dominated by considerations which may change rapidly during the progress of Statistical Science. In problems of Distribution relatively little progress has hitherto been made, these problems still affording a field for valuable enquiry by highly trained mathematicians. The principal purpose of this paper is to put forward a general solution of problems of Estimation.

Of the criteria used in problems of Estimation only the criterion of Consistency has hitherto been widely applied; in Section 5 are given examples of the adequate and inadequate application of this criterion. The criterion of Efficiency is shown to be a special but important case of the criterion of Sufficiency, which latter requires that the whole of the relevant information supplied by a sample shall be contained in the statistics calculated.

In order to make clear the nature of the general method of satisfying the criterion of Sufficiency, which is here put forward, it has been thought necessary to reconsider Bayes' problem in the light of the more recent criticisms to which the idea of "inverse probability" has been exposed. The conclusion is drawn that two radically distinct concepts, both of importance in influencing our judgment, have been confused under the single name of *probability*. It is proposed to use the term *likelihood* to designate the state of our information with respect to the parameters of hypothetical populations, and it is shown that the quantitative measure of likelihood does not obey the mathematical laws of probability.

A proof is given in Section 7 that the criterion of Sufficiency is satisfied by that set of values for the parameters of which the likelihood is a maximum, and that the same function may be used to calculate the efficiency of any other statistics, or, in other words, the percentage of the total available information which is made use of by such statistics.

This quantitative treatment of the information supplied by a sample is illustrated by an investigation of the efficiency of the method of moments in fitting the Pearsonian curves of Type III.

Section 9 treats of the location and scaling of Error Curves in general, and contains definitions and illustrations of the *intrinsic accuracy*, and of the *centre of location* of such curves.

In Section 10 the efficiency of the method of moments in fitting the general Pearsonian curves is tested and discussed. High efficiency is only found in the neighbourhood of the normal point. The two causes of failure of the method of moments in locating these curves are discussed and illustrated. The special cause is discovered for the high efficiency of the third and fourth moments in the neighbourhood of the normal point.

It is to be understood that the low efficiency of the moments of a sample in estimating the form of these curves does not at all diminish the value of the notation of moments as a means of the comparative specification of the form of such curves as have finite moment coefficients.

Section 12 illustrates the application of the method of maximum likelihood to discontinuous distributions. The Poisson series is shown to be sufficiently fitted by the mean. In the case of grouped normal data, the Sheppard correction of the crude moments is shown to have a very high efficiency, as compared to recent attempts to improve such fits by making χ^2 a minimum; the reason being that χ^2 is an expression only approximate to a true value derivable from likelihood. As a final illustration of the scope of the new process, the theory of the estimation of micro-organisms by the dilution method is investigated.

Finally it is a pleasure to thank Miss W.A. Mackenzie, for her valuable assistance in the preparation of the diagrams.

References

(1) K. Pearson (1920). "The Fundamental Problem of Practical Statistics," 'Biom.,' xiii., pp. 1–16.
(2) F.Y. Edgeworth (1921) "Molecular Statistics," 'J.R.S.S.,' lxxxiv., p. 83.
(3) G.U. Yule (1912). "On the Methods of Measuring Association between two Attributes." 'J.R.S.S.,' lxxv., p. 587.
(4) Student (1908). "The Probable Error of a Mean," 'Biom.,' vi., p. 1.
(5) R.A. Fisher (1915). "Frequency Distribution of the Values of the Correlation Coefficient in Samples from an Indefinitely Large Population," 'Biom.,' x., 507.
(6) R.A. Fisher (1921). "On the 'Probable Error' of a Coefficient of Correlation deduced from a Small Sample," 'Metron.,' i., pt. iv., p. 82.

(7) R.A. Fisher (1920). "A Mathematical Examination of the Methods of Determining the Accuracy of an Observation by the Mean Error and by the Mean Square Error," 'Monthly Notices of R.A.S.,' lxxx., 758.

(8) E. Pairman and K. Pearson (1919). "On Corrections for the Moment Coefficients of Limited Range Frequency Distributions when there are finite or infinite Ordinates and any Slopes at the Terminals of the Range," 'Biom.,' xii., p. 231.

(9) R.A. Fisher (1912). "On an Absolute Criterion for Fitting Frequency Curves," 'Messenger of Mathematics,' xli., p. 155.

(10) Bayes (1763). "An Essay towards Solving a Problem in the Doctrine of Chances," 'Phil. Trans.,' liii., p. 370.

(11) K. Pearson (1919). "Tracts for Computers. No. 1: Tables of the Digamma and Trigamma Functions," By E. Pairman, Camb. Univ. Press.

(12) K. Smith (1916). "On the 'best' Values of the Constants in Frequency Distributions," 'Biom.,' xi., p. 262.

(13) K. Pearson (1914). "Tables for Statisticians and Biometricians," Camb. Univ. Press.

(14) G. Vega (1764). "Thesaurus Logarithmorum Completus," p. 643.

(15) K. Pearson (1900). "On the Criterion that a given System of Deviations from the Probable in the case of a Correlated System of Variables is such that it can be reasonably supposed to have arisen from Random Sampling," 'Phil. Mag.,' l., p. 157.

(16) K. Pearson and L.N.G. Filon (1898). "Mathematical Contributions to the Theory of Evolution. IV.—On the Probable Errors of Frequency Constants, and on the influence of Random Selection on Variation and Correlation," 'Phil. Trans.," cxci., p. 229.

(17) R.A. Fisher (1922). "The Interpretation of χ^2 from Contingency Tables, and the Calculation of P," 'J.R.S.S.,' lxxxv., pp. 87–94.

(18) K. Pearson (1915). "On the General Theory of Multiple Contingency, with special reference to Partial Contingency," 'Biom.,' xi., p. 145.

(19) K. Pearson (1903). "On the Probable Errors of Frequency Constants," 'Biom.,' ii., p. 273, Editorial.

(20) W.F. Sheppard (1898). "On the Application of the Theory of Error to Cases of Normal Distribution and Correlations," 'Phil. Trans.,' A., cxcii., p. 101.

(21) J. M. Keynes (1921). "A Treatise on Probability," Macmillan & Co., London.

Introduction to
Hotelling (1931) The Generalization of Student's Ratio

T.W. Anderson
Stanford University

1. Hotelling's T^2

Perhaps the most frequently used statistic in statistical inference is the Student t-statistic. If $X_1, \ldots, X_N$ are N observations from the univariate normal distribution with mean μ and variance σ^2, the Student-t statistic is $t = \sqrt{N}\bar{x}/s$, where $\bar{x} = \sum_{\alpha=1}^{N} X_\alpha/N$, the sample mean, and $s = [\sum_{\alpha=1}^{N} (X_\alpha - \bar{x})^2/n]^{1/2}$, the sample standard deviation; here $n = N - 1$, the number of degrees of freedom.[*] Student (1908) argued that the density of t is

$$\text{const.}\left(1 + \frac{t^2}{n}\right)^{-(n+1)/2}.$$

His "proof" was incomplete; Fisher (1925) gave an explicit rigorous proof. This t-statistic is used, for example, to test the null hypothesis that the mean μ of a normal distribution $N(\mu, \sigma^2)$ is 0. A variant of the Student t-statistic can be used for the two-sample problem of testing whether the means of two normal populations with common variance are equal; a t-statistic can also be formulated to test a hypothesis about a regression coefficient.

In this paper, Hotelling generalizes the square of t to the multivariate case when $X_1, \ldots, X_N$ are p-component vectors and the normal distribution has a vector mean μ and a covariance matrix Σ with the sample mean defined as $\bar{x} = (1/N)\sum_{\alpha=1}^{N} X_\alpha$ and the sample covariance matrix as[†]

---

[*] The notation in this introduction conforms as closely as possible to Hotelling's notation although it may differ from modern usage.

[†] Hotelling's notation is A for the sample covariance matrix, while the current conventional notation is S.

$$S = \frac{1}{n} \sum_{\alpha=1}^{n} (X_\alpha - \bar{x})(X_\alpha - \bar{x})'.$$

Hotelling's generalized T^2 is given as

$$T^2 = N\bar{x}'S^{-1}\bar{x}.$$

The analogy of T^2 to $t^2 = N\bar{x}^2/s^2$ is obvious in this vector notation. Hotelling actually carried out his exposition entirely in terms of the components of X_α, $\bar{x}$, S, etc., although now that seems very cumbersome.

The t-statistic, $t = \sqrt{N}\bar{x}/s$, is scale-invariant; that is, if each observation is multiplied by a positive constant c, the sample mean $\bar{x}$ and standard deviation s are both multiplied by c, but t is unaffected. For example, if X_α represents the length in feet of the αth object and $c = 12$ in. per foot, X_α^* is the length measured in inches. The t-statistic does not depend on the units of measurement. Hotelling points out that $T^2 = N\bar{x}'S^{-1}\bar{x}$ is invariant under linear transformations $X_\alpha^* = CX_\alpha$, where C is a nonsingular matrix because $\bar{x}$ is replaced by $\bar{x}^* = C\bar{x}$ and S is replaced by $S^* = CSC'$, leaving T^2 unchanged. In particular, T^2 does not depend on the scale for each component of X. In the univariate case, the real line has a positive and a negative direction, but in the multivariate case, no direction has special meaning. Hotelling's T^2 really corresponds to Student's t^2 since t^2 is invariant with respect to multiplication of X_α by any real constant different from 0.

The statistic T^2 is the only invariant function of the sufficient statistics, $\bar{x}$ and S; that is, any function of $\bar{x}$ and S that is invariant is a function of T^2. An important use of the T^2-statistic is to test the null hypothesis that $\mu = 0$ in $N(\mu, \Sigma)$; the hypothesis is rejected if T^2 is greater than a number preassigned to attain a desired significance level. This test is the uniformly most powerful invariant test. [See Sect. 5.3 of Anderson (1984), for example.] The only invariant of the parameters, μ and Σ, is $\mu'\Sigma^{-1}\mu$. The power of the T^2-test is, therefore, a function of this invariant of the parameters.

Hotelling approached the distribution of T^2 when $\mu = 0$ by a geometric method similar to that introduced by Fisher (1915) in finding the distribution of the Pearson correlation coefficient. Since T^2 is invariant with respect to nonsingular transformation $X_\alpha^* = CX_\alpha$, in the case of $\mu = 0$ the distribution of T^2 is the same as the distribution of $T^{*2} = N\bar{x}^{*\prime}(S^*)^{-1}\bar{x}^*$. If X_α has the distribution $N(0, \Sigma)$, then X_α^* has the distribution $N(0, C\Sigma C')$. Inasmuch as C can be chosen so that $C\Sigma C' = I$, Hotelling assumed $\Sigma = I$ when he derived the distribution of T^2 under the null hypothesis $\mu = 0$. In this case, the observed components $x_{i\alpha}$, $i = 1, \ldots, p$, $\alpha = 1, \ldots, N$, are independent and each has the standard normal distribution $N(0, 1)$; the vectors $\mathbf{x}_i = (x_{i1}, \ldots, x_{iN})'$, $i = 1, \ldots, p$, are independently distributed $N(0, I_N)$. The critical feature of $N(0, I_N)$ is that it is a spherical distribution.

Consider $\mathbf{X} = (\mathbf{x}_1, \ldots, \mathbf{x}_p)$ as spanning a p-dimensional linear space V_p and let $\zeta = \mathbf{e} = (1, \ldots, 1)'$ be an N-dimensional vector. Then the point in V_p closest to $\zeta = \mathbf{e}$ is (by usual least squares)

$$\hat{\mathbf{e}} = \mathbf{X}(\mathbf{X}'\mathbf{X})^{-1}\mathbf{X}'\mathbf{e} = N\mathbf{X}(\mathbf{X}'\mathbf{X})^{-1}\bar{x}.$$

The length of this vector is

$$\hat{\mathbf{e}}'\hat{\mathbf{e}} = N^2\bar{x}'(\mathbf{X}'\mathbf{X})^{-1}\bar{x}.$$

The squared distance of $\mathbf{e}$ from V_p is

$$(\mathbf{e} - \hat{\mathbf{e}})'(\mathbf{e} - \hat{\mathbf{e}}) = N - N^2\bar{x}'(\mathbf{X}'\mathbf{X})^{-1}\bar{x}.$$

The cotangent of the angle θ between $\mathbf{e}$ and V_p is given by

$$\cot^2 \theta = \frac{\|\hat{\mathbf{e}}\|^2}{\|\mathbf{e} - \hat{\mathbf{e}}\|^2} = \frac{N\bar{x}'(\mathbf{X}'\mathbf{X})^{-1}\bar{x}}{1 - N\bar{x}'(\mathbf{X}'\mathbf{X})^{-1}\bar{x}} = \frac{N\bar{x}'(nS + N\bar{x}\bar{x}')^{-1}\bar{x}}{1 - N\bar{x}'(nS_N + \bar{x}\bar{x}')^{-1}\bar{x}}.$$

A little algebra shows that this is $T^2/n = N\bar{x}'(nS)^{-1}\bar{x}$.

The distribution of T^2 when $\mu = 0$ is generated by the vectors $\mathbf{x}_1, \ldots, \mathbf{x}_p$; the direction of each of these is random in the sense that its projection on the unit sphere has a uniform distribution. It follows that the distribution of the angle between $\mathbf{e}$ and V_p is the same as the distribution of the angle between an arbitrary vector $\mathbf{y}$ and V_p. In fact, $\mathbf{y}$ can be assigned a distribution $N(0, I_N)$ independent of $\mathbf{x}_1, \ldots, \mathbf{x}_p$. The cotangent squared of the angle θ_y between $\mathbf{y}$ and V_p is

$$\cot^2 \theta_y = \frac{\mathbf{y}'\mathbf{X}(\mathbf{X}'\mathbf{X})^{-1}\mathbf{X}'\mathbf{y}}{\mathbf{y}'\mathbf{y} - \mathbf{y}'\mathbf{X}(\mathbf{X}'\mathbf{X})^{-1}\mathbf{X}'\mathbf{y}}.$$

Since $\mathbf{y}$ has a spherical distribution, the distribution of $\cot^2 \theta_y$ does not depend on $\mathbf{X}$, that is, V_p. Hotelling then found the distribution of $\cot^2 \theta_y$ by representing the projection of $\mathbf{y}$ on the unit sphere in polar coordinates and performing integration. Of course, from normal regression theory, we know that the numerator of $\cot^2 \theta_y$ has a χ^2-distribution with p degrees of freedom and the denominator has a χ^2-distribution with $N - p$ degrees of freedom independent of the numerator. Hence, $(N - p) \cot^2 \theta_y/p$ has the F-distribution with p and $N - p$ degrees of freedom.

A form of the T^2-statistic can be used to test the hypothesis $\mu = \mu'$ on the basis of observations $X_1, \ldots, X_{N_1}$ from $N(\mu, \Sigma)$ and $X'_1, \ldots, X'_{N_2}$ from $N(\mu', \Sigma)$. Hotelling defined

$$\xi = \sqrt{\frac{N_1 N_2}{N_1 + N_2}}(\bar{x} - \bar{x}'),$$

which has the distribution $N(0, \Sigma)$ under the null hypothesis, and

$$nS = \sum_{\alpha=1}^{N_1}(X_\alpha - \bar{x})(X_\alpha - \bar{x})' + \sum_{\alpha=1}^{N_2}(X'_\alpha - \bar{x}')(X'_\alpha - \bar{x}')',$$

where $n = N_1 + N_2 - 2$. The matrix nS has the distribution of $\sum_{\alpha=1}^{N_1+N_2-2} Z_\alpha Z'_\alpha$, where the Z_α's are independent and Z_α has the distribution $N(0, \Sigma)$. Then

$$T^2 = \xi'S^{-1}\xi = \frac{N_1 N_2}{N_1 + N_2}(\bar{x} - \bar{x}')'S^{-1}(\bar{x} - \bar{x}').$$

When $\mu = \mu'$, this T^2 times $(N_1 + N_2 - p - 1)$ and divided by p has the $F_{p, N_1 + N_2 - p - 1}$-distribution.

Fisher (1936) found this same statistic when he derived the linear discriminant function and related it to Hotelling's T^2 [Fisher (1938)]. This is a sample analog of

$$p\Delta^2 = (\mu - \mu')'\Sigma - 1(\mu - \mu'),$$

known as Mahalanobis squared distance [Mahalanobis (1930)]. The distribution of the "studentized" version of this quantity was published by Bose and Roy (1938).

2. The Life and Works of Harold Hotelling

Harold Hotelling was a kind of Renaissance man; his interests were catholic. His career "progressed" from journalist to mathematician to economist to statistician. (Not completely coincidentally, his colleague Abraham Wald followed a similar progression, without preliminary journalistic experience; the difference showed up in their expositions.) Hotelling wrote in an almost encylopedic style; no detail was omitted.

Hotelling was born on September 29, 1895, in Fulda, Minnesota. When he was nine years old, his family moved to Seattle, Washington, where he attended high school and the University of Seattle. (Around Columbia University, there was an apocryphal story that at a faculty meeting he argued that the university should move out of the surrounding costly urban area; when asked where he thought the university should settle, he said he considered Seattle a favorable location.)

After he received an undergraduate degree in journalism in 1919, Hotelling got a position with the *Washington Standard* in Olympia (near Seattle). He wrote later that "journalism had seemed to offer both a means of livelihood and an opportunity to stimulate proper action on public matters; later I concluded that it had been overrated in both respects." He proceeded to study mathematics at the University of Washington and then Princeton (after being refused a fellowship in economics at Columbia University). He studied topology, although his primary interest was economics, acquiring his Ph.D. in mathematics in 1924.

His interest in economics was called upon in his next position at the Food Research Institute at Stanford University. Since Hotelling was expected to help in estimating crop yields and food requirements, it was essential that he acquire a sound knowledge of statistical methods. He started by reading Fisher. This led to correspondence with Fisher and his spending the academic year 1929–30 at the Rothamstead Experimental Station. His contact with Fisher had a profound impact on Hotelling's views and work in statistics. In fact, the two planned a book together, but soon found that their views of statistics were not entirely in harmony.

After seven years at Stanford, moving gradually into the department of mathematics, Hotelling accepted the position of professor of mathematical economics at Columbia University, succeeding H.L. Moore. Although Columbia had perhaps the most outstanding economics department, it was actually not very friendly to the mathematical approach. During the period 1925–1943 Hotelling wrote 11 seminal and penetrating papers in economics. These were reprinted in *The Collected Economics Articles by Harold Hotelling*, edited by Adrian C. Darnell; the volumes also include a biographical article by the editor.

Hotelling's interests turned more toward mathematical statistics. During Hotelling's first year at Columbia, S.S. Wilks held a National Research Fellowship, followed by J.L. Doob. Soon Hotelling attracted large numbers of graduate students, some such as Madow and Girshick working in multivariate analysis. These 15 years at Columbia were very productive in mathematical statistics as well as in mathematical economics. Incidentally, Hotelling continued to work on a comprehensive exposition of statistics, but never was satisfied enough to let the manuscript pass beyond the hands of his students.

After World War II, during which he served as Operations Director of Columbia's Statistical Research Group, Hotelling moved to the University of North Carolina, where an Institute of Statistics had been established. The Institute with Gertrude Cox as director consisted of a department of applied statistics at North Carolina State College in Raleigh and a department of mathematical statistics at Chapel Hill headed by Hotelling. In his later years, he devoted himself primarily to teaching, passing away in 1973.

3. This Paper in the Context of Current Statistical Research

In his 1915 paper, Fisher derived the exact distribution of the Pearson correlation coefficient by a rigorous method (as contrasted to Student's 1908 paper). He subsequently found the distributions of other statistics by his geometric method. In particular, the distribution of the multiple correlation coefficient was derived by Fisher (1924) when the population multiple correlation was 0 and by Fisher (1928) when the population coefficient was different from 0. Hotelling points out the "affinity" of T^2 with the multiple correlation.

In this 1915 paper, Fisher actually derived the distribution of the sample covariance matrix S for $p = 2$. In Wishart (1928), deriving the distribution for arbitrary p, the author acknowledges the help of Fisher; the method was Fisher's geometric one. Later Wishart and Bartlett (1933) proved the result by characteristic functions.

Following "The generalization of Student's ratio" came many generalizations of univariate statistics. Wilks, who had spent the academic year 1931–32 with Hotelling, published "Certain generalizations in the analysis of variance" (1932). While Hotelling generalized Student's t, Wilks generalized the F-statistic basic to the analysis of variance; one generalization is the likelihood ratio criterion, often called "Wilks' lambda." If $X_\alpha^{(i)}$, $\alpha = 1, \ldots, N_i$, $i = 1, \ldots, q$, is the αth observation from the ith distribution $N(\mu_i, \Sigma)$, we define the "between" covariance matrix as

$$S_1 = \frac{1}{q-1} \sum_{i=1}^{q} [X_\alpha^{(i)} - \bar{X}^{(i)}][X_\alpha^{(i)} - \bar{X}^{(i)}]'$$

and the "within" covariance matrix as

$$S_2 = \frac{1}{\sum_{i=1}^{q} N_i - q} \sum_{i=1}^{q} \sum_{\alpha=1}^{N_i} [X_\alpha^{(i)} - \bar{X}^{(i)}][X_\alpha^{(i)} - \bar{X}^{(i)}]'.$$

In the scalar case ($p = 1$), the F-statistic is $F = S_1/S_2$. In the multivariate case, Wilks' lambda is $N/2$ times

$$\Lambda = \frac{|(\sum_{i=1}^{q} N_i - q)S_2|}{|(q-1)S_1 + (\sum_{i=1}^{q} N_i - q)S_2|},$$

which for $p = 1$ reduces to

$$\frac{1}{(q-1)F/(\sum_{i=1}^{q} N_i - q) + 1}.$$

Wilks found the moments of Λ and the distribution in some special cases. In a paper written the next year (when Wilks was in London and Cambridge), E.S. Pearson and Wilks (1933) treated a more general problem when $p = 2$, testing the homogeneity of covariance matrices as well as of means in the case of $N(\mu_i, \Sigma_i)$, $i = 1, \ldots, q$. Wilks continued his research in multivariate statistics; see S.S. Wilks, *Collected Papers* (1967).

Later Hotelling (1947, 1951) proposed a "generalized T test" for the analysis of variance. In the preceding notation, it was tr $S_1 S_2^{-1}$, known also as T_0^2, to test the hypothesis $\mu_1 = \cdots = u_q$. However, Lawley (1938) had already made this generalization.

Further study was done on the T^2-statistics. Hsu (1938) found the distribution of $T^2 = N\bar{x}'S\bar{x}$ when $\mu \neq 0$. This leads to the power function of a T^2-test. Simaika (1941) proved that of all tests of $H : \mu = 0$ with the power depending only on $N\mu'\Sigma^{-1}\mu$, the T^2-test is uniformily most powerful. Hsu (1945) proved an optimal property of the t^2-test that involves averaging the power over μ and Σ.

Bartlett, who was at Cambridge University with Wishart, developed much of the theory of the multivariate generalization. In his paper (1934), he developed further the geometric approach to multivariate analysis.

The study of properties of the T^2-test, alternatives to the test, and adaptations continue. A key paper was Stein (1956), in which it was shown that the T^2-test is admissible within the class of all tests of $H : \mu = 0$ (not just invariant tests). The proof depended on a more general theorem concerning exponential distributions and closed convex acceptance regions. An alternative proof of the admissibility of the T^2-test is to show that it is a proper Bayes procedure [Kiefer and Schwartz (1965)]. A different kind of test procedure of testing $H : \mu = 0$ is a step-down procedure. Marden and Perlman (1990) showed that a step-down procedure is admissible only if it is trivial, that is, has no step.

For references up to 1966, see Anderson, Das Gupta, and Styan (1972). About 125 papers are listed under the category "Tests of hypotheses about one or two mean vectors of multivariate normal distributions and Hotelling's T^2."

4. Comments

For a modern reader, this paper has the disadvantage of being written explicitly in terms of the components of constituent vectors and matrices. Linear operations and inverses do not seem as natural as in matrix notation. We are accustomed to defining A^{-1} the inverse to a nonsingular matrix A, as the (unique) matrix satisfying $A(A^{-1}) = I$. In keeping with usage at that time, Hotelling defined a as the determinant of (a_{ij}) and an element of the symmetric inverse as

$$A_{ij} = A_{ji} = \frac{\text{cofactor of } a_{ij} \text{ in } a}{a}.$$

The exposition of regression (pages 57–58) is particularly opaque. Let the observation matrix be $X = (x_{i\alpha})$, $i = 1, \ldots, p$, $\alpha = 1, \ldots, n$. Then $X = H + E$, where $\mathscr{E}E = 0$. The model can be written

$$\underset{p \times N}{H} = \underset{p \times q}{Z} \ \underset{q \times N}{G'},$$

where $Z = (\zeta_{is})$ is the matrix of parameters and $G = (g_{\alpha s})$ the matrix of independent variables. The matrix of regression coefficients Z is estimated by minimizing each diagonal element of

$$[X - ZG'][X - ZG']'$$

with respect to the elements of that row of Z.

Hotelling found the distribution of T^2 under the null hypothesis $\mu = 0$, but his approach can be developed to obtain the distribution when $\mu \neq 0$; the distribution is the noncentral F-distribution with the same number of degrees of freedom and noncentrality parameter $N\mu'\Sigma^{-1}\mu$.

Hotelling was clearly very familiar with Fisher's work, in particular, his obtaining distributions of the various correlation coefficients. However, Hotelling seemed not to realize that finding the distribution of T^2 is exactly the same as finding the distribution of $R^2/(1 - R^2)$, when R is the multiple correlation coefficient between a vector $y \sim N(WB, \sigma^2 I_N)$ and the columns of W, $N \times p$. In fact, when the notation is translated, Hotelling's derivation is almost the same as Fisher's, except that Fisher relied entirely on geometry (and did not include as much detail). The relationship between these two problems is spelled out in Problems 2 to 4 and alternatively 5 to 10 of Chap. 5 of Anderson (1984).

References

Anderson, T.W. (1984). *An Introduction to Multivariate Statistical Analysis*, 2nd Ed. Wiley, New York.

Anderson, T.W., Das Gupta, S., and Styan, G.P.H. (1972). *A Bibliography of Multivariate Analysis*. Oliver & Boyd, Edinburgh (reprinted by Robert E. Kreiger, Malabar, Florida, 1977).

Bartlett, M.S. (1934). The vector representation of a sample. *Proc. Cambridge Philos. Soc.*, **30**, 327–340.

Bose, R.C., and Roy, S.N. (1938). The distribution of the studentized D^2-statistic, *Sankhyā*, **4**, 19–38.

Darnell, A.C. (1990). *The Collected Economics Articles of Harold Hotelling*. Springer-Verlag, New York.

Fisher, R.A. (1915). Frequency distribution of the values of the correlation coefficient in samples from an indefinitely large population, *Biometrika*, **10**, 507–521.

Fisher, R.A. (1924). The influence of rainfall on the yield of wheat at Rothamstead, *Philos. Trans. Roy. Soc. London, Ser. B*, **213**, 89–142.

Fisher, R.A. (1925). Applications of Student's distribution, *Metron*, **5**, 90–104.

Fisher, R.A. (1928). The general sampling distribution of the multiple correlation coefficient, *Proc. Roy. Soc. London, Ser. A*, **121**, 654–673.

Fisher, R.A. (1936). The statistical use of multiple measurements in taxonomic problems, *Ann. of Eugenics*, **7**, 179–188.

Fisher, R.A. (1938). The statistical utilization of multiple measurements, *Ann. Eugenics*, **8**, 376–386.

Hotelling, H. (1947). Multivariate quality control, illustrated by the air testing of sample bombsights, in *Techniques of Statistical Analysis* (C. Eisenhart, M. Hastay, and W.A. Wallis, eds.). McGraw-Hill, New York, pp. 111–184.

Hotelling, H. (1951). A generalized T test and measure of multivariate dispersion, in *Proceedings of the Second Berkeley Symposium on Mathematical Statistics and Probability* (J. Neyman, ed.). University of California, Los Angeles and Berkeley, pp. 23–41.

Hsu, P.L. (1938). Notes on Hotelling's generalized T, *Ann. Math. Statist.*, **9**, 231–243.

Hsu, P.L. (1945). On the power functions for the E^2-test and the T^2-test, *Ann. Math. Statist.*, **16**, 278–286.

Kiefer, J., and Schwartz, R. (1965). Admissible Bayes character of T^2-, R^2-, and other fully invariant tests for classical multivariate normal problems, *Ann. Math. Statist.*, **36**, 747–770

Lawley, D.N. (1938). A generalization of Fisher's z test, *Biometrika*, **30**, 180–187.

Mahalanobis, P.C. (1930). On tests and measures of group divergence. Part I. Theoretical formulae, *J. and Proc. Asiatic Soc. Bengal*, **26**, 541–588.

Marden, J.I., and Perlman, M.D. (1990). On the inadmissibility of step-down procedures for the Hotelling T^2 problem, *Ann. Statist.*, **18**, 172–190.

Pearson, E.S., and Wilks, S.S. (1933). Methods of analysis appropriate for k samples of two variables, *Biometrika*, **25**, 353–378.

Simaika, J.B. (1941). On an optimum property of two important statistical tests, *Biometrika*, **32**, 70–80.

Stein, C. (1956). The admissibility of Hotelling's T^2-test, *Ann. Math. Statist.*, **27**, 616–623.

Student (W.S. Gosset) (1908). The probable error of a mean, *Biometrika*, **6**, 1–25.

Wilks, S.S. (1932). Certain generalizations in the analysis of variance, *Biometrika*, **24**, 471–494.

Wilks, S.S. (1967). *Collected Papers: Contributions to Mathematical Statistics* (T.W. Anderson, ed.), Wiley, New York.

Wishart, J. (1928). The generalized product moment distribution in samples from a multivariate normal distribution, *Biometrika*, **20A**, 32–52.

Wishart, J., and Bartlett, M.S. (1933). The generalized product moment calculation in a normal system, *Proc. Cambridge Philos. Soc.*, **29**, 260–270.

The Generalization of Student's Ratio*

Harold Hotelling
Columbia University

The accuracy of an estimate of a normally distributed quantity is judged by reference to its variance, or rather, to an estimate of the variance based on the available sample. In 1908 "Student" examined the ratio of the mean to the standard deviation of a sample.[1] The distribution at which he arrived was obtained in a more rigorous manner in 1925 by R.A. Fisher,[2] who at the same time showed how to extend the application of the distribution beyond the problem of the significance of means, which had been its original object, and applied it to examine regression coefficients and other quantities obtained by least squares, testing not only the deviation of a statistic from a hypothetical value but also the difference between two statistics.

Let ξ be any linear function of normally and independently distributed observations of equal variance, and let s be the estimate of the standard error of ξ derived by the method of maximum likelihood. If we let t be the ratio to s of the deviation of ξ from its mathematical expectation, Fisher's result is that the probability that t lies between t_1 and t_2 is

$$\frac{1}{\sqrt{\pi n}} \frac{\Gamma\left(\dfrac{n+1}{2}\right)}{\Gamma\left(\dfrac{n}{2}\right)} \int_{t_1}^{t_2} \frac{dt}{\left(1 + \dfrac{t^2}{n}\right)^{(n+1)/2}} \tag{1}$$

where n is the number of degrees of freedom involved in the estimate s.

---

* Presented at the meeting of the American Mathematical Society at Berkeley, April 11, 1931.

[1] Biometrika, vol. 6 (1908), p. 1.

[2] Applications of Student's Distribution, Metron, vol. 5 (1925), p. 90.

It is easy to see how this result may be extended to cases in which the variances of the observations are not equal but have known ratios and in which, instead of independence among the observations, we have a known system of intercorrelations. Indeed, we have only to replace the observations by a set of linear functions of them which are independently distributed with equal variance. By way of further extension beyond the cases discussed by Fisher, it may be remarked that the estimate of variance s^2 may be based on a body of data not involved in the calculation of ξ. Thus the accuracy of a physical measurement may be estimated by means of the dispersion among similar measurements on a different quantity.

A generalization of quite a different order is needed to test the simultaneous deviations of several quantities. This problem was raised by Karl Pearson in connection with the determination whether two groups of individuals do or do not belong to the same race, measurements of a number of organs or characters having been obtained for all the individuals. Several "coefficients of racial likeness" have been suggested by Pearson and by V. Romanovsky with a view to such biological uses. Romanovsky has made a careful study[1] of the sampling distributions, assuming in each case that the variates are independently and normally distributed. One of Romanovsky's most important results is the exact sampling distribution of L, a constant multiple of the sum of the squares of the values of t for the different variates. This distribution function is given by a somewhat complex infinite series. For large samples and numerous variates it slowly approximates to the normal form; for 500 individuals, Romanovsky considers that an adequate approach to normality requires that no fewer than 62 characters be measured in each individual. When it is remembered that all these characters must be entirely independent, and that it is usually hard to find as many as three independent characters, the difficulties in application will be apparent. To avoid these troubles, Romanovsky proposes a new coefficient of racial likeness, H, the average of the ratios of variances in the two samples for the several characters. He obtains the exact distribution of H, again as an infinite series, though it approaches normality more rapidly than the distribution of L. But H does not satisfy the need for a comparison between magnitudes of characters, since it concerns only their variabilities.

Joint comparisons of correlated variates, and variates of unknown correlations and standard deviations, are required not only for biologic purposes, but in a great variety of subjects. The eclipse and comparison star plates used in testing the Einstein deflection of light show deviations in right ascension and in declination; an exact calculation of probability combining the two least-square solutions is desirable. The comparison of the prices of a list of

[1] V. Romanovsky, On the criteria that two given samples belong to the same normal population (on the different coefficients of racial likeness), Metron, vol. 7 (1928), no. 3, pp. 3–46; K. Pearson, On the coefficient of racial likeness, Biometrika, vol. 18 (1926), pp. 105–118.

commodities at two times, with a view to discovering whether the changes are more than can reasonably be ascribed to ordinary fluctuation, is a problem dealt with only very crudely by means of index numbers, and is one of many examples of the need for such a coefficient as is now proposed. We shall generalize Student's distribution to take account of such cases.

We consider p variates $x_1, x_2, \ldots, x_p$, each of which is measured for N individuals, and denote by $X_{i\alpha}$ the value of x_i for the αth individual. Taking first the problem of the significance of the deviations from a hypothetical set of mean values $m_1, m_2, \ldots, m_p$, we calculate the means $\bar{x}_1, \bar{x}_2, \ldots, \bar{x}_p$, of the samples, and put

$$\xi_i = (\bar{x}_i - m_i)\sqrt{N}.$$

Then the mean values of the ξ_i will all be zero, and the variances and co-variances will be the same as for the corresponding x_i, since the individuals are supposed chosen independently from an infinite population.[1] In order to estimate them with the help of the deviations

$$x_i = X_{i\alpha} - \bar{x}_i$$

from the respective means, we call $n = N - 1$ the number of degrees of freedom and take as the estimates of the variances and covariances,

$$a_{ji} = a_{ij} = \frac{1}{n} \sum_{\alpha=1}^{N} x_{i\alpha} x_{j\alpha} \tag{2}$$

We next put:

$$a = \begin{vmatrix} a_{11} & a_{12} & \cdots & a_{1p} \\ a_{21} & a_{22} & \cdots & a_{2p} \\ \cdots\cdots\cdots\cdots\cdots\cdots \\ a_{p1} & a_{p2} & \cdots & a_{pp} \end{vmatrix}$$

$$A_{ij} = A_{ji} = \frac{\text{cofactor of } a_{ij} \text{ in } a}{a}. \tag{3}$$

The measure of simultaneous deviations which we shall employ is

$$T^2 = \sum_{i=1}^{p} \sum_{j=1}^{p} A_{ij} \xi_i \xi_j. \tag{4}$$

For a single variate it is natural to take $A_{11} = 1/a_{11}$; then T reduces to t, the ordinary "critical ratio" of a deviation in a mean to its estimated standard error, a ratio which has "Student's distribution," (1). For examining the deviations from zero of two variates x and y,

----

[1] "Mean Value" is used in the sense of mathematical expectation; the variance of a quantity whose mean value is zero is defined as the expectation of its squares; the covariance of two such quantities is the expectation of their product. Thus the correlation of the two in a hypothetical infinite population is the ratio of their covariance to the geometric mean of the variances.

$$T = \frac{N}{L - r^2} \left\{ \frac{\bar{x}^2}{s_1^2} - \frac{2r\bar{x}\bar{y}}{s_1 s_2} + \frac{\bar{y}^2}{s_2^2} \right\},$$

where

$$s_1^2 = \frac{\sum (X - \bar{x})^2}{N - 1}, \qquad s_2^2 = \frac{\sum (Y - \bar{y})^2}{N - 1},$$

$$r = \frac{\sum (X - \bar{x})(Y - \bar{y})}{\sqrt{\sum (X - \bar{x})^2 \sum (Y - \bar{y})^2}}.$$

For comparing the means of two samples, one of N_1 and the other of N_2 individuals, we distinguish symbols pertaining to the second sample by primes, and write

$$\xi_i = \frac{\bar{x}_i - \bar{x}_i'}{\sqrt{1/N_1 + 1/N_2}} \tag{5}$$

$$n = N_1 + N_2 - 2,$$

$$a = \frac{1}{n} [\sum (X_{i\alpha} - \bar{x}_i)(X_{j\alpha} - \bar{x}_j) + \sum (X_{i\alpha}' - \bar{x}_i')(X_{j\alpha}' - \bar{x}_j')]$$

$$= \frac{1}{n} [\sum X_{i\alpha} X_{j\alpha} - N_1 \bar{x}_i \bar{x}_j + \sum X_{i\alpha} X_{j\alpha} - N_2 \bar{x}_i' \bar{x}_j'] \tag{6}$$

and take as our "coefficients of racial likeness" the value (4) of T^2, in which the ξ_i are calculated from (5) and the A_{ij} from (6) and (3).

Other situations to which the measure T^2 of simultaneous deviations can be applied include comparisons of regression coefficients and slopes of lines of secular trend, comparisons which for single variates have been explained by R.A. Fisher.[1] In each case we deal for each variate with a linear function ξ_i of the observed values, such that the sum of the squares of the coefficients is unity, so that the variance is the same as for a single observation, and such that the expectation of ξ_i is, on the hypothesis to be tested, zero. Deviations $x_{i\alpha}$ of the observations from means, or from trend lines or other such estimates, are used to provide the estimated variances and covariances a_{ij} by (2). The number of degrees of freedom n is the difference between the number N of individuals and the number q of independent linear relations which must be satisfied by the quantities $x_{i1}, x_{i2}, \ldots, x_{iN}$ on account of their method of derivation. For all the variates, these relations and n must be the same.

The general procedure is to set up what may be called normal values $\bar{x}_{i\alpha}$ for the respective $X_{i\alpha}$, putting

$$x_{i\alpha} = X_{i\alpha} - \bar{x}_{i\alpha}. \tag{7}$$

The underlying assumption is that $X_{i\alpha}$ is composed of two parts, of which one,

[1] Metron, loc. cit., and Statistical Methods for Research Workers, Oliver and Boyd, third edition (1928).

$\varepsilon_{i\alpha}$, is normally and independently distributed about zero with variance σ_i^2 which is the same for all the observations on x_i. The other component is determined by the time, place, or other circumstances of the αth observation in some regular manner, the same for all the variates. Denoting this part by $\eta_{i\alpha}$, we have

$$X_{i\alpha} = \eta_{i\alpha} + \varepsilon_{i\alpha}.$$

Specifically, we take $\eta_{i\alpha}$ to be a linear function, with known coefficients $g_{\alpha s}$, of q unknown parameters $\zeta_{i1}, \ldots, \zeta_{iq}$ where $q < N$:

$$\eta_{i\alpha} = \sum_{s=1}^{q} g_{\alpha s}\zeta_{is}. \tag{8}$$

Thus in dealing with a secular trend representable by a polynomial in the time, we may take the g's as powers of the time-variable, the ζ's as the coefficients. For differences of means, the g's are 0's and 1's, and the ζ's the true means.

We estimate the ζ's by minimizing

$$2V_i = \sum_{\alpha=1}^{N} \varepsilon_{i\alpha}^2 = \sum_{\alpha=1}^{N} (X_{i\alpha} - \eta_{i\alpha})^2. \tag{9}$$

Substituting from (8), differentiating with respect to ζ_{is}, and replacing $\eta_{i\alpha}$ by $\bar{x}_{i\alpha}$ for the minimizing value, we obtain:

$$\sum_{\alpha=1}^{N} g_{\alpha s}(X_{i\alpha} - \bar{x}_{i\alpha}) = 0 \qquad (s = 1, 2, \ldots, q) \tag{10}$$

or by (7),

$$\sum_{\alpha=1}^{N} g_{\alpha s}x_{i\alpha} = 0 \qquad (s = 1, 2, \ldots, q). \tag{11}$$

Denoting also the minimizing values of ζ_{is} by z_{is}, we have made from (8),

$$\bar{x}_{i\alpha} = \sum_{s=1}^{q} g_{\alpha s}z_{is}.$$

Subtracting (8),

$$\bar{x}_{i\alpha} - \eta_{i\alpha} = \sum_{s=1}^{q} g_{\alpha s}(z_{is} - \zeta_{is}). \tag{12}$$

From (9),

$$2V = \sum_{\alpha=1}^{N} [(X_{i\alpha} - \bar{x}_{i\alpha}) + (\bar{x}_{i\alpha} - \eta_{i\alpha})]^2$$

$$= \sum_{\alpha=1}^{N} (X_{i\alpha} - \bar{x}_{i\alpha})^2 + 2 \sum_{\alpha=1}^{N} (X_{i\alpha} - \bar{x}_{i\alpha})(\bar{x}_{i\alpha} - \eta_{i\alpha})$$

$$+ \sum_{\alpha=1}^{N} (\bar{x}_{i\alpha} - \eta_{i\alpha})^2. \tag{13}$$

The middle term, by (12), equals

$$2 \sum_{\alpha=1}^{N} \sum_{s=1}^{q} g_{\alpha s}(X_{i\alpha} - \bar{x}_{i\alpha})(z_{is} - \zeta_{is}),$$

this, by (10), is zero. Hence, by (7) and (13),

$$U_i = V_i + W_i,$$

where

$$2V_i = \sum_{\alpha=1}^{N} x_{i\alpha}^2$$

$$2W_i = \sum_{\alpha=1}^{N} (\bar{x}_{i\alpha} - \eta_{i\alpha})^2.$$

If the q equations (10) be solved for $\bar{x}_{i1}, \bar{x}_{i2}, \ldots, \bar{x}_{iN}$, the values of these quantities will be found to be homogeneous linear functions of the observations $X_{i\alpha}$. By (7), therefore, the quantities

$$\bar{x}_{i1}, \bar{x}_{i2}, \ldots, \bar{x}_{iN}$$

are homogeneous linear functions of the $X_{i\alpha}$. But they are not linearly independent functions, since they are connected by the q relations (11). Hence V is a quadratic form of rank

$$n = N - q.$$

Since V_i, by (9), is of rank N, W is of rank q.

This shows that Np new quantities $x'_{i\alpha}$, given by equations of the form

$$x'_{i\alpha} = \sum_{\beta=1}^{N} c_{\alpha\beta} x_{i\beta} = \sum_{\beta=1}^{N} c_{\alpha\beta} X_{i\beta}, \qquad (\alpha = 1, 2, \ldots, n)$$

$$x'_{i\alpha} = \sum_{\beta=1}^{N} c_{\alpha\beta}(\bar{x}_{i\beta} - \eta_{i\beta}) = \sum_{\beta=1}^{N} (c_{\alpha\beta} X_{i\beta} - c_{\alpha\beta} \eta_{i\beta}), \qquad (\alpha = n+1, \ldots, N)$$

(14)

can be found such that

$$2V_i = \sum_{\alpha=1}^{N} x_{i\alpha}^2 = \sum_{\alpha=1}^{N} x_{i\alpha}'^2,$$

$$2W_i = \sum_{\alpha=n+1}^{N} x_{i\alpha}'^2,$$

(15)

and therefore

$$2U_i = \sum_{\alpha=1}^{N} x_{i\alpha}'^2.$$

(16)

Substituting (14) in (15) and equating like coefficients,

$$\sum_{\alpha=1}^{n} c_{\alpha\beta} c_{\alpha\gamma} = \delta_{\beta\gamma}$$

(17)

where $\delta_{\beta\gamma}$ is the Kronecker delta, equal to 1 if $\beta = \gamma$, to 0 if $\beta \neq \gamma$.

The coefficients $c_{\alpha\beta}$ depend only on the g_{as}, which have been assumed to be the same for all the p variates. Thus (14) may be written

$$x'_{j\alpha} = \sum_{\gamma=1}^{N} c_{\alpha\gamma} x_{j\gamma}.$$

Multiplying by (14), summing with respect to α from 1 to n, and using (17),

$$\sum_{\alpha=1}^{n} x'_{i\alpha} x'_{j\alpha} = \sum_{\alpha=1}^{n} \sum_{\beta=1}^{N} \sum_{\gamma=1}^{N} c_{\alpha\beta} c_{\alpha\gamma} x_{i\beta} x_{j\gamma}$$

$$= \sum_{\beta=1}^{N} \sum_{\gamma=1}^{N} \delta_{\beta\gamma} x_{i\beta} x_{j\gamma} = \sum_{\beta=1}^{N} x_{i\beta} x_{j\beta}. \qquad (18)$$

Just as in (2), we define a_{ij} in this generalized case by

$$a_{ij} = \frac{1}{n} \sum_{\alpha=1}^{N} x_{i\alpha} x_{j\alpha}. \qquad (19)$$

Then by (18),

$$a_{ij} = \frac{1}{n} \sum_{\alpha=1}^{N} x'_{i\alpha} x'_{j\alpha}. \qquad (20)$$

Of the last equation, (6) is a special case.

The random parts $\varepsilon_{i\alpha}$ of the observations on x_i have by hypothesis the distribution

$$\frac{1}{(\sigma_i \sqrt{2\pi})^N} e^{-U_i/2\sigma_i^2} d\varepsilon_{i1} \, d\varepsilon_{i2}, \ldots, d\varepsilon_{iN},$$

where V_i is given by (9). From what has been shown, it is clear that this may be transformed into

$$\frac{1}{(\sigma_i \sqrt{2\pi})^N} e^{-(x'^2_{i1} + x'^2_{i2} + \cdots + x'^2_{iN})/2\sigma_i^2} dx'_{i1}, \ldots, dx'_{iN},$$

showing that $x'_{i1}, \ldots, x'_{iN}$ are normally and independently distributed with equal variance σ_i^2.

The statistic ξ_i must be independent of the quantities $x'_{i1}, x'_{i2}, \ldots, x'_{in}$ entering into (20), its mean value must be zero, and its variance must be σ_i^2. These conditions are satisfied in the cases which have been mentioned, and are satisfied in general if ξ_i is a linear homogeneous function of $x'_{i,n+1}, \ldots, x'_{iN}$ with the sum of the squares of the coefficients equal to unity.

The measure of simultaneous discrepancy is

$$T^2 = \sum_{i=1}^{p} \sum_{j=1}^{p} A_{ij} \xi_i \xi_j,$$

A_{ij} being defined by (3) on the basis of (19). It is evident that

$$T^2 = -\frac{\begin{vmatrix} 0 & \xi_1 & \xi_2 & \cdots & \xi_p \\ \xi_1 & a_{11} & a_{12} & \cdots & a_{1p} \\ \xi_2 & a_{21} & a_{22} & \cdots & a_{2p} \\ & \cdots & a & \cdots & \\ \xi_p & a_{p1} & a_{p2} & \cdots & a_{pp} \end{vmatrix}}{\begin{vmatrix} a_{11} & a_{12} & \cdots & a_{1p} \\ a_{21} & a_{22} & \cdots & a_{2p} \\ & \cdots & & \\ a_{p1} & a_{p2} & \cdots & a_{pp} \end{vmatrix}} \tag{21}$$

as appears when the numerator is expanded by the first row, and the resulting determinants by their first columns.

A most important property of T is that it is an absolute invariant under all homogeneous linear transformations of the variates $x_i, \ldots, x_p$. This may be seen most simply by tensor analysis; for ξ_i is covariant of the first order and A_{ij} is contravariant of the second order.

The invariance of T shows that in seeking its sampling distribution we may, without loss of generality, assume that the variates $x_1, \ldots, x_p$ have, in the normal population, zero correlations and equal variances for they may always by a linear transformation be replaced by such variates.

Let us now take

$$\xi_i, x'_{i1}, x'_{i2}, \ldots, x'_{in}$$

as rectangular coordinates of a point P_i in space V_{n+1} of $n + 1$ dimensions. Since these quantities are normally and independently distributed with equal variance about zero, the probability density for P_i has spherical symmetry about the origin. Indefinite repetition of the sampling would result in a globular cluster of representative points for each variate. Actually the sample in hand fixes the points $P_1, P_2, \ldots, P_p$, which may be regarded as taken independently.

We shall now show that T is a function of the angle θ between the ξ-axis and the flat space V_p containing the points $P_1, P_2, \ldots, P_p$ and the origin 0. We shall denote by A the point on the ξ-axis of coordinates $1, 0, 0, \ldots, 0$, and by V_n the flat space containing the remaining axes. Since in V_{n+1} one equation specifies V_n and $n + 1 - p$ equations V_p, the intersection of V_n and V_p is specified by all these $n + 2 - p$ equations, and is therefore of $p - 1$ dimensions. Call it V_{p-1}.

If $P_1, P_2, \ldots, P_p$ be moved about in V_p, θ will not change, and neither will T, since T is invariant under linear transformations, equivalent to such motions of the P_i. Hence T always has the value which it takes if all the lines OP_1, $OP_2, \ldots, OP_p$ are perpendicular, with the last $p - 1$ of these lines lying in V_{p-1}. In this case the angle AOP_1 equals θ. Applying to the coordinates of A and of P_1 the formula for the cosine of an angle at the origin of lines to $(x_1, x_2, \ldots)$ and $(y_1, y_2, \ldots)$, namely,

$$\cos \theta = \frac{\sum xy}{\sqrt{\sum x^2 \sum y^2}}. \tag{22}$$

We obtain

$$\cos \theta = \frac{\xi}{\sqrt{\xi_1^2 + x_{11}'^2 + \cdots + x_{1n}'^2}}.$$

Since $x_{11}'^2 + \cdots + x_{1n}'^2 = na_{11}$, it follows that

$$n \cot^2 \theta = \xi_1^2/a_{11}. \tag{23}$$

The fact that $P_2, P_3, \ldots, P_p$ lie in V_{p-1}, and therefore in V_n, shows that in this case

$$\xi_2 = \xi_3 = \cdots = \xi_p = 0.$$

Because $OP_1, OP_2, \ldots, OP_p$ are mutually perpendicular, (20) and (22) show that $a_{ij} = 0$ whenever $i \neq j$. Hence, by (21) and (23),

$$T = \xi_1/a_{11} = \sqrt{n} \cot \theta. \tag{24}$$

By this result the problem of the sampling distribution of T is reduced to that of the angle θ between a line OA in V_{n+1} and the flat space V_p containing p other lines drawn independently through the origin. The distribution will be unaffected if we suppose V_p fixed and OA drawn at random, with spherical symmetry for the points A.[1] Let us then, abandoning the coordinates hitherto used, take new axes of rectangular coordinates $y_1, y_2, \ldots, y_{n+1}$, of which the first p lie in V_p. A unit hypersphere about 0 is defined in terms of the generalized latitude-longitude parameters $\phi_1, \ldots, \phi_n$ if we put

$$y_1 = \sin \phi_1 \sin \phi_2 \sin \phi_3 \ldots \sin \phi_{p-1} \cos \phi_p$$

$$y_2 = \cos \phi_1 \sin \phi_2 \sin \phi_3 \ldots \sin \phi_{p-1} \cos \phi_p$$

$$y_3 = \qquad \cos \phi_2 \sin \phi_3 \ldots \sin \phi_{p-1} \cos \phi_p$$

$$y_4 = \qquad\qquad \cos \phi_3 \ldots \sin \phi_{p-1} \cos \phi_p$$

. . . .

[1] This geometrical interpretation of T shows its affinity with the multiple correlation coefficient, whose interpretation as the cosine of an angle of a random line with a V_p enabled R.A. Fisher to obtain its exact distribution (Phil. Trans., vol. 213B, 1924, p. 91; and Proc. Roy. Soc., vol. 121A, 1928, p. 654). The omitted steps in Fisher's argument may be supplied with the help of generalized polar coordinates as in the text. Other examples of the use of these coordinates in statistics have been given by the author in The Distribution of Correlation Ratios Calculated from Random Data, Proc. Nat. Acad. Sci., vol. 11 (1925), p. 657, and in The Physical State of Protoplasm, Koninklijke Akademie van Wetenschappen te Amsterdam, verhandlingen, vol. 25 (1928), no. 5, pp. 28–31.

$$y_p = \cos \phi_{p-1} \cos \phi_p$$

$$y_{p+1} = \sin \phi_p \cos \phi_{p+1}$$

$$\cdots\cdots\cdots\cdots\cdots\cdots\cdots\cdots\cdots\cdots\cdots\cdots\cdots\cdots\cdots$$

$$y_n = \sin \phi_p \sin \phi_{p+1} \ldots \cos \phi_n$$

$$y_{n+1} = \sin \phi_p \sin \phi_{p+1} \ldots \sin \phi_n,$$

for the sum of the squares is unity. Since

$$y_{p+1}^2 + \cdots + y_{n+1}^2 = \sin^2 \phi_p$$

we have

$$\phi_p = \theta.$$

The element of probability is proportional to the element of generalized area, which is given by

$$\sqrt{D} \, d\phi_1 \, d\phi_2, \ldots, d\phi_n,$$

where D is an n-rowed determinant in which the element in the ith row and jth column is

$$\sum_{k=1}^{n+1} \frac{\partial y_k}{\partial \phi_i} \frac{\partial y_k}{\partial \phi_j}.$$

For $i \neq j$, this is zero. Of the diagonal elements, the first $p - 1$ contain the factor $\cos^2 \phi_p$; the pth is unity; and the remaining $n - p$ elements contain the factor $\sin^2 \phi_p$. Since ϕ is not otherwise involved, the element of area is the product of

$$\cos^{p-1} \phi_p \sin^{n-p} \phi_p \, d\phi_p$$

by factors independent of ϕ_p. The distribution function of θ is obtained by replacing ϕ_p by θ and integrating with respect to the other parameters. Since θ lies between 0 and $\pi/2$, we divide by the integral between these limits and obtain for the frequency element,

$$\frac{2\Gamma\left(\dfrac{n+1}{2}\right)}{\Gamma\left(\dfrac{p}{2}\right)\Gamma\left(\dfrac{n-p+1}{2}\right)} \cos^{p-1} \theta \sin^{n-p} \theta \, d\theta.$$

Substituting from (24) we have as the distribution of T:

$$\frac{2\Gamma\left(\dfrac{n+1}{2}\right)}{\Gamma\left(\dfrac{p}{2}\right)\Gamma\left(\dfrac{n-p+1}{2}\right) n^{p/2}} \frac{T^{p-1} \, dT}{\left(1 + \dfrac{T^2}{n}\right)^{(n+1)/2}}. \tag{25}$$

For $p = 1$ this reduces to the form of Student's distribution given by Fisher and tabulated in the issue of Metron cited; however, as T may be negative as well as positive in this case, Fisher omits the factor 2.

For $p = 2$ the distribution becomes

$$\frac{n-1}{n} \frac{T \, dT}{\left(1 + \dfrac{T^2}{n}\right)^{(n+1)/2}}.$$

From this it is easy to calculate as the probability that a given value of T will be exceeded by chance,

$$P = \frac{1}{\left(1 + \dfrac{T^2}{n}\right)^{(n-1)/2}}, \tag{26}$$

a very convenient expression.

The probability integral for higher values of p may be calculated in various ways, the most direct being successive integration by parts, giving a series of terms analogous to (26) to which, if p is odd, is added an integral which may be evaluated with the help of the tables of Student's distribution. If p is large, this process is laborious; but other methods are available.

The probability integral is reduced to the incomplete beta function if we put

$$x = (1 + T^2/n)^{-1},$$

for then the integral of (25) from T to infinity becomes

$$P = I_x\left(\frac{n - p + 1}{2}, \ \frac{p}{2}\right),$$

the notation being

$$B_x(p, q) = \int_0^x x^{p-1}(1 - x)^{q-1} \, dx,$$

$$B(p, q) = \int_0^1 x^{p-1}(1 - x)^{q-1} \, dx,$$

$$I_x(p, q) = \frac{B_x(p, q)}{B(p, q)}.$$

Many methods of calculation have been discussed by H.E. Soper[1] and by V. Romanovsky.[2] An extensive table of the incomplete beta function being

[1] Tracts for Computers, no. 7 (1921).

[2] On certain expansions in series of polynomials of incomplete B-functions (in English), Recueil Math. de la Soc. de Moscou, vol. 33 (1926), pp. 207–229.

prepared under the supervision of Professor Karl Pearson has not yet been published.

Perhaps the most generally useful method now available is to make the substitution

$$z = \tfrac{1}{2} \log_e (n - p + 1) T^2 - \tfrac{1}{2} \log_e np,$$

$$n_1 = p,$$

$$n_2 = n - p + 1,$$

reducing (25) to a form considered by Fisher. Table VI in his book, "Statistical Methods for Research Workers," gives the values of z which will be exceeded by chance in 5 per cent and in 1 per cent of cases. If the value of z obtained from the data is greater than that in Fisher's table, the indication is that the deviations measured are real.

If the variances and covariances are known a priori, they are to be used instead of the a_{ij}; the resulting expression T has the well known distribution of χ, with p degrees of freedom. For very large samples the estimates of the covariances from the sample are sufficiently accurate to permit the use of the χ distribution for T. This is well shown by (25), in which, as n increases, the factor involving T approaches

$$T^{p-1} e^{-T^2/2} dT,$$

which is proportional to the frequency element for χ when χ is put for T.

As Pearson pointed out, the labor of calculating χ, which we replace by T, is prohibitive when forty or fifty characters are measured on each individual. With two, three, or four characters, however, the labor is very moderate, and the results far more accurate than any attainable with the Pearson coefficient. The great advantage of using T is the simplicity of its distribution, with its complete independence of any correlations among the variates which may exist in the population.

To means of a single variate it is customary to attach a "probable error," with the assumption that the difference between the true and calculated values is almost certainly less than a certain multiple of the probable error. A more precise way to follow out this assumption would be to adopt some definite level of probability, say $P = .05$, of a greater discrepancy, and to determine from a table of Student's distribution the corresponding value of t, which will depend on n; adding and subtracting the product of this value of t by the estimated standard error would give upper and lower limits between which the true values may with the given degree of confidence be said to lie. With T an exactly analogous procedure may be followed, resulting in the determination of an ellipse or ellipsoid centered at the point $\xi_1, \xi_2, \ldots, \xi_p$. Confidence corresponding to the adopted probability P may then be placed in the proposition that the set of true values is represented by a point within this boundary.

Introduction to
Neyman and Pearson (1933) On the Problem of the Most Efficient Tests of Statistical Hypotheses

E.L. Lehmann
University of California at Berkeley

Hypothesis testing throughout the 19th century was sporadic and was (1) based on large sample approximations to the distributions of test statistics that were (2) chosen on intuitive grounds.

In contrast to (1), a small-sample approach was initiated in Student's paper of 1908. Between 1915 and 1925, small-sample distribution theory underwent spectacular development at the hand of Fisher who derived the exact distributions of t, χ^2, F and various correlation coefficients in normal models and applied the results to a variety of testing problems. By establishing a collection of useful tests, he incidentally drew attention to hypothesis testing as an important statistical tool and thereby provided the impetus for a more systematic study of the subject. In particular, regarding (2), a logical foundation for the choice of a test statistic more solid than mere intuitive appeal, E.S. Pearson writes (1966), "In the middle of the 1920's the development of small-sample distribution theory ... drew more than one statistician into examining the philosophy of choice among statistical techniques."

One may wonder why an interest in this question with regard to hypothesis testing had to wait for the work of Student and Fisher. One reason is that the large-sample distribution of statistics such as Student's t requires no assumptions concerning the form of the underlying distribution. On the other hand, such assumptions are at the very heart of the small-sample approach. Since the optimal choice of a test statistic depends on the underlying distribution, the choice question became meaningful only after this change of viewpoint had occurred.

Another factor has already been implied. Nineteenth century hypothesis testing tended to be informal and to occur primarily in the context of specific applications. It was only with the work of Student and Fisher (and somewhat earlier that of Karl Pearson) that the spotlight fell on hypothesis testing as a general methodology. As E.S. Pearson (1966) writes, "The need for critical discussion [now] became apparent and the illustrative material on which to base discussion had become ample enough to make it worthwhile."

In an earlier paper, after explaining how recent developments had set the stage for the choice question to be asked, Pearson proceeds with an account of the steps that led to the Neyman–Pearson theory.

> I had been trying to discover some principle beyond that of practical expediency which would justify the use of "Student's" ratio $z = (\bar{x} - m)/s$ in testing the hypothesis that the mean of the sample population was at m. Gosset's reply [to the letter in which Pearson (1939) had raised the question] had a tremendous influence on the direction of my subsequent work, for the first paragraph contains the germ of that idea which has formed the basis of all the later joint researches of Neyman and myself. It is the simple suggestion that the only valid reason for rejecting a statistical hypothesis is that some alternative hypothesis explains the observed events with a greater degree of probability.

The following is the passage in Student's letter to which Pearson refers:

> It doesn't in itself necessarily prove that the sample is not drawn randomly from the [hypothetical] population even if the chance is very small, say .00001: what it does is to show that if there is any alternative hypothesis which will explain the occurrence of the sample with a more reasonable probability, say .05 (such as that it belongs to a different population or that the sample wasn't random or whatever will do the trick) you will be very much more inclined to consider that the original hypothesis is not true.

Student's suggestion led Pearson to propose to Neyman the likelihood ratio criterion that rejects the hypothesis when the maximum likelihood of the observed sample under the alternative is sufficiently large compared to its value under the hypothesis. Neyman and Pearson's joint work on this approach culminated in a 100-page paper (1928) in which they set out their ideas and then applied them successfully to many specific testing problems.

Pearson felt that the likelihood ratio provided the unified method for which they had been looking, but Neyman was not yet satisfied. It seemed to him that the likelihood ratio principle itself was somewhat ad hoc and was lacking a fully logical basis. His search for a firmer foundation, which constitutes the third of the three steps, eventually led him to a new formulation: The most desirable test would be obtained by maximizing the power of the test, subject to the condition that under the hypothesis, the rejection probability has a preassigned value, the level of the test. In March 1930, he announced to Pearson that he had found a general solution of this problem in the case where both the hypothesis and alternative are simple. The solution was what became known as the Neyman–Pearson fundamental lemma. [A more detailed account of this history, based on the unpublished correspondence between Neyman and Pearson, can be found in Reid's book (1982). An earlier account is provided by Pearson (1966).]

In the course of working out these ideas, Neyman and Pearson were led from the case of testing a simple hypothesis against a simple alternative to the concepts of uniformly most powerful (UMP) tests and, in the presence of nuisance parameters, to UMP similar tests. They published the results of these investigations together with their application to a number of important examples in the paper reprinted here, which laid the foundations of the theory of hypothesis testing.

Later developments of the NP theory include the concepts of UMP unbiased and invariant tests, tests that maximize the minimum power, extensions to sequential tests and to approximate hypotheses. An important offshoot is Neyman's theory of confidence sets. [For a book-length treatment of the whole subject, see Lehmann (1986).]

The emphasis of the paper and some of its elaborations on the second kind of error and therefore on power led to the gradual realization of the importance of power both for measuring the ability of a test to detect significant treatment effects or other deviations from the hypothesis and for determining the sample size required to achieve adequate power. These have become central concerns of statistical practice and are treated fully in the books by Kraemer and Thiemann (1987) and Cohen (1988). [See also Chap. 14 of Bailar and Mosteller (1986).]

With their formulation, Neyman and Pearson ushered in a new paradigm that became enormously influential: a small-sample theory in which statistical procedures are derived according to some optimality criteria and hence as solutions to clearly stated mathematical problems. To assess the performance of a test on which to base the definition of optimality, Neyman and Pearson (1933) suggested that "we may search for rules to govern our behavior ..., in following which we insure that in the long run of experience we shall not be too often wrong." The idea of viewing statistical procedures as a guide to behavior became a cornerstone of Neyman's statistical philosophy [see, e.g., Neyman (1957)] and is the direct forerunner of Wald's general decision theory. This idea was strongly criticized by Fisher (1955, 1973) who argued that such an approach was suitable for the repetitive situations encountered in acceptance sampling but was inappropriate for scientific inference.

The NP theory was criticized not only on philosophical grounds. In order to discuss some other criticisms,* which in turn led to various extensions and modifications, let us summarize the principal steps required when applying the theory to specific cases.

(i) The starting point is the specification of a model, which Neyman and Pearson assumed to be a parametric family of distributions from one of which the observations are drawn.

(ii) Within this model, the question under investigation is assumed to specify a hypothesis concerning the parameter(s) of interest, say, $H : \theta = \theta_0$,

---

* We shall not discuss here criticism emanating from the Bayesian school which has a completely different basis from that adopted by NP. For a critical discussion of the NP theory from a Bayesian perspective, see, e.g., Berger (1985).

and a class K of alternatives, e.g., one-sided, say, $\theta > \theta_0$ or two-sided $\theta \neq \theta_0$.

(iii) Next a significance level α is specified, which represents the maximum permissible probability for an error of the first kind (i.e., falsely rejecting H).

(iv) Within the framework (i–iii), an optimal procedure is derived for testing H against K at the given level.

(v) A crucial assumption of the theory is that all the steps are to be taken before any observations have been seen.

The basic objection to this program is that it is too rigid and as a result far from statistical practice. The following are more specific objections to some of the steps:

(1) In practice, the models are often selected in the light of the data, which violates assumption (v).

(2) Frequently many questions are asked of the same data rather than the single question stipulated by Neyman and Pearson.

(3) In many situations, a two-decision rule based on a fixed level may not be appropriate.

In response to (1), a new methodology developed, not envisaged by Neyman and Pearson but in the spirit of their theory. It is the theory of model selection, in which the data are used to select a "best" model from a large class of candidates. Similarly, (2) has led to an extension of hypothesis testing to multiple comparisons or more generally simultaneous inference. [A result of this theory is the optimality under appropriate conditions of either Scheffé's S-method or Tukey's T-method; see Wijsman (1979) and Lehmann (1986).] Finally with respect to (3), a popular alternative to the use of a fixed level is to calculate the so-called p-value, the smallest level at which the data would reject the hypothesis. As was recently pointed out by Schweder (1988), the NP theory translates readily into a corresponding theory of p-values. In this connection, it should be mentioned that the frequently criticized use of conventional levels such as $\alpha = .01$ or $.05$ was not recommended by Neyman and Pearson (1933). On the contrary, they say that in any given case, the determination of "just how the balance [between the two kinds of errors, i.e. between level and power] should be struck, must be left to the investigator."

The availability of the options mentioned in the preceding paragraph narrows the gap between theory and practice, but does not close it. The problem of models (or hypotheses) suggested by the data can, of course, be handled by collecting the data in two stages or dividing them into two parts. The first part is used in a freewheeling data analysis to decide on the model and the hypothesis to be tested; once the problem has been formulated, an NP analysis can then be performed on the second part of the data. (This is Tukey's distinction between exploratory and confirmatory analysis.) Such a two-stage procedure will often appear to be wasteful of the data, but it does provide a possible compromise between the too stringent formalism of the NP approach and a set of ad hoc tests not following a predetermined protocol and therefore without any error control.

A difficulty inherent in the application of any model-based theory is that models are never reliable and even at best provide only approximations. The resulting concerns about the sensitivity of the level of a test to model assumptions have led to a large body of robustness studies. A particularly sensitive feature tends to be the optimality of a procedure. This realization has led to the development of tests whose efficiency properties are more robust than those of the normal theory tests (e.g., certain rank tests) and to the theory of adaptive tests, which adapt themselves to the true distribution and thereby become asymptotically efficient for a large class of models.

The great popularity and somewhat indiscriminate use of hypothesis testing as a preferred statistical procedure have led some statisticians to take a strong stand against all testing and in favor of estimation. When the hypothesis specifies the value θ_0 of a parameter θ, the totality of tests for varying θ_0 is, of course, equivalent to confidence sets for θ, so that the difference is then largely a matter of presentation. Confidence intervals for θ have the advantage of providing more information than recording only whether the hypothesis was accepted or rejected.

In addition, there are however many situations (e.g, nonparametric ones), in which no clearly specified alternatives are available and there is no parameter to be estimated. As in many such ideological controversies, there is no universally correct approach: What is most appropriate depends on the particular situation. [A classification of situations into those in which testing or estimation is more appropriate is given by Anscombe (1956).]

Tests that are optimal in the sense of the NP theory, say, UMP, UMP unbiased, or UMP invariant, have been derived for many important problems including univariate and certain multivariate linear hypotheses. However, the existence of such optimal tests is limited. Basically, it is restricted to problems involving exponential or certain group families. Tests maximizing the minimum power against a separated class of alternatives will exist fairly generally, but may be complicated and unappealing. These limitations and the criticisms mentioned earlier have led to some disillusionment with small-sample optimality theory.

Nevertheless, despite their shortcomings, the new paradigm formulated in the 1933 paper and the many developments carried out within its framework continue to play a central role in both the theory and practice of statistics and can be expected to do so in the forseeable future.

References

Anscombe, F.J. (1956). Discussion of David, F.N., and Johnson, N.L., *Some tests of significance with ordered variables, J. Roy. Statist. Soc.*, **18**, 24–27

Bailar, J.C., and Mosteller, F. (1986). *Medical Uses of Statistics*. NEJM Books, Waltham, Mass.

Berger, J.O. (1985). *Statistical Decision Theory and Bayesian Analysis*, 2nd ed., Springer-Verlag, New York.

Cohen, J. (1988). *Statistical Power Analysis for the Behavioral Sciences*, 2nd ed. Lawrence Erlbaum, Hillsdale, N.J.

Fisher, R.A. (1955). Statistical methods and scientific induction, *J. Roy. Statist. Soc.*, **17**, 69–78.

Fisher, R.A. (1973). *Statistical Methods and Scientific Inference*, 3rd ed. Hafner Press, New York.

Kraemer, H.C., and Thiemann, S. (1987). *How Many Subjects? Statistical Power Analysis in Research*. Sage Publications, Newbury Park, Calif.

Lehmann, E.L. (1986). *Testing Statistical Hypotheses*, 2nd ed. Wiley, New York, (1991 Wadsworth & Brooks/Cole, Pacific Grove.)

Neyman, J. (1957). "Inductive behavior" as a basic concept of philosophy of science, *Rev. Int. Statist. Inst.*, **25**, 7–22.

Neyman, J., and Pearson, E.S. (1928). On the use and interpretation of certain test criteria for purposes of statistical inference, *Biometrika*, **20A**, 175–240, 263–295.

Neyman, J., and Pearson, E.S. (1933). On the problem of the most efficient tests of statistical hypotheses, *Phil. Trans. R. Soc., Ser. A*, **231**, 289–337.

Pearson, E.S. (1939). William Sealy Gosset, 1876–1937, *Biometrika*, **30**, 205–250.

Pearson, E.S. (1966). The Neyman-Pearson story: 1926–34, in *Research Papers in Statistical: Festschrift for J. Neyman* (David, F.N. Ed.). Wiley, New York.

Reid, C. (1982). *Neyman—From Life*. Springer-Verlag, New York.

Schweder, T. (1988). A significance version of the basic Neyman-Pearson theory for scientific hypothesis testing, *Scand. J. Statist.*, **15**, 225–242.

Student (1908). On the probable error of a mean, *Biometrika*, **6**, 151–164.

Wijsman, R.A. (1979). Constructing all smallest simultaneous confidence sets in a given class, with applications to MANOVA, *Ann. Statist.*, **7**, 1003–1018.

On the Problem of the Most Efficient Tests of Statistical Hypotheses

J. Neyman
Nencki Institute
Warsaw, Poland

E.S. Pearson
Department of Applied Statistics,
University College
London, United Kingdom

I. Introductory

The problem of testing statistical hypotheses is an old one. Its origins·are usually connected with the name of Thomas Bayes, who gave the well-known theorem on the probabilities *a posteriori* of the possible "causes" of a given event.* Since then it has been discussed by many writers of whom we shall here mention two only, Bertrand† and Borel,‡ whose differing views serve well to illustrate the point from which we shall approach the subject.

Bertrand put into statistical form a variety of hypotheses, as for example the hypothesis that a given group of stars with relatively small angular distances between them as seen from the earth, form a "system" or group in space. His method of attack, which is that in common use, consisted essentially in calculating the probability, P, that a certain character, x, of the observed facts would arise if the hypothesis tested were true. If P were very small, this would generally be considered as an indication that the hypothesis, H, was probably false, and *vice versa*. Bertrand expressed the pessimistic view that no test of this kind could give reliable results.

Borel, however, in a later discussion, considered that the method described could be applied with success provided that the character, x, of the observed facts were properly chosen—were, in fact, a character which he terms "en quelque sorte remarquable".

We appear to find disagreement here, but are inclined to think that, as is so often the case, the difference arises because the two writers are not really

---

* *Phil. Trans.* vol. 53, p. 370 (1763).

† *Calcul des Probabilités*, Paris (1907).

‡ *Le Hasard*, Paris (1920).

considering precisely the same problem. In general terms the problem is this: Is it possible that there are any efficient tests of hypotheses based upon the theory of probability, and if so, what is their nature? Before trying to answer this question, we must attempt to get closer to its exact meaning. In the first place, it is evident that the hypotheses to be tested by means of the theory of probability must concern in some way the probabilities of the different kinds of results of certain trials. That is to say, they must be of statistical nature, or as we shall say later on, they must be statistical hypotheses.

Now what is the precise meaning of the words "an efficient test of a hypothesis?" There may be several meanings. For example, we may consider some specified hypothesis, as that concerning the group of stars, and look for a method which we should hope to tell us, with regard to a particular group of stars, whether they form a system, or are grouped "by chance", their mutual distances apart being enormous and their relative movements unrelated.

If this were what is required of "an efficient test", we should agree with Bertrand in his pessimistic view. For however small be the probability that a particular grouping of a number of stars is due to "chance", does this in itself provide any evidence of another "cause" for this grouping but "chance?" "Comment définir, d'ailleurs, la singularité dont on juge le hasard incapable?"[*] Indeed, if x is a continuous variable—as for example is the angular distance between two stars—then any value of x is a singularity of relative probability equal to zero. We are inclined to think that as far as a particular hypothesis is concerned, no test based upon the theory of probability[†] can by itself provide any valuable evidence of the truth or falsehood of that hypothesis.

But we may look at the purpose of tests from another view-point. Without hoping to know whether each separate hypothesis is true or false, we may search for rules to govern our behaviour with regard to them, in following which we insure that, in the long run of experience, we shall not be too often wrong. Here, for example, would be such a "rule of behaviour": to decide whether a hypothesis, H, of a given type be rejected or not, calculate a specified character, x, of the observed facts; if $x > x_0$ reject H, if $x \leqslant x_0$ accept H. Such a rule tells us nothing as to whether in a particular case H is true when $x \leqslant x_0$ or false when $x > x_0$. But it may often be proved that if we behave according to such a rule, then in the long run we shall reject H when it is true not more, say, than once in a hundred times, and in addition we may have evidence that we shall reject H sufficiently often when it is false.

If we accept the words, "an efficient test of the hypothesis H" to mean simply such a rule of behaviour as just described, then we agree with Borel that efficient tests are possible. We agree also that not any character, x, what-

---

[*] Bertrand, *loc. cit.*, p. 165.

[†] Cases will of course, arise where the verdict of a test is based on certainty. The question "Is there a black ball in this bag?" may be answered with certainty if we find one in a sample from the bag.

ever is equally suitable to be a basis for an efficient test,* and the main purpose of the present paper is to find a general method of determining tests, which, from the above point of view would be the most efficient.

In common statistical practice, when the observed facts are described as "samples", and the hypotheses concern the "populations" from which the samples have been drawn, the characters of the samples, or as we shall term them criteria, which have been used for testing hypotheses, appear often to have been fixed by a happy intuition. They are generally functions of the moment coefficients of the sample, and as long as the variation among the observations is approximately represented by the normal frequency law, moments appear to be the most appropriate sample measures that we can use. But as Fisher† has pointed out in the closely allied problem of Estimation, the moments cease to be efficient measures when the variation departs widely from normality. Further, even though the moments are efficient, there is considerable choice in the particular function of these moments that is most appropriate to test a given hypothesis, and statistical literature is full of examples of confusion of thought on this choice.

A blind adoption of the rule,

Standard Error of $(x - y)$

$$= \sqrt{[(\text{Standard Error of } x)^2 + (\text{Standard Error of } y)^2]}, \qquad (1)$$

has led to frequent inconsistencies. Consider, for example, the problem of testing the significance of a difference between two percentages or proportions; a sample of n_1 contains t_1 individuals with a given character and an independent sample of n_2 contains t_2. Following the rule (1), the standard error of the difference $d = t_1/n_1 - t_2/n_2$ is often given as

$$\sigma_d = \sqrt{\left[\frac{t_1}{n_1^2}\left(1 - \frac{t_1}{n_1}\right) + \frac{t_2}{n_2^2}\left(1 - \frac{t_2}{n_2}\right)\right]}. \qquad (2)$$

But in using

$$\frac{t_1}{n_1^2}\left(1 - \frac{t_1}{n_1}\right) \quad \text{and} \quad \frac{t_2}{n_2^2}\left(1 - \frac{t_2}{n_2}\right)$$

as the squares of the estimates of the two standard errors, we are proceeding on the supposition that sample estimates must be made of two different population proportions p_1 and p_2. Actually, it is desired to test the hypothesis that $p_1 = p_2 = p$, and it follows that the best estimate of p is obtained by combining together the two samples, whence we obtain

---

* This point has been discussed in earlier papers. *See* (a) Neyman and Pearson. *Biometrika*, vol. 20A, pp. 175 and 263 (1928); (b) Neyman. *C.R. Premier Congrès Math., Pays Slaves, Warsaw*, p. 355. (1929); (c) Pearson and Neyman. *Bull. Acad. Polonaise Sci. Lettres*, Série A, p. 73 (1930).

† *Phil. Trans.* vol. 222, A, p. 326 (1921).

Estimate of

$$\sigma_d = \sqrt{\left[\frac{t_1 + t_2}{n_1 + n_2} \left(1 - \frac{t_1 + t_2}{n_1 + n_2} \right) \left(\frac{1}{n_1} + \frac{1}{n_2} \right) \right]}. \tag{3}$$

A rather similar situation arises in the case of the standard error of the difference between two means. In the case of large samples there are two forms of estimate*

$$\sigma_d = \sqrt{\left(\frac{s_1^2}{n_1} + \frac{s_2^2}{n_2} \right)}, \tag{4}$$

$$\sigma_d' = \sqrt{\left(\frac{s_1^2}{n_2} + \frac{s_2^2}{n_1} \right)}, \tag{5}$$

The use of the first form is justified if we believe that in the populations sampled there are different standard deviations, σ_1 and σ_2; while if $\sigma_1 = \sigma_2$ the second form should be taken. The hypothesis concerning the two means has not, in fact, always to be tested under the same conditions, and which form of the criterion is most appropriate is a matter for judgement based upon the evidence available regarding those conditions.

The role of sound judgement in statistical analysis is of great importance and in a large number of problems common sense may alone suffice to determine the appropriate method of attack. But there is ample evidence to show that this has not and cannot always be enough, and that it is therefore essential that the ideas involved in the process of testing hypotheses should be more clearly understood.

In earlier papers we have suggested that the criterion appropriate for testing a given hypothesis could be obtained by applying the principle of likelihood.† This principle was put forward on intuitive grounds after the consideration of a variety of simple cases. It was subsequently found to link together a great number of statistical tests already in use, besides suggesting certain new methods of attack. It was clear, however, in using it that we were still handling a tool not fully understood, and it is the purpose of the present investigation to widen, and we believe simplify, certain of the conceptions previously introduced.

We shall proceed to give a review of some of the more important aspects of the subject as a preliminary to a more formal treatment.

---

* s_1 and s_2 are observed standard deviations in independent samples of n_1 and n_2. The expression (5) is the limiting form of

$$\sqrt{\left[\frac{n_1 s_1^2 + n_2 s_2^2}{n_1 + n_2 - 2} \left(\frac{1}{n_1} + \frac{1}{n_2} \right) \right]}$$

when n_1 and n_2 are large, based upon the common estimate of the unknown population variance σ^2.

† *Biometrika*, vol. 20A.

II. Outline of General Theory

Suppose that the nature of an event, E, is exactly described by the values of n variates,

$$x_1, x_2, \ldots, x_n. \tag{6}$$

For example, the series (6) may represent the value of a certain character observed in a sample of n individuals drawn at random from a given population. Or again, the x's may be the proportions or frequencies of individuals in a random sample falling into n out of the $n + 1$ categories into which the sampled population is divided. In any case, the event, E, may be represented by a point in a space of n dimensions having (6) as its co-ordinates; such a point might be termed an Event Point, but we shall here speak of it in statistical terms as a Sample Point, and the space in which it lies as the Sample Space. Suppose now that there exists a certain hypothesis, H_0, concerning the origin of the event which is such as to determine the probability of occurrence of every possible event E. Let

$$p_0 = p_0(x_1, x_2, \ldots, x_n) \tag{7}$$

be this probability—or if the sample point can vary continuously, the elementary probability of such a point. To obtain the probability that the event will give a sample point falling into a particular region, say w, we shall have either to take the sum of (7) over all sample points included in w, or to calculate the integral

$$P_0(w) = \int \cdots \int_w p_0(x_1, x_2, \ldots, x_n)\, dx_1\, dx_2 \ldots dx_n. \tag{8}$$

The two cases are quite analogous as far as the general argument is concerned, and we shall consider only the latter. That is to say, we shall assume that the sample points may fall anywhere within a continuous sample space (which may be limited or not), which we shall denote by W. It will follow that

$$P_0(W) = 1. \tag{9}$$

We shall be concerned with two types of hypotheses, (a) simple and (b) composite. The hypothesis that an event E has occurred subject to a completely specified probability law $p_0(x_1, x_2, \ldots, x_n)$ is a simple one; while if the functional form of p_0 is given, though it depends upon the value of c unspecified parameters, H_0 will be called a composite hypothesis with c degrees of freedom.* The distinction may be illustrated in the case where H_0 concerns the population Π from which a sample, Σ has been drawn at random. For example, the normal frequency law,

---

* The idea of degrees of freedom as defined above, though clearly analogous, is not to be confused with that introduced by Fisher.

$$p(x) = \frac{1}{\sigma\sqrt{(2\pi)}}e^{-(1/2)(x-a)^2/\sigma^2}, \tag{10}$$

represents an infinite set of population distributions. A simple hypothesis is that Σ has been sampled from a definite member of this set for which $a = a_0$, $\sigma = \sigma_0$. A composite hypothesis with one degree of freedom is that the sampled population, II, is one of the sub-set for which $a = a_0$ but for which σ may have any value whatever. "Student's" original problem consisted in testing this composite hypothesis.*

The practice of using observational data to test a composite hypothesis is a familiar one. We ask whether the variation in a simple character may be considered as following the normal law; whether two samples are likely to have come from a common population; whether regression is linear; whether the variance in a number of samples differs significantly. In these cases we are not concerned with the exact value of particular parameters, but seek for information regarding the conditions and factors controlling the events.

It is clear that besides H_0 in which we are particularly interested, there will exist certain admissible alternative hypotheses. Denote by Ω the set of all simple hypotheses which in a particular problem we consider as admissible. If H_0 is a simple hypothesis, it will clearly belong to Ω. If H_0 is a composite hypothesis, then it will be possible to specify a part of the set Ω, say ω, such that every simple hypothesis belonging to the sub-set ω will be a particular case of the composite hypothesis H_0. We could say also that the simple hypotheses belonging to the sub-set ω, may be obtained from H_0 by means of some additional conditions specifying the parameters of the function (7) which are not specified by the hypothesis H_0.

In many statistical problems the hypotheses concern different populations from which the sample, Σ, may have been drawn. Therefore, instead of speaking of the sets Ω or ω of simple hypothesis, it will be sometimes convenient to speak of the sets Ω or ω of populations. A composite hypothesis, H_0, will then refer to populations belonging to the sub-set ω, of the set Ω. Every test of a statistical hypothesis in the sense described above, consists in a rule of rejecting the hypothesis when a specified character, x, of the sample lies within certain critical limits, and accepting it or remaining in doubt in all other cases. In the n-dimensional sample space, W, the critical limits for x will correspond to a certain critical region w, and when the sample point falls within this region we reject the hypothesis. If there are two alternative tests for the same hypothesis, the difference between them consists in the difference in critical regions.

We can now state briefly how the criterion of likelihood is obtained. Take any sample point, Σ, with co-ordinates $(x_1, x_2, \ldots, x_n)$ and consider the set A_Σ of probabilities $p_H(x_1, x_2, \ldots, x_n)$ corresponding to this sample point and determined by different simple hypotheses belonging to Ω. We shall sup-

---

* *Biometrika*, vol. 6, p. 1 (1908).

pose that whatever be the sample point the set A_Σ is bounded. Denote by $p_\Omega(x_1, x_2, \ldots, x_n)$ the upper bound of the set A_Σ, then if H_0 is a simple hypothesis, determining the elementary probability $p_0(x_1, x_2, \ldots, x_n)$, we have defined its likelihood to be

$$\lambda = \frac{p_0(x_1, x_2, \ldots, x_n)}{p_\Omega(x_1, x_2, \ldots, x_n)}. \tag{11}$$

If H_0 is a composite hypothesis, denote by $A_\Sigma(\omega)$ the sub-set of A_Σ corresponding to the set ω of simple hypotheses belonging to H_0 and by $p_\omega(x_1, x_2, \ldots, x_n)$ the upper bound of $A_\Sigma(\omega)$. The likelihood of the composite hypothesis is then

$$\lambda = \frac{p_\omega(x_1, x_2, \ldots, x_n)}{p_\Omega(x_1, x_2, \ldots, x_n)}. \tag{12}$$

In most cases met with in practice, the elementary probabilities, corresponding to different simple hypotheses of the set Ω are continuous and differentiable functions

$$p(a_1, \alpha_2, \ldots, \alpha_c, \alpha_{c+1}, \ldots, \alpha_k; x_1, x_2, \ldots, x_n), \tag{13}$$

of the certain number, k, of parameters $\alpha_1, \alpha_2, \ldots, \alpha_c, \alpha_{c+1}, \ldots, \alpha_k$; and each simple hypothesis specifies the values of these parameters. Under these conditions the upper bound, $p_\Omega(x_1, x_2, \ldots, x_n)$, is often a maximum of (13) (for fixed values of the x's) with regard to all possible systems of the α's. If H_0 is a composite hypothesis with c degrees of freedom, it specifies the values of $k - c$ parameters, say $\alpha_{c+1}, \alpha_{c+2}, \ldots, \alpha_k$ and leaves the others unspecified. Then $p_\omega(x_1, x_2, \ldots, x_n)$ is often a maximum of (13) (for fixed values of the x's and of $\alpha_{c+1}, \alpha_{c+2}, \ldots, \alpha_k$) with regard to all possible values of $\alpha_1, \alpha_2, \ldots, \ldots, \alpha_c$.

The use of the principle of likelihood in testing hypotheses, consists in accepting for critical regions those determined by the inequality

$$\lambda \leqslant C = \text{const.} \tag{14}$$

Let us now for a moment consider the form in which judgements are made in practical experience. We may accept or we may reject a hypothesis with varying degrees of confidence; or we may decide to remain in doubt. But whatever conclusion is reached the following position must be recognized. If we reject H_0, we may reject it when it is true; if we accept H_0, we may be accepting it when it is false, that is to say, when really some alternative H_i is true. These two sources of error can rarely be eliminated completely; in some cases it will be more important to avoid the first, in others the second. We are reminded of the old problem considered by Laplace of the number of votes in a court of judges that should be needed to convict a prisoner. Is it more serious to convict an innocent man or to acquit a guilty? That will depend upon the consequences of the error; is the punishment death or fine; what is the danger to the community of released criminals; what are the current ethical views on punishment? From the point of view of mathematical theory all

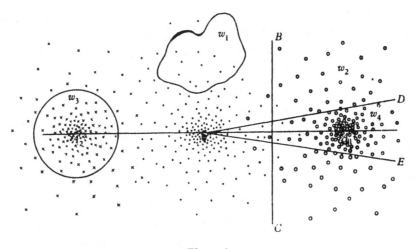

Figure 1

that we can do is to show how the risk of the errors may be controlled and minimized. The use of these statistical tools in any given case, in determining just how the balance should be struck, must be left to the investigator.

The principle upon which the choice of the critical region is determined so that the two sources of errors may be controlled is of first importance. Suppose for simplicity that the sample space is of two dimensions, so that the sample points lie on a plane. Suppose further that besides the hypothesis H_0 to be tested, there are only two alternatives H_1 and H_2. The situation is illustrated in fig. 1, where the cluster of spots round the point O, of circles round A_1, and of crosses round A_2 may be taken to represent the probability or density field appropriate to the three hypotheses. The spots, circles and crosses in the figure suggest diagrammatically the behaviour of the functions $p_i(x_1, x_2)$, $i = 0, 1, 2$, in the sense that the number of spots included in any region w is proportional to the integral of $p_0(x_1, x_2)$ taken over this region, etc. Looking at the diagram we see that, if the process of sampling was repeated many times, then, were the hypothesis H_0 true, most sample points would lie somewhere near the point O. On the contrary, if H_1 or H_2 were true, the sample points would be close to O in comparatively rare cases only.

In trying to choose a proper critical region, we notice at once that it is very easy to control errors of the first kind referred to above. In fact, the chance of rejecting the hypothesis H_0 when it is true may be reduced to as low a level as we please. For if w is any region in the sample space, which we intend to use as a critical region, the chance, $P_0(w)$, of rejecting the hypothesis H_0 when it is true, is merely the chance determined by H_0 of having a sample point inside of w, and is equal either to the sum (when the sample points do not form a continuous region) or to the integral (when they do) of $p_0(x_1, x_2)$ taken over the region w. It may be easily made $\leqslant \varepsilon$, by choosing w sufficiently small.

Four possible regions are suggested on the figure; (1) w_1; (2) w_2, i.e. the region to the right of the line BC; (3) w_3, the region within the circle centred at A_2; (4) w_4, the region between the straight lines OD, OE.

If the integrals of $p_0(x_1, x_2)$ over these regions, or the numbers of spots included in them, are equal, we know that they are all of equal value in regard to the first source of error; for as far as our judgment on the truth or falsehood of H_0 is concerned, if an error cannot be avoided it does not matter on which sample we make it.* It is the frequency of these errors that matters, and this—for errors of the first kind—is equal in all four cases.

It is when we turn to consider the second source of error—that of accepting H_0 when it is false—that we see the importance of distinguishing between different critical regions. If H_1 were the only admissible alternative to H_0, it is evident that we should choose from w_1, w_2, w_3 and w_4 that region in which the chance of a sample point falling, if H_1 were true, is greatest; that is to say the region in the diagram containing the greatest number of the small circles forming the cluster round A_1. This would be the region w_2, because for example,

$$P_1(w_2) > P_1(w_1) \qquad \text{or} \qquad P_1(W - w_2) < P_1(W - w_1).$$

This we do since in accepting H_0 when the sample point lies in $(W - w_2)$, we shall be accepting it when H_1 is, in fact, true, less often than if we used w_1. *We need indeed to pick out from all possible regions for which $P_0(w) = \varepsilon$, that region, w_0, for which $P_1(w_0)$ is a maximum and $P_1(W - w_0)$ consequently a minimum;* this region (or regions if more than one satisfy the condition) we shall term the Best Critical Region for H_0 with regard to H_1. There will be a family of such regions, each member corresponding to a different value of ε. The conception is simple but fundamental.

It is clear that in the situation presented in the diagram the best critical region with regard to H_1 will not be the best critical region with regard to H_2. While the first may be w_2, the second may be w_3. But it will be shown below that in certain problems there is a common family of best critical regions for H_0 with regard to the whole class of admissible alternative hypotheses Ω.† In these problems we have found that the regions are also those given by using the principle of likelihood, although a general proof of this result has not so far been obtained, when H_0 is composite.

In the problems where there is a *different* best critical region for H_0 with regard to each of the alternatives constituting Ω, some further principle must be introduced in fixing what may be termed a Good Critical Region with regard to the set Ω. We have found here that the region picked out by the likelihood method is the envelope of the best critical regions with regard to

---

* If the samples for which H_0 is accepted are to be used for some purpose and those for which it is rejected to be discarded, it is possible that other conceptions of relative value may be introduced. But the problem is then no longer the simple one of discriminating between hypotheses.

† Again as above, each member of the family is determined by a different value of ε.

the individual hypotheses of the set. This region appears to satisfy our intuitive requirements for a good critical region, but we are not clear that it has the unique status of the common best critical region of the former case.

We have referred in an earlier section to the distinction between simple and composite hypotheses, and it will be shown that the best critical regions may be found in both cases, although in the latter case they must satisfy certain additional conditions. If, for example in fig. 1, H_0 were a composite hypothesis with one degree of freedom such that while the centre of the cluster of spots were fixed at O, the scale or measure of radial expansion were unspecified, it is clear that w_4 could be used as a critical region, since $P_0(w_4) = \varepsilon$ would remain constant for any radial expansion or contraction of the field. Neither w_1, w_2 nor w_3 satisfy this condition. "Student's" test is a case in which a hyperconical region of this type is used.

III. Simple Hypotheses

(a) General Theory

We shall now consider how to find the best critical region for H_0 with regard to a single alternative H_1; this will be the region w_0 for which $P_1(w_0)$ is a maximum subject to the condition that

$$P_0(w_0) = \varepsilon. \tag{15}$$

We shall suppose that the probability laws for H_0 and H_1, namely,

$$p_0(x_1, x_2, \ldots, x_n) \quad \text{and} \quad p_1(x_1, x_2, \ldots, x_n),$$

exist, are continuous and not negative throughout the whole sample space W, further that

$$P_0(W) = P_1(W) = 1. \tag{16}$$

Following the ordinary method of the Calculus of Variations, the problem will consist in finding an unconditioned minimum of the expression

$$P_0(w_0) - kP_1(w_0) = \int \cdots \int_{w_0} \{p_0(x_1, x_2, \ldots, x_n) \\ - kp_1, (x_1, x_2, \ldots, x_n)\} \, dx_1 \ldots dx_n, \tag{17}$$

k being a constant afterwards to be determined by the condition (15). Suppose that the region w_0 has been determined and that S is the hypersurface limiting it. Let s_1 be any portion of S such that every straight line parallel to the axis Ox_n cuts s_1 not more than once; denote by σ_1 the orthogonal projection of s_1 on the prime or hyperplane $x_n = 0$. Clearly σ_1 will be a region in the $(n-1)$ dimensioned space of

$$x_1, x_2, \ldots, x_{n-1}, \tag{18}$$

for which there will exist a uniform function

$$_{s_1}x_n = \psi(x_1, \ldots, x_{n-1}), \tag{19}$$

whose value together with the corresponding values of (18) will determine the points on the hypersurface s_1.

Consider now a region $w_0(\alpha)$ bounded by

(1) The hypersurface s_2 with equation

$$_{s_2}x_n = \psi(n_1, x_2, \ldots, x_{n-1}) + \alpha\theta(x_1, x_2, \ldots, x_{n-1}), \tag{20}$$

where α is a parameter independent of the x's, and θ any uniform and continuous function.

(2) By the hypercylinder projecting s_1 on to σ_1.

(3) By any hypersurface s_3 with equation,

$$_{s_3}x_n = \phi(x_1, \ldots, x_{n-1}), \tag{21}$$

having the following properties:

(a) It is not cut more than once by any straight line parallel to the axis Ox_n.

(b) It lies entirely inside the region w_0.

(c) Every point on a straight line parallel to Ox_n lying between the points of its intersection with s_1 and s_3 belongs to w_0.

For the case of $n = 2$, the situation can be pictured in two dimensions as shown in fig. 2. Here $w_0(\alpha)$ is the shaded region bounded by the curves s_2 and s_3, and the two vertical lines through A and B. σ_1 is the portion AB of the axis Ox_1. In general if the region w_0 is such as to give a minimum value to the expression (17), it follows that

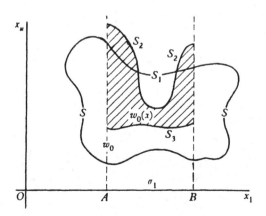

Figure 2

$$I(\alpha) = P_0(w_0(\alpha)) - kP_1(w_0(\alpha))$$

$$= \int \cdots \int_{\sigma_1} dx_1 \dots dx_{n-1} \int_{s_3 x_n}^{s_2 x_n} (p_0 - kp_1)\, dx_n, \qquad (22)$$

considered as a function of the varying parameter α must be a minimum for $\alpha = 0$. Hence differentiating

$$\frac{dI}{d\alpha} = \pm \int \cdots \int_{\sigma_1} \theta\{p_0(x_1, \dots, x_{n-1}, s_2 x_n) - kp_1(x_1, x_2, \dots, x_{n-1}, s_2 x_n)\}$$

$$\times\, dx_1 \dots dx_{n-1} = 0, \qquad (23)$$

whatever be the form of the function θ. This is known to be possible only if the expression within curled brackets on the right hand side of (23) is identically zero. It follows that if w_0 is the best critical region for H_0 with regard to H_1, then at every point on the hypersurface s_1 and consequently at every point on the complete boundary S, we must have

$$p_0(x_1, x_2, \dots, x_n) = kp_1(x_1, x_2, \dots, x_n), \qquad (24)$$

k being a constant. This result gives the necessary boundary condition. We shall now show that the necessary and sufficient condition for a region w_0, being the best critical region for H_0 with regard to the alternative hypothesis, H_1, consists in the fulfilment of the inequality $p_0(x_1, x_2, \dots, x_n) > kp_1(x_1, x_2, \dots, x_n)$, k being a constant, at any point outside w_0; that is to say that w_0 is defined by the inequality

$$p_0(x_1, x_2, \dots, x_n) \leqslant kp_1(x_1, x_2, \dots, x_n). \qquad (25)$$

Denote by w_0 the region defined by (25) and let w_1 be any other region satisfying the condition $P_0(w_1) = P_0(w_0) = \varepsilon$ (say). These regions may have a common part, w_{01}. The situation is represented diagrammatically in fig. 3.

It will follow that,

$$P_0(w_0 - w_{01}) = \varepsilon - P_0(w_{01}) = P_0(w_1 - w_{01}), \qquad (26)$$

and consequently

$$kP_1(w_0 - w_{0i}) \geqslant P_0(w_0 - w_{01}) = P_0(w_1 - w_{01}) \geqslant kP_1(w_1 - w_{01}). \qquad (27)$$

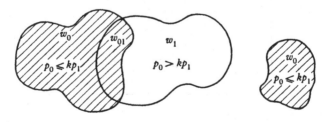

Figure 3

If we add $kP_1(w_{01})$ to both sides of the inequality, we obtain

$$kP_1(w_0) \geqslant kP_1(w_1)$$

or

$$P_1(w_0) \geqslant P_1(w_1). \tag{28}$$

From the considerations advanced above it follows that w_1 is less satisfactory as a critical region than w_0. That is to say, of the regions w, for which $P_0(w) = \varepsilon$, satisfying the boundary condition (24), the region w_0 defined by the inequality (25) is the best critical region with regard to the alternative H_1. There will be a family of such best critical regions, each member of which corresponds to a different value of ε.

As will appear below when discussing illustrative examples, in certain cases the family of best critical regions is not the same for each of the admissible alternatives $H_1, H_2, \ldots$; while in other cases a single common family exists for the whole set of alternatives. In the latter event the basis of the test is remarkably simple. If we reject H_0 when the sample point, Σ, falls into w_0, the chance of rejecting it when it is true is ε, and the risk involved can be controlled by choosing from the family of best critical regions to which w_0 belongs, a region for which ε is as small as we please. On the other hand, if we accept H_0 when Σ falls outside w_0, we shall sometimes be doing this when some H_t of the set of alternatives is really true. But we know that whatever be H_t, the region w_0 has been so chosen as to reduce risk to a minimum. In this case even if we had precise information as to the *a priori* probabilities of the alternatives H_1, $H_2, \ldots$ we could not obtain any improved test.*

It is now possible to see the relation between best critical regions and the region defined by the principle of likelihood described above. Suppose that for a hypothesis H_t belonging to the set of alternatives Ω, the probability law for a given sample is defined by

(1) an expression of given functional type $p(x_1, x_2, \ldots, x_n)$
(2) the values of c parameters contained in this expression, say

$$\alpha_t^{(1)}, \alpha_t^{(2)}, \ldots, \alpha_t^{(c)}. \tag{29}$$

This law for H_t may be written as $p_t = p_t(x_1, x_2, \ldots, x_n)$. The hypothesis of maximum likelihood, H, (Ω max.), is obtained by maximizing p_t with regard to these c parameters, or in fact from a solution of the equations,

$$\frac{\partial p}{\partial \alpha^{(i)}} = 0 \quad (i = 1, 2, \ldots, c). \tag{30}$$

The values of the α's so obtained are then substituted into p to give $p(\Omega$ max.). Then the family of surfaces of constant likelihood, λ, appropriate for testing

---

* For properties of critical regions given by the principle of likelihood from the point of view of probabilities *a posteriori*, see Neyman. "Contribution to the Theory of Certain Test Criteria," *Bull. Inst. int. Statist.* vol. 24, pp. 44 (1928).

a simple hypothesis H_0 is defined by

$$p_0 = \lambda p(\Omega \text{ max}). \tag{31}$$

It will be seen that the members of this family are identical with the envelopes of the family

$$p_0 = k p_t, \tag{32}$$

which bound the best critical regions. From this it follows that,

(a) If for a given ε a common best critical region exists with regard to the whole set of alternatives, it will correspond to its envelope with regard to these alternatives, and it will therefore be identical with a region bounded by a surface (31). Further, in this case, the region in which $\lambda \leqslant \lambda_0 = \text{const.}$ will correspond to the region in which $p_0 \leqslant \lambda_0 p_t$. The test based upon the principle of likelihood leads, in fact, to the use of best critical regions.

(b) If there is not a common best critical region, the likelihood of H_0 with regard to a particular alternative H_t will equal the constant, k, of equation (32). It follows that the surface (31) upon which the likelihood of H_0 with regard to the whole set of alternatives is constant, will be the envelope of (32) for which $\lambda = k$. The interpretation of this result will be seen more clearly in some of the examples which follow.

(b) Illustrative Examples

(1) *Sample Space Unlimited; Case of the Normal Population*

EXAMPLE (1). Suppose that it is known that a sample of n individuals, $x_1, x_2, \ldots, x_n$ has been drawn randomly from *some* normally distributed population with standard deviation $\sigma = \sigma_0$, but it is desired to test the hypothesis H_0 that the mean in the sampled population is $a = a_0$. Then the admissible hypotheses concern the set of populations for which

$$p(x) = \frac{1}{\sigma\sqrt{(2\pi)}} e^{-(1/2)(x-a)^2/\sigma^2}, \tag{33}$$

the mean, a, being unspecified, but σ always equal to σ_0. Let H_1 relate to the member of this set for which $a = a_1$. Let $\bar{x}$ and s be the mean and standard deviation of the sample. The probabilities of its occurrence determined by H_0 and by H_1 will then be

$$p_0(x_1, \ldots, x_n) = \left(\frac{1}{\sigma_0\sqrt{(2\pi)}}\right)^n \exp - \left\{ n\frac{(\bar{x} - a_0)^2 + s^2}{2\sigma_0^2} \right\}, \tag{34*}$$

$$p_1(x_1, \ldots, x_n) = \left(\frac{1}{\sigma_0\sqrt{(2\pi)}}\right)^n \exp - \left\{ n\frac{(\bar{x} - a_1)^2 + s^2}{2\sigma_0^2} \right\}, \tag{35}$$

* *Editor's note*: The exp notation is introduced here for typographical convenience. The symbol e is used in the original paper.

and the equation (24) becomes

$$\frac{p_0}{p_1} = \exp - \left\{ n \frac{(\bar{x} - a_0)^2 - (\bar{x} - a_1)^2}{2\sigma_0^2} \right\} = k. \tag{36}$$

From this it follows that the best critical region for H_0 with regard to H_1, defined by the inequality (25), becomes

$$(a_0 - a_1)\bar{x} \leqslant \frac{1}{2}(a_0^2 - a_1^2) + \frac{\sigma_0^2}{n} \log k = (a_0 - a_1)\bar{x}_0 \quad \text{(say).} \tag{37}$$

Two cases will now arise,

(a) $a_1 < a_0$, then the region is defined by

$$\bar{x} = \frac{1}{n} \sum_{i=1}^{n} (x_i) \leqslant \bar{x}_0. \tag{38}$$

(b) $a_1 > a_0$, then the region is defined by

$$\bar{x} \geqslant \bar{x}_0. \tag{39}$$

We see that whatever be H_1 and a_1, the family of hypersurfaces corresponding to different values of k, bounding the best critical region, will be the same namely,

$$\bar{x} = \frac{1}{n} \sum_{i=1}^{n} (x_i) = \bar{x}_0. \tag{40}$$

These are primes in the n-dimensional space lying at right angles to the line,

$$x_1 = x_2 = \cdots = x_n. \tag{41}$$

If, however, the class of admissible alternatives includes both those for which $a < a_0$ and $a > a_0$, there will not be a single best critical region; for the first it will be defined by $\bar{x} \leqslant \bar{x}_0$ and for the second by $\bar{x} \geqslant \bar{x}_0$, where $\bar{x}_0$ is to be chosen so that $p_0(\bar{x} \leqslant \bar{x}_0) = \varepsilon.$* This situation will not present any difficulty in practice. Suppose $\bar{x} > a_0$ as in fig. 4. We deal first with the class of alternatives for which $a > a_0$. If $\varepsilon = 0.05$; $\bar{x}_0 = a_0 + 1.6449\sigma_0/\sqrt{n}$, and if $\bar{x} < \bar{x}_0$, we shall probably decide to accept the hypothesis H_0 as far as this class of alternatives is concerned. That being so, we shall certainly not reject H_0 in favour of the class for which $a < a_0$, for the risk of rejection when H_0 were true would be too great.

* In this example

$$P_0(\bar{x} \geqslant \bar{x}_0) = \frac{1}{\sigma_0} \sqrt{\frac{n}{2\pi}} \int_{-\infty}^{x_0} \exp - \left\{ n \frac{(\bar{x} - a_0)^2}{2\sigma_0^2} \right\} d\bar{x},$$

$$P_0(\bar{x} \geqslant \bar{x}_0) = \frac{1}{\sigma_0} \sqrt{\frac{n}{2\pi}} \int_{x_0}^{+\infty} \exp - \left\{ n \frac{(\bar{x} - a_0)^2}{2\sigma_0^2} \right\} d\bar{x}.$$

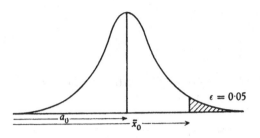

Figure 4

The test obtained by finding the best critical region is in fact the ordinary test for the significance of a variation in the mean of a sample; but the method of approach helps to bring out clearly the relation of the two critical regions $\bar{x} \leqslant \bar{x}_0$ and $\bar{x} \geqslant \bar{x}_0$. Further, it has been established that starting from the same information, the test of this hypothesis could not be improved by using any other form of criterion or critical region.

EXAMPLE (2). The admissible hypotheses are as before given by (33), but in this case the means are known to have a given common value a_0, while σ is unspecified. We may suppose the origin to be taken at the common mean, so that $a = a_0 = 0$. H_0 is the hypothesis that $\sigma = \sigma_0$, and an alternative H_1 is that $\sigma = \sigma_1$. In this case it is easy to show that the best critical region with regard to H_1 is defined by the inequality,

$$\frac{1}{n}\sum_{i=1}^{n}(x_i^2)(\sigma_0^2 - \sigma_1^2) = (\bar{x}^2 + s^2)(\sigma_0^2 - \sigma_1^2) \leqslant v^2(\sigma_0^2 - \sigma_1^2), \qquad (42)$$

where v is a constant depending only on ε, σ_0, σ_1. Again two cases will arise,

(a) $\sigma_1 < \sigma_0$; then the region is defined by

$$\bar{x}^2 + s^2 \leqslant v^2. \qquad (43)$$

(b) $\sigma_1 > \sigma_0$ when it is defined by

$$\bar{x}^2 + s^2 \geqslant v^2. \qquad (44)$$

The best critical regions in the n-dimensioned space are therefore the regions (a) inside and (b) outside hyperspheres of radius $v\sqrt{n}$ whose centres are at the origin of coordinates. This family of hyperspheres will be the same whatever be the alternative value σ_1; there will be a common family of best critical regions for the class of alternatives $\sigma_1 < \sigma_0$, and another common family for the class $\sigma_1 > \sigma_0$.

It will be seen that the criterion is the second moment coefficient of the sample about the known population mean,

$$m_2' = \bar{x}^2 + s^2, \qquad (45)$$

and not the sample variance s^2. Although a little reflexion might have suggested this result as intuitively sound, it is probable that s^2 has often been used as the criterion in cases where the mean is known. The probability integral of the sampling distributions of m'_2 and s^2 may be obtained from the distribution of $\psi = \chi^2$, namely,

$$p(\psi) = c\psi^{(1/2)f-1}e^{-(1/2)\psi}, \tag{46}$$

by writing

$$m'_2 = \sigma_0^2\psi/n, \qquad f = n, \tag{47}$$

and

$$s^2 = \sigma_0^2\psi/n, \qquad f = n - 1. \tag{48}$$

It is of interest to compare the relative efficiency of the criteria m'_2 and s^2 in avoiding errors of the second type, that is of accepting H_0 when it is false. If it is false, suppose the true hypothesis to be H_1 relating to a population in which

$$\sigma_1 = h\sigma_0 > \sigma_0. \tag{49}$$

In testing H_0 with regard to the class of alternatives for which $\sigma > \sigma_0$, we should determine the critical value of ψ_0 so that

$$P_0(\psi \geqslant \psi_0) = \int_{\psi_0}^{+\infty} p(\psi)\,d\psi = \varepsilon, \tag{50}$$

and would accept H_0 if $\psi < \psi_0$. But if H_1 is true, the chance of finding $\psi < \psi_0$, ψ_0 being determined from (50), that is of accepting H_0 (though it is false), will be

$$P_1(\psi \leqslant \psi_0) = \int_0^{\psi_0 h^{-2}} p(\psi)\,d\psi. \tag{51}$$

The position is shown in fig. 5. Suppose that for the purpose of illustration we take $\varepsilon = 0.01$ and $n = 5$.

(a) Using m'_2 and thus the best critical region, we shall put $f = 5$ in (46), and from (50) entering the tables of the χ^2 integral with 5 degrees of freedom,

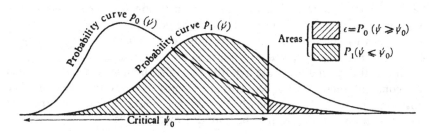

Figure 5

find that $\psi_0 = 15.086$. Hence from (51),

$$\begin{cases} \text{if} \quad h = 2, \quad (\sigma_1 = 2\sigma_0), \quad P_1(\psi \leqslant \psi_0) = 0.42 \\ \text{if} \quad h = 3, \quad (\sigma_1 = 3\sigma_0), \quad P_1(\psi \leqslant \psi_0) = 0.11. \end{cases}$$

(b) On the other hand, if the variance, s^2, is used as criterion, we must put $f = 4$ in (46) and find that $\psi_0 = 13.277$. Hence

$$\begin{cases} \text{if} \quad h = 2, \quad (\sigma_1 = 2\sigma_0), \quad P_1(\psi \leqslant \psi_0) = 0.49 \\ \text{if} \quad h = 3, \quad (\sigma_1 = 3\sigma_0), \quad P_1(\psi \leqslant \psi_0) = 0.17. \end{cases}$$

In fact for $h = 2, 3$ or any other value, it is found that the second test has less power of discrimination between the false and the true than the test associated with the best critical region.

EXAMPLE (3). The admissible hypotheses are given by (33), both a and σ being in this case unspecified. We have to test the simple hypothesis H_0, that $a = a_0$, $\sigma = \sigma_0$. The best critical region with regard to a single alternative H_1, with $a = a_1, \sigma = \sigma_1$, will be defined by

$$\frac{p_0}{p_1} = \left(\frac{\sigma_1}{\sigma_0}\right)^n \exp -\left[\frac{1}{2}\sum_{i=1}^{n}\left\{\left(\frac{x_i - a_0}{\sigma_0}\right)^2 - \left(\frac{x_i - a_1}{\sigma_1}\right)^2\right\}\right] \leqslant k. \qquad (52)$$

This inequality may be shown to result in the following

(a) If $\sigma_1 < \sigma_0$

$$\frac{1}{n}\sum_{i=1}^{n}(x_i - \alpha)^2 = (\bar{x} - \alpha)^2 + s^2 \leqslant v^2. \qquad (53)$$

(b) If $\sigma_1 > \sigma_0$

$$\frac{1}{n}\sum_{i=1}^{n}(x_i - \alpha)^2 = (\bar{x} - \alpha)^2 + s^2 \geqslant v^2. \qquad (54)$$

where

$$\alpha = \frac{a_0\sigma_1^2 - a_1\sigma_0^2}{\sigma_1^2 - \sigma_0^2}, \qquad (55)$$

and v is a constant, whose value will depend upon $a_0, a_1, \sigma_0, \sigma_1$ and ε. It will be seen that a best critical region in the n-dimensioned space is bounded by a hypersphere of radius $v\sqrt{n}$ with centre at the point $(x_1 = x_2 = \cdots = x_n = \alpha)$. The region will be the space inside or outside the hypersphere according as $\sigma_1 < \sigma_0$ or $\sigma_1 > \sigma_0$. If $a_1 = a_0 = 0$ the case becomes that of example (2).

Unless the set of admissible hypotheses can be limited to those for which $\alpha = $ constant, there will not be a common family of best critical regions. The position can be seen most clearly by taking $\bar{x}$ and s as variables; the best critical regions are then seen to be bounded by the circles

$$(\bar{x} - \alpha)^2 + s^2 = v^2. \qquad (56)$$

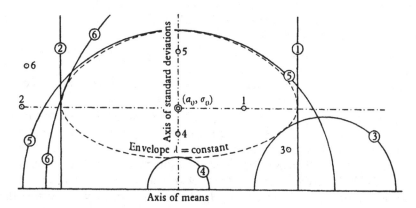

Figure 6. N.B. The same number is assigned to a point (a, σ) and to the boundary of the corresponding best critical region.

If $p_0(v)$ be the probability law for v, then the relation between ε and v_0, the radius of the limiting circles is given by

$$\int_0^{v_0} p_0(v)\, dv = \varepsilon \qquad \text{if} \qquad \sigma_1 < \sigma_0, \tag{57}$$

and

$$\int_{v_0}^{+\infty} p_0(v)\, dv = \varepsilon \qquad \text{if} \qquad \sigma_1 > \sigma_0. \tag{58}$$

By applying the transformation

$$\bar{x} = \alpha + v \cos \phi, \qquad s = v \sin \phi, \tag{59}$$

to

$$p_0(\bar{x}, s) = cs^{n-2} \exp\{-\tfrac{1}{2}n(\bar{x}^2 + s^2)\}, \tag{60}$$

it will be found that

$$p_0(v) = ce^{-(1/2)n\alpha^2} v^{n-1} e^{-(1/2)nv^2} \int_0^{\pi} e^{-n\alpha v \cos \phi} \sin^{n-2} \phi\, d\phi. \tag{61}$$

This integral may be expressed as a series in ascending powers of v,* but no simple method of finding v_0 for a given value of ε has been evolved.

The relation between certain population points (a, σ), and the associated best critical regions is shown in fig. 6. A single curve of the family bounding the best critical regions is shown in each case.

Cases (1) and (2). $\sigma_1 = \sigma_0$, then $\alpha = \pm\infty$. The B.C.R. (best critical region) will be to the right of straight line (1), $(a_1 > a_0)$, or to the left of straight line (2), $(a_1 < a_0)$.

* It is a series containing a finite number of terms if n be odd, and an infinite series if n be even.

Case (3). $\sigma_1 < \sigma_0$. Suppose $\sigma_1 = \frac{1}{2}\sigma_0$, then $\alpha = a_0 + \frac{4}{3}(a_1 - a_0)$ and the B.C.R. lies inside the semi-circle (3).

Case (4). $\sigma_1 < \sigma_0$ and $a_1 = a_0$. $\alpha = a_0$. The B.C.R. lies inside the semi-circle (4).

Case (5). $\sigma_1 > \sigma_0$ and $a_1 = a_0$. $\alpha = a_0$. The B.C.R. lies outside the semi-circle (5).

Case (6). $\sigma_1 > \sigma_0$. Suppose $\sigma_1 = \frac{3}{2}\sigma_0$, then $\alpha = a_0 - \frac{4}{5}(a_1 - a_0)$, and for $a_1 < a_0$, $\alpha > a_0$. In the diagram the B.C.R. lies outside the large semi-circle, part of which is shown as curve (6).

It is evident that there is no approach to a common best critical region with regard to all the alternatives H_t, of the set Ω represented by equation (33). If $w_0(t)$ is the best critical region for H_t, then $W - w_0(t)$ may be termed the region of acceptance of H_0 with regard to H_t. The diagram shows how these regions of acceptance will have a large common part, namely, the central space around the point $a = a_0, \sigma = \sigma_0$. This is the region of acceptance picked out by the criterion of likelihood. It has been pointed out above that if λ be the likelihood of H_0 with regard to the set Ω, then the hypersurfaces $\lambda = k$ are the envelopes of the hypersurfaces $p_0/p_t = k = \lambda$ considered as varying with regard to a_t and σ_t. The equation of these envelopes we have shown elsewhere to be,*

$$\left(\frac{\bar{x} - a_0}{\sigma_0}\right)^2 - \log\left(\frac{s}{\sigma_0}\right)^2 = 1 - \frac{2}{n}\log\lambda. \tag{62}$$

The dotted curve shown in fig. 6 represents one such envelope. The region in the $(\bar{x}, s)$ plane outside this curve and the corresponding region in the n-dimensioned-space may be termed good critical regions, but have not the unique status of the best critical region common for all H_t. Such a region is essentially one of compromise, since it includes a part of the best critical regions with regard to each of the admissible alternatives.

It is also clear that considerations of *a priori* probability may now need to be taken into account in testing H_0. If a certain group of alternatives were more probable than others *a priori*, we might be inclined to choose a critical region more in accordance with the best critical regions associated with the hypotheses of that group than the λ region. Occasionally it happens that *a priori* probabilities can be expressed in exact numerical form,† and if this is so, it would at any rate be possible theoretically to pick out the region w_0 for which $P_0(w_0) = \varepsilon$, such that the chance of accepting H_0 when one of the weighted alternatives H_t is true is a minimum. But in general, we are doubtful of the value of attempts to combine measures of the probability of an event if

---

* *Biometrika*, vol. 20A, p. 188 (1928). The ratio p_0/p_t is given by equation (52) if we write a_t and σ_t for a_1 and σ_1. It should be noted that the envelope is obtained by keeping $\lambda = k = $ constant, and since k is a function of a_t and σ_t, this will not mean that $\varepsilon = $ constant for the members of the system giving the envelope.

† As for example in certain Mendelian problems.

a hypothesis be true, with measures of the *a priori* probability of that hypothesis. The difficulty seems to vanish in this as in the other cases, if we regard the λ surfaces as providing (1) a control by the choice of ε of the first source of error (the rejection of H_0 when true); and (2) a good compromise in the control of the second source of error (the acceptance of H_0 when some H_t is true). The vague *a priori* grounds on which we are intuitively more confident in some alternatives than in others must be taken into account in the final judgement, but cannot be introduced into the test to give a single probability measure.*

(2) *The Sample Space Limited*; *Case of the Rectangular Population*

Hitherto we have supposed that there is a common sample space, W, for all admissible hypotheses, and in the previous examples this has been the unlimited n-dimensional space. We must, however, consider the case in which the space W_0, in which $p_0 > 0$, associated with H_0, does not correspond exactly with the space W_1, associated with an alternative H_1 where $p_1 > 0$. Should W_0 and W_1 have no common part, then we are able to discriminate absolutely between H_0 and H_1. Such would be the case for example if $p_t(x) = 0$ when $x < a_t$ or $x > b_t$, and it happened that $a_1 > b_0$. But more often W_0 and W_1 will have a common part, say W_{01}. Then it is clear that $W_1 - W_{01}$ should be *included* in the best critical region for H_0 with regard to H_1. If this were the whole critical region, w_0, we should never reject H_0 when it is true, for $P_0(w_0) = 0$, but it is possible that we should accept H_0 too often when H_1 is true. Consequently we may wish to make up w_0 by adding to $W_1 - W_{01}$ a region w_{00} which is a part of W_{01} for which $P_0(w_{00}) = P_0(w_0) = \varepsilon$. The method of choosing the appropriate w_{00} with regard to H_1 will be as before, except that the sample space for which it may be chosen is now limited to W_{01}. If, however, a class of alternatives exists for which the space W_{0t} varies with t, there will probably be no common best critical region. The position may be illustrated in the case of the so-called rectangular distribution, for which the probability law can be written

$$\left.\begin{array}{llll} p(x) = 1/b & \text{for} & a - \tfrac{1}{2}b \leqslant x \leqslant a + \tfrac{1}{2}b \\ p(x) = 0 & \text{for} & x < a - \tfrac{1}{2}b \ \text{ and } \ x > a + \tfrac{1}{2}b. \end{array}\right\} \tag{63}$$

a will be termed the mid-point and b the range of the distribution.

EXAMPLE (4). Suppose that a sample of n individuals $x_1, x_2, \ldots, x_n$ is known to have been drawn at random for *some* population with distribution following (63), in which $b = b_0$, and it is wished to test the simple hypothesis H_0 that in the sampled population, $a = a_0$. For the admissible set of alternatives, $b = b_0$,

---

* Tables and diagrams to assist in using this λ-test have been given in *Biometrika*, vol. 20A, p. 233 (1928), and are reproduced in *Tables for Statisticians and Biometricians*, Part II.

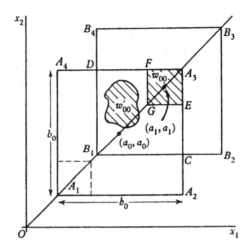

Figure 7

but a is unspecified. For H_0 the sample space W_0 is the region within the hypercube, defined by

$$a_0 - \tfrac{1}{2}b_0 \leqslant x_i \leqslant a_0 + \tfrac{1}{2}b_0. \tag{64}$$

If H_1 be a member of the set of alternatives for which $a = a_1$, then

$$p_0(x_1, x_2, \ldots, x_n) = p_1(x_1, x_2, \ldots, x_n) = \frac{1}{b^n}, \tag{65}$$

provided the sample point lies within W_{01}. It follows that at every point in W_{01} $p_0/p_1 = k = 1$, and that $P_0(w_0) = \varepsilon$ for any region whatsoever within W_{01}, the content of which equals ε times the content, b_0^n, of the hypercube W_0.

There is in fact no single best critical region with regard to H_1. Fig. 7 illustrates the position for the case of samples of 2. The sample spaces W_0 and W_1 are the squares $A_1 A_2 A_3 A_4$ and $B_1 B_2 B_3 B_4$ respectively. A critical region for H_0 with regard to H_1 will consist of

(1) The space $W_1 - W_{01} = A_3 C B_2 B_3 B_4 D$;
(2) Any region such as w_{00}' lying wholly inside the common square

$$W_{01} = B_1 C A_3 D,$$

containing an area εb_0^2.

The value of ε is at our choice and may range from O to $(a_0 - a_1 + b_0)^2$, according to the balance it is wished to strike between the two kinds of error. We shall not allow any part of w_0 to lie outside $B_1 C A_3 D$ in the space $W_0 - W_{01}$, for this would lead to the rejection of H_0 in cases where the alternative H_1 could not be true.

For different alternatives, H_t, of the set, the mid-point of the square

$B_1 B_2 B_3 B_4$ will shift along the diagonal $OA_1 A_3$. For a fixed ε we cannot find a region that will be included in W_{0t} for every H_t, but we shall achieve this result as nearly as possible if we can divide the alternatives into two classes—

(a) $a_1 > a_0$. Take w_{00} as the square $GEA_3 F$ with length of side $= b_0\sqrt{\varepsilon}$ lying in the upper left-hand corner of W_0.

(b) $a_1 < a_0$. Take a similar square with corner at A_1.

In both cases the whole space outside W_0 must be added to make up the critical region w_0. In the general case of samples of n, the region w_{00} will be a hypercube with length of side $b_0 \sqrt[n]{\varepsilon}$ fitting into one or other of the two corners of the hypercube of W_0 which lie on the axis $x_1 = x_2 = \cdots = x_n$. The whole of the space outside W_0 within which sample points can fall will also be added to w_{00} to make up w_0.[*]

EXAMPLE (5). Suppose that the set of alternatives consists of distributions of form (63), for all of which $a = a_0$, but b may vary. H_0 is the hypothesis that $b = b_0$. The sample spaces, W_t, are now hypercubes of varying size all centred at the point $(x_1 = x_2 = \cdots = x_n = a_0)$. A little consideration suggests that we should make the critical region w_0 consist of—

(1) The whole space outside the hypercube W_0.
(2) The region w_{00} inside a hypercube with centre at $(x_1 = x_2 = \cdots = x_n = a_0)$, sides parallel to the co-ordinate axes and of volume εb_0^n. This region w_{00} is chosen because it will lie completely within the sample space W_{0t} common to H_0 and H_1 for a larger number of the set of alternatives than any other region of equal content.

EXAMPLE (6). H_0 is the hypothesis that $a = a_0$, $b = b_0$, and the set of admissible alternatives is given by (63) in which both α and b are now unspecified. Both the mid-point $(x_1 = x_2 = \cdots = x_n = a_t)$ and the length of side, b_t, of the alternative sample spaces W_t can therefore vary. Clearly we shall again include in w_0 the whole space outside W_0, but there can be no common region w_{00} within W_0.

Fig. 8(a) represents the position for $n = 2$. Four squares W_1, W_2, W_3, and W_4 correspond to the sample spaces of possible alternatives H_1, H_2, H_3, and H_4, and the smaller shaded squares w_1, w_2, w_3, and w_4 represent possible critical regions for H_0 with regard to these. What compromise shall we make in choosing a critical region with regard to the whole set Ω As we have shown elsewhere[†] the method of likelihood fixes for the critical region that part of the space that represents samples for which the range (the difference between extreme variates) is less than a given value, say $l \leqslant l_0$. For samples of 2,

---

[*] If the set is limited to distributions for which $b = b_0$, no sample point can lie outside the envelope of hypercubes whose centres lie on the axis $x_1 = x_2 = \cdots = x_n$.

[†] *Biometrika*, vol. 20A, p. 208 (1928). Section on 'Samples from a Rectangular Population'.

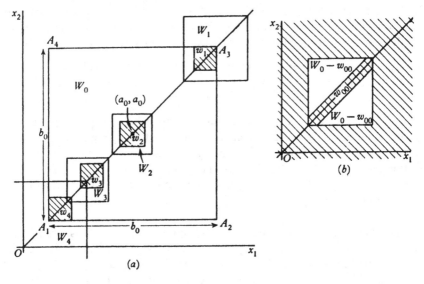

Figure 8

$l = x_1 - x_2$ if $x_1 > x_2$, and $x_2 - x_1$ if $x_1 < x_2$, and the critical region w_{00} will therefore lie between two straight lines parallel to and equidistant from the axis $x_1 = x_2$. A pair of such lines will be the envelope of the small squares w_1, w_2, etc., of fig. 8(a). In fact, the complete critical region will be as shown in fig. 8(b), the belt w_{00} being chosen so that its area is εb_0^2.

For $n = 3$ the surface $l = l_0$ is a prism of hexagonal cross-section, whose generating lines are parallel to the axis $x_1 = x_2 = x_3$. The space w_{00}, within this and the whole space outside the cube W_0 will form the critical region w_0. In general for samples of n the critical region of the likelihood method will consist of the space outside the hypercube W_0, and the space of content εb_0^n within the envelope of hypercubes having centres on the axis $x_1 = x_2 = \cdots = x_n$, and edges parallel to the axes of co-ordinates.

It will have been noted that a correspondence exists between the hypotheses tested in examples (1) and (4), (2) and (5), (3) and (6), and between the resulting critical regions. Consider for instance the position for $n = 3$ in example (3); the boundary of the critical region may be obtained by rotating fig. 6 in 3-dimensioned space about the axis of means. The region of acceptance of H_0 is then bounded by a surface analogous to an anchor ring surrounding the axis $x_1 = x_2 = x_3$, traced out by the rotation of the dotted curve $\lambda = $ constant. Its counterpart in example (6) is the region inside a cube from which the hexagonal sectioned prism w_{00} surrounding the diagonal $x_1 = x_2 = x_3$ has been removed. A similar correspondence may be traced in the case of sampling from a distribution following the exponential law. It continues to hold in the higher dimensioned spaces with $n > 3$.

The difference between the normal, rectangular and exponential laws is of course, very great, but the question of what may be termed the stability in form of best critical regions for smaller changes in the frequency law,

$$p(x_1, x_2, \ldots, x_n),$$

is of considerable practical importance.

IV. Composite Hypotheses

(a) Introductory

In the present investigation we shall suppose that the set Ω of admissible hypotheses defines the functional form of the probability law for a given sample, namely—

$$p(x_1, x_2, \ldots, x_n), \tag{66}$$

but that this law is dependent upon the values of $c + d$ parameters

$$\alpha^{(1)}, \alpha^{(2)}, \ldots, \alpha^{(c)}; \quad \alpha^{(c+1)}, \ldots, \alpha^{(c+d)}. \tag{67}$$

A composite hypothesis, H_0', of c degrees of freedom is one for which the values of d of these parameters are specified and c unspecified. We shall denote these parameters by

$$\alpha^{(1)}, \alpha^{(2)}, \ldots, \alpha^{(c)}; \quad \alpha^{(c+1)}, \ldots, \alpha^{(c+d)}. \tag{68}$$

This composite hypothesis consists of a sub-set ω (of the set Ω) of simple hypotheses. We shall denote the probability law for H_0' by

$$p_0 = p_0(x_1, x_2, \ldots, x_n), \tag{69}$$

associating with (69) in any given case the series (68). An alternative simple hypothesis which is definitely specified will be written as H_t, and with this will be associated

(1) a probability law

$$p_t = p_t(x_1, x_2, \ldots, x_n); \tag{70}$$

(2) a series of parameters

$$\alpha_t^{(1)}, \alpha_t^{(2)}, \ldots, \alpha_t^{(c+d)}. \tag{71}$$

We shall suppose that there is a common sample space W for any admissible hypothesis H_t, although its probability law p_t may be zero in some parts of W.

As when dealing with simple hypotheses we must now determine a family of critical regions in the sample space, W, having regard to the two sources of error in judgment. In the first place it is evident that a necessary condition for a critical region, w, suitable for testing H'_0 is that

$$P_0(w) = \int\int \cdots \int_w p_0(x_1, x_2, \ldots, x_n) \, dx_1 \, dx_2 \ldots dx_n = \text{constant} = \varepsilon \quad (72)$$

for every simple hypothesis of the sub-set ω. That is to say, it is necessary for $P_0(w)$ to be independent of the values of $\alpha^{(1)}, \alpha^{(2)}, \ldots, \alpha^{(c)}$. If this condition is satisfied we shall speak of w as a region of "size" ε, similar to W with regard to the c parameters $\alpha^{(1)}, \alpha^{(2)}, \ldots, \alpha^{(c)}$.

Our first problem is to express the condition for similarity in analytical form. Afterwards it will be necessary to pick out from the regions satisfying this condition that one which reduces to a minimum the chance of accepting H'_0 when a simple alternative hypothesis H_t is true. If this region is the same for all the alternatives H_t of the set Ω, then we shall have a common best critical region for H'_0 with regard to the whole set of alternatives. The fundamental position from which we start should be noted at this point. It is assumed that the only possible critical regions that can be used are similar regions; that is to say regions such that $P(w) = \varepsilon$ for every simple hypothesis of the sub-set ω. It is clear that were it possible to assign differing measures of *a priori* probability to these simple hypotheses, a principle might be laid down for determining critical regions, w, for which $P(w)$ would vary from one simple hypothesis to another. But it would seem hardly possible to put such a test into working form.

We have, in fact, no hesitation in preferring to retain the simple conception of control of the first source of error (rejection of H'_0 when it is true) by the choice of ε, which follows from the use of similar regions. This course seems necessary as a matter of practical policy, apart from any theoretical objections to the introduction of measures of *a priori* probability.

(b) Similar Regions for Case in Which H'_0 Has One Degree of Freedom

We shall commence with this simple case for which the series (68) becomes

$$\alpha^{(1)}; \alpha_0^{(2)}, \alpha_0^{(3)}, \ldots, \alpha_0^{(1+d)}. \quad (73)$$

We have been able to solve the problem of similar regions only under very limiting conditions concerning p_0. These are as follows:

(a) p_0 is indefinitely differentiable with regard to $\alpha^{(1)}$ for all values of $\alpha^{(1)}$ and in every point of W, except perhaps in points forming a set of measure zero. That is to say, we suppose that $\partial^k p_0 / \partial (\alpha^{(1)})^k$ exists for any $k = 1, 2, \ldots$ and is integrable over the region W.

Denote by

$$\phi = \frac{\partial \log p_0}{\partial \alpha^{(1)}} = \frac{1}{p_0} \frac{\partial p_0}{\partial \alpha^{(1)}}; \qquad \phi' = \frac{\partial \phi}{\partial \alpha^{(1)}}. \qquad (74)$$

(b) The function p_0 satisfies the equation

$$\phi' = A + B\phi, \qquad (75)$$

where the coefficients A and B are functions of $\alpha^{(1)}$ but are independent of $x_1, x_2, \ldots, x_n$.

This last condition could be somewhat generalized by adding the term $C\phi^2$ to the right-hand side of (75), but this would introduce some complication and we have not found any practical case in which p_0 does satisfy (75) in this more general form and does not in the simple form. We have, however, met instances in which neither of the two forms of the condition (b) is satisfied by p_0.

If the probability law p_0 satisfies the two conditions (a) and (b), then it follows that a necessary and sufficient condition for w to be similar to W with regard to $\alpha^{(1)}$ is that

$$\frac{\partial^k P_0(w)}{\partial (\alpha^{(1)})^k} = \int\!\!\int \cdots \int_w \frac{\partial^k p_0}{\partial (\alpha^{(1)})^k} dx_1\, dx_2, \ldots, dx_n = 0 \qquad (k = 1, 2, \ldots). \quad (76)$$

Taking in (76) $k = 1$ and 2 and writing

$$\frac{\partial p_0}{\partial \alpha^{(1)}} = p_0 \phi, \qquad (77)$$

$$\frac{\partial^2 p_0}{\partial (\alpha^{(1)})^2} = \frac{\partial}{\partial \alpha^{(1)}} (p_0 \phi) = p_0(\phi^2 + \phi'), \qquad (78)$$

it will be found that

$$\frac{\partial P_0(w)}{\partial \alpha^{(1)}} = \int\!\!\int \cdots \int_w p_0 \phi\, dx_1\, dx_2 \ldots dx_n = 0, \qquad (79)$$

$$\frac{\partial^2 P_0(w)}{\partial (\alpha^{(1)})^2} = \int\!\!\int \cdots \int_w p_0(\phi^2 + \phi')\, dx_1\, dx_2 \ldots dx_n = 0. \qquad (80)$$

Using (75) we may transform this last equation into the following

$$\frac{\partial^2 P_0(w)}{\partial (\alpha^{(1)})^2} = \int\!\!\int \cdots \int_w p_0(\phi^2 + A + B\phi)\, dx_1\, dx_2 \ldots dx_n = 0. \qquad (81)$$

Having regard to (72) and (79) it follows from (81) that

$$\int\!\!\int \cdots \int_w p_0 \phi^2\, dx_1 \ldots dx_n = -A\varepsilon = \varepsilon \psi_2(\alpha^{(1)}) \qquad \text{(say)}. \qquad (82)$$

The condition (76) for $k = 3$ may now be obtained by differentiating (81). We shall have

$$\frac{\partial^3 P_0(w)}{\partial (\alpha^{(1)})^3} = \int\!\!\int \cdots \int_w p_0(\phi^3 + 3B\phi^2 + (3A + B^2 + B')\phi + A' + AB)\, dx_1 \ldots dx_n$$

$$= 0, \tag{83}$$

which, owing to (72), (79) and (82) is equivalent to the condition

$$\int\!\!\int \cdots \int_w p_0\phi^3 \, dx_1 \ldots dx_n = (3AB - A' - AB)\varepsilon = \varepsilon\psi_3(\alpha^{(1)}) \qquad \text{(say)}. \tag{84}$$

As it is easy to show, using the method of induction, this process may be continued indefinitely and we shall find

$$\int\!\!\int \cdots \int_w p_0\phi^k \, dx_1 \, dx_2 \ldots dx_n = \varepsilon\psi_k(\alpha^{(1)}) \quad (k = 1, 2), \tag{85}$$

where $\psi_k(\alpha^{(1)})$ is a function of $\alpha^{(1)}$ but independent of the sample x's, since the quantities A, B and their derivatives with regard to $\alpha^{(1)}$ are independent of the x's. $\psi_k(\alpha^{(1)})$ is also independent of the region w, and it follows that whatever be w, and its size ε, if it be similar to W with regard to $\alpha^{(1)}$, the equation (85) must hold true for every value of k, i.e., 1, 2, ... Since the complete sample space W is clearly similar to W and of size unity, it follows that

$$\frac{1}{\varepsilon}\int\!\!\int \cdots \int_w p_0\phi^k \, dx_1 \, dx_2 \ldots dx_n = \int\!\!\int \cdots \int_W p_0\phi^k \, dx_1 \, dx_2 \ldots dx_n \quad (k = 1, 2, 3). \tag{86}$$

Now $p_0(x_1, x_2, \ldots, x_n)$ is a probability law of n variates $x_1, x_2, \ldots, x_n$, defined in the region W; similarly $(1/\varepsilon)p(x_1, x_2, \ldots, x_n)$ may be considered as a probability law for the same variates under the condition that their variation is limited to the region w. We may regard ϕ as a dependent variate which is a known function of the n independent variates x_i. The integral on the right-hand side of (86) is the kth moment coefficient of this variate ϕ obtained on the assumption that the variation in the sample point $x_1, x_2, \ldots, x_n$ is limited to the region W, while the integral on the left-hand side is the same moment coefficient obtained for variation of the sample point, limited to the region w. Denoting these moment coefficients $\mu_k(W)$ and $\mu_k(w)$, we may rewrite (86) in the form:

$$\mu_k(w) = \mu_k(W) \quad (k = 1, 2, 3). \tag{87}$$

It is known that if the set of moment coefficients satisfy certain conditions, the corresponding frequency distribution is completely defined.* Such, for instance, is the case when the series $\Sigma\{\mu_k(it)^k/k!\}$ is convergent, and it then represents the characteristic function of the distribution.

We do not, however, propose to go more closely into this question, and shall consider only the cases in which the moment coefficients of ϕ satisfy the

---

* Hamburger; *Math Ann.*, vol. 81, p. 4 (1920).

conditions of H. Hamburger*. In these cases, to which the theory developed below only applies†, it follows from (87) that when ϕ, which is related to $p_0(x_1, x_2, \ldots, x_n)$ by (74) is such as to satisfy the equation (75), the identity of the two distributions of ϕ is the necessary (and clearly also sufficient) condition for w being similar to W with regard to the parameter $\alpha^{(1)}$.

The significance of this result may be grasped more clearly from the following consideration. Every point of the sample space W will fall on to one or other of the family of hypersurfaces

$$\phi = \text{constant} = \phi_1. \tag{88}$$

Then if

$$P_0(w(\phi)) = \int\int \cdots \int_{w(\phi_1)} p_0 \, dw(\phi_1), \tag{89}$$

$$P_0(W(\phi)) = \int\int \cdots \int_{W(\phi_1)} p_0 \, dW(\phi_1) \tag{90}$$

represent the integral of p_0 taken over the common parts, $w(\phi)$ and $W(\phi)$ of $\phi = \phi_1$ and w and W respectively, it follows that if w be similar to W and of size ε,

$$P_0(w(\phi)) = \varepsilon P_0(W(\phi)), \tag{91}$$

whatever be ϕ_1.

Whatever be ε, a similar region is, in fact, built up of pieces of the hypersurfaces (88) for which (91) is true.

We shall give at this stage only a single example of this result, which will be illustrated more fully when dealing with the best critical regions.

EXAMPLE (7) (A Single Sample of n from a Normal Population; σ Unspecified).

$$p(x_1, x_2, \ldots, x_n) = \left(\frac{1}{\sigma\sqrt{2\pi}}\right)^n \exp - \left\{n\frac{(\bar{x} - a)^2 + s^2}{2\sigma^2}\right\}. \tag{92}$$

For H_0'

$$\alpha^{(1)} = \sigma, \qquad \alpha_0^{(2)} = \alpha_0, \tag{93}$$

$$\phi = \frac{\partial \log p_0}{\partial \sigma} = -\frac{n}{\sigma} + n\frac{(\bar{x} - a_0)^2 + s^2}{4\sigma^3}, \tag{94}$$

$$\phi' = \frac{n}{\sigma^2} - 3n\frac{(\bar{x} - a_0)^2}{\sigma^4} = -\frac{2n}{\sigma^2} - \frac{3}{\sigma}\phi. \tag{95}$$

---

* We are indebted to Dr R.A. Fisher for kindly calling our attention to the fact that we had originally omitted to refer to this restriction.

† It may easily be proved that these conditions are satisfied in the case of examples (7), (8), (9), (10) and (11) discussed below.

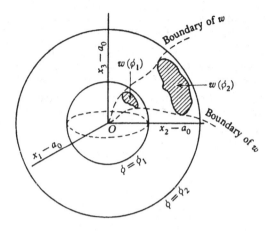

Figure 9

Equation (95) shows that the condition (75) is satisfied. Further, ϕ is constant on any one of the family of hypersurfaces

$$n\{(\bar{x} - a_0)^2 + s^2\} = \sum_{i=1}^{n} (x_i - a_0)^2 = \text{const.} \tag{96}$$

Consequently the most general region w similar to W (which in this case is the whole n-dimensioned space of the x's) is built up of pieces of the hyperspheres (96) which satisfy the relation (91). Since $p_0(x_1, x_2, \ldots, x_n)$ is constant upon each hypersphere, the content of the "piece" $w(\phi)$ must be in a constant proportion, $\varepsilon : 1$ to the content of the complete hyperspherical shell $W(\phi)$. The possible similar regions may be of infinite variety in form. They need not be hypercones, but may be of irregular shape as suggested in fig. 9 for the case $n = 3$. It is out of these possible forms that the best critical region has to be chosen.

(c) Choice of the Best Critical Region

Let H_t be an alternative simple hypothesis defined by the relations (70) and (71). We shall assume that regions similar to W with regard to $\alpha^{(1)}$ do exist. Then w_0, the best critical region for H_0 with regard to H_t, must be determined to maximise

$$P_t(w_0) = \iint \cdots \int_w p_t(x_1, x_2, \ldots, x_n) \, dx_1 \ldots dx_n, \tag{97}$$

subject to the condition (91) holding for all values of ϕ, which implies the condition (72). We shall now prove that if w_0 is chosen to maximise $P_t(w_0)$ under the condition (72), then except perhaps for a set of values of ϕ of measure zero, the region $w_0(\phi)$ will maximise

$$P_t(w(\phi)) = \int \cdots \int_{w(\phi)} p_t(x_1, x_2, \ldots, x_n) \, dw(\phi), \tag{98}$$

under the condition (91). That is to say, we shall prove that whatever be the $(n-1)$-dimensioned region, say $v(\phi)$, being a part of the hypersurface $\phi = $ const. and satisfying the condition

$$P_0(v(\phi)) = \varepsilon P_0(W(\phi)), \tag{99}$$

we should have

$$P_t(v(\phi)) \leqslant P_t(w_0(\phi)), \tag{100}$$

except perhaps for a set of values of ϕ of measure zero.

Suppose in fact that the proposition is not true and that there exists a set E of values of ϕ of positive measure for which it is possible to define the regions $v(\phi)$ satisfying (99), and such that

$$P_t(v(\phi)) > P_t(w_0(\phi)). \tag{101}$$

Denote by CE the set of values of ϕ complementary to E. We shall now define a region, say v, which will be similar to W with regard to $\alpha^{(1)}$ and such that

$$P_t(v) \geqslant P_t(w_0), \tag{102}$$

which will contradict the assumption that w_0 is the best critical region with regard to H_t.

The region v will consist of parts of hypersurfaces $\phi = $ const. For ϕ's included in CE, these parts, $v(\phi)$, will be identical with $w_0(\phi)$ and for ϕ's belonging to E, they will be $v(\phi)$ satisfying (101). Now,

$$P_t(v) = \int_{E+CE} P_t(v(\phi)) \, d\phi,$$

$$P_t(w_0) = \int_{E+CE} P_t(w_0(\phi)) \, d\phi,$$

and, owing to the properties of v,

$$P_t(v) - P_t(w_0) = \int_E (P_t(v(\phi)) - P_t(w_0(\phi)) \, d\phi > 0. \tag{103}$$

It follows that if w_0 is the best critical region, then (101) may be true at most for a set of ϕ's of measure zero. It follows also that if (100) be true for every ϕ and every $v(\phi)$ satisfying (99), then the region w_0, built up of parts of hypersurfaces $\phi = $ const. satisfying (91), is the best critical region required.

Having established this result the problem of finding the best critical region, w_0, is reduced to that of finding parts, $w_0(\phi)$, of $W(\phi)$, which will maximise $P(w(\phi))$ subject to the condition

$$P_0(w_0(\phi)) = \varepsilon P_0(W(\phi)) \tag{104}$$

where ϕ is fixed. This is the same problem that we have treated already when dealing with the case of a simple hypothesis (see pp. 82–86), except that instead of the regions w_0 and W, we have the regions $w_0(\phi)$ and $W(\phi)$, and a space of one dimension less. The inequality

$$p_t \geqslant k(\phi)p_0 \tag{105}$$

will therefore determine the region $w_0(\phi)$, where $k(\phi)$ is a constant (whose value may depend upon ϕ) chosen subject to the condition (104).

The examples which follow illustrate the way in which the relations (104) and (105) combine to give the best critical region. It will be noted that if the family of surfaces bounding the pieces $w_0(\phi)$ conditioned by (105) is independent of the parameters $\alpha_t^{(1)}, \alpha_t^{(2)}, \ldots, \alpha_t^{(1+d)}$, then a common best critical region will exist for H_0 with regard to all hypotheses H_t of the set Ω.

(d) Illustrative Examples

(1) EXAMPLE (8) (The Hypothesis Concerning the Population Mean ("Student's" Problem)). A sample of n has been drawn at random from some normal population, and H_0' is the composite hypothesis that the mean in this population is $a = a_0$, σ being unspecified. We have already discussed the problem of determining similar regions for H_0' in example (7). H_t is an alternative for which

$$\alpha_t^{(1)} = \sigma_t, \qquad \alpha_t^{(2)} = \alpha_t. \tag{106}$$

The family of hypersurfaces, $\phi = $ constant, in the n-dimensioned space are hyperspheres (96) centred at $(x_1 = x_2 = \cdots = x_n = a_0)$; we must determine the nature of the pieces defined by condition (105), to be taken from these to build up the best critical region for H_0' with regard to H_t.

Using (92), it is seen that the condition $p_t \geqslant kp_0$ becomes

$$\frac{1}{\sigma_t^n} \exp - \left\{ \frac{n[(\bar{x} - a_t)^2 + s^2]}{2\sigma_t^2} \right\} \geqslant k \frac{1}{\sigma^n} \exp - \left\{ \frac{n[(\bar{x} - a_t)^2 + s^2]}{2\sigma^2} \right\}. \tag{107}$$

As we are dealing with regions similar with regard to $\alpha^{(1)}$, that is, essentially independent of the value of the parameter $\alpha^{(1)} = \sigma$, we may put $\sigma = \sigma_t$ and find that (107) reduces to,

$$\bar{x}(a_t - a_0) \geqslant \frac{1}{n}\sigma_t^2 \log k + \frac{1}{2}(a_t^2 - a_0^2) = (a_t - a_0)k_1(\phi) \quad \text{(say)}. \tag{108}$$

Two cases must be distinguished in determining $w_0(\phi)$—

$$\text{(a)} \quad a_t > a_0, \qquad \text{then} \qquad \bar{x} \geqslant k_1(\phi), \tag{109}$$

$$\text{(b)} \quad a_t < a_0, \qquad \text{then} \qquad \bar{x} \leqslant k_1(\phi), \tag{110}$$

where $k_1(\phi)$ has to be chosen so that (91) is satisfied. Conditions (109) and

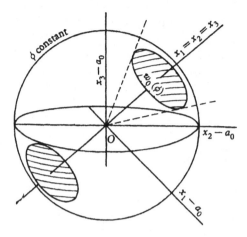

Figure 10

(110) will determine the pieces of the hyperspheres to be used. In the case $n = 3$, $\bar{x} = \frac{1}{3}(x_1 + x_2 + x_3)$ is a plane perpendicular to the axis $x_1 = x_2 = x_3$, and it follows that $w_0(\phi)$ will be a "polar cap" on the surface of the sphere surrounding this axis. The pole is determined by the condition $a_t > a_0$ or $a_t < a_0$. The condition (91) implies that the area of this cap must be ε times the surface area of the whole sphere. The position is indicated in fig. 10. For all values of ϕ, that is to say, for all the concentric spherical shells making up the complete space, these caps must be subtended a constant angle at the centre. Hence the pieces, $w_0(\phi)$, will build up into a cone of circular cross-section, with vertex at (a_0, a_0, a_0) and axis $x_1 = x_2 = x_3$. For each value of ε there will be a cone of different vertical angle. There will be two families of these cones containing the best critical regions:

(a) For the class of hypotheses $a_t > a_0$; the cones will lie in the quadrant of positive x's.
(b) For the class of hypotheses $a_t < a_0$; the cones will lie in the quadrant of negative x's.

It is of interest to compare the general type of similar region suggested in fig. 9 with the special best critical region of fig. 10.

For the cases $n > 3$ we may either appeal to the geometry of multiple space, or proceed analytically as follows.

If $m_2' = (\bar{x} - a_0)^2 + s^2$, then it can be deduced from the probability law (92) that

$$p_0(\bar{x}, m_2') = c_1 \sigma^{-n}\{m_2' - (\bar{x} - a_0)^2\}^{(1/2)(n-3)} \exp - \left\{\frac{nm_2'}{2\sigma^2}\right\}, \qquad (111)$$

$$p_0(m_2') = c_2 \sigma^{-n}\{m_2'\}^{(1/2)(n-2)} \exp - \left\{\frac{nm_2'}{2\sigma^2}\right\}, \qquad (112)$$

where c_1 and c_2 are constants depending on n only. Taking the class of alternatives $a_t > a_0$, $w_0(\phi)$ is that portion of the hypersphere on which $m_2' = $ constant, for which $\bar{x} \geqslant k_1(\phi)$. Consequently the expression (91) becomes

$$\int_{k_1}^{a_0 + \sqrt{m_2'}} p_0(\bar{x}, m_2') \, d\bar{x} = \varepsilon p_0(m_2'), \tag{113}$$

or

$$\int_{k_1}^{a_0 + \sqrt{m_2'}} \{m_2' - (\bar{x} - a_0)^2\}^{(1/2)(n-3)} \, d\bar{x} = \varepsilon \frac{c_2}{c_1} (m_2')^{(1/2)(n-2)}. \tag{114}$$

Make now the transformation

$$\bar{x} - a_0 = \frac{z \sqrt{m_2'}}{\sqrt{(1 + z^2)}}, \tag{115}$$

from which it follows that

$$d\bar{x} = \frac{\sqrt{m_2'}}{\sqrt{[(1 + z^2)^3]}} \, dz, \tag{116}$$

and the relation (114) becomes

$$\int_{z_0}^{+\infty} (1 + z^2)^{-(1/2)n} \, dz = \varepsilon \sqrt{\pi} \frac{\Gamma(\frac{1}{2}[n - 1])}{\Gamma(\frac{1}{2}n)}, \tag{117}$$

the constant multiplying ε necessarily assuming this value so that $\varepsilon = 1$ when $z_0 = -\infty$. But it is seen from (115) that $z = (\bar{x} - a_0)/s$; consequently the boundary of the partial region $w_0(\phi)$ lies on the intersection of the hypersphere, $m_2' = $ constant, and the hypercone, $(\bar{x} - a_0)/s = z_0$. This is independent of ϕ; its axis is the line $x_1 = x_2 = \cdots = x_n$ and its vertical angle is $2\theta = 2 \cot^{-1} z_0$.

If the admissible alternatives are divided into two classes, there will therefore be for each a common best critical region of size ε.

(a)   *Class $a_t > a_0$*;   region w_0 defined by $z = \dfrac{\bar{x} - a_0}{s} \geqslant z_0$. $\qquad$ (118)

(b)   *Class $a_t < a_0$*;   region w_0 defined by $z = \dfrac{\bar{x} - a_0}{s} \leqslant z_0' = -z_0$,   (119)

where z_0 is related to ε by (117), and z_0' by a similar expression in which the limits of the integral are $-\infty$ and $z_0' = -z_0$.

This is "Student's" test.* It is also the test reached by using the principle of likelihood. Further, it has now been shown that starting with information in the form supposed, there can be no better test for the hypothesis under consideration.

* *Biometrika*, vol. 6, p. 1 (1908).

(2) EXAMPLE (9) (The Hypothesis Concerning the Variance in the Sampled Population). The sample has been drawn from *some* normal population and H_0' is the hypothesis that $\sigma = \sigma_0$, the mean α being unspecified. We shall have for H_0'

$$\alpha^{(1)} = a; \qquad \alpha_0^{(2)} = \sigma_0, \tag{120}$$

while for an alternative H_t the parameters are as in (106).
    Further

$$\phi = \frac{\partial \log p_0}{\partial a} = \frac{n(\bar{x} - a)}{\sigma_0^2}, \tag{121}$$

$$\phi' = -n/\sigma_0^2, \tag{122}$$

satisfying the condition (75) with $B = 0$. We must therefore determine on each of the family of hypersurfaces $\phi = \phi_1$ (that is, from (121), $\bar{x} = $ constant) regions $w_0(\phi)$ within which $p_t \geqslant k(\phi_1)p_0$, where $k(\phi_1)$ is chosen so that

$$P_0(w_0(\phi_1)) = \varepsilon P_0(W(\phi_1)). \tag{123}$$

Since we are dealing with regions similar with regard to the mean a, we may put $a = a_t$, and consequently find that

$$s^2(\sigma_0^2 - \sigma_t^2) \leqslant -(\bar{x} - a_t)^2(\sigma_0^2 - \sigma_t^2) + 2\sigma_0^2\sigma_t^2 \left\{ \log \frac{\sigma_0}{\sigma_t} - \frac{1}{n} \log k \right\}$$

$$= (\sigma_0^2 - \sigma_t^2)k'(\phi_1) \quad \text{(say)}. \tag{124}$$

The admissible alternatives must again be broken into two classes according as $\sigma_t > \sigma_0$ or $< \sigma_0$, and since $\bar{x}$ is constant on $\phi = \phi_1$, the regions $w_0(\phi)$ will be given by the following inequalities:

(a)   Case $\sigma_t > \sigma_0$, $s^2 \geqslant k'(\phi)$. $\qquad\qquad$ (125)

(b)   Case $\sigma_t < \sigma_0$, $s^2 \leqslant k'(\phi)$. $\qquad\qquad$ (126)

But since for samples from a normal distribution $\bar{x}$ and s^2 are completely independent the values of $k'(\phi)$ that determine the regions $w_0(\phi)$ so as to satisfy (123), will be functions of ε and n only. It follows that the best critical regions, w_0, for H_0' will be:

(a)   for the class of alternatives $\sigma_t > \sigma_0$,   defined by $s^2 \geqslant s_0^2$, (127)

(b)   for the class of alternatives $\sigma_t < \sigma_0$,   defined by $s^2 \leqslant s_0'^2$. (128)

These regions lie respectively outside and inside hypercylinders in the n-dimensioned space. The relation between ε and the critical values s_0^2 and $s_0'^2$ may be found from equations (46), (48) and (50) of example 2.* ...

---

* The difference between the two cases should be noted: In example (2) the population mean is specified, H_0 is a simple hypothesis and m_2' is the criterion. In example (9) the mean is not specified, H_0' is composite and the criterion is s^2.

VI. Summary of Results

1. A new basis has been introduced for choosing among criteria suitable for testing any given statistical hypothesis, H_0, with regard to an alternative H_t. If θ_1 and θ_2 are two such possible criteria and if in using them there is the same chance, ε, of rejecting H_0 when it is in fact true, we should choose that one of the two which assures the minimum chance of accepting H_0 when the true hypothesis is H_t.

2. Starting from this point of view, since the choice of a criterion is equivalent to the choice of a critical region in multiple space, it was possible to introduce the conception of the best critical region with regard to the alternative hypothesis H_t. This is the region, the use of which, for a fixed value of ε, assures the minimum chance of accepting H_0 when the true hypothesis is H_t. The criterion, based on the best critical region, may be referred to as to the most efficient criterion with regard to the alternative H_t.

3. It has been shown that the choice of the most efficient criterion, or of the best critical region, is equivalent to the solution of a problem in the Calculus of Variations. We give the solution of this problem for the case of testing a simple hypothesis.

To solve the same problem in the case where the hypothesis tested is composite, the solution of a further problem is required; this consists in determining what has been called a region similar to the sample space with regard to a parameter.

We have been able to solve this auxiliary problem only under certain limiting conditions; at present, therefore, these conditions also restrict the generality of the solution given to the problem of the best critical region for testing composite hypotheses.

4. An important case arises, when the best critical regions are identical with regard to a certain class of alternatives, which may be considered to include all admissible hypotheses. In this case—which, as has been shown by several examples, is not an uncommon one—unless we are in a position to assign precise measures of a priori probability to the simple hypotheses contained in the composite H_0, it appears that no more efficient test than that given by the best critical region can be devised.

5. The question of the choice of a "good critical region" for testing a hypothesis, when there is no common best critical region with regard to every alternative admissible hypothesis, remains open. It has, however, been shown that the critical region based on the principle of likelihood satisfies our intuitive requirements of a "good critical region".

6. The method of finding best critical regions for testing both simple and composite hypotheses has been illustrated for several important problems commonly met in statistical analysis. Owing to the considerable size which the paper has already reached, the solution of the same problem for other important types of hypotheses must be left for separate publication.

Introduction to
Bartlett (1937) Properties of Sufficiency and Statistical Tests

D.A.S. Fraser
York University

1. Overview

Fisher's immense contribution to statistics in the 1920s and 1930s arose in sharp contrast to the general statistics of the time and the developing Neyman–Pearson theory; for an overview, see the published version of the first R.A. Fisher lecture [Bartlett (1965)]. With very strong credentials in the background statistics of the time, Bartlett in this paper and its predecessor [Bartlett (1936)], examines analytically and seriously many of the Fisher concepts against the background theory. From the perspective of the 1965 review paper and even that of the present time, this paper stands as a major initiative in the development of statistics.

2. Background and Contents

This study of sufficiency and statistical tests appeared at a time when statistics had a strong basis in the biological sciences with a developing concern for experimentation, particularly in agricultural science. The contemporary statistical forces were in conflicting directions: Fisher (1922, 1925, 1930, 1934, 1935; also 1950) proposing the concepts of sufficiency, information, ancillarity, likelihood, and fiducial; Neyman and Pearson (1933, 1936a, 1936b) developed the formal accept–reject theory of hypothesis testing. Rather than on estimation or testing, this paper focuses on the "structure of small-sample" procedures, on the distributions for inference concerning one parameter in the presence of "irrelevant unknown parameters" and "variation" (error with a known distribution). It thus provides the seeds for statistical inference as a distinct area of study, largely delineated following the publication of the books by Fisher (1956) and Savage (1954).

The paper uses concepts and theory from both the Fisher and Neyman–Pearson schools and proposes statistical theory for inference in a manner that might now be called unified.

In Sec. 5 exact tests are discussed; these relate to the similar tests of the Neyman–Pearson school and, in part, address the larger issue of the distribution for inference concerning an interest parameter in the presence of "irrelevant unknown" (nuisance) parameters.

As a part of this, Bartlett "state(s) as a general principle that all exact tests of composite hypotheses are equivalent to tests of simple hypotheses for conditional samples." This focuses on what would now be expressed by the factorization

$$f(y; \psi, \lambda) = f(y_2|y_1; \psi)f(y_1; \psi, \lambda), \tag{1}$$

with ψ as the interest and λ as the nuisance parameter. In this, y_1 is sufficient for λ, given any particular value for ψ, and is perhaps slightly more general than the prescription: "for the variation in the (response variable) to be independent of (the nuisance parameter λ), a sufficient ... statistic(.) must exist for (λ)." The discussion uses the term "variation" in the context of inference, thus anticipating the current separation of *variation* from *effect* in general statistical inference, distinct from the special case in the analysis of variance as proposed by Fisher.

As a first example, Bartlett discusses briefly the conditional analysis of independence in the 2×2 contingency table which has a long and continuing history of proponents and detractors. For some recent views, see Yates (1984).

As a second example, Bartlett considers in Sec. 6 a likelihood ratio test of the homogeneity of sample variances. The starting point, however, is the conditional (also marginal) model given the sufficient statistic for regression parameters; this avoids the usual degrees-of-freedom problem commonly illustrated by the many-means problem [Neyman and Scott, (1948)]. As part of approximating the (conditional) likelihood ratio chi-square statistic, he derives corrections of a type now generally called Bartlett corrections; see, e.g., McCullagh (1987).

In the preamble to the example, Bartlett proposes that a procedure "be based directly on the conditional likelihood (for the interest parameter)." This notion of conditional likelihood has only recently been pursued generally, e.g., Cox and Reid (1987); Fraser and Reid (1989), although closely related marginal likelihoods have had longer attention; Fraser (1967, 1968); Kalbfleisch and Sprott (1970); Fraser and Reid (1989).

In Sec. 7, Bartlett discusses exact tests of fit, examining initially the normal model. For this, he notes that the conditional distribution of the response $(y_1, \ldots, y_n)$ given the sufficient statistic $(\bar{y}, s_y^2)$ is free of the parameters (μ, σ^2)

and is thus available for model testing. This procedure anticipates much contemporary theory for model testing and for conditional inference.

At the end of this section, Bartlett briefly mentions a corresponding procedure for the location Cauchy distribution and notes the goodness of fit would be based on the *marginal* distribution of the configuration statistic; see Fisher (1934). The procedure would now be expressed by the factorization

$$f(y; \psi, \lambda) = f(y_2; \psi)f(y_1|y_2; \psi, \lambda), \tag{2}$$

in which a marginal variable y_2 is used for inference concerning ψ. It should be noted that this exact test procedure would contradict Bartlett's general principle cited before Eq. (1). In fact, exact tests can come from marginal (2) as well as conditional (1) models concerning an interest parameter, see, e.g., Fraser and Reid, (1989).

Sections 9 and 10 consider many discrete and contingency table examples of *conditional inference for an interest parameter in the presence of a nuisance parameter*; this would be the current language although Bartlett mainly used the term "exact tests."

Bartlett's paper initiates many of the methods of conditional and marginal inference. The conditional methods apply widely to component parameters in exponential family models, and to generalizations using the exponential as a pattern. The marginal methods apply widely to component parameters in transformation parameter models, and to generalizations using the transformation model as a pattern. For an overview, see Fraser and Reid (1990). The paper also initiates the study of distributional corrections for the likelihood ratio statistic, such as the Bartlett and mean-variance corrections.

3. Personal Background

Maurice S. Bartlett was born in London, June 18, 1910. After studying at Cambridge, he took an appointment at University College, London. In 1934, he left to work at a research station of Imperial Chemical Industries, but returned to the academic world in 1938 as a lecturer in mathematics at Cambridge. In 1947, he took the chair in mathematical statistics at the University of Manchester, in 1960, became professor of statistics at University College, London, and from 1967 to his retirement in 1975, was professor of biostatistics at the University of Oxford. For a detailed interview, see Olkin (1989).

Bartlett's strengths range from mathematics and the foundations through to the realities of statistics in application. His departure from University College to go to Imperial Chemical Industries in 1934 was triggered by the need to teach statistics with an understanding of its applications. He thus had the background and concerns for this early and penetrating investigation toward unity in statistical inference.

References

Bartlett, M.S. (1965). R.A. Fisher and the last fifty years of statistical methodology, *J. Amer. Statist. Assoc.*, **60**, 395–409.

Bartlett, M.S. (1936). Statistical information and properties of sufficiency, *Proc. Roy. Soc., London, Ser. A*, **154**, 124–137.

Cox, D.R., and Reid, N. (1987). Parameter orthogonality and approximate conditional inference, *J. Roy. Statist. Soc., Ser. B*, **49**, 1–39.

Fisher, R.A. (1922). On the mathematical foundations of theoretical statistics, *Phil. Trans. Roy. Soc. London, Ser. A*, **222**, 309–68 [also as paper 10 in Fisher (1950)].

Fisher, R.A. (1925). Theory of statistical estimation, *Proc. Cambridge Phil. Soc.*, **22**, 700–725 [also as paper 11 in Fisher (1950)].

Fisher, R.A. (1930). Inverse probability, *Proc. Cambridge Phil. Soc.*, **26**, 528–535 [also as paper 22 in Fisher (1950)].

Fisher, R.A. (1934). Two new properties of mathematical likelihood, *Proc. Roy. Soc. London, Ser. A*, **144**, 285–307 [also as paper 24 in Fisher (1950)].

Fisher, R.A. (1935). The fiducial argument in statistical inference, *Ann. Eugenics*, **6**, 391–398. [also as paper 25 in Fisher (1950)].

Fisher, R.A. (1950). *Contributions to Mathematical Statistics*. Wiley, New York.

Fisher, R.A. (1956). *Statistical Methods and Scientific Inference*. Oliver and Boyd, Edinburgh.

Fraser, D.A.S. (1967). Data transformations and the linear model, *Ann. Math. Statist.*, **38**, 1456–1465.

Fraser, D.A.S. (1968). *Inference and Linear Models*. McGraw-Hill, New York.

Fraser, D.A.S., and Reid, N. (1989). Adjustments to profile likelihood, *Biometrika*, **76**, 477–488.

Fraser, D.A.S., and Reid, N. (1990). Statistical inference: Some theoretical methods and directions, *Environmetrics*, **1**, 21–36.

Kalbfleisch, J.D., and Sprott, D.A. (1970). Application of likelihood methods to models involving large numbers of parameters, *J. Roy. Statist. Soc., Ser. B*, **32**, 175–208.

McCullagh, P. (1987). *Tensor Methods in Statistics*. Chapman and Hall, London.

Neyman, J., and Pearson, E.S. (1933). On the problem of the most efficient tests of statistical hypotheses, *Phil. Trans. Roy. Soc., Ser. A*, **231**, 28–337.

Neyman, J., and Pearson, E.S. (1936a). Contributions to the theory of testing statistical hypotheses. I. Unbiased critical regions of type A and type A1, *Statist. Res. Mem.*, **1**, 1–37.

Neyman, J., and Pearson, E.S. (1936b). Sufficient statistics and uniformly most powerful tests of statistical hypotheses, *Statist. Res. Mem.*, **1**, 113–137.

Neyman, J., and Scott, E.L. (1948). Consistent estimates based on partially consistent observations, *Econometrica*, **16**, 1–32.

Olkin, I. (1989). A conversation with Maurice Bartlett, *Statist. Sci.*, **4**, 151–163.

Savage, L.J. (1954). *The Foundations of Statistics*. Wiley, New York.

Yates, F. (1984). Tests of significance for 2 × 2 contingency tables, *J. Roy. Statist. Soc. Ser. A*, **147**, 426–463.

Properties of Sufficiency and Statistical Tests

M.S. Bartlett
Imperial Chemical Industries, Ltd.,
United Kingdom

Introduction

1—In a previous paper*, dealing with the importance of properties of sufficiency in the statistical theory of small samples, attention was mainly confined to the theory of estimation. In the present paper the structure of small sample tests, whether these are related to problems of estimation and fiducial distributions, or are of the nature of tests of goodness of fit, is considered further.

The notation $a|b$ implies as before that the variate a is conditioned by† a given value of b. The fixed variate b may be denoted by $|b$, and analogously if b is clear from the context, $a|b$ may be written simply as $a|$. Corresponding to the idea of ancillary information introduced by Fisher for the case of a single unknown θ, where auxiliary statistics control the accuracy of our estimate, I have termed a conditional statistic of the form $T|$, quasi-sufficient, if its distribution satisfies the "sufficiency" property and contains all the information on θ. In the more general case of other unknowns, such a statistic may contain all the *available* information on θ.

---

* Bartlett (1936a): I have noticed an error on p. 128 of this paper which I will take this opportunity to correct. In the example the order of magnitude of the two observations was lost sight of. The information in one observation, *if it be recorded whether it is the greater or smaller*, is found to be 1·386, and is thus more than that in the mean.

† With this notation and phraseology, b is in general a known statistic. Inside a probability bracket, it may sometimes be necessary to stress that the distribution depends on an unknown parameter θ, and the usual notation is then adopted of writing $p(a)$ more fully as $p(a|\theta)$, and $p(a|b)$ as $p(a|b, \theta)$.

Sufficient Statistics and Fiducial Distributions

2—It has been noted (Bartlett 1936*a*) that if our information on a population parameter θ can be confined to a single degree of freedom, a fiducial distribution for θ can be expected to follow, and possible sufficiency properties that would achieve this result have been enumerated. A corresponding classification of fiducial distributions is possible.

Since recently Fisher (1935) has put forward the idea of a simultaneous fiducial distribution, it is important to notice that the sufficient set of statistics $\bar{x}$ and s^2 obtained from a sample drawn from a normal population (usual notation) do not at once determine fiducial distributions for the mean m and variance σ^2. That for σ^2 follows at once from the relation

$$p(\bar{x}, s^2 | m, \sigma^2) = p(\bar{x}|m, \sigma^2)p(s^2|\sigma^2), \tag{1}$$

but that for $\bar{x}$ depends on the possibility of the alternative division

$$p\{\Sigma(x - m)^2 | \sigma^2\}p(t), \tag{2}$$

where t depends only on the unknown quantity m. No justification has yet been given that because the above relations are equivalent respectively to fiducial distributions denoted by $fp(m|\sigma^2)fp(\sigma^2)$ and $fp(\sigma^2|m)fp(m)$, and hence symbolically to $fp(m, \sigma^2)$, that the idea of a simultaneous fiducial distribution, and hence by integration the fiducial distribution of either of the two parameters, is valid when *both* relations of form (1) and (2) do not exist (Bartlett 1936*b*). Moreover, even in the above example, the simultaneous distribution is only to be regarded as a symbolic one, for there is no reason to suppose that from it we may infer the fiducial distribution of, say, $m + \sigma$.

3—In certain cases where a fiducial distribution exists for a population parameter, it will similarly exist for the corresponding statistic in an unknown sample. If, for example, a sufficient statistic T_1 exists for θ in the known sample S_1, we shall have a corresponding unknown statistic T_2 in an unknown sample S_2, and an unknown statistic T for the joint sample S. If we write

$$p(T_1|\theta)p(T_2|\theta) = p(T|\theta)p(T_1, T_2|T), \tag{3}$$

then $p(T_1, T_2|T)$ depends only on T_1 and the unknown T_2 (for which T_1 may be regarded as a sufficient statistic), and will lead to a fiducial distribution for T_2. Alternatively, if the unknown sample S_2 is merely the remainder of a "sample" from which, in order to infer its contents, a subsample S_1 has been drawn, we may obtain the fiducial distribution of T. If T_2 or T is an unbiased estimate of θ, we obtain the fiducial distribution of θ by letting the size of sample S_2 tend to infinity.

No corresponding fiducial distribution for T_2 (or T) exists if these statistics are only quasi-sufficient, since the configuration of the second sample will be unknown. T_2 has not then the same claim to summarize the contents of sample S_2.

For similar inferences on both $\bar{x}$ and s^2 (or $\bar{x}_2$ and s_2^2) in normal theory, the relevant probability distribution will be

$$p(\bar{x}_1, \bar{x}_2, s_1^2, s_2^2 | \bar{x}, s^2), \tag{4}$$

which is necessarily independent of m and σ^2. This distribution can be split up into two factors in three ways, corresponding to the association of s_1^2, s_2^2 or $(n_2 - 1)s_1^2 + (n_2 - 1)s_2^2$ with the t-factor. We have

$$p\left(\frac{\bar{x}_1 - \bar{x}_2}{s_1}\right)p\left(\frac{s_2^2}{s^2}\right) \tag{5}$$

$$= p\left(\frac{\bar{x}_1 - \bar{x}_2}{s_2}\right)p\left(\frac{s_1^2}{s^2}\right) \tag{6}$$

$$= p\left(\frac{\bar{x}_1 - \bar{x}_2}{s}\right)p\left(\frac{s_1^2}{s_2^2}\right). \tag{7}$$

Since (5) is equivalent to $fp(\bar{x}_2)fp(s_2^2|\bar{x}_2)$, and (7) to $fp(\bar{x}_2|s_2^2)fp(s_2^2)$, it is consistent to speak of the simultaneous distribution $fp(\bar{x}_2|s_2^2)$. But while (5) is also equivalent to $fp(\bar{x})fp(s^2|\bar{x})$, $fp(\bar{x}|s^2)$ is obtained from the first factor, and $fp(s^2)$ from the second factor, of (6), so that $fp(\bar{x}, s^2)$ also exists, but by virtue of a *different* factorization (cf. Fisher 1935).

For discontinuous variation, a relation (3) may similarly hold. While a fiducial distribution (in Fisher's sense) will no longer exist, the probability distribution $p(T_1, T_2|T)$ will still be available for inferences on T_2 or T. Thus if S_1 contains r_1 members out of n_1 with an attribute A, etc., we obtain

$$p(r_1, r_2|r) = p(r_1)p(r_2)/p(r)$$

$$= \frac{n_1!n_2!(n - r)!r!}{(n_1 - r_1)!(n_2 - r_2)!r_1!r_2!}, \tag{8}$$

which will determine the chance, say, of obtaining as few as r_1 members in S_1 when S contains r such members, or S_2 at least r_2 such members.*

4—The equivalence of a sufficient statistic (or, when relevant, the fiducial distribution derived from it) to the original data implies that when it exists it should lead to the most efficient test. It does not follow that a uniformly most powerful test, as defined by Neyman and Pearson (1933), will necessarily exist; but if the probability (or fiducial probability) distribution is known, the consequences of any procedure based on it will also be known.

The converse principle, that the existence of a uniformly more powerful test must depend on the necessary sufficiency properties being present, and is, moreover, only possible for the testing of a single unknown parameter, has been denied by Neyman and Pearson (1936b); but while agreeing that the examples they give are formal exceptions, I think it is worth while examining

---

* For approximate methods of using equation (8), see Bartlett (1937).

their examples further, since they could reasonably be regarded as bearing out the principle to which formally they are anomalous. It seems to me more valuable to recognize the generality of the principle that a test of a single parameter should be most sensitive to variations in it than to reject the principle because of apparent exceptions.

In example I of their paper, the distribution

$$p(x) = \beta e^{-\beta(x-\gamma)}, \quad (x \geqslant \gamma), \tag{9}$$

is considered. It is required to test whether $\gamma \leqslant \gamma_0$ and/or $\beta \geqslant \beta_0$. Since if any observation occurs that is less than γ_0 no statistical test is necessary, we are effectively concerned only with samples for which all observations are greater than γ_0. *For such observations*, the distribution law is

$$p(x) = \beta e^{-\beta(x-\gamma)}/e^{-\beta(\gamma_0-\gamma)}, \quad (x \geqslant \gamma_0),$$

$$= \beta e^{-\beta(x-\gamma_0)}, \tag{10}$$

and is independent of γ. The sufficient statistic for β is $\bar{x}$, so that we are merely testing one independent parameter β for which a sufficient statistic $\bar{x}$ exists.

Example II is merely a special case of this with $\beta = 1 + \gamma^2$, $(\gamma_0 = 0)$, and again $\bar{x}$ is the sufficient statistic for β and hence for γ, $(\gamma \leqslant 0)$.

The above examples remind us, however, that a fiducial distribution is of more limited application than the sufficient statistic from which it is derived, and if restrictions are placed on the possible values of an unknown parameter, may become irrelevant. If the restriction is on an eliminated unknown, it might prove more profitable to use an inequality for a test of significance than an exact test. Thus if in normal theory it were known that $\sigma^2 \leqslant \sigma_0^2$, a test based on $p(\bar{x}|m, \sigma_0^2)$ might be more useful than one based on $p(t)$, though the possibility of using exact information on the range of other parameters in this way is in practice rare.

Conditional Variation and Exact Tests of Significance

5—By exact tests will be meant tests depending on a known probability distribution; that is, independent of irrelevant unknown parameters. It is assumed that no certain information is available on the range of these extra parameters, so that their complete elimination from our distributions is desirable.

In order for the variation in the sample S to be independent of irrelevant unknowns ϕ, a sufficient set of statistics U must exist for ϕ. All exact tests of significance which are to be independent of ϕ must be based on the calculable *conditional variation* of the sample $S|U$. We may in fact state as a general principle that *all exact tests of composite hypotheses are equivalent to tests of simple hypotheses for conditional samples.* For this principle to be general, conditional variation is understood to include theoretical conditional variation; for we have seen that in certain cases allied to problems in estimation, the set U may be functions of a primary unknown θ.

A useful illustration of the principle is given by the known exact test (Fisher 1934, p. 99) for the 2×2 contingency table (observed frequencies n_{11}, n_{12}, n_{21}, n_{22}). The sufficient statistics U for the unknown probabilities of independent attributes A and B are $n_{1.}|n$ and $n_{.1}|n$, where $n_{1.} = n_{11} + n_{12}$, etc. Hence any exact test of independence *must* be based on the variation $S|U$, which has one degree of freedom, and a distribution

$$p(S|U) = p(S)/p(n_{1.}|n)p(n_{.1}|n)$$

$$= \frac{n_{1.}!n_{2.}!n_{.1}!n_{.2}!}{n_{11}!n_{12}!n_{21}!n_{22}!n!}. \tag{11}$$

6—It is of some importance to consider the relation of tests dependent on this principle of conditional variation with those obtained by the likelihood criterion introduced by Neyman and Pearson. Suppose there is only one degree of freedom after elimination of irrelevant unknowns, as in quasi-sufficient solutions of estimation problems; and suppose further the relation exists,

$$p(T_1|\theta_1)p(T_2|\theta_2) = p(T_1, T_2|T)p(T|\theta), \tag{12}$$

when $\theta_1 = \theta_2 = \theta$. We have $p(T_1, T_2|T) \equiv p(T_1|T)$ and $T_1|T$ is the statistic for testing discrepancies between θ_1 and θ_2. By the likelihood criterion, however, the appropriate variate will be of the form

$$\lambda = \frac{p(T_1|T)p(T|\hat{\theta})}{p(T_1|\hat{\theta}_1)p(T_2|\hat{\theta}_2)},$$

where $\hat{\theta}$ is the maximum likelihood estimate of θ from the distribution of T, etc., whence

$$\lambda = \frac{f(T_1|T)F(T)}{f_1(T_1)f_2(T_2)}, \tag{13}$$

say, which, for variation of S, will not necessarily be equivalent to $T_1|T$, or independent of θ. The condition for λ to provide the same test as $T_1|T$ appears to be that $T_1|T$ should be equivalent to a function $\psi(T_1, T_2)$ independent of T, and that $F(T)/f_1(T_1)f_2(T_2)$ should be an appropriate function $\phi(\psi)$. This holds when λ provides the proper test in normal theory, but it clearly must fail when only quasi-sufficiency (not convertible into pure sufficiency) properties exist.

A modification of the criterion when statistics U exist is proposed here. For a comprehensive test of "goodness of fit" involving all the remaining degrees of freedom of the sample, the test may be based directly on the *conditional likelihood* of $S|$. For the joint testing of definite unknowns, the conditional likelihood of the relevant statistics $T|$ would be considered. A criterion of this kind, if it differs from λ, is denoted subsequently by μ.

The mathematical definition of likelihood adopted is not separated from the more fundamental conception of the chance $p(S)$ of the observed data. For discontinuous data* the two are identical, so that the logarithm L is $\log p(S)$. For continuous variation it is sometimes convenient to drop the infinitesimal elements dx in $p(S)$, but some caution is necessary, this becoming more apparent when the likelihood of derived statistics is to be considered. Thus for s^2 (n degrees of freedom) in normal theory, $L(s^2)$ must, like $p(s^2)$, be invariant for a simultaneous change in scale in both s^2 and σ^2, and is defined to be

$$C + n \log s - n \log \sigma - \frac{n}{2}\left(\frac{s^2}{\sigma^2} - 1\right), \tag{14}$$

it being permissible to drop the term $d \log s^2$ (but not ds^2).

As an example we shall derive the μ appropriate for testing discrepancies among several variances. It is assumed that means and other regression parameters have already been eliminated, the statistic U being the pooled residual variance s^2, and the T the k individual variances s_r^2. Then

$$L(T) = L(T|U) + L(U),$$

or

$$L(T|U) = C' + \Sigma n_r \log s_r - n \log s.$$

For convenience $L(T|U)$ is measured from its maximum value C', so that

$$-2 \log \mu = n \log s^2 - \Sigma n_r \log s_r^2. \tag{15}$$

This criterion is not identical with that proposed by Neyman and Pearson, the weights being the number of degrees of freedom and not the number of observations. It is, however, put forward as an improvement, and a practicable method of using it derived below. With the original criterion λ it would be possible, if several regression parameters were eliminated from samples of unequal size, for fluctuations of a variance reduced to one or two degrees of freedom to mask real discrepancies in more stable samples; this effect is corrected when the weight for such a variance is reduced correspondingly.

If any likelihood with f degrees of freedom tends to a limiting normal form $C \exp\{-\frac{1}{2}A(x, x)\}$, then $-2 \log \lambda$ will tend to be distributed as χ^2 with f degrees of freedom. This, apart from its practical importance, is a useful reminder of the goodness of fit, or test of homogeneity, character of such tests, and should warn us against pooling together components which there is reason to separate.

To obtain more precisely the value of the χ^2 approximation in the present problem, consider first μ from (14) (that is, $k = 2$, $n_1 = n$, $n_2 = \infty$). From the known form (14) of the distribution of s^2, we readily obtain by integration the characteristic function of $-2 \log \mu$ for this case; the expected value

---

* Variation in which only discrete values of the variate are possible is specified for convenience by the term "discontinuous variation".

Table I.

n_1	$n_1 = n_2$					$n_2 = \infty$			
	1	2	3	6	12	2	4	9	∞
C	1.5	1.25	1.167	1.083	1.042	1.167	1.083	1.037	1
$P = 0.10$	2.48	2.66	2.69	2.70	2.70	2.68	2.70	2.70	2.706
$P = 0.02$	4.61	5.16	5.31	5.39	5.41	5.28	5.38	5.41	5.412

$$M \equiv E(\mu)^{-2t}$$

$$= \frac{\Gamma\left\{\frac{n}{2}(1 - 2t)\right\}\left(\frac{n}{2}\right)^{nt} e^{-nt}}{(1 - 2t)^{n/2}\Gamma\left(\frac{n}{2}\right)}, \tag{16}$$

$$K \equiv \log M = t\left(1 + \frac{1}{3n} + \cdots\right) + \frac{t^2}{2!}\left(2 + \frac{4}{3n} + \cdots\right)$$

$$+ \frac{t^3}{3!}\left(8 + \frac{24}{3n} + \cdots\right) + \cdots$$

$$= t\left(1 + \frac{1}{3n}\right) + 2\frac{t^2}{2!}\left(1 + \frac{1}{3n}\right)^2 + 8\frac{t^3}{3!}\left(1 + \frac{1}{3n}\right)^3 + \cdots \tag{17}$$

approximately. If we call the exact function $K(n)$, we have for the general equation (15), owing to the equivalence of the statistics $s_r^2 | s^2$ to angle variables independent of s^2,

$$K = \Sigma K(n_r) - K(n), \tag{18}$$

or neglecting the effect of terms of $O\left(\frac{1}{n_r^2}\right)$, as in (17), we write

$$-\frac{2\log\mu}{C} = \chi^2 \tag{19}$$

with $k - 1$ degrees of freedom, where

$$C = 1 + \frac{1}{3(k - 1)}\left\{\Sigma\frac{1}{n_r} - \frac{1}{n}\right\}. \tag{20}$$

The practical value of (19) was checked by means of the special case $k = 2$, representative values of (19) being given in Table I* for $P = 0.10$ and $P =$

* The values for $n_1 = n_2$ were obtained by means of Fisher's z table. When $n_2 = \infty$, the values for $n_1 = 2$ and 4 were obtained by an iterative method. It was subsequently noticed that the case $n_1 = 2$ could be checked from Table I of Neyman and Pearson's paper (1936a), from which the values for $n_1 = 9$ were added.

0.02, corresponding to the correct values of χ^2 (one degree of freedom) of 2.706 and 5.412 respectively. The use of C tends to over-correct in very small samples the otherwise exaggerated significance levels, but it greatly increases the value of the approximation, and serves also as a gauge of its closeness.

Continuous Variation—Normal Theory

7—For continuous variation, such as in normal theory, exact tests of significance have often been obtained owing to the readiness of the sample to factorize into independent statistics. Thus, for all inferences on the normality of a sample the relevant distribution

$$p(S|\bar{x}, s^2)$$

is expressible as a product of t distributions, and the usual statistics g_1 and g_2 (or β_1 and β_2) for testing for non-normality are independent of $\bar{x}$ and s^2.

The usual χ^2 goodness of fit test is not, but since the expected frequencies corresponding to $p(S|)$ would in any case only be used in a "large sample" test, which is an approximation, the alternative use of the estimated normal distribution, $m = \bar{x}$, $\sigma^2 = s^2$, will be legitimate. We may appeal to Fisher's proof that the distribution of χ^2 when m and σ^2 are efficiently estimated follows the well-known form with $f - 3$ degrees of freedom, where f is the number of cells.

But it is of theoretical interest to note that the *true* expected values for $S|$ could be found. Since the expected frequencies would then have three *fixed* linear conditions, for n, $\bar{x}$ and s^2, the number of degrees of freedom for χ^2 could, from this point of view, never have been questioned. $S|$ implies a sample point distributed over the $n - 2$-dimensional surface of a hypersphere of radius $s\sqrt{(n-1)}$ in the "plane" $\Sigma x = n\bar{x}$, but the *expected* distribution of the n variates is more simply expressed by the distribution (n times that for any one variate)

$$E(S|) = \frac{n\Gamma\left(\dfrac{n}{2}\right)}{\Gamma(\tfrac{1}{2})\Gamma\left(\dfrac{n-1}{2}\right)} \left\{1 - \frac{(x-\bar{x})^2}{(n-1)s^2}\right\}^{(1/2)(n-3)} \frac{dx}{s\sqrt{(n-1)}}. \tag{21}$$

There is here a distinction between an exact test of goodness of fit for the normal law (which does not imply fitting a *normal* distribution at all), and the estimation of the normal law, which may be taken to be $m = \bar{x}$, $\sigma^2 = s^2$ (or $(n-1)s^2/n$).

Similarly for the exponential distribution with known origin 0 but unknown scale, the sufficient statistic for the scale is $\bar{x}$, the geometrical sample point is distributed at random over the "plane" $\Sigma x = n\bar{x}$, $(x > 0)$, and the expected distribution is

$$E(S|) = \frac{n}{\bar{x}} \left(1 - \frac{x}{n\bar{x}} \right)^{n-2} dx. \tag{22}$$

For the distribution $p = dx/\pi\{1 + (x - m)^2\}$, for which

$$p(S|m) = p(\bar{x}|C, m)p(C) \tag{23}$$

(where C is the configuration), the goodness of fit will be based on C and the estimation problem has entirely disappeared.

8—For two correlated variates x_1 and x_2, no function of the estimated correlation coefficients r and r' from two samples S and S' is a sufficient statistic for the true coefficient ρ. Hence no "best" test for a discrepancy in correlation coefficients is possible.

If, however, the degree of association between x_1 and x_2 were to be compared in the two samples on the basis of two populations of the same variability, an appropriate distribution is

$$p(v_{12}, v'_{12}|V_{11}, V_{12}, V_{22}), \tag{24}$$

where v_{12} is the sample covariance of S, V_{12} that of $S + S'$ (with elimination of both sample means), etc. This distribution, which is necessarily independent of σ_1^2, σ_2^2 and ρ, is thus a valid test of the difference between two covariances, although owing to the conditional nature of the distribution, the test would be rather an impracticable one even if the mathematical form of the distribution were known.

Discontinuous Variation—Poisson and Binomial Theory

9—For discontinuous variation, as for continuous variation which has been grouped, it is expedient for all but very small samples to be treated by approximate tests, but it is still important to consider the exact tests when they exist, not only for use with very small samples, but so that the basis of the approximate tests may be more clearly realized.

Consider first the Poisson distribution. For two samples with observed frequencies r_1 and r_2, the distribution of $(r_1, r_2|r)$, where $r = r_1 + r_2$, is simply a partition or configuration distribution giving the number of ways of dividing r between the two samples, and is

$$p(r_1, r_2|r) = \frac{r!}{2^r r_1! r_2!}, \tag{25}$$

or the terms of the binomial distribution $(\frac{1}{2} + \frac{1}{2})^r$. This will be the distribution for testing a discrepancy between the observed values of two Poisson samples.

For several samples, we have similarly a distribution depending on the multinomial distribution

$$\left(\frac{1}{n} + \frac{1}{n} + \cdots + \frac{1}{n}\right)^r.$$ (26)

The χ^2 test for dispersion is

$$\chi^2 = \Sigma(r_i - k_1)^2/k_1 = (n - 1)k_2/k_1$$ (27)

(where the usual semi-invariant notation is used, so that $k_1 = r/n$). Before an exact test is possible, a suitable criterion must be adopted. χ^2 and the μ criterion will no longer be completely equivalent in small samples, but since for distributions near to the Poisson the ratio k_2/k_1 may be defined as an index of dispersion, it seems in practice convenient still to consider χ^2, or equivalently (since k_1 is fixed and is the *true* expected value in (27)) the variance k_2.

The moments of $k_2|k_1$ are of some interest; they would be obtained from the identity

$$k_2 = \frac{1}{n-1}\{\Sigma r_i^2 - nk_1^2\}.$$

Thus, after some algebra with factorial moments, it is found that

$$\kappa_1(k_2|k_1) = k_1,$$ (28)

$$\kappa_2(k_2|k_1) = \frac{2k_1(nk_1 - 1)}{n(n-1)}.$$ (29)

These results may be compared with

$$\kappa_1(k_2) = m,$$ (30)

$$\kappa_2(k_2 - k_1) = \frac{2m^2}{(n-1)},$$ (31)

and the approximate solution from (27),

$$\kappa_2(k_2|k_1) \sim \frac{2k_1^2}{n-1}.$$ (32)

When a large number of samples are available, a Poisson distribution may be fitted. Analogously to the goodness of fit tests for the normal and other continuous distributions, the *true* expected frequencies for observed values 0, 1, 2, etc.; could be found from the distribution of $S|$, but as a good enough approximation we fit the estimated Poisson distribution, $m = k_1$. The true expected distribution for $S|$ is the binomial

$$n\left(\frac{n-1}{n} + \frac{1}{n}\right)^r.$$ (33)

10—Similar, though somewhat more complicated, properties hold for the binomial distribution. For two samples, the exact distribution for inferring the contents of a second sample was given by (8), and this distribution may similarly be used for comparing two known samples. The problem is a special case of a 2×2 contingency table where the marginal frequencies one way, corresponding to the sizes of the two samples, are absolutely fixed.

For l samples, we have similarly

$$p(S|) = \frac{(N - R)!R!}{N!} \prod_{i=1}^{l} \frac{n_i!}{(n_i - r_i)!r_i!} \tag{34}$$

(N and R referring to the total sample $S = \Sigma S_i$).

For l samples with more than a two-way classification of the contents, $R_1 \ldots R_m$ being the total observed numbers in the m groups,

$$p(S|) = \frac{R_1! \ldots R_m!}{N!} \prod_{i=1}^{l} \frac{n_i!}{r_{i1}! \ldots r_{im}!} \tag{35}$$

This corresponds also to an $l \times m$ contingency table.

For testing the dispersion among l binomial samples of equal size n, the usual test is

$$\chi^2 = \frac{N\Sigma(r_i - k_1)^2}{k_1(N - lk_1)} = \frac{N(l - 1)k_2}{k_1(N - lk_1)}, \tag{36}$$

with $l - 1$ degrees of freedom. The exact distribution of $k_2|k_1$ could always be investigated, if necessary, from (34). The alternative use of the μ criterion is considered in section 12.

11—The moments of $k_2|k_1$ could also be found*, using factorial moments of r_i. For example,

$$\kappa_1(k_2|k_1) = \frac{nR(N - R)}{N(N - 1)}, \tag{37}$$

where $k_1 = R/l$ (cf. equation (46), of which this is a special case).

It might be noticed that the factorial moment-generating function for (33) is either the coefficient of x^R in

$$\frac{(N - R)!R!}{N!}(1 + x + xt)^{n_1}(1 + x)^{n_2}, \tag{38}$$

or the coefficient of x^{n_1} in

$$\frac{n_1!n_2!}{N!}(1 + x + xt)^R(1 + x)^{N-R}. \tag{39}$$

The expression (39) is most readily generalized for classification of individuals into more than two *groups*, and becomes

* *Note added in proof*, 23 March 1937. Compare Cochran's treatment (1936). The exact value for the variance κ_2 appears somewhat complicated, but for large l it becomes approximately $2n^3R^2(N - R)^2/\{N^4(l - 1)(n - 1)\}$, which checks with Cochran's result.

$$\frac{n_1!n_2!}{N!} \prod_{j=1}^{m} (1 + x + xt_j)^{R_j}, \tag{40}$$

while (38) is most easily generalized for the case of more than two *samples*, and becomes

$$\frac{(N - R)!R!}{N!} \prod_{i=1}^{l} (1 + x + xt_i)^{n_i}. \tag{41}$$

The general case (35), corresponding to the $l \times m$ contingency table, is more complicated. Its generating function can be regarded as the coefficient of $x_1^{R_1}x_2^{R_2}\ldots x_m^{R_m}$ in

$$\frac{R_1!R_2!\ldots R_m!}{N!} \prod_{i=1}^{l} \left\{\prod_{j=1}^{m} x_j(1 + t_{ij})\right\}^{n_i}. \tag{42}$$

For a large number of equal binomial samples, the expected distribution of $S|$ is the hypergeometric distribution

$$E(S|) = \frac{l \cdot n!(N - n)!R!(N - R)!}{(n - r)!r!(N - n - R + r)!(R - r)!N!}. \tag{43}$$

12—The 2×2 contingency table has been already mentioned. The exact solution for testing interactions in a $2 \times 2 \times 2$ table is also known (Bartlett 1935), but the immediate derivation of the probability of any partition, complete with constant, is no longer possible, owing to the complication which lack of independence introduces. Thus the number of ways of filling up a 2×2 table is the coefficient of $x^{n_1 \cdot} y^{n_{\cdot 1}}$ in

$$(1 + x + y + xy)^n = (1 + x)^n(1 + y)^n$$

or

$$\frac{n_{1\cdot}!n_{2\cdot}!n_{\cdot 1}!n_{\cdot 2}!}{n!n!},$$

but the number of ways of filling up a $2 \times 2 \times 2$ table *when testing the interaction* is the coefficient of

$$x_{11}^{n_{11}}x_{12}^{n_{12}}x_{21}^{n_{21}}x_{22}^{n_{22}}y_{11}^{n_{11}}y_{12}^{n_{12}}y_{21}^{n_{21}}y_{22}^{n_{22}}z_{11}^{n_{11}}z_{12}^{n_{12}}z_{21}^{n_{21}}z_{22}^{n_{22}},$$

in

$$(x_{11}y_{11}z_{11} + x_{12}y_{12}z_{11} + x_{21}y_{11}z_{12} + x_{22}y_{12}z_{12}$$
$$+ x_{11}y_{21}z_{21} + x_{21}y_{22}z_{21} + x_{21}y_{21}z_{22} + x_{22}y_{22}z_{22})^n, \tag{44}$$

this last expression no longer factorizing.*

---

* The symbols x_{11}, x_{12}, x_{21} and x_{22} represent four parallel edges of a cube, the y's four other parallel edges, and the z's the remaining four. Each observed frequency n_{ijk}, $(i, j, k = 1, 2)$, corresponds to a corner of the cube, and hence to the three edges which intersect there. The sum of the frequencies at the end of every edge is fixed.

The expected value in the cell of a 2×2 table is the first factorial moment. For example,

$$E(n_{11}) = \frac{n_1 . n_{.1}}{n}. \tag{45}$$

While this result is evident, it should be noted that the expected values in other χ^2 problems have not remained unaltered when S is modified to $S|$; χ^2 for a contingency table appears to have a slight theoretical advantage here when approximations to small sample theory are being considered.

Since the expected value corresponding to (34) must also be expressible as a rational fraction, the solution in terms of a cubic equation (Bartlett 1935) is an approximation, valid for large sample theory.

For the $l \times m$ table the second factorial moment for any cell is

$$\frac{n_{i.}(n_{i.} - 1)n_{.j}(n_{.j} - 1)}{n(n - 1)},$$

whence the expected value of χ^2 itself is readily shown to be

$$E(\chi^2) = \frac{n}{n - 1}(l - 1)(m - 1), \tag{46}$$

so that the bias of χ^2 is small and unimportant in comparison with the more general effects of discontinuity on its distribution (Yates 1934).

Since for an $l \times m$ contingency table to be tested for independence, $p(S|)$ is given by (35), the μ criterion is

$$\mu = \prod_{i=1}^{l} \prod_{j=1}^{m} \frac{n'_{ij}!}{n_{ij}!}, \tag{47}$$

where the n'_{ij} are the values of n_{ij} maximizing $p(S|)$. If this criterion were used, the values n'_{ij} must be found by inspection, though they will be near the expected values of n_{ij}. Equation (47) may be contrasted with the λ criterion given by Wilks (1935).

From (47) a small sample test is always possible. Thus for three binomial samples of 20 (equivalent to a 2×3 contingency table), with numbers

$$2 \quad 0 \quad 7$$

the exact significance level corresponding to μ (and also in this instance to χ^2) is found to be 0.007.

For large samples $-2 \log \mu$ will, like $-2 \log \lambda$, be distributed like χ^2, the three tests becoming equivalent. For medium to small samples, for which an exact test is too laborious, it is doubtful whether the usual χ^2 test can be bettered, for μ is not easier to compute, and its approximate distribution is only known in so far as μ tends to the form $\exp(-\frac{1}{2}\chi^2)$.

13—If a test of significance is to be independent of the particular *kind of population* from which the sample values were obtained, the whole set S of sample values must be regarded as fixed. This might seem to imply that noth-

ing is left to vary; but permutations of order are still possible. The relation of
the sample values x with different groups or treatments, or with values of a
second variate y, leads to tests for the significance of differences in means, or
for the significance of apparent association. Thus in an experiment the assignment of treatments is at our choice; randomization ensures the validity of a
test along these lines, this test tending to the usual test for a reasonable number of replications.

Summary

Properties of sufficiency must necessarily be considered for all small sample
tests of significance, whether these are related to problems of estimation and
fiducial distributions, or are of the nature of tests of goodness of fit.

The idea of "conditional variation" is developed, and its bearing on common
tests, depending either on continuous or discontinuous variation, is shown.
In particular, the use of χ^2 and other likelihood criteria is re-examined; and
a new application of χ^2 proposed for testing the homogeneity of a set of
variances.

References

Bartlett 1935 *J.R. Statist. Soc.* (Suppl.), **2**, 248.
———— 1936a *Proc. Roy. Soc.* A, **154**, 124.
———— 1936b *Proc. Camb. Phil. Soc.* **32**, 560.
———— 1937 *J.R. Statist. Soc.* (Suppl.), **4**.
Cochran 1936 *Ann. Eugen.* **7**, 207.
Fisher, R.A. 1934 "Statistical Methods for Research Workers," 5th ed.
———— 1935 *Ann. Eugen.* **6**, 391.
Neyman and Pearson 1933 *Philos. Trans.* A, **231**, 289.
———— 1936a *Statistical Research Memoirs*, **1**, 1.
———— 1936b *Statistical Research Memoirs*, **1**, 113.
Wilks 1935 *Ann. Math. Statist.* **6**, 190.
Yates 1934 *J.R. Statist. Soc.* (Suppl.), **1**, 217.

Introduction to
de Finetti (1937) Foresight: Its Logical Laws, Its Subjective Sources

R.E. Barlow
University of California at Berkeley

1. Importance of the Paper

De Finetti's paper, "La prévision ..." has had a major impact not only for statisticians, but also for mathematicians and philosophers. It is a fundamental paper in statistics and one of the few which possess this multidisciplinary character.

There are several reasons for its importance. The paper presented for the first time a rigorous and systematic treatment of the concept of *exchangeability* together with the fundamental result which became known as "de Finetti's representation theorem". Among the theorem's implications discussed in the paper, we find the connection between the concept of probability, which de Finetti only understands as subjective, and of frequencies, which are at the core of the so-called frequentist or classical school of statistics. The paper illuminates the conditions under which frequencies may be related to subjective probabilities and also formalizes this connection. It replaces the classical notion of observations assumed to be "independent and identically distributed with unknown distribution" by the concept of exchangeable observations. The strong law of large numbers, which is often used to justify the frequentist approach, is also given a subjective interpretation by de Finetti.

De Finetti's presentation of the connection between known frequencies and random quantities, if exchangeable, may be seen as a solution for the philosophical problem of induction. This problem was also discussed by David Hume; a philosopher greatly admired by de Finetti.

"La Prévision ..." was the first important paper of de Finetti written in a "foreign" language (i. e. French) and marks the beginning of the re-birth of Bayesian statistics and of the subjectivistic approach in science and statistics.

2. Prehistory of the paper

Until the first half of the last century, the Bayesian or *inverse probability* approach to statistics was dominant. Bayes' original paper on the Bayesian approach was published in 1763 after his death in 1761. Later, Laplace in France used a similar approach. At the beginning of the 20th century, however, this approach gradually lost its importance due to a variety of causes: the widespread assumption of uniform prior probabilities, the rivalry between mathematicians on both sides of the English Channel and the "objectivistic" trend in science connected to positivism and related ideas. The British statisticians eventually succeeded in imposing a new paradigm and when R. A. Fisher made his far-reaching contributions, the Bayesian approach was virtually dead. Jerzy Neyman at Berkeley built an alternative school of statistics which became famous as the "Neyman-Pearson school of statistics". The "Fisher-Neyman-Pearson" approach became dominant in the United States in the field of statistics. It was against this scenario that de Finetti struggled all his life. His writings did not receive much attention in the English-speaking world since they were usually written in Italian or French. In the 1950's, Jimmy Savage became influenced by "La prévision ..." and in his 1954 book on foundations, Savage refers to this paper by de Finetti as providing the "personalistic view of probability" upon which his book was based. The publication of "La prévision ..." was the first major event that provoked the rebirth of Bayesian statistics. It was first translated into English by Henry Kyburg, Jr. in 1964. A corrected English version appears in a revised 1980 version of the book *Studies in Subjective Probability*, edited by Kyburg and Smokler.

3. Main Points of the Paper

The theme of de Finetti's paper is the relation between probabilities, defined as degrees of belief, and observed frequencies. Contrary to widely accepted belief, probability and frequency are completely different concepts. His subjectivistic explanation of this relation was the subject of many of his papers and conversations even until the end of his life in July 1985. The problem that worried him is that of induction: How can we pass from observed things to anticipate future things? This was also essentially the problem posed by Thomas Bayes in his posthumous 1763 paper, namely:

> Given the number times in which an unknown event has happened and failed: Required the chance that the probability of its happening in a single trial lies somewhere between any two degrees of probability that can be named.

Both Bayes and de Finetti define probability as a degree of belief. However, Bayes defined "chance" to be the same as "probability". Hence Bayes is asking for a probability of a probability while for de Finetti this is an operationally meaningless question.

De Finetti was influenced by Bridgman's (1927) book, *The Logic of Modern Physics*, and the need for operationally verifiable definitions. He very often emphasized that an "event" is a unique fact and that what are sometimes called repetitions or trials of the same event are really distinct events although they may have common characteristics or symmetries. Hence Bayes' first sentence is not well defined and needs clarification.

De Finetti cast Bayes' problem in a modern framework by considering a conceptually infinite sequence of events and gave the first rigorous solution, based on the concept of exchangeability and his famous 'representation theorem' [equation (19)]. (There is no evidence, however, that de Finetti ever read Bayes' original paper before 1937. Bayes never considered an infinite sequence, and this together with exchangeability is the crux of what is missing in Bayes' argument.) Random quantities $x_1, x_2, \ldots, x_n$ are exchangeable if and only if their joint probability measure is invariant relative to their order. This idea relative to finite sequences was first announced by Jules Haag at the Toronto meeting of the International Congress of Mathematicians in 1924. (The Cambridge philosopher W.E. Johnson also invented the concept of exchangeability which he called the "permutation postulate" circa 1924.) De Finetti, on the other hand, first announced his results on exchangeability at the 1928 Congress of Mathematicians. Haag also discussed the case of infinite sequences, but does not have the representation theorem. In his original paper, de Finetti used the word "equivalent" instead of exchangeable. The term "exchangeable" was proposed by Fréchet and has been used by most English speaking writers since 1957.

Chapter III of de Finetti's paper is concerned with a conceptually infinite sequence of exchangeable events. In this same chapter he solves Bayes' original problem using his famous representation theorem. Let f_N be the relative frequency of the number of occurrences of the first N exchangeable events, E_1, $E_2, \ldots, E_N$, in a conceptually infinite sequence. Then your probability for f_N is

$$P\left(f_N = \frac{k}{N}\right) = \binom{N}{k} \int_0^1 \theta^k (1 - \theta)^{N-k} \, dF(\theta), \qquad k = 0, 1, \ldots, N,$$

where F is the limiting distribution function of f_M when M becomes infinite. Now, Bayes' original problem can be solved. Given $f_N = \frac{k}{N}$, the probability that another exchangeable event E_{N+1} (or any other specified exchangeable event) occurs is

$$P\left(E_{N+1} = 1 | f_N = \frac{k}{N}\right) = \int_0^1 \theta \, dF\left(\theta | f_N = \frac{k}{N}\right),$$

where

$$F\left(\theta | f_N = \frac{k}{N}\right) = \frac{\int_0^\theta u^k (1 - u)^{N-k} \, dF(u)}{\int_0^1 u^k (1 - u)^{N-k} \, dF(u)}. \tag{1}$$

Although Bayes argued that F should be the uniform distribution, there is no reason, in general, why this should be so. Of course, eq. (1) is the posterior distribution for the limiting frequency given $f_N = \dfrac{k}{N}$. Also (1) follows from de Finetti's representation theorem and agrees with Bayes' original formula.

Chapter II of his paper is concerned with the evaluation of probabilities. Probability and frequency are indirectly related by the fact that the relative frequency is the average of the probabilities of the single events envisaged, and the probability may (under ordinary assumptions, but not necessarily) be expected to be roughly close to the relative frequency.

To clarify this statement, let $E_1, E_2, \ldots, E_n$ be n exchangeable events and $p_1, p_2, \ldots, p_n$ your probabilities for their respective occurrence. Also, let w_i be your probability that exactly i out of n of the events occur. Then by the laws of probability

$$p_1 + p_2 + \cdots + p_n = 0w_0 + 1w_1 + \cdots + mw_m + \cdots + nw_n$$

is always valid, whatever your state of knowledge. Also suppose that in your judgment $p_1 = p_2 = \cdots = p_n = p$. Now suppose you know that exactly m of the n possible events occur, but you do not know which of the events actually occurred. Hence, w_m is now equal to 1 for you. Let $p' = P(E_i = 1 | m$ out of n events occur). Then

$$p' = \frac{p' + p' + \cdots + p'}{n} = \frac{mw_m}{n} = \frac{m}{n},$$

so that your evaluation of the probability that E_i occurred given that m out of n events actually occurred is, in this case, the observed frequency.

In introducing the concept of exchangeability, de Finetti is able to prove both the weak and strong laws of large numbers under the judgment of exchangeability and finite first and second cross moments of the exchangeable random quantities. He proves that if $f_N = f$ and E_i is any event such that $i > N + 1$, then

$$P(E_i = 1 | f_N = f) \to f$$

as N becomes infinite. Thus he has provided the subjective basis and circumstances which may justify adopting a frequency for your probability. Bernoulli's 1713 weak law of large numbers was conditional on a parameter, say, p, while both Bayes and de Finetti were interested in the problem when p is unknown.

I have deliberately put off discussing de Finetti's definition of probabilities, their assessment and his derivation of the laws of probability. Of course neither the consistency (or better, coherency) of the laws of probability nor their subjectivity is in dispute, but only their assessment.

Like Bayes, de Finetti begins his paper with a definition of probability and a derivation of the calculus of probability. Their definitions of an individual's probability for the occurrence of a single event E is virtually the same. As de Finetti states,

> Let us suppose that an individual is obliged to evaluate the rate p at which he would be ready to exchange the possession of an arbitrary sum S (positive or

negative) dependent on the occurrence of a given event E, for the possession of the sum pS; we will say by definition that this number p is the measure of the degree of probability attributed by the individual considered to the event $E\ldots$

This definition is still controversial since a person's utility for money need not be linear in money. In a footnote added in 1964 to the English translation of his paper, de Finetti attempts to address this problem by considering only small stakes. Rubin (1987) introduces a weak system of axioms for utility based on "rational" behavior that lead to the existence of utility but concludes that, in general, utility cannot be separated from probability. Rubin says "I came to question the comparability of the utility scales for different states of nature. That this is a problem was already noted by Ramsey; in this article the attempt by de Finetti, and also by Savage to get around this problem is questioned, and *I reject their solution.*" [Emphasis added by this author.] However, Rubin goes on to say: "This does not mean that one is forbidden to use a separate utility function and prior, or forbidden to add axioms requiring this…"

De Finetti's influence on contemporary research is partly documented in the *Proceedings of the International Conference on Exchangeability in Probability and Statistics* held in Rome in 1981. (See Koch and Spizzichino (1982).) A purely mathematical treatment of exchangeability can be found in Aldous (1983). A recent discussion of partial exchangeability and sufficiency in the language of Bayesian statistics can be found in Diaconis and Freedman (1984).

Applications of de Finetti's ideas have appeared in many papers; e.g. Lindley and Novick (1981). Textbooks have been published based on his approach to probability; e.g. Daboni and Wedlin (1982). Despite the growing number of papers and books which have been influenced by de Finetti, his over all influence is still minimal.

Why is this? Perhaps one reason is communication. De Finetti's style of writing is difficult to understand, even for Italian mathematicians and few English-speaking mathematicians have really tried to interpret what he has written.

De Finetti was very much interested in applications even to the extent of setting up an unofficial "Operations Research Laboratory" in Rome not far from the University of Rome, La Sapienza. It might not be surprising if engineers and scientists in the future become the greatest beneficiaries of his ideas. In a sense this may already be happening. Examples are Lewis's (1990) book on *Technological Risk* that is de Finettian as well as a recent MIT Ph.D. thesis by Mendel (1989).

4. Misunderstandings

De Finetti's paper has sometimes been misunderstood. De Finetti was always very explicit in stating that probability is always subjective. In the preface to his 1974 book on the theory of probability, he states emphatically, "*Probabili-*

ty Does not Exist," by which he means that probability has no objective existence, only a subjective one. In his famous representation theorem, the mixing measure is subjective. It is not the "de Finetti measure" as some mathematicians have interpreted it. The mixing measure is a subjective probability which describes one's personal opinion about the related random quantities.

De Finetti also had strong views concerning the role of parameters in probability models. He emphasized that probability models should be formulated relative to *observable* random quantities. Parameters should be functions of (or the limit of functions of) observable random quantities.

5. Personal Background

Bruno de Finetti was born in Innsbruck (Austria) on June 13, 1906 of an Italian family. He spent his childhood and adolescence in Trentino. This is a northern region of Italy which belonged to the Austro-Hungarian empire until the end of World War I. He was a student at the technical high school in Milan and then at Milan University as an undergraduate, where he took his first degree in mathematics. His graduation thesis was on affine geometry. His first research work dealt with mathematical biology and was published in 1926 when he was still an undergraduate. After graduation and up to 1931, he worked in the mathematical office of the Central Italian Agency for Statistics (ISTAT). From 1931–46, de Finetti worked in Trieste at Assicurazioni Generali, one of the most important insurance companies in Italy. In the same period, he lectured at the University of Trieste and the University of Padua.

From 1946, he devoted himself full-time to teaching at Trieste University although he had won a competition for a chair as a full professor in 1939. In 1954 he moved to the faculty of economics at the University of Rome. In 1961, he changed to the faculty of sciences in Rome where he was professor of the calculus of probability. He died in Rome on July 20, 1985.

It is impossible to summarize in a few paragraphs the scientific activity of de Finetti in the different fields of mathematics (probability), measure theory, analysis, geometry, mathematics of finance, economics, the social sciences, teaching, computer science, and biomathematics or to describe his generous and complex personality as a scientist and a humanitarian. De Finetti discussed his own life in a book edited by Gani (1982). See also the article by Lindley (1989).

Acknowledgement

I would like to acknowledge many fruitful discussions with Fabio Spizzichino and Sergio Wechsler on this subject.

References

Aldous, D.J. (1983). Exchangeability and related topics, in Ecole d'Ete de Probabilites de Saint-Flour XIII—edited by A. Dold and B. Eckman; Springer-Verlag, *Lecture Notes in Mathematics*, New York, pp. 1–198.

Bayes, T. (1763). An essay towards solving a problem in the doctrine of chances, *Philos. Trans. R. Soc. London*, **53**, 370–418. Reprinted in *Biometrika*, **45**, 1958, 293–315.

Bridgman, P. (1927). *The Logic of Modern Physics*, New York, The Macmillan Co.

Daboni, L. and Wedlin, A. (1982). *Statistica, un'introduzione all'impostazione neo-bayesiana*, Unione Tipografico-Editrice Torinese corso Raffaello, 28—10125, Torino.

de Finetti, B. (1937). "Foresight: Its logical laws, its subjective sources" in *Annales de l'Institut Henri Poincaré*, **7**, 1–68. (English translation by H.E. Kyburg, Jr. and H.E. Smokler, eds. in *Studies in Subjective Probability*. (1964, 2nd ed. 1980), Robert E. Krieger, Huntington, New York.)

de Finetti, B. (1970). *Teoria delle probabilità*. Einaudi, Torino. English translation: *Theory of probability* (1974–75), 2 vols. Wiley and Sons, New York.

de Finetti, B. (1982). Probability and my life, in J. Gani, ed. *The Making of Statisticians*. Springer-Verlag, New York.

Diaconis, P. and Freedman, D. (1984). Partial exchangeability and sufficiency. *Proceedings of the Indian Statistical Institute Golden Jubilee International Conference on Statistics: Applications and New Directions*. J.K. Ghosh and J. Roy (eds.), Indian Statistical Institute, Calcutta, pp. 205–236.

Haag, J. (1928). Sur un problème général de probabilités et ses diverses applications. *Proceedings of the International Congress of Mathematicians; Toronto 1924*, pp. 658–674, University Press, Toronto.

Johnson, W.E. (1924). *Logic; Part III. The Logical Foundations of Science*, Cambridge University Press, New York. (Reprinted in 1964, Dover, New York).

Koch, G. and Spizzichino, F. eds. (1982). *Exchangeability in Probability and Statistics*. North-Holland, Amsterdam.

Lewis, H. (1990). *Technological Risk*. W.W. Norton, New York.

Lindley, D.V. and Novick, M.R. (1981). The Role of Exchangeability in Inference. *Ann. of Statistics*, **9**, 45–58.

Lindley, D.V. (1989). Bruno de Finetti. Supplement to the *Encyclopedia of Statistical Sciences*, S. Kotz, N.L. Johnson, and C.B. Read, eds., Wiley, New York.

Mendel, M.B. (1989). *Development of Bayesian parametric theory with applications to control*. Ph.D. thesis, MIT, Department of Mechanical Engineering, Cambridge, Mass.

Rubin, H. (1987). A weak system of axioms for "rational" behavior and the non-separability of utility from prior. *Statistics & Decisions*, **5**, 47–58.

Savage, L.J. (1954). *The Foundations of Statistics*. Wiley and Sons, New York.

Foresight: Its Logical Laws,
Its Subjective Sources

Bruno de Finetti
(Translated from the French by Henry E. Kyberg, Jr.)

Words

The word "equivalent" of the original has been translated throughout as "exchangeable." The original term (used also by Khinchin) and even the term "symmetric" (used by Savage and Hewitt) appear to admit ambiguity. The word "exchangeable," proposed by Fréchet, seems expressive and unambiguous and has been adopted and recommended by most authors, including de Finetti.

The word "subjectiv" was used ambiguously in the original paper, both in the sense of "subjective" or "personal," as in "subjective probability," and in the sense of "subjectivistic," as in "the subjectivistic theory of probability," where "subjectiv" does not mean subjective (personal, private) at all. The distinction between the two concepts is made throughout the translation; the word "subjectivist" is reserved to mean "one who holds a subjectivistic theory."

"Cohérence" has been translated "coherent" following the usage of Shimony, Kemeny, and others. "Consistency" is used by some English and American authors, and is perfectly acceptable to de Finetti, but it is ambiguous (from the logician's point of view) because, applied to beliefs, it has another very precise and explicit meaning in formal logic. As the words are used in this translation, to say that a body of beliefs is "consistent" is to say (as in logic) that it contains no two beliefs that are contradictory. To say that in addition the body of beliefs is "coherent" is to say that the *degrees* of belief satisfy certain further conditions.

"Nombre aléatoire" has been translated as "random quantity." Although the phrase "random variable" is far more familiar to English-speaking mathe-

* Chapters I–IV only of this paper are reproduced here.

maticians and philosophers, there are excellent reasons, as de Finetti points out, for making this substitution. I shall quote two of these reasons from de Finetti's correspondence. The first reason is that emphasized repeatedly in connection with the word "event." "While frequentists speak of an event as something admitting repeated 'trials,' for those who take a subjectivistic (or logical) view of probability, any trial is a different 'event.' Likewise, for frequentists, a random variable X is something assuming different values in repeated 'trials,' and only with this interpretation is the word 'variable' proper. For me any single trial gives a *random quantity*; there is nothing *variable*: the value is univocally indicated; it is only *unknown*; there is only *uncertainty* (for me, for somebody) about the unique value it will exhibit." The second objection de Finetti raises to the phrase "random variable" is one that is quite independent of any particular point of view with respect to probability. "Even with the statistical conception of probability, it is unjustifiably asymmetric to speak of random points, random functions, random vectors, etc., and of random variables when the 'variable' is a number or quantity; it would be consistent to say 'random variable' *always*, specifying, if necessary, 'random variable numbers,' 'random variable points,' 'random variable vectors,' 'random variable functions,' etc., as particular kinds of random variables."

"Loi" is used in the text both in the sense of "*theorem*" (as in "the law of large numbers") and in the sense of "distribution" (as in "normal law"). This is conventional French usage, and to some extent English and American usage has followed the French in this respect. But de Finetti himself now avoids the ambiguity by reserving the word "law" for the first sense (theoremhood) only, and by introducing the term "distribution" in a general sense to serve the function of the word "law" in its second sense. "Distribution" in this general sense may refer to specific distribution functions (as in "normal distribution"), the additive function of events P(E), or distributions that are not indicated by particular functions at all. I have attempted, with de Finetti's advice and suggestions, to introduce this distinction in translation.

Notation

The original notation has been followed closely, with the single exception of that for the "conditional event," E given A, which is written in the (currently) usual way, E|A. In the original this is written $\frac{E}{A}$. I have also substituted the conventional " $\vee$ " for the original " $+$ " in forming the expression denoting the disjunction of two events.

Foreword

In the lectures which I had the honor to give at the Institut Henri Poincaré the second, third, eighth, ninth, and tenth of May 1935, the text of which is reproduced in the pages that follow, I attempted to give a general view of two subjects which particularly interest me, and to clarify the delicate relationship that unites them. There is the question, on the one hand, of the definition of probability (which I consider a purely subjective entity) and of the meaning of its laws, and, on the other hand, of the concepts and of the theory of "exchangeable" events and random quantities; the link between the two subjects lies in the fact that the latter theory provides the solution of the problem of inductive reasoning for the most typical case, according to the subjectivistic conception of probability (and thus clarifies, in general, the way in which the problem of induction is posed). Besides, even if this were not so, that is to say, even if the subjective point of view which we have adopted were not accepted, this theory would have no less validity and would still be an interesting chapter in the theory of probability.

The exposition is divided into six chapters, of which the first two deal with the first question, the following two with the second, and of which the last two examine the conclusions that can be drawn. The majority of the questions treated here have been dealt with, sometimes in detail, sometimes briefly, but always in a fragmentary way,[1] in my earlier works. Among these, those which treat questions studied or touched upon in these lectures are indicated in the bibliography.[2]

For more complete details concerning the material in each of these chapters, I refer the reader to the following publications.

Chapter I. The logic of the probable: [26], [34].
     II. The evaluation of probability: [49], [63], [70].
     III. Exchangeable events: [29], [40].
     IV. Exchangeable random quantities: [46], [47], [48].
     V. Reflections on the notion of exchangeability: [51], [62].
     VI. Observation and prediction: [32], [36], [62].

Each of these chapters constitutes one of the five lectures,[3] with the exception of Chapters IV and V, which correspond to the fourth, in which the text has been amplified in order to clarify the notion used there of integration in function space. The text of the other lectures has not undergone any essential modifications beyond a few improvements, for example, at the beginning of Chapter III, where, for greater clarity, the text has been completely revised. For these revisions, I have profited from the valuable advice of MM. Fréchet and Darmois, who consented to help with the lectures, and of M. Castelnuovo, who read the manuscript and its successive modifications several times; the editing of the text has been reviewed by my colleague M.V. Carmona and by

---

[1] A more complete statement of my point of view, in the form of a purely critical and philosophical essay, without formulas, is to be found in [32].

[2] The numbers in boldface type refer always to this list (roman numerals for the works of other authors; arabic numerals for my own, arranged by general chronological order).

[3] Their titles are those of the six chapters, with the exception of Chapter V.

M. Al. Proca, who suggested to me a number of stylistic changes. For their kind help I wish to express here my sincere appreciation. Finally, I cannot end these remarks without again thanking the director and the members of the governing committee of the Institut Henri Poincaré for the great honor they have done me by inviting me to give these lectures in Paris.

Trieste, December 19, 1936

Introduction

Henri Poincaré, the immortal scientist whose name this institute honors, and who brought to life with his ingenious ideas so many branches of mathematics, is without doubt also the thinker who attributed the greatest domain of application to the theory of probability and gave it a completely essential role in scientific philosophy. "Predictions," he said, "can only be probable. However solidly founded a prediction may appear to us, we are never absolutely sure that experience will not refute it." The calculus of probability rests on "an obscure instinct, which we cannot do without; without it science would be impossible, without it we could neither discover a law nor apply it." "On this account all the sciences would be but unconscious applications of the calculus of probability; to condemn this calculus would be to condemn science entirely."[1]

Thus questions of principle relating to the significance and value of probability cease to be isolated in a particular branch of mathematics and take on the importance of fundamental epistemological problems.

Such questions evidently admit as many different answers as there are different philosophical attitudes; to give one answer does not mean to say something that can convince and satisfy everybody, but familiarity with one particular point of view can nevertheless be interesting and useful even to those who are not able to share it. The point of view I have the honor of presenting here may be considered the extreme of subjectivistic solutions; the link uniting the diverse researches that I propose to summarize is in fact the principal common goal which is pursued in all of them, beyond other, more immediate and concrete objectives; this goal is that of bringing into the framework of the subjectivistic conception and of explaining even the problems that seem to refute it and are currently invoked against it. The aim of the first lecture will be to show how the logical laws of the theory of probability can be rigorously established within the subjectivistic point of view; in the others it will be seen how, while refusing to admit the existence of an objective meaning and value for probabilities, one can get a clear idea of the reasons, themselves subjective, for which in a host of problems the subjective judgments of diverse normal individuals not only do not differ essentially from each other, but even coincide exactly. The simplest cases will be the subject

---

[1] [XXVIII], p. 183, 186.

of the second lecture; the following lectures will be devoted to the most delicate question of this study: that of understanding the subjectivistic explanation of the use we make of the results of observation, of past experience, in
our predictions of the future.

This point of view is only one of the possible points of view, but I would
not be completely honest if I did not add that it is the only one that is not in
conflict with the logical demands of my mind. If I do not wish to conclude
from this that it is "true," it is because I know very well that, as paradoxical
as it seems, nothing is more subjective and personal than this "instinct of that
which is logical" which each mathematician has, when it comes to the matter
of applying it to questions of principle.

Chapter I: The Logic of the Probable

Let us consider the notion of probability as it is conceived by all of us in
everyday life. Let us consider a well-defined event and suppose that we do not
know in advance whether it will occur or not; the doubt about its occurrence
to which we are subject lends itself to comparison, and, consequently, to
gradation. If we acknowledge only, first, that one uncertain event can only
appear to us (a) equally probable, (b) more probable, or (c) less probable
than another; second, that an uncertain event always seems to us more probable than an impossible event and less probable than a necessary event; and
finally, third, that when we judge an event E' more probable than an event E,
which is itself judged more probable than an event E'', the event E' can only
appear more probable than E'' (transitive property), it will suffice to add to
these three evidently trivial axioms a fourth, itself of a purely qualitative nature, in order to construct rigorously the whole theory of probability. This
fourth axiom tells us that inequalities are preserved in logical sums: if E is
incompatible with E_1 and with E_2, then $E_1 \vee E$ will be more or less probable
than $E_2 \vee E$, or they will be equally probable, according to whether E_1 is
more or less probable than E_2, or they are equally probable. More generally,
it may be deduced from this[2] that two inequalities, such as

$$E_1 \text{ is more probable than } E_2,$$

$$E_1' \text{ is more probable than } E_2',$$

can be added to give

$$E_1 \vee E_1' \text{ is more probable than } E_2 \vee E_2',$$

provided that the events added are incompatible with each other (E_1 with
E_1', E_2 with E_2'). It can then be shown that when we have events for which we

---

[2] See [34], p. 321, note 1.

know a subdivision into possible cases that we judge to be equally probable, the comparison between their probabilities can be reduced to the purely arithmetic comparison of the ratio between the number of favorable cases and the number of possible cases (not because the judgment then has an objective value, but because everything substantial and thus subjective is already included in the judgment that the cases constituting the division are equally probable). This ratio can then be chosen as the appropriate index to measure a probability, and applied in general, even in cases other than those in which one can effectively employ the criterion that governs us there. In these other cases one can evaluate this index by comparison: it will be in fact a number, uniquely determined, such that to numbers greater or less than that number will correspond events respectively more probable or less probable than the event considered. Thus, while starting out from a purely qualitative system of axioms, one arrives at a quantitative measure of probability, and then at the theorem of total probability which permits the construction of the whole calculus of probabilities (for conditional probabilities, however, it is necessary to introduce a fifth axiom: see note 8).

One can, however, also give a direct, quantitative, numerical definition of the degree of probability attributed by a given individual to a given event, in such a fashion that the whole theory of probability can be deduced immediately from a very natural condition having an obvious meaning. It is a question simply of making mathematically precise the trivial and obvious idea that the degree of probability attributed by an individual to a given event is revealed by the conditions under which he would be disposed to bet on that event.[3] The axiomatization whose general outline we have just indicated above has the advantage of permitting a deeper and more detailed analysis, of starting out with only qualitative notions, and of eliminating the notion of "money," foreign to the question of probability, but which is required to talk of stakes; however, once it has been shown that one can overcome the distrust that is born of the somewhat too concrete and perhaps artificial nature of the definition based on bets, the second procedure is preferable, above all for its clarity.

Let us suppose that an individual is obliged to evaluate the rate p at which he would be ready to exchange the possession of an arbitrary sum S (positive or negative) dependent on the occurrence of a given event E, for the possession of the sum pS; we will say by definition that this number p is the measure of the degree of probability attributed by the individual considered to the event E, or, more simply, that p is the probability of E (according to the indi-

---

[3] Bertrand ([1], p. 24) beginning with this observation, gave several examples of subjective probabilities, but only for the purpose of contrasting them with "objective probabilities." The subjectivistic theory has been developed according to the scheme of bets in the exposition (Chap. I and II) in my first paper of 1928 on this subject. This was not published in its original form, but was summarized or partially developed in [27], [34], [35], etc.

vidual considered; this specification can be implicit if there is no ambiguity.)[a]

Let us further specify that, in the terminology that I believe is suitable to follow, an event is always a singular fact; if one has to consider several trials, we will never say "trials of the same event" but "trials of the same phenomenon" and each "trial" will be one "event." The point is obviously not the choice of terms: it is a question of making precise that, according to us, one has no right to speak of the "probability of an event" if one understands by "event" that which we have called a "phenomenon"; one can only do this if it is a question of one specific "trial."[4]

This being granted, once an individual has evaluated the probabilities of certain events, two cases can present themselves: either it is possible to bet with him in such a way as to be assured of gaining, or else this possibility does not exist. In the first case one clearly should say that the evaluation of the probabilities given by this individual contains an incoherence, an intrinsic contradiction; in the other case we will say that the individual is coherent.[b] It is precisely this condition of coherence which constitutes the sole principle from which one can deduce the whole calculus of probability: this calculus then appears as a set of rules to which the subjective evaluation of probability of various events by the same individual ought to conform if there is not to be a fundamental contradiction among them.

Let us see how to demonstrate, on this view, the theorem of total probability: it is an important result in itself, and also will clarify the point of view followed. Let $E_1, E_2, \ldots, E_n$ be incompatible events, of which one (and one

---

[a] Such a formulation could better, like Ramsey's, deal with expected *utilities*; I did not know of Ramsey's work before 1937, but I was aware of the difficulty of money bets. I preferred to get around it by considering sufficiently small stakes, rather than to build up a complex theory to deal with it. I do not remember whether I failed to mention this limitation to small amounts inadvertently or for some reason, for instance considering the difficulty overcome in the artificial situation of compulsory choice.

Another shortcoming of the definition—or of the device for making it operational—is the possibility that people accepting bets against our individual have better information than he has (or know the outcome of the event considered). This would bring us to game-theoretic situations.

Of course, a device is always imperfect, and we must be content with an idealization. A better device (in this regard) is that mentioned in B. de Finetti and L.J. Savage, "Sul modo di scegliere le probabilità iniziali," *Biblioteca del Metron*, S. C. Vol. 1, pp. 81–147 (English summary pp. 148–151), and with some more detail in B. de Finetti, "Does it make sense to speak of 'good probability appraisers'?" *The Scientist Speculates: An anthology of partly-baked ideas*, Gen. Ed. I.J. Good, Heinemann, London, 1962. This device will be fully presented by the same authors in a paper in preparation.

[4] This same point of view has been taken by von Kries [XIX]; see [65]. [70], and, for the contrary point of view, see [XXV].

[b] To speak of coherent or incoherent (consistent or inconsistent) individuals has been interpreted as a criticism of people who do not accept a specific behavior rule. Needless to say, this is meant only as a technical distinction. At any rate, it is better to speak of coherence (consistency) of probability evaluations rather than of individuals, not only to avoid this charge, but because the notion belongs strictly to the evaluations and only indirectly to the individuals. Of course, an individual may make mistakes sometimes, often without meriting contempt.

only) must occur (we shall say: a *complete* class of incompatible events), and let $p_1, p_2, \ldots, p_n$ be their probabilities evaluated by a given individual; if one fixes the stakes (positive or negative) $S_1, S_2, \ldots, S_n$, the gains in the n possible cases will be the difference between the stake of the bet won and the sum of the n paid outlays.

$$G_h = S_h - \sum_1^n {}_i p_i S_i$$

By considering the S_h as unknowns, one obtains a system of linear equations with the determinant

$$\begin{vmatrix} 1 - p_1 & -p_2 & \cdots & -p_n \\ -p_1 & 1 - p_2 & \cdots & -p_n \\ \cdots & \cdots & \cdots & \cdots \\ -p_1 & -p_2 & \cdots & 1 - p_n \end{vmatrix} = 1 - (p_1 + p_2 + \cdots + p_n);$$

if this determinant is not zero, one can fix the S_h in such a way that the G_h have arbitrary values, in particular, all positive, contrary to the condition of coherence; consequently coherence obliges us to impose the condition $p_1 + p_2 \cdots + p_n = 1.$[c] This necessary condition for coherence is also sufficient because, if it is satisfied, one has identically (whatever be the stakes S_h)

$$\sum_1^n {}_h p_h G_h = 0$$

and the G_h can never, in consequence, all be positive.

Thus one has the theorem of total probabilities in the following form: *in a complete class of incompatible events, the sum of the probabilities must be equal to 1.* The more general form, *the probability of the logical sum of n incompatible events is the sum of their probabilities,* is only an immediate corollary.

However, we have added that the condition is also sufficient; it is useful to make the sense of this assertion a little clearer, for in a concrete case one can throw into clear relief the distinction, fundamental from this point of view, between the logic of the probable and judgments of probability. In saying that the condition is sufficient, we mean that, a complete class of incompatible events $E_1, E_2, \ldots, E_n$ being given, all the assignments of probability that attribute to $p_1, p_2, \ldots, p_n$ any values whatever, which are non-negative and have a sum equal to unity, are admissible assignments: each of these evaluations corresponds to a coherent opinion, to an opinion legitimate in itself, and every individual is free to adopt that one of these opinions which he prefers, or, to put it more plainly, that which he *feels.* The best example is that of a championship where the spectator attributes to each team a greater or smaller probability of winning according to his own judgment; the theory

[c] Of course the proof might have been presented in an easier form by considering simply the case of $S_1 = S_2 = \cdots = S_n = S$ (as I did in earlier papers). On this occasion I preferred a different proof which perhaps gives deeper insight.

cannot reject *a priori* any of these judgments unless the sum of the probabilities attributed to each team is not equal to unity. This arbitrariness, which any one would admit in the above case, exists also, according to the conception which we are maintaining, in all other domains, including those more or less vaguely defined domains in which the various objective conceptions are asserted to be valid.

Because of this arbitrariness, the subject of the calculus of probabilities is no longer a single function $\mathbf{P}(E)$ of events E, that is to say, their probability considered as something objectively determined, but the set of all functions $\mathbf{P}(E)$ corresponding to admissible opinions. And when a *calculation* of the probability $\mathbf{P}(E)$ of an event E is wanted, the statement of the problem is to be made precise in this sense: calculate the value that one is obliged to attribute to the event E if one wants to remain in the domain of coherence, after having assigned definite probabilities to the events constituting a certain class $\mathscr{E}$. Mathematically the function $\mathbf{P}$ is adopted over the set $\mathscr{E}$, and one asks what unique value or what set of values can be attributed to $\mathbf{P}(E)$ without this extension of $\mathbf{P}$ making an incoherence appear.

It is interesting to pose the following general question: what are the events E for which the probability is determined by the knowledge of the probabilities attributed to the events of a given class $\mathscr{E}$? We are thus led to introduce the notion (which I believe novel) of "linearly independent events" [26]. Let $E_1, E_2, \ldots, E_n$ be the events of $\mathscr{E}$. Of these n events some will occur, others will not; there being 2^n subclasses of a class of n elements (including the whole class $\mathscr{E}$ and the empty class), there will be at most 2^n possible cases $C_1, C_2, \ldots, C_s$ ($s \leqslant 2^n$) which we call, after Boole, "constituents." ("At most," since a certain number of combinations may be impossible.)[5] Formally, the C_h are the events obtained by starting with the logical product $E_1 \cdot E_2 \cdot \ldots \cdot E_n$ and replacing any group of E_i by the contrary events (negations) $\sim E_i$ (or, in brief notation, $\bar{E_i}$). The constituents form a complete class of incompatible events; the E_i are logical sums of constituents, and the events which are the sums of constituents are the only events logically dependent on the E_i, that is, such that one can always say whether they are true or false when one knows, for each event $E_1, \ldots, E_n$, if it is true or false.

To give the probability of an event E_i means to give the sum of the probabilities of its constituents

$$c_{i_1} + c_{i_2} + \cdots + c_{i_h} = p_i;$$

the probabilities of $E_1, \ldots, E_n$ being fixed, one obtains n equations of this type, which form, with the equation $c_1 + c_2 + \cdots + c_s = 1$, a system of $n + 1$ linear equations relating the probabilities c_h of the constituents. It may be seen that, E being an event logically dependent on $E_1, \ldots, E_n$, and thus a logical sum of constituents $E = C_{h_1} \vee C_{h_2} \vee \cdots \vee C_{h_k}$, its probability

$$p = c_{h_1} + c_{h_2} + \cdots + c_{h_k}$$

[5] These notions are applied to the calculus of probability in Medolaghi [XXIV].

is uniquely determined when this equation is linearly dependent on the preceding system of equations. Observe that this fact does not depend on the function **P**, but only on the class $\mathscr{E}$ and the event E and can be expressed by saying that E is *linearly dependent on $\mathscr{E}$*, or—what comes to the same thing if the E_i are linearly independent—that $E_1, E_2, \ldots, E_n$ and E are linearly related among themselves.

The notion of linear independence thus defined for events is perfectly analogous to the well-known geometrical notion, and enjoys the same properties; instead of this fact being demonstrated directly, it can quickly be made obvious by introducing a geometrical representation which makes a point correspond to each event, and the notion of geometrical "linear independence" correspond to the notion of logical "linear independence." The representation is as follows: the constituents C_h are represented by the Apexes A_h of a simplex in a space of $s - 1$ dimensions, the event which is the sum of k constituents by the center of gravity of the k corresponding apexes given a mass k, and finally, the certain event (the logical sum of all the s constituents) by the center 0 of the simplex, given a mass s.

This geometric representation allows us to characterize by means of a model the set of all possible assignments of probability. We have seen that a probability function **P**(E) is completely determined when one gives the relative values of the constituents, $c_1 = \mathbf{P}(C_1), c_2 = \mathbf{P}(C_2), \ldots, c_s = \mathbf{P}(C_s)$, values which must be non-negative and have a sum equal to unity. Let us now consider the linear function f which takes the values c_h on the apexes A_h; at the point A, the center of gravity of $A_{h_1}, A_{h_2}, \ldots, A_{h_k}$, it obviously takes the value $f(A) = (1/k)(c_{h_1} + c_{h_2} + \cdots + c_{h_k})$, while the probability **P**(E) of the event E, the logical sum of the constituents $C_{h_1}, C_{h_2}, \ldots, C_{h_k}$ will be $c_{h_1} + c_{h_2} + \cdots + c_{h_k}$. We have, then, in general, $\mathbf{P}(E) = k \cdot f(A)$: the probability of an event E is the value of f at its representative point A, multiplied by the mass k; one could say that it is given as the value of f for the point A endowed with a mass k, writing $\mathbf{P}(E) = f(k \cdot A)$.[d] The center 0 corresponding to the certain event, one has in particular $1 = f(s \cdot 0) = s \cdot f(0)$, that is, $f(0) = (1/s)$.

It is immediately seen that the possible assignments of probability correspond to all the linear functions of the space that are non-negative on the simplex and have the value $1/s$ at the origin; such a function f being characterized by the hyperplane $f = 0$, assignments of probability correspond biunivocally to the hyperplanes which do not cut the simplex. It may be seen that the probability $\mathbf{P}(E) = f(k \cdot A)$ is the moment of the given mass point kA (distance × mass) relative to the hyperplane $f = 0$ (taking as unity the moment of $s0$). If, in particular, the s constituents are equally probable, the hyperplane goes to infinity.

By giving the value that it takes on a certain group of points, a linear

---

[d] The notion of "weighted point," or "geometrical formation of the first kind," belongs to the geometrical approach and notations of Grassmann-Peano, to which the Italian school of vector calculus adheres.

function f is defined for all those points linearly dependent on them, but it remains undetermined for linearly independent points: the comparison with the above definition of linearly dependent events thus shows, as we have said, that the linear dependence and independence of events means dependence and independence of the corresponding points in the geometric representation. The two following criteria characterizing the linear dependence of events can now be deduced in a manner more intuitive than the direct way. In the system of barycentric coordinates, where $x_i = 1$, $x_j = 0$ $(j \neq i)$ represents the point A_i, the coordinates of the center of gravity of A_{h_1}, A_{h_2}, ..., A_{h_k} having a mass k will be

$$x_{h_1} = x_{h_2} = \cdots = x_{h_k} = 1, \qquad x_j = 0 \quad (j \neq h_1, h_2, \ldots, h_k);$$

the sum of the constituents can thus be represented by a symbol of s digits, 1 or 0 (for example, the sum $C_1 \vee C_3$ by $10100 \cdots 0$). Events are linearly dependent when the matrix of the coordinates of the corresponding points and of the center 0 is of less than maximum rank, the rows of this matrix being the expressions described above corresponding to the events in question and—for the last line which consists only of 1's—the certain event. The other condition is that the events are linearly dependent when a coefficient can be assigned to each of them in such a way that in every possible case the sum of the coefficients of the events that occur always has the same value. If, in fact, the points corresponding to the given events and the point 0 are linearly dependent, it is possible to express 0 by a linear combination of the others, and this means that there exists a combination of bets on these events equivalent to a bet on the certain event.

An assignment of probability can be represented not only by the hyperplane $f = 0$ but also by a point not exterior to the simplex, conjugate to the hyperplane,[6] and defined as the center of gravity of s points having masses proportional to the probabilities of the events (constituents) that they represent. This representation is useful because the simplex gives an intuitive image of the space of probability laws, and above all because linear relations are conserved. The ∞^{s-1} admissible assignments of probability can in fact be combined linearly: if P_1, P_2, ..., P_m are probability functions, $P = \Sigma \lambda_i P_i$ $\lambda_i \geq 0$, $\Sigma \lambda_i = 1$ is also, and the point representing P is given by the same relation i.e., it is the center gravity of the representative points of P_1, ..., P_m with masses $\lambda_1, \ldots, \lambda_m$; the admissible assignments of probability constitute then, as do the non-exterior points of the simplex, a closed, convex set. This simple remark allows us to complete our results quickly, by specifying the

---

[6] In the polarity $f(^x_y) = \Sigma x_i y_i = 0$ (barycentric coordinates). It is convenient here, having to employ metric notions, to consider the simplex to be equilateral. It can be specified, then, that it is a question of the polarity relative to the imaginary hypersphere $\Sigma x_i^2 = 0$, and that it makes correspond to any point A whatever the hyperplane orthogonal to the line AO passing through the point A' corresponding to A in an inversion about the center O. In vectorial notation, the hyperplane is the locus of all points Q such that the scalar product $(A - O) \cdot (Q - O)$ gives $-R^2$, where $R = l/\sqrt{2s}$, l being the length of each edge of the simplex.

lack of determination of the probability of an event which remains when the event is linearly independent of certain others after the probability of the others has been fixed. It suffices to note that by fixing the value of the probability of certain events, one imposes linear conditions on the function **P**; the functions **P** that are still admissible also constitute a closed, convex set. From this one arrives immediately at the important conclusion that when the probability of an event E is not uniquely determined by those probabilities given, the admissible numbers are all those numbers in a closed interval $p' \leqslant p \leqslant p''$. If E' and E'' are respectively the sum of all the constituents contained in E or compatible with E, p' will be the smallest value admissible for the probability of E' and p'' the greatest for E''.

When the events considered are infinite in number, our definition introduces no new difficulty: **P** is a probability function for the infinite class of events $\mathscr{E}$ when it is a probability function for all finite subclasses of $\mathscr{E}$. This conclusion implies that the theorem of total probability cannot be extended to the case of an infinite or even denumerable number of events[7]; a discussion of this subject would carry us too far afield.

We have yet to consider the definition of conditional probabilities and the demonstration of the multiplication theorem for probabilities. Let there be two events E' and E''; we can bet on E' and condition this bet on E'': if E'' does not occur, the bet will be annulled; if E'' does occur, it will be won or lost according to whether E' does or does not occur. One can consider, then, the "conditional events" (or "tri-events"), which are the events of a three-valued logic: this "tri-event," "E' conditioned on E''," E'|E'', is the logical entity capable of having three values: *true* if E'' and E' are true: *false* if E'' is true and E' false; *void* if E'' is false. It is clear that two tri-events $E'_1|E''_1$ and $E'_2|E''_2$ are equal if $E''_1 = E''_2$ and $E'_1 E''_1 = E'_2 E''_2$; we will say that E'|E'' is written in *normal* form if $E' \rightarrow E''$, and it may be seen that any tri-event can be written in a single way in normal form: $E'E''|E''$. We could establish for the tri-events a three-valued logic perfectly analogous to ordinary logic [64], but this is not necessary for the goal we are pursuing.

Let us define the probability p of E' conditioned on E'' by the same condition relative to bets, but in this case we make the convention that the bet is to be called off if E'' does not happen. The bet can then give three different results: if S is the stake, outlay paid will be pS, and the gain $(1 - p)S$, $-pS$, or 0 according to whether E'|E'' will be true, false, or void, for in the first case one gains the stake and loses the outlay, in the second one loses the outlay, and in the last the outlay is returned (if $S < 0$ these considerations remain unchanged; we need only to change the terminology of debit and credit). Let us suppose that $E' \rightarrow E''$, and let p' and p'' be the probabilities of E' and E'': we will show that for coherence we must have $p' = p \cdot p''$. If we make three bets: one on E' with the stake S', one on E'' with the stake S'', and one on E'|E'' with the stake S, the total gain corresponds, in the three possible cases, to

---

[7] See [16], [24], [X], [28], [XI], [64].

E': $\qquad G_1 = (1 - p') \cdot S' + (1 - p'') \cdot S'' + (1 - p)S;$

E'' and not E': $\qquad G_2 = -p'S' + (1 - p'')S'' - pS;$

not E'': $\qquad G_3 = -p'S' - p''S''.$

If the determinant

$$\begin{vmatrix} 1 - p' & 1 - p'' & 1 - p \\ -p' & 1 - p'' & -p \\ -p' & -p'' & 0 \end{vmatrix} = p' - pp''$$

is not zero, one can fix S, S', and S'' in such a way that the G's have arbitrary values, in particular, all positive, and that implies a lack of coherence. Therefore $p' = pp''$, and, in general, if E' does not imply E'', this will still be true if we consider E'E'' rather than E': we thus have the multiplication theorem for probabilities[8]

$$P(E' \cdot E'') = P(E') \cdot P(E''|E'). \qquad (1)$$

The condition is not only necessary, but also sufficient, in the same sense as in the case of the theorem of total probability. According to whether an individual evaluates $P(E'|E'')$ as greater than, smaller than, or equal to $P(E')$, we will say that he judges the two events to be in a positive or negative correlation, or as independent: it follows that the notion of independence or dependence of two events has itself only a subjective meaning, relative to the particular function P which represents the opinion of a given individual.

We will say that $E_1, E_2, \ldots, E_n$ constitute a class of independent events if each of them is independent of any product whatever of several others of these events (pairwise independence, naturally, does not suffice); in this case the probability of a logical product is the product of the probabilities, and, the constituents themselves being logical products, the probability of any event whatever logically dependent on $E_1, \ldots, E_n$ will be given by an algebraic function of $p_1, p_2, \ldots, p_n$.

We obtain as an immediate corollary of (1), Bayes's theorem, in the form[9]

$$P(E''|E') = \frac{P(E'') \cdot P(E'|E'')}{P(E')}, \qquad (2)$$

which can be formulated in the following particularly meaningful way: The probability of E', relative to E'', is modified in the same sense and in the same measure as the probability of E'' relative to E'.

---

[8] This result, which, in the scheme of bets, can be deduced as we have seen from the definition of coherence, may also be expressed in a purely qualitative form, such as the following, which may be added as a fifth axiom to the preceding four: If E' and E'' are contained in E, E'|E is more or less probable than (or is equal in probability to) E''|E according to whether E' is more or less probable than (or equal in probability to) E''.

[9] It is also found expressed in this form in Kolmogorov [XVII].

In what precedes I have only summarized in a quick and incomplete way some ideas and some results with the object of clarifying what ought to be understood, from the subjectivistic point of view, by "logical laws of probability" and the way in which they can be proved. These laws are the conditions which characterize coherent opinions (that is, opinions admissible in their own right) and which distinguish them from others that are intrinsically contradictory. The choice of one of these admissible opinions from among all the others is not objective at all and does not enter into the logic of the probable; we shall concern ourselves with this problem in the following chapters.

Chapter II: The Evaluation of a Probability

The notion of probability which we have described is without doubt the closest to that of "the man in the street"; better yet, it is that which he applies every day in practical judgments. Why should science repudiate it? What more adequate meaning could be discovered for the notion?

It could be maintained, from the very outset, that in its usual sense probability cannot be the object of a mathematical theory. However, we have seen that the rules of the calculus of probability, conceived as conditions necessary to ensure coherence among the assignments of probability of a given individual, can, on the contrary, be developed and demonstrated rigorously. They constitute, in fact, only the precise expression of the rules of the logic of the probable which are applied in an unconscious manner, qualitatively if not numerically, by all men in all the circumstances of life.[e]

It can still be doubted whether this conception, which leaves each individual free to evaluate probabilities as he sees fit, provided only that the condition of coherence be satisfied, suffices to account for the more or less strict agreement which is observed among the judgments of diverse individuals, as well as between predictions and observed results. Is there, then, among the infinity of evaluations that are perfectly admissible in themselves, one particular evaluation which we can qualify, in a sense as yet unknown, as *objectively correct*? Or, at least, can we ask if a given evaluation is better than another?

There are two procedures that have been thought to provide an objective meaning for probability: the scheme of equally probable cases, and the consideration of frequencies. Indeed it is on these two procedures that the evalua-

---

[e] Such a statement is misleading if, as unfortunately has sometimes happened, it is taken too seriously. It cannot be said that people compute according to arithmetic or think according to logic, unless it is understood that mistakes in arithmetic or in logic are very natural for all of us. It is still more natural that mistakes are common in the more complex realm of probability; nevertheless it seems correct to say that, fundamentally, people behave according to the rules of coherence even though they frequently violate them (just as it may be said that they accept arithmetic and logic). But in order to avoid frequent misunderstandings it is essential to point out that probability theory is not an attempt to describe actual behavior; its subject is coherent behavior, and the fact that people are only more or less coherent is inessential.

tion of probability generally rests in the cases where normally the opinions of most individuals coincide. However, these same procedures do not oblige us at all to admit the existence of an objective probability; on the contrary, if one wants to stretch their significance to arrive at such a conclusion, one encounters well-known difficulties, which disappear when one becomes a little less demanding, that is to say, when one seeks not to eliminate but to make more precise the subjective element in all this. In other words, it is a question of considering the coincidence of opinions as a psychological fact; the reasons for this fact can then retain their subjective nature, which cannot be left aside without raising a host of questions of which even the sense is not clear. Thus in the case of games of chance, in which the calculus of probability originated, there is no difficulty in understanding or finding very natural the fact that people are generally agreed in assigning equal probabilities to the various possible cases, through more or less precise, but doubtless very spontaneous, considerations of symmetry. Thus the classical definition of probability, based on the relation of the number of favorable cases to the number of possible cases, can be justified immediately: indeed, if there is a complete class of n incompatible events, and if they are judged equally probable, then by virtue of the theorem of total probability each of them will necessarily have the probability $p = 1/n$ and the sum of m of them the probability m/n. A powerful and convenient criterion is thus obtained: not only because it gives us a way of calculating the probability easily when a subdivision into cases that are judged equally probable is found, but also because it furnishes a general method for evaluating by comparison any probability whatever, by basing the quantitative evaluation on purely qualitative judgments (equality or inequality of two probabilities). However this criterion is only applicable on the hypothesis that the individual who evaluates the probabilities judges the cases considered equally probable; this is again due to a subjective judgment for which the habitual considerations of symmetry which we have recalled can furnish psychological reasons, but which cannot be transformed by them into anything objective. If, for example, one wants to demonstrate that the evaluation in which all the probabilities are judged equal is alone "right," and that if an individual does not begin from it he is "mistaken," one ought to begin by explaining what is meant by saying that an individual who evaluates a probability judges "right" or that he is "mistaken." Then one must show that the conditions of symmetry cited imply necessarily that one must accept the hypothesis of equal probability if one does not want to be "mistaken." But any event whatever can only happen or not happen, and neither in one case nor in the other can one decide what would be the degree of doubt with which it would be "reasonable" or "right" to expect the event before knowing whether it has occurred or not.

Let us now consider the other criterion, that of frequencies. Here the problem is to explain its value from the subjectivistic point of view and to show precisely how its content is preserved. Like the preceding criterion, and like all possible criteria, it is incapable of leading us outside the field of subjective judgments; it can offer us only a more extended psychological analysis. In the case of frequencies this analysis is divided into two parts: an elementary part comprised of the relations between evaluations of probabilities and predictions of future frequencies, and a second, more delicate part concerning the relation between the observation of past frequencies and the prediction of future frequencies. For the moment we will limit ourselves to the first question, while admitting as a known psychological fact, whose reasons will be analyzed later, that one generally predicts frequencies close to those that have been observed.

The relation we are looking for between the evaluation of probabilities and the prediction of frequencies is given by the following theorem. Let E_1, E_2, ..., E_n be any events whatever.[1] Let us assign the values p_1, p_2, ..., p_n to their probabilities and the values ω_0, ω_1, ..., ω_n, to the probabilities that zero, or only one, or two, etc., or finally, all these events will occur (clearly $\omega_0 + \omega_1 + \omega_2 + \cdots + \omega_n = 1$). For coherence, we must have:

$$p_1 + p_2 + \cdots + p_n = 0 \times \omega_0 + 1 \times \omega_1 + 2 \times \omega_2 + \cdots + n \times \omega_n$$

or simply

$$\bar{p} = \bar{f} \tag{3}$$

where $\bar{p}$ indicates the arithmetic mean of the p_i, and $\bar{f}$ the mathematical expectation of the frequency (that is to say of the random quantity which takes the values $0/n$, $1/n$, $2/n$, ..., n/n according to whether 0, 1, 2, ..., n of the E_i occur); we note that in this respect the notion of mathematical expectation has itself a subjective meaning, since it is defined only in relation to the given judgment which assigns to the $n + 1$ possible cases the probabilities ω_h.

This relation can be further simplified in some particular cases: if the frequency is known, the second member simply represents that value of the frequency; if one judges that the n events are equally probable, the first member is nothing but the common value of the probability. Let us begin with the case in which both simplifying assumptions are correct: there are n events, m are known to have occurred or to be going to occur, but we are ignorant of which, and it is judged equally probable that any one of the events should occur. The only possible evaluation of the probability in this case leads to the

[1] In order to avoid a possible misunderstanding due to the divergence of our conception from some commonly accepted ones, it will be useful to recall that, in our terminology, an "event" is always a determinate singular fact. What are sometimes called *repetitions* or *trials* of the same event are for us distinct events. They have, in general, some common characteristics or symmetries which make it natural to attribute to them equal probabilities, but we do not admit any *a priori* reason which prevents us in principle from attributing to each of these trials E_1, ..., E_n some different and absolutely arbitrary probabilities p_1, ..., p_n. In principle there is no difference for us between this case and the case of n events which are not analogous to each other; the analogy which suggests the name "trials of the same event" (we would say "of the same phenomenon") is not at all essential, but, at the most, valuable because of the influence it can exert on our psychological judgment in the sense of making us attribute equal or very nearly equal probabilities to the different events.

value $p = m/n$. If $m = 1$, this reduces to the case of n equally probable, incompatible possibilities.

If, in the case where the frequency is known in advance, our judgment is not so simple, the relation is still very useful to us for evaluating the n probabilities, for by knowing what their arithmetic mean has to be, we have a gross indication of their general order of magnitude, and we need only arrange to augment certain terms and diminish others until the relation between the various probabilities corresponds to our subjective judgment or the inequality of their respective chances. As a typical example, consider a secret ballot: one knows that among the n voters A_1, A_2, ..., A_n, one has m favorable ballots; one can then evaluate the probabilities $p_1, p_2, ..., p_n$ that the different voters have given a favorable vote, according to the idea one has of their opinions; in any case this evaluation must be made in such a way that the arithmetic mean of the p_i will be m/n.

When the frequency is not known, the equation relates two terms which both depend on a judgment of probability: the evaluation of the probabilities p_i is no longer bound by their average to something given objectively, but to the evaluation of other probabilities, the probabilities ω_h of the various frequencies. Still, it is an advantage not to have to evaluate exactly all the ω_h in order to apply the given relation to the evaluation of the probabilities p_i; a very vague estimation of a qualitative nature suffices, in fact, to evaluate $\bar{f}$ with enough precision. It suffices, for example, to judge as "not very probable" that the frequency differs noticeably from a certain value a, which is tantamount to estimating as very small the sum of all the ω_h for which $|h/n - a|$ is not small, to give approximately $\bar{f} = a$.

Once $\bar{f}$ has been evaluated, nothing is changed of what we said earlier concerning the case where the frequency is known: if the n events are judged equally probable, their common probability is $p = \bar{f}$; if that is not the case, then certain probabilities will be augmented or diminished in order that their arithmetic mean will be $\bar{f}$.

It is thus that one readily evaluates probabilities in most practical problems, for example, the probability that a given individual, let us say Mr. A, will die in the course of the year. If it is desired to estimate directly under these conditions what stakes (or insurance, as one would prefer to say in this case) seem to be equitable, this evaluation would seem to us to be affected with great uncertainty; the application of the criterion described above facilitates the estimation greatly. For this one must consider other events, for example, the deaths, during the year, of individuals of the same age and living in the same country as Mr. A. Let us suppose that among these individuals about 13 out of 1000 will die in a year; if, in particular, all the probabilities are judged equal, their common value is $p = 0.013$, and the probability of death for Mr. A is 0.013; if in general there are reasons which make the chances we attribute to their deaths vary from one individual to another, this average value of 0.013 at least gives us a base from which we can deviate in one direction or the other in taking account of the characteristics which differentiate Mr. A from other individuals.

This procedure has three distinct and successive phases: the first consists of the choice of a class of events including that which we want to consider; the second is the prediction of the frequency; the third is the comparison between the average probability of the single events and that of the event in question. Some observations in this regard are necessary in order to clarify the significance and value that are attributed to these considerations by subjectivists' points of view, and to indicate how these views differ from current opinion. Indeed, it is only the necessity of providing some clarification about these points before continuing that makes it indispensable to spend some little time on such an elementary question.

The choice of a class of events is in itself arbitrary; if one chooses "similar" events, it is only to make the application of the procedure easier, that is to say, to make the prediction of the frequency and the comparison of the various probabilities easier: but this restriction is not at all essential, and even if one admits it, its meaning is still very vague. In the preceding example, one could consider, not individuals of the same age and the same country, but those of the same profession, of the same height, of the same profession and town, etc., and in all these cases one could observe a noticeable enough similarity. Nothing prevents a priori the grouping of the event which interests us with any other events whatever. One can consider, for example, the death of Mr. A during the year as a claim in relation to all the policies of the company by which he is insured, comprising fire insurance, transport insurance, and others; from a certain point of view, one can still maintain that these events are "similar."

This is why we avoid expressions like "trials of the same event," "events which can be repeated," etc., and, in general, all the frequency considerations which presuppose a classification of events, conceived as rigid and essential, into classes or collections or series. All classifications of this sort have only an auxiliary function and an arbitrary value.

The prediction of the frequency is based generally on the hypothesis that its value remains nearly constant: in our example, the conviction that the proportion of deaths is 13 per 1000 can have its origin in the observation that in the course of some years past the mortality of individuals of the same kind was in the neighborhood of 13/1000. The reasons which justify this way of predicting could be analyzed further; for the moment it suffices to assume that in effect our intuition leads us to judge thus. Let us remark that such a prediction is generally the more difficult the narrower the class considered.

On the other hand, the comparison of the different probabilities is more difficult in the same proportion the events are more numerous and less homogeneous: the difficulty is clearly reduced to a minimum when the events appear to us equally probable. In practice one must attempt to reconcile as best one can these opposing demands, in order to achieve the best application of the two parts of the procedure: it is only as a function of these demands that the class of events considered can be chosen in a more or less appropriate fashion.

An illustration will render these considerations still clearer. If one must give an estimate of the thickness of a sheet of paper, he can very easily arrive at it by estimating first the thickness of a packet of n sheets in which it is inserted, and then by estimating the degree to which the various sheets have the same thickness. The thickness can be evaluated the more easily the larger the packet; the difficulty of the subsequent comparison of the sheets is on the contrary diminished if one makes the packet thinner by saving only those sheets judged to have about the same thickness as the sheet that interests us.

Thus the criterion based on the notion of frequency is reduced, like that based on equiprobable events, to a practical method for linking certain subjective evaluations of probability to other evaluations, themselves subjective, but preferable either because more accessible to direct estimation, or because a rougher estimate or even one of a purely qualitative nature suffices for the expected conclusions. A priori, when one accepts the subjectivistic point of view, such ought to be the effective meaning and the value of any criterion at all.

In the case of predictions of frequencies, one only relates the evaluation of p_i to that of the ω_h and to a comparison between the p_i; the estimation of the ω_h does not need to come up to more than a rough approximation, such as suffices to determine the p_i closely enough. It must be remarked nevertheless that this prediction of the frequency is nothing else than an evaluation of the ω_h; it is not a prophecy which one can call correct if the frequency is equal or close to $\bar{f}$, and false in the contrary case. All the frequencies $0/n$, $1/n$, $2/n, \ldots, n/n$ are possible, and whatever the realized frequency may be, nothing can make us right or wrong if our actual judgment is to attribute to these $n + 1$ cases the probabilities ω_h, leading to a certain value

$$\bar{p} = \bar{f} = \frac{\omega_1 + 2\omega_2 + 3\omega_3 + \cdots + n\omega_n}{n}. \tag{3}$$

It is often thought that these objections may be escaped by observing that the impossibility of making the relations between probabilities and frequencies precise is analogous to the practical impossibility that is encountered in all the experimental sciences of relating exactly the abstract notions of the theory and the empirical realities.[2] The analogy is, in my view, illusory: in the other sciences one has a theory which asserts and predicts with certainty and exactitude what would happen if the theory were completely exact; in the calculus of probability it is the theory itself which obliges us to admit the possibility of all frequencies. In the other sciences the uncertainty flows indeed from the imperfect connection between the theory and the facts; in our case,

---

[2] This point of view is maintained with more or less important variations in most modern treatises, among others those of Castelnuovo [VI], Fréchet-Halbwachs [XII], Lévy [XX], von Mises [XXV].

on the contrary, it does not have its origin in this link, but in the body of the theory itself [32], [65], [IX]. No relation between probabilities and frequencies has an empirical character, for the observed frequency, whatever it may be, is always compatible with all the opinions concerning the respective probabilities; these opinions, in consequence, can be neither confirmed nor refuted, once it is admitted that they contain no categorical assertion such as: such and such an event *must* occur or *can not* occur.

This last consideration may seem rather strange if one reflects that the prediction of a future frequency is generally based on the observation of those past; one says, "we will correct" our initial opinions if "experience refutes them." Then isn't this instinctive and natural procedure justified? Yes; but the way in which it is formulated is not exact, or more precisely, is not meaningful. It is not a question of "correcting" some opinions which have been "refuted"; it is simply a question of substituting for the initial evaluation of the probability the value of the probability which is conditioned on the occurrence of facts which have already been observed; this probability is a completely different thing from the other, and their values can very well not coincide without this non-coincidence having to be interpreted as the "correction of a refuted opinion."

The explanation of the influence exercised by experience on our future predictions, developed according to the ideas that I have just expounded, constitutes the point that we have left aside in the analysis of the criterion based on frequencies. This development will be the subject of the following chapters, in which we will make a more detailed study of the most typical case in this connection: the case of exchangeable events, and, in general, of any exchangeable random quantities or elements whatever. This study is important for the development of the subjectivistic conception, but I hope that the mathematical aspect will be of some interest in itself, independently of the philosophical interpretation; in fact, exchangeable random quantities and exchangeable events are characterized by simple and significant conditions which can justify by themselves a deep study of the problems that arise in connection with them.

Chapter III: Exchangeable Events

Why are we obliged in the majority of problems to evaluate a probability according to the observation of a frequency? This is a question of the relations between the observation of past frequencies and the prediction of future frequencies which we have left hanging, but which presents itself anew under a somewhat modified form when we ask ourselves if a prediction of frequency can be in a certain sense confirmed or refuted by experience. The question we pose ourselves now includes in reality the problem of reasoning by induction. Can this essential problem, which has never received a satisfactory solution up to now, receive one if we employ the conception of subjective probability and the theory which we have sketched?

In order to fix our ideas better, let us imagine a concrete example, or rather a concrete interpretation of the problem, which does not restrict its generality at all. Let us suppose that the game of heads or tails is played with a coin of irregular appearance. The probabilities of obtaining "heads" on the first, the second, the hth toss, that is to say, the probabilities $P(E_1), P(E_2), \ldots, P(E_h), \ldots$ of the events $E_1, E_2, \ldots, E_h, \ldots$ consisting of the occurrence of heads on the different tosses, can only be evaluated by calculating a priori the effect of the apparent irregularity of the coin.

It will be objected, no doubt, that in order to get to this point, that is to say, to obtain the "correct" probabilities of future trials, we can utilize the results obtained in the previous trials: it is indeed in this sense that—according to the current interpretation—we "correct" the evaluation of $P(E_{n+1})$ after the observation of the trials which have, or have not, brought about E_1, $E_2, \ldots, E_n$. Such an interpretation seems to us unacceptable, not only because it presupposes the objective existence of unknown probabilities, but also because it cannot even be formulated correctly: indeed the probability of E_{n+1} evaluated with the knowledge of a certain result, A, of the n preceding trials is no longer $P(E_{n+1})$ but $P(E_{n+1}|A)$. To be exact, we will have

$$A = E_{i_1} E_{i_2} \ldots E_{i_r} \bar{E}_{j_1} \bar{E}_{j_2} \ldots \bar{E}_{j_s} \quad (r + s = n),$$

the result A consisting of the r throws $i_1, i_2, \ldots, i_r$ giving "heads" and the other s throws $j_1, j_2, \ldots, j_s$ giving tails: A is then one of the constituents formed with $E_1, E_2, \ldots, E_n$. But then, if it is a question of a conditional probability, we can apply the theorem of compound probability, and the interpretation of the results which flow from this will constitute our justification of inductive reasoning.

In general, we have

$$P(E_{n+1}|A) = \frac{P(A \cdot E_{n+1})}{P(A)}; \tag{4}$$

our explanation of inductive reasoning is nothing else, at bottom, than the knowledge of what this formula expresses: the probability of E_{n+1} evaluated when the result A of $E_1, \ldots, E_n$ is known, is not something of an essentially novel nature (justifying the introduction of a new term like "statistical" or "a posteriori" probability). This probability is not independent of the "a priori probability" and does not replace it; it flows in fact from the same a priori judgment by subtracting, so to speak, the components of doubt associated with the trials whose results have been obtained.[f]

[f] This terminology derives from the time when a philosophical distinction was made between probabilities evaluated by considerations of symmetry (a priori probabilities), and those justified statistically (a posteriori probabilities); this dualistic view is now rejected not only in the subjectivistic theory maintained here, but also by most authors of other theories. With reference to current views, it is proper to speak simply of initial and final probabilities (the difference being relative to a particular problem where one has to deal with evaluations at different times, before and after some specific additional information has been obtained); the terminology has not been modernized here because the passage makes reference to the older views.

In order to avoid erroneous interpretations of what follows, it is best at the outset to recall once more the sense which we attribute to a certain number of terms in this work. Let us consider, to begin with, a class of events (as, for example, the various tosses of a coin). We will say sometimes that they constitute the *trials* of a given phenomenon; this will serve to remind us that we are almost always interested in applying the reasoning that follows to the case where the events considered are events *of the same type*, or which have *analogous* characteristics, without attaching an intrinsic significance or a precise value to these exterior characteristics whose definition is largely arbitrary. Our reasoning will only bring in the events, that is to say, the trials, each taken individually; the analogy of the events does not enter into the chain of reasoning in its own right but only to the degree and in the sense that it can influence in some way the judgment of an individual on the probabilities in question.

It is evident that by posing the problem as we have, it will be impossible for us to *demonstrate* the validity of the principle of induction, that is to say, the principle according to which the probability ought to be close to the observed frequency—for example, in the preceding case: $P(E_{n+1}|A) \cong r/n$. That this principle can only be justified in particular cases is not due to an insufficiency of the method followed, but corresponds logically and necessarily to the essential demands of our point of view. Indeed, probability being purely subjective, nothing obliges us to choose it close to the frequency; all that can be shown is that such an evaluation follows in a coherent manner from our initial judgment when the latter satisfies certain perfectly clear and natural conditions.

We will limit ourselves in what follows to the simplest conditions which define the events which we call exchangeable, and to fix our ideas we will exhibit these conditions in the example already mentioned; our results will nevertheless be completely general

The problem is to evaluate the probabilities of all the possible results of the n first trials (for any n). These possible results are 2^n in number, of which $\binom{n}{n} = 1$ consist of the repetition of "heads" n times, $\binom{n}{n-1} = n$ of $n-1$ occurrences of "heads" and one occurrence of "tails," ..., and in general $\binom{n}{r}$ of r occurrences of "heads" and $n-r$ occurrences of "tails." If we designate by $\omega_r^{(n)}$ the probability that one obtains in n tosses, in any order whatever, r occurrences of "heads" and $n-r$ occurrences of "tails," $\omega_r^{(n)}$ will be the sum of the probabilities of the $\binom{n}{r}$ distinct ways in which one can obtain this result; the average of these probabilities will then be $\omega_r^{(n)} \big/ \binom{n}{r}$. Having grouped the 2^n results in this way, we can distinguish usefully, though arbitrarily, two kinds of variation in the probabilities: to begin with we have an average probability which is greater or smaller for each frequency, and then we have a more or less uniform subdivision of the probabilities $\omega_r^{(n)}$ among the various results of equal frequency that only differ from one another in the order of succession of favorable and unfavorable trials. In general, different probabilities will be assigned, depending on the order, whether it is supposed that one toss has an influence on the one which follows it immediately, or whether the exterior circumstances are supposed to vary, etc.; nevertheless it is particularly interesting to study the case where the probability does not depend on the order of the trials. In this case every result having the same frequency r/n on n trials has the same probability, which is $\omega_r^{(n)} \Big/ \binom{n}{r}$; if this condition is satisfied, we will say that the events of the class being considered, e.g., the different tosses in the example of tossing coins, are *exchangeable* (in relation to our judgment of probability). We will see better how simple this condition is and the extent to which its significance is natural, when we have expressed it in other forms, some of which will at first seem more general, and others more restrictive.

It is almost obvious that the definition of exchangeability leads to the following result: the probability that n determinate trials will all have a favorable result is always the same, whatever the n-tuple chosen: this probability will be equal to $\omega_n = \omega_n^{(n)}$, since the first n cases constitute a particular n-tuple. Conversely, if the probabilities of the events have this property, the events are exchangeable, for, as will be shown a little later, it follows from this property that all the results having r favorable and s unfavorable results out of n trials have the same probability, that is:

$$\frac{\omega_r^{(n)}}{\binom{n}{r}} = (-1)^s \Delta^s \omega_r \tag{5}$$

Another conclusion has already been obtained: the probability that r trials will be favorable and s unfavorable will always be $\omega_r^{(n)} \Big/ \binom{n}{r}$ (with $n = r + s$), not only when it is a question of the first n trials in the original order, but also in the case of any trials whatever.

Another condition, equivalent to the original definition, can be stated: the probability of any trial E whatever, conditional on the hypothesis A that there have been r favorable and s unfavorable results on the other specific trials, does not depend on the events chosen, but simply on r and s (or on r and $n = r + s$).[8] If

$$P(A) = \frac{\omega_r^{(n)}}{\binom{n}{r}} \quad \text{and} \quad P(A \cdot E) = \frac{\omega_{r+1}^{(n+1)}}{\binom{n+1}{r+1}}$$

[8] This may also be expressed by saying that the observed frequency r/n and n give a *sufficient statistic*, or that the likelihood is only a function of r/n and n.

then we will have

$$\mathbf{P}(E|A) = \frac{r+1}{n+1}\left(\frac{\omega_{r+1}^{(n+1)}}{\omega_r^{(n)}}\right) = p_r^{(n)} \tag{6}$$

a function of n and r only; if, on the other hand, one supposes that $\mathbf{P}(E|A) = p_r^{(n)}$, a function of n and r only, it follows clearly that for every n-tuple the probability that all the trials will be favorable is

$$\omega_n = p_0^{(0)} \cdot p_1^{(1)} \ldots p_{n-1}^{(n-1)}. \tag{7}$$

In general it may easily be seen that in the case of exchangeable events, the whole problem of probabilities concerning $E_{i_1}, E_{i_2}, \ldots, E_{i_n}$ does not depend on the choice of the (distinct) indices $i_1, \ldots, i_n$, but only on the probabilities $\omega_0, \omega_1, \ldots, \omega_n$. This fact justifies the name of "exchangeable events" that we have introduced: when the indicated condition is satisfied, any problem is perfectly well determined if it is stated for *generic* events. This same fact makes it very natural to extend the notion of exchangeability to the larger domain of random quantities: We shall say that $X_1, X_2, \ldots, X_n, \ldots$ are exchangeable random quantities if they play a symmetrical role in relation to all problems of probability, or, in other words, if the probability that $X_{i_1}, X_{i_2}, \ldots, X_{i_n}$ satisfy a given condition is always the same however the distinct indices $i_1 \ldots i_n$ are chosen. As in the case for exchangeable *events*, any problem of probability is perfectly determined when it has been stated for *generic* random quantities; in particular if $X_1, X_2, \ldots, X_n, \ldots$ are exchangeable random quantities, the events $E_i = (X_i \leqslant x)$ (where x is any fixed number) or more generally $E_i = (X_i \in I)$ (I being any set of numbers) are exchangeable. This property will be very useful to us, as in the following case: the mathematical expectation of any function of n exchangeable random quantities does not change when we change the n-tuple chosen; in particular there will be values $m_1, m_2, \ldots, m_k, \ldots$ such that $\mathcal{M}(X_i) = m_1$, whatever i may be; $\mathcal{M}(X_i X_j) = m_2$, whatever be i and j ($i \neq j$), and in general $\mathcal{M}(X_{i_1} X_{i_2} \ldots X_{i_k}) = m_k$ whatever be the distinct $i_1, i_2, \ldots, i_k$. This observation has been made by Kinchin[1] who has used it to simplify the proofs of some of the results that I have established for exchangeable events. I have used this idea in the study of exchangeable random quantities, and I will avail myself of it equally in this account.

One can, indeed, treat the study of exchangeable events as a special case of the study of exchangeable random quantities, by observing that the events E_i are exchangeable only if that is also true of their "indicators," that is to say, the random quantities X_i such that $X_i = 1$ or $X_i = 0$ according to whether E_i occurs or not. We mention in connection with these "indicators" some of the simple properties which explain their usefulness.

The indicator of $\bar{E}_i$ is $1 - X_i$; that of $E_i E_j$ is $X_i X_j$; that of $E_i \vee E_j$ is $1 - (1 - X_i)(1 - X_j) = X_i + X_j - X_i X_j$—it is not, as it is in the case of incom-

---

[1] [XV]; also see [XVI].

patible events where $X_i X_j = 0$, simply $X_i + X_j$. The indicator of $E_{i_1} E_{i_2} \cdots E_{i_r} \bar{E}_{j_1} \bar{E}_{j_2} \cdots \bar{E}_{j_s}$ is then

$$X_{i_1} X_{i_2} \ldots X_{i_r} (1 - X_{j_1})(1 - X_{j_2}) \ldots (1 - X_{j_s})$$

$$= X_{i_1} X_{i_2} \ldots X_{i_r} - \sum_{h=1}^{s} X_{i_1} X_{i_2} \ldots X_{i_r} X_{j_h}$$

$$+ \sum_{k,h=1}^{s} X_{i_1} X_{i_2} \ldots X_{i_r} X_{j_h} X_{j_k} - \cdots \pm X_1 X_2 \ldots X_n$$

The mathematical expectation of the indicator is only the probability of the corresponding event; thus the possibility of transforming the logical operations on the events into arithmetical operations on the indicators greatly facilitates the solution of a certain number of problems. One infers immediately, in particular, the formula (5) stated for $\omega_r^{(n)}$ in the case of exchangeable events: if the product of h trials always has the probability ω_h, then the probability $\omega_r^{(n)} \Big/ \binom{n}{r}$ of $E_{i_1} E_{i_2} \ldots E_{i_r} \bar{E}_{j_1} \bar{E}_{j_2} \ldots \bar{E}_{j_s}$ is deduced from the above development of the indicator of this event and one obtains

$$\frac{\omega_r^{(n)}}{\binom{n}{r}} = \omega_r - \binom{s}{1}\omega_{r+1} + \binom{s}{2}\omega_{r+2} - \cdots (-1)^s \omega_{r+s} = (-1)^s \Delta^s \omega_r. \qquad (5)$$

Putting $\omega_0 = 1$, the formula remains true for $r = 0$.

Leaving aside for the moment the philosophical question of the principles which have guided us here, we will now develop the study of exchangeable events and exchangeable random quantities, showing first that the law of large numbers and even the strong law of large numbers are valid for exchangeable random quantities X_i, and that the probability distribution of the average Y_n of n of the random quantities X_i tends toward a limiting distribution when n increases indefinitely. It suffices even, in the demonstration, to suppose

$$\mathcal{M}(X_i) = m_1, \qquad \mathcal{M}(X_i^2) = \mu_2, \qquad \mathcal{M}(X_i X_j) = m_2$$

for all i and j ($i \neq j$), a condition which is much less restrictive than that of exchangeability. We remark again that it suffices to consider explicitly random quantities, the case of events being included by the consideration of "indicators"; an average Y_n is identical, in this case, with the frequency on n trials.

The "law of large numbers" consists of the following property: *if Y_h and Y_k are respectively the averages of h and of k random quantities X_i (the two averages may or may not contain some terms in common), the probability that $|Y_h - Y_k| > \varepsilon$ ($\varepsilon > 0$) may be made as small as we wish by taking h and k sufficiently large*; this follows immediately from the calculation of the mathematical expectation of $(Y_h - Y_k)^2$:

$$\mathcal{M}(Y_h - Y_k)^2 = \frac{h + k - 2r}{hk}(\mu_2 - m_2)$$

$$= \left(\frac{1}{h} + \frac{1}{k} - \frac{2r}{hk}\right)(\mu_2 - m_2) \leqslant \left(\frac{1}{h} + \frac{1}{k}\right)(\mu_2 - m_2), \qquad (8)$$

where r is the number of common terms, i.e., the X_i that occur in Y_h as well as in Y_k. In particular, if it is a question of "successive" averages, that is to say, if all the terms in the first expression appear also in the other, as for example if

$$Y_h = (1/h)(X_1 + X_2 + \cdots + X_h), \qquad Y_k = (1/k)(X_1 + X_2 + \cdots + X_k)(h < k)$$

we will have $r = h$, and

$$\mathcal{M}(Y_h - Y_k)^2 = \left(\frac{1}{h} - \frac{1}{k}\right)(\mu_2 - m_2). \qquad (9)$$

When successive averages are considered, we have in addition the following result, which constitutes the strong law of large numbers: *ε and θ being given, it suffices to choose h sufficiently great in order that the probability of finding the successive averages* Y_{h+1}, Y_{h+2}, ..., Y_{h+q} *all between* $Y_h - ε$ *and* $Y_h + ε$ *differs from unity by a quantity smaller than θ, q being as great as one wants.* If one admits that the probability that all the inequalities

$$|Y_h - Y_{h+i}| < ε \quad (i = 1, 2, 3, \ldots)$$

are true is equal to the limit of the analogous probability of $i = 1, 2, \ldots, q$, when $q \to \infty$, then one can say that *all* the averages Y_{h+i} $(i = 1, 2, \ldots)$ fall between $Y_h - ε$ and $Y_h + ε$, excepting in a case whose probability is less than $θ$; I prefer however to avoid this sort of statement, for it presupposes essentially the extension of the theorem of total probabilities to the case of a denumerably infinite number of events, and this extension is not admissible, at least according to my point of view.

The proof of the strong law of large numbers can be obtained easily, by considering the variation among the terms Y_{h+i} with the index $(h + i)$ square, and then the variation in the segments between two successive square indices. If the Y's with square indices do not differ among themselves by more than $ε/3$, and the Y's with indices falling between two successive square indices do not differ from each other by more than $ε/3$, the deviations among the Y_{h+i} obviously cannot exceed $ε$. But it suffices to apply the Bienaymé-Tchebycheff inequality to succeed in overestimating the probability of an exception to one of these partial limitations,[2] and the corresponding probabilities come out

---

[2] The formula $\mathcal{M}(Y_h - Y_k)^2 = \frac{k - h}{hk}(\mu_2 - m_2)$ gives, by the Bienaymé-Tchebycheff inequality,

$P(|Y_h - Y_k| > ε) < \frac{1}{ε^2}(\mu_2 - m_2)\frac{k - h}{hk}$; applying the theorem of total probabilities in the manner indicated in my note [47], it is possible to draw from this inequality the conclusions that follow in the text.

less than $36(\mu_2 - m_2)\varepsilon^{-2}\sum_{i=s}^{\infty} i^{-2}$ (s = the integral part of $\sqrt{h}$); the probability of an exception to one or the other of the partial limitations cannot therefore exceed

$$72(\mu_2 - m_2)\varepsilon^{-2}\sum_{i=s}^{\infty} i^{-2}.$$

This value does not depend on q, and tends toward zero when $s \to \infty$ (that is to say, when $h \to \infty$); the strong law of large numbers is thus demonstrated.

From the fact that the law of large numbers holds, the other result stated follows easily; the *distribution* $\Phi_n(\xi) = P(Y_n \leqslant \xi)$ *tends to a limit as* $n \to \infty$. If the probability that $|Y_h - Y_k| > \varepsilon$ is smaller than θ, the probability that $Y_h \leqslant \xi$ and $Y_k > \xi + \varepsilon$ will *a fortiori* be smaller than θ, and one will have $\Phi_h(\xi) < \Phi_k(\xi + \varepsilon) + \theta$, and similarly $\Phi_h(\xi) > \Phi_k(\xi - \varepsilon) - \theta$. As ε and θ can be chosen as small as we wish, it follows that there exists a limiting distribution $\Phi(\xi)$ such that $\lim_{n\to\infty} \Phi_n(\xi) = \Phi(\xi)$ except perhaps for points of discontinuity.[3]

If, in particular, the random quantities $X_1, X_2, \ldots, X_n, \ldots$ are the indicators of exchangeable trials of a given phenomenon, that is to say, if they correspond to the exchangeable events $E_1, E_2, \ldots, E_n, \ldots$, the hypothesis will be satisfied; it would suffice even that the events be equally probable $[P(E_i) = \mathscr{M}(X_i) = m_1 = \omega_1]$ and have the same two-by-two correlation $[P(E_iE_j) = \mathscr{M}(X_iX_j) = m_2 = \omega_2]$. We remark that for the indicators one has $X^2 = X$ (since $0^2 = 0$ and $1^2 = 1$) so that $\mu_2 = m_1 = \omega_1$. For Y_h, the frequency on h trials, we then have

$$\mathscr{M}(Y_h) = \omega_1, \qquad \mathscr{M}(Y_h - Y_k)^2 = (1/h + 1/k - 2r/hk)(\omega_1 - \omega_2) \quad (10)$$

and the demonstrated results show simply that the frequencies of two sufficiently numerous groups of trials are, almost surely, very close [even if it is a question of disjoint groups ($r = 0$); if there are some common events ($r > 0$), so much the better]. The same results further signify that the successive frequencies in the same sequence of experiments oscillate almost surely with a quantity less than a given ε, beginning from a rank h sufficiently large, whatever be the number of subsequent events; and finally that there exists a probability distribution $\Phi(\xi)$ differing only slightly from that of a frequency Y_h for very large h.

In order to determine completely the limiting distribution $\Phi(\xi)$, the knowledge of m_1, m_2, μ_2, is evidently no longer sufficient, except in the limiting case where there is no two-by-two correlation and $m_2 = m_1^2$; here $\Phi(\xi)$ is degenerate and reduces to the distribution where the probability is concentrated in

---

[3] We remark that if the X_i are exchangeable, the distribution $\Phi_n(\xi)$ is the same for all the averages Y_n of n terms; one then has a sequence of functions Φ_n depending solely on n and tending toward Φ; with a less restrictive hypothesis than the demonstration assumes, two averages Y_n' and Y_n'' formed from distinct terms can have two different distributions Φ_n' and Φ_n'', but the result will still hold in the sense that all the $\Phi_n(\xi)$ concerning the average of any n terms whatever will differ very little from $\Phi(\xi)$ (and thus from one another) if n is sufficiently large.

one point $\xi = m_1$. In this case the law of large numbers and the strong law of large numbers reduce to the laws of Bernoulli and Cantelli [III], [V], according to which the deviation between Y_h and the value m_1, fixed in advance, tends stochastically toward zero in a "strong" way. In the general case of a class of exchangeable random quantities, Φ is determined by the knowledge of the complete sequence $m_1, m_2, \ldots, m_n, \ldots$, for these values are the *moments* relative to the distribution Φ:

$$m_n = \int_0^1 \xi^n \, d\Phi(\xi) \tag{11}$$

and then

$$\psi(t) = \sum_{0=n}^{\infty} \frac{i^n t^n}{n!} m_n \tag{12}$$

is the characteristic function of Φ.

Indeed,

$$Y_h^n = \frac{1}{h^n}(X_1 + X_2 + \cdots + X_n)^n = \frac{1}{h^n} \sum X_{i_1} X_{i_2} \ldots X_{i_n};$$

among the h^n products there are $h(h-1)(h-2)\ldots(h-n+1)$ that are formed from distinct factors; the products containing the same term more than one time constitute a more and more negligible fraction as h is increased, so that

$$\mathscr{M}(Y_h^n) = \frac{h(h-1)\ldots(h-n+1)}{h^n} m_n + \mathcal{O}\left(\frac{1}{h}\right) \to m_n \quad (h \to \infty). \tag{13}$$

If, in particular, the X_i are the indicators of exchangeable trials of a phenomenon, Y_h the frequency on h trials, and m_n is the probability ω_n that n trials will all have a favorable result, then (13) evaluates the mean of the nth power of the relative frequency on a large number of trials. The characteristic function of $\Phi(\xi)$ is

$$\psi(t) = \sum_{n=0}^{\infty} \frac{i^n t^n}{n!} \omega_n \tag{14}$$

and then we have

$$\Phi(\xi) = \frac{1}{2\pi} \int_{-\infty}^{\infty} \frac{e^{it} - e^{-it\xi}}{it} \psi(t) \, dt \tag{15}$$

for—the Y_h signifying frequencies—the probability distribution can only fall between 0 and 1, and thus $\Phi(-1) = 0$. The characteristic function of $\Phi_h(\xi)$ is

$$\psi_h(t) = \Omega_h(e^{i(t/h)}), \tag{16}$$

where Ω_h is the polynomial

$$\Omega_h(z) = \sum_{k=0}^{h} \binom{h}{k} \omega_k (z-1)^k, \tag{17}$$

and $\Omega_h(t)$ converges uniformly to $\psi(t)$. This fact can be proved directly; it is from this standpoint that I developed systematically the study of exchangeable events in my first works [29], [40], and demonstrated the existence of the limiting distribution Φ, and of ψ, which I call the "characteristic function of the phenomenon."[4]

To give the limiting distribution Φ, or the characteristic function ψ, is, as we have seen, equivalent to giving the sequence ω_n; it follows that this suffices to determine the probability for any problem definable in terms of exchangeable events. All such problems lead, indeed, in the case of exchangeable events, to the probabilities $\omega_r^{(n)}$ that on n trials, a number r will be favorable; we have (putting $s = n - r$)

$$\omega_r^{(n)} = (-1)^s \binom{n}{r} \Delta^s \omega_r = \binom{n}{r} \int_0^1 \xi^r (1 - \xi)^s \, d\Phi(\xi), \tag{18}$$

and an analogous formula having the same significance is valid for the general case. Indeed, let $\mathbf{P}_\xi(E)$ be the probability attributed to the generic event E when the events $E_1, E_2, \ldots, E_n, \ldots$ are considered independent and equally probable with probability ξ; the probability $\mathbf{P}(E)$ of the same generic event, the E_i being exchangeable events with the limiting distribution $\Phi(\xi)$, is

$$\mathbf{P}(E) = \int_0^1 \mathbf{P}_\xi(E) \, d\Phi(\xi).[5] \tag{19}$$

This fact can be expressed by saying that the probability distributions $\mathbf{P}$ corresponding to the case of exchangeable events are linear combinations of the distributions $\mathbf{P}_\xi$ corresponding to the case of independent equiprobable events, the weights in the linear combination being expressed by $\varphi(\xi)$.

This conclusion exhibits an interesting fact which brings our case into agreement with a well known scheme, with which it even coincides from a formal point of view. If one has a phenomenon of exchangeable trials, and if Φ is the limiting distribution of the frequencies, a scheme can easily be imagined which gives for every problem concerning this phenomenon the same probabilities; it suffices to consider a random quantity X whose probability distribution is Φ and events which, conforming to the hypothesis $X = \xi$ (ξ any value between 0 and 1), are independent and have a probability $p = \xi$; the trials of a phenomenon constructed thus are always exchangeable events. Further on, we will analyze the meaning of this result, after having examined its extension to exchangeable random quantities. For the moment, we will limit ourselves to deducing the following result: in order that Φ may represent the limiting distribution corresponding to a class of exchangeable events, it is

---

[4] I had then reserved the name "phenomenon" for the case of exchangeable trials; I now believe it preferable to use this word in the sense which is commonly given to it, and to specify, if it should be the case, that it is a question of a phenomenon whose trials are judged exchangeable.

[5] It is clear that the particular case just mentioned—formula (18)—is obtained by putting $E = $ "on n (given) trials, r results are favorable"; then, indeed,

$$\mathbf{P}_\xi(E) = \binom{n}{r} \xi^r (1 - \xi)^{n-r}, \qquad \mathbf{P}(E) = \omega_r^{(n)}.$$

necessary and sufficient that the distribution be limited to values between 0 and 1 [so that $\Phi(-\varepsilon) = 0$, $\Phi(1 + \varepsilon) = 1$ when $\varepsilon > 0$]; in other words it is necessary that the ω_h be the moments of a distribution taking values between 0 and 1, or again that $(-1)^s \Delta^s \omega_r \geqslant 0$ $(r, s = 1, 2, \ldots)$, as results from the expression for $\omega_n^{(r)}$.

If only the probabilities of the various frequencies on n trials, $\omega_0^{(n)}$, $\omega_1^{(n)}$, $\omega_2^{(n)}, \ldots, \omega_n^{(n)}$, are known, the condition under which there can exist a phenomenon consisting of exchangeable trials for which the $\omega_r^{(n)}$ have the given values, will clearly be that the corresponding $\omega_1, \omega_2, \ldots, \omega_n$ be the first n moments of a distribution on $(0, 1)$; these ω_h can be calculated as a function of the $\omega_r^{(n)}$ by the formula

$$\omega_h = \sum_{r=h}^{n} \omega_r^{(n)} \frac{r!(n-h)!}{n!(r-h)!}; \tag{20}$$

finally, the condition that $\omega_1, \ldots, \omega_n$ be the first n moments of a distribution on $(0, 1)$ is that all the roots of the polynomial

$$f(\xi) = \begin{vmatrix} 1 & \xi & \xi^2 & \ldots & \xi^k \\ \omega_0 & \omega_1 & \omega_2 & \ldots & \omega_k \\ \omega_1 & \omega_2 & \omega_3 & \ldots & \omega_{k+1} \\ \ldots & \ldots & \ldots & \ldots & \ldots \\ \ldots & \ldots & \ldots & \ldots & \ldots \\ \omega_{k-1} & \omega_k & \omega_{k+1} & \ldots & \omega_{2k-1} \end{vmatrix} \quad \text{if} \quad n = 2k - 1. \tag{21}$$

$$f(\xi) = \begin{vmatrix} 1 & \xi & \xi^2 & \ldots & \xi^k \\ \omega_1 & \omega_2 & \omega_3 & \ldots & \omega_{k+1} \\ \omega_2 & \omega_3 & \omega_4 & \ldots & \omega_{k+2} \\ \ldots & \ldots & \ldots & \ldots & \ldots \\ \ldots & \ldots & \ldots & \ldots & \ldots \\ \omega_k & \omega_{k+1} & \omega_{k+2} & \ldots & \omega_{2k} \end{vmatrix} \quad \text{if} \quad n = 2k. \tag{22}$$

fall in the interval $(0, 1)$, including the endpoints.[6]

Chapter IV: Exchangeable Random Quantities

Thus, as we have seen, in any problem at all concerning the exchangeable events $E_1, E_2, \ldots, E_n$, the probability will be completely determined either by the sequence of probabilities ω_n or by the limiting distribution of the frequency $\Phi(\xi)$ [or, what amounts to the same thing, by the corresponding characteristic function $\psi(t)$]. We have thus completely characterized the families of exchangeable events, and we have, in particular, elucidated the essential significance of $\Phi(\xi)$ connected with the fundamental result we have

[6] This result follows from Castelnuovo [VII] (see also [VIII]), as we have noted in [29].

demonstrated: the probability distributions **P**, corresponding to the case of exchangeability, are linear combinations of the probability distributions $\mathbf{P}_\xi$ corresponding to the case of independence and equiprobability (probability = ξ). We have indeed,

$$\mathbf{P}(E) = \int \mathbf{P}_\xi(E)\, d\Phi(\xi) \tag{19}$$

where $d\Phi(\xi)$ represents the distribution of weights in the linear combination.

We are going to extend this fundamental result to the case of exchangeable random quantities for which, up to now, we have only demonstrated the preliminary theorems, which we have used to establish certain results concerning the events themselves, rather than to solve the analogous problem, i.e. to characterize completely families of exchangeable random quantities.

Let us now consider the case of exchangeable random quantities and let us take an example to fix our ideas. In the study of exchangeable events, we have taken as an example the case of a game of heads or tails; let us now suppose that $X_1, X_2, \ldots, X_n$ represent measurements of the same magnitude; it suffices that the conditions under which the measurements are made do not present any apparent asymmetry which could justify an asymmetry in our evaluation of the probabilities, in order that we be able to consider them as exchangeable random quantities.

The extension of our earlier conclusions to this case will clearly be less easy than in the case of events, a *random quantity* being no longer characterized, from the probabilistic point of view, by a number (probability) as are the events, but by a function (for example, a distribution function or a characteristic function, etc.). Here the case of independence and equiprobability corresponds to the hypothesis of the independence of the random quantities X_i and the existence of a general distribution function $V(x)$; by calling $\mathbf{P}_v(E)$ the probability attributable to a generic event E, when the X_i are considered to be independent and to have the same distribution function V, the linear combinations will be distributions of the type

$$\mathbf{P}(E) = \sum c_i \mathbf{P}_{v_i}(E)$$

(with the weights $c_i > 0$, $\sum c_i = 1$); in the limit

$$\mathbf{P}(E) = \int \mathbf{P}_v(E)\, d\mathscr{F}(V), \tag{23}$$

the integral being extended over the function space of distribution functions, and the distribution of weights being characterized by the functional $\mathscr{F}(V)$, in a manner which will be made precise in what follows. Even before knowing the exact meaning of this integration, one is led to notice immediately that if $\mathbf{P}(E)$ is a linear combination of the $\mathbf{P}_v(E)$ one has the case of exchangeability: it suffices to observe that each $\mathbf{P}_v(E)$ giving the same probability to the events defined in a symmetrical[1] fashion in relation to $X_1, \ldots, X_n, \ldots$, the same condition will necessarily be satisfied by every linear combination $P(E)$; it is a question then only of proving the inverse, i.e. of showing that, in the case of exchangeability, $P(E)$ is necessarily of the form $\int P_v(E) \, d\mathscr{F}(V)$.[2]

The definition of the integral

$$\int f(V) \, d\mathscr{F}(V)$$

that we must introduce over the function space is only a completely natural generalization of the Stieltjes-Riemann integral:[3] by subdividing the space of distribution functions into a finite number of partial domains in any way whatever, we consider the expressions $\sum \bar{f}_i c_i$ and $\sum \underline{f}_i c_i$ where c_i is the weight of a generic element of these parts, and $\bar{f}_i$ and $\underline{f}_i$ are respectively the upper and lower bounds of the values taken by the function f in these domains. The lower bound of $\sum \bar{f}_i c_i$ and the upper bound of $\sum \underline{f}_i c_i$, when the subdivision is changed in all possible ways, are respectively the superior and inferior integral of f, extended to the function space of distribution functions in relation to the distribution of weights $\mathscr{F}$; when they coincide, their common value is precisely the integral $\int f(V) \, d\mathscr{F}(V)$ that we are going to examine more closely.

We are going to show that, in the circumstances that interest us, this integral exists, and that in order to determine its value, it suffices to know the

---

[1] Symmetric in the sense that, for example, the event E = "the point determined by the coordinates $X_1, X_2, \ldots, X_n$ will fall in the domain D" (in Euclidean space of n dimensions) is symmetrical to the events consisting in the same eventuality for one of the $n!$ points $X_{i_1}, X_{i_2}, \ldots, X_{i_n}$ corresponding to the $n!$ permutations of the coordinates. In particular:

(for D rectangular):

$$E = \text{``} a_h < X_h < b_h \qquad (h = 1, 2, \ldots, n)\text{''}$$

and

$$\text{``} a_h < X_{i_h} < b_h \qquad (h = 1, 2, \ldots, n)\text{''};$$

(for D spherical):

$$E = \text{``} \sum (X_h - a_h)^2 < \rho^2 \text{''}$$

and

$$\text{``} \sum (X_{i_h} - a_h)^2 < \rho^2 \text{''};$$

(for D a half-space):

$$E = \text{``} \sum a_h X_h > a \text{''} \qquad \text{and} \qquad \text{``} \sum a_h X_{i_h} > a, \text{''} \ldots$$

[2] One can accept this result and omit the following developments which are devoted to proving it and making it precise (toward the end of Chap. IV), without prejudice to an overall view of the thesis maintained in these lectures.

[3] For the reasons which make us regard the Stieltjes-Riemann integral as more appropriate to the calculus of probability, see [58] and [64].

weight for some very simple functional domains of distribution functions. Suppose to begin with the $f(V)$ depends only on the values

$$y_1 = V(x_1), \qquad y_2 = V(x_2), \ldots, \qquad y_s = V(x_s)$$

which the function V takes on a finite given set of abscissas $x_1, x_2, \ldots, x_s$; $f(V)$ is thus the probability that n random variables following the distribution V will all fall in a rectangular domain D, the first falling between x_1 and x_1', the second between x_2 and $x_2' \ldots$, the last between x_n and x_n'. This probability is[4]

$$f(V) = [V(x_1') - V(x_1)][V(x_2') - V(x_2)] \ldots [V(x_n') - V(x_n)]$$
$$= (y_1' - y_1)(y_2' - y_2) \ldots (y_n' - y_n) \quad (s = 2n) \tag{24}$$

It is clear that in order to evaluate the integral of such a function, it is sufficient to know the weights of the functional domains defined only by the ordinates $y_1, \ldots, y_s$ corresponding to the abscissas $x_1, \ldots, x_s$, i.e. the weights of the domains of the space of s dimensions defined by $y_1, \ldots, y_s$; if f is a continuous function of the y_i it will suffice to know the weights of the domains defined by the inequalities $y_i < a_i$ $(i = 1, 2, \ldots, s)$. The significance of these domains is the following: they comprise the distribution functions V whose representative curve $y = V(x)$ remains below each of the s points (x_i, a_i). Let $\Phi(x)$ be the stepwise curve of which the points (x_i, a_i) are the lower corners; the above condition can now be expressed by $V(x) < \Phi(x)$ [for all x],[5] and the weights of the set of distribution functions V such that $V(x) < \Phi(x)$ will be designated by $\mathscr{F}(\Phi)$; thus we give a concrete meaning to $\mathscr{F}$ which until now has represented a distribution of weights in a purely symbolic way. In this case the integral $\int f(V) \, d\mathscr{F}(V)$ is only the ordinary Stieltjes-Riemann integral in the space of s dimensions. If $f(V)$ does not depend solely on the ordinates of $V(x)$ for a finite set of abscissas $x_1, \ldots, x_s$ we will consider the case where it is possible to approach $f(V)$, from above and below, by means of functions of the preceding type, in such a way that the value of the integral will be uniquely determined by the values approached from above and below. In other words, it will be necessary that, for an arbitrary ε, one be able to find two functions $f'(V)$ and $f''(V)$ depending on a finite number of values $V(x_i)$, such that

---

[4] It is not necessary to be particularly concerned with the discontinuity points: indeed, a determinate function V is continuous almost everywhere (better: everywhere except, at most, on a denumerable set of points), and likewise in the integration, the weight of the set of distribution functions having x as a point of discontinuity is always zero, except, at most, for a denumerable set of points x; it suffices to observe that these points are the points of discontinuity of $\Phi(x) = \int V(x) \, d\mathscr{F}(V)$, and that $\Phi(x)$ is a distribution function.

[5] It is always understood that an inequality like $f(x) < g(x)$ between two functions means that it holds for all x (unless one has explicitly a particular case in mind when it is a question of a determinate value x).

$$f'(V) \leqslant f(V) \leqslant f''(V) \quad \text{and} \quad \int f'(V) \, d\mathscr{F}(V) > \int f''(V) \, d\mathscr{F}(V) - \varepsilon.$$

We return to the case of n independent random quantities having the distribution $V(x)$: if $f(V)$ is the probability that the point $(X_1, X_2, \ldots, X_n)$ falls in a domain D which is not reducible to a sum of rectangular domains, f' and f'' can represent the analogous probabilities for the domains D' contained in D, and D'' containing D, each formed from a sum of rectangular domains.

We have no need to pursue to the end the analysis of the conditions of integrability; we will content ourselves with having shown that they are satisfied in some sufficiently general conditions which contain all the interesting cases. We now return to the problem concerning the exchangeable random quantities $X_1, X_2, \ldots, X_n, \ldots$ in order to show the existence of the functional $\mathscr{F}$ having a meaning analogous to $\Phi(\xi)$ for exchangeable events. Let V be a stepwise function of which the lower corners are the s points

$$(x_i, y_i)(i = 1, 2, \ldots, s; x_{i+1} > x_i; y_{i+1} > y_i);$$

we will designate by $\mathscr{F}_h(V)$ the probability that, of h numbers $X_1, X_2, \ldots, X_h$, hy_1 at the most will exceed x_1, hy_2 at the most will exceed x_1, hy_2 at the most will exceed x_s, or, in other words, the probability

$$\mathbf{P}\{G_h(x) \leqslant V(x)\}$$

that the distribution function $G_h(x)$ of the values of $X_1, X_2, \ldots, X_h$ never exceeds $V(x)$. More precisely, the function $G_h(x)$ is the "observed distribution function" resulting from the observation of $X_1, \ldots, X_h$; it represents the stepwise curve of which the ordinate is zero to the left of the smallest of the h numbers $X_1, \ldots, X_h$, equal to $1/h$ between the smallest and the second, equal afterwards to $2/h$, $3/h$, $\ldots$, $(h - 1)/h$, and finally equal to unity to the right of the largest of the h numbers considered. The steps of $G_h(x)$ are placed on the points of the axes of abscissas which correspond to the values X_i; before knowing these values, $G_h(x)$ is a random function, since these abscissas are random quantities.

It is easy to show, by extending a theorem given by Glivenko[6] for the case of independent random quantities to the case of exchangeable random quantities, that it is very probable that for h and k sufficiently large, $G_h(x)$ and $G_k(x)$ differ very little, and, in the case of a set of *successive* averages $G_h(x)$, $G_{h+1}(x)$, $\ldots$, we have a strong stochastic convergence. By dividing the x axis into a sufficiently large finite number of points $x_1, \ldots, x_s$ the proof can be based on that given for the analogous properties in the case of exchangeable events. For a given x, $G_h(x)$ and $G_k(x)$ give respectively the frequencies Y_h and Y_k for the h and k trials of the set of exchangeable events $E_i = (X_i < x)$; the difference between $G_h(x)$ and $G_k(x)$ then has standard deviation less than

---

[6] [XIII], see also Kolmogorov [XVIII], and [45].

$\sqrt{\dfrac{1}{h} + \dfrac{1}{k}}$ [see formula (10)], and the probability that it exceeds ε can be made as small as one wishes by choosing h and k larger than a sufficiently large number N. By taking N so that the probability of a difference greater than ε is less than θ/s for each of the abscissas

$$x = x_1, x_2, \ldots, x_s,$$

we see that, except in a case whose total probability is less than θ, the two functions $G_h(x)$ and $G_k(x)$ will not differ by more than ε for any of the abscissas $x_1, \ldots, x_s$.

Under these conditions, the probability $\mathscr{F}_k(V - \varepsilon)$ that $G_k(x)$ will not exceed the stepwise curve $V(x) - \varepsilon$ for any x, which is to say the probability of having

$$G_k(x_i) < y_i - \varepsilon \quad (i = 1, 2, \ldots, s)$$

cannot be more than $\mathscr{F}_h(V) + \theta$, for, in order to satisfy the imposed conditions, it is necessary either that $G_h(x)$ not exceed $V(x)$ for any x, or that we have $G_h(x) - G_k(x) > \varepsilon$ for at least one of the abscissas $x_1 \ldots x_s$. We thus have

$$\mathscr{F}_k(V - \varepsilon) - \theta \leqslant \mathscr{F}_h(V) \leqslant \mathscr{F}_k(V + \varepsilon) + \theta \tag{25}$$

(the second inequality can be proved in the same way); by defining convergence in an appropriate way (as for distribution functions[7]), one concludes that $\mathscr{F}_h \to \mathscr{F}$; it is the functional $\mathscr{F}$ which allows us to characterize the family of exchangeable random quantities we have in mind.

To prove the fundamental formula

$$\mathbf{P}(E) = \int \mathbf{P}_V(E) \, d\mathscr{F}(V) \tag{23}$$

we remark that we have, for all h,

$$\mathbf{P}(E) = \int \mathbf{P}_{h, V}(E) \, d\mathscr{F}(V) \tag{26}$$

where $\mathbf{P}_{h, V}(E)$ is the probability of E, given the hypothesis

$$G_h(x) = V(x).$$

If the event E depends on the n first random quantities $X_1, \ldots, X_n$ (to fix our ideas by a simple example, let us imagine that the event E consists of X_1 falling between a_1 and b_1, X_2 between a_2 and b_2, $\ldots$, X_n between a_n and b_n), it will naturally be necessary to suppose $h \geqslant n$; if h is very large in relation to n, it is clear that $\mathbf{P}_{h, V}(E) \cong \mathbf{P}_V(E)$, for the probability $\mathbf{P}_{h, V}(E)$ is obtained by supposing $X_1, \ldots, X_n$ chosen by chance, simultaneously (that is, without repetition) from among the h values where $G_h = V$ is discontinuous, whereas $\mathbf{P}_V(E)$ is

---

[7] See Lévy [XX], p. 194.

the analogous probability obtained by considering all the combinations possible on the supposition of independent choices. The fact of including or excluding repetitions has a more and more negligible influence as $h \to \infty$; thus $P_{h,V}(E) \to P_V(E)$. This relation and the relation $\mathscr{F}_h(V) \to \mathscr{F}(V)$ provide the proof that

$$P(E) = \int P_{h,V}(E) \, d\mathscr{F}_h(V) = \int P_V(E) \, d\mathscr{F}(V).$$

We shall consider a particular type of event E, which will permit us to analyze the relation between the functional distribution given by $\mathscr{F}$, relative to the exchangeable random quantities X_i, and the linear distributions $\Phi_x(\xi)$, that is to say, the limiting distributions $\Phi(\xi)$, related to the events $E_i = (X_i < x)$. An event E will belong to the particular type envisaged if it expresses a condition depending solely on the fact that certain random quantities X_1, ..., X_n are less than or greater than[8] a unique given number x. For example, $E = "X_1, X_3, X_8$ are $> x$, X_2 and X_7 are $< x"$; $E = "$among the numbers X_2, X_3, X_9, X_{12} there are three which are $> x$ and one $< x"$; $E = "$in the sequence $X_1 X_2 \ldots X_{100}$ there are no more than three consecutive numbers $> x"$; etc. In other words, the event E is a logical combination of the $E_i = (X_i < x)$ for a unique given x.

The theory of exchangeable events tells us that the probability of any event E of this type is completely determined by the knowledge of $\Phi_x(\xi)$, and we can express this probability with the aid of $\mathscr{F}(V)$; we can then express $\Phi_x(\xi)$ by means of $\mathscr{F}(V)$, and we then have precisely

$$\Phi_x(\xi) = \int_{V(x)<\xi} d\mathscr{F}(V); \qquad d\Phi_x(\xi) = \int_{\xi<V(x)<\xi+d\xi} d\mathscr{F}(V). \qquad (27)$$

Indeed, let $E^{(n)}$ be the event consisting in this; that the frequency of values $< x$ on the first n trials $X_1, \ldots, X_n$ not exceed ξ; by definition $\Phi_x(\xi) = \lim P(E^{(n)})$, and moreover $P_V(E^{(n)})$, for n very large, is very close[9] to 0, if $V(\xi) > \xi$, or to 1, if $V(x) < \xi$; we will have, then,

$$\Phi_x(\xi) = \lim \int P_V(E^{(n)}) \, d\mathscr{F}(V)$$

$$= \int_{V(x)<\xi} 1 \cdot d\mathscr{F}(V) + \int_{V(x)>\xi} 0 \cdot d\mathscr{F}(V) = \int_{V(x)<\xi} d\mathscr{F}(V). \qquad (27)$$

[8] This case of equality has a zero probability, except for particular values of x finite or at most denumerable in number; we neglect this case for simplicity, observing, besides, that it does not entail any essential difficulty, but only the consideration of two distinct values of the distribution function to the left and right of x.

[9] We recall that $P_V(E^{(n)})$ is simply the probability of a frequency $< \xi$ on n independent trials, with constant probability $p = V(x)$, and therefore has the value

$$\sum_{v<\xi n} \binom{n}{v} p^v q^{n-v} \quad [q = 1 - p = 1 - V(x)].$$

One can deduce from this (or, better, obtain directly) the following result: $\omega_r^{(n)}(x)$, the probability that r out of n random quantities $X_{i_1}, \ldots, X_{i_n}$, chosen in advance, should be $< x$, is equal to

$$\omega_r^{(n)}(x) = \binom{n}{r} \int [V(x)]^r [1 - V(x)]^{n-r} \, d\mathscr{F}(V), \tag{28}$$

and in particular for $r = n$:

$$\omega_n(x) = \int [V(x)]^n \, d\mathscr{F}(V), \qquad \omega_1(x) = \int V(x) \, d\mathscr{F}(V); \tag{29}$$

this last formula giving, in particular, the probability that a general fixed number X_i is less than x; and this is the distribution function attributed to each of the X_i before beginning the trials.

Up to now, in $\Phi_x(\xi)$, $\omega_r^{(n)}(x)$, $\omega_n(x)$, we have considered x only as a parameter which determines the events $E_i = (X_i < x)$, but which does not vary; if, on the contrary, these expressions are considered as functions of x, certain remarks can be made which throw a new light on them. Let us consider n of the given random quantities: $X_1, X_2, \ldots, X_n$; $\omega_n^{(n)}$ is the probability that none of these numbers exceed x, and thus constitutes the distribution function of the maximum value among $X_1, \ldots, X_n$; $\omega_n^{(n)}(x) + \omega_{n-1}^{(n)}(x)$ is in an analogous way the distribution function of the next-to-largest of the numbers $X_1, \ldots, X_n$ arranged in increasing order; $\omega_n^{(n)}(x) + \omega_{n-1}^{(n)}(x) + \cdots + \omega_r^{(n)}$ that of the rth; and finally $\omega_n^{(n)}(x) + \cdots + \omega_1^{(n)}(x) = 1 - \omega_0^{(n)}(x)$ that of the smallest of the X_i. As the identity

$$\omega_1^{(n)}(x) + 2\omega_2^{(n)}(x) + 3\omega_3^{(n)}(x) + \cdots + n\omega_n^{(n)}(x) = n\omega_1(x) \tag{30}$$

shows, the average of these n distribution functions is $\omega_1(x)$, that is to say, the distribution function of any one whatever of the X_i: this fact is very natural, for, according to the definition of exchangeability, each number X_i has the same probability $1/n$ of being the smallest, or the second, $\ldots$, or the greatest, and, in general, all the permutations of the X_i have the same probability $1/n!$ of being disposed in increasing order (if there exists a probability different from zero that the n values are not distinct, the modifications to make in these statements are obvious).

There exists a close relation between the distribution functions of a random quantity of determinate rank and the function $\Phi_x(\xi)$: by definition, $\Phi_x(\xi)$ is the limiting value, for $n \to \infty$, of the probability that of n random quantities $X_1, \ldots, X_n$ there will be at most ξn which are $< x$; this probability is equal to $\sum_r \omega_r^{(n)}(x)$, the sum being extended over the indices $r < \xi n$. But this sum is the distribution function of those of the numbers $X_1, \ldots, X_n$ which occupy the rank "whole part of ξn" where the random quantities are arranged in order of increasing magnitude: by considering ξ fixed, $\Phi_x(\xi)$ is (as a function of x) the distribution function of the number of rank $\cong \xi n$ on a very large number n of given random quantities.

It is easily seen that $\Phi_x(\xi)$ is a never-decreasing function of ξ and x, that $\Phi = 0$ if $\xi < 0$, and $\Phi = 1$ if $\xi > 1$ (Φ is thus defined substantially only on the interval $0 \leqslant \xi \leqslant 1$), and finally that $\Phi \to 0$ and $\Phi \to 1$ respectively for $x \to -\infty$ and $x \to +\infty$. Conversely, each function $\Phi_x(\xi)$ having these properties can be associated in an infinite number of ways with a probability distribution for exchangeable random quantities; one such function $\Phi_x(\xi)$ being given, one can always construct a distribution of weights $\mathscr{F}(V)$ in function space, such that formula (27) holds. The simplest way of doing this is the following: let $V_\lambda(x) = \xi$ be the explicit equation of the contour line $\Phi_x(\xi) = \lambda$, which represents, due to the properties of $\Phi_x(\xi)$, a distribution function, and defines the distribution $\mathscr{F}(V)$ by attributing the weights $\lambda' - \lambda(\lambda' > \lambda)$ to the set of $V(x)$ such that

$$V_\lambda(x) < V(x) < V_{\lambda'}(x)$$

for all x; in this way the integration in function space is reduced to a simple integral:

$$\int f(V)\, d\mathscr{F}(V) = \int_0^1 f(V_\lambda)\, d\lambda. \tag{31}$$

We have, for example,

$$\omega_n(x) = \int_0^1 [V_\lambda(x)]^n\, d\lambda = \int_0^1 \xi^n\, d\Phi_x(\xi); \tag{32}$$

this suffices to show that the distribution we have obtained satisfies the desired condition; it results directly from the calculation of

$$\Phi_x(\xi) = \int_{V(x)<\xi} d\mathscr{F}(V)$$

$$= \int_{V\lambda(x)<\xi} d\lambda = \{\text{the value of } \lambda \text{ for which } V_\lambda(x) = \xi\} \tag{33}$$

However, there always exists an infinity of other distributions $\mathscr{F}(V)$ corresponding to the same function $\Phi_x(\xi)$: it suffices to observe, for example, that if one puts in any way at all

$$\Phi_x(\xi) = c_1\Phi_x^{(1)}(\xi) + c_2\Phi_x^{(2)}(\xi) + \cdots + c_k\Phi_x^{(k)}(\xi)$$

with $c_i > 0$, $\sum c_i = 1$, the $\Phi_x^{(i)}(\xi)$ satisfying the same conditions as Φ, and if one introduces the corresponding $V_\lambda^{(i)}(x)$, one will always have

$$\omega_n(x) = \sum c_i \int_0^1 [V^{(i)}(x)]^n\, d\lambda. \tag{34}$$

The function $\Phi_x(\xi)$ thus characterizes neatly all the families of exchangeable events $E_i = (X_i < x)$ for any x whatever, but this does not suffice in problems where the interdependence of these various families comes into play: complete knowledge of the distribution $\mathscr{F}(V)$ in function space is then indispensable.

It should be noted once more that if one were to consider exchangeable random elements of any space whatever, one would arrive at perfectly analogous results: implicitly, we have already indeed considered some exchangeable random functions, since, for example, the $G_h(x)$, the distribution functions of $X_{i_1}, X_{i_2}, \ldots, X_{i_h}$ constitute a family of exchangeable random functions, when all possible groups $i_1, i_2, \ldots, i_h$ are considered. The general result which has been established for events and for random quantities, and which could be demonstrated for random elements of any space whatever, may be expressed by saying that *the probability distribution of classes of exchangeable random elements are "averages" of the probability distributions of classes of independent random elements.* . . . .

Bibliography

[I]	Bertrand (J.)	*Calcul des probabilités*, Paris, 1889.
[II]	Bridgman (P.)	*The Logic of Modern Physics*, New York, 1927.
[III]	Cantelli (F.P.)	"Sulla probabilità come limite della frequenza," *Rendiconti Reale Accademia Nazionale dei Lincei*, Series 5, Vol. XXVI, 1st Semester (1917).
[IV]	——	"Una teoria astratta del calcolo della probabilità," *Giornale dell'Istituto Italiano degli Attuari*, Vol. III-2 (1932).
[V]	——	"Considérations sur la convergence dans le calcul des probabilités," *Annales de l'Institut Henri Poincaré*, Vol. V, fasc. 1 (1935).
[VI]	Castelnuovo (G.)	*Calcolo delle probabilità*, Zanichelli, Bologna, 1925.
[VII]	——	"Sul problema dei momenti," *Gior. Ist. Ital. Attuari*, Vol. I-2 (1930).
[VIII]	——	"Sur quelques problèmes se rattachant au calcul des probabilités," *Ann. Inst. H. Poincaré*, Vol. III, fasc. 4 (1933).
[IX]	Dörge (K.)	"Ueber das Anwendung der Wahrscheinlichkeitsrechnung und das Induktionsproblem," *Deutsche Mathematik*, 1 (1936).
[X], [XI]	Fréchet (R.M.)	"Sur l'extension du théorème des probabilités totales au cas d'une suite infinie d'événements," *Rendiconti Reale Istituto Lombardo di Scienze Lettere ed Arte*, Series 2, Vol. LXIII, first note, fasc. 11-15, second note, fasc. 16-18 (1930).
[XII]	Fréchet (R.M.) and Halbwachs (M.)	*Le calcul des probabilités à la portée de tous*, Dunod, Paris, 1925.
[XIII]	Glivenko (V.)	"Sulla determinazione empirica delle leggi di probabilità," *Gior. Ist. Ital Attuari*, Vol. IV-1 (1933).
[XIV]	Hosiasson (J.)	"La théorie des probabilités est-elle une logique généralisée?" Vol. IV of *Actes du Congrès International de Philosophie Scientifique* (1935), Act. Scient. Indust. #391, Hermann, Paris, 1936.
[XV]	Khinchin (A.)	"Sur les classes d'événements équivalentes," *Matematičeskii Sbornik, Recueil. Math. Moscou*, 39-3 (1932).

[XVI]   ———   "Remarques sur les suites d'événements obéissant à la loi des grandes nombres," *Ibid.*

[XVII]   Kolmogorov (A.)   *Foundations of the Theory of Probability*, 2nd English edition (translation edited by N. Morrison), New York, Chelsea, 1956.

[XVIII]   ———   "Sulla determinazione empirica di una legge di distribuzione," *Gior. Ist. Ital. Attuari*, Vol. IV-1 (1933).

[XIX]   von Kries (J.)   *Die Prinzipien der Wahrscheinlichkeitsrechnung*, Freiburg i.b., 1886.

[XX]   Lévy (P.)   *Calcul des probabilités*, Paris, Gauthier-Villars, 1925.

[XXI]   Lomnicki (A.)   "Nouveaux fondements de la théorie des probabilités," *Fundamenta Mathematica*, Vol. 4 (1923)

[XXII]   Lukasiewicz (J.)   "Philosophische Bemerkungen zu mehrwertigen Systemen des Aussagenkalküls," *Comptes Rendus Soc. Sci. Lett. Varsovie*, Classe III, Vol. XXIII (1930).

[XXIII]   Mazurkiewicz (S.) "Zur Axiomatik der Wahrscheinlichkeitsrechnung," *C. R. Soc. Sci. Lett. Varsovie*, Classe III, Vol. XXV (1932).

[XXIV]   Medolaghi (P.)   "La logica mathematica e il calcolo delle probabilità," *Bolletino Assoc. Ital. Attuari*, **18** (1907).

[XXV]   von Mises (R.)   *Probability, Statistics and Truth*, 2nd revised English edition (prepared by H. Geringer), New York, Macmillan, 1957.

[XXVI]   ———   "Théorie des probabilités: fondements et applications," *Ann. Inst. H. Poincaré*, Vol. III, fasc. 2 (1932).

[XXVII]   Pareto (V.)   *Manuel d'économie politique*, Paris, Giard et Brière, 1909.

[XXVIII]   Poincaré (H.)   *Science and Hypothesis* (translated by W.J. Greenstreet), Chapter: The Calculus of Probability, New York, Dover, 1952.

[XXIX]   Reichenbach (H.)   *The Theory of Probability* (English edition translated by Ernest Hutten and Maria Reichenbach), Berkeley, 1949.

[XXX]   ———   "Die Induktion als Methode der wissenschaftliche Erkenntnis," Vol. IV of *Actes du Congrès Int. de Phil. Scient.* (1935), Act Scient. Indust. #391, Hermann, Paris, 1936.

[XXXI]   ———   "Wahrscheinlichkeitslogik als form wissenschaftlichen Denkens," *Ibid.*

[16]   "Sui passagi al limite nel calcolo della probabilità," *Rend. R. Ist. Lombardo*, Series 2, Vol. LXIII, fasc. 2–5 (1930).

[24]   "A proposito dell'estensione del teorema delle probabilità totali alle classi numerabili," *ibid.*, fasc. 11–15 (1930).

[26]   "Problemi determinati e indeterminati nel calcolo delle probabilità," *Rend. R. Acc. Naz. Lincei*, Series 6, Vol. XII, fasc. 9 (Nov., 1930).

[27]   "Fondamenti logici del ragionamento probabilistico," *Boll. Un. Mat. Ital.*, Vol. IX-5 (Dec., 1930).

[28]   "Ancora sull'estensione alle classi numerabili del teorema delle probabilità totali," *Rend. R. Ist. Lombardo*, Series 2, Vol. LXIII, fasc. 1–18 (1930).

[29]   "Funzione caratteristica di un fenomeno aleatorio," *Memorie R. Acc. Naz. Lincei*, Series 6, Vol. IV, fasc. 5 (1930).

[32]   "Probabilismo: Saggio critico sulla teoria delle probabilità e sul valore
            della scienze," *Biblioteca di Filosofia diretta da Antonio Aliotta*, Naples,
            Perrella (1931).
[34]   "Sul significato soggettivo della probabilità," *Fund. Math.*, Vol. 17, War-
            saw (1931).
[35]   "Sui fondamenti logici del ragionamento probabilistico," *Atti. Soc. Ital.
            Progr. Scienze*, Riunione Bolzano-Trento del 1930, Vol. 11, Rome (1931).
[36]   "Le leggi differenziali e la rinunzia al determinismo," *Rend. Semin. Mat.
            R. Univ. Roma*, Series 2, Vol. VII (1931).
[38]   "Probabilità fuori dagli schemi di urne," *Period di Mat.*, Series 4, Vol.
            XII-1 (1932).
[40]   "Funzione caratteristica di un fenomeno aleatorio," *Att. Congr. Int.
            Mathem.*, Bologna, 1928, Vol. 6, Bologna, Zanichelli ed. (1932).
[45]   "Sull'approssimazione empirica di una legge di probabilità," *Gior. Ist.
            Ital. Attuari*, Vol. 4-3 (1933).
[46 to 48] "Classi di numeri aleatori equivalenti. La legge dei grandi numeri nel caso
            dei numeri aleatori equivalenti. Sulla legge di distrubuzione dei valori in
            una successione di numeri aleatori equivalenti," *Rend. R. Acc. Naz. Lin-
            cei*, Series 6, Vol. XVIII, 2e sem., fasc. 3–4 (1933).
[49]   "Sul concetto di probabilità," *Riv. Ital. Statist.*, Vol. 1 (1933).
[51]   "Indipendenza stocastica ed equivalenza stocastica," *Atti. Soc. Ital. Prog.
            Scienze*, Riunione Bari del 1933, Vol. II, Rome (1934).
[R. 2.]   Review of Reichenbach, H. [XXIX] *Zentralblatt für Mathematik und ihre
            Grenzegebiete*, 10-8 (1935).
[56]   "Il problema della perequazione," *Atti. Soc. Ital. Progr. Scienze*, Riunione
            Napoli del 1934, Vol. II, Rome (1935).
[58]   (In collaboration with M.M. Jacob). "Sull'integrale di Stieltjes-Riemann,"
            Gior. Ist. Itol. Attuari, Vol. VI-4 (1935).
[62]   "La logique de la probabilité" (Communication to the Congrès Int. de
            Philosophie Scient., Paris, 1935), Act. Scient. Indus. #391, Hermann,
            1936.
[64]   "Les probabilités nulles," *Bull. Sci. Math.*, Second Series, **60** (1936).
[65]   "Statistica e probabilità nella concezione di R. von Mises," *Suppl. Statis.
            Nuovi Probl.*, Vol. II-3 (1936).
[70]   "Riflessioni teoriche sulle assicurazioni elementari," Vol. 2 of *Comptes
            Rendus du Onzième Congrès International d'Actuaires* (Paris, June 17–24,
            1936), Gauthier-Villars, Paris, 1937.

Introduction to
Cramér (1942) On Harmonic Analysis in Certain Functional Spaces

M.R. Leadbetter
University of North Carolina at Chapel Hill

Harald Cramér was born in 1893 in Stockholm, Sweden. He attended Stockholm University in 1912 and graduated with a Ph.D. degree in mathematics in 1917. He became interested in actuarial mathematics (applications of mathematics to insurance) and was appointed to a chair in actuarial mathematics and mathematical statistics in 1929. Details of his life and work may be found, for example, in the obituary by Leadbetter (1988) and references therein. Here we indicate only some influences leading to the present paper.

His work in insurance applications led Cramér naturally to the interest in probability and statistics from which his notable contributions over half a century emanated. His early work in central limit theory involved an essential use of characteristic functions, of which the Fourier representation (1) is an immediate relative. At the same time in the 1920s and 1930s, Cramér worked increasingly with stochastic processes (again arising initially from insurance applications). In hindsight, at least, it therefore seems almost inevitable that his interests should then turn to a combination of these areas. It was this focus on the development of Fourier theory for stochastic processes that led to the formulation of the elegant and important result (2), which we know as the spectral representation for stationary processes.

It may seem unlikely that a paper that (1) does not explicitly discuss the topic for which it is remembered and (2) is mathematically equivalent to work published eight years previously would appear in a "breakthrough" volume. In this introduction to Cramér's paper, we discuss (from an admitted time distance) something of the circumstances surrounding it and the significant impacts of the work.

So-called "spectral theory" for stationary processes must surely be regarded as one of the most significant historical developments in stochastic process theory. Aside from its important theoretical uses, potential applications abound, and the theory pervades much of subsequent activity in modern time-series analysis.

The use of Fourier theory for deterministic functions has, of course, a long and venerable history, and, in some contexts, it has been almost automatic to split a function of time into its harmonic components before thinking further. This may be done for its own sake, as in illustrative examples for mathematics training, showing how unlikely periodic functions may be regarded as superpositions of sine and cosine waves. It may be done to solve specific problems such as the description of the harmonic content of music or speech. Or, perhaps most important, its purpose may be to describe the output of a linear system in terms of its input by recognizing that such systems transmit harmonic waves with potentially altered amplitude and phase but leaving frequency unchanged.

It is certainly not unnatural that workers in Fourier theory would in due course wish to consider families of functions, and that probabilists would wish to use it for *random* functions, i.e., stochastic processes. Cramér had both (essentially equivalent) viewpoints in mind, as will be evident more explicitly below. His paper is written in the former context—indeed, with deliberate avoidance of the stochastic process language in which the work has had its truly significant impact, being recognized by its inclusion in this volume. If not already well known to a reader, that significance should be evident from these introductory comments.

In addition to the interest in Fourier theory and characteristic functions, the decade of the 1930s saw important developments in correlation theory and foundations for stationary processes, in, e.g., the work of Wiener (1934) and especially Khintchine (1934). In particular, Khintchine had obtained the spectral representation

$$R(\tau) = \int_{-\infty}^{\infty} e^{i\tau\lambda} \, dF(\lambda), \tag{1}$$

for the correlation function $R(\tau) = \mathrm{corr}[X(t), X(t + \tau)]$ of a (mean square continuous) stationary process $X(t)$, in terms of the so-called "spectral distribution function" F. This representation for the correlation between values of $X(t)$ at time points separated by τ has, of course, been basic to much statistical study of the correlation properties of stationary processes.

The time around 1940, although complicated by World War II, was therefore ripe in terms of interest and activity for the consideration of even more basic questions concerning corresponding representations for a stationary process $X(t)$ itself. In 1941, Cramér announced the representation

$$X(t) = \int_{-\infty}^{\infty} e^{it\lambda} \, dZ(\lambda), \tag{2}$$

where Z has orthogonal increments and $\mathscr{E}[Z(\lambda_2) - Z(\lambda_1)]^2 = F(\lambda_2) - F(\lambda_1)$ for $\lambda_1 < \lambda_2$, F being the spectral distribution function, as above in Eq. (1). Put intuitively, this means that $X(t)$ is regarded as a superposition of pure harmonic waves with angular frequency λ having, loosely speaking, the random amplitudes $|dZ(\lambda)|$ whose variance is $dF(\lambda)$. Thus, F provides an important measure of the "amount of frequency content up to λ" present in the process. The usefulness of Eq. (2) is well exemplified by the fact that if the random waveform $X(t)$ is the input to a linear system, the output $Y(t)$ is given by

$$Y(t) = \int G(\lambda)e^{it\lambda}\, dZ(\lambda)$$

with spectral distribution function $\int_{-\infty}^{\lambda} |G(x)|^2\, dF(x)$, where G is the system "gain" or "transfer function."

The representation (2) was presented by Cramér at a conference on mathematical probability in Stockholm in February 1941, at which attendance was limited, even from Scandinavian countries, due to the restrictions imposed by the war. In 1942, the present paper was published in *Arkiv för Matematik, Astronomi och Fysik*. As noted above (and may be seen from Sec. 2 of the paper) except for a reference the language of probability is totally and deliberately avoided, it being Cramér's intention (stated in Sec. 3) to write a subsequent paper discussing further properties including probabilistic interpretations. Evidently this intention was modified and the explicit representation (2) was stated in the survey paper [Cramér (1946)] given at the 10th Scandinavian Mathematics Congress held in Copenhagen and published in the 1946 proceedings of that conference.

In spite of the absence of a specific statement in stochastic form, the present paper is certainly widely regarded as the significant source of the representation (2). As will be seen, the paper is not a long one. The proof, while omitting some details, is also not long; it involves a direct definition and use of stochastic Fourier integrals and is still instructive [see Doob (1953)]. On the other hand, modern proofs [see, e.g. Cramér and Leadbetter (1967)] are very simple applications of Hilbert space theory by the use of an isomorphism of the space $L_2(\{X_t\})$ with $L_2(dF)$ that maps X_t into the function $e^{i\lambda t}$. Cramér (1976) notes that "The fundamental importance of this (i.e., Hilbert space) theory for the study of stochastic processes did not become known to us until after the war." At the same time, interesting historical notes given by Doob indicate that the Russian school was aware of the identity of certain probability and Hilbert space problems, and used Hilbert space methods for obtaining spectral representations at this time [Doob (1953), p. 636].

In fact, in one sense the entire result was already available in 1932, in a Hilbert space context, with the appearance of the paper of Stone (1930) on representations of unitary groups in Hilbert spaces. As subsequently noted by Cramér (1976), the representation (2) is a probabilistic version of Stone's result. But this had little or no probabilistic impact at the time, being couched in terms of apparently quite unrelated abstract mathematical areas. This may seem curious to today's probabilist, totally accustomed to unitary operators in Hilbert spaces, but was very natural at this time of developing theory. At any rate, the direct attack on the problem certainly provided a needed "break-through" for stochastic process theory that would subsequently be unified with the abstract theory and Hilbert space methods, thereby enriching both fields.

Finally, without diminishing admiration for Cramér's contribution, it should be repeated that the time was right for the consideration of such problems and that other workers produced significant results along similar lines. For example, Loève apparently obtained the representation (2) at about the same time [see Lévy (1948) & also Cramér (1976)]. Contributions of others (e.g., von Neumann, Obukhoff) are described in the historical notes of Doob (1953) referred to above.

References

Cramér, H. (1942). On harmonic analysis in certain functional spaces, *Arkiv Mat. Astron. Fysik*, **28B**.

Cramér, H. (1946). On the theory of stochastic processes, in *Proceedings of the 10th Scandinavian Congress of Mathematics*. Copenhagen, pp. 28–39.

Cramér, H. (1976). Half a century with probability theory: Some personal recollections, *Ann. Prob.*, **4**, 509–546.

Cramér, H., and Leadbetter, M.R. (1967). *Stationary and related Stochastic Processes*. Wiley, New York.

Doob, J.L. (1953). *Stochastic Processes*. Wiley, New York.

Khintchine, Y.A. (1934). Korrelationstheorie der stationären stochastischen Prozesse, *Math. Ann.*, **109**, 604–615.

Leadbetter, M.R. (1988). Harald Cramér 1893–1985, *Internat. Statist. Ins. Rev.*, **56**, 89–97.

Lévy, P. (1948). *Processus stochastiques et mouvement Brownien*, Gauthier-Villars Paris. 299–352.

Stone, M.H. (1930). Linear transformation in Hilbert Space III. Operational methods and group theory, *Proc. Nat. Acad. Sci. USA*, **16**, 172–175.

Wiener, N. (1934). Generalizedären stochastischen Prozesse, *Math. Ann.*, **109**, 604–615.

On Harmonic Analysis in Certain Functional Spaces

Harald Cramér

1. In generalized harmonic analysis as developed by Wiener [7, 8] and Bochner [1] we are concerned with a measurable complex-valued function $f(t)$ of the real variable t (which may be thought of as representing time), and it is assumed that the limit

$$\varphi(t) = \lim_{T \to \infty} \frac{1}{2T} \int_{-T}^{T} f(\tau + t)\overline{f(\tau)}\, d\tau \tag{1}$$

exists for all real t. If, in addition, $\varphi(t)$ is continuous at the particular point $t = 0$, it is continuous for all real t and may be represented by a Fourier-Stieltjes integral

$$\varphi(t) = \int_{-\infty}^{\infty} e^{itx}\, d\Phi(x), \tag{2}$$

where $\Phi(x)$ is real, bounded and never decreasing.

Denoting by x_0 a fixed continuity point of $\Phi(x)$, the infinite integral

$$F(x) = -\frac{1}{2\pi i} \int_{-\infty}^{\infty} \frac{e^{-itx} - e^{-itx_0}}{t} f(t)\, dt \tag{3}$$

exists, by the Plancherel theorem, as a »limit in the mean» over $-\infty < x < \infty$. Conversely, we have for almost all real t

$$f(t) = \lim_{\varepsilon \to 0} \int_{-\infty}^{\infty} e^{itx} \frac{F(x + \varepsilon) - F(x - \varepsilon)}{2\varepsilon}\, dx, \tag{4}$$

where the integral should be similarly interpreted as in (3). In the particular case when $F(x)$ as defined by (3) is of bounded variation in $(-\infty, \infty)$, the expression (4) reduces to an ordinary Fourier-Stieltjes integral:

$$f(t) = \int_{-\infty}^{\infty} e^{itx}\, dF(x). \tag{4a}$$

In the general case, (4) may be regarded as a *generalized form* of the Fourier-Stieltjes representation (4a).

Further, we have for any real numbers a and b, which are continuity points of $\Phi(x)$,

$$\Phi(b) - \Phi(a) = \lim_{\varepsilon \to 0} \int_a^b \frac{|F(x + \varepsilon) - F(x - \varepsilon)|^2}{2\varepsilon}\, dx. \tag{5}$$

2. All the results so far quoted are concerned with the harmonic analysis of an individual function $f(t)$. On the other hand, in many important applications it is required to deal simultaneously with an infinite class of functions $f(t)$ and to deduce results, analogous to those quoted above, which are true *almost always*, with reference to some appropriate measure defined in the class of functions under consideration.

In this respect, a particular interest is attached to the case when the measure considered is *invariant for translations in time*, in the sense that whenever S is a measurable set of functions $f(t)$, then the set S_h obtained by substituting $f(t + h)$ for $f(t)$ is also measurable and has the same measure for any real h. Wiener [9] has already obtained certain results concerning this case, and it is the purpose of the present note to give a further contribution to the study of the problem.

The subject-matter of this note is intimately related to the theory of *stationary random processes* developed by Khintchine [3] and other authors, and indeed the theorem given below could be stated in the terminology of this theory. In this preliminary note we shall, however, avoid every reference to the theory of probability, reserving a further discussion of these questions, as well as of various generalizations and applications, for a later publication.

3. In the sequel, we shall consider the same $\mathfrak{S}$ of all finite complex-valued functions $f(t) = g(t) + ih(t)$ of the real variable t. A set of functions $f(t)$ defined by a finite number of pairs of inequalities of the form

$$\alpha_v < g(t_v) \leqq \beta_v,$$

$$\gamma_v < h(t_v) \leqq \delta_v,$$

will be called an *interval* in $\mathfrak{S}$. Let $\mathfrak{C}$ denote the smallest additive class of sets in $\mathfrak{S}$ containing all intervals, and let $P(S)$ denote a non-negative measure defined for all sets S belonging to $\mathfrak{C}$, and such that $P(\mathfrak{S})$ is finite. According to a theorem of Kolmogoroff [4, p. 27], $P(S)$ is uniquely defined by its values on all intervals.

For any functional $G(f)$ which is measurable ($\mathfrak{S}$), the Lebesgue-Stieltjes integral with respect to the measure $P(S)$ over a measurable set S_0 will be denoted by[1]

---

[1] For the definition of this integral, cf. Saks [5], Ch. I.

$$\int_{S_0} G(f) \, dP.$$

If t_0 is any given real number, $G(f) = |f(t_0)|^2$ is a measurable functional, and we shall always assume that the integral

$$\int_{\mathfrak{S}} |f(t_0)|^2 \, dP$$

is finite.

If the measure $P(S)$ is invariant for translations in time in the sense defined above, we shall then obviously have

$$\begin{cases} \displaystyle\int_{\mathfrak{S}} f(t_0) \, dP = K, \\[2mm] \displaystyle\int_{\mathfrak{S}} f(t_0 + t) f(t_0) \, dP = \varphi(t), \end{cases} \tag{6}$$

where K is a constant and $\varphi(t)$ is a function of t only[2]. Conversely, we may say that the relations (6) imply *invariance of the measure P for time transla-tions, as far as moments of the first and second orders are concerned.*

If $\varphi(t)$ is continuous at the particular point $t = 0$, it can be shown[1] that $\varphi(t)$ is uniformly continuous for all real t and representable in the form (2), with a real, bounded and never decreasing $\Phi(x)$.

Whenever a measure $P(S)$ satisfies the relations (6) with a continuous $\varphi(t)$, we shall say, in accordance with a definition introduced by Khintchine[3], that $P(S)$ is attached to a stationary and continuous distribution in function space $\mathfrak{S}$.

4. *A net N on the real axis* will be defined as a sequence of divisions D_1, $D_2, \ldots$ of that axis, such that D_n consists of a finite number k_n of points

$$\alpha_1^{(n)} < \alpha_2^{(n)} < \cdots < \alpha_{k_n}^{(n)}.$$

Denoting by δ_n the maximum length of the intervals $\alpha_{v+1}^{(n)} - \alpha_v^{(n)}$ occurring in D_n, we shall require that

$$\alpha_1^{(n)} \to -\infty, \qquad \alpha_{k_n}^{(n)} \to +\infty, \qquad \delta_n \to 0,$$

as $n \to \infty$.

If, for a given function $F(x)$ and a given real t, the limit

$$\Lambda = \lim_{n \to \infty} \sum_{v=1}^{k_n - 1} e^{it\alpha_v^{(n)}} [F(\alpha_{v+1}^{(n)}) - F(\alpha_v^{(n)})]$$

[2] In order to emphasize the analogy with ordinary harmonic analysis as reviewed in section 1, we use systematically the same letters to denote corresponding functions. It is, however, obvious that the integrals occurring e.g. in (1) and (6) are fundamentally different (1) defines an average over the time axis for an individual function $f(t)$, while (6) deals with averages over function space $\mathfrak{S}$.

[3] Khintchine [3]. Cf. also Cramér [2].

exists, we shall call this limit a *generalized* Fourier-Stieltjes *integral associated with the net N*, and we shall write

$$\Lambda = \int_{-\infty}^{\infty} e^{itx} \, d^{(N)}F(x).$$

We shall finally say that a net N is *efficient with respect to the measure* $P(S)$, if I) every $\alpha_v^{(n)}$ is a point of continuity of $\Phi(x)$, and II) the three series with positive terms

$$\sum_n [\Phi(\alpha_1^{(n)}) - \Phi(-\infty)], \qquad \sum_n [\Phi(\infty) - \Phi(\alpha_{k_n}^{(n)})], \qquad \sum_n \delta_n^2$$

are all convergent.

5. After these preliminaries, we can now state the following theorem, where the expression »almost all functions $f(t)$« will have to be interpreted with respect to P-measure, while the expression »almost all real t« refers to ordinary Lebesgue measure.

Theorem. *Let $P(S)$ be a measure attached to a stationary and continuous distribution in $\mathfrak{S}$. It is then possible to find a transformation*

$$T[f] = F,$$

by which to every element $f(t)$ of $\mathfrak{S}$ corresponds another uniquely defined element $F(t)$ of the same space, so that the following properties hold:

I) *If N is any net efficient with respect to $P(S)$, then almost all functions $f(t)$ satisfy the relation*

$$f(t) = \int_{-\infty}^{\infty} e^{itx} \, d^{(N)}F(x) \tag{7}$$

for almost all real t.

II) *For any real numbers $a < b$ which are continuity points of $\Phi(x)$, we have*

$$\Phi(b) - \Phi(a) = \int_{\mathfrak{S}} |F(b) - F(a)|^2 \, dP. \tag{8}$$

The analogy between the relations (7) and (8) of the theorem and the corresponding relations (4), (4a) and (5) of ordinary harmonic analysis is obvious. A proposition recently given by Slutsky [6] is contained in the particular case of the theorem when $\varphi(t)$ is an almost periodic functions.—Owing to restrictions of space, only the main lines of the proof can be indicated here. Denoting by x_0 a fixed continuity point of $\Phi(x)$, the expression

$$F_{m,n}[x, f(t)] = -\frac{1}{2\pi i} \sum_{v=-mn}^{mn} \frac{e^{-(vix)/n} - e^{-(vix_0)/n}}{v} f\left(\frac{v}{n}\right), \quad (v \neq 0), \tag{9}$$

is, for any fixed integers m, n and any fixed real x, a $\mathbb{C}$-measurable function in $\mathfrak{S}$. Obviously $|F_{m,n}|^2$ has a finite integral with respect to P over $\mathfrak{S}$. The structural analogy between (9) and (3) is evident.

Now, let E denote the enumerable set of discontinuity points of $\Phi(x)$. For any x which does not belong to E, it can be shown that the repeated limit in the mean over $\mathfrak{S}$

$$F[x, f(t)] = \underset{m \to \infty}{\text{l.i.m.}} \left\{ \underset{n \to \infty}{\text{l.i.m.}} F_{m,n}[x, f(t)] \right\}$$

exists. In other words, there are functionals $F_m[x, f(t)]$ and $F[x, f(t)]$ such that

$$\lim_{n \to \infty} \int_{\mathfrak{S}} |F_{m,n} - F_m|^2 \, dP = 0,$$

$$\lim_{m \to \infty} \int_{\mathfrak{S}} |F_m - F|^2 \, dP = 0.$$

For any x which does not belong to E, and for almost all $f(t)$, the functional $F[x, f(t)]$ is determined as the limit of a convergent sub-sequence of $F_{m,n}$. In all other cases, we put $F = 0$, so that F is completely defined for all real x and all functions $f(t)$ in $\mathfrak{S}$. The integral of $|F|^2$ with respect to P over $\mathfrak{S}$ is finite for every fixed x.

We now define the transformation $T[f]$ by putting for any $f(t)$ in $\mathfrak{S}$

$$T[f] = F(x) = F[x, f(t)].$$

It can then be shown without difficulty that this function $F(x)$ satisfies part II of the theorem.

On the other hand, if we consider the expression

$$h_n[t, f] = \sum_v e^{it\alpha_v^{(n)}} [F(\alpha_{v+1}^{(n)}) - F(\alpha_v^{(n)})],$$

where the $\alpha_v^{(n)}$ belong to an efficient net N, we obtain for any fixed n (integer) and t (real)

$$\int_{\mathfrak{S}} |f - h_n|^2 \, dP < t^2 \delta_n^2 + \Phi(\infty) - \Phi(\alpha_{k_n}^{(n)}) + \Phi(\alpha_1^{(n)}) - \Phi(-\infty).$$

Hence, considering the product space $\mathfrak{T}\mathfrak{S}$, where $\mathfrak{T}$ denotes the real t-axis, it follows from the properties of an efficient net that almost all functions $f(t)$ satisfy the relation

$$f(t) = \lim_{n \to \infty} h_n[t, f]$$

for almost all real t. According to our definition of a generalized Fourier-Stieltjes integral, this proves the remaining part of the theorem.

References

1) Bochner, S., Monotone Funktionen, Stieltjessche Integrale und harmonische Analyse. Math. Ann. 108 (1933), p. 378.
2) Cramér, H., On the theory of stationary random processes. Ann. of Math., 41 (1940), p. 215.
3) Khintchine, A., Korrelationstheorie der stationären stochastischen Prozesse. Math. Ann. 109 (1934), p. 604.
4) Kolmogoroff, A., Grundbegriffe der Wahrscheinlichkeitsrechnung. Berlin 1933.
5) Saks, S., Theory of the integral, 2:d ed., Warszawa-Lwów 1937.
6) Slutsky, E., Sur les fonctions aléatoires presque périodiques etc. Actualités scientifiques et industr., 738 (1938), p. 33.
7) Wiener, N., Generalized harmonic analysis. Acta Math. 55 (1930), p. 117.
8) ———, The Fourier integral. Cambridge 1933.
9) ———, The homogeneous chaos. American J. of Math., 60 (1938), p. 897.

Introduction to
Gnedenko (1943) On the Limiting Distribution of the Maximum Term in a Random Series

Richard L. Smith
University of Surrey and
University of North Carolina at Chapel Hill

1. General Comments

Gnedenko's paper was the first mathematically rigorous treatment of the fundamental limit theorems of extreme value theory. In its influence on the probabilistic theory of extremes, the paper set the agenda for the next 30 years. Its influence on statistical methodology was less direct, although many of the more recent statistical developments have exploited the probabilistic theory that Gnedenko's paper spawned.

The main contributions of the paper were the complete characterization of the "three types" of limiting distributions. and the development of necessary and sufficient conditions for a distribution function to be in the domain of attraction of a given extreme value limit. In two of the three cases, Gnedenko's solution was complete. In the third case, however, Gnedenko's necessary and sufficient condition involved the existence of a function that was not explicitly defined, so this left open the task of determining a directly checkable condition. The latter problem gave rise to a series of papers, culminating in de Haan's work in the early 1970s.

Apart from these most famous contributions of the paper, it also established weak laws of large numbers for extreme values and gave some results on the most important class of distributions for which the three-types theory does not apply, namely, discrete distributions. Both of these led to some substantial developments later on.

Despite the extensive nature of Gnedenko's contribution, he was not the first to discuss the three-types theory. Fisher and Tippett (1928) formulated the three types and claimed (with some supporting arguments but no rigorous proofs) that these were the only extreme value limit distributions, whereas von Mises (1936) gave sufficient but not necessary conditions for their domains of attraction. Both of these were acknowledged by Gnedenko, although in a footnote he claimed to be unaware of the Fisher–Tippett work when he originally derived the results.

2. Statement of the Problem and Earlier Work

Suppose $X_1, X_2, \ldots$ denote independent random variables with a common distribution function F; let $M_n = \max(X_1, \ldots, X_n)$. Throughout this introduction, I use $x_F = \sup\{x : F(x) < 1\}$ to denote the upper bound of the distribution, which may be finite or infinite. The distribution of M_n is given by

$$\Pr\{M_n \leq x\} = F^n(x).$$

The main problem of interest is to determine conditions under which there exist constants $a_n > 0$, b_n and a nondegenerate distribution function Φ such that

$$\lim_{n \to \infty} \Pr\left\{\frac{M_n - b_n}{a_n} \leq x\right\} = \lim_{n \to \infty} F^n(a_n x + b_n) = \Phi(x). \qquad (1)$$

When (1) holds we say that Φ is an *extreme value limiting distribution function* and F is in the *domain of attraction* of Φ. An entirely analogous definition exists for sample minima instead of sample maxima, but we do not consider this separately as it is a straightforward transformation of the maximum case.

The most striking result of the whole theory is that there are only three possible types for Φ. Two distribution functions F and G are said to be *of the same type* if $F(x) = G(Ax + B)$ for constants $A > 0$ and B. It is obvious that if one Φ can arise as a limit in (1), so can any other distribution function of the same type as Φ. Conversely, if two different distributions Φ and Φ^* arise (under different a_n, b_n sequences) in (1), then Φ and Φ^* must be of the same type. Hence, it is natural to talk of limiting types rather than limiting distribution functions.

The main result is that there are only three possible types of Φ. These are

$$\Phi_\alpha(x) = \begin{cases} 0, & x \leq 0, \\ \exp(-x^{-\alpha}), & x > 0, \end{cases}$$

$$\Psi_\alpha(x) = \begin{cases} \exp\{-(-x)^\alpha\}, & x < 0, \\ 1, & x \geq 0, \end{cases}$$

$$\Lambda(x) = \exp(-e^{-x}), \qquad -\infty < x < \infty.$$

In the case of Φ_α and Ψ_α, $0 < \alpha < \infty$.

This result was essentially stated by Fisher and Tippett (1928) but without rigorous proof. This seems to have been one of many instances in which Fisher's marvellous intuition led him to the right result without him being able to provide a proof acceptable to mathematicians. The elegance of Gnedenko's eventual proof shows that this was by no means a trivial task.

Independently of Fisher and Tippett, Fréchet (1927) considered the corresponding problem with $b_n \equiv 0$. In this case, the only limit is Φ_α, though Fréchet experimented with a number of different formulations of the problem before arriving at this result. Although Fréchet's paper tends to be overlooked in the light of the more general results of Fisher–Tippett and Gnedenko, it deserves to be remembered as a significant contribution in its own right, one which foreshadowed Gnedenko's in a number of respects.

The other major precedent of Gnedenko was von Mises (1936). Assuming the Fisher–Tippett result, von Mises established sufficient conditions for F to be in a domain of attraction. These conditions are not necessary, but are good enough for nearly all practical applications. In particular, von Mises' sufficient condition for F to be in the domain of attraction of Λ is that

$$\lim_{x \to \infty} \frac{d}{dx} \left\{ \frac{1 - F(x)}{F'(x)} \right\} = 0, \tag{2}$$

where $F' = dF/dx$ and it is assumed that $x_F = +\infty$.

An interesting side issue is why the formulation (1) was adopted at all, with its affine normalization of M_n. Fisher and Tippett did not explain this, whereas Gnedenko offered only the analogy with stable distribution theory for sums, which seems to be begging the question. Perhaps the real explanation is that no one came up with an alternative formulation that led to interesting results. The same explanation is still valid today.

3. Summary of the Paper

Section 1 is something of a preliminary to the main part of the paper. Defining M_n as above, Gnedenko is interested in results of the form $M_n - A_n \to 0$ or $M_n/B_n \to 1$, where A_n, B_n are constants and the convergence is in probability. Gnedenko calls the first type of result a *law of large numbers*, while in the second case the sequence of maxima is called *relatively stable*. Theorems 1 and 2 give necessary and sufficient conditions for the law of large numbers and relative stability, respectively. It is obvious that the one result is directly obtainable from the other through an exponential or logarithmic transformation, but Gnedenko does not remark on this, preferring to give both proofs in full. One wonders whether any modern author would take this luxury, or any modern journal editor permit it!

Section 2 is concerned with the main result about the three limiting types of the extreme value distribution, Theorem 3 establishing the anticipated result that the three limiting types of Fisher and Tippett are the only possibilities. In other words, assuming (1), he wants to show that Φ is necessarily of the same type as one of Φ_α, Ψ_α, Λ. The first step is to show that, given (1), there must exist sequences $\alpha_k > 0$ and β_k such that

$$\Phi^k(\alpha_k x + \beta_k) = \Phi(x), \qquad \text{any } k \geq 1.$$

From here the proof splits into three separate cases according to the constants α_k, $k \geq 1$. Case 1 is when $\alpha_k < 1$ for at least one k. The first thing that Gnedenko shows in this case is that $x_\Phi = \beta_k/(1 - \alpha_k)$ and so, in particular, x_Φ is finite. The next step is to show that $\alpha_k < 1$ for all k. Suppose this is false, i.e., there exists r such that $\alpha_r \geq 1$. Splitting this further into the cases $\alpha_r = 1$, $\alpha_r > 1$, he shows that in either case, Φ must be degenerate (Gnedenko uses the word improper or unitary). Since this is a contradiction, we deduce that $\alpha_k < 1$ for all k and hence, since $x_\Phi = \beta_k/(1 - \alpha_k)$ for one k, it must be true for all k. Defining $\bar{\Phi}(z) = \Phi(z + x_\Phi)$, Gnedenko writes down a functional equation for $\bar{\Phi}$ and deduces that Φ is of the type Ψ_α for some $\alpha > 0$.

Now suppose $\alpha_k > 1$ for some k. By essentially the same argument (no tedious repetition this time!), it follows that $\alpha_k > 1$ for all k and Φ is of type Φ_α. This leaves only the case $\alpha_k = 1$ for all k, and in this case, Gnedenko is able to establish quickly that Φ satisfies a functional equation whose only solution is of the same type as Λ. These are the main steps leading to the desired conclusion.

Section 3 is concerned with some technical lemmas connected with convergence in distribution. Lemma 1 asserts that if F_n is a sequence of distribution functions and $\{a_n, b_n\}$ and $\{\alpha_n, \beta_n\}$, two sets of normalizing constants such that

$$F_n(a_n x + b_n) \to \Phi(x), \quad F_n(\alpha_n x + \beta_n) \to \Phi(x) \tag{3}$$

both hold with nondegenerate Φ, then

$$\frac{\alpha_n}{a_n} \to 1, \qquad \frac{(b_n - \beta_n)}{a_n} \to 0. \tag{4}$$

Lemma 2 establishes the converse: If (4) holds and one of the convergences in (3), then so does the other. Lemma 3 applies this to a specific case in which F_n is either F^n or F^{n+1} to deduce that if $F^n(a_n x + b_n)$ converges, we must have $a_n/a_{n+1} \to 1$ and $(b_n - b_{n+1})/a_n \to 0$. Lemma 4 asserts that $F^n(a_n x + b_n) \to \Phi(x)$ if and only if $n\{1 - F(a_n x + b_n)\} \to -\log \Phi(x)$. It is not clear why he included a proof of this obvious statement.

The paper turns in Sec. 4 to the domain of attraction of Φ_α for given $\alpha > 0$. The main result (Theorem 4) is elegant and precise: In modern terminology, it is necessary and sufficient that the function $1 - F(x)$ be regularly varying at infinity, with index $-\alpha$. In 1943, the concept of regular variation was not nearly as widely known as it is today, and Gnedenko's results, on both this and the problem of convergence to stable and infinitely divisible laws (see Sec. 6), were among the first major applications of it. The proof splits into two halves. First, assume $1 - F$ is regularly varying. Defining a_n such that $1 - F(a_n - 0) \leq 1/n \leq 1 - F(a_n + 0)$, he deduces that $n\{1 - F(a_n x)\} \sim \{1 -$

$F(a_n x)\}/\{1 - F(a_n)\}$, which converges to $x^{-\alpha}$ by regular variation. The sufficiency part of the result now follows from Lemma 4. The necessity (i.e., convergence of F^n implies regular variation) would be just as easy if n could be regarded as a continuous variable, but since it is not, there is some slightly delicate approximation necessary (see Eqs. 44 and 45) to obtain the desired result. The argument given here is repeated, in various guises, several times during the remainder of the paper.

Section 5 and Theorem 5 prove rather similar results about the domain of attraction of Ψ_α. Then in Sec. 6, Gnedenko comes to the hardest case, the domain of attraction of Λ. The main result is Theorem 7, stating that a necessary and sufficient condition for F to be in the domain of attraction of Λ is that there exists a function $A(z)$ such that $A(z) \to 0$ as $z \to x_F \le \infty$ and

$$\lim_{z \uparrow x_F} \frac{1 - F[z + zA(z)x]}{1 - F(z)} = e^{-x}. \tag{3}$$

This is followed by two corollaries, the second of which establishes that a wide class of discrete random variables (the Poisson is cited as an explicit example) do not belong to any domain of attraction.

In the remaining part of the paper, Theorem 8 seems of little independent interest, but the final few lines of the paper rederive von Mises' sufficient condition (2) and point out that it also holds when $x_F < \infty$, provided the limit $x \to \infty$ is replaced by $x \uparrow x_F$.

4. Subsequent Developments

As Gnedenko made clear in his preliminary discussion, he did not regard the result for Λ as definitive: Although condition (3) is necessary and sufficient, no prescription is given for constructing the function $A(z)$. This part of the work was completed, in stages, by Mejzler (1949), Marcus and Pinsky (1969), and de Haan (1970, 1971).

In the notation of de Haan, Mejzler's contribution was to show that if the inverse function U is defined by

$$U(x) = \inf\{y : 1 - F(y) \le x\},$$

then a necessary and sufficient condition for F to be in the domain of attraction of Λ is that

$$\lim_{t \downarrow 0} \frac{U(tx) - U(t)}{U(ty) - U(t)} = \frac{\log x}{\log y} \tag{4}$$

for all $x > 0$, $y > 0$. This improves on Gnedenko in that it is an explicit condition, but one would still like the condition to be defined directly in terms of F rather than U.

Marcus and Pinsky, apparently unaware of Mejzler, rederived (4) and gave a representation for U, reminiscent of Karamata's representation for regularly varying functions.

de Haan (1970, 1971) completed this line of development by developing a detailed theory for functions satisfying (4). Among his results, de Haan showed that Gnedenko's function $A(z)$ may, in general, be given by

$$A(z) = \frac{\int_z^{x_F} \{1 - F(t)\}\, dt}{z\{1 - F(z)\}},$$

and that $1 - F$ satisfies the Karamata-like representation

$$1 - F(x) = c(x) \exp\left\{-\int_{x_1}^x \frac{a(t)}{f(t)}\, dt\right\}, \tag{5}$$

where $c(x) > 0$, $c(x) \to c_1 > 0$ as $x \uparrow x_F$, $a(x) \to 1$ as $x \uparrow x_F$, $f(x) > 0$ for all x and $f'(x) \to 0$ as $x \uparrow x_F$. Balkema and de Haan (1972) showed that no generality is lost by taking $a(t) \equiv 1$ in (5).

Although this ends the sequence of developments that may be said to stem directly from Gnedenko's paper, some subsequent ones are worth noting. de Haan (1976) showed that, by formulating the whole problem in terms of U instead of F, one can obtain a much shorter proof of the three-types theorem. This is the proof now preferred for textbook treatment, e.g., Leadbetter, Lindgren, and Rootzén (1983). Another short proof, based on quite different principles, was given by Weissman (1975). Finally, a number of authors have extended the notion of domain of attraction by adding to (1) the condition that either the first or second derivatives of $F^n(a_n x + b_n)$ should converge to the corresponding derivatives of Φ. One consequence of this is a renewed appreciation of the importance of von Mises' conditions. See de Haan and Resnick (1982), Sweeting (1985), and Pickands (1986).

Another line of development is the characterization of convergence in terms of moments. Clough and Kotz (1965) suggested using the mean and standard deviation of M_n as scaling constants in place of b_n and a_n. This is valid, provided the rescaled mean and variance themselves converge to the mean and variance of the limiting distribution. Conditions for that were given by Pickands (1968).

Two other problems, mentioned in the summary above, have also led to interesting subsequent developments. Gnedenko's results about the weak law of large numbers pose the obvious corresponding questions about the strong law of large numbers. The most substantial contributions to this question were due to Geffroy (1958) and Barndorff–Nielsen (1961, 1963). The study of discrete random variables also led to subsequent work, in particular, that of Anderson (1970) who gave both laws of large numbers and, in lieu of limiting distributions, lim-sup and lim-inf results for renormalized maxima. However, there are numerous unexpected results in this area; for an example, see Kimber (1983).

5. The Broader Influence of Gnedenko's Paper

Gnedenko's paper influenced subsequent developments in probability theory, analysis, and statistics. The most obvious probabilistic development is to relax the requirement that $X_1, X_2, \ldots$ be i.i.d. The first case considered was that in which the random variables are independent but not identically distributed. This problem was solved, in part, by Juncosa (1949), and much more completely by Mejzler in a series of papers in the 1940s and 1950s. Galambos (1987) has a very detailed account of these results. Interestingly, there are still three classes of solution, but they are much broader than Gnedenko's parametric types, essentially involving concavity conditions on log Φ.

The literature on extremes in dependent sequences is far too large for us to summarise here. Early references were Watson (1954) on m-dependent sequences, Berman (1962) on exchangeable sequences, Berman (1964) on stationary Gaussian sequences, and Loynes (1965) on mixing sequences. A prevailing theme through much of this literature is that the three limiting types of Gnedenko are still valid in a variety of dependent cases. For instance, if $\{X_n\}$ is a stationary sequence with marginal distribution F such that $F^n(a_n x + b_n) \to \Phi(x)$ and if certain mixing conditions hold, then it is true under quite general conditions that $\Pr\{(M_n - b_n)/a_n) \le x\} \to \Phi^\theta(x)$, where $\theta \in [0, 1]$ is a parameter that Leadbetter (1983) termed the *extremal index*. Except when $\theta = 0$, it is immediate that Φ^θ is of the same type as Φ and hence that the limiting distribution in the dependent case is of the same type as the independent case. In other cases such as exchangeable sequences, the limiting distributions are different from the Gnedenko types but nevertheless related to them (by a mixing operation, in the case of exchangeability). Thus, the Gnedenko paper may be said to be an important precursor of the modern theory of extremes in stochastic processes. For recent accounts, see the texts by Leadbetter, Lindgren, and Rootzén (1983) and Galambos (1987), and the review article by Leadbetter and Rootzén (1988).

Two other developments suggested by Gnedenko's paper are toward extremes of vector rather than scalar random variables (multivariate extreme value theory), and the joint distribution of several high-order statistics from a univariate sample. For an indication of how far these topics have developed, see the monographs by Resnick (1987) and Reiss (1989), respectively.

It would be too much to suggest that all these developments have stemmed directly from Gnedenko's paper, but to the extent that his paper was the first to establish a rigorous asymptotic theory for extremes, it can be identified as the key foundational paper that got the subject going.

Gnedenko's paper also had an influence on analysis. Although his domain of attraction conditions were not the first probabilistic applications of regular variation, they were important and served to draw attention to this area of interaction between probabilists and analysts. Feller (1971) gave an alternative account of the application of regularly varying functions to extreme value theory. The theory of functions satisfying (4), developed at length by de Haan, led to a theory of *extended regular variation* that, in turn, found other probabilistic and analytic applications. The monograph by Bingham, Goldie, and Teugels (1987) is a rich source of information about this.

On the other hand, the *statistical* influence of Gnedenko's paper was less direct. The statistical application of the three types was pioneered by Gumbel (1958), but Gumbel was not a mathematician and seems not to have appreciated the importance of Gnedenko's contribution as compared with those of Fréchet and Fisher–Tippett. On the other hand, recent statistical work on extremes has heavily exploited the asymptotic distribution of excedances over a high threshold, and it is hard to believe that this would have been possible without the rigorous foundations laid by Gnedenko. The papers by Pickands (1975) and Weissman (1978) were among those that started this line of research, and Davison and Smith (1990) is just one of many papers to represent recent developments.

6. The Author

Boris Vladimirovich Gnedenko was born January 1, 1912 in Simbirsk, now (1991) called Ul'yanov. He was a student at the University of Saratov from 1927–30 and then taught at the Ivanova Textile Institute. From 1934–37, he was a graduate student under Khinchin in Moscow, and from 1938, was a lecturer at Moscow University. He received his doctorate in 1942. During the period 1937–41, he did what is probably his best known work, a complete characterization of infinitely divisible laws and their domains of attraction. He also worked in this period on stable laws, and his work in these areas eventually led to the famous monograph by Gnedenko and Kolmogorov, *Limit Distributions for Sums of Independent Random Variables*, the first edition of which was published in 1949. The paper on extreme value theory was, then, part of a general interest on limit laws and their domains of attraction, and although the extreme values problem is simpler than the corresponding problem involving sums of independent random variables, there is some overlap in the techniques employed, particularly the use of regular variation.

Why did the paper appear in French in an American journal? Apparently, the paper was one of several papers by Soviet authors, invited by the editors of *Annals of Mathematics* as a gesture of friendship toward the United States' new wartime ally. Gnedenko knew no English but was given permission to write in French.

In 1945, Gnedenko moved to the University of Lvov in the Ukraine. In 1945, he was elected Corresponding Member, and in 1948, Academician of the Ukrainian Academy of Sciences. From 1950, he was at the University of Kiev, and from 1960, back at Moscow University. Apart from his work on limit theorems, he made extensive contributions to mathematical statistics, reliability, and queueing theory and wrote an intermediate textbook on probability that went through at least five editions and was translated into English, German, and Polish. He is a Member of the International Statistical Institute, a Fellow of the Institute of Mathematical Statistics, and an Honorary Fellow of the Royal Statistical Society.

Acknowledgments. I thank Norman Johnson, Sam Kotz, and Ishay Weissman for comments about the first draft. The translation of Gnedenko's paper into English was completed by Norman Johnson. The biographical information in Sec. 6 is based partly on the 50th, 60th, and 70th birthday tributes published in *Uspekhi Math Nauk* (in English, *Russian Mathematical Surveys*).

References

Anderson, C.W. (1970). Extreme value theory for a class of discrete distributions with applications to some stochastic processes, *J. Appl. Prob.*, **7**, 99–113.

Balkema, A.A., and de Haan, L. (1972). On R. von Mises' condition for the domain of attraction of $\exp(-e^{-x})$, *Ann. Math. Statist.*, **43**, 1352–1354.

Barndorff-Nielsen, O. (1961). On the rate of growth of the partial maxima of a sequence of independent identically distributed random variables, *Math. Scand.*, **9**, 383–394.

Barndorff-Nielsen, O. (1963). On the limit behaviour of extreme order statistics, *Ann. Math. Statist.*, **34**, 992–1002.

Berman, S.M. (1962). Limiting distribution of the maximum term in a sequence of dependent random variables, *Ann. Math. Statist.*, **33**, 894–908.

Berman, S.M. (1964). Limit theorems for the maximum term in stationary sequences, *Ann. Math. Statist.*, **35**, 502–516.

Bingham, N.H., Goldie, C.M., and Teugels, J.L. (1987). *Regular Variation*. Cambridge University Press, New York.

Clough, D.J., and Kotz, S. (1965). Extreme-value distributions with a special queueing model application, *CORS J.*, **3**, 96–109.

Davison, A.C., and Smith, R.L. (1989). Models for excedances over high thresholds (with discussion), *J. Roy. Statist. Soc., Ser. B*, **52**, 393–442.

de Haan, L. (1970). *On Regular Variation and Its Application to the Weak Convergence of Sample Extremes*. Mathematical Centre Tracts No. 32, Amsterdam.

de Haan, L. (1971). A form of regular variation and its application to the domain of attraction of the double exponential, *Z. Wahrsch. v. Geb.*, **17**, 241–258.

de Haan, L. (1976). Sample extremes: An elementary introduction, *Statist. Neerlandica*, **30**, 161–172.

de Haan, L., and Resnick, S. (1982). Local limit theorems for sample extremes, *Ann. Prob.*, **10**, 396–413.

Feller, W. (1971). *An Introduction to Probability Theory and Its Applications*, Vol. II, 2nd ed. Wiley, New York.

Fisher, R.A., and Tippett, L.H.C. (1928). Limiting forms of the frequency distributions of the largest or smallest member of a sample, *Proc. Cambridge Philos. Soc.*, **24**, 180–190.

Fréchet, M. (1927). Sur la loi de probabilité de l'écart maximum, *Ann. Soc. Polonaise Math.* (Cracow), **6**, 93.

Galambos, J. (1987). *The Asymptotic Theory of Extreme Order Statistics*, 2nd ed. Krieger, Melbourne, Fla. (1st ed. published in 1978 by Wiley, New York).

Geffroy, J. (1958, 1959). Contributions à la théorie des valeurs extrêmes, *Publ. Inst. Statist. Univ. Paris*, **7/8**, 37–185.

Gumbel, E.J. (1958). *Statistics of Extremes*. Columbia University Press, New York.

Juncosa, M.L. (1949). On the distribution of the minimum in a sequence of mutually independent random variables, *Duke Math. J.*, **16**, 609–618.

Kimber, A.C. (1983). A note on Poisson maxima, *Z. Wahrsch. v. Geb.*, **63**, 551–552.

Leadbetter, M.R (1983). Extremes and local dependence in stationary sequences, *Z. Wahrsch. v. Geb.*, **65**, 291–306.

Leadbetter, M.R., Lindgren, G., and Rootzén, H. (1983). *Extremes and Related Properties of Random Sequences and Series*. Springer-Verlag, New York.

Leadbetter, M.R., and Rootzén, H. (1988). Extremal theory for stochastic processes, *Ann. Prob.*, **16**, 431–478.

Loynes, R.M. (1965). Extreme values in uniformly mixing stationary stochastic processes, *Ann. Math. Statist.*, **36**, 993–999.

Marcus, M.B., and Pinsky, M. (1969). On the domain of attraction of $\exp(-e^{-x})$, *J. Math. Anal. Appl.*, **28**, 440–449.

Mejzler, D.G. (1949). On a theorem of B.V. Gnedenko, *Sb. Trudov Inst. Mat. Akad. Nauk. Ukrain. SSR*, **12**, 31–35 (in Russian).

Pickands, J. (1968). Moment convergence of sample extremes, *Ann. Math. Statist.*, **39**, 881–889.

Pickands, J. (1975). Statistical inference using extreme order statistics, *Ann. Statist.*, **3**, 119–131.

Pickands, J. (1986). The continuous and differentiable domains of attraction of the extreme value distributions, *Ann. Prob.*, **14**, 996–1004.

Reiss, R.D. (1989). *Approximate Distributions of Order Statistics*. Springer-Verlag, New York.

Resnick, S. (1987). *Extreme Values, Point Processes and Regular Variation*. Springer-Verlag, New York.

Sweeting, T.J. (1985). On domains of uniform local attraction in extreme value theory, *Ann. Prob.*, **13**, 196–205.

von Mises, R. (1936). La distribution de la plus grande de *n* valeurs, reprinted in *Selected Papers II*. American Mathematical Society, Providence, R.I. (1954), pp. 271–294.

Watson, G.S. (1954). Extreme values in samples from *m*-dependent stationary stochastic processes, *Ann. Math. Statist.*, **25**, 798–800.

Weissman, I. (1975). On location and scale functions of a class of limiting processes with application to extreme value theory, *Ann. Prob.*, **3**, 178–181.

Weissman, I. (1978). Estimation of parameters and large quantiles based on the *k* largest observations, *J. Amer. Statist. Assoc.*, **73**, 812–815.

On the Limiting Distribution of the Maximum Term in a Random Series

B.V. Gnedenko
(Translated from the French by Norman L. Johnson)

Let us consider a sequence

$$x_1, x_2, \ldots, x_n, \ldots$$

of mutually independent random variables with a common distribution law $F(x)$. We construct another sequence of variables

$$\xi_1, \xi_2, \ldots, \xi_n, \ldots,$$

putting $\xi_n = \max(x_1, x_2, \ldots, x_n)$.

It is easy to see that the distribution function of ξ_n is

$$F_n(x) = P\{\xi_n < x\} = F^n(x).$$

The study of the function $F_n(x)$ for large values of n is of great interest. Many studies have been devoted to this problem. In particular, M. Fréchet [1] has discovered those laws which can be limits of $F_n(a_n x)$ for suitable choice of the positive constants a_n.

This class of limiting laws consists of laws of the following types[1]:

$$\Phi_\alpha(x) = \begin{cases} 0 & \text{for } x \le 0 \\ e^{-x^{-\alpha}} & \text{for } x > 0 \end{cases}$$

and

$$\Psi_\alpha(x) = \begin{cases} e^{-(-x)^\alpha} & \text{for } x \le 0 \\ 1 & \text{for } x > 0 \end{cases}$$

where α denotes a positive constant.

[1] Following A. Khintchine and P. Lévy we will call the type of the law $\Phi(x)$ to be the set of laws $\Phi(ax + b)$ for all possible choices of the real constants $a > 0$ and b.

In 1928, R.A. Fisher and L.H.C. Tippett [2] showed that limiting laws of $F_n(a_n x + b_n)$, where $a_n > 0$ and b_n are suitably chosen real numbers is composed of the types $\Phi_\alpha(x)$, $\Psi_\alpha(x)$ and the law

$$\Lambda(x) = e^{-e^{-x}}.$$

R. Misès [3], who began a systematic study of limit laws for the maximum term about 1936, has found many sufficient conditions for the convergence of the laws $F_n(a_n x + b_n)$ to each of these types, for a given fixed choice of values of the constants $a_n > 0$ and b_n; at the conclusion of the final paragraph of the present paper we will formulate the sufficient condition for convergence to the law $\Lambda(x)$ found by R. Misès.

However, these papers do not solve the most fundamental problem concerning the limiting distribution of the maximum of a random series, namely, the problem of determining the domain of attraction of each proper[2] limit law $\Phi(x)$. That is, we want to determine the set of all distribution functions $F(x)$ such that, for a suitable choice of constants $a_n > 0$ and b_n, we have

$$\lim_{n \to \infty} F_n(a_n x_n + b_n) = \Phi(x).$$

It is interesting to note that not only the problems posed on the distributions of maximum terms, but also as we shall see, the results obtained are markedly analogous with the corresponding problems and results in the theory of stable laws for the sums of independent random variables. (See, for example, Chapter V of [4] and [5].)

In the present paper it will be shown that the class of limit laws for maximum terms contains no other types beyond those already mentioned. We will also give necessary and sufficient conditions for the domains of attraction of each of the possible limit types. However, although the results found for the laws $\Phi_\alpha(x)$ and $\Psi_\alpha(x)$ can be regarded as definitive, this is not so for the conditions regarding the law $\Lambda(x)$; in our opinion, these conditions have not yet attained a well-established form which is easily applied. In §1 necessary and sufficient conditions for the law of large numbers and for relative stability of maxima are given. It is worth noting that lemmas 1 and 2 appear to be of interest on their own account and can be used in research on other limit problems.

It is easy to see that the results of the paper can be extended to the distribution of the minimum term of a random series. It is only necessary to note that if

$$\eta_n = \min(x_1 x_2, \ldots, x_n)$$

---

[2] A distribution function is called *improper* or *unitary* if it belongs to the type

$$\varepsilon(x) = \begin{cases} 0 & \text{for } x < 0 \\ 1 & \text{for } x > 0 \end{cases}$$

then

$$-\eta_n = \max(-x_1, -x_2, \ldots, -x_n).$$

1. The Law of Large Numbers

We will say that the sequence of maxima

$$\xi_1, \xi_2, \ldots, \xi_n, \ldots \tag{1}$$

of a series of mutually independent random variables

$$x_1, x_2, \ldots, x_n, \ldots \tag{2}$$

is subject to the law of large numbers if there exist constants A_n such that

$$P\{|\xi_n - A_n| < \varepsilon\} \to 1 \tag{3}$$

as $n \to \infty$ for all fixed $\varepsilon > 0$.

The sequence of maxima (1) will be called relatively stable if, for suitable choice of positive constants B_n,

$$\lim_{n \to \infty} P\left\{ \left| \frac{\xi_n}{B_n} - 1 \right| < \varepsilon \right\} = 1 \tag{4}$$

is true for all $\varepsilon > 0$.

If the distribution function $F(x)$ of the random variables in the series (2) is such that there exists a number x_0 for which

$$F(x_0) = 1 \qquad \text{and} \qquad F(x_0 - \varepsilon) < 1 \tag{5}$$

for all $\varepsilon > 0$, then the sequence (1) is subject to the law of large numbers.

Indeed, if conditions (5) are satisfied, we have

$$P\{|\xi_n - x_0| < \varepsilon\} = 1 - F_n(x_0 - \varepsilon) = 1 - F^n(x_0 - \varepsilon).$$

Since for all $\varepsilon > 0$, we have

$$\lim_{n \to \infty} F^n(x_0 - \varepsilon) = 0,$$

it follows that

$$\lim_{n \to \infty} P\{|\xi_n - x_0| < \varepsilon\} = 1.$$

Analogously if conditions (5) hold, and $x_0 > 0$, then the sequence (1) is relatively stable.

In this case we have

$$P\left\{ \left| \frac{\xi_n}{x_0} - 1 \right| < \varepsilon \right\} = 1 - F^n(x_0(1 - \varepsilon))$$

and since from (5), as $n \to \infty$

$$F^n(x_0(1 - \varepsilon)) \to 0$$

we obtain, for $n \to \infty$

$$P\left\{ \left| \frac{\xi_n}{x_0} - 1 \right| < \varepsilon \right\} \to 1.$$

Clearly, if $x_0 \leq 0$, relative stability cannot hold.

We thus see that the only difficulty in determining conditions for the law of large numbers and relative stability arises from distributions for which $F(x) < 1$ for all finite values of x.

Theorem 1. *For the sequence* (1) *to be subject to the law of large numbers, if $F(x) < 1$ for all finite values of x, it is necessary and sufficient that*

$$\lim_{x \to \infty} \frac{1 - F(x + \varepsilon)}{1 - F(x)} = 0 \qquad \text{for all} \qquad \varepsilon > 0. \tag{6}$$

PROOF. From the obvious equality

$$P\{|\xi_n - A_n| < \varepsilon\} = F^n(A_n + \varepsilon) - F^n(A_n - \varepsilon)$$

the conditions for the law of large numbers can be expressed in the following form: for all $\varepsilon > 0$,

$$F^n(A_n + \varepsilon) \to 1 \qquad \text{and} \qquad F^n(A_n - \varepsilon) \to 0$$

as $n \to \infty$.

From the first of these relations it follows that, taking account of the condition of the theorem, $A_n \to \infty$ for $n \to \infty$.

These relations are equivalent to the following conditions:

$$n \log F(A_n + \varepsilon) \to 0$$

and

$$n \log F(A_n - \varepsilon) \to -\infty$$

as $n \to \infty$; now, since $1 - F(x) \to 0$ as $x \to \infty$, and under this condition

$$\log F(x) = \log(1 - (1 - F(x))) = -(1 - F(x)) - \tfrac{1}{2}(1 - F(x))^2 - \cdots$$
$$= -(1 - F(x))(1 + 0(1)),$$

the conditions are equivalent to

$$n(1 - F(A_n + \varepsilon)) \to 0$$
$$n(1 - F(A_n - \varepsilon)) \to -\infty \tag{7}$$

as $n \to \infty$.

We now suppose that the conditions of the theorem are satisfied and demonstrate that the law of large numbers holds. To do this, we define the constant A_n as the smallest value of x for which the inequalities

$$F(x - 0) \le 1 - \frac{1}{n} \le F(x + 0) \tag{8}$$

hold. Because of the hypotheses on $F(x)$ in the statement of the theorem, it is clear that $A_n \to \infty$ as $n \to \infty$. From the condition in the theorem, we have, for all ε and η $(\varepsilon > \eta > 0)$

$$\frac{1 - F(A_n + \varepsilon)}{1 - F(A_n + \eta)} \to 0, \qquad \frac{1 - F(A_n + \varepsilon)}{1 - F(A_n - \eta)} \to 0 \qquad \text{as } n \to \infty;$$

now, since $\eta > 0$ is arbitrary, we conclude that for $n \to \infty$

$$\frac{1 - F(A_n + \varepsilon)}{1 - F(A_n + 0)} \to 0, \qquad \frac{1 - F(A_n + \varepsilon)}{1 - F(A_n - 0)} \to 0. \tag{9}$$

From (8), it follows that

$$\frac{1 - F(A_n + \varepsilon)}{1 - F(A_n - 0)} \le n(1 - F(A_n + \varepsilon)) \le \frac{1 - F(A_n + \varepsilon)}{1 - F(A_n + 0)}$$

and consequently, from (9) we have, as $n \to \infty$

$$n(1 - F(A_n + \varepsilon)) \to 0. \tag{10}$$

From the condition of the theorem it follows that, for all $\varepsilon > 0$

$$\lim_{x \to \infty} \frac{1 - F(x)}{1 - F(x - \varepsilon)} = 0;$$

and by analogous arguments we then see that

$$n(1 - F(A_n - \varepsilon)) \to \infty \qquad \text{as} \qquad n \to \infty. \tag{11}$$

So we have seen that the relation (3) follows from (10) and (11).

Now we suppose that the law of large numbers holds, that is, there is a sequence of constants A_n such that the conditions (10) and (11) are satisfied for all $\varepsilon > 0$. We proceed to show that equation (6) holds also.

Clearly, from (10) we have $A_n \to \infty$ as $n \to \infty$, and we can assume that the A_n's are nondecreasing. For each sufficiently large value of x, we can find a number n such that

$$A_{n-1} \le x \le A_n.$$

Clearly the inequalities

$$1 - F(A_{n-1} - \eta) \ge 1 - F(x - \eta) \ge 1 - F(A_n - \eta)$$

$$1 - F(A_{n-1} + \eta) \ge 1 - F(x + \eta) \ge 1 - F(A_n + \eta)$$

hold for all $\eta > 0$, as also do the inequalities

$$\frac{1 - F(A_{n-1} + \eta)}{1 - F(A_n - \eta)} \geq \frac{1 - F(x + \eta)}{1 - F(x - \eta)} \geq \frac{1 - F(A_n + \eta)}{1 - F(A_{n-1} - \eta)}.$$

From (10) and (11) it follows that

$$\lim_{x \to \infty} \frac{1 - F(x + \eta)}{1 - F(x - \eta)} = 0.$$

Replacing $x - \eta$ by x and 2η by ε we obtain the condition stated in the theorem.   □

Theorem 2. *For the sequence* (1) *to be relatively stable, if* $F(x) < 1$ *for all finite values of* x, *it is necessary and sufficient that*

$$\lim_{x \to \infty} \frac{1 - F(kx)}{1 - F(x)} = 0 \tag{12}$$

for all $k > 1$.

PROOF. Noting the obvious equality

$$P\left\{ \left| \frac{\xi_n}{B_n} - 1 \right| < \varepsilon \right\} = F^n(B_n(1 + \varepsilon)) - F^n(B_n(1 - \varepsilon))$$

we can write the stability conditions in the following way:

$$F^n(B_n(1 + \varepsilon)) \to 1, \qquad F^n(B_n(1 - \varepsilon)) \to 0$$

as $n \to \infty$.

By arguments analogous to those used in the proof of the preceding theorem, we see that these conditions are equivalent to the following:

$$n(1 - F(B_n(1 + \varepsilon))) \to 0 \tag{13}$$

and

$$n(1 - F(B_n(1 + \varepsilon))) \to \infty \tag{14}$$

as $n \to \infty$.

First, suppose that the condition stated in the theorem holds. We define B_n as the smallest value of x for which

$$F(x(1 - 0)) \leq 1 - \frac{1}{n} \leq F(x(1 + 0)). \tag{15}$$

From the assumption about the function $F(x)$, we conclude that $B_n \to \infty$ as $n \to \infty$.

From (12), for all $\varepsilon > \eta > 0$, we have

$$\frac{1 - F(B_n(1 + \varepsilon))}{1 - F(B_n(1 + \eta))} \to 0, \qquad \frac{1 - F(B_n(1 + \varepsilon))}{1 - F(B_n(1 - \eta))} \to 0$$

as $n \to \infty$. Now, since $\varepsilon > 0$, we obtain

$$\frac{1 - F(B_n(1 + \varepsilon))}{1 - F(B_n(1 + 0))} \to 0, \qquad \frac{1 - F(B_n(1 + \varepsilon))}{1 - F(B_n(1 - 0))} \to 0. \tag{16}$$

From the inequality (15) we conclude that

$$\frac{1 - F(B_n(1 + \varepsilon))}{1 - F(B_n(1 - 0))} \le n(1 - F(B_n(1 + \varepsilon))) \le \frac{1 - F(B_n(1 + \varepsilon))}{1 - F(B_n(1 + 0))},$$

and hence, from (16)

$$n(1 - F(B_n(1 + \varepsilon))) \to 0 \tag{17}$$

as $n \to \infty$. Now, since it follows from the condition stated in the theorem, that for all $\varepsilon > 0$, we have

$$\lim_{x \to \infty} \frac{1 - F(x(1 - \varepsilon))}{1 - F(x)} = \infty,$$

we obtain by analogous arguments

$$n(1 - F(B_n(1 - \varepsilon))) \to \infty \qquad \text{as} \qquad n \to \infty. \tag{18}$$

But, as has already been shown, relative stability of the maxima follows from (17) and (18).

Now assume that the maxima are relatively stable, and, hence, (13) and (14) hold. We will now demonstrate that (12) also holds. Since $F(x) < 1$ for all finite x, it follows from (13) that

$$B_n \to \infty \qquad \text{for} \qquad n \to \infty.$$

Clearly, we can assume that the B_n's are non-decreasing. For all sufficiently large values of x we can find an integer, n, such that

$$B_{n-1} \le x \le B_n.$$

Clearly, for all $\varepsilon > 0$ and $\eta > 0$ we have

$$1 - F(B_{n-1}(1 - \eta)) \ge 1 - F(x(1 - \eta)) \ge 1 - F(B_n(1 - \eta))$$

and

$$1 - F(B_{n-1}(1 + \varepsilon)) \ge 1 - F(x(1 + \varepsilon)) \ge 1 - F(B_n(1 + \varepsilon)),$$

from which it is seen that

$$\frac{1 - F(B_{n-1}(1 + \varepsilon))}{1 - F(B_n(1 - \eta))} \ge \frac{1 - F(x(1 + \varepsilon))}{1 - F(x(1 - \eta))} \ge \frac{1 - F(B_n(1 + \varepsilon))}{1 - F(B_{n-1}(1 - \eta))}.$$

From (13) and (14), for all $\varepsilon > 0$ and $\eta > 0$ we have

$$\lim_{x \to \infty} \frac{1 - F(x(1 + \varepsilon))}{1 - F(x(1 - \eta))} = 0.$$

Putting $X = x(1 - \eta)$, $k = \dfrac{1 + \varepsilon}{1 - \eta}$ we obtain (12). $\qquad\qquad \square$

By way of example, consider the following distribution functions:

$$F_1(x) = \begin{cases} 0 & \text{for } x \leq 1 \\ 1 - \dfrac{1}{x^\alpha} & \text{for } x > 1 \end{cases} \tag{19}$$

$$F_2(x) = \begin{cases} 0 & \text{for } x \leq 0 \\ 1 - e^{-x^\alpha} & \text{for } x > 0 \end{cases}. \tag{20}$$

Since

$$\lim_{x \to \infty} \frac{1 - F_1(x + \varepsilon)}{1 - F_1(x)} = 1$$

and

$$\lim_{x \to \infty} \frac{1 - F_1(kx)}{1 - F_1(x)} = \frac{1}{k^\alpha},$$

we see that for the distribution function (19) the maxima satisfy neither the law of large numbers, nor relative stability. However, since, for any $\alpha > 0$

$$\lim_{x \to 0} \frac{1 - F_2(x + \varepsilon)}{1 - F_2(x)} = \begin{cases} 0 & \text{for } \alpha > 1 \\ e^{-\varepsilon} & \text{for } \alpha = 1. \\ 1 & \text{for } \alpha < 1 \end{cases}$$

and

$$\lim_{x \to 0} \frac{1 - F_2(kx)}{1 - F_2(x)} = 0 \quad (k > 1),$$

we see that for laws (20), relative stability holds for all $\alpha > 0$, while they are subject to the law of large numbers only if $\alpha > 1$.

It can easily be verified that

1) for the Poisson distribution the maxima are relatively stable but do not satisfy the law of large numbers, and
2) for Gaussian distributions with standard deviation one and expected value zero we have

$$P\left\{ \left| \frac{\xi_n}{\sqrt{2 \log n}} - 1 \right| < \varepsilon \right\} \to 1$$

and

$$P\{|\xi_n - \sqrt{2 \log n}| < \varepsilon\} \to 1$$

as $n \to \infty$.

In 1932, Bruno de Finetti [6] gave some conditions for applicability of the law of large numbers. Finetti considered random variables having densities

$f(x) = F'(x)$, and subject to certain further conditions; the sufficient condition found by Finetti is expressible as the equality

$$\lim_{z \to \infty} \frac{f(z + \varepsilon)}{f(z)} = 0$$

for all $\varepsilon > 0$. Finetti's condition follows easily from Theorem 1 (without any further condition being imposed on the random variables). In fact, if the derivative $f(x) = F'(x)$ exists for all values of x, l'Hospital's rule gives

$$\lim_{x \to \infty} \frac{1 - F(x + \varepsilon)}{1 - F(x)} = \lim_{z \to \infty} \frac{f(z + \varepsilon)}{f(z)}$$

from which it follows, from Theorem 1 of the present paper, that if the limit

$$\lim_{z \to \infty} \frac{f(z + \varepsilon)}{f(z)}$$

exists for all $\varepsilon > 0$, Finetti's condition is necessary and sufficient.

Obviously, an analogous condition can be established for relative stability of the minima.

2. The Class of Limit Laws

Theorem 3. *The class of limit laws for $F_n(a_n x + b_n)$, where $a_n > 0$ and b_n are suitably chosen constants, contains only laws of types $\Phi_\alpha(x)$, $\Psi_\alpha(x)$ and $\Lambda(x)$.*

PROOF. Suppose that $F_n(a_n x + b_n) = F^n(a_n x + b_n) \to \Phi(x)$ as $n \to \infty$. Then the equality

$$\lim_{n \to \infty} [F^n(a_{nk} x + b_{nk})]^k = \Phi(x) \tag{21}$$

holds for all integers $k > 0$. It follows that with k fixed and $n \to \infty$, the sequence of functions $F^n(a_{nk} x + b_{nk})$ tends to a limit function. From a theorem of A. Khintchine ([4], theorem 43), this limit function must belong to the same type as $\Phi(x)$, so that we must have

$$\lim_{n \to \infty} F^n(a_{nk} x + b_{nk}) = \Phi(\alpha_k x + \beta_k) \tag{22}$$

where $\alpha_k > 0$ and β_k are constants.

---

[3] When I first obtained this theorem, the results of Fisher & Tippett [2] were unknown to me. Since the proofs given by these authors are not, in my opinion, sufficiently detailed, and appeal to the unnecessary hypothesis of analyticity of the quantities a_n and b_n as functions of n, I thought it would be worthwhile setting out, in the present paper, the results of this section with all necessary developments.

From (21) and (22), for any natural number k, the limit law satisfies

$$\Phi^k(\alpha_k x + \beta_k) = \Phi(x). \tag{23}$$

We now consider three separate cases.

1) $\alpha_k < 1$ for some $k > 1$

Then for

$$x \geq \frac{\beta_k}{1 - \alpha_k}$$

we have

$$a_k x + \beta_k \leq x.$$

Since the function $\Phi(x)$ is monotone, we can write

$$\Phi(\alpha_k x + \beta_k) \leq \Phi(x).$$

Hence, for the distribution function $\Phi(x)$, equation (23) cannot be satisfied unless

$$\Phi(x) = 1 \qquad \text{for} \qquad x \geq \frac{\beta_k}{1 - \alpha_k}.$$

We now show that for $x < \dfrac{\beta_k}{1 - \alpha_k}$ we must have $\Phi(x) < 1$. Suppose this is not true, that is there exists a value $x_0 < \dfrac{\beta_k}{1 - \alpha_k}$ giving

$$\Phi(x_0) = 1. \tag{24}$$

Clearly, it is always possible to choose, for all $x \leq x_0$, an integer n such that

$$x_0 \leq \alpha_k^n x + \beta_k(1 + \alpha_k + \cdots + \alpha_k^{n-1}).$$

Now, from (24) we must have

$$\Phi(\alpha_k^n x + \beta_k(1 + \alpha_k + \cdots + \alpha_k^{n-1})) = 1.$$

Hence it follows from (23) that

$$\begin{aligned}
\Phi^{k^n}(\alpha_k^n x &+ \beta_k(1 + \alpha_k + \cdots + \alpha_k^{n-1})) \\
&= [\Phi^k(\alpha_k\{\alpha_k^{n-1}x + \beta_k(1 + \alpha_k + \cdots + \alpha_k^{n-2})\} + \beta_k]^{k^{n-1}} \\
&= \Phi^{k^{n-1}}(\alpha_k^{n-1}x + \beta_k(1 + \alpha_k + \cdots + \alpha_k^{n-2})) \\
&\quad \vdots \\
&= \Phi^k(\alpha_k x + \beta_k),
\end{aligned} \tag{25}$$

that is,

$$\Phi(x) = 1$$

for all x, which is impossible.

We see, therefore, that $\Phi(x) \to 1$ for $x \geq \beta_k/(1 - \alpha_k)$ and $\Phi(x) < 1$ for $x < \beta_k/(1 - \alpha_k)$.

We now show that if $\Phi(x)$ is a proper law, and if $\alpha_k < 1$ for some value of k, this inequality holds for all values of k. Let us suppose that there is an $r > 1$ such that $\alpha_r \geq 1$.

If $\alpha_r = 1$, then for all values of x,

$$\Phi^r(x + \beta_r) = \Phi(x)$$

and so

$$\Phi^r(x) = \Phi(x - \beta_r).$$

In particular,

$$\left.\begin{array}{l}
\Phi^r\left(\dfrac{\beta_k}{1 - \alpha_k} + \beta_r\right) = \Phi\left(\dfrac{\beta_k}{1 - \alpha_k}\right) = 1, \\[12pt]
\Phi\left(\dfrac{\beta_k}{1 - \alpha_k} - \beta_r\right) = \Phi^r\left(\dfrac{\beta_k}{1 - \alpha_k}\right) = 1.
\end{array}\right\} \tag{26}$$

If $\beta_r \neq 0$ we have

$$\min(\beta_k/(1 - \alpha_k)) + \beta_r : \beta_k/(1 - \alpha_k) - \beta_r) < \beta_k/(1 - \alpha_k)$$

and it follows from (26) by an obvious argument that $\Phi(x) \equiv 1$; if $\beta_r = 0$, then we have

$$\Phi^r(x) = \Phi(x)$$

and $\Phi(x) \equiv 1$, or $\Phi(x) = 1$ for $x \geq \beta_k/(1 - \alpha_k)$ while $\Phi(x) = 0$ for $x < \beta_k/(1 - \alpha_k)$. Hence, since $\Phi(x)$ is a proper law and α_k satisfies the inequality $\alpha_k < 1$, we see that $\alpha_r \neq 1$ for all r.

If $\alpha_r > 1$, we have, for $x \leq \beta_r/(1 - \alpha_r)$

$$\alpha_r x + \beta_r \leq x$$

and hence

$$\Phi(\alpha_r x + \beta_r) \leq \Phi(x).$$

Remembering (23) we obtain

$$\Phi(x) = 0 \quad \text{for} \quad x \leq \frac{\beta_r}{1 - \alpha_r}. \tag{27}$$

Take $x < \beta_k/(1 - \alpha_k)$. For all $\varepsilon > 0$ an integer n can be found such that

$$\frac{\beta_k}{1 - \alpha_k} - \varepsilon < \alpha_k^n x + \beta_k(1 + \alpha_k + \cdots + \alpha_k^{n-1}) = z.$$

As in (25) and (27), we obtain

$$\Phi^{k^n}(\alpha_k^n x + \beta_k(1 + \alpha_k + \cdots + \alpha_k^{n-1})) = \Phi(x) = 0,$$

so that for all $z < \beta_k/(1 - \alpha_k)$ we have

$$\Phi(z) = 0.$$

The law $\Phi(x)$ is therefore improper, contradicting our initial hypothesis.

Hence the function $\Phi(x)$ is equal to 1 for $x \geq \beta_k(1 - \alpha_k) = x_0$ and less than 1 for $x < \beta_k/(1 - \alpha_k) = x_0$. The value x_0 being clearly independent of k, we have

$$\frac{\beta_k}{1 - \alpha_k} = \frac{\beta_n}{1 - \alpha_n}$$

for all values of k and n.

Put

$$\bar{\Phi}(z) = \Phi\left(z + \frac{\beta_k}{1 - \alpha_k}\right)$$

(corresponding to moving the origin to the point $\beta_k/(1 - \alpha_k)$).

Clearly,

$$\bar{\Phi}(\alpha_k z) = \Phi\left(\alpha_k z + \frac{\beta_k}{1 - \alpha_k}\right). \tag{28}$$

From (23), the function $\bar{\Phi}(z)$ satisfies the equation

$$\bar{\Phi}^k(\alpha_k z) = \bar{\Phi}(z)$$

for all positive integers k. The solution of this functional equation is well-known (see, for example, [4], page 95). The only distribution function satisfying the equation (28), subject to the condition $\bar{\Phi}(z) = 1$ for $z \geq 0$, is the function $\Psi_\alpha(x)$.

2) $\alpha_k > 1$ for some value of k

From the arguments already developed it follows that $\alpha_k > 1$ for all values of k. We have already seen (27) that

$$\Phi(x) = 0 \qquad \text{for} \qquad x \leq \frac{\beta_k}{1 - \alpha_k}.$$

Demonstration of the inequality $\Phi(x) > 0$ for $x > \dfrac{\beta_k}{1 - \alpha_k}$ follows from the equation

$$\Phi^{k^n}(x) = \Phi\left[\frac{x}{\alpha_k^n} - \beta_k\left(\frac{1}{\alpha_k} + \frac{1}{\alpha_k^2} + \cdots + \frac{1}{\alpha_k^n}\right)\right]$$

which follows, in turn, from (23). It follows from this same inequality $\Phi(x) < 1$ for all $x > \beta_k/(1 - \alpha_k)$. Similarly we see that

$$\frac{\beta_k}{1 - \alpha_k} = \frac{\beta_n}{1 - \alpha_n} \qquad \text{for all values of } k \text{ and } n,$$

and that the function

$$\bar{\Phi}(z) = \Phi\left(z + \frac{\beta_k}{1 - \alpha_k}\right)$$

satisfies equation (28) for all $k > 0$.

The only distribution function satisfying this equation and also the condition $\bar{\Phi}(z) = 0$ for $z < 0$ is the function $\Phi_a(x)$.

3) $\alpha_k = 1$ for Some k

From our preceding analysis, this implies $\alpha_k = 1$ for all k. Making the change of variables

$$z = e^x; \quad \text{with} \quad \beta_k = e^{c_k}, \quad \bar{\Phi}(z) = \begin{cases} \Phi(\log z) & \text{for } z > 0 \\ 0 & \text{for } z \leq 0 \end{cases}$$

we reduce equation (23) to the form

$$\bar{\Phi}^k(c_k z) = \bar{\Phi}(z).$$

The only distribution function satisfying this equation and the condition $\bar{\Phi}(0) = 0$ is the function $\Phi_a(z)$. Thus, we have

$$\Phi(x) = e^{-e^{-ax}}.$$

This function is of type $\Lambda(x)$, which we have described in the Introduction.

3. Auxiliary Propositions

Lemma 1. *Let $F_n(x)$ and $\Phi(x)$ be distribution functions, $\Phi(x)$ not being unitary. If, for some sequences of real numbers $a_n > 0$ and b_n, $\alpha_n > 0$ and β_n we have*

$$F_n(a_n x + b_n) \to \Phi(x)$$

and

$$F_n(\alpha_n x + \beta_n) \to \Phi(x).$$

As $n \to \infty$, we will also have

$$\frac{a_n}{\alpha_n} \to 1 \quad \text{and} \quad \frac{b_n - \beta_n}{a_n} \to 0.$$

PROOF. For brevity, put $V_n(x) = F_n(a_n x + b_n)$.

From the conditions stated in the lemma,

$$V_n(x) \to \Phi(x)$$

and

$$V_n(A_n x + B_n) \to \Phi(x)$$

and $n \to \infty$, where

$$A_n = \alpha_n/a_n; \qquad B_n = (\beta_n - b_n)/a_n = 0.$$

Define a sequence of indices $n_1 < n_2 < \cdots < n_k < \cdots$ such that the limits

$$\lim_{k \to \infty} A_{n_k} = A, \qquad \lim_{k \to \infty} B_{n_k} = B \quad (0 \le A \le +\infty; \ -\infty \le B \le +\infty)$$

exist. We will show that $A < +\infty$. Supposing the contrary, that is, $A = +\infty$, denote by x_0 the upper bound of numbers x for which

$$\overline{\lim_{k \to \infty}} \ (A_{n_k} x + B_{n_k}) < +\infty.$$

It is clear that for all $x > x_0$,

$$\overline{\lim_{k \to \infty}} \ (A_{n_k} x + B_{n_k}) = +\infty,$$

while for all $x < x_0$,

$$\overline{\lim_{k \to \infty}} \ (A_{n_k} x + B_{n_k}) = -\infty.$$

We deduce that

$$\Phi(x) = \begin{cases} 0 & \text{for } x < x_0 \\ 1 & \text{for } x > x_0 \end{cases}.$$

But since this is excluded by the conditions of the lemma, we must have $A < +\infty$. It follows that B, also, is finite, since the hypothesis $B = -\infty$ would lead to

$$\overline{\lim_{k \to \infty}} \ (A_{n_k} x + B_{n_k}) = -\infty$$

for all x, which would imply, in turn, that $\Phi(x) \equiv 0$; in the same way, the hypothesis $B = +\infty$ would lead to

$$\lim_{k \to \infty} (A_{n_k} x + B_{n_k}) = +\infty$$

from which it would follow that $\Phi(x) \equiv 1$. It is apparent that $A > 0$, since the α_n's and a_n's play the same role, and if $A = 0$ we would have

$$\overline{\lim_{k \to \infty}} \ (a_{n_k}/\alpha_{n_k}) = +\infty,$$

the impossibility of which has just been demonstrated.

Let x be a value such that, at the points x and $Ax + B$, the function $\Phi(x)$

is continuous. It is clear that

$$\Phi(x) = \lim_{k \to \infty} V_{n_k}(A_{n_k} x + B_{n_k}) = \Phi(Ax + B). \tag{29}$$

Since this equality must hold for all x, except for points of discontinuity, it follows that $A = 1$ and $B = 0$. In fact, suppose that this is not so, and let us consider the possible cases that might arise.

For $A < 1$, iterating (29) we obtain for all x, and n an arbitrary positive integer

$$\Phi(x) = \Phi(A^n x + B(1 + A + \cdots + A^{n-1})).$$

Since $A^n x$ can be made as small as desired by taking n to be sufficiently large, we have, for all x

$$\Phi(x) = \lim_{n \to \infty} \Phi(A^n x + B(1 + A + \cdots + A^{n-1})) = \Phi\left(\frac{B}{1 - A}\right).$$

The function $\Phi(x)$ is therefore not a distribution function.

For $A > 1$, we write (29) in the form:

$$\Phi(x) = \Phi\left(\frac{x}{A} - \frac{B}{A}\right)$$

and by arguments analogous to earlier ones we arrive at the equation

$$\Phi(x) = \Phi\left(\frac{B}{1 - A}\right)$$

which must hold for all values of x, which proves that $\Phi(x)$ cannot be a distribution function. Finally, we will prove the impossibility of the case $A = 1$, $B \neq 0$. In fact, from (29) we see that for all integers n,

$$\Phi(x) = \Phi(x + nB).$$

If, $B > 0$ $(B < 0)$, then for all x and for $n \to \infty$ $(n \to -\infty)$

$$\Phi(x) = \Phi(x + nB) \to \Phi(+\infty),$$

on the other hand, for $n \to -\infty$ $(n \to +\infty)$

$$\Phi(x) = \Phi(x + nB) \to \Phi(-\infty)$$

which is evidently only possible if $\Phi(x)$ is constant, and consequently, is not a distribution function. Thus we see that when there is a sequence of indices $\{n_k\}$ such that the limits

$$\lim_{k \to \infty} A_{n_k} = A, \qquad \lim_{k \to \infty} B_{n_k} = B$$

exist, then necessarily $A = 1$, $B = 0$. Now, we have to show that

$$\lim_{k \to \infty} A_n = 1, \qquad \lim_{k \to \infty} B_n = 0.$$

If one or other of these relations do not hold, there would clearly be a sequence of indices $\{n_k\}$ and a number $\varepsilon > 0$ such that one, at least, of the inequalities

$$\lim_{k \to \infty} |A_{n_k} - 1| \geq \varepsilon, \qquad \lim_{k \to \infty} |B_{n_k}| \geq \varepsilon \tag{30}$$

would be satisfied. Further, the sequence $\{n_k\}$ can be chosen so that A_{n_k} and B_{n_k} tend to fixed limits as $k \to \infty$. Now, these limits can only be 1 and 0 respectively, as we have just demonstrated, which contradicts (30) and so the lemma is proved.   □

Later we shall also need to use the converse proposition.

Lemma 2. *If $F_n(x)$ is a sequence of functions for which*

$$\lim_{n \to \infty} F_n(a_n x + b_n) = \Phi(x)$$

holds, for a suitable choice of constants $a_n > 0$ and b_n, and for all values of x, then for two sequences of constants $\alpha_n > 0$ and β_n, such that as $n \to \infty$

$$\frac{a_n}{\alpha_n} \to 1, \qquad \frac{b_n - \beta_n}{\alpha_n} \to 0 \tag{31}$$

we have

$$F_n(\alpha_n x + \beta_n) \to \Phi(x)$$

as $n \to \infty$, for all values of x.

PROOF. Let x_1, x and x_2 $(x_1 < x < x_2)$ be points of continuity of the function $\Phi(x)$. From our hypothesis, we have, for n sufficiently large

$$x_1 < \frac{\alpha_n}{a_n} x + \frac{\beta_n - b_n}{a_n} < x_2.$$

Since

$$\alpha_n x + \beta_n = a_n \left(\frac{\alpha_n}{a_n} x + \frac{\beta_n - b_n}{a_n} \right) + b_n$$

we have, for n sufficiently large

$$a_n x_1 + b_n < \alpha_n x + \beta_n < a_n x_2 + b_n$$

and hence

$$F_n(a_n x_1 + b_n) \leq F_n(\alpha_n x + \beta_n) \leq F(a_n x_2 + b_n).$$

As x_1 and x_2 tend to x, we have

$$\Phi(x_1) \to \Phi(x), \qquad \Phi(x_2) \to \Phi(x),$$

x being assumed to be a point of continuity of $\Phi(x)$. We therefore have

$$\Phi(x) \leq \varliminf_{n \to \infty} F_n(\alpha_n x + \beta_n) \leq \varlimsup_{n \to \infty} F_n(\alpha_n x + \beta_n) \leq \Phi(x)$$

i.e., $\lim F_n(\alpha_n x + \beta_n) = \Phi(x)$. $\square$

Lemma 3. *If $F(x)$ is a distribution function, and if, for suitable choice of constants $a_n > 0$ and b_n,*

$$F^n(a_n x + b_n) \to \Phi(x) \tag{32}$$

as $n \to \infty$, for all values of x, where $\Phi(x)$ is a proper distribution function, then, as $n \to \infty$

$$\frac{a_n}{a_{n+1}} \to 1, \qquad \frac{b_n - b_{n+1}}{a_n} \to 0.$$

PROOF. In fact, if relation (32) holds, then, for all values of x such that $\Phi(x) \neq 0$, we have

$$\lim_{n \to \infty} F(a_n x + b_n) = 1,$$

whence

$$F^{n+1}(a_n x + b_n) \to \Phi(x)$$

as $n \to \infty$.

We are now in the situation of Lemma 1 with $\alpha_n = a_{n-1}$, $\beta_n = b_{n-1}$, and this demonstrates the validity of Lemma 3.

Lemma 4. *In order to have*

$$F^n(a_n x + b_n) \to \Phi(x) \tag{33}$$

as $n \to \infty$, for all values of x, it is necessary and sufficient that

$$n[1 - F(a_n x + b_n)] \to -\log \Phi(x) \tag{34}$$

as $n \to \infty$ for all values of x for which $\Phi(x) \neq 0$.

PROOF. Assume that relation (33) holds; then for all values of x such that $\Phi(x) \neq 0$ we have

$$\lim_{n \to \infty} F(a_n x + b_n) = 1. \tag{35}$$

Clearly, for these values of x the condition (33) is equivalent to the following:

$$n \log F(a_n x + b_n) \to \log \Phi(x) \quad (\Phi(x) \neq 0) \tag{36}$$

as $n \to \infty$. Now, from (35) we have

$$\log F(a_n x + b_n) = -(1 - F(a_n x + b_n)) - \tfrac{1}{2}(1 - F(a_n x + b_n))^2$$
$$= -(1 - F(a_n x + b_n)) \ (1 + o(1)). \tag{37}$$

Since (33) holds, condition (34) is necessarily satisfied. Conversely, if condition

(34) is satisfied, then so is (35) because from (37)

$$-n[1 - F(a_n x + b_n)] = \{n \log F(a_n x + b_n)\} \quad (1 + o(1)).$$

Hence from (34), (36) holds, and, consequently, so does (33). □

4. The Domain of Attraction of the Law $\Phi_\alpha(x)$

Theorem 4. *For a distribution function $F(x)$ to belong to the domain of attraction of the law $\Phi_\alpha(x)$ it is necessary and sufficient that*

$$\lim_{x \to \infty} \frac{1 - F(x)}{1 - F(kx)} = k^\alpha \tag{38}$$

for all values of $k > 0$.

PROOF. First, we will suppose that condition (38) is satisfied, and show that $F(x)$ belongs to the domain of attraction of the law $\Phi_\alpha(x)$.

From (38) it is clear that $F(x) < 1$ for all values of x. Hence, for sufficiently large n, the values of x for which

$$1 - F(x) \le \frac{1}{n}$$

are all positive.

Define a_n as the smallest x satisfying the inequalities

$$1 - F(x(1 + 0)) \le \frac{1}{n} \le 1 - F(x(1 - 0)). \tag{39}$$

It follows from the preceding analysis that $a_n \to \infty$ as $n \to \infty$. From (38) we have

$$\frac{1 - F(a_n x)}{1 - F(a_n(1 + \varepsilon))} \to \left(\frac{1 + \varepsilon}{x}\right)^\alpha \quad \text{and} \quad \frac{1 - F(a_n x)}{1 - F(a_n(1 - \varepsilon))} \to \left(\frac{1 - \varepsilon}{x}\right)^\alpha$$

for all values of x and ε $(0 < \varepsilon < 1)$, as $n \to \infty$.

As the left-hand members of these relations are monotonic functions of ε, and the right-hand members are continuous functions of ε, the convergence is uniform, which allows us to write, as $n \to \infty$

$$\frac{1 - F(a_n x)}{1 - F(a_n(1 + 0))} \to \frac{1}{x^\alpha} \quad \text{and} \quad \frac{1 - F(a_n x)}{1 - F(a_n(1 - 0))} \to \frac{1}{x^\alpha}.$$

Now, since we have, from (39),

$$\frac{1 - F(a_n x)}{1 - F(a_n(1 - 0))} \le n(1 - F(a_n x)) \le \frac{1 - F(a_n x)}{1 - F(a_n(1 + 0))}$$

we see that, for all $x > 0$,

$$n(1 - F(a_n x)) \to x^{-\alpha} \qquad \text{as} \qquad n \to \infty.$$

From Lemma 4 of the preceding paragraph

$$F^n(a_n x) \to \Phi_\alpha(x) \qquad \text{as} \qquad n \to \infty \quad (-\infty < x < +\infty).$$

Now suppose that $F(x)$ belongs to the domain of attraction of the law $\Phi_\alpha(x)$, that is, suppose that, for a suitable choice of constants $a_n > 0$ and b_n, the relation

$$n(1 - F(a_n x + b_n)) \to x^{-\alpha} \qquad \text{as} \qquad n \to \infty \tag{40}$$

holds for all $x > 0$. We will show that as a consequence, (38) holds.

For any constant $\beta > 1$ we have, as $n \to \infty$

$$n\beta(1 - F(a_n x + b_n)) \to \beta x^{-\alpha}.$$

Since it follows from (40) that for $x > 0$ and $n \to \infty$

$$1 - F(a_n x + b_n) \to 0,$$

we see that, as $n \to \infty$ we must have

$$[n\beta](1 - F(a_n x + b_n)) \to \beta x^{-\alpha},$$

where $[n\beta]$ denotes the integer part of $n\beta$.

Applying the change of variable $x = z\beta^{1/\alpha}$, this relation takes the form

$$[n\beta](1 - F(a_n \beta^{1/\alpha} z + b_n)) \to z^{-\alpha} \tag{41}$$

as $n \to \infty$, for all $x < 0$.

From (41) it follows that

$$[n\beta](1 - F(a_{[n\beta]} x + b_{[n\beta]})) \to x^{-\alpha} \tag{42}$$

as $n \to \infty$.

From Lemmas 4 and 1, and taking account of (4) and (5), we conclude that the relations

$$\frac{a_{[n\beta]}}{a_n} \to \beta^{1/\alpha} \qquad \text{and} \qquad \frac{b_{[n\beta]} - b_n}{a_{[n\beta]}} \to 0$$

as $n \to \infty$, must hold. From Lemma 2, the relation (42) is unchanged if we put

$$a_{[n\beta]} = a_n \beta^{1/\alpha}, \qquad b_{[n\beta]} = b_n. \tag{43}$$

For a fixed integer n, define $n_1 = [n\beta]$, $n_s = [n_{s-1}\beta]$ for $s > 1$. It follows from (43) that, for all natural numbers s, we have

$$a_{n_s} = a_n \beta^{s/\alpha}, \qquad b_{n_s} = b_n.$$

Hence, as $s \to \infty$

$$b_{n_s}/a_{n_s} \to 0$$

and consequently, by Lemma 2

$$n_s(1 - F(a_{n_s} x)) \to x^{-\alpha} \qquad \text{as} \qquad s \to \infty. \tag{44}$$

Suppose that $y \to \infty$; for all sufficiently large values of y we can find a value of s such that

$$a_{n_s} x \leq y \leq a_{n_{s+1}} x$$

and so

$$1 - F(a_{n_{s+1}} x) \leq 1 - F(y) \leq 1 - F(a_{n_s} x)$$

and, for $k > 0$

$$1 - F(a_{n_{s+1}} kx) \leq 1 - F(ky) \leq 1 - F(a_{n_s} kx).$$

From this we obtain the inequality

$$\frac{1 - F(a_{n_{s+1}} x)}{1 - F(a_{n_s} kx)} \leq \frac{1 - F(y)}{1 - F(ky)} \leq \frac{1 - F(a_{n_s} x)}{1 - F(a_{n_{s+1}} kx)}. \tag{45}$$

We note that

$$\frac{n_{s+1}}{n_s} = \frac{n_s \beta - \theta_s}{n_s}$$

where $0 \leq \theta_s < 1$; therefore as $s \to \infty$ we have

$$n_{s+1}/n_s \to \beta.$$

From (44) and (45) we conclude that

$$\frac{1}{\beta} k^\alpha \leq \lim_{y \to \infty} \frac{1 - F(y)}{1 - F(ky)} \leq \beta k^\alpha;$$

and since β can be chosen as close to unity as desired, condition (38) is established. $\square$

We note that from the preceding analysis it follows that any distribution $F(x)$, belonging to the domain of attraction of the law $\Phi_\alpha(x)$, approaches the limit $\Phi_\alpha(x)$ in a very special manner, as follows: for appropriate choice of constants a_n,

$$\lim_{n \to \infty} F^n(a_n x) = \Phi_\alpha(x).$$

5. The Domain of Attraction of the Law $\Psi_\alpha(x)$

Theorem 5. *For a distribution function $F(x)$ to belong to the domain of attraction of the law $\Psi_\alpha(d)$, it is necessary and sufficient that:*

1. *There is an x_0 such that*

$$F(x_0) = 1 \qquad and \qquad F(x_0 - \varepsilon) < 1$$

for all $\varepsilon > 0$;

2.
$$\lim_{x \to -0} \frac{1 - F(kx + x_0)}{1 - F(x + x_0)} = k^\alpha$$

for all $k > 0$.

PROOF. We suppose that the conditions of the theorem are satisfied and will show that the function $f(x)$ belongs to the domain of attraction of the law $\Psi_\alpha(x)$. To do this, we define a_n as the smallest value of $x > 0$, for which the inequalities

$$1 - F(-x(1 - 0) + x_0) \le n^{-1} \le 1 - F(-x(1 + 0) + x_0) \tag{46}$$

From the first condition of the theorem

$$a_n \to 0 \quad \text{as} \quad n \to \infty$$

The second condition of the theorem gives

$$\frac{1 - F(a_n x + x_0)}{1 - F(-a_n(1 + \varepsilon) + x_0)} \to \left(-\frac{x}{1 + \varepsilon}\right)^\alpha$$

and

$$\frac{1 - F(a_n x + x_0)}{1 - F(-a_n(1 + \varepsilon) + x_0)} \to \left(-\frac{x}{1 - \varepsilon}\right)^\alpha$$

as $n \to \infty$ for all $\varepsilon > 0$ and $x < 0$.

The left-hand expressions of these two relations are monotonic functions of ε and the right-hand sides are continuous functions of ε; therefore the convergence is uniform and we can write

$$\frac{1 - F(a_n x + x_0)}{1 - F(-a_n(1 + 0) + x_0)} \to (-x)^\alpha$$

and

$$\frac{1 - F(a_n x + x_0)}{1 - F(-a_n(1 - 0) + x_0)} \to (-x)^\alpha.$$

Now, since we have, from (1),

$$\frac{1 - F(a_n x + x_0)}{1 - F(-a_n(1 + 0) + x_0)} \le n(1 - F(a_n x + x_0)) \le \frac{1 - F(a_n x + x_0)}{1 - F(-a_n(1 - 0) + x_0)}$$

we can state that

$$n(1 - F(a_n x + x_0)) \to (-x)^\alpha$$

as $n \to \infty$, for all $x < 0$, so that from Lemma 4, we see that the function $F(x)$ belongs to the domain of attraction of the law $\Psi_\alpha(x)$.

Now we suppose that $F(x)$ belongs to the domain of attraction of the law $\Psi_\alpha(x)$, which means that for all values of x, and appropriate choice of $a_n > 0$ and b_n,

$$F^n(a_n x + b_n) \to \Psi_\alpha(x), \tag{47}$$

as $n \to \infty$. From this we draw the conclusion

$$F^{2n}(a_n x + b_n) \to \Psi_\alpha^2(x) = \Psi_\alpha(\gamma x) \tag{48}$$

where

$$\gamma = 2^{1/\alpha}.$$

Replacing γx in (48) by x, we see that, as $n \to \infty$, we have

$$F^{2n}\left(\frac{a_n}{\gamma} x + b_n\right) \to \Psi_\alpha(x). \tag{49}$$

Comparing (47) and (49), we conclude, using Lemmas 1 and 2, that the a_n's and b_n's can be chosen in such a way that

$$a_{2n} = a_n/\gamma, \qquad b_{2n} = b_n.$$

It follows that we can always make this choice so as to have, for all k,

$$b_{2^k n} = b_{2^{(k-1)}n} = b_n \tag{50}$$

and $a_n \to 0$ as $n \to \infty$.

If the inequality

$$F(x) < 1$$

holds for all values of x, we deduce from (47), by putting $x = 0$, that $b_n \to \infty$ as $n \to \infty$, which contradicts (50). We have thus demonstrated the necessity of the first condition of the theorem. If relation (47) holds, one must choose b_n so as to have

$$F^n(b_n) \to \Psi_\alpha(0) = 1$$

that is to say, one must have $b_n \to x_0$.

From Lemma 2 and (50), we can make this choice by putting

$$b_n - x_0. \tag{51}$$

From Lemma 4 and (51), the relation (47) is equivalent to the following:

$$n(1 - F(a_n x + x_0)) \to (-x)^\alpha \tag{52}$$

as $n \to \infty$.

From this relation we immediately obtain the equation $\lim_{n\to\infty} a_n = 0$. Indeed, we have $a_n x + x_0 < x_0$ for $x < 0$, and for the left-hand side of (52) to tend to a finite limit it is necessary that the equality $\lim_{n\to\infty} (a_n x + x_0) = x_0$ be satisfied for all $x < 0$.

Now suppose that $y \to -0$. For all sufficiently small y it is possible to find a value of n sufficiently large to have $-a_n \leq y \leq -a_{n-1}$, if $a_{n+1} \leq a_n$, or $-a_{n+1} \leq y \leq -a_n$ if $a_n \leq a_{n+1}$.

In the first case we see that

$$1 - F(-a_{n+1} + x_0) \leq 1 - F(y + x_0) \leq 1 - F(-a_n + x_0)$$

and, for all $k > 0$

$$1 - F(-a_{n+1}k + x_0) \leq 1 - F(ky + x_0) \leq 1 - F(-a_n k + x_0),$$

whence

$$\frac{1 - F(-a_{n+1}k + x_0)}{1 - F(-a_n + x_0)} \leq \frac{1 - F(ky + x_0)}{1 - F(y + x_0)} \leq \frac{1 - F(-a_n k + x_0)}{1 - F(-a_{n+1} + x_0)}.$$

In the second case we obtain, analogously, the inequalities

$$\frac{1 - F(-a_n k + x_0)}{1 - F(-a_{n+1} + x_0)} \leq \frac{1 - F(ky + x_0)}{1 - F(y + x_0)} \leq \frac{1 - F(-a_{n+1}k + x_0)}{1 - F(-a_n + x_0)}.$$

Since, in both cases, from (52) the two extreme limits of the inequalities tend to k^α as $n \to \infty$, we are led to the second condition of the theorem. $\quad\square$

6. The Domain of Attraction of the Law $\Lambda(x)$

In the preceding paragraphs we have seen that laws of type $\Phi_\alpha(x)$ are distribution functions only if $F(x) < 1$ for all x, and laws of type $\Psi_\alpha(x)$ are distribution functions only if $F(x_0) = 1$ for a finite value of x_0 and $F(x_0 - \varepsilon) < 1$ for all $\varepsilon > 0$. It is easy to see that the law $\Lambda(x)$ attracts functions of both of these kinds. We give some examples.

EXAMPLE 1. Let

$$F(x) = \begin{cases} 0 & \text{for } x < 0 \\ 1 - e^{-x^\alpha} & \text{for } x > 0 \end{cases}$$

where $\alpha > 0$ is constant. Without difficulty we find that

$$F^n(a_n x + b_n) \to e^{-e^{-x}},$$

where

$$a_n = \frac{1}{\alpha}(\log n)^{(\alpha-1)/\alpha}; \qquad b_n = (\log n)^{1/\alpha}.$$

EXAMPLE 2. Let

$$F(x) = \begin{cases} 0 & \text{for } x \leq 0 \\ 1 - e^{-x/(1-x)} & \text{for } 0 < x \leq 1 \\ 1 & \text{for } x > 1 \end{cases}$$

Here also, as $n \to \infty$

$$F^n(a_n x + b_n) \to e^{-e^{-x}}$$

if

$$a_n = \log^{-2} n, \qquad b_n = \frac{\log n}{1 + \log n}.$$

Lemma 5. *If, for some a_n and b_n, and for all values of x, we have*

$$\lim_{n \to \infty} n(1 - F(a_n x + b_n)) = e^{-x} \tag{53}$$

then

$$\frac{a_n}{b_n} \to 0 \qquad as \qquad n \to \infty. \tag{54}$$

PROOF. From (53), we conclude that, for all x

$$a_n x + b_n \to \infty \qquad as \qquad n \to \infty \tag{55}$$

if $F(x) < 1$ for all values of x, and that

$$a_n x + b_n \to x_0 \tag{56}$$

if $F(x_0) = 1$ while $F(x_0 - \varepsilon) < 1$ for all $\varepsilon > 0$. Putting $x = -A$ ($A > 0$) it is, then, clear from (55) and (56) that in the case $x_0 > 0$, we have, for all sufficiently large n, and all A

$$a_n A < b_n,$$

that is

$$\frac{a_n}{b_n} < \frac{1}{A}.$$

Since A is arbitrary, relation (54) follows.

If $x_0 < 0$, we derive from (56), by putting $x = 0$ that

$$b_n \to x_0 \qquad as \qquad n \to \infty.$$

From the same relation (56), putting $x = 1$, we obtain

$$a_n + b_n \to x_0 \qquad as \qquad n \to \infty$$

whence $a_n \to 0$ as $n \to \infty$, and consequently $a_n/b_n \to 0$ as $n \to \infty$.

In the case $x_0 = 0$, then, although $b_n \to x_0 = 0$ as $n \to \infty$, we have $b_n < 0$ for all sufficiently large n, because if this were not so we would have, for $x > 0$, $a_n x + b_n > 0$ which would lead to the equation

$$n(1 - F(a_n x + b_n)) = 0$$

which contradicts condition (53) of the lemma.

From (53), for sufficiently large n and for all x, we have

$$a_n x + b_n < 0;$$

therefore, for $x > 0$

$$\frac{a_n}{-b_n} < \frac{1}{x}.$$

Since this inequality holds for all x, relation (54) follows.   □

Theorem 6. *A necessary and sufficient condition for a distribution function $F(x)$ to belong to the domain of attraction of the law $\Lambda(x)$ is that the relation*

$$\lim_{n \to \infty} n(1 - F(a_n x + b_n)) = e^- x \tag{57}$$

holds for all values of x, where the constants b_n are defined as the smallest values of x satisfying the inequalities

$$F(x - 0) \leq 1 - \frac{1}{n} \leq F(x + 0) \tag{58}$$

and the constants a_n are the smallest values of x satisfying the inequalities

$$F(x(1 - 0) + b_n) \leq 1 - \frac{1}{ne} \leq F(x(1 + 0) + b_n). \tag{59}$$

PROOF. From Lemma 4, the stated conditions are sufficient for a distribution function $F(x)$ to belong to the domain of attraction of the law $\Lambda(x)$.

Conversely, if $F(x)$ belongs to the domain of attraction of the law $\Lambda(x)$, then, from Lemma 4, we must have, for some α_n and β_n

$$n(1 - F(\alpha_n x + \beta_n)) \to e^{-x} \tag{60}$$

as $n \to \infty$ for all x.

For all $\varepsilon > 0$, from (60), for sufficiently large n, the inequalities

$$n(1 - F(\alpha_n \varepsilon + \beta_n) + \eta < 1 < n(1 - F(-\alpha_n \varepsilon + \beta_n) - \eta$$

and

$$n(1 - F(\alpha_n(1 + \varepsilon) + \beta_n) + \eta < \frac{1}{\varepsilon} < n(1 - F(\alpha_n(1 - \varepsilon) + \beta_n) - \eta$$

must hold, with $\eta = \frac{1}{2e}(1 - e^- \varepsilon)$.

From these inequalities, together with (58) and (59), we derive for all ε, the inequalities

$$-\alpha_n \varepsilon + \beta_n \leq b_n \leq \alpha_n \varepsilon + \beta_n,$$

$$\alpha_n(1 - \varepsilon) + \beta_n \leq a_n + b_n \leq \alpha_n(1 + \varepsilon) + \beta_n.$$

Since these inequalities hold for all $\varepsilon > 0$, and all sufficiently large n, we can choose a sequence $\varepsilon_n > 0$ ($\varepsilon_n \to 0$ as $n \to \infty$) in such a way as to have

$$-\alpha_n \varepsilon + \beta_n \le b_n \le \alpha_n \varepsilon_n + \beta_n,$$

$$\alpha_n(1 - \varepsilon) + \beta_n \le a_n + b_n \le \alpha_n(1 + \varepsilon_n) + \beta_n.$$

The first of these inequalities provides us with the inequality

$$\left| \frac{b_n - \beta_n}{\alpha_n} \right| \le \varepsilon_n$$

and the second, together with the one just presented, leads to the inequality

$$\left| \frac{\alpha_n}{\alpha_n} - 1 \right| \le 2\varepsilon_n.$$

Hence, from Lemma 2, we can state that whenever the relation (60) is satisfied for a certain choice of α_n and β_n, so also is (57), when a_n and b_n are defined by (58) and (59). The theorem is therefore proved. $\qquad\square$

Theorem 7. *In order for a distribution function $F(x)$ to belong to the domain of attraction of the law $\Lambda(x)$, it is necessary and sufficient that there exist a continuous function $A(z)$ such that $A(z) \to 0$ as $z \to z_0$, and that for all values of x*

$$\lim_{z \to x_0 - 0} \frac{1 - F(z(1 + A(z)x))}{1 - F(z)} = e^{-x}, \tag{61}$$

the number $x_0 \le \infty$ being determined by the relations $F(x_0) = 1$, $F(x) < 1$ for $x < x_0$.

PROOF. First let us suppose that $F(x)$ belongs to the domain of attraction of the law $\Lambda(x)$. Then for a certain choice of the constants $a_n > 0$ and b_n we have

$$\lim_{n \to \infty} n(1 - F(a_n x + b_n)) = e^{-x} \tag{62}$$

for all values of x. It follows that

$$\lim_{n \to \infty} n(1 - F(b_n)) = 1. \tag{63}$$

Clearly the equality $a_n = 0$ can only hold for a finite number of values of n; we can therefore always assume that $a_n > 0$ for all values of n. From Theorem 6, we can take the b_n's to be nondecreasing functions of n.

Let us put $A(b_n) = a_n/b_n$ for all values of n, and, for $b_{n-1} \le z \le b_n$, define $A(z)$ so that it is a continuous and monotonic function of z. Clearly, for all sufficiently great $z < x_0$ it is always possible to find an integer n such that $b_{n-1} \le z \le b_n$. Hence

$$1 - F(b_n) \le 1 - F(z) \le 1 - F(b_{n-1}). \tag{64}$$

From the definition of the function $A(z)$ we must have either

$$\frac{a_{n-1}}{b_{n-1}} \le A(z) \le \frac{a_n}{b_n}$$

or

$$\frac{a_n}{b_n} \le A(z) \le \frac{a_{n-1}}{b_{n-1}}$$

In the first case we see that for $x > 0$

$$a_{n-1}x + b_{n-1} \le z(1 + A(z)x) \le a_n x + b_n$$

and for $x < 0$

$$a_n x + b_{n-1} \le z(1 + A(z)x) \le a_{n-1}x + b_n.$$

Therefore, in the first case we have, for $x > 0$

$$1 - F(a_n x + b_n) \le 1 - F(z(1 + A(z)x)) \le 1 - F(a_{n-1}x + b_{n-1}) \quad (65)$$

and, for $x < 0$

$$1 - F(a_{n-1}x + b_n) \le 1 - F(z(1 + A(z)x)) \le 1 - F(a_n x + b_{n-1}). \quad (66)$$

From (63), (64) and (65) we obtain, for $x > 0$, the inequalities

$$(n - 1)[1 - F(a_n x + b_n)] \le \frac{1 - F(z(1 + A(z)x))}{1 - F(z)} \le n[1 - F(a_{n-1}x + b_{n-1})].$$

By virtue of Lemmas 3 and 2 and the relation (62), these inequalities imply, for $x > 0$, equation (61).

We obtain (61), starting from (66), in a similar way. In the case not yet considered $((a_n/b_n) \le A(z) \le (a_{n-1}/b_{n-1}))$ we can reach (61) again by analogous arguments. This completes the proof of the necessity of condition (61).

As for the proof of sufficiency of this condition, we first of all note that we obtain from (61) the equation,

$$\lim_{z \to x_0 - 0} \frac{1 - F(z + 0)}{1 - F(z)} = 1. \quad (67)$$

In fact, since we have, for all $x > 0$,

$$F(z(1 + A(z)x)) \ge F(z + 0)$$

we can write

$$1 \ge \frac{1 - F(z + 0)}{1 - F(z)} \ge \frac{1 - F(z(1 + A(z)x))}{1 - F(y)}.$$

Hence

$$1 \ge \varlimsup_{z \to x_0 - 0} \frac{1 - F(z + 0)}{1 - F(z)} \ge \varliminf_{z \to x_0 - 0} \frac{1 - F(z + 0)}{1 - F(z)} \ge e^{-x}.$$

Since these inequalities hold for all values of x, they also hold for the limit as $x \to 0$.

Now we suppose that the conditions of the theorem hold and we will show that, then, $F(x)$ belongs to the domain of attraction of the law $\Lambda(x)$. To this end, we define b_n to be the smallest value of x satisfying

$$1 - F(x + 0) \leq \frac{1}{n} \leq 1 - F(x - 0) = 1 - F(x).$$

From this we obtain

$$\frac{1 - F(b_n(1 + A(b_n)x))}{1 - F(b_n - 0)} \leq n(1 - F(b_n(1 + A(b_n)x))) \leq \frac{1 - F(b_n(1 - A(b_n)x))}{1 - F(b_n + 0)}.$$

From (61) and (67) we see that

$$\lim_{n \to \infty} n(1 - F(b_n(1 + A(b_n)x)) = e^{-x}.$$

Putting $a_n = b_n A(b_n)$, we obtain (62). The theorem is proved. From this theorem there results the following corollary. $\square$

Corollary 1. *Suppose that the distribution function $F(x)$ is such that $F(x) < 1$ for all x. Then, for the function $F(x)$ to belong to the domain of attraction of the law $\Lambda(x)$, it is necessary that*

$$\frac{1 - F(ky)}{1 - F(y)} \to 0$$

for all positive constant k, as $y \to \infty$, that is to say, that the sequence of maxima is relatively stable.

PROOF. Put

$$\Phi_z(x) = \frac{1 - F(z(1 + A(z)x))}{1 - F(z)}; \tag{68}$$

the functions $\Phi_z(x)$ are nonincreasing with respect to x. On several occasions we have made use of the remark that if a sequence of monotone functions converges, at all points, to a continuous function, the convergence is uniform. From the uniform convergence of $\Phi_z(x)$ to e^{-x}, if x_z is a sequence tending to infinity as $z \to \infty$, we must have

$$\lim_{z \to \infty} \Phi_z(x_z) = \lim_{z \to \infty} e^{-x_z} = 0. \tag{69}$$

Take $\alpha > 0$ and put $x_z = \alpha/A(z)$.

By the definition of the function $A(z)$, we have, for $z \to \infty$

$$\lim_{z \to \infty} A(z) = \infty.$$

From (68) and (69)

$$\lim_{z \to \infty} \frac{1 - F((1 + \alpha)z)}{1 - F(z)} = 0.$$

It is easy to see that the necessary condition we have just found is by no means sufficient. In order to show this, we will consider a distribution function defined as follows:

$$F(x) = \begin{cases} 0 & \text{for } x < 0 \\ 1 - e^{-[x]} & \text{for } x > 0 \end{cases}$$

where $[x]$ denotes the integer part of x, and demonstrate that it is impossible to choose a_n's and b_n's such that

$$\lim_{n \to \infty} n(1 - F(a_n x + b_n)) = e^{-x}$$

for all values of x.

In other words, we will show that there cannot exist constants a_n and b_n for which

$$\lim_{n \to \infty} n e^{-[a_n x + b_n]} = e^{-x}$$

for all x, or equivalently

$$\lim_{n \to \infty} (\log n - [a_n x + b_n]) = -x$$

for all values of x. We will consider the subsequence $n_k = [e^{k+0.5}]$ where k is an integer. As $k \to \infty$, $\log n_k - k \to 0.5$, and so there cannot exist b_{n_k}'s such that

$$\lim_{k \to \infty} (\log n_k - [a_n \cdot 0 + b_{n_k}]) = \lim_{k \to \infty} (\log n_k - [b_{n_k}]) = 0.$$

This demonstrates the corollary.

Corollary 2. *Let $F(x)$ be a distribution function. If there exists a sequence $x_1 < x_2 < \cdots < x_k < \cdots$, with $\lim_{k \to \infty} x_k = x_0{}^4$ ($x_0 \le +\infty$) such that*

$$\frac{1 - F(x_k - 0)}{1 - F(x_k + 0)} \ge 1 + \beta \tag{70}$$

and the constant β is positive, then the function $F(y)$ cannot belong to the domain of attraction of $\Lambda(x)$.

PROOF. Inequality (70) is incompatible with the equality (67), which follows from (61). □

Remark. It follows from Theorems 4 and 5 that if condition (70) holds $F(x)$ cannot belong to the domains of attraction of the laws $\Phi_\alpha(x)$ and $\Psi_\alpha(x)$.

EXAMPLE. The Poisson law

$$F(x) = \begin{cases} 0 & \text{for } x < 0 \\ \displaystyle\sum_{0 \le k < x} e^{-\lambda} \frac{\lambda^k}{k!} & \text{for } x > 0 \end{cases}$$

is not attracted to any of our three limit laws. In fact, putting $x_k = k$, we see

---

[4] Here x_0 is defined in the same way as in Theorem 7.

that

$$\frac{1 - F(k - 0)}{1 - F(k + 0)} = \frac{\sum_{s \geq k} \dfrac{\lambda^s}{s!}}{\sum_{s > k} \dfrac{\lambda^s}{s!}} \geq \frac{k + 1}{\lambda};$$

therefore condition (70) holds for $k + 1 > \lambda$.

The following theorem gives a simple necessary and sufficient condition for convergence to the law $\Lambda(x)$ with a special choice of values of the constants a_n.

Theorem 8. *A necessary and sufficient condition that the distribution function* $F(x)$ *satisfies*

$$F^n(ax + b_n) \to \Lambda(x) \tag{71}$$

as $n \to \infty$, *for a suitable choice of values of* a *and* b_n, *is that*

$$\frac{1 - F(\log x)}{1 - F(\log kx)} \to k^\alpha \qquad as \qquad x \to \infty \tag{72}$$

for all constant $k > 0$, *where* $\alpha a = 1$.

PROOF. If condition (71) holds, we have, from Lemma 5, the inequality $F(x) < 1$ for all values of x, and we see that $b_n \to \infty$ as $n \to \infty$.

Therefore, it is obvious that determining whether (71) holds is equivalent to establishing conditions under which

$$\lim_{n \to \infty} F^n(x + b_n) = e^{-e^{-ax}}. \tag{73}$$

Put $x = \log z$, $b_n = \log \beta_n$; $F_1(z) = F(\log z)$.

It is clear that $F_1(z)$ is a distribution function. In these circumstances the determination of conditions under which (73) holds, comes down to the same question for

$$F_1^n(\beta_n z) = F^n(\log \beta_n z) \to e^{-z^{-\alpha}}. \tag{74}$$

Necessary and sufficient conditions for (74) to hold have been found in §3; they are equivalent to condition (72).

A convenient criterion for applications, to see whether the law $F(x)$, satisfying the condition $F(x) < 1$ for all values of x, belongs to the domain of attraction of the law $\Lambda(x)$, has been stated by Mises in his cited work. Mises' condition is:

Let $F(x)$ be a function having derivatives of first two orders for all x greater than a certain value x_0. Put

$$f(x) = \frac{F'(x)}{1 - F(x)};$$

then if

$$\lim_{x \to \infty} \frac{d}{dx}\left[\frac{1}{f(x)}\right] = 0,$$

we have

$$\lim_{n \to \infty} F^n\left(x_n + \frac{x}{nF'(x_n)}\right) = e^{-e^{-x}}$$

where x_n is the smallest root of the equation $1 - F(x) = \frac{1}{n}$. A similar condition can be derived for a function not satisfying the condition $F(x) < 1$ for all x. If there exists a value x_0 such that, for all $\varepsilon > 0$

$$F(x_0 - \varepsilon) < 1, \qquad F(x_0) = 1$$

and, for some $x < x_0$, $F(x)$ has first and second derivatives, with

$$\lim_{x \to x_0 - 0} \frac{d}{dx}\left[\frac{1}{f(x)}\right] = 0$$

then $F(x)$ belongs to the domain of attraction of the law $\Lambda(x)$.   □

References

1. M. Fréchet. *Sur la loi de probabilité de l'écart maximum.* Annales de la société polonaise de Mathématiques, t. VI, p. 93, Cracovie, 1927.
2. R.A. Fisher and L.H.C. Tippett. *Limiting forms of the frequency distribution of the largest or smallest member of a sample.* Proceedings Cambridge Philos. Soc. V, XXIV, part II, pp. 180–190, 1928.
3. R. de Misès. *La distribution de la plus grande de n valeurs.* Revue Mathématique de l'union interbalkanique, t. I. f. 1, p. 141–160, 1939.
4. A. Khintchine. Limit theorems for sums of independent random variables (In Russian), Moscow, 1938.
5. B. Gnedenko. *To the theory of the domains of attraction of stable laws,* Uchenye Zapiski of the Moscow State University, t. 30, p.p. 61–81 (1939).
6. B. de Finetti. *Sulla legge di probabilità degli estremi,* Metron v. IX, No. 3–9, p.p. 127–138 (1932).

Introduction to
Rao (1945) Information and the Accuracy Attainable in the Estimation of Statistical Parameters

P.K. Pathak
University of New Mexico

1. General Remarks

The object of this introduction is to present a brief account of a paper that remains an unbroken link in the continuing evolution of modern statistics. In it are contained some of the fundamental paradigms of the foundations of statistics, e.g., unbiased minimum variance estimation, the Cramér–Rao inequality, the Rao–Blackwellization, and the Rao space of probability distributions furnished with a Riemannian quadratic differential metric and the associated geodesic distance (called the Rao distance). The paper furnishes some of the basic theoretical tools for the discussion of a variety of problems in statistical inference.

The concept of averaging is now widely recognized as an indispensable tool in probability, statistics, and other disciplines. In probability, its great potential was noticed by J.L. Doob (1953) in the defining property of martingales, whereas in statistics its use in the form of Rao–Blackwellization was initiated by C.R. Rao. Both of these paradigms, based on ideas of conditioning, have now become standard tools for statisticians. The Cramér–Rao inequality, the finite-sample analogoue of the Fisher information inequality, is now widely used in diverse areas such as signal processing and optics [see Cedarquist et al. (1986) and Fossi et al. (1989)]. Rao's distance (or divergence between probability distributions) has become a basic tool in constructing tests for comparing populations and in cluster analysis.

2. Brief Review

Although the problem of estimation of parameters in the formal sense began in the middle of the 19th century with the method of least squares due to

Gauss and Markoff, it was not until the first half of this century that the theory was put on a sound theoretical basis in the pioneering papers of Fisher (1922) and Rao (1945). This development was made possible to a great extent because of their rigorous mathematical treatment of the underlying statistical subject matter.

Rao's paper is divided into eight sections. Of these, the first outlines the Fisherian concepts of consistency, efficiency, and sufficiency, as well as Aitken's (1941) model for unbiased minimum variance estimation. The second section notes that the minimum mean square error, or for that matter the criterion of minimum variance for unbiased estimators, is necessary if estimators under study are to have the maximum concentration of probability around the parameter of interest. The section concludes with a note of caution about the somewhat arbitrariness of the requirement of unbiasedness as follows:

> Until a unified solution of the problem of estimation is set forth, we have to subject the estimating functions to a critical examination as to its bias, variance and the frequency of a given amount of departure of the estimating function from the parameter before utilising it.

Section 3 investigates the problem of estimation of a single real-valued parameter. In it are presented the two time-honored theorems of statistical inference, namely, the Cramér–Rao inequality and the Rao–Blackwell theorem. The proof of the Cramér–Rao inequality is based on a simple yet ingenious application of the Cauchy–Schwarz inequality, whereas the proof of the Rao–Blackwell theorem is a consequence of the transformation of variables and the decomposition of the variance into two orthogonal components. It is stated that if T is sufficient for θ and t is such that $E(t) = \theta$, then $f(T) = E(t|T)$ is unbiased for θ with a variance not greater than that of t. An illustrative example based on the Koopman one-parameter exponential family is presented.

Section 4 is devoted to the case of several parameters, and the multivariate analogue of the Cramér–Rao inequality is derived. It is shown that for the Koopman multiparameter exponential family, these bounds are attained.

Section 5 is devoted to a study of the loss of information when the sample data are replaced by a set of statistics. Among other things, it is shown that the score function based on a statistic is simply the Rao–Blackwellization of the score function based on the sample given this statistic. Fisher's information inequality is then an immediate consequence of the Rao–Blackwell theorem. The multivariate analogue of Fisher's information inequality is established.

Sections 6 and 7 are devoted to a differential geometric approach to the study of neighborhoods of probability distributions indexed by a vector-valued parameter. It is shown that the geodesic distance based on the element of length given by $ds^2 = \sum I_{ij} \, d\theta_i \theta_j$, in which I_{ij} denotes the ijth element of the information matrix, is a suitable measure of the distance between two probability distributions.

Section 8 outlines a general procedure for tests of significance and classification based on a given measure of distance between two populations. The possibility of using Rao's geodesic distance for this purpose is explored.

The theory and practice of statistics underwent a revolutionary change in the first half of the present century. The use of standard errors in estimation and testing developed by Karl Pearson is appropriate only for large samples, and there was a pressing need to develop exact theory for sample sizes that often arise in practical work. Student's seminal work on the t-distribution provided a breakthrough; it led to the development by R.A. Fisher of exact small-sample tests in a variety of situations.

In his fundamental paper, "Mathematical Foundations of Theoretical Statistics," Fisher (1922) introduced the concepts of consistency, efficiency, sufficiency, and loss of information in the context of estimation. Of these, consistency and efficiency (in the sense of asymptotic variance) are essentially large-sample concepts and their applicability in small samples remained to be explored. Rao's paper (1945) provided the answer for small samples by replacing consistency by unbiasedness and asymptotic variance by exact variance. All the results of Fisher on the asymptotic behavior of efficient estimators found their counterpart in Rao's small-sample theory.

Fisher's concept of sufficiency as introduced in his 1922 paper and developed by him in later papers had an intuitive appeal due to the interesting property that a sufficient statistic has the same information (defined in a certain sense) as the whole sample. The first concrete demonstration of the deep significance of sufficiency appears in Rao's paper as the process of Rao–Blackwellization. Thus, Rao's paper (1945) complements the fundamental ideas of Fisher's paper (1922) and forms an important link in the transition from large-sample theory to exact small-sample theory with a rigorous mathematical foundation.

3. Impact on Current Research

The object of this section is to briefly review the impact of the paper of C.R. Rao on current research in statistics and more specifically in estimation theory. Over the last four decades, research in the theory of estimation has continued at a remarkable pace. A significant amount of this work has been either a direct extension of the results contained in C.R. Rao's paper or motivated by the ideas outlined in it.

3.1. The Rao–Blackwell Theorem

The Rao–Blackwell theorem under the L_2-metric was first established by Rao (1945). Under the same metric, it was discovered *independently* by Blackwell (1947) who extended its applicability to unbiased estimation under optional stopping rules (sequential procedures). Blackwell's method is an ingenius classical approach which has been extensively used in statistical inference for the last 10 years. It was later extended to the L_p-metric by Barankin (1950) and for convex functions by Hodges and Lehmann (1950). Bahadur (1957) established the converse of the Rao–Blackwell theorem when the underlying family of probability measures is dominated by a σ-finite measure. More recently, Torgersen (1988) has shown that this converse goes through for any family of probability measures that is dominated by a nonnegative measure. In the context of size-biased sampling, Pathak (1964a, 1965) showed that the Rao–Blackwell theorem goes through for certain conditional sampling schemes as well as for ratio estimators of population ratios commonly employed in practice. Some of the major implications of the Rao–Blackwell theorem are as follows:

For convex loss function, randomization is irrelevant in the sense that given an estimator, its Rao–Blackwellization provides a more efficient estimator that is free of randomization.

Suppose that an efficient estimator of $\pi(\theta)$ is desired. Let $d(X)$ be an unbiased estimator of $\pi(\theta)$, but one known to be inefficient. Typically in applications, $d(X)$ is based on only a few observations and consequently inadmissible. Then the Rao–Blackwellization of $d(X)$ provides a simple procedure of constructing an efficient estimator of $\pi(\theta)$. This technique has been used effectively for the construction of efficient unbiased estimators in sampling theory by Basu (1958), Pathak (1976), Mitra and Pathak (1984), Kremers (1986), and others.

Let T be a complete sufficient statistic. Then a uniformly minimum variance unbiased estimator of $\pi(\theta)$ exists if and only if the system of equations

$$E_\theta d(T) = \pi(\theta)$$

for all θ has a solution; such a solution must necessarily be unique. Thus, if the parametric function $\pi(\theta)$ admits an unbiased estimator $d(T)$, then $d(T)$ is the uniformly minimum variance unbiased estimator of $\pi(\theta)$. It may be noted that the notion of completeness of a sufficient statistic T in the sense that $Ef(T) = 0$ for all values of the underlying parameter implies that $f(T) = 0$ was introduced in a later paper by Rao (1948). This led to the notion of the uniqueness of the Rao–Blackwellized estimator. For certain important applications of these ideas, we refer the reader to Kolmogorov (1950), Doob (1953), Lehmann and Scheffé (1950, 1955, 1956), Pathak (1964b, 1975), and others.

3.2. The Cramér–Rao Inequality

During a lecture at Calcutta University in 1943, C.R. Rao, who was then hardly 23 years old, proved in his class a result first obtained by R.A. Fisher regarding the lower bound for the variance of an estimator for large samples. When a student asked, "Why don't you prove it for finite samples?", Rao went back home, worked all night, and the next day proved what is now known as the Cramér–Rao inequality for finite samples. Work on this inequality has continued unabated ever since. The inequality has been strengthened and generalized from numerous angles. Some of the major contributions on this inequality have been made by Bhattacharya (1946), Blyth (1974), Blyth and Roberts (1972), Chapman and Robbins (1951), Fabian and Hannan (1977), Hammersley (1950), Simons (1980), Simons and Woodroofe (1983), Wolfowitz (1947), and others. In addition to its applications in establishing that in certain standard applications the inequality is attained, its most useful applications in statistics have been in the area of decision theory and probability. Hodges and Lehmann (1951) use the Cramér–Rao inequality in establishing the admissibility of certain statistical procedures. For applications in proving limit theorems in probability, reference may be made to Brown (1982) and Mayer–Wolf (1988). In the latter paper, the concept of the Cramér–Rao functional is developed and used as a variational tool.

In a follow-up paper, Rao (1946) considered the extension of his results to the multiparameter case in some detail. Independently Cramér (1945, 1946) discovered the univariate parameter result (1945) and extended it (1946) to the multiparameter case. It is also interesting to note that the elegant proof of the Cramér–Rao inequality was used by Kallianpur and Rao (1955) in establishing Fisher's inequality for an asymptotic variance, under Fisher consistency, in a rigorous manner. All these earlier results paved the way for Rao's fundamental work on second-order efficiency published in 1961 and 1963.

3.3. Differential Geometry, Riemannian Space, and Geodesics

Over 45 years ago, when Rao introduced differential geometric methods in statistical inference, statisticians were not conversant with this branch of mathematics. His work remained unnoticed for a long time, and it is only recently that systematic research on the application of differential-geometric methods in statistical inference has started. [A principal figure in this area is now Shun-ichi Amari, who refers to Rao's pioneering work in his valuable monograph, *Differential-Geometric Methods in Statistics* (1985).]

Among recent developments may be mentioned the work of Efron (1975) on statistical curvature and the interpretation of the Fisher–Rao measure of second-order efficiency of an estimator, of Amari (1985) on affine connections in Riemannian geometry and their use in the higher-order asymptotic theory of statistical inference, and of Burbea and Oller (1989) on the use of Rao's geodesic distance as a test criterion for comparing different population distributions, as an alternative to the likelihood ratio criterion.

4. Biography

Calyampudi Radhakrishna Rao was born on September 10, 1920 in Hadagali, Karnataka State, India. He received an M.A. in mathematics in 1940 from Andhra University and an M.A. in statistics in 1943 from Calcutta University. From Cambridge University, he received a Ph.D. in 1948 completing a thesis entitled "Statistical Problems of Biological Classification" under R.A. Fisher and an Sc.D. in 1965 for significant contributions to statistical theory and applications. Dr. Rao was elected a Fellow of the Royal Society of U.K. in 1967, and has received 14 honorary doctorate degrees from universities all over the world. Dr. Rao became a professor at the Indian Statistical Institute in 1949 at the early age of 29, the Director of the Research and Training School of the Indian Statistical Institute in 1964, Secretary and Director of the Institute in 1972, and a Jawaharlal Nehru Professor in 1976. He was a professor at the University of Pittsburgh during 1979–88. He currently holds the Eberly Chair in Statistics at Pennsylvania State University and the title of National Professor, awarded to him by the government of India for his outstanding contributions to science.

Some of the well-known results in statistics that bear his name are the Rao–Blackwell theorem, the Cramér–Rao inequality, the Fisher–Rao theorem [on second-order efficiency; see Efron (1975)], Rao's U-test (in multivariate analysis), Rao's score test, the Neyman–Rao test [see Hall and Mathiason (1990)], Rao's distance, Rao's MINIQUE (for the estimation of variance components), Rao's orthogonal arrays (in combinatorics), the Rao–Hamming bound (in coding theory), Rao's quadratic entropy, the Kagan–Linnik–Rao theorem, the Rao–Rubin theorem, etc. To these may be added Rao's theory of the generalized inverse of matrices, which provided a valuable matrix operator for developing a statistical methodology in linear models.

One of the reasons for Dr. Rao's exceptional success undoubtedly lies in the counsel he received from his mentor R.A. Fisher in developing statistical methodology from given data rather than pursuing a mathematical type of research following the work of others. In his statistical career, Dr. Rao has adhered steadfastly to the Fisherian approach to statistics, although he has drawn inspiration from diverse applications. He has collaborated effectively with anthropologists, biologists, geologists, national planners, economists, and demographers.

Any account of Dr. Rao's biography would be incomplete if it made no mention of his superb human qualities and his personal life that has been just as exemplary as his professional life. In his own anonymous ways, he has been aiding numerous individuals in need of help throughout the world. His exceptional organizing skills and personal dedication for half a century have shaped the Indian Statistical Institute to what it is today, from a humble beginning to a large and prestigious international institute of higher learning.

References

Aitken, A.C. (1941). On the estimation of statistical parameters, *Proc. Roy. Soc. Edin.*, **61**, 56–62.

Amari, S. (1985). *Differential-Geometric Methods in Statistics*. Springer-Verlag, New York.

Bahadur, R.R. (1957). On unbiased estimates of uniformly minimum variance, *Sankhyā*, **18**, 211–224.

Barankin, E.W. (1950). Extension of a theorem of Blackwell, *Ann. Math. Statist.* **21**, 280–284.

Basu, D. (1958). On sampling with and without replacement, *Sankhyā*, **20**, 287–294.

Bhattacharya, A. (1946). On some analogues to the amount of information and their uses in statistical estimation, *Sankhyā*, **8**, 1–14, 201–208.

Blackwell, D. (1947). Conditional expectation and unbiased sequential estimation, *Ann. Math. Statist.*, **18**, 105–110.

Blyth, C.R. (1974). Necessary and sufficient conditions for inequalities of Cramér–Rao type, *Ann. Statist.*, **2**, 464–473.

Blyth, C.R., and Roberts, D.M. (1972). On inequalities of Cramér–Rao type and admissibility proofs, in *Proceedings of 6th Berkeley Symposium on Mathematical Statistics and Probability*. University of California Press, Vol. 1, pp. 17–30.

Brown, L.D. (1982). A proof of the central limit theorem motivated by the Cramér–Rao inequality, in *Statistics and Probability: Essays in Honor of C.R. Rao* (Kallianpur et al., eds.). North-Holland, Amsterdam, pp. 141–148.

Burbea, J., and Oller, J.M. (1989). On Rao distance asymptotic distribution. *Mathematics Reprint Series*, No. 67, University of Barcelona, Spain.

Cedarquist, J., Robinson, S.R., and Kryskowski, D. (1986). Cramér–Rao lower bound on wavefront sensor error, *Opt. Eng.*, **25**, 586–592.

Chapman, D.C., and Robbins, H. (1951). Minimum variance estimation without regularity assumptions, *Ann. Math. Statist.*, **22**, 581–586.

Cramér, H. (1945). *Mathematical Methods of Statistics*, Almqvist and Wiksell, Uppsala, Sweden.

Cramér, H. (1946). Contributions to the theory of statistical estimation, *Skand. Aktuarietidsk*, **29**, 85–94.

Doob, J.L. (1953). *Stochastic Processes*. Wiley, New York.

Efron, B. (1975). Defining the curvature of a statistical problem (with applications to second order efficiency), *Ann. Statist.*, **3**, 1189–1242.

Fabian, V., and Hannan, J. (1977). On the Cramér–Rao inequality, *Ann. Statist.*, **5**, 197–205.

Fisher, R.A. (1922). On the mathematical foundations of theoretical statistics, *Philos. Trans. Roy. Soc. London, Ser. A*, **222**, 309–368.

Fossi, M., Giuli, D., and Gherardelli, M. (1989). Cramér–Rao bounds and maximum-likelihood estimation of Doppler frequency of signals received by a polarimetric radar, *IEEE Proc., Part F, Radar and Signal Processing*, **136**, 175–184.

Hall, W.J., and Mathiason, D.J. (1990). On large-sample estimation and testing in parametric models, *Int. Statist. Rev.*, **58**, 77–97.

Hammersley, J.M. (1950). On estimating restricted parameters, *J. Roy. Statist. Soc., Ser. B*, **12**, 192–240.

Hodges, J.L., Jr., and Lehmann, E.L. (1950). Some problems in minimax point estimation, *Ann. Math. Statist.*, **21**, 182–197.

Hodges, J.L., Jr., and Lehmann, E.L. (1951). Some applications of the Cramér–Rao inequality, in *Proceedings of 2nd Berkeley Symposium on Mathematical Statistics and Probability*. University of California Press, pp. 13–22.

Kallianpur, G., and Rao, C.R. (1955). On Fisher's lower bound to asymptotic variance of a consistent estimate, *Sankhyā*, **17**, 105–114.

Kolmogorov, A.N. (1950). Unbiased estimates, *Izvestia Akad. Nauk SSSR, Ser. Math.*, **14**, 303–326 (American Mathematical Society Translation No. **98**).

Kremers, W.K. (1986). Completeness and unbiased estimation for sum-quota sampling, *J. Amer. Statist. Assoc.*, **81**, 1070–1073.

Lehmann, E.L., and Scheffé, H. (1950, 1955, 1956). Completeness, similar regions, and unbiased estimation, *Sankhyā*, **10**, 305–340; **15**, 219–236; **17**, 250.

Mayer-Wolf, E. (1988). The Cramér-Rao functional and limit laws. *Mimeo Series*, No. **1773**, Department of Statistics, University of North Carolina.

Mitra, S.K., and Pathak, P.K. (1984). The nature of simple random sampling, *Ann. Statist.*, **12**, 1536–1542.

Pathak, P.K. (1964a). On sampling schemes providing unbiased ratio estimators, *Ann. Math. Statist.*, **35**, 222–231.

Pathak, P.K. (1964b). On inverse sampling with unequal probabilities, *Biometrika*, **51**, 185–193.

Pathak, P.K. (1965). Estimating population parameters from conditional sampling schemes. 35th Session of the International Statistical Institute, Belgrade, Yugoslavia.

Pathak, P.K. (1975). An extension of a theorem of Hoeffding, *Studia Sci. Math. Hung.*, **10**, 73–74.

Pathak, P.K. (1976). Unbiased estimation in a fixed cost sequential sampling scheme, *Ann. Statist.*, **4**, 1012–1017.

Rao, C.R. (1946). Minimum variance and the estimation of several parameters, *Proc. Cambridge Philos. Soc.*, **43**, 280–283.

Rao, C.R. (1948). Sufficient statistics and minimum variance estimation, *Proc. Cambridge Philos. Soc.*, **45**, 215–218.

Rao, C.R. (1961). Asymptotic efficiency and limiting information, in *Proceedings of 4th Berkeley Symposium on Mathematical Statistics and Probability*, Vol. 1. University of California Press, pp. 531–546.

Rao, C.R. (1963). Criteria of estimation in large samples, *Sankhyā*, **25**, 189–206.

Simons, G. (1980). Sequential estimators and the Cramér–Rao lower bound, *J. Statist. Planning and Inf.*, **4**, 67–74.

Simons, G., and Woodroofe, M. (1983). The Cramér–Rao inequality holds almost everywhere, in *Recent Advances in Statistics* (papers in honor of Herman Chernoff on his 60th birthday) (Rizvi et al., eds.). Academic Press, New York, pp. 69–93.

Torgersen, E. (1988). On Bahadur's converse of the Rao–Blackwell theorem. Extension to majorized experiments, *Scand. J. Statist.*, **15**, 273–280.

Wolfowitz, J. (1947). The efficiency of sequential estimates and Wald's equation for sequential processes, *Ann. Math. Statist.*, **18**, 215–230.

Information and the Accuracy Attainable in the Estimation of Statistical Parameters

C. Radhakrishna Rao

Introduction

The earliest method of estimation of statistical parameters is the method of least squares due to Markoff. A set of observations whose expectations are linear functions of a number of unknown parameters being given, the problem which Markoff posed for solution is to find out a linear function of observations whose expectation is an assigned linear function of the unknown parameters and whose variance is a minimum. There is no assumption about the distribution of the observations except that each has a finite variance.

A significant advance in the theory of estimation is due to Fisher (1921) who introduced the concepts of *consistency*, *efficiency* and *sufficiency* of estimating functions and advocated the use of the maximum likelihood method. The principle accepts as the estimate of an unknown parameter θ, in a probability function $\phi(\theta)$ of an assigned type, that function $t(x_1, \ldots, x_n)$ of the sampled observations which makes the probability density a maximum. The validity of this principle arises from the fact that out of a large class of unbiassed estimating functions following the normal distribution the function given by maximising the probability density has the least variance. Even when the distribution of t is not normal the property of minimum variance tends to hold as the size of the sample is increased.

Taking the analogue of Markoff's set up Aitken (1941) proceeded to find a function $t(x_1, \ldots, x_n)$ such that

$$\int t\phi(\theta)\pi \, dx_i = \theta$$

and

$$\int (t - \theta)^2 \phi(\theta)\pi \, dx_i \qquad \text{is minimum.}$$

Estimation by this method was possible only for a class of distribution functions which admit sufficient statistics. Some simple conditions under which the maximum likelihood provides an estimate accurately possessing the minimum variance, even though the sample is finite and the distribution of the estimating function is not normal, have emerged.

The object of the paper is to derive certain inequality relations connecting the elements of the *Information Matrix* as defined by Fisher (1921) and the variances and covariances of the estimating functions. A class of distribution functions which admit estimation of parameters with the minimum possible variance has been discussed.

The concept of distance between populations of a given type has been developed starting from a quadratic differential metric defining the element of length.

Estimation by Minimising Variance

Let the probability density $\phi(x_1, \ldots, x_n; \theta)$ for a sample of n observations contain a parameter θ which is to be estimated by a function $t = f(x_1, \ldots, x_n)$ of the observations. This estimate may be considered to be the best, if with respect to any other function t', independent of θ, the probabilities satisfy the inequality

$$P(\theta - \lambda_1 < t < \theta + \lambda_2) \not< P(\theta - \lambda_1 < t' < \theta + \lambda_2) \qquad (2.1)$$

for all positive λ_1 and λ_2 in an interval $(0, \lambda)$. The choice of the interval may be fixed by other considerations depending on the frequency and magnitude of the departure of t from θ. If we replace the condition (2.1) by a less stringent one that (2.1) should be satisfied for all λ we get as a necessary condition that

$$E(t - \theta)^2 \not> E(t' - \theta)^2, \qquad (2.2)$$

where E stands for the mathematical expectation. We may further assume the property of unbiassedness of the estimating functions *viz.*, $E(t) = \theta$, in which case the function t has to be determined subject to the conditions $E(t) = \theta$ and $E(t - \theta)^2$ is minimum.

As no simple solution exists satisfying the postulate (2.1) the inevitable arbitrariness of these postulates of unbiassedness and minimum variance needs no emphasis. The only justification for selecting an estimate with minimum variance from a class of unbiassed estimates is that a necessary condition for (2.1) with the further requirement that $E(t) = \theta$ is ensured. The condition of unbiassedness is particularly defective in that many biassed estimates with smaller variances lose their claims as estimating functions when compared with unbiassed estimates with greater variances. There are, however, numerous examples where a slightly biassed estimate is preferred to an unbiassed estimate with a greater variance. Until a unified solution of the problem of estimation is set forth we have to subject the estimating functions to a critical examination as to its bias, variance and the frequency of a given amount of departure of the estimating function from the parameter before utilising it.

Single Parameter and the Efficiency Attainable

Let $\phi(x_1, \ldots, x_n)$ be the probability density of the observations $x_1, x_2, \ldots, x_n$, and $t(x_1, \ldots, x_n)$ be an unbiassed estimate of θ. Then

$$\int \cdots \int t\phi\pi \, dx_i = \theta. \tag{3.1}$$

Differentiating with respect to θ under the integral sign, we get

$$\int \cdots \int t\frac{d\phi}{d\theta}\pi \, dx_i = 1 \tag{3.2}$$

if the integral exists, which shows that the covariance of t and $\dfrac{1}{\phi}\dfrac{d\phi}{d\theta}$ is unity.

Since the square of the covariance of two variates is not greater than the product of the variances of the variates we get using V and C for variance and covariance

$$V(t)V\left(\frac{1}{\phi}\frac{d\phi}{d\theta}\right) \not< \left\{C\left(t, \frac{1}{\phi}\frac{d\phi}{d\theta}\right)\right\}^2 \tag{3.3}$$

which gives that

$$V(t) \not< 1/I$$

where

$$I = V\left(\frac{1}{\phi}\frac{d\phi}{d\theta}\right) = E\left\{-\frac{d^2 \log \phi}{d\theta^2}\right\} \tag{3.4}$$

is the intrinsic accuracy defined by Fisher (1921). This shows that *the variance of any unbiassed estimate of θ is greater than the inverse of I which is defined independently of any method of estimation*. The assumption of the normality of the distribution function of the estimate is not necessary.

If instead of θ we are estimating $f(\theta)$, a function of θ, then

$$V(t) \not< \{f'(\theta)\}^2/I. \tag{3.5}$$

If there exists a sufficient statistic T for θ then the necessary and sufficient condition is that $\phi(x; \theta)$ the probability density of the sample observations satisfies the equality

$$\phi(x; \theta) = \Phi(T, \theta)\psi(x_1, \ldots, x_n), \tag{3.6}$$

where ψ does not involve θ and $\Phi(T, \theta)$ is the probability density of T. If t is an unbiassed estimate of θ then

$$\theta = \int t\phi\pi \, dx_i = \int f(T)\Phi(T, \theta) \, dT \tag{3.7}$$

which shows that there exists a function $f(T)$ of T, independent of θ and is an unbiassed estimate of θ. Also

$$\int (t - \theta^2)\phi\pi \, dx_i = \int [t - f(T)]^2\phi\pi \, dx_i + \int [f(T) - \theta]^2\Phi(T, \theta) \, dT$$

$$\geq \int [f(T) - \theta]^2\Phi(T, \theta) \, dT, \tag{3.8}$$

which shows that

$$V[f(T)] \ngtr V(t) \tag{3.9}$$

and hence we get the result that *if a sufficient statistic and an unbiassed estimate exist for θ, then the best unbiassed estimate of θ is an explicit function of the sufficient statistic.* It usually happens that instead of θ, a certain function of θ can be estimated by this method for a function of θ may admit an unbiassed estimate.

It also follows that if T is a sufficient statistic for θ and $E(T) = f(\theta)$, then there exists no other statistic whose expectation is $f(\theta)$ with the property that its variance is smaller than that of T.

It has been shown by Koopman (1936) that under certain conditions, the distribution function $\phi(x, \theta)$ admitting a sufficient statistic can be expressed as

$$\phi(x, \theta) = \exp(\Theta_1 X_1 + \Theta_2 + X_2), \tag{3.10}$$

where X_1 and X_2 are functions of $x_1, x_2, \ldots, x_n$ only and Θ_1 and Θ_2 are functions of θ only. Making use of the relation

$$\int \exp(\Theta_1 X_1 + \Theta_2 + X_2)\pi \, dx_i = 1, \tag{3.11}$$

we get

$$E(X_1) = -\frac{d\Theta_2}{d\Theta_1} \quad \text{and} \quad V(X_1) = -\frac{d^2\Theta_2}{d\Theta_1^2}. \tag{3.12}$$

If we choose $-\dfrac{d\Theta_2}{d\Theta_1}$ as the parameter to be estimated we get the minimum variance attainable is by (3.5)

$$\left\{ \frac{d}{d\theta} \frac{d\Theta_2}{d\Theta_1} \right\}^2 \bigg/ \left\{ \frac{d^2\Theta_2}{d\Theta_1^2} \frac{d\Theta_1}{d\theta} \right\} = -\frac{d^2\Theta_2}{d\Theta_1^2} = V(X_1). \tag{3.13}$$

Hence X_1 is the best unbiassed estimate of $-\dfrac{d\Theta_2}{d\Theta_1}$. Thus for *the distributions of the type* (3.10), *there exists a function of the observations which has the maximum precision as an estimate of a function of* θ.

Case of Several Parameters

Let $\theta_1, \theta_2, \ldots, \theta_q$ be q unknown parameters occurring in the probability density $\phi(x_1, \ldots, x_n; \theta_1, \theta_2, \ldots, \theta_q)$ and $t_1, t_2, \ldots, t_q$ be q functions independent of $\theta_1, \theta_2, \ldots, \theta_q$ such that

$$\int \cdots \int t_i \phi \pi \, dx_j = \theta_i. \tag{4.1}$$

Differentiating under the integral sign with respect to θ_i and θ_j, we get, if the following integrals exist,

$$\int \cdots \int t_i \frac{\partial \phi}{\partial \theta_i} \pi \, dx_k = 1, \tag{4.2}$$

and

$$\int \cdots \int t_i \frac{\partial \phi}{\partial \theta_j} \pi \, dx_k = 0. \tag{4.4}$$

Defining

$$E\left[-\frac{\partial^2 \log \phi}{\partial \theta_i \partial \theta_j} \right] = I_{ij}, \tag{4.4}$$

and

$$E(t_i - \theta_i)(t_j - \theta_j) = V_{ij}, \tag{4.5}$$

we get the result that the matrix of the determinant

$$\begin{vmatrix} V_{ii} & 0 & \cdots & 1 & \cdots & 0 \\ 0 & I_{11} & \cdots & I_{1i} & \cdots & I_{1q} \\ \cdots & \cdots & \cdots & \cdots & \cdots & \cdots \\ 1 & I_{i1} & \cdots & I_{ii} & \cdots & I_{iq} \\ \cdots & \cdots & \cdots & \cdots & \cdots & \cdots \\ 0 & I_{q1} & \cdots & I_{qi} & \cdots & I_{qq} \end{vmatrix} \tag{4.6}$$

being the dispersion matrix of the stochastic variates t_i and $\dfrac{1}{\phi} \dfrac{\partial \phi}{\partial \theta_j}$ $(j = 1,$

$2, \ldots, q$) is positive definite or semi-definite. If we assume that there is no linear relationship of the type

$$\sum \lambda_j \frac{1}{\phi} \frac{\partial \phi}{\partial \theta_j} = 0 \tag{4.7}$$

among the variables $\frac{1}{\phi} \frac{\partial \phi}{\partial \theta_j}$ ($j = 1, 2, \ldots, q$) then the matrix $\|I_{ij}\|$, which is known as the information matrix due to $\theta_1, \theta_2, \ldots, \theta_q$, is positive definite in which case there exists a matrix $\|I^{ij}\|$ inverse to $\|I_{ij}\|$. From (4.6) we derive that

$$V_{ii} - I^{ii} \geq 0 \tag{4.8}$$

which shows that minimum variance attainable for the estimating function of θ_i when $\theta_1, \theta_2, \ldots, \theta_q$ are not known is I^{ii}, the element in the i-th row and the i-th column of the matrix $\|I^{ij}\|$ inverse to the information matrix $\|I_{ij}\|$.

The equality is attained when

$$t_i - \theta_i = \sum \mu_j \frac{1}{\phi} \frac{\partial \phi}{\partial \theta_j}. \tag{4.9}$$

We can obtain a generalisation of (4.8) by considering the dispersion matrix of $t_1, t_2, \ldots, t_i$ and $\frac{1}{\phi} \frac{\partial \phi}{\partial \theta_r}$ ($r = 1, 2, \ldots, q$)

$$\begin{vmatrix} V_{11} & \cdots & V_{1i} & 1 & 0 & \cdots & 0 & \cdots & 0 \\ V_{21} & \cdots & V_{2i} & 0 & 1 & \cdots & 0 & \cdots & 0 \\ \cdots & \cdots & \cdots & \cdots & \cdots & \cdots & \cdots & \cdots & \cdots \\ V_{i1} & \cdots & V_{ii} & 0 & 0 & \cdots & 1 & \cdots & 0 \\ 1 & \cdots & 0 & I_{11} & I_{12} & \cdots & I_{1i} & \cdots & I_{1q} \\ \cdots & \cdots & \cdots & \cdots & \cdots & \cdots & \cdots & \cdots & \cdots \\ 0 & \cdots & 0 & I_{q1} & I_{q2} & \cdots & I_{qi} & \cdots & I_{qq} \end{vmatrix} \tag{4.10}$$

This being positive definite or semi-definite we get the result that the determinant

$$|V_{rs} - I^{rs}| \geq 0, \quad (r, s = 1, 2, \ldots, i) \tag{4.11}$$

for $i = 1, 2, \ldots, q$. The above inequality is evidently independent of the order of the elements so that, in particular, we get that the determinant

$$\begin{vmatrix} V_{ii} - I^{ii}, & V_{ij} - I^{ij} \\ V_{ji} - I^{ji}, & V_{jj} - I^{jj} \end{vmatrix} \geq 0, \tag{4.12}$$

which gives the result that if $V_{ii} = I^{ii}$, so that maximum precision is attainable for the estimation of θ_i, then $V_{ij} = I^{ij}$ for ($j = 1, 2, \ldots, q$).

In the case of the normal distribution

$$\phi(x; m, \sigma) = \text{const. exp} - \tfrac{1}{2}\{\sum (x_i - m)^2/\sigma^2\}, \tag{4.13}$$

we have

$$I_{mm} = n/\sigma^2, \quad I_{m\sigma} = 0, \quad I_{\sigma\sigma} = 2n/\sigma^2. \tag{4.14}$$

Since the mean of observations $(x_1 + x_2 + \cdots + x_n)/n$ is the best unbiassed estimate of the parameter m and the maximum precision is attainable *viz.*, $V_{mm} = I^{mm}$, it follows that any unbiassed estimate of the parameter σ is uncorrelated with the mean of observations for $V_{m\sigma} = I^{m\sigma} = 0$. *Thus in the case of the univariate normal distribution any function of the observations whose expectation is a function of σ and independent of m is uncorrelated with the mean of the observations.* This can be extended to the case of multivariate normal populations where any unbiassed estimates of the variances and covariances are uncorrelated with the means of the observations for the several variates.

If there exists no functional relationships among the estimating functions $t_1, t_2, \ldots, t_q$ then $\|V^{ij}\|$ the inverse of the matrix $\|V_{ij}\|$ exists in which case we get that the determinant

$$|V^{rs} - I_{rs}|, \quad (r, s = 1, 2, \ldots, i) \tag{4.15}$$

is greater than or equal to zero for $i = 1, 2, \ldots, q$, which is analogous to (4.11).

If a sufficient set of statistics $T_1, T_2, \ldots, T_q$ exist for $\theta_1, \theta_2, \ldots, \theta_q$ then we can show as in the case of a single parameter that the best estimating functions of the parameters or functions of parameters are explicit functions of the sufficient set of statistics.

Koopman (1936) has shown that under some conditions the distribution function $\phi(x_1, x_2, \ldots, x_n; \theta_1, \theta_2, \ldots, \theta_q)$ admitting a set of statistics $T_1, T_2, \ldots, T_q$ sufficient for $\theta_1, \theta_2, \ldots, \theta_q$ can be expressed in the form

$$\phi = \exp(\Theta_1 X_1 + \Theta_2 X_2 + \cdots + \Theta_q X_q + \Theta + X) \tag{4.16}$$

where X's are independent of θ's and Θ's are independent of x's. Making use of the relation

$$\int \phi \, dv = 1, \tag{4.17}$$

we get

$$\left. \begin{aligned} E(X_i) &= -\frac{\partial \Theta}{\partial \Theta_i}, \\[2mm] V(X_i) &= -\frac{\partial^2 \Theta}{\partial \Theta_i^2}, \\[2mm] \mathrm{cov}(X_i X_j) &= -\frac{\partial^2 \Theta}{\partial \Theta_i \partial \Theta_j}. \end{aligned} \right\} \tag{4.18}$$

This being the maximum precision available we get that for this class of distribution laws there exist functions of observations which are the best possible estimates of functions of parameters.

Loss of Information

If $t_1, t_2, \ldots, t_q$, the estimates of $\theta_1, \theta_2, \ldots, \theta_q$, have the joint distribution $\Phi(t_1, t_2, \ldots, t_q; \theta_1, \theta_2, \ldots, \theta_q)$ then the information matrix on $\theta_1, \theta_2, \ldots, \theta_q$ due to $t_1, t_2, \ldots, t_q$ is $\|F_{ij}\|$ where

$$F_{ij} = E\left\{-\frac{\partial^2 \log \Phi}{\partial\theta_i\partial\theta_j}\right\}. \tag{5.1}$$

The equality

$$I_{ij} = (I_{ij} - F_{ij}) + F_{ij} \tag{5.2}$$

effects a partition of the covariance between $\dfrac{1}{\phi}\dfrac{\partial\phi}{\partial\theta_i}$ and $\dfrac{1}{\phi}\dfrac{\partial\phi}{\partial\theta_j}$ as within and between the regions formed by the intersection of the surfaces for constant values of $t_1, t_2, \ldots, t_q$. Hence we get that the matrices

$$\|I_{ij} - F_{ij}\| \qquad \text{and} \qquad \|F_{ij}\| \tag{5.3}$$

which may be defined as the dispersion matrices of the quantities $\dfrac{1}{\phi}\dfrac{\partial\phi}{\partial\theta_i}$ $(i = 1, 2, \ldots, q)$ within and between the meshes formed by the surfaces of constant values of $t_1, t_2, \ldots, t_q$, is positive definite or semidefinite. This may be considered as a generalisation of Fisher's inequality $I_{ii} \geq F_{ii}$ in the case of a single parameter.

If $I_{ii} = F_{ii}$, then it follows that $I_{ij} = F_{ij}$ for all j for otherwise the determinant

$$\begin{vmatrix} I_{ii} - F_{ii} & I_{ij} - F_{ij} \\ I_{ij} - F_{ij} & I_{jj} - F_{jj} \end{vmatrix} < 0. \tag{5.4}$$

If in the determinant

$$|I_{ij} - F_{ij}|, \quad (i, j = 1, 2, \ldots, q), \tag{5.5}$$

the zero rows and columns are omitted, the resulting determinant will be positive and less than the determinant obtained by omitting the corresponding rows and columns in $|I_{ij}|$. If we represent the resulting determinants by dashes, we may define the loss of information in using the statistics $t_1, t_2, \ldots, t_q$ as

$$|I_{ij} - F_{ij}|'/|I_{ij}|'. \tag{5.6}$$

If Φ is the joint distribution of $t_1, t_2, \ldots, t_q$ the estimates of $\theta_1, \theta_2, \ldots, \theta_q$ with the dispersion matrix $\|V_{ij}\|$ then we have the relations analogous to (4.11) and (4.15) connecting the elements of $\|V_{ij}\|$ and $\|F_{ij}\|$ defined above. Proceeding as before we get that the determinants

$$|V_{rs} - F^{rs}| \qquad \text{and} \qquad |F_{rs} - V^{rs}|, \quad (r, s = 1, 2, \ldots, i), \tag{5.7}$$

are greater than or equal to zero for all $i = 1, 2, \ldots, q$.

The Population Space

Let the distribution of a certain number of characters in a population be characterised by the probability differential

$$\phi(x, \theta_1, \ldots, \theta_q) \, dv. \tag{6.1}$$

The quantities θ_1, θ_2, ..., θ_q are called population parameters. Given the functional form in x's as in (6.1) which determines the type of the distribution function, we can generate different populations by varying θ_1, θ_2, ..., θ_q. If these quantities are represented in a space of q dimensions, then a population may be identified by a point in this space which may be defined as the population space (P.S).

Let θ_1, θ_2, ..., θ_q and $\theta_1 + d\theta_1$, $\theta_2 + d\theta_2$, ..., $\theta_q + d\theta_q$ be two contiguous points in (P.S). At any assigned value of the characters of the populations corresponding to these contiguous points, the probability densities differ by

$$d\phi(\theta_1, \theta_2, \ldots, \theta_q) \tag{6.2}$$

retaining only first order differentials. It is a matter of importance to consider the relative discrepancy $d\phi/\phi$ rather than the actual discrepancy. The distribution of this quantity over the x's summarises the consequences of replacing $\theta_1, \theta_2, \ldots, \theta_q$ by $\theta_1 + d\theta_1, \ldots, \theta_q + d\theta_q$. The variance of this distribution or the expectation of the square of this relative discrepancy comes out as the positive definite quadratic differential form

$$ds^2 = \sum \sum g_{ij} \, d\theta_i \, d\theta_j, \tag{6.3}$$

where

$$g_{ij} = E\left(\frac{1}{\phi} \frac{\partial \phi}{\partial \theta_i}\right)\left(\frac{1}{\phi} \frac{\partial \phi}{\partial \theta_j}\right). \tag{6.4}$$

Since the quadratic form is invariant for transformations in (P.S) it follows that g_{ij} form the components of a covariant tensor of the second order and is also symmetric for $g_{ij} = g_{ji}$ by definition. This quadratic differential form with its *fundamental tensor* as the elements of the *Information matrix* may be used as a suitable measure of divergence between two populations defined by two contiguous points. The properties of (P.S) may be studied with this as the *quadratic differential metric* defining the element of length. The space based on such a metric is called the Riemannian space and the geometry associated with this is the Riemannian geometry with its definitions of distances and angles.

The Distance Between Two Populations

If two populations are represented by two points A and B in (P.S) then we can find the distance between A and B by integrating along a geodesic using the element of length

$$ds^2 = \sum \sum g_{ij} \, d\theta_i \, d\theta_j. \tag{7.1}$$

If the equations to the geodesic are

$$\theta_i = f_i(t), \tag{7.2}$$

where t is a parameter, then the functions f_i are derivable from the set of differential equations

$$\sum_1^q {}_j g_{jk} \frac{d^2\theta_j}{dt^2} + \sum_1^q {}_{jl} [jl, k] \frac{d\theta_j}{dt} \frac{d\theta_l}{dt} = 0, \tag{7.3}$$

where $[jl, k]$ is the Christoffel symbol defined by

$$[jl, k] = \frac{1}{2} \left[\frac{\partial g_{jk}}{\partial \theta_l} + \frac{\partial g_{lk}}{\partial \theta_j} + \frac{\partial g_{jl}}{\partial \theta_k} \right]. \tag{7.4}$$

The estimation of distance, however, present some difficulty. If the two samples from two populations are large then the best estimate of distance can be found by substituting the maximum likelihood estimates of the parameters in the above expression for distance. In the case of small samples we can get the fiducial limits only in a limited number of cases.

We apply the metric (7.1) to find the distance between two normal populations defined by (m_1, σ_1) and (m_2, σ_2) the distribution being of the type

$$\phi(x, m, \sigma) = \frac{1}{\sqrt{(2\pi\sigma^2)}} \exp. -\frac{1}{2} \frac{(x - m)^2}{\sigma^2}. \tag{7.5}$$

The quantities g_{ij} defined above have the values

$$g_{11} = 1/\sigma^2, \qquad g_{12} = 0, \qquad g_{22} = 2/\sigma^2, \tag{7.6}$$

so that the element of length is obtained from

$$ds^2 = \frac{(dm)^2}{\sigma^2} + \frac{2}{\sigma^2}(d\sigma)^2. \tag{7.7}$$

If $m_1 \neq m_2$ and $\sigma_1 \neq \sigma_2$ then the distance comes out as

$$D_{AB} = \sqrt{2} \log \frac{\tan \theta_1/2}{\tan \theta_2/2} \tag{7.8}$$

where

$$\theta_i = \sin^{-1} \sigma_i/\beta \quad \text{and} \quad \beta = \sigma_1^2 + [(m_1 - m_2)^2 - 2(\sigma_2^2 - \sigma_1^2)]/8(m_1 - m_2)^2. \tag{7.9}$$

If $m_1 = m_2$ and $\sigma_1 \neq \sigma_2$

$$D_{AB} = \sqrt{2} \log(\sigma_2/\sigma_1). \tag{7.10}$$

If $m_1 \neq m_2$ and $\sigma_1 = \sigma_2$

$$D_{AB} = \frac{m_1 - m_2}{\sigma}. \tag{7.11}$$

Distance in Tests of Significance and Classification

The necessity for the introduction of a suitable measure of distance between two populations arises when the position of a population with respect to an assigned set of characteristics of a given population or with respect to a number of populations has to be studied. The first problem leads to tests of significance and the second to the problem of classification. Thus if the assigned values of parameters which define some characteristics in a population are $\bar{\theta}_1$, $\bar{\theta}_2, \ldots, \bar{\theta}_q$ represented by the point O, and the true values are $\theta_1, \theta_2, \ldots, \theta_q$ represented by the point A, then we can define the divergence from the assigned sets of parameters by D_{AO}, the distance defined before in the (P.S.). The testing of the hypothesis

$$\bar{\theta}_i = \theta_i, \quad (i = 1, 2, \ldots, q), \tag{8.1}$$

may be made equivalent to the test for the significance of the estimated distance D_{AO} on the large sample assumption. If $D_{AO} = \psi(\theta_1, \ldots, \theta_q; \bar{\theta}_1, \ldots, \bar{\theta}_q)$ and the maximum likelihood estimates of $\theta_1, \theta_2, \ldots, \theta_q$ are $\hat{\theta}_1, \hat{\theta}_2, \ldots, \hat{\theta}_q$, then the estimate of D_{AO} is given by

$$\hat{D}_{AO} = \psi(\hat{\theta}_1, \ldots, \hat{\theta}_q; \bar{\theta}_1, \ldots, \bar{\theta}_q). \tag{8.2}$$

The covariances between the maximum likelihood estimates being given by the elements of the information matrix, we can calculate the large sample approximation to the variance of the estimate of D_{AO} by the following formula

$$V(\hat{D}_{AO}) = \sum \sum \frac{\partial \psi}{\partial \theta_i} \frac{\partial \psi}{\partial \theta_j} \operatorname{cov}(\hat{\theta}_i \hat{\theta}_j) \tag{8.3}$$

We can substitute the maximum likelihood estimates of $\theta_1, \theta_2, \ldots, \theta_q$ in the expression for variance. The statistic

$$w = \frac{\hat{D}_{AO}}{[V(\hat{D}_{AO})]^{1/2}} \tag{8.4}$$

can be used as a normal variate with zero mean and unit variance to test the hypothesis (8.1).

If the hypothesis is that two populations have the same set of parameters then the statistic

$$w = \frac{\hat{D}_{AB}}{[V(\hat{D}_{AB})]^{1/2}}, \tag{8.5}$$

where $\hat{D}_{AB}$ is the estimate of the distance between two populations defined by two points A and B in (P.S) can be used as (8.4). The expression for variance has to be calculated by the usual large sample assumption.

If the sample is small the appropriate test will be to find out a suitable region in the sample space which affords the greatest average power over the surfaces in the (P.S) defined by constant values of distances. The appropriate methods for this purpose are under consideration and will be dealt with in a future communication.

The estimated distances can also be used in the problem of classification. It usually becomes necessary to know whether a certain population is closer to one of a number of given populations when it is known that populations are all different from one another. In this case the distances among the populations taken two by two settle the question. We take that population whose distance from a given population is significantly the least as the one closest to the given population.

This general concept of distance between two statistical populations (as different from tests of significance) was first developed by Prof. P.C. Mahalanobis. The generalised distance defined by him (Mahalanobis, 1936) has become a powerful tool in biological and anthropological research. A perfectly general measure of divergence has been developed by Bhattacharya (1942) who defines the distance between populations as the angular distance between two points representing the populations on a unit sphere. If π_1, $\pi_2, \ldots, \pi_k$ are the proportions in a population consisting of k classes then the population can be represented by a point with coordinates $\sqrt{\pi_1}, \sqrt{\pi_2}, \ldots, \sqrt{\pi_k}$ on a unit sphere in a space of k dimensions. If two populations have the proportions $\pi_1, \pi_2, \ldots, \pi_k$ and $\pi_1', \pi_2', \ldots, \pi_k'$ the points representing them have the co-ordinates $\sqrt{\pi_1}, \sqrt{\pi_2}, \ldots, \sqrt{\pi_k}$ and $\sqrt{\pi_1'}, \sqrt{\pi_2'}, \ldots, \sqrt{\pi_k'}$. The distance between them is given by

$$\cos^{-1}\{\sqrt{(\pi_1\pi_1')} + \sqrt{(\pi_2\pi_2')} + \cdots + \sqrt{(\pi_k\pi_k')}\}. \tag{8.6}$$

If the populations are continuous with probability densities $\phi(x)$ and $\psi(x)$ the distance is given by

$$\cos^{-1}\int \sqrt{\{\phi(x)\psi(x)\}}\,dx. \tag{8.7}$$

The representation of a population as a point on a unit sphere as given by Bhattacharya (1942) throws the quadratic differential metric (7.1) in an interesting light. By changing $\theta_1, \theta_2, \ldots, \theta_q$ the parameters occurring in the probability density, the points representing the corresponding populations describe a surface on the unit sphere. It is easy to verify that the element of length ds connecting two points corresponding to $\theta_1, \theta_2, \ldots, \theta_q$ and $\theta_1 + d\theta_1, \ldots, \theta_q + d\theta_q$ on this is given by

$$ds^2 = \sum (d\phi)^2/\phi = \sum\sum g_{ij}\,d\theta_i\,d\theta_j, \tag{8.8}$$

where g_{ij} are the same as the elements of the quadratic differential metric defined in (7.1).

Further aspects of the problems of distance will be dealt with in an extensive paper to be published shortly.

References

Aitken, A.C., (1941), On the Estimation of Statistical Parameters. *Proc. Roy. Soc. Edin.*, **61**, 56–62.

Bhattacharya, A., (1942), On Discrimination and Divergence. *Proc. Sc. Cong.*
Fisher, R.A., (1921), On the Mathematical Foundations of Theoretical Statistics. *Phil. Trans. Roy. Soc. A*, **222**, 309–368.
Koopman, B.O., (1936), On Distributions Admitting Sufficient Statistics. *Trans. Am. Math. Soc.*, **39**, 399–409.
Mahalanobis, P.C., (1936), On the Generalised Distance in Statistics. *Proc. Nat. Inst. Sc. Ind.*, **2**, 49–55.

Introduction to
Wald (1945) Sequential Tests of Statistical Hypotheses

B.K. Ghosh
Lehigh University

Probability theory came of age with the advent of Kolmogorov's axiomatics in 1933 and the subsequent developments in limit theorems and stochastic processes. Statistical inference came of age with the advent of the Neyman–Pearson theory in 1933 and the subsequent formalization of hypothesis testing, estimation, and decision theory. In the present paper, Wald unified the two seemingly dissimilar areas most elegantly. To probabilists, he offered gems of new results in random walks, martingales, stochastic processes, and limit theorems. He also pointed out indirectly how they can seek probability questions from statistical inference and use the latter area as a testing ground for their abstract theorems. To statisticians, he showed that statistical inference is not just the analysis or "significance" of an existing body of data; it also entails their entry into the very process of experimentation and a continual analysis of the data as they become available. The sequential probability ratio test embodies this aspect. Wald also showed them how results from "pure" areas of mathematics, particularly probability theory, can be adapted to put statistical inference, in general, on stronger footing without losing sight of its practical nature.

An account by L. Weiss of Abraham Wald's career can be found in the introduction to Wald (1949). [See also Morgenstern (1951), Menger (1952), Tintner (1952), and Wolfowitz (1952)]. Some specific details of the paper will now be taken up.

1. The Genesis

As Wallis (1980) relates, the Applied Mathematics Panel of the U.S. Office of Scientific Research and Development established a Statistical Research Group at Columbia University on July 1, 1942. The purpose of the group was to advise the Defense Department on statistical methods for experiments that were being conducted on a vast scale in war production. Allen Wallis headed the group with Harold Hotelling and Jacob Wolfowitz as its founding members. Wald, Harold Freeman, and Milton Friedman, among others, joined a few months later.

In late March 1943, Wallis discussed the following problem with Friedman. If $x_1, \ldots, x_N$ are independent observations with a common distribution f and one is testing the hypothesis $H_0: f = f_0$ against $H_1: f = f_1$, then the Neyman–Pearson theory leads to a most powerful test that accepts H_0 if the likelihood ratio λ_N is small, $\leq c$, say, or rejects H_0 if $\lambda_N > c$. They noticed that in some problems the inequality $\lambda_N \leq c$ reduces to $\psi(x_1) + \cdots + \psi(x_N) \leq C_N$ for some nonnegative function ψ. In such cases, one can obviously stop with the partial sample $(x_1, \ldots, x_m)$ if $\psi(x_1) + \cdots + \psi(x_m) > C_N$ and reject H_0, thereby saving the cost of the redundant observations $x_{m+1}, \ldots, x_N$, which would not have changed the final decision anyway. The economy is achieved here without sacrificing the power of the most powerful test. The phenomenon occurs, in particular, when one is testing the normal standard deviation σ, the binomial proportion p, or the Poisson mean μ, which were some of the major problems in those days.

The discussion led Wallis and Friedman to the following conjecture: It may be more advantageous to observe $x_1, x_2, \ldots$ one by one and then accept H_0, reject H_0, or *continue sampling* depending on the information provided by the successive observations. They argued intuitively that the advantage would come from a reduction in the average sample size vis-à-vis the sample size of the most powerful test, and this could more than offset any loss in power. They presented their ideas to Wolfowitz due to the latter's famed mathematical skills. However, Wolfowitz was quite skeptical about the existence of a procedure that could improve the most powerful test. Unconvinced, Wallis and Friedman finally approached Wald in early April 1943. Wald was initially unenthusiastic, but phoned them two days later to admit that their conjecture was, indeed, correct. At a later meeting, he explained the details of the *sequential probability ratio test* (*SPRT*). Apparently, Hotelling coined the term *sequential* to characterize Wald's procedure.

Wald's (1943) theoretical results appeared in a "restricted" report of the group in September 1943, and a companion report on applications by Freeman (1944) appeared in July 1944. The term restricted meant that the Defense Department considered the results significantly important for the war effort and, therefore, decided not to disseminate them to the general public for a while. Nevertheless, a 10-page synopsis of Wald's results was allowed to be circulated as a memorandum (1944) in April 1944 to a select group of individuals including some statisticians who are well known today: R.E. Bechhofer, J. Berkson, G.W. Brown, J.H. Curtiss, G.B. Dantzig, W.E. Deming, H.F. Dodge, J.L. Doob, B. Epstein, W. Feller, F.E. Grubbs, S. Kullback, J. Neyman, E.G. Olds, H. Scheffé, J.W. Tukey, J. von Neumann, and S.S. Wilks in the United States, and G.A. Barnard, M.S. Bartlett, R.A. Fisher, J.O. Irwin, E.S.

Pearson, E.J.G. Pitman, R.L. Plackett, and F. Yates in Great Britain. The reports of the group were "declassified" in early 1945, thereby allowing the present paper to appear in June of that year. Interestingly enough, Wald (1944) had already published many of the mathematical foundations of sequential procedures without any mention of the word sequential or any reference to his 1943 report, as the protocol called for.

To all intents and purposes, the present paper is an amalgamation of Wald's 1943 report and 1944 paper, and Freeman's 1944 report. Wald's (1947) famous text, except for Chapters 10 and 11, is an outgrowth of this paper and its expository companion (1945).

2. The Content

Sequential analysis is concerned with probabilistic and statistical techniques for analyzing data in which the number of observations (or events) may be random. Up until Wald (1943), there had been a general awareness that such methods might be useful in some statistical investigations, but actual attempts in this direction were rare and mostly heuristic. The only formal methods were the double-sampling plan by Dodge and Romig (1929) and its extension, the multiple-sampling plan, by Bartky (1943) for testing $p = p_0$ against $p = p_1$ about a binomial proportion p. Barnard (1944) submitted a report on sequential tests for the binomial p and the Poisson mean μ to the British Ministry of Supply. However, his approach and methods are different from Wald's and do not lend themselves to any generalization to other situations [see Barnard (1946), p. 2 and Wolfowitz (1952), p. 10 for some comments]. With this preamble, Wald proceeds to the heart of the paper.

Wald starts (Sec. 1) with a review of the Neyman–Pearson theory of tests (nonrandomized) of $H_0: f = f_0$ against $H_1: f = f_1$ based on a fixed number N of observations $x_1, \ldots, x_N$ that need not be independent. In Sec. 2, the description of an arbitrary sequential test is given, along with its stopping time n, the decision $D(= H_0, H_1$ or $n = \infty)$, and the error probabilities $\alpha = P_0(D = H_1)$ and $\beta = P_1(D = H_0)$. The concept of admissibility and various notions of optimality of a test are also introduced. In Sec. 3, an SPRT is defined by the stopping time $n^* = $ first $m \geq 1$ such that $\lambda_m \leq B$ or $\lambda_m \geq A$, and the decision rules are $D = H_0$ (i.e., accept f_0) if $\lambda_{n^*} \leq B$ and $D = H_1$ (i.e., reject f_0) if $\lambda_{n^*} \geq A$. Here, λ_m denotes the likelihood ratio p_{1m}/p_{0m}, $p_{im} = f_i(x_1, \ldots, x_m)$, and $A > B > 0$ are constants (the paper uses the same symbol n to denote an arbitrary stopping time and the SPRT n^*). He also gives a Bayesian interpretation of the SPRT. He then shows that $A \leq (1 - \beta)/\alpha$ and $B \geq \beta/(1 - \alpha)$, and argues that if the *excesses* $\lambda_{n^*} - A$ (when $D = H_1$) and $B - \lambda_{n^*}$ (when $D = H_0$) are small, one can use the *Wald boundaries* $A = (1 - \beta_0)/\alpha_0$ and $B = \beta_0/(1 - \alpha_0)$ in order to satisfy the practical requirements $\alpha \approx \alpha_0$ and $\beta \approx \beta_0$ for the SPRT. In order to develop further properties of an SPRT, Wald assumes that the random variables $\log(\lambda_i/\lambda_{i-1})$, $i \geq 1$ ($\lambda_0 = 1$), are i.i.d. (independent and identically distributed).

Two general results for stopping times are obtained in Sec. 3 and 4 (and in the 1944 paper) in the following framework. Let $z_1, z_2, \ldots$ be i.i.d. and non-degenerate random variables with $E(z_1) = \mu$, $E(\exp[tz_1]) = \varphi(t)$, and let $Z_m = z_1 + \cdots + z_m$ for $m \geq 1$. Then, for an arbitrary stopping time n, one arrives at Wald's identity $E(\varphi^{-n}(t) \exp[tZ_n]) = 1$ if $P(n < \infty) = 1$ and Wald's equation (or lemma) $E(Z_n) = \mu E(n)$ if $E(n) < \infty$. In Sec. 3, Wald takes $z_i = \log(\lambda_i/\lambda_{i-1})$, $a = \log A > 0 > b = \log B$, and $n = n^*$ to show that $P_f(n^* < \infty) = 1$ as long as $P_f(z_1 = 0) < 1$ under an arbitrary f. He then uses his identity to derive an approximation and bounds for the *operating characteristic* or *boundary crossing probability* $P_f(Z_{n^*} \leq b)$. In Sec. 4, he uses his lemma to obtain an approximation and bounds for the *average sample number* $E_f(n^*)$ as long as $E_f(z_1) \neq 0$. Some results for the characteristic function of n^* are also given. Computational details for three specific situations are worked out in Sec. 3–5: When one is testing $p = p_0$ against $p = p_1$ for a binomial proportion, $\theta = \theta_0$ against $\theta = \theta_1$ for a normal mean (known variance), and $u = u_0$ against $u = u_1$, $u = p_2(1 - p_1)/p_1(1 - p_2)$, about two binomial proportions p_1 and p_2. In these situations, the operating characteristic function is monotonic in p, θ, or u, and therefore, the same SPRT can be used for testing one-sided hypotheses against one-sided alternatives (e.g., $p \leq p_0$ against $p \geq p_1$).

The well-known *optimum property* of the SPRT, as conjectured by Wald on p. 298, states that an SPRT minimizes $E_0(n)$ and $E_1(n)$ among all tests whose error probabilities are at most those of the SPRT. In Sec. 4.7, Wald gives a simple proof of this property when Z_{n^*} hits a or b with probability 1, which happens to be true in the binomial case and, as discovered later, in certain continuous time stochastic processes (e.g., a Wiener process with drift μ). In particular, he shows that the SPRT requires an average of 33 to 67% of the sample size needed by the most powerful test.

The main purpose of the last section (Sec. 6), is to formulate certain questions on sequential tests for H_0: $\theta \in \omega$ against H_1: $\theta \in \bar{\omega}$ when θ is a vector-valued parameter and ω and $\bar{\omega}$ are disjoint sets containing more than one point. Wald proposes a modification of the SPRT in which one uses *weight functions* on ω and $\bar{\omega}$. He works out the details of the so-called sequential t^2-test when $\theta = (\mu, \sigma)$ in a normal distribution and H_0: $\mu = 0$, H_1: $|\mu| = d_0\sigma$. The main difficulty with this approach is that there is no unique way to choose the weight functions.

3. The Breakthroughs and Their Influence

The most profound impact of the paper on statistical inference is the way experimental scientists and statisticians started collaborating with each other thereafter. In the Precambrian era of scientific and engineering experiments, statistical thinking was a nonentity. In the Paleozoic period, experimentalists realized that their results were often influenced by chance factors and therefore sought advice from "mathematicians." However, they retained the prerogative of designing their own experiments. In the Mesozoic period, Fisher and others developed statistics into a bona fide discipline and introduced statistical considerations into the advance planning of experiments. However, it still remained a situation in which the observed results would not be continually analyzed between the onset and conclusion of an experiment. Finally, in the current Cenozoic era, Wald's sequential analysis injected statistical considerations into the very process of experimentation itself. He provided a concrete tool, the SPRT, to analyze the data on a continual basis, which, apart from its inherent naturalness, often resulted in a speedier conclusion of the experiment. If the accumulated data at any given stage are insufficient, the statisticians no longer feel compelled to make a final decision at that stage.

At a time when Kolmogorov's axiomatics and the Neyman–Pearson theory were only ten years old, Wald had to develop new statistical concepts and improvise mathematical tools to crystallize his theory of the SPRT. These concepts and tools influenced future research in statistics in several directions. First, a huge number of papers have appeared since 1945 on the exact and approximate behavior of α, β, $P_f(Z_{n^*} \leq b)$, $E_f(n^*)$, and the characteristic function of n^* [see Ghosh (1970), Woodroofe (1982), Siegmund (1985)]. Second, people soon realized that the SPRT was no panacea; Wald himself recognized this in Sec. 6 for composite hypotheses and in his discussion of the truncated SPRT in Sec. 4.6 for simple hypotheses. This stimulated the search for a wide range of sequential tests to serve special purposes; the generalized SPRT [Weiss (1953), Anderson (1960)], the invariant SPRT [Cox (1952), Hall, Wijsman, and Ghosh (1965)], and repeated significance tests [Sen (1981), Woodroofe (1982), Siegmund (1985)] are some of these. Third, Wald's conjecture on the optimality of SPRT aroused interest in finding a general proof of this property [Wald and Wolfowitz (1948), Burkholder and Wijsman (1963), Irle and Schmitz (1984)]. Moreover, his various concepts of optimality in Sec. 2.2 and the eventual proof by him and Wolfowitz of a special optimality for the SPRT led to similar questions for other sequential tests [Kiefer and Weiss (1957), Schwarz (1962), Lai (1981)]. Fourth, owing to its easy adaptability, the SPRT made the application of statistical principles more widespread in different scientific and engineering disciplines. The Shewhart chart transformed to Page's (1954) CUSUM chart and Robbins' (1970) test with power one in process control, while a whole new field of sequential medical trials opened up [Armitage (1975), Siegmund (1985)]. Finally, the economic aspect of the SPRT raised the natural question whether sequential methods can be profitably used also in statistical estimation and decision theory. The relevant issues were raised by Wald (1947, 1950) himself and are being pursued by numerous reseachers today.

In probability theory, it soon became evident that $Z_m - m\mu$ of the SPRT is a martingale and the SPRT itself is a random walk. Sequential analysis thus became a favorite topic from statistics that is often cited by probabilists as an application of their results in random walks and martingales under optional stopping [Feller (1950, 1966), Doob (1953)]. On a theoretical level, Wald's equation $E(Z_n - n\mu) = 0$ and its companion $E(Z_n - n\mu)^2 = E(n)\,\mathrm{var}(z_1)$ became important enough to be generalized to other martingales as well as when the z_i's are not i.i.d. [Wolfowitz (1947), Doob (1953), Feller (1966), Chow, Robbins, and Siegmund (1971)]. Similarly, Wald's identity also evolved into several generalizations [Doob (1953), Bellman (1957), Bahadur (1958), Tweedie (1960)]. Finally, the exact and asymptotic tools provided by Wald in Sec. 3.4 generated a constant flow of boundary-crossing problems in the vast probability literature.

As a final note, it would be inconceivable without this paper to have a separate journal on *Sequential Analysis*.

Reviewer's Note

Only a part of Wald's (1945) original article is reproduced in the following pages due to space limitations of the present volume. The original article in *The Annals of Mathematical Statistics* also contains a *Table of Contents* (p. 117) and a *Part II. Sequential Test of a Simple or Composite Hypothesis Against a Set of Alternatives* (pp. 158–186), which are not reproduced here. However, the preceding introduction reviews the entire article.

References

Anderson, T.W. (1960). A modification of the sequential probability ratio test to reduce the sample size, *Ann. Math. Statist.*, **31**, 165–197.

Armitage P. (1975). *Sequential Medical Trials*. Wiley, New York.

Bahadur, R.R. (1958). A note on the fundamental identity of sequential analysis, *Ann. Math. Statist.*, **29**, 534–543.

Barnard, G.A. (1944). *Economy in Sampling with Special Reference to Engineering Experimentation*. British Ministry of Supply.

Barnard, G.A. (1946). Sequential tests in industrial statistics, *J. Roy. Statist. Soc.*, *Supplement* **8**, 1–26.

Bartky, W. (1943). Multiple sampling with constant probability, *Ann. Math. Statist.*, **14**, 363–377.

Bellman, R. (1957). On a generalization of the fundamental identity of Wald, *Proc. Cambridge Philos. Soc.*, **53**, 257–259.

Burkholder, D.L., and Wijsman, R.A. (1963). Optimum properties and admissibility of sequential tests, *Ann. Math. Statist.*, **34**, 1–17.

Chow, Y.S., Robbins, H., and Siegmund, D. (1971). *Great Expectations: The Theory of Optimal Stopping*. Houghton-Mifflin, Boston.

Cox, D.R. (1952). Sequential tests for composite hypotheses, *Proc. Cambridge Philos. Soc.*, **48**, 290–299.

Dodge, H.F., and Romig, H.G. (1929). A method of sampling inspection, *Bell Syst. Tech. J.*, **8**, 613–631.

Doob, J.L. (1953). *Stochastic Processes*. Wiley, New York.

Feller, W. (1950, 1966). *An Introduction to Probability Theory and Its Applications*, Vols. I, II. Wiley, New York.

Freeman, H. (1944). *Sequential Analysis of Statistical Data: Applications.* Statistical Research Group, Columbia University.

Ghosh, B.K. (1970). *Sequential Tests of Statistical Hypotheses.* Addison-Wesley, Reading, Mass.

Hall, W.J., Wijsman, R.A., and Ghosh, J.K. (1965). The relationship between sufficiency and invariance with applications to sequential analysis, *Ann. Math. Statist.*, **36**, 575–614.

Irle, A., and Schmitz, N. (1984). On the optimality of the SPRT for processes with continuous time parameter, *Math. Oper. u. Statist.*, **15**, 91–104.

Kiefer, J., and Weiss, L. (1957). Some properties of generalized sequential probability ratio tests, *Ann. Math. Statist.*, **28**, 57–74.

Lai, T.L. (1981). Asymptotic optimality of invariant sequential probability ratio tests, *Ann. Statist.*, **9**, 318–333.

Memorandum (1944). *Sequential tests of statistical significance.* Statistical Research Group, Memo 180 (AMP Memo 30.1M), Columbia University.

Menger, K. (1952). The formative years of Abraham Wald and his work in geometry, *Ann. Math. Statist.*, **23**, 14–20.

Morgenstern, O. (1951). Abraham Wald, 1902–50, *Econometrica*, **19**, 361–367.

Page, E.S. (1954). Continuous inspection schemes, *Biometrika*, **41**, 100–114.

Robbins, H. (1970). Statistical methods related to the law of iterated logarithm, *Ann. Math. Statist.*, **41**, 1397–1409.

Schwarz, G. (1962). Asymptotic shapes of Bayes sequential testing regions, *Ann. Math. Statist.*, **33**, 224–236.

Sen, P.K. (1981). *Sequential Nonparametrics.* Wiley, New York.

Siegmund, D. (1985). *Sequential Analysis: Tests and Confidence Intervals.* Springer-Verlag, New York.

Tintner, G. (1952). Abraham Wald's contributions to econometrics, *Ann. Math. Statist.*, **23**, 21–28.

Tweedie, M.C.K. (1960). Generalization of Wald's fundamental identity of sequential analysis to Markov chains, *Proc. Cambridge Philos. Soc.*, **56**, 205–214.

Wald, A. (1943). *Sequential Analysis of Statistical Data: Theory.* Statistical Research Group, Columbia University.

Wald, A. (1944). On cumulative sums of random variables, *Ann. Math. Statist.*, **15**, 283–296.

Wald, A. (1945). Sequential method of sampling for deciding between two courses of action, *J. Amer. Statist. Assoc.*, **40**, 277–306.

Wald, A. (1947). *Sequential Analysis.* Wiley, New York.

Wald, A. (1950). *Statistical Decision Functions.* Wiley, New York.

Wald, A., and Wolfowitz, J. (1948). Optimum character of the sequential probability ratio test, *Ann. Math. Statist.*, **19**, 326–339.

Wallis, W.A. (1980). The Statistical Research Group, 1942–1945, *J. Amer. Statist. Assoc.*, **75**, 320–335.

Weiss, L. (1953). Testing one simple hypothesis against another, *Ann. Math. Statist.*, **24**, 273–281.

Wolfowitz, J. (1947). The efficiency of sequential estimates and Wald's equation for sequential processes, *Ann. Math. Statist.*, **18**, 215–230.

Wolfowitz, J. (1952). Abraham Wald, 1902–1950, *Ann. Math. Statist.*, **23**, 1–13.

Woodroofe, M. (1982). *Nonlinear Renewal Theory in Sequential Analysis.* SIAM, Philadelphia, Pa.

Sequential Tests of Statistical Hypotheses

A. Wald
Columbia University

A. Introduction

By a sequential test of a statistical hypothesis is meant any statistical test procedure which gives a specific rule, at any stage of the experiment (at the n-th trial for each integral value of n), for making one of the following three decisions: (1) to accept the hypothesis being tested (null hypothesis), (2) to reject the null hypothesis, (3) to continue the experiment by making an additional observation. Thus, such a test procedure is carried out sequentially. On the basis of the first trial, one of the three decisions mentioned above is made. If the first or the second decision is made, the process is terminated. If the third decision is made, a second trial is performed. Again on the basis of the first two trials one of the three decisions is made and if the third decision is reached a third trial is performed, etc. This process is continued until either the first or the second decision is made.

An essential feature of the sequential test, as distinguished from the current test procedure, is that the number of observations required by the sequential test is not predetermined, but is a random variable due to the fact that at any stage of the experiment the decision of terminating the process depends on the results of the observations previously made. The current test procedure may be considered a limiting case of a sequential test in the following sense: For any positive integer n less than some fixed positive integer N, the third decision is always taken at the n-th trial irrespective of the results of these first n trials. At the N-th trial either the first or the second decision is taken. Which decision is taken will depend, of course, on the results of the N trials.

In a sequential test, as well as in the current test procedure, we may commit two kinds of errors. We may reject the null hypothesis when it is true (error of the first kind), or we may accept the null hypothesis when some alternative hypothesis is true (error of the second kind). Suppose that we wish to test the null hypothesis H_0 against a single alternative hypothesis H_1, and that we want the test procedure to be such that the probability of making an error of the first kind (rejecting H_0 when H_0 is true) does not exceed a preassigned value α, and the probability of making an error of the second kind (accepting H_0 when H_1 is true) does not exceed a preassigned value β. Using the current test procedure, i.e., a most powerful test for testing H_0 against H_1 in the sense of the Neyman-Pearson theory, the minimum number of observations required by the test can be determined as follows: For any given number N of observations a most powerful test is considered for which the probability of an error of the first kind is equal to α. Let $\beta(N)$ denote the probability of an error of the second kind for this test procedure. Then the minimum number of observations is equal to the smallest positive integer N for which $\beta(N) \leq \beta$.

In this paper a particular test procedure, called the sequential probability ratio test, is devised and shown to have certain optimum properties (see section 4.7). The sequential probability ratio test in general requires an expected number of observations considerably smaller than the fixed number of observations needed by the current most powerful test which controls the errors of the first and second kinds to exactly the same extent (has the same α and β) as the sequential test. The sequential probability ratio test frequently results in a saving of about 50% in the number of observations as compared with the current most powerful test. Another surprising feature of the sequential probability ratio test is that the test can be carried out without determining any probability distributions whatsoever. In the current procedure the test can be carried out only if the probability distribution of the statistic on which the test is based is known. This is not necessary in the application of the sequential probability ratio test, and only simple algebraic operations are needed for carrying it out. Distribution problems arise in connection with the sequential probability ratio test only if we want to make statements about the probability distribution of the number of observations required by the test.

This paper consists of two parts. Part I deals with the theory of sequential tests for testing a simple hypothesis against a single alternative. In Part II a theory of sequential tests for testing simple or composite hypotheses against infinite sets of alternatives is outlined. The extension of the probability ratio test to the case of testing a simple hypothesis against a set of one-sided alternatives is straight forward and does not present any difficulty. Applications to testing the means of binomial and normal distributions, as well as to testing double dichotomies are given. The theory of sequential tests of hypotheses with no restrictions on the possible values of the unknown parameters is, however, not as simple. There are several unsolved problems in this case and it is hoped that the general ideas outlined in Part II will stimulate further research.

Sections 5.2, 5.3 and 5.4 in Part II deal with the applications of the sequential probability ratio test to binomial distributions, double dichotomies and normal distributions. These sections are nearly self-contained and can be understood without reading the rest of the paper. Thus, readers who are primarily interested in these special cases of the sequential probability ratio test rather than in the general theory, may profitably read only the above mentioned sections. For the benefit of readers who lack a sufficient background in the mathematical theory of statistics the exposition in sections 5.2, 5.3 and 5.4 is kept on a fairly elementary level.

It should be pointed out that whenever the number of observations on which the test is based is for some reason determined in advance, for instance, if certain data are available from past history and no additional data can be obtained, then the current most powerful test procedure is preferable. The superiority of the sequential probability ratio test is due to the fact that it requires a smaller expected number of observations than the current most powerful test. This feature of the sequential probability ratio test is, however, of no value if the number of observations is for some reason determined in advance.

B. Historical Note

To the best of the author's knowledge the first idea of a sequential test, i.e., a test where the number of observations is not predetermined but is dependent on the outcome of the observations, goes back to H.F. Dodge and H.G. Romig who proposed a double sampling inspection procedure [1]. In this double sampling scheme the decision whether a second sample should be drawn or not depends on the outcome of the observations in the first sample. The reason for introducing a double sampling method was, of course, the recognition of the fact that double sampling results in a reduction of the amount of inspection as compared with "single" sampling.

The double sampling method does not fully take advantage of sequential analysis, since it does not allow for more than two samples. A multiple sampling scheme for the particular case of testing the mean of a binomial distribution was proposed and discussed by Walter Bartky [2]. His procedure is closely related to the test which results from the application of the sequential probability ratio test to testing the mean of a binomial distribution. Bartky clearly recognized the fact that multiple sampling results in a considerable reduction of the average amount of inspection.

The idea of chain experiments discussed briefly by Harold Hotelling [3] is also somewhat related to our notion of sequential analysis. An interesting example of such a chain of experiments is the series of sample censuses of area of jute in Bengal carried out under the direction of P. C. Mahalanobis [6]. The successive preliminary censuses, steadily increasing in size, were primarily designed to obtain some information as to the parameters to be estimated so that an efficient design could be set up for the final sampling of the whole immense jute area in the province.

In March 1943, the problem of sequential analysis arose in the Statistical Research Group, Columbia University,[1] in connection with a specific question posed by Captain G. L. Schuyler of the Bureau of Ordnance, Navy Department. It was pointed out by Milton Friedman and W. Allen Wallis that the mere notion of sequential analysis could slightly improve the efficiency of some current most powerful tests. This can be seen as follows: Suppose that N is the planned number of trials and W_N is a most powerful critical region based on N observations. If it happens that on the basis of the first n trials ($n < N$) it is already certain that the completed set of N trials must lead to a rejection of the null hypothesis, we can terminate the experiment at the n-th trial and thus save some observations. For instance, if W_N is defined by the inequality $x_1^2 + \cdots + x_N^2 \geq c$, and if for some $n < N$ we find that $x_1^2 + \cdots + x_n^2 \geq c$, we can terminate the process at this stage. Realization of this naturally led Friedman and Wallis to the conjecture that modifications of current tests may exist which take advantage of sequential procedure and effect substantial improvements. More specifically, Friedman and Wallis conjectured that a sequential test may exist that controls the errors of the first and second kinds to exactly the same extent as the current most powerful test, and at the same time requires an expected number of observations substantially smaller than the number of observations required by the current most powerful test.[2]

It was at this stage that the problem was called to the attention of the author of the present paper. Since infinitely many sequential test procedures exist, the first and basic problem was, of course, to find the particular sequential test procedure which is most efficient, i.e., which effects the greatest possible saving in the expected number of observations as compared with any other (sequential or non-sequential) test. In April, 1943 the author devised such a test, called the sequential probability ratio test, which for all practical purposes is most efficient when used for testing a simple hypothesis H_0 against a single alternative H_1.

Because of the substantial savings in the expected number of observations effected by the sequential probability ratio test, and because of the simplicity of this test procedure in practical applications, the National Defense Research Committee considered these developments sufficiently useful for the war effort to make it desirable to keep the results out of the reach of the enemy, at least for a certain period of time. The author was, therefore, requested to submit his findings in a restricted report [7] which was dated September, 1943.[3] In this report the sequential probability ratio test is devised and its

---

[1] The Statistical Research Group operates under a contract with the Office of Scientific Research and Development and is directed by the Applied Mathematics Panel of the National Defense Research Committee.

[2] Bartky's multiple sampling scheme [2] for testing the mean of a binomial distribution provides, of course, an example of such a sequential test (see, for example, the remarks on p. 377 in [2]). Bartky's results were not known to us at that time, since they were published nearly a year later.

[3] The material was recently released making the present publication possible.

mathematical theory is developed. In July 1944 a second report [8] was issued by the Statistical Research Group which gives an elementary non-mathematical exposition of the applications of the sequential probability ratio test, together with charts, tables and computational simplifications to facilitate applications.

Independently of the developments here, G.A. Barnard [9] recognized the merits of a sequential method of testing, i.e., the possibility of a saving in the number of observations as compared with the current most powerful test. He also devised an interesting sequential test for testing double dichotomies, which differs from the one obtained by applying the sequential probability ratio test.

Some further developments in the theory of the sequential probability ratio test took place in 1944. Extending the methods used in [7], C.M. Stockman [10] found the operating characteristic curve of the sequential probability ratio test applied to a binomial distribution. Independently of Stockman, Milton Friedman and George W. Brown (independently of each other) obtained the same result which can be extended to the normal distribution and a few other specific distributions, but is not applicable to more general distributions. The general operating characteristic curve for any sequential probability ratio test was derived by the author [11]. A few months later the author developed a general theory of cumulative sums [4] which gives not only the operating characteristic curve for any sequential probability ratio test but also the characteristic function of the number of observations required by the test.

The theory of the sequential probability ratio test as given in the present paper differs considerably from the exposition given in [7], since the new developments in [4] have been taken into account. However, some tables and a few sections of the original report [7] are included in the present paper without any substantial changes.

Part I. Sequential Test of a Simple Hypothesis Against a Single Alternative

1. The Current Test Procedure

Let X be a random variable. In what follows in this and the subsequent sections it will be assumed that the random variable X has either a continuous probability density function or a discrete distribution. Accordingly, by the probability distribution $f(x)$ of a random variable X we shall mean either the probability density function of X or the probability that $X = x$, depending upon whether X is a continuous or a discrete variable. Let the hypothesis H_0 to be tested (null hypothesis) be the statement that the distribution of X

is $f_0(x)$. Suppose that H_0 is to be tested against the single alternative hypothesis H_1 that the distribution of X is given by $f_1(x)$.

According to the Neyman-Pearson theory of testing hypotheses a most powerful critical region W_N for testing H_0 against H_1 on the basis of N independent observations $x_1, \ldots, x_N$ on X is given by the set of all sample points $(x_1, \ldots, x_N)$ for which the inequality

$$\frac{f_1(x_1)f_1(x_2)\ldots f_1(x_N)}{f_0(x_1)f_0(x_2)\ldots f_0(x_N)} \geq k \tag{1.1}$$

is fulfilled. The quantity k on the right hand side of (1.1) is a constant and is chosen so that the size of the critical region, i.e., the probability of an error of the first kind should have the required value α.

For a fixed sample size N the probability β of an error of the second kind is a single valued function of α, say $\beta_N(\alpha)$, if a most powerful critical region is used. Thus, if in addition to fixing the value of α it is required that the probability of an error of the second kind should have a preassigned value β, or at least it should not exceed a preassigned value β, we are no longer free to choose the sample size N. The minimum number of observations required by the test satisfying these conditions is equal to the smallest integral value of N for which $\beta_N(\alpha) \leq \beta$.

Thus, the current most powerful test procedure for testing H_0 against H_1 can be briefly stated as follows: We choose as critical region the region defined by (1.1) where the constant k is determined so that the probability of an error of the first kind should have a preassigned value α and N is equal to the smallest integer for which the probability of an error of the second kind does not exceed a preassigned value β.

2. The Sequential Test Procedure: General Definitions

2.1. Notion of a Sequential Test

In current tests of hypotheses the number of observations is treated as a constant for any particular problem. In sequential tests the number of observations is no longer a constant, but a random variable. In what follows the symbol n is used for the number of observations required by a sequential test and the symbol N is used when the number of observations is treated as a constant.

Sequential tests can be described as follows: For each positive integer m the m-dimensional sample space M_m is subdivided into three mutually exclusive parts R_m^0, R_m^1 and R_m. After the first observation x_1 has been drawn H_0 is accepted if x_1 lies in R_1^0, H_0 is rejected (i.e., H_1 is accepted) if x_1 lies in R_1^1, or a second observation is drawn if x_1 lies in R_1. If the third decision is reached and a second observation x_2 drawn, H_0 is accepted, H_1 is accepted, or a third observation is drawn according as the point (x_1, x_2) lies in R_2^0, R_2^1 or in R_2. If (x_1, x_2) lies in R_2, a third observation x_3 is drawn and one of the three decisions is made according as (x_1, x_2, x_3) lies in R_3^0, R_3^1 or in R_3, etc. This process is stopped when, and only when, either the first decision or the second decision is reached. Let n be the number of observations at which the process is terminated. Then n is a random variable, since the value of n depends on the outcome of the observations. (It will be seen later that the probability is one that the sequential process will be terminated at some finite stage.)

We shall denote by $E_0(n)$ the expected value of n if H_0 is true and by $E_1(n)$ the expected value of n if H_1 is true. These expected values, of course, depend on the sequential test used. In order to put this dependence in evidence, we shall occasionally use the symbols $E_0(n|S)$ and $E_1(n|S)$ to denote the values $E_0(n)$ and $E_1(n)$, respectively, when the sequential test S is applied.

2.2. *Efficiency of a Sequential Test*

As in the current test procedure, errors of two kinds may be committed in sequential analysis. We may reject H_0 when it is true (error of the first kind), or we may accept H_0 when H_1 is true (error of the second kind). With any sequential test there will be associated two numbers α and β between 0 and 1 such that if H_0 is true the probability is α that we shall commit an error of the first kind and if H_1 is true, the probability is β that we shall commit an error of the second kind. We shall say that two sequential tests S and S' are of equal strength if the values α and β associated with S are equal to the corresponding values α' and β' associated with S'. If $\alpha < \alpha'$ and $\beta \leq \beta'$, or if $\alpha \leq \alpha'$ and $\beta < \beta'$, we shall say that S is stronger than S' (S' is weaker than S). If $\alpha > \alpha'$ and $\beta < \beta'$, or if $\alpha < \alpha'$ and $\beta > \beta'$, we shall say that the strength of S is not comparable with that of S'.

Restricting ourselves to sequential tests of a given strength, we want to make the number of observations necessary for reaching a final decision as small as possible. If S and S' are two sequential tests of equal strength we shall say that S' is better than S if either $E_0(n|S') < E_0(n|S)$ and $E_1(n|S') \leq E_1(n|S)$, or $E_0(n|S') \leq E_0(n|S)$ and $E_1(n|S') < E_1(n|S)$. A sequential test will be said to be an admissible test if no better test of equal strength exists. If a sequential test S satisfies both inequalities $E_0(n|S) \leq E_0(n|S')$ and $E_1(n|S) \leq E_1(n|S')$ for any sequential test S' of strength equal to that of S, then the test S can be considered to be a best sequential test. That such tests exist, i.e., that it is possible to minimize $E_0(n)$ and $E_1(n)$ simultaneously, is not proved here; but it is shown later (section 4.7) that for the so called sequential probability ratio test defined in section 3.1 both $E_0(n)$ and $E_1(n)$ are very nearly minimized.[4]

---

[4] The author conjectures that $E_0(n)$ and $E_1(n)$ are exactly minimized for the sequential probability ratio test, but he did not succeed in proving this, except for a special class of problems (see section 4.7).

Thus, for all practical purposes the sequential probability ratio test can be considered best.

Since it is unknown that a sequential test always exists for which both $E_0(n)$ and $E_1(n)$ are exactly minimized, we need a substitute definition of an optimum test. Several substitute definitions are possible. We could, for example, require that the test be admissible and the maximum of the two values $E_0(n)$ and $E_1(n)$ be minimized, or that the mean $\dfrac{E_0(n) + E_1(n)}{2}$, or some other weighted average be minimized. All these definitions are equivalent if a sequential test exists for which both $E_0(n)$ and $E_1(n)$ are minimized; but if they cannot be minimized simultaneously the definitions differ. Which of them is chosen is of no significance for the purpose of this paper, since for the sequential probability ratio test proposed later both expected values $E_0(n)$ and $E_1(n)$ are, if not exactly, very nearly minimized. If we had a priori knowledge as to how frequently H_0 and how frequently H_1 will be true in the long run, it would be most reasonable to minimize a weighted average (weighted by the frequencies of H_0 and H_1, respectively) of $E_0(n)$ and $E_1(n)$. However, when such knowledge is absent, as is usually the case in practical applications, it is perhaps more reasonable to minimize the maximum of $E_0(n)$ and $E_1(n)$ than to minimize some weighted average of $E_0(n)$ and $E_1(n)$. Hence the following definition is introduced.

A sequential test S is said to be an optimum test if S is admissible and Max $[E_0(n|S), E_1(n|S)] \leq$ Max $[E_0(n|S'), E_1(n|S')]$ for all sequential tests S' of strength equal to that of S.

By the efficiency of a sequential test S is meant the value of the ratio[5]

$$\frac{\text{Max}[E_0(n|S^*), E_1(n|S^*)]}{\text{Max}[E_0(n|S), E_1(n|S)]},$$

where S^* is an optimum sequential test of strength equal to that of S.

2.3. Efficiency of the Current Procedure, Viewed as a Particular Case of a Sequential Test

The current test procedure can be considered as a particular case of a sequential test. In fact, let N be the size of the sample used in the current procedure and let W_N be the critical region on which the test is based. Then the current procedure can be considered as a sequential test defined as follows: For all $m < N$, the regions R_m^0, R_m^1 are the empty subsets of the m-dimensional sample space M_m, and $R_m = M_m$. For $m = N$, R_N^1 is equal to W_N, R_N^0 is equal to the complement $\overline{W}_N$ of W_N and R_N is the empty set. Thus, for the current procedure we have $E_0(n) = E_1(n) = N$.

---

[5] The existence of an optimum sequential test is not essential for the definition of efficiency, since Max $[E_0(n|S^*), E_1(n|S^*)]$ could be replaced by the greatest lower bound of Max $[E_0(n|S'), E_1(n|S')]$ with respect to all sequential tests S' of strength equal to that of S.

It will be seen later that the efficiency of the current test based on the most powerful critical region is rather low. Frequently it is below $\frac{1}{2}$. In other words, an optimum sequential test can attain the same α and β as the current most powerful test on the basis of an expected number of observations much smaller than the fixed number of observations needed for the current most powerful test.

In the next section we shall propose a simple sequential test procedure, called the sequential probability ratio test, which for all practical purposes can be considered an optimum sequential test. It will be seen that these sequential tests usually lead to average savings of about 50% in the number of trials as compared with the current most powerful test.

3. Sequential Probability Ratio Test

3.1. *Definition of the Sequential Probability Ratio Test*

We have seen in section 2.1 that the sequential test procedure is defined by subdividing the m-dimensional sample space M_m ($m = 1, 2, \ldots$, ad inf.) into three mutually exclusive parts R_m^0, R_m^1 and R_m. The sequential process is terminated at the smallest value n of m for which the sample point lies either in R_n^0 or in R_n^1. If the sample point lies in R_n^0 we accept H_0 and if it lies in R_n^1 we accept H_1.

An indication as to the proper choice of the regions R_m^0, R_m^1 and R_m can be obtained from the following considerations: Suppose that before the sample is drawn there exists an a priori probability that H_0 is true and the value of this probability is known. Denote this a priori probability by g_0. Then the a priori probability that H_1 is true is given by $g_1 = 1 - g_0$, since it is assumed that the hypotheses H_0 and H_1 exhaust all possibilities. After a number of observations have been made we gain additional information which will affect the probability that H_i ($i = 0, 1$) is true. Let g_{0m} be the a posteriori probability that H_0 is true and g_{1m} the a posteriori probability that H_1 is true after m observations have been made. Then according to the well known formula of Bayes we have

$$g_{0m} = \frac{g_0 p_{0m}(x_1, \ldots, x_m)}{g_0 p_{0m}(x_1, \ldots, x_m) + g_1 p_{1m}(x_1, \ldots, x_m)} \tag{3.1}$$

and

$$g_{1m} = \frac{g_1 p_{1m}(x_1, \ldots, x_m)}{g_0 p_{0m}(x_1, \ldots, x_m) + g_1 p_{1m}(x_1, \ldots, x_m)}, \tag{3.2}$$

where $p_{im}(x_1, \ldots, x_m)$ denotes the probability density in the m-dimensional sample space calculated under the hypothesis $H_i(i = 0, 1)$.[6] As an abbreviation for $p_{im}(x_1, \ldots, x_m)$ we shall use simply p_{im}.

Let d_0 and d_1 be two positive numbers less than 1 and greater than $\frac{1}{2}$. Suppose that we want to construct a sequential test such that the conditional probability of a correct decision under the condition that H_0 is accepted is greater than or equal to d_0, and the conditional probability of a correct decision under the condition that H_1 is accepted is greater than or equal to d_1.[7] Then the following sequential process seems reasonable: At each stage calculate g_{0m} and g_{1m}. If $g_{1m} \geq d_1$, accept H_1. If $g_{0m} \geq d_0$, accept H_0. If $g_{1m} < d_1$ and $g_{0m} < d_0$, draw an additional observation. R_m^0 in this sequential process is thus defined by the inequality $g_{0m} \geq d_0$, R_m^1 by the inequality $g_{1m} \geq d_1$, and R_m by the simultaneous inequalities $g_{1m} < d_1$ and $g_{0m} < d_0$. It is necessary that the sets R_m^0, R_m^1 and R_m be mutually exclusive and exhaustive. For this it suffices that the inequalities

$$g_{1m} = \frac{g_1 p_{1m}}{g_0 p_{0m} + g_1 p_{1m}} \geq d_1 \tag{3.3}$$

and

$$g_{0m} = \frac{g_0 p_{0m}}{g_0 p_{0m} + g_1 p_{1m}} \geq d_0 \tag{3.4}$$

be not fulfilled simultaneously. To show that (3.3) and (3.4) are incompatible, we shall assume that they are simultaneously fulfilled and derive a contradiction from this assumption. The two inequalities sum to

$$g_{1m} + g_{0m} \geq d_1 + d_0. \tag{3.5}$$

Since $g_{0m} + g_{1m} = 1$, we have

$$1 \geq d_1 + d_0$$

which is impossible, since by assumption $d_i > \frac{1}{2}(i = 0, 1)$. Hence it is proved that the sets R_m^0, R_m^1 and R_m are mutually exclusive and exhaustive.

The inequalities (3.3) and (3.4) are equivalent to the following inequalities, respectively:

$$\frac{p_{1m}}{p_{0m}} \geq \frac{g_0}{g_1} \frac{d_1}{1 - d_1} \tag{3.6}$$

and

$$\frac{p_{1m}}{p_{0m}} \leq \frac{g_0}{g_1} \frac{1 - d_0}{d_0}. \tag{3.7}$$

---

[6] If the probability distribution is discrete $p_{im}(x_1, \ldots, x_m)$ denotes the probability that the sample point $(x_1, \ldots, x_m)$ will be obtained.

[7] The restriction $d_0 > 1/2$ and $d_1 > 1/2$ are imposed because otherwise it might happen that the hypothesis with the smaller a posteriori probability will be accepted.

The constants on the right hand sides of (3.6) and (3.7) do not depend on m.

If an a priori probability of H_0 does not exist, or if it is unknown, the inequalities (3.6) and (3.7) suggest the use of the following sequential test: At each stage calculate p_{1m}/p_{0m}. If $p_{1m} = p_{0m} = 0$, the value of the ratio p_{1m}/p_{0m} is defined to be equal to 1. Accept H_1 if

$$\frac{p_{1m}}{p_{0m}} \geq A. \tag{3.8}$$

Accept H_0 if

$$\frac{p_{1m}}{p_{0m}} \leq B. \tag{3.9}$$

Take an additional observation if

$$B < \frac{p_{1m}}{p_{0m}} < A. \tag{3.10}$$

Thus, the number n of observations required by the test is the smallest integral value of m for which either (3.8) or (3.9) holds. The constants A and B are chosen so that $0 < B < A$ and the sequential test has the desired value α of the probability of an error of the first kind and the desired value β of the probability of an error of the second kind. We shall call the test procedure defined by (3.8), (3.9) and (3.10), a sequential probability ratio test.

The sequential test procedure given by (3.8), (3.9) and (3.10) has been justified here merely on an intuitive basis. Section 4.7, however, shows that for this sequential test the expected values $E_0(n)$ and $E_1(n)$ are very nearly minimized.[8] Thus, for practical purposes this test can be considered an optimum test.

3.2. Fundamental Relations Among the Quantities α, β, A and B

In this section the quantities α, β, A and B will be related by certain inequalities which are of basic importance for the sequential analysis.

Let $\{x_m\}$ ($m = 1, 2, \ldots$, ad inf.) be an infinite sequence of observations. The set of all possible infinite sequences $\{x_m\}$ is called the infinite dimensional sample space. It will be denoted by M_∞. Any particular infinite sequence $\{x_m\}$ is called a point of M_∞. For any set of n given real numbers $a_1, \ldots, a_n$ we shall denote by $C(a_1, \ldots, a_n)$ the subset of M_∞ which consists of all points (infinite sequences) $\{x_m\}$ ($m = 1, 2, \ldots$, ad inf.) for which $x_1 = a_1, \ldots, x_n = a_n$. For any values $a_1, \ldots, a_n$ the set $C(a_1, \ldots, a_n)$ will be called a cylindric point of order n. A subset S of M_∞ will be called a cylindric point, if there exists a positive

---

[8] It seems likely to the author that $E_0(n)$ and $E_1(n)$ are exactly minimized for the sequential probability ratio test. However, he did not succeed in proving it, except for a special class of problems (see section 4.7).

integer n for which S is a cylindric point of order n. Thus, a cylindric point may be a cylindric point of order 1, or of order 2, etc. A cylindric point $C(a_1, \ldots, a_n)$ will be said to be of type 1 if

$$\frac{p_{1n}}{p_{0n}} = \frac{f_1(a_1)f_1(a_2)\ldots f_1(a_n)}{f_0(a_1)f_0(a_2)\ldots f_0(a_n)} \geq A$$

and

$$B < \frac{p_{1m}}{p_{0m}} = \frac{f_1(a_1)\ldots f_1(a_m)}{f_0(a_1)\ldots f_0(a_m)} < A \quad (m = 1, \ldots, n-1).$$

A cylindric point $C(a_1, \ldots, a_n)$ will be said to be of type 0 if

$$\frac{p_{1n}}{p_{0n}} = \frac{f_1(a_1)\ldots f_1(a_n)}{f_0(a_1)\ldots f_0(a_n)} \leq B$$

and

$$B < \frac{p_{1m}}{p_{0m}} = \frac{f_1(a_1)\ldots f_1(a_m)}{f_0(a_1)\ldots f_0(a_m)} < A \quad (m = 1, \ldots, n-1).$$

Thus, if a sample $(x_1, \ldots, x_n)$ is observed for which $C(x_1, \ldots, x_n)$ is a cylindric point of type i, the sequential test defined by (3.8), (3.9) and (3.10) leads to the acceptance of H_i $(i = 0, 1)$.

Let Q_i be the sum of all cylindric points of type i $(i = 0, 1)$. For any subset M of M_∞ we shall denote by $P_i(M)$ the probability of M calculated under the assumption that H_i is true $(i = 0, 1)$. Now we shall prove that

$$P_i(Q_0 + Q_1) = 1 \quad (i = 0, 1). \tag{3.11}$$

This equation means that the probability is equal to one that the sequential process will eventually terminate. To prove (3.11) we shall denote the variate $\log \dfrac{f_1(x_i)}{f_0(x_i)}$ by z_i and $z_1 + \cdots + z_m$ by Z_m $(i, m = 1, 2, \ldots,$ ad inf.$)$. Furthermore, denote by n the smallest integer for which either $Z_n \geq \log A$ or $Z_n \leq \log B$. If no such finite integer n exists we shall say that $n = \infty$. Clearly, n is the number of observations required by the sequential test and (3.11) is proved if we show that the probability that $n = \infty$ is zero. But the latter statement was proved by the author elsewhere (see Lemma 1 in [4]). Hence equation (3.11) is proved.

With the help of (3.11) we shall be able to derive some important inequalities satisfied by the quantities α, β, A and B. Since for each sample $(x_1, \ldots, x_n)$ for which $C(x_1, \ldots, x_n)$ is an element of Q_1 the inequality $p_{1n}/p_{0n} \geq A$ holds, we see that

$$P_1(Q_1) \geq A P_0(Q_1). \tag{3.12}$$

Similarly, for each sample $(x_1, \ldots, x_n)$ for which $C(x_1, \ldots, x_n)$ is a point of Q_0 the inequality $p_{1n}/p_{0n} \leq B$ holds. Hence

$$P_1(Q_0) \leq B P_0(Q_0). \tag{3.13}$$

But $P_0(Q_1)$ is the probability of committing an error of the first kind and $P_1(Q_0)$ is the probability of making an error of the second kind. Thus, we have

$$P_0(Q_1) = \alpha; \qquad P_1(Q_0) = \beta. \tag{3.14}$$

Since Q_0 and Q_1 are disjoint, it follows from (3.11) that

$$P_0(Q_0) = 1 - \alpha; \qquad P_1(Q_1) = 1 - \beta. \tag{3.15}$$

From the relations (3.12)–(3.15) we obtain the important inequalities

$$1 - \beta \geq A\alpha \tag{3.16}$$

and

$$\beta \leq B(1 - \alpha). \tag{3.17}$$

These inequalities can be written as

$$\frac{\alpha}{1 - \beta} \leq \frac{1}{A} \tag{3.18}$$

and

$$\frac{\beta}{1 - \alpha} \leq B. \tag{3.19}$$

The above inequalities are of great value in practical applications, since they supply upper limits for α and β when A and B are given. For instance, it follows immediately from (3.18) and (3.19), and the fact that $0 < \alpha < 1$, $0 < \beta < 1$ that

$$\alpha \leq \frac{1}{A} \tag{3.20}$$

and

$$\beta \leq B. \tag{3.21}$$

A pair of values α and β can be represented by a point in the plane with the coordinates α and β. It is of interest to determine the set of all points (α, β) which satisfy the inequalities (3.18) and (3.19) for given values of A and B. Consider the straight lines L_1 and L_2 in the plane given by the equations

$$A\alpha = 1 - \beta \tag{3.22}$$

and

$$\beta = B(1 - \alpha), \tag{3.23}$$

respectively. The line L_1 intersects the abscissa axis at $\alpha = \dfrac{1}{A}$ and the ordinate axis at $\beta = 1$. The line L_2 intersects the abscissa axis at $\alpha = 1$ and the ordinate

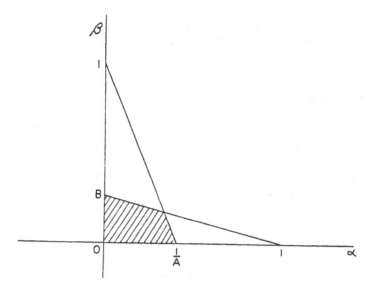

Figure 1

axis at $\beta = B$. The set of all points (α, β) which satisfy the inequalities (3.18) and (3.19) is the interior and the boundary of the quadrilateral determined by the lines L_1, L_2 and the coordinate axes. This set is represented by the shaded area in figure 1.

The fundamental inequalities (3.18) and (3.19) were derived under the assumption that $x_1, x_2, \ldots$, ad inf. are independent observations on the same random variable X. The independence of the observations is, however, not necessary for the validity of (3.18) and (3.19). In fact, the independence of the observations was used merely to show the validity of (3.11). But (3.11) can be shown to hold also for dependent observations under very general conditions. Hence, if H_i states that the joint distribution of $x_1, x_2, \ldots, x_m$ is given by the joint probability density function $p_{im}(x_1, \ldots, x_m)$[9] $(i = 0, 1; m = 1, 2, \ldots,$ ad inf.) and if (3.11) holds, then for the sequential test of H_0 against H_1, as defined by (3.8), (3.9) and (3.10), the inequalities (3.18) and (3.19) remain valid. For instance, let λ_0 and λ_1 be two different positive values < 1 and let $H_i(i = 0, 1)$ be the hypothesis that the joint probability density function of $x_1, \ldots, x_m$ is given by

$$p_{im}(x_1, \ldots, x_m) = \frac{1}{(2\pi)^{m/2}} e^{-(1/2)x_1^2 - (1/2)\sum_{j=2}^{m}(x_j - \lambda_i x_{j-1})^2} \quad (i = 0, 1),$$

i.e., that x_1 and $(x_j - \lambda_i x_{j-1})(j = 2, 3, \ldots,$ ad inf.) are normally and indepen-

---

[9] Of course, for any positive integers m and m' with $m < m'$ the marginal distribution of $x_1, \ldots, x_m$ determined on the basis of the joint distribution $P_{im'}(x_1, \ldots, x_{m'})$ must be equal to $P_{im}(x_1, \ldots, x_m)$

dently distributed with zero means and unit variances, then the inequalities (3.18) and (3.19) will hold for the sequential test defined by (3.8), (3.9) and (3.10).

3.3. Determination of the Values A and B in Practice

Suppose that we wish to have a sequential test such that the probability of an error of the first kind is equal to α and the probability of an error of the second kind is equal to β. Denote by $A(\alpha, \beta)$ and $B(\alpha, \beta)$ the values of A and B for which the probabilities of the errors of the first and second kinds will take the desired values α and β. The exact determination of the values $A(\alpha, \beta)$ and $B(\alpha, \beta)$ is rather laborious, as will be seen in Section 3.4. The inequalities at our disposal, however, permit the problem to be solved satisfactorily for practical purposes. From (3.18) and (3.19) it follows that

$$A(\alpha, \beta) \leq \frac{1 - \beta}{\alpha} \tag{3.24}$$

and

$$B(\alpha, \beta) \geq \frac{\beta}{1 - \alpha}. \tag{3.25}$$

Suppose we put $A = \dfrac{1 - \beta}{\alpha} = a(\alpha, \beta)$ (say), and $B = \dfrac{\beta}{1 - \alpha} = b(\alpha, \beta)$ (say). Then A is greater than or equal to the exact value $A(\alpha, \beta)$, and B is less than or equal to the exact value $B(\alpha, \beta)$. This procedure, of course, changes the probabilities of errors of the first and second kind. If we were to use the exact value of B and a value of A which is greater than the exact value, then evidently we would lower the value of α, but slightly increase the value of β. Similarly, if we were to use the exact value of A and a value of B which is below the exact value, then we would lower the value of β, but slightly increase the value of α. Thus, it is not clear what will be the resulting effect on α and β if a value of A is used which is higher than the exact value, and a value of B is used which is lower than the exact value. Denote by α' and β' the resulting probabilities of errors of the first and second kind, respectively, if we put $A = \dfrac{1 - \beta}{\alpha}$ and $B = \dfrac{\beta}{1 - \alpha}$.

We now derive inequalities satisfied by the quantities α', β', α and β. Substituting $a(\alpha, \beta)$ for A, $b(\alpha, \beta)$ for B, α' for α and β' for β we obtain from (3.18) and (3.19)

$$\frac{\alpha'}{1 - \beta'} \leq \frac{1}{a(\alpha, \beta)} = \frac{\alpha}{1 - \beta} \tag{3.26}$$

and

$$\frac{\beta'}{1 - \alpha'} \leq b(\alpha, \beta) = \frac{\beta}{1 - \alpha}. \tag{3.27}$$

From these inequalities it follows that

$$\alpha' \leq \frac{\alpha}{1 - \beta} \tag{3.28}$$

and

$$\beta' \leq \frac{\beta}{1 - \alpha}. \tag{3.29}$$

Multiplying (3.26) by $(1 - \beta)(1 - \beta')$ and (3.27) by $(1 - \alpha)(1 - \alpha')$ and adding the two resulting inequalities, we have

$$\alpha' + \beta' \leq \alpha + \beta. \tag{3.30}$$

Thus, we see that at least one of the inequalities $\alpha' \leq \alpha$ and $\beta' \leq \beta$ must hold. In other words, by using $a(\alpha, \beta)$ and $b(\alpha, \beta)$ instead of $A(\alpha, \beta)$ and $B(\alpha, \beta)$, respectively, at most one of the probabilities α and β may be increased.

If α and β are small (say less than .05), as they frequently will be in practical applications, $\frac{\alpha}{1 - \beta}$ and $\frac{\beta}{1 - \alpha}$ are nearly equal to α and β, respectively. Thus, we see from (3.28) and (3.29) that the quantity by which α' can possibly exceed α, or β' can exceed β, must be small. Section 3.4 contains further inequalities which show that the amount by which $\alpha'(\beta')$ can possibly exceed $\alpha(\beta)$ is indeed extremely small. Thus, for all practical purposes $\alpha' \leq \alpha$ and $\beta' \leq \beta$.

If $f_1(x)$ (the distribution under the alternative hypothesis) is sufficiently near $f_0(x)$ (the distribution under the null hypothesis), $A(\alpha, \beta)$ and $B(\alpha, \beta)$ will be nearly equal to $\frac{1 - \beta}{\alpha}$ and $\frac{\beta}{1 - \alpha}$, respectively; and consequently α' and β' are also very nearly equal to α and β respectively. The reason that (3.18) and (3.19) and therefore also (3.24) and (3.25) are inequalities instead of equalities is that the sequential process may terminate with $\frac{p_{1n}}{p_{0n}} > A$ or $\frac{p_{1n}}{p_{0n}} < B$. If at the final stage $\frac{p_{1n}}{p_{0n}}$ were exactly equal to A or B, then $A(\alpha, \beta)$ and $B(\alpha, \beta)$ would be exactly $\frac{1 - \beta}{\alpha}$ and $\frac{\beta}{1 - \alpha}$, respectively. If $f_1(x)$ is near $f_0(x)$, it is almost certain that the value of $\frac{p_{1n}}{p_{0n}}$ is changed only slightly by one additional observation. Thus, at the final stage $\frac{p_{1n}}{p_{0n}}$ be only slightly above A, or slightly below B and consequently $A(\alpha, \beta)$ and $B(\alpha, \beta)$ will be nearly equal to $\frac{1 - \beta}{\alpha}$ and $\frac{\beta}{1 - \alpha}$, respectively. If fractional observations were possible, that is to say, if the number of observations were a continuous variable, $\dfrac{p_{1m}}{p_{0m}}$ would also be a continuous function of m and consequently $A(\alpha, \beta)$ and $B(\alpha, \beta)$ would be exactly equal to $\dfrac{1 - \beta}{\alpha}$ and $\dfrac{\beta}{1 - \alpha}$, respectively. Thus, we have inequalities in (3.24) and (3.25) instead of equalities merely on account of the fact that the number m of observations is discontinuous, i.e., m can take only integral values.

Hence for all practical purposes the following procedure can be adopted: *To construct a sequential test such that the probability of an error of the first kind does not exceed α and the probability of an error of the second kind does not exceed β, put $A = \dfrac{1 - \beta}{\alpha}$ and $B = \dfrac{\beta}{1 - \alpha}$ and carry out the sequential test as defined by the inequalities* (3.8), (3.9) *and* (3.10).

In most practical cases the calculation of the exact values $A(\alpha, \beta)$ and $B(\alpha, \beta)$ will be of little interest for the following reasons: when $A = a(\alpha, \beta) = \dfrac{1 - \beta}{\alpha}$ and $B = b(\alpha, \beta) = \dfrac{\beta}{1 - \alpha}$, the probability α' of an error of the first kind cannot exceed α and the probability β' of an error of the second kind cannot exceed β, except by a very small quantity which can be neglected for practical purposes. Thus, for all practical purposes the use of $a(\alpha, \beta)$ and $b(\alpha, \beta)$ instead of $A(\alpha, \beta)$ and $B(\alpha, \beta)$ will not decrease the strength of the sequential test. The only possible disadvantage from the substitution is that it may increase the expected number of trials necessary for a decision. Since the discrepancy between $A(\alpha, \beta)$ and $B(\alpha, \beta)$ on the one hand and $a(\alpha, \beta)$ and $b(\alpha, \beta)$ on the other, arises only from the discontinuity of the number m of observations, it is clear that the increase in the expected number of trials caused by the use of $a(\alpha, \beta)$ and $b(\alpha, \beta)$ will be slight. This slight increase, however, cannot be considered entirely a loss for the following reason: if $a(\alpha, \beta) > A(\alpha, \beta)$ or $b(\alpha, \beta) < B(\alpha, \beta)$, then we can sharpen the inequality (3.30) to $\alpha' + \beta' < \alpha + \beta$. Hence by using $a(\alpha, \beta)$ and $b(\alpha, \beta)$ we gain in strength.

The fact that for practical purposes we may put $A = a(\alpha, \beta)$ and $B = b(\alpha, \beta)$ brings out a surprising feature of the sequential test as compared with current tests. While current tests cannot be carried out without finding the probability distribution of the statistic on which the test is based, there are no distribution problems in connection with sequential tests. In fact, $a(\alpha, \beta)$ and $b(\alpha, \beta)$ depend on α and β only, and the ratio $\dfrac{p_{1m}}{p_{0m}}$ can be calculated from the data of the problem without solving any distribution problems. Distribution problems arise in connection with the sequential process only if it is desired to find the probability distribution of the number of trials necessary for reaching a final decision. (This subject is discussed later.) But this is of secondary importance as long as we know that the sequential test on the average leads to a saving in the number of trials.

3.4. Probability of Accepting H_0 (or H_1) When Some Third Hypothesis H Is True

In Section 3.2 we were concerned with the probability that the sequential probability ratio test will lead to the acceptance of H_0 (or H_1) when H_0 or H_1 is true. Since in Part II we shall admit an infinite set of alternatives, and since this is the practically important case, it is of interest to study the probability of accepting H_0 (or H_1) when any third hypothesis H, not necessarily equal to H_0 or H_1, is true. Let H be the hypothesis that the distribution of X is given by $f(x)$. If $f(x)$ is equal to $f_0(x)$ or $f_1(x)$ we have the special case discussed in Section 3.2. In what follows in this and the subsequent sections any probability relationship will be stated on the assumption that H is true, unless a statement to the contrary is explicitly made. Denote by γ the probability that the sequential probability ratio test will lead to the acceptance of H_1.[10] Clearly, if $H = H_0$, then $\gamma = \alpha$ and if $H = H_1$, then $\gamma = 1 - \beta$.

The probability γ can readily be derived on the basis of the general theory of cumulative sums given in [4]. Denote $\log \dfrac{f_1(x_i)}{f_0(x_i)}$ by z_i. Then $\{z_i\}$ $(i = 2, \ldots,$ ad inf.) is a sequence of independent random variables each having the same distribution. Denote by Z_j the sum of the first j elements of the sequence $\{z_i\}$ i.e.,

$$Z_j = z_1 + \cdots + z_j \quad (j = 1, 2, \ldots, \text{ad inf.}). \tag{3.31}$$

For any relation R we shall denote by $P(R)$ the probability that R holds. For any random variable Y the symbol EY will denote the expected value of Y. Let n be the smallest positive integer for which either $Z_n \geq \log A$ or $Z_n \leq \log B$ holds. If $\log B < Z_m < \log A$ holds for $m = 1, 2, \ldots,$ ad inf., we shall say that $n = \infty$. Obviously, n is the number of observations required by the sequential probability ratio test. As we have seen in Section 3.3, in practice we shall put $A = a(\alpha, \beta) = \dfrac{1 - \beta}{\alpha}$ and $B = b(\alpha, \beta) = \dfrac{\beta}{1 - \alpha}$. Since B must be less than A, we shall consider only values α and β for which $\dfrac{1 - \beta}{\alpha} > \dfrac{\beta}{1 - \alpha}$. This inequality is equivalent to $\alpha + \beta < 1$, which in turn implies that $B < 1$ and $A > 1$. Thus, in all that follows it will be assumed that $A > 1$ and $B < 1$. We shall also assume that the variance of z_i is not zero.

According to Lemma 1 in [4] the relation $P(n = \infty) = 0$ holds. Hence, the probability is equal to one that the sequential process will eventually terminate. This implies that the probability of accepting H_0 is equal to $1 - \gamma$.

Let z be a random variable whose distribution is equal to the common distribution of the variates z_i $(i = 1, 2, \ldots,$ ad inf.). Denote by $\varphi(t)$ the moment generating function of z, i.e.,

---

[10] The probability that H_0 will be accepted is equal to $1 - \gamma$, as will be seen later.

$$\varphi(t) = Ee^{zt}.$$

It was shown in [4] that under very mild restrictions on the distribution of z there exists exactly one real value h such that $h \neq 0$ and $\varphi(h) = 1$. Furthermore, it was shown in [4] (see equation (16) in [4]) that

$$Ee^{Z_n h} = 1. \qquad (3.32)$$

Let E^* be the conditional expected value of $e^{Z_n h}$ under the restriction that H_0 is accepted, i.e., that $Z_n \leq \log B$, and let E^{**} be the conditional expected value of $e^{Z_n h}$ under the restriction that H_1 is accepted, i.e., that $Z_n \geq \log A$. Then we obtain from (3.32)

$$(1 - \gamma)E^* + \gamma E^{**} = 1. \qquad (3.33)$$

Solving for γ we obtain

$$\gamma = \frac{1 - E^*}{E^{**} - E^*}. \qquad (3.34)$$

If both the absolute value of Ez and the variance of z are small, which will be the case when $f_1(x)$ is near $f_0(x)$, E^* and E^{**} will be nearly equal to B^h and A^h, respectively. Hence, in this case a good approximation to γ is given by the expression

$$\bar{\gamma} = \frac{1 - B^h}{A^h - B^h}. \qquad (3.35)$$

It is easy to verify that $h = 1$ if $H = H_0$, and $h = -1$ if $H = H_1$. The difference $\bar{\gamma} - \gamma$ approaches zero if both the mean and the variance of z converge to zero.

To judge the goodness of the approximation given by $\bar{\gamma}$, it is desirable to derive lower and upper limits for γ. Such limits for γ can be obtained by deriving lower and upper limits for E^* and E^{**}. First we consider the case when $h > 0$. Let ζ be a real variable restricted to values > 1, and let ρ be a positive variable restricted to values < 1. For any random variable Y and any relationship R we shall denote by $E(Y|R)$ the conditional expected value of Y under the restriction that R holds. It was shown in [4] that the following inequalities hold:[11]

$$B^h \left\{ \underset{\zeta}{\text{g.l.b.}}\ \zeta E\left(e^{hz} | e^{hz} \leq \frac{1}{\zeta} \right) \right\} \leq E^* \leq B^h \quad (h > 0) \qquad (3.36)$$

and

$$A^h \leq E^{**} \leq A^h \left\{ \underset{\rho}{\text{l.u.b.}}\ \rho E\left(e^{hz} | e^{hz} \geq \frac{1}{\rho} \right) \right\} \quad (h > 0). \qquad (3.37)$$

The symbol g.l.b. stands for the greatest lower bound with respect to ζ, and

---

[11] See relations (23) and (26) in [4]. The notation used here is somewhat different from that in [4].

the symbol l.u.b. stands for least upper bound with respect to ρ. Putting
$$\text{g.l.b.}_{\zeta}\ \zeta E\left(e^{hz}|e^{hz} \le \frac{1}{\zeta}\right) = \eta \tag{3.38}$$
and
$$\text{l.u.b.}_{\rho}\ \rho E\left(e^{hz}|e^{hz} \ge \frac{1}{\rho}\right) = \delta, \tag{3.39}$$

the inequalities (3.36) and (3.37) can be written as
$$B^h\eta \le E^* \le B^h \quad (h > 0) \tag{3.40}$$
and
$$A^h \le E^{**} \le A^h\delta \quad (h > 0). \tag{3.41}$$

Since $B < 1$ and $A > 1$, we see that $E^* < 1$ and $E^{**} > 1$ if $h > 0$. From this and the relations (3.34), (3.40) and (3.41) it follows easily that
$$\frac{1 - B^h}{\delta A^h - B^h} \le \gamma \le \frac{1 - \eta B^h}{A^h - \eta B^h} \quad (h > 0). \tag{3.42}$$

If $h < 0$, limits for γ can be obtained as follows: Let $z' = -z$, $A' = \frac{1}{B}$, $B' = \frac{1}{A}$. Then $h' = -h > 0$ and $\gamma' = 1 - \gamma$. Thus, according to (3.42) we have
$$\frac{1 - (B')^{h'}}{\delta'(A')^{h'} - (B')^{h'}} \le \gamma' \le \frac{1 - \eta'(B')^{h'}}{(A')^{h'} - \eta'(B')^{h'}}, \tag{3.43}$$

where δ' and η' are equal to the expressions we obtain from (3.38) and (3.39), respectively, by substituting h' for h and z' for z. Since η and δ depend only on the product $hz = h'z'$, we see that $\delta' = \delta$ and $\eta' = \eta$. Hence, we obtain from (3.43)
$$\frac{1 - A^h}{\delta B^h - A^h} \le 1 - \gamma \le \frac{1 - \eta A^h}{B^h - \eta A^h} \quad (h < 0), \tag{3.44}$$

where δ and η are given by (3.38) and (3.39), respectively.

In Section 3.5 we shall calculate the value of η and δ for binomial and normal distributions. If the limits of γ, as given in (3.42) and (3.44), are too far apart, it may be desirable to determine the exact value of γ, or at least to find a closer approximation to γ than that given in (3.35). A solution of this problem is given in [4] (see section 7 of that paper). There the exact value of γ is derived when z can take only a finite number of integral multiples of a constant d. If z does not have this property, arbitrarily fine approximation to the value of γ can be obtained, since the distribution of z can be approximated to any desired degree by a discrete distribution of the type mentioned before if the constant d is chosen sufficiently small. The results obtained in [4] can be stated as follows: There is no loss of generality in assuming that $d = 1$, since the quantity d can be chosen as the unit of measurement. Thus, we shall assume that z takes only a finite number of integral values. Let g_1 and g_2 be two positive integers such that $P(z = -g_1)$ and $P(z = g_2)$ are positive and z can take only integral values $\geq -g_1$ and $\leq g_2$. Denote $P(z = i)$ by h_i. Then the moment generating function of z is given by

$$\varphi(t) = \sum_{i=-g_1}^{g_2} h_i e^{it}.$$

Put $u = e^t$ and let $u_1, \ldots u_g$ be the $g = g_1 + g_2$ roots of the equation of g-th degree

$$\sum_{i=-g_1}^{g_2} h_i u^i = 1. \tag{3.45}$$

Denote by $[a]$ the smallest integer $\geq \log A$, and by $[b]$ the largest integer $\leq \log B$. Then Z_n can take only the values

$$[b] - g_1 + 1, [b] - g_1 + 2, \ldots, [b], [a], [a] + 1, \ldots, [a] + g_2 - 1. \tag{3.46}$$

Denote the g different integers in (3.46) by $c_1, \ldots, c_g$, respectively. Let Δ be the determinant value of the matrix $\|u_i^{c_j}\|$ $(i, j = 1, \ldots, g)$ and let Δ_j be the determinant we obtain from Δ by substituting 1 for the elements in the j-th column. Then, if $\Delta \neq 0$, the probability that $Z_n = c_j$ is given by

$$P(Z_n = c_j) = \frac{\Delta_j}{\Delta}. \tag{3.47}$$

Hence

$$\gamma = P(Z_n \geq [a]) = \sum_j \frac{\Delta_j}{\Delta}, \tag{3.48}$$

where the summation is to be taken over all vaues of j for which $c_j \geq [a]$.

3.5. Calculation of δ and η for Binomial and Normal Distributions

Let X be a random variable which can take only the values 0 and 1. Let the probability that $X = 1$ be p_i if H_i is true $(i = 0, 1)$, and p if H is true. Denote $1 - p$ by q and $1 - p_i$ by q_i $(i = 0, 1)$. Then $f_i(1) = p_i$; $f_i(0) = q_i$, $f(1) = p$ and $f(0) = q$. It can be assumed without loss of generality that $p_1 > p_0$. The moment generating function of $z = \log \dfrac{f_1(x)}{f_0(x)}$ is given by

$$\varphi(t) = E\left(\frac{f_1(x)}{f_0(x)}\right)^t = p\left(\frac{p_1}{p_0}\right)^t + q\left(\frac{q_1}{q_0}\right)^t.$$

Let $h \neq 0$ be the value of t for which $\varphi(h) = 1$, i.e.,

$$p\left(\frac{p_1}{p_0}\right)^h + q\left(\frac{q_1}{q_0}\right)^h = 1.$$

First we consider the case when $h > 0$. It is clear that $e^{zh} = \left(\dfrac{f_1(x)}{f_0(x)}\right)^h > 1$ implies that $x = 1$. Hence $e^{zh} > 1$ implies that $e^{zh} = \left(\dfrac{f_1(1)}{f_0(1)}\right)^h = \left(\dfrac{p_1}{p_0}\right)^h$. From this and the definition of δ given in (3.39) it follows that

$$\delta = \left(\frac{p_1}{p_0}\right)^h \quad (h > 0). \tag{3.49}$$

Similarly, the inequality $e^{zh} < 1$ implies that $e^{zh} = \left(\dfrac{q_1}{q_0}\right)^h$. From this and the definition of η given in (3.38) it follows that

$$\eta = \left(\frac{q_1}{q_0}\right)^h \quad (h > 0). \tag{3.50}$$

If $h < 0$, it can be shown in a similar way that

$$\delta = \left(\frac{q_1}{q_0}\right)^h \quad (h < 0) \tag{3.51}$$

and

$$\eta = \left(\frac{p_1}{p_0}\right)^h \quad (h < 0). \tag{3.52}$$

Now we shall calculate the values of δ and η if X is normally distributed. Let

$$f_i(x) = \frac{1}{\sqrt{2\pi}} e^{-(1/2)(x-\theta_i)^2} \quad (i = 0, 1) \tag{3.53}$$

and

$$f(x) = \frac{1}{\sqrt{2\pi}} e^{-(1/2)(x-\theta)^2}. \tag{3.54}$$

We can assume without loss of generality that $\theta_0 = -\Delta$ and $\theta_1 = \Delta$ where $\Delta > 0$, since this can always be achieved by a translation. Then

$$z = \log \frac{f_1(x)}{f_0(x)} = 2\Delta x. \tag{3.55}$$

The moment generating function of z is given by

$$\varphi(t) = e^{2\Delta\theta t + 2\Delta^2 t^2}. \tag{3.56}$$

Hence

$$h = -\frac{\theta}{\Delta}. \tag{3.57}$$

Substituting this value of h in (3.38) and (3.39) we obtain

$$\delta = \text{l.u.b.} \; \rho E\left(e^{-2\theta x}|e^{-2\theta x} \geq \frac{1}{\rho}\right) \tag{3.58}$$

and

$$\eta = \text{g.l.b.} \; \zeta E\left(e^{-2\theta x}|e^{-2\theta x} \leq \frac{1}{\zeta}\right). \tag{3.59}$$

For any relation R let $P^*(R)$ denote the probability that the relation R holds calculated under the assumption that the distribution of x is normal with mean θ and variance unity. Furthermore, let $P^{**}(R)$ denote the probability that R holds if the distribution of x is normal with mean $-\theta$ and variance unity. Since $e^{-2\theta x}$ is equal to the ratio of the normal probability density function with mean $-\theta$ and variance unity to the normal probability density function with mean θ and variance unity, we see that

$$E\left(e^{-2\theta x}|e^{-2\theta x} \geq \frac{1}{\rho}\right) = \frac{P^{**}\left(e^{-2\theta x} \geq \frac{1}{\rho}\right)}{P^*\left(e^{-2\theta x} \geq \frac{1}{\rho}\right)} \tag{3.60}$$

and

$$E\left(e^{-2\theta x}|e^{-2\theta x} \leq \frac{1}{\zeta}\right) = \frac{P^{**}\left(e^{-2\theta x} \leq \frac{1}{\zeta}\right)}{P^*\left(e^{-2\theta x} \leq \frac{1}{\zeta}\right)}. \tag{3.61}$$

It can easily be verified that the right hand side expressions in (3.60) and (3.61) have the same values for $\theta = \lambda$ as for $\theta = -\lambda$. Thus, also δ and η have the same values for $\theta = \lambda$ as for $\theta = -\lambda$. It will be, therefore, sufficient to compute δ and η for negative values of θ. Let $\theta = -\lambda$ where $\lambda > 0$. First we show that $\eta = \frac{1}{\delta}$. Clearly

$$\frac{\zeta P^{**}\left(e^{2\lambda x} \leq \frac{1}{\zeta}\right)}{P^*\left(e^{2\lambda x} \leq \frac{1}{\zeta}\right)} = \frac{\zeta P^{**}(e^{-2\lambda x} \geq \zeta)}{P^*(e^{-2\lambda x} \geq \zeta)} \quad (1 \leq \zeta < \infty). \tag{3.62}$$

Putting $\zeta = \frac{1}{\rho} (0 < \rho \leq 1)$ in (3.62) gives

$$\frac{\zeta P^{**}\left(e^{2\lambda x} \leq \frac{1}{\zeta}\right)}{P^*\left(e^{2\lambda x} \leq \frac{1}{\zeta}\right)} = \frac{P^{**}\left(e^{-2\lambda x} \geq \frac{1}{\rho}\right)}{\rho P^*\left(e^{-2\lambda x} \geq \frac{1}{\rho}\right)}. \tag{3.63}$$

Hence

$$\eta = \text{g.l.b.}_\zeta \left\{ \frac{\zeta P^{**}\left(e^{2\lambda x} \le \frac{1}{\zeta}\right)}{P^*\left(e^{2\lambda x} \le \frac{1}{\zeta}\right)} \right\} = \frac{1}{\text{l.u.b.}_\rho \left\{ \frac{\rho P^*\left(e^{-2\lambda x} \ge \frac{1}{\rho}\right)}{P^{**}\left(e^{-2\lambda x} \ge \frac{1}{\rho}\right)} \right\}}. \tag{3.64}$$

Because of the symmetry of the normal distribution, it is easily seen that

$$\text{l.u.b.}_\rho \left\{ \frac{\rho P^*\left(e^{-2\lambda x} \ge \frac{1}{\rho}\right)}{P^{**}\left(e^{-2\lambda x} \ge \frac{1}{\rho}\right)} \right\} = \text{l.u.b.}_\rho \left\{ \frac{\rho P^{**}\left(e^{2\lambda x} \ge \frac{1}{\rho}\right)}{P^*\left(e^{2\lambda x} \ge \frac{1}{\rho}\right)} \right\} = \delta.$$

Hence

$$\eta = \frac{1}{\delta}. \tag{3.65}$$

Now we shall calculate the value of δ. Denote $\dfrac{1}{\sqrt{2\pi}} \displaystyle\int_x^\infty e^{-t^2/2}\, dt$ by $G(x)$. Then

$$P^{**}\left(e^{2\lambda x} \ge \frac{1}{\rho}\right) = P^{**}\left(2\lambda x \ge \log \frac{1}{\rho}\right)$$

$$= P^{**}\left(x \ge \frac{1}{2\lambda}\log\frac{1}{\rho}\right) = G\left(\frac{1}{2\lambda}\log\frac{1}{\rho} - \lambda\right).$$

Similarly

$$P^*\left(e^{2\lambda x} \ge \frac{1}{\rho}\right) = P^*\left(x \ge \frac{1}{2\lambda}\log\frac{1}{\rho}\right) = G\left(\frac{1}{2\lambda}\log\frac{1}{\rho} + \lambda\right).$$

Denote $\dfrac{1}{2\lambda}\log\dfrac{1}{\rho}$ by u. Since ρ can vary from 0 to 1, u can take any value from 0 to ∞. Since $\rho = e^{-2\lambda u}$, we have

$$\delta = \text{l.u.b.}_\rho \left\{ \frac{\rho P^{**}\left(e^{2\lambda x} \ge \frac{1}{\rho}\right)}{P^*\left(e^{2\lambda x} \ge \frac{1}{\rho}\right)} \right\} = \text{l.u.b.}_u \left\{ e^{-2\lambda u}\frac{G(u-\lambda)}{G(u+\lambda)} \right\} \quad (0 \le u < \infty). \tag{3.66}$$

We shall prove that

$$e^{-2u\lambda}\frac{G(u-\lambda)}{G(u+\lambda)} = \chi(u) \quad \text{(say)} \tag{3.67}$$

is a monotonically decreasing function of u and consequently the maximum is at $u = 0$. For this purpose it suffices to show that the derivative of $\log \chi(u)$ is never positive. Now

$$\log \chi(u) = \log G(u - \lambda) - \log G(u + \lambda) - 2\lambda u. \tag{3.68}$$

Denote $\dfrac{1}{\sqrt{2\pi}} e^{-(1/2)x^2}$ by $\Phi(x)$. Since $\dfrac{d}{du} G(u) = -\Phi(u)$ it follows from (3.68) that

$$\frac{d}{du} \log \chi(u) = -\frac{\Phi(u - \lambda)}{G(u - \lambda)} + \frac{\Phi(u + \lambda)}{G(u + \lambda)} - 2\lambda. \tag{3.69}$$

It follows from the mean value theorem that the right hand side of (3.69) is never positive if $\dfrac{d}{du}\left(\dfrac{\Phi(u)}{G(u)}\right)$ is equal to or less than 1 for all values of u. Thus, we need merely to show that

$$\frac{d}{du}\left(\frac{\Phi(u)}{G(u)}\right) = \frac{\Phi'(u)G(u) - G'(u)\Phi(u)}{G^2(u)}$$

$$= \frac{\Phi'(u)G(u) + \Phi^2(u)}{G^2(u)} = \frac{\Phi^2(u)}{G^2(u)} - u\frac{\Phi(u)}{G(u)} \leq 1. \tag{3.70}$$

Denote $\dfrac{\Phi(u)}{G(u)}$ by y. The roots of the equation $y^2 - uy - 1 = 0$ are

$$y = \frac{u \pm \sqrt{u^2 + 4}}{2}.$$

Hence the inequality $y^2 - uy - 1 \leq 0$ holds if and only if

$$\frac{u - \sqrt{u^2 + 4}}{2} \leq y \leq \frac{u + \sqrt{u^2 + 4}}{2}.$$

Since y cannot be negative, this inequality is equivalent to

$$\frac{\Phi(u)}{G(u)} = y \leq \frac{u + \sqrt{u^2 + 4}}{2}. \tag{3.71}$$

Thus we have merely to prove (3.71). We shall show that (3.71) holds for all real values of u. Birnbaum has shown [5] that for $u > 0$

$$\frac{\sqrt{u^2 + 4} - u}{2}\Phi(u) \leq G(u). \tag{3.72}$$

Hence

$$\frac{\Phi(u)}{G(u)} \leq \frac{2}{\sqrt{u^2 + 4} - u} = \frac{\sqrt{u^2 + 4} + u}{2} \quad (u > 0), \tag{3.73}$$

which proves (3.71) for $u > 0$. Now we prove (3.71) for $u < 0$. Let $u = -v$ where $v > 0$. Then it follows from (3.73) that

$$\frac{\Phi(v)}{G(v)} \le \frac{2}{\sqrt{4 + v^2} - v}. \tag{3.74}$$

Taking reciprocals, we obtain from (3.74)

$$\frac{G(v)}{\Phi(v)} \ge \frac{\sqrt{4 + v^2} - v}{2}. \tag{3.75}$$

Since

$$\frac{G(u)}{\Phi(u)} \ge \frac{G(v) + 2v\Phi(v)}{\Phi(v)} = \frac{G(v)}{\Phi(v)} + 2v,$$

we obtain from (3.75)

$$\frac{G(u)}{\Phi(u)} \ge \frac{\sqrt{v^2 + 4} + 3v}{2} \ge \frac{\sqrt{v^2 + 4} + v}{2}. \tag{3.76}$$

Taking reciprocals, we obtain

$$\frac{\Phi(u)}{G(u)} \le \frac{2}{\sqrt{v^2 + 4} + v} = \frac{\sqrt{v^2 + 4} - v}{2} = \frac{\sqrt{u^2 + 4} + u}{2}.$$

Hence (3.71) is proved for all values of u and consequently δ is equal to the value of the expression (3.67) if we substitute 0 for u. Thus,

$$\delta = \frac{G(-\lambda)}{G(\lambda)}. \tag{3.77}$$

4. The Number of Observations Required by the Sequential Probability Ratio Test

4.1. *Expected Number of Observations Necessary for Reaching a Decision*

As before, let

$$z = \log \frac{f_1(x)}{f_0(x)}, \qquad z_i = \log \frac{f_1(x_i)}{f_0(x_i)} \quad (i = 1, 2, \ldots, \text{ad inf.})$$

and let n be the number of observations required by the sequential test, i.e., n is the smallest integer for which $Z_n = z_1 + \cdots + z_n$ is either $\ge \log A$ or $\le \log B$. To determine the expected value $E(n)$ of n under any hypothesis H we shall consider a fixed positive integer N. The sum $Z_N = z_1 + \cdots + z_N$ can be split in two parts as follows

$$Z_N = Z_n + Z'_n \tag{4.1}$$

where $Z'_n = z_{n+1} + \cdots + z_N$ if $n \le N$ and $Z'_n = Z_N - Z_n$ if $n > N$. Taking expected values on both sides of (4.1) we obtain

$$NEz = EZ_n + EZ'_n. \tag{4.2}$$

Since the probability that $n > N$ converges to zero as $N \to \infty$, and since $|Z'_n| < 2(\log A + |\log B|)$ if $n > N$, it can be seen that

$$\lim_{N=\infty} [EZ'_n - E(N - n)Ez] = 0. \tag{4.3}$$

From (4.2) and (4.3) it follows that

$$EZ_n = EnEz. \tag{4.4}$$

Hence

$$En = \frac{EZ_n}{Ez}. \tag{4.5}$$

Let E^*Z_n be the conditional expected value of Z_n under the restriction that the sequential analysis leads to the acceptance of H_0, i.e. that $Z_n \le \log B$. Similarly, let $E^{**}Z_n$ be the conditional expected value of Z_n under the restriction that H_1 is accepted, i.e., that $Z_n \ge \log A$. Since γ is the probability that $Z_n \ge \log A$, we have

$$EZ_n = (1 - \gamma)E^*Z_n + \gamma E^{**}Z_n. \tag{4.6}$$

From (4.5) and (4.6) we obtain

$$En = \frac{(1 - \gamma)E^*Z_n + \gamma E^{**}Z_n}{Ez}. \tag{4.7}$$

The exact value of EZ_n, and therefore also the exact value of En, can be computed if z can take only integral multiples of a constant d, since in this case the exact probability distribution of Z_n was obtained (see equation (3.47)). If z does not satisfy the above restriction, it is still possible to obtain arbitrarily fine approximations to the value of EZ_n, since the distribution of z can be approximated to any desired degree by a discrete distribution of the type mentioned above if the constant d is chosen sufficiently small.

If both $|Ez|$ and the standard deviation of z are small, E^*Z_n is very nearly equal to $\log B$ and $E^{**}Z_n$ is very nearly equal to $\log A$. Hence in this case we can write

$$En \sim \frac{(1 - \gamma) \log B + \gamma \log A}{Ez}. \tag{4.8}$$

To judge the goodness of the approximation given in (4.8) we shall derive lower and upper limits for En by deriving lower and upper limits for E^*Z_n and $E^{**}Z_n$. Let r be a non-negative variable and let

$$\xi = \operatorname*{Max}_{r} E(z - r | z \geq r) \quad (r \geq 0) \tag{4.9}$$

and

$$\xi' = \operatorname*{Min}_{r} E(z + r | z + r \leq 0) \quad (r \geq 0). \tag{4.10}$$

It is easy to see that

$$\log A \leq E^{**} Z_n \leq \log A + \xi \tag{4.11}$$

and

$$\log B + \xi' \leq E^* Z_n \leq \log B. \tag{4.12}$$

We obtain from (4.7), (4.11) and (4.12)

$$\frac{(1 - \gamma)(\log B + \xi') + \gamma \log A}{Ez} \leq En \leq \frac{(1 - \gamma) \log B + \gamma(\log A + \xi)}{Ez} \quad \text{if } Ez > 0 \tag{4.13}$$

and

$$\frac{(1 - \gamma) \log B + \gamma(\log A + \xi)}{Ez} \leq En \leq \frac{(1 - \gamma)(\log B + \xi') + \gamma \log A}{Ez} \quad \text{if } Ez < 0. \tag{4.14}$$

4.2. Calculation of the Quantities ξ and ξ' for Binomial and Normal Distributions

Let X be a random variable which can take only the values 0 and 1. Let the probability that $X = 1$ be p_i if H_i is true ($i = 0, 1$), and p if H is true. Denote $1 - p$ by q and $1 - p_i$ by q_i ($i = 0, 1$). Then $f_i(1) = p_i$, $f_i(0) = q_i$, $f(1) = p$ and $f(0) = q$. It can be assumed without loss of generality that $p_1 > p_0$. It is clear that $\log \dfrac{f_1(x)}{f_0(x)} > 0$ implies that $x = 1$ and consequently $\log \dfrac{f_1(x)}{f_0(x)} = \log \dfrac{f_1(1)}{f_0(1)} = \log \dfrac{p_1}{p_0}$. Hence

$$\xi = \operatorname*{Max}_{r} E(z - r | z \geq r) = \log \frac{p_1}{p_0}. \tag{4.15}$$

Since $\log \dfrac{f_1(x)}{f_0(x)} \leq 0$ implies that $x = 0$, we have

$$\xi' = \operatorname*{Min}_{r} E(z + r | z + r \leq 0) = \log \frac{q_1}{q_0}. \tag{4.16}$$

Now we shall calculate the values ξ and ξ' if X is normally distributed. Let

$$f_i(x) = \frac{1}{\sqrt{2\pi}} e^{-(x-\theta_i)^2/2} \quad (i = 0, 1)(\theta_1 > \theta_0)$$

and

$$f(x) = \frac{1}{\sqrt{2\pi}} e^{-(x-\theta)^2/2}.$$

We may assume without loss of generality that $\theta_0 = -\Delta$ and $\theta_1 = \Delta$ where $\Delta > 0$, since this can always be achieved by a translation. Then

$$z = \log \frac{f_1(x)}{f_0(x)} = 2\Delta x. \tag{4.17}$$

Denote $\frac{1}{\sqrt{2\pi}} e^{-(1/2)x^2}$ by $\Phi(x)$ and $\frac{1}{\sqrt{2\pi}} \int_x^\infty e^{-(1/2)t^2} dt$ by $G(x)$. Let $t = x - \theta$. Then $z = 2\Delta(t + \theta)$ and

$$E(z - r|z - r \geq 0) = 2\Delta E\left(t + \theta - \frac{r}{2\Delta} \middle| t + \theta - \frac{r}{2\Delta} \geq 0 \right)$$

$$= \frac{2\Delta}{G(t_0)} \int_{t_0}^\infty (t - t_0)\Phi(t) \, dt = \frac{2\Delta}{G(t_0)} [-t_0 G(t_0) + \Phi(t_0)], \tag{4.18}$$

where

$$t_0 = \frac{r}{2\Delta} - \theta. \tag{4.19}$$

In section 3.5 (see equation (3.70)) it was proved that $\frac{\Phi(t_0)}{G(t_0)} - t_0$ is a monotonically decreasing function of t_0. Hence the maximum of $E(z - r|z - r \geq 0)$ is reached for $r = 0$ and consequently

$$\xi = \frac{2\Delta}{G(-\theta)} [\theta G(-\theta) + \Phi(-\theta)] = 2\Delta\left[\theta + \frac{\Phi(-\theta)}{G(-\theta)} \right]. \tag{4.20}$$

Now we shall calculate ξ'. We have

$$\xi' = \underset{r}{\text{Min}}\, E(z + r|z + r \leq 0) = -\underset{r}{\text{Max}}\, E(-z - r|-z - r \geq 0)$$

$$= -2\Delta \underset{r}{\text{Max}}\, E\left(-x - \frac{r}{2\Delta} \middle| -x - \frac{r}{2\Delta} \geq 0 \right). \tag{4.21}$$

Let $t = -x + \theta$ and $t_0 = \frac{r}{2\Delta} + \theta$. Then

$$E\left(-x - \frac{r}{2\Delta} \middle| -x - \frac{r}{2\Delta} \geq 0 \right) = E(t - t_0|t - t_0 \geq 0)$$

$$= \frac{1}{G(t_0)} \int_{t_0}^\infty (t - t_0)\Phi(t) \, dt = \frac{\Phi(t_0)}{G(t_0)} - t_0. \tag{4.22}$$

Since this is a monotonically decreasing function of t_0, we have

$$\text{Max } E\left(-x - \frac{r}{2\Delta}\left|-x - \frac{r}{2\Delta} \geq 0\right.\right) = \frac{\Phi(\theta)}{G(\theta)} - \theta. \qquad (4.23)$$

From (4.21) and (4.23) we obtain

$$\xi' = -2\Delta\left[\frac{\Phi(\theta)}{G(\theta)} - \theta\right]. \qquad (4.24)$$

4.3. Saving in the Number of Observations as Compared with the Current Test Procedure

We consider the case of a normally distributed variate, such that

$$f_0(x) = \frac{1}{\sqrt{2\pi}}e^{-(1/2)(x-\theta_0)^2}$$

and

$$f_1(x) = \frac{1}{\sqrt{2\pi}}e^{-(1/2)(x-\theta_1)^2} \quad (\theta_1 \neq \theta_0).$$

Denote by $n(\alpha, \beta)$ the minimum number of observations necessary in the current most powerful test for the probabilities of errors of the first and second kinds to be α and β, respectively, or less.

We shall calculate the number of observations required by the most powerful test. It can be assumed without loss of generality that $\theta_0 \leq \theta_1$. According to the current most powerful test procedure the hypothesis H_0 is accepted if $\bar{x} \leq d$ and the hypothesis H_1 is accepted if $\bar{x} > d$, where $\bar{x}$ is the arithmetic mean of the observations and d is a properly chosen constant. The probability of an error of the first kind is given by $G[\sqrt{n}(d - \theta_0)]$ and the probability of an error of the second kind is given by $1 - G[\sqrt{n}(d - \theta_1)]$ where $G(t) = \frac{1}{\sqrt{2\pi}}\int_t^\infty e^{-x^2/2}\,dx$. To equate these probabilities to α and β, respectively, the quantities d and n must satisfy

$$G[\sqrt{n}(d - \theta_0)] = \alpha \qquad (4.25)$$

and

$$1 - G[\sqrt{n}(d - \theta_1)] = \beta. \qquad (4.26)$$

Denote by λ_0 and λ_1 the values for which $G(\lambda_0) = \alpha$ and $G(\lambda_1) = 1 - \beta$. Then we have

$$\sqrt{n}(d - \theta_0) = \lambda_0 \qquad (4.27)$$

and

$$\sqrt{n}(d - \theta_1) = \lambda_1. \qquad (4.28)$$

Subtracting (4.27) from (4.28) we obtain

$$\sqrt{n}(\theta_0 - \theta_1) = \lambda_1 - \lambda_0. \tag{4.29}$$

From (4.29)

$$n = n(\alpha, \beta) = \frac{(\lambda_1 - \lambda_0)^2}{(\theta_0 - \theta_1)^2}. \tag{4.30}$$

If the expression on the right hand side of (4.30) is not an integer, $n(\alpha, \beta)$ is the smallest integer in excess.

In the sequential probability ratio test we put $A = a(\alpha, \beta) = \dfrac{1 - \beta}{\alpha}$ and $B = b(\alpha, \beta) = \dfrac{\beta}{1 - \alpha}$. Then the probability of an error of the first (second) kind cannot exceed $\alpha(\beta)$ except by a negligible amount. Let $A(\alpha, \beta)$ and $B(\alpha, \beta)$ be the values of A and B for which the probabilities of errors of the first and second kinds become exactly equal to α and β, respectively. It has been shown in section 3.2 that $A(\alpha, \beta) \leq a(\alpha, \beta)$ and $B(\alpha, \beta) \geq b(\alpha, \beta)$. Thus, the expected values $E_1(n)$ and $E_0(n)$ are only increased by putting $A = a(\alpha, \beta)$ and $B = b(\alpha, \beta)$ instead of $A = A(\alpha, \beta)$ and $B = B(\alpha, \beta)$.

Consider the case where $|\theta_1 - \theta_0|$ is small so that the quantities ξ and ξ' can be neglected. Thus, we shall use the approximation (4.8). Since $\gamma = \alpha$ if $H = H_0$ and $\gamma = 1 - \beta$ if $H = H_1$, we obtain from (4.8)

$$E_1(n) = \frac{a^*}{E_1(z)} - \beta \frac{a^* + |b^*|}{E_1(z)} \tag{4.31}$$

and

$$E_0(n) = \frac{-b^*}{E_0(-z)} - \alpha \frac{-b^* + a^*}{E_0(-z)}, \tag{4.32}$$

where $a^* = \log a(\alpha, \beta) = \log \dfrac{1 - \beta}{\alpha}$ and $b^* = \log b(\alpha, \beta) = \log \dfrac{\beta}{1 - \alpha}$. Since

$$E_1(z) = \tfrac{1}{2}(\theta_0 - \theta_1)^2 \tag{4.33}$$

and

$$E_0(-z) = \tfrac{1}{2}(\theta_0 - \theta_1)^2, \tag{4.34}$$

it follows from (4.30), (4.31) and (4.32) that $\dfrac{E_1(n)}{n(\alpha, \beta)}$ and $\dfrac{E_0(n)}{n(\alpha, \beta)}$ are independent of the parameters θ_0 and θ_1.

The average saving of the sequential analysis as compared with the current method is $100\left(1 - \dfrac{E_1(n)}{n(\alpha, \beta)}\right)$ per cent if H_1 is true, and $100\left(1 - \dfrac{E_0(n)}{n(\alpha, \beta)}\right)$ per cent if H_0 is true. In Table 1 the expression $100\left(1 - \dfrac{E_1(n)}{n(\alpha, \beta)}\right)$ is shown in

Table 1. Average Percentage Saving of Sequential Analysis, as Compared with Current Most Powerful Test for Testing Mean of a Normally Distributed Variate.

A. When alternative hypothesis is true:

β \ α	.01	.02	.03	.04	.05
.01	58	60	61	62	63
.02	54	56	57	58	59
.03	51	53	54	55	55
.04	49	50	51	52	53
.05	47	49	50	50	51

B. When null hypothesis is true:

β \ α	.01	.02	.03	.04	.05
.01	58	54	51	49	47
.02	60	56	53	50	49
.03	61	57	54	51	50
.04	62	58	55	52	50
.05	63	59	55	53	51

Panel A, and the expression $100\left(1 - \dfrac{E_0(n)}{n(\alpha, \beta)}\right)$ in Panel B, for several values of α and β. Because of the symmetry of the normal distribution, Panel B is obtained from Panel A simply by interchanging α and β.

As can be seen from the table, for the range of α and α from .01 to .05 (the range most frequently employed), the sequential process leads to an average saving of at least 47 per cent in the necessary number of observations as compared with the current procedure. The true saving is slightly greater than shown in the table, since $E_i(n)$ calculated under the condition that $A = a(\alpha, \beta)$ and $B = b(\alpha, \beta)$ is greater than $E_i(n)$ calculated under the condition that $A = A(\alpha, \beta)$ and $B = B(\alpha, \beta)$.

4.4. The Characteristic Function, the Moments and the Distribution of the Number of Observations Necessary for Reaching a Decision

It was shown in [4] (see equation (15) in [4]) that the following fundamental identity holds

$$E\{e^{Z_n t}[\varphi(t)]^{-n}\} = 1 \quad (\varphi(t) = E e^{z t}) \tag{4.35}$$

for all points t of the complex plane for which $\varphi(t)$ exists and $|\varphi(t)| \geq 1$. The symbol n denotes the number of observations required by the sequential test, i.e., n is the smallest positive integer for which Z_n is either $\geq \log A$ or $\leq \log B$, and $\varphi(t)$ denotes the moment generating function of z.

On the basis of the identity (4.35) the exact characteristic function of n is derived in section 7 of [4] in the case when z can take only integral multiples of a constant. If the number of different values which Z_n can take is large, the calculation of the exact characteristic function is cumbersome, because a large number of simultaneous linear equations have to be solved. However, if $|Ez|$ and σ_z are small so that $|Z_n - \log A|$ (when $Z_n \geq \log A$) and $|Z_n - \log B|$ (when $Z_n \geq \log B$) can be neglected, the calculation of the characteristic function is much simpler, as was shown in [4]. We shall briefly state the results obtained in [4]. Let h be the real value $\neq 0$ for which $\varphi(h) = 1$. Furthermore let $t = t_1(\tau)$ and $t = t_2(\tau)$ be the roots of the equation in t

$$-\log \varphi(t) = \tau$$

such that $\lim_{\tau=0} t_1(\tau) = 0$ and $\lim_{\tau=0} t_2(\tau) = h$. Finally, let $\psi_1(\tau)$ be the characteristic function of the conditional distribution of n under the restriction that $Z_n \geq \log A$, and $\psi_2(\tau)$ the characteristic function of the conditional distribution of n under the restriction that $Z_n \leq \log B$. Then, if $|Z_n - \log A|$ (when $Z_n \geq \log A$) and $|Z_n - \log B|$ (when $Z_n \leq \log B$) can be neglected, $\psi_1(\tau)$ and $\psi_2(\tau)$ are the solutions of the linear equations

$$\gamma\psi_1(\tau)A^{t_1(\tau)} + (1 - \gamma)\psi_2(\tau)B^{t_1(\tau)} = 1 \tag{4.36}$$

and

$$\gamma\psi_1(\tau)A^{t_2(\tau)} + (1 - \gamma)\psi_2(\tau)B^{t_2(\tau)} = 1, \tag{4.37}$$

where

$$\gamma = P(Z_n \geq \log A) = \frac{1 - B^h}{A^h - B^h}.$$

The characteristic function of the unconditional distribution of n is

$$\psi(\tau) = \gamma\psi_1(\tau) + (1 - \gamma)\psi_2(\tau). \tag{4.38}$$

As an illustration we shall determine $\psi_1(\tau)$, $\psi_2(\tau)$ and $\psi(\tau)$ when z has a normal distribution. Then we have

$$-\log \varphi(t) = -(Ez)t - \frac{\sigma_z^2}{2}t^2 = \tau.$$

Hence

$$h = -\frac{2Ez}{\sigma_z^2}, \tag{4.39}$$

$$t_1(\tau) = \frac{1}{\sigma_z^2}(-Ez + \sqrt{(Ez)^2 - 2\sigma_z^2\tau}),$$

$$t_2(\tau) = \frac{1}{\sigma_z^2}(-Ez - \sqrt{(Ez)^2 - 2\sigma_z^2\tau}). \tag{4.40}$$

From (4.36), (4.37) and (4.38) we obtain

$$\gamma\psi_1(\tau) = \frac{B^{g_2} - B^{g_1}}{A^{g_1}B^{g_2} - A^{g_2}B^{g_1}}, \tag{4.41}$$

$$(1 - \gamma)\psi_2(\tau) = \frac{A^{g_1} - A^{g_2}}{A^{g_1}B^{g_2} - A^{g_2}B^{g_1}}, \tag{4.42}$$

and

$$\psi(\tau) = \frac{A^{g_1} + B^{g_2} - A^{g_2} - B^{g_1}}{A^{g_1}B^{g_2} - A^{g_2}B^{g_1}}, \tag{4.43}$$

where

$$g_1 = \frac{1}{\sigma_z^2}(-Ez + \sqrt{(Ez)^2 - 2\sigma_z^2\tau}) \tag{4.44}$$

and

$$g_2 = \frac{1}{\sigma_z^2}(-Ez - \sqrt{(Ez)^2 - 2\sigma_z^2\tau}). \tag{4.45}$$

For any positive integer r the r-th moment of n i.e., $E(n^r)$ is equal to the r-th derivative of $\psi(\tau)$ taken at $\tau = 0$. Let $E^*(n^r)$ be the conditional expected value of n^r under the restriction that $Z_n \leq \log B$, and let $E^{**}(n^r)$ be the conditional expected value of n^r under the restriction that $Z_n \geq \log A$. Then

$$E^*(n^r) = \frac{d^r\psi_2(\tau)}{d\tau^r}\bigg|_{\tau=0} \quad \text{and} \quad E^{**}(n^r) = \frac{d^r\psi_1(\tau)}{d\tau^r}\bigg|_{\tau=0}. \tag{4.46}$$

It may be of interest to note that $\dfrac{d^r\psi_k(\tau)}{d\tau^r}\bigg|_{\tau=0}$ $(k = 1, 2)$ and therefore also the moments of n can be obtained from the identity (4.35) directly by successive differentiation. In fact, the identity (4.35) can be written as (neglecting the excess of Z_n over the boundaries $\log A$ and $\log B$)

$$\gamma A'\psi_1[-\log \varphi(t)] + (1 - \gamma)B'\psi_2[-\log \varphi(t)] = 1. \tag{4.47}$$

Taking the first r derivatives of (4.47) with respect to t at $t = 0$ and $t = h$ we obtain a system of $2r$ linear equations in the $2r$ unknowns $\dfrac{d^j\psi_k(\tau)}{d\tau^j}\bigg|_{\tau=0}$ $(k = 1, 2; j = 1, \ldots, r)$ from which these unknowns can be determined. For example, $\dfrac{d\psi_k(\tau)}{d\tau}\bigg|_{\tau=0}$ $(k = 1, 2)$ can be determined as follows: Taking the first

derivative of (4.47) with respect to t and denoting $\dfrac{d^r\psi_k(\tau)}{d\tau^r}$ by $\psi_k^{(r)}(\tau)$ we obtain

$$\gamma(\log A)A^t\psi_1[-\log \varphi(t)] - \gamma A^t\frac{\varphi'(t)}{\varphi(t)}\psi_1^{(1)}[-\log \varphi(t)]$$

$$+ (1 - \gamma)(\log B)B^t\psi_2[-\log \varphi(t)] - (1 - \gamma)B^t\frac{\varphi'(t)}{\varphi(t)}\psi_2^{(1)}[-\log \varphi(t)] = 0.$$

$$(4.48)$$

Putting $t = 0$ and $t = h$ we obtain the equations

$$\gamma \log A - \gamma\frac{\varphi'(0)}{\varphi(0)}\psi_1^{(1)}(0) + (1 - \gamma) \log B - (1 - \gamma)\frac{\varphi'(0)}{\varphi(0)}\psi_2^{(1)}(0) = 0 \quad (4.49)$$

and

$$\gamma(\log A)A^h - \gamma A^h\frac{\varphi'(h)}{\varphi(h)}\psi_1^{(1)}(0) + (1 - \gamma)(\log B)B^h - (1 - \gamma)B^h\frac{\varphi'(h)}{\varphi(h)}\psi_2^{(1)}(0)$$

$$= 0 \quad (4.50)$$

from which $\psi_1^{(1)}(0)$ and $\psi_2^{(1)}(0)$ can be determined.

The distribution of n can be obtained by inverting the characteristic function of $\psi(\tau)$. This was done in [4] (neglecting the excess of Z_n over $\log A$ and $\log B$) in the case when z is normally distributed. The results obtained in [4] can be briefly stated as follows: If $B = 0$, or if $B > 0$ and $A = \infty$, the distribution of n is a simple elementary function. If $B = 0$ and $Ez > 0$, the distribution of $m = \dfrac{1}{2\sigma_z^2}(Ez)^2 n$ is given by

$$F(m) \, dm = \frac{c}{2\Gamma(\frac{1}{2})m^{3/2}}e^{-c^2/4m-m+c} \, dm \quad (0 \le m < \infty), \quad (4.51)$$

where

$$c = \frac{1}{\sigma_z^2}(Ez) \log A. \quad (4.52)$$

If $B > 0$, $A = \infty$ and $Ez < 0$ the distribution of $m = \dfrac{1}{2\sigma_z^2}(Ez)^2 n$ is given by the

expression we obtain from (4.51) if we substitute $\dfrac{1}{\sigma_z^2}(Ez) \log B$ for c.

If $B > 0$ and $A < \infty$, the distribution of m is given by an infinite series where each term is of the form (4.51) (see equation (76) in [4]).

Since m is a discrete variable, it may seem paradoxical that we obtained a probability density function for m. However, the explanation lies in the fact that we neglected the excess of Z_n over log A and log B which is zero only in the limiting case when Ez and σ_z approach zero.

The distribution of m given in (4.51) can be used as a good approximation to the exact distribution of m even if $B > 0$, provided that the probability that $Z_n \geq$ log A is nearly equal to 1.

It was pointed out in [4] that if $|Ez|$ and σ_z are sufficiently small, the distribution of n determined under the assumption that z is normally distributed will be a good approximation to the exact distribution of n even if z is not normally distributed.

4.5. Lower Limit of the Probability That the Sequential Process Will Terminate with a Number of Trials Less than or Equal to a Given Number

Let $P_i(n_0)$ be the probability that the sequential process will terminate at a value $n \leq n_0$, calculated under H_i ($i = 0, 1$). Let

$$\bar{P}_0(n_0) = P_0 \left[\sum_{\alpha=1}^{n_0} z_\alpha \leq \log B \right] \tag{4.53}$$

and

$$\bar{P}_1(n_0) = P_1 \left[\sum_{\alpha=1}^{n_0} z_\alpha \geq \log A \right]. \tag{4.54}$$

It is clear that

$$\bar{P}_i(n_0) \leq P_i(n_0) \quad (i = 0, 1). \tag{4.55}$$

For calculating $\bar{P}_i(n_0)$ we shall assume that n_0 is sufficiently large so that $\sum_{\alpha=1}^{n_0} z_\alpha$ can be regarded as normally distributed. Let $G(\lambda)$ be defined by

$$G(\lambda) = \frac{1}{\sqrt{2\pi}} \int_\lambda^\infty e^{-(1/2)t^2} \, dt. \tag{4.56}$$

Furthermore, let

$$\lambda_1(n_0) = \frac{\log A - n_0 E_1(z)}{\sqrt{n_0}\sigma_1(z)} \tag{4.57}$$

and

$$\lambda_0(n_0) = \frac{\log B - n_0 E_0(z)}{\sqrt{n_0}\sigma_0(z)}, \tag{4.58}$$

where $\sigma_i(z)$ is the standard deviation of z under H_i. Then

$$\bar{P}_1(n_0) = G[\lambda_1(n_0)] \tag{4.59}$$

and

Table 2. Lower Bound of the Probability* That a Sequential Analysis Will Terminate Within Various Numbers of Trials, When the Most Powerful Current Test Requires Exactly 1000 Trials.

Number of trials	$\alpha = .01$ and $\beta = .01$		$\alpha = .01$ and $\beta = .05$		$\alpha = .05$ and $\beta = .05$	
	Alternative hypothesis true	Null hypothesis true	Alternative hypothesis true	Null hypothesis true	Alternative hypothesis true	Null hypothesis true
1000	.910	.910	.799	.891	.773	.773
1200	.950	.950	.871	.932	.837	.837
1400	.972	.972	.916	.957	.883	.883
1600	.985	.985	.946	.972	.915	.915
1800	.991	.991	.965	.982	.938	.938
2000	.995	.995	.977	.989	.955	.955
2200	.997	.997	.985	.993	.967	.967
2400	.999	.999	.990	.995	.976	.976
2600	.999	.999	.994	.997	.982	.982
2800	1.00	1.00	.996	.998	.987	.987
3000	1.00	1.00	.997	.999	.990	.990

* The probabilities given are lower bounds for the true probabilities. They relate to a test of the mean of a normally distributed variate, the difference between the null and alternative hypothesis being adjusted for each pair of values of α and β so that the number of trials required under the most powerful current test is exactly 1000.

$$\bar{P}_0(n_0) = 1 - G[\lambda_0(n_0)]. \tag{4.60}$$

Hence we have the inequalities

$$P_1(n_0) \geq G[\lambda_1(n_0)] \tag{4.61}$$

and

$$P_0(n_0) \geq 1 - G[\lambda_0(n_0)]. \tag{4.62}$$

Putting $\log A = \log \dfrac{1 - \beta}{\alpha}$ and $\log B = \log \dfrac{\beta}{1 - \alpha}$, Table 2 shows the values of $\bar{P}_1(n_0)$ and $\bar{P}_0(n_0)$ corresponding to different pairs (α, β) and different values of n_0. In these calculations it has been assumed that the distribution under H_0 is a normal distribution with mean zero and unit variance, and the distribution under H_1 is a normal distribution with mean θ and unit variance. For each pair (α, β) the value of θ was determined so that the number of observations required by the current most powerful test of strength (α, β) is equal to 1000.

4.6. Truncated Sequential Analysis

In some applications a definite upper bound for the number of observations may be desirable. Thus, a certain integer n_0 is chosen so that if the sequential process does not lead to a final decision for $n \leq n_0$, a new rule is given for the acceptance or rejection of H_0 at the stage $n = n_0$.

A simple and reasonable rule for the acceptance or rejection of H_0 at the stage $n = n_0$ can be given as follows: If $\sum_{\alpha=1}^{n_0} z_\alpha \leq 0$ accept H_0 and if $\sum_{\alpha=1}^{n_0} z_\alpha > 0$ we accept H_1. By thus truncating the sequential process we change, however, the probabilities of errors of the first and second kinds. Let α and β be the probabilities of errors of the first and second kinds, respectively, if the sequential test is not truncated. Let $\alpha(n_0)$ and $\beta(n_0)$ be the probabilities of errors of the first and second kinds if the test is truncated at $n = n_0$. We shall derive upper bounds for $\alpha(n_0)$ and $\beta(n_0)$.

First we shall derive an upper bound for $\alpha(n_0)$. Let $\rho_0(n_0)$ be the probability (under the null hypothesis) that the following three conditions are simultaneously fulfilled:

(i) $$\log B < \sum_{\alpha=1}^{n} z_\alpha < \log A \qquad \text{for } n = 1, \ldots, n_0 - 1,$$

(ii) $$0 < \sum_{\alpha=1}^{n_0} z_\alpha < \log A,$$

(iii) continuing the sequential process beyond n_0, it terminates with the acceptance of H_0.

It is clear that

$$\alpha(n_0) \leq \alpha + \rho_0(n_0). \tag{4.63}$$

Let $\bar{\rho}_0(n_0)$ be the probability (under the null hypothesis) that $0 < \sum_{\alpha=1}^{n_0} z_\alpha < \log A$. Then obviously

$$\rho_0(n_0) \leq \bar{\rho}_0(n_0)$$

and consequently

$$\alpha(n_0) \leq \alpha + \bar{\rho}_0(n_0). \tag{4.64}$$

Let $\rho_1(n_0)$ be the probability under the alternative hypothesis that the following three conditions are simultaneously fulfilled:

(i) $$\log B < \sum_{\alpha=1}^{n} z_\alpha < \log A \qquad \text{for } n = 1, \ldots, n_0 - 1,$$

(ii) $$\log B < \sum_{\alpha=1}^{n_0} z_\alpha \leq 0,$$

(iii) continuing the sequential process beyond n_0, it terminates with the acceptance of H_1.

It is clear that

$$\beta(n_0) \leq \beta + \rho_1(n_0). \tag{4.65}$$

Let $\bar{\rho}_1(n_0)$ be the probability (under the alternative hypothesis) that $\log B <$

$\sum_{\alpha=1}^{n_0} z_\alpha \le 0$. Then $\rho_1(n_0) \le \bar{\rho}_1(n_0)$ and consequently

$$\beta(n_0) \le \beta + \bar{\rho}_1(n_0). \tag{4.66}$$

Let

$$v_1 = \frac{-n_0 E_0(z)}{\sqrt{n_0 \sigma_0(z)}}, \qquad v_2 = \frac{\log A - n_0 E_0(z)}{\sqrt{n_0 \sigma_0(z)}},$$

$$v_3 = \frac{-n_0 E_1(z)}{\sqrt{n_0 \sigma_1(z)}}, \qquad v_4 = \frac{\log B - n_0 E_1(z)}{\sqrt{n_0 \sigma_1(z)}},$$

where $\sigma_i(z)$ is the standard deviation of z under H_i $(i = 0, 1)$. Then

$$\bar{\rho}_0(n_0) = G(v_1) - G(v_2) \tag{4.67}$$

and

$$\bar{\rho}_1(n_0) = G(v_4) - G(v_3). \tag{4.68}$$

From (4.64), (4.66), (4.67) and (4.68) we obtain

$$\alpha(n_0) \le \alpha + G(v_1) - G(v_2) \tag{4.69}$$

and

$$\beta(n_0) \le \beta + G(v_4) - G(v_3). \tag{4.70}$$

Table 3. Effect on Risks of Error of Truncating* a Sequential Analysis at a Predetermined Number of Trials.

Number of trials	$\alpha = .01$ and $\beta = .01$		$\alpha = .01$ and $\beta = .05$		$\alpha = .05$ and $\beta = .05$	
	Upper bound of effective α	Upper bound of effective β	Upper bound of effective α	Upper bound of effective β	Upper bound of effective α	Upper bound of effective β
1000	.020	.020	.033	.070	.095	.095
1200	.015	.015	.024	.063	.082	.082
1400	.013	.013	.019	.058	.072	.072
1600	.012	.012	.016	.055	.066	.066
1800	.011	.011	.014	.053	.062	.062
2000	.010	.010	.012	.052	.058	.058
2200	.010	.010	.012	.051	.056	.056
2400	.010	.010	.011	.051	.055	.055
2600	.010	.010	.011	.051	.053	.053
2800	.010	.010	.010	.050	.053	.053
3000	.010	.010	.010	.050	.052	.052

* If the sequential analysis is based on the values α and β shown, but a decision is made at n_0 trials even when the normal sequential criteria would require a continuation of the process, the realized values of α and β will not exceed the tabular entries. The table relates to a test of the mean of a normally distributed variate, the difference between the null and alternative hypotheses being adjusted for each pair (α, β) so that the number of trials required by the current test is 1000.

The upper bounds given in (4.69) and (4.70) may considerably exceed $\alpha(n_0)$ and $\beta(n_0)$, respectively. It would be desirable to find closer limits.

Table 3 shows the values of the upper bounds of $\alpha(n_0)$ and $\beta(n_0)$ given by formulas (4.69) and (4.70) corresponding to different pairs (α, β) and different values of n_0. In these calculations we have put $\log A = \log \dfrac{1 - \beta}{\alpha}$, $\log B = \log \dfrac{\beta}{1 - \alpha}$ and assumed that the distribution under H_0 is a normal distribution with mean zero and unit variance, and the distribution under H_1 is a normal distribution with mean θ and unit variance. For each pair (α, β) the value of θ has been determined so that the number of observations required by the current most powerful test of strength (α, β) is equal to 1000.

It seems to the author that the upper limits given in (4.69) and (4.70) are considerably above the true $\alpha(n_0)$ and $\beta(n_0)$ respectively, when n_0 is not much higher than the value of n needed for the current most powerful test.

4.7. Efficiency of the Sequential Probability Ratio Test

Let S be any sequential test for which the probability of an error of the first kind is α, the probability of an error of the second kind is β and the probability that the test procedure will eventually terminate is one. Let S' be the sequential probability ratio test whose strength is equal to that of S. We shall prove that the sequential probability ratio test is an optimum test, i.e., that $E_i(n|S) \geq E_i(n|S')$ $(i = 0, 1)$, if for S' the excess of Z_n over $\log A$ and $\log B$ can be neglected. This excess is exactly zero if z can take only the values d and $-d$ and if $\log A$ and $\log B$ are integral multiples of d. In any other case the excess will not be identically zero. However, if $|Ez|$ and σ_z are sufficiently small, the excess of Z_n over $\log A$ and $\log B$ is negligible.

For any random variable u we shall denote by $E_i^*(u|S)$ the conditional expected value of u under the hypothesis H_i $(i = 0, 1)$ and under the restriction that H_0 is accepted. Similarly, let $E_i^{**}(u|S)$ be the conditional expected value of u under the hypothesis H_i $(i = 0, 1)$ and under the restriction that H_1 is accepted. In the notations for these expected values the symbol S stands for the sequential test used. Denote by $Q_i(S)$ the totality of all samples for which the test S leads to the acceptance of H_i. Then we have

$$E_0^* \left(\frac{p_{1n}}{p_{0n}} \Big| S \right) = \frac{P_1[Q_0(S)]}{P_0[Q_0(S)]} = \frac{\beta}{1 - \alpha}, \tag{4.71}$$

$$E_0^{**} \left(\frac{p_{1n}}{p_{0n}} \Big| S \right) = \frac{P_1[Q_1(S)]}{P_0[Q_1(S)]} = \frac{1 - \beta}{\alpha}, \tag{4.72}$$

$$E_1^* \left(\frac{p_{0n}}{p_{1n}} \Big| S \right) = \frac{P_0[Q_0(S)]}{P_1[Q_0(S)]} = \frac{1 - \alpha}{\beta}, \tag{4.73}$$

and

$$E_1^{**} \left(\frac{p_{0n}}{p_{1n}} \Big| S \right) = \frac{P_0[Q_1(S)]}{P_1[Q_1(S)]} = \frac{\alpha}{1 - \beta}. \tag{4.74}$$

To prove the efficiency of the sequential probability ratio test, we shall first derive two lemmas.

Lemma 1. *For any random variable u the inequality*

$$e^{Eu} \le E e^u \tag{4.75}$$

holds.

PROOF. Inequality (4.75) can be written as

$$1 \le E e^{u'} \tag{4.76}$$

where $u' = u - Eu$. Lemma 1 is proved if we show that (4.76) holds for any random variable u' with zero mean. Expanding $e^{u'}$ in a Taylor series around $u' = 0$, we obtain

$$e^{u'} = 1 + u' + \tfrac{1}{2} u'^2 e^{\xi(u')} \quad \text{where} \quad 0 \le \xi(u') \le u'. \tag{4.77}$$

Hence

$$E e^{u'} = 1 + \tfrac{1}{2} E[u'^2 e^{\xi(u')}] \ge 1 \tag{4.78}$$

and Lemma 1 is proved. □

Lemma 2. *Let S be a sequential test such that there exists a finite integer N with the property that the number n of observations required for the test is $\le N$. Then*

$$E_i(n|S) = \frac{E_i\left(\log \dfrac{p_{1n}}{p_{0n}} \Big| S \right)}{E_i(z)} \qquad (i = 0, 1). \tag{4.79}$$

The proof is omitted, since it is essentially the same as that of equation (4.5) for the sequential probability ratio test

On the basis of Lemmas 1 and 2 we shall be able to derive the following theorem.

Theorem. *Let S be any sequential test for which the probability of an error of the first kind is α, the probability of an error of the second kind is β and the probability that the test procedure will eventually terminate is equal to one. Then*

$$E_0(n|S) \ge \frac{1}{E_0(z)}\left[(1 - \alpha) \log \frac{\beta}{1 - \alpha} + \alpha \log \frac{1 - \beta}{\alpha} \right] \tag{4.80}$$

and

$$E_1(n|S) \ge \frac{1}{E_1(z)}\left[\beta \log \frac{\beta}{1 - \alpha} + (1 - \beta) \log \frac{1 - \beta}{\alpha} \right]. \tag{4.81}$$

PROOF. First we shall prove the theorem in the case when there exists a finite integer N such that n never exceeds N. According to Lemma 2 we have

$$E_0(n|S) = \frac{1}{E_0(z)} E_0\left(\log \frac{p_{1n}}{p_{0n}} | S\right)$$

$$= \frac{1}{E_0(z)}\left[(1-\alpha)E_0^*\left(\log \frac{p_{1n}}{p_{0n}}|S\right) + \alpha E_0^{**}\left(\log \frac{p_{1n}}{p_{0n}}|S\right)\right] \quad (4.82)$$

and

$$E_1(n|S) = \frac{1}{E_1(z)} E_1\left(\log \frac{p_{1n}}{p_{0n}} | S\right)$$

$$= \frac{1}{E_1(z)}\left[\beta E_1^*\left(\log \frac{p_{1n}}{p_{0n}}|S\right) + (1-\beta)E_1^{**}\left(\log \frac{p_{1n}}{p_{0n}}|S\right)\right]. \quad (4.82)$$

From equations (4.71)–(4.74) and Lemma 1 we obtain the inequalities

$$E_0^*\left(\log \frac{p_{1n}}{p_{0n}}|S\right) \le \log \frac{\beta}{1-\alpha}, \quad (4.84)$$

$$E_0^{**}\left(\log \frac{p_{1n}}{p_{0n}}|S\right) \le \log \frac{1-\beta}{\alpha}, \quad (4.85)$$

$$E_1^*\left(\log \frac{p_{0n}}{p_{1n}}|S\right) = -E_1^*\left(\log \frac{p_{1n}}{p_{0n}}|S\right) \le \log \frac{1-\alpha}{\beta}, \quad (4.86)$$

and

$$E_1^{**}\left(\log \frac{p_{0n}}{p_{1n}}|S\right) = -E_1^{**}\left(\log \frac{p_{1n}}{p_{0n}}|S\right) \le \log \frac{\alpha}{1-\beta}. \quad (4.87)$$

Since $E_0(z) < 0$, (4.80) follows from (4.82), (4.84) and (4.85). Similarly, since $E_1(z) > 0$, (4.81) follows from (4.83), (4.86) and (4.87). This proves the theorem when there exists a finite integer N such that $n \le N$.

To prove the theorem for any sequential test S of strength (α, β), for any positive integer N let S_N be the sequential test we obtain by truncating S at the N-th observation if no decision is reached before the N-th observation. Let (α_N, β_N) be the strength of S_N. Then we have

$$E_0(n|S) \ge E_0(n|S_N) \ge \frac{1}{E_0(z)}\left[(1-\alpha_N)\log \frac{\beta_N}{1-\alpha_N} + \alpha_N \log \frac{1-\beta_N}{\alpha_N}\right] \quad (4.88)$$

and

$$E_1(n|S) \ge E_1(n|S_N) \ge \frac{1}{E_1(z)}\left[\beta_N \log \frac{\beta_N}{1-\alpha_N} + (1-\beta_N)\log \frac{1-\beta_N}{\alpha_N}\right]. \quad (4.89)$$

Since $\lim_{N=\infty} \alpha_N = \alpha$ and $\lim_{N=\infty} \beta_N = \beta$, inequalities (4.80) and (4.81) follow from (4.88) and (4.89). Hence the proof of the theorem is completed.

If for the sequential probability ratio test S' the excess of the cumulative sum Z_n over the boundaries $\log A$ and $\log B$ is zero, $E_0(n|S')$ is exactly equal to the right hand side member of (4.80) and $E_1(n|S')$ is exactly equal to the right hand side member of (4.81). Hence, in this case S' is exactly an optimum test. If both $|Ez|$ and σ_z are small, also the expected value of the excess over the boundaries will be small and, therefore, $E_0(n|S')$ and $E_1(n|S')$ will be only slightly larger than the right hand members of (4.80) and (4.81), respectively. Thus, in such a case the sequential probability ratio test is, if not exactly, very nearly an optimum test.[12]

References

[1] H.F. Dodge and H.G. Romig, "A method of sampling inspection," *The Bell System Tech. Jour.*, Vol. 8 (1929), pp. 613–631.

[2] Walter Bartky, "Multiple sampling with constant probability", *Annals of Math. Stat.*, Vol. 14 (1943), pp. 363–377.

[3] Harold Hotelling, "Experimental determination of the maximum of a function", *Annals of Math. Stat.*, Vol. 12 (1941).

[4] Abraham Wald, "On cumulative sums of random variables", *Annals of Math. Stat.*, Vol. 15 (1944).

[5] Z.W. Birnbaum, "An inequality for Mill's ratio", *Annals of Math. Stat.*, Vol. 13 (1942).

[6] P.C. Mahalanobis, "A sample survey of the acreage under jute in Bengal, with discussion on planning of experiments," Proc. 2nd Ind. Stat. Conf., Calcutta, Statistical Publishing Soc. (1940).

[7] Abraham Wald, *Sequential Analysis of Statistical Data: Theory.* A report submitted by the Statistical Research Group, Columbia University to the Applied Mathematics Panel, National Defense Research Committee, Sept. 1943.

[8] Harold Freeman, *Sequential Analysis of Statistical Data: Applications.* A Report submitted by the Statistical Research Group, Columbia University to the Applied Mathematics Panel, National Defense Research Committee, July 1944.

[9] G.A. Barnard, M.A., *Economy in Sampling with Reference to Engineering Experimentation*, (British) Ministry of Supply, Advisory Service on Statistical Method and Quality Control, Technical Report, Series 'R', No. Q.C./R/7 Part 1.

[10] C.M. Stockman, *A Method of Obtaining an Approximation for the Operating Characteristic of a Wald Sequential Probability Ratio Test Applied to a Binomial Distribution*, (British) Ministry of Supply, Advisory Service on Statistical Method and Quality Control, Technical Report, Series 'R' No. Q.C./R/19.

[11] Abraham Wald, *A General Method of Deriving the Operating Characteristics of any Sequential Probability Ratio Test.* A Memorandum submitted to the Statistical Research Group, Columbia University, April 1944.

---

[12] The author conjectures that the sequential probability ratio test is exactly an optimum test even if the excess of Z_n over the boundaries is not zero. However, he did not succeed in proving this.

Introduction to
Hoeffding (1948) A Class of Statistics with Asymptotically Normal Distribution

P.K. Sen
University of North Carolina at Chapel Hill

Wassily Hoeffding was born on June 12, 1914 in Mustamaki, Finland, near St. Petersburg (now Leningrad), USSR. His parents were of Danish origin; his father was an economist and his mother had studied medicine. Although at that time, Finland was a part of the Russian Empire, the Bolshevik movement was quite intense and consolidated under the Lenin dictatorship in the Civil War of 1918–20. The Hoeffding family left Russia for Denmark in 1920, and four years later, they moved on to Berlin. In 1933, Wassily finished high school and went on to college to study economics. However, a year later, he gave up economics and entered Berlin University to study mathematics. He earned a Ph.D. degree from Berlin University in 1940 with a dissertation in correlation theory, which dealt with some properties of bivariate distributions that are invariant under arbitrary monotone transformations of the margins. In this context, he studied some (mostly, descriptive) aspects of rank correlations, and a few years later, while investigating the sampling aspects of such measures, he formulated in a remarkably general and highly original form the general distribution theory of symmetric, unbiased estimators (which he termed U-statistics). This is depicted in this outstanding article (under commentary). During World War II, Wassily worked in Berlin partly as an editorial assistant and partly as a research assistant in actuarial science. In September 1946, he was able to immigrate to the United States, and there he started attending lectures at Columbia University in New York. In 1947, he was invited by Harold Hotelling to join the newly established department of Statistics at the University of North Carolina at Chapel Hill. Since then Wassily had been in Chapel Hill with occasional visits to other campuses (Columbia University and Cornell University in New York, the Steklov Institute in Leningrad, USSR, and the Indian Statistical Institute in Calcutta, among others). In 1979, he retired from active service. He died on February 28, 1991 at Chapel Hill.

Wassily Hoeffding had a variety of research interests in statistics, ranging from correlation theory through U-statistics, sequential analysis, probability inequalities, large deviation probabilities to general asymptotic methods. In each of these areas, his contributions are top-rate and highly significant. Although he received his basic education in Berlin, Hoeffding was never very happy about that. He felt that "probability and statistics were very poorly represented in Berlin at that time," and only after his arrival in America, was he was able to appreciate fully the statistical aspects of his study. This led him to writing the outstanding paper on U-statistics that follows. He had indeed a great deal of affection for this paper. He writes [Gani (1982)], "I like to think of this paper as my 'real' Ph.D. dissertation." Reading in between the lines of his biography [Gani (1982)], I have no doubt in my mind that had it not been for World War II, he would have made this accomplishment even earlier. During the peak of his research career (1947–66), he made outstanding research contributions in diverse areas, and undoubtedly, this U-statistics paper is the "jewel in the crown" of Wassily's creativity and ingenuity in research.

To appreciate fully the impact of this paper in statistics and to characterize its true breakthrough nature, let us ponder the state of the art of nonparametric inference before the appearence of this article. Prior to 1947, estimation theory in a nonparametric fashion was practically in a dormant state, with the scanty literature on nonparametric tests having predominantly a "randomization" flavor. Although justifications for the use of symmetric and unbiased estimators were partially made earlier by Halmos (1946) and others, their treatment lacked the complete generality of the subject matter as treated in this paper of Hoeffding's. In a parametric setup, parameters are algebraic constants appearing in the known (or assumed) functional forms of distribution functions governing the random variables under consideration. In a nonparametric setup, this functional form is not known, and it is only assumed that the unknown distribution belongs to a general class. As such, statistical parameters are expressed as functionals of the underlying distribution functions (and termed regular functionals or estimable parameters). This generality of the setup invariantly calls for more general forms of statistics or estimators that may not be linear in the sample observations (or suitable transformations of them). As such, the usual formulation of linear estimators may no longer be valid here. Moreover, the distribution theory (even in an asymptotic setup) of linear estimators may not be applicable for such statistical functionals. This outstanding paper by Hoeffding deals with the basic formulation of statistical parameters in a nonparametric setup, the construction of suitable unbiased and symmetric estimators of such functionals (termed U-statistics), and a thorough and most unified treatment of their (sampling) distributional properties with easy access to the related asymptotic theory. In fact, the simple projection result Hoeffding considered in this respect was by far the most novel and natural means to approximate plausible nonlinear statistics by a sum of independent random variables, and this alone qualifies this paper for the coveted breakthrough category. Although the letter "U" used by Wassily for unbiasedness, the symmetric nature of U-statistics plays a more important role in the characterization of their optimality properties, and this aspect has also been thoroughly probed in this paper. A year earlier, von Mises (1947) considered a class of (differentiable) statistical functions (now referred to as V-statistics). Hoeffding's U-statistics and von Mises' V-statistics are indeed vey close to each other, although V-statistics are not generally unbiased for the functionals they estimate. von Mises' approach hinges on the asymptotic behavior of the usual empirical distributional processes and some differentiability properties of such functionals (which were not properly known at that time), and it requires more stringent regularity conditions than U-statistics. In this context too, Wassily was able to establish the close affinity of U- and V-statistics under quite simple regularity conditions, and this provided easy access for the study of the asymptotic properties of V-statistics via U-statistics. Prior to von Mises and Hoeffding, there was some scattered (and piecemeal) work on U-statistics, such as Gini's mean difference, sample moments, k-statistics, Kendall's tau statistics, Spearman's rank correlation coefficient and some other ad hoc nonparametric test statistics that are expressible as U-statistics. However, these were dealt with on an individual (and often with an inadequate theoretical) basis, and therefore, a genuine need existed for a unified treatment to cover all these specific cases: A complete treatise on this indeed appears in this paper of Hoeffding's. All these specific cases have been incorporated in the form of examples in the concluding section of this paper, where the unified theory developed in earlier sections has been used to provide a sound basis for the general methodology underlying the finite sample theory as well as the asymptotic ones.

To appreciate fully the novelty of the Hoeffding approach (1948) and its impact on mathematical statistics and probability theory, let us start with the definition of a U-statistic. Consider an estimable parameter $\theta(F) = \int \cdots \int g(x_1, \ldots, x_m) \, dF(x_1) \ldots dF(x_m)$ of degree $m(\geq 1)$, where the kernel $g(x_1, \ldots, x_m)$ is assumed to be a symmetric function of its m arguments. $\theta(F)$ has a domain F, a subset of the class of all distributions F. Then, for a sample $X_1, \ldots, X_n$ of size $n(\geq m)$, the U-statistic corresponding to $\theta(F)$ is defined by

$$U_n = \binom{n}{m}^{-1} \sum_{\{1 \leq i_1 < \cdots < i_m \leq n\}} g(X_{i_1}, \ldots, X_{i_m}). \tag{1}$$

For $m = 1$, U_n reduces to an average over independent and identically distributed random variables (i.i.d.r.v.), so that the standard results in probability theory may be borrowed to study its properties. The picture becomes quite different when $m \geq 2$. Although U_n remains an unbiased and symmetric estimator, the summands in (1) are no longer all independent. This creates some difficulties in incorporating the standard central limit theorems for the study of the asymptotic normality of $n^{1/2}[U_n - \theta(F)]$. Also, the variance of U_n is no longer n^{-1} times the variance of $g(X_1, \ldots, X_m)$, and covariance terms for these summands lead to a more complicated expression. Hoeffding (1948) succeeded in deriving a neat expression for this variance term as a linear function of m estimable (variance) parameters (say, $\zeta_1, \ldots, \zeta_m$) depicting the order of the dependence. These have decreasing weights. This enabled him to show that for every finite n, $n^{1/2}[U_n - \theta(F)]$ has a variance bounded from below by $m^2\zeta_1$ and this lower bound is asymptotically attained. {Actually, he showed that the variance of $n^{1/2}[U_n - \theta(F)]$ is monotonically nonincreasing in n; its upper bound $m\zeta_m$ is attained for $n = m$ and the lower bound for $n \to \infty$.} The formulation of the parameters $\zeta_1, \ldots, \zeta_m$ (which represent the variances of successive order projections of the kernel) was certainly a very clever idea. What was more is that in this process of taking appropriate conditional expectations, he managed to obtain a projection of U_n as a sum of i.i.d.r.v.'s, and this opened up a new and unified approach to studying the asymptotic normality and other properties of nonlinear estimators. This is by far the most significant contribution of this outstanding paper.

To illustrate this point, let us denote

$$g_1(x_1) = E_F[g(x_1, X_2, \ldots, X_m)] \qquad \text{(so that } \zeta_1 = \text{var}[g_1(X_1)]). \qquad (2)$$

Then, Hoeffding (1948) showed that

$$[U_n - \theta(F)] = mn^{-1}[g_1(X_1) + \cdots + g_1(X_n) - n\theta(F)] + R_n, \qquad (3)$$

where $E(R_n^2) = 0(n^{-2})$ and the sum $\sum_{i=1}^n[g_1(X_i) - \theta(F)]$ involves independent summands for which various standard asymptotic results can be established under appropriate regularity conditions. For example, the law of large numbers under the (Khintchine) first moment condition or the classical central limit theorem under the second moment condition, and in general, related asymptotic theory under appropriate moment conditions on $g_1(X_1)$ can be established by direct adaptation of standard theory as available prior to Hoeffding (1948). Thus, the crux of the problem lies in the handling of the remainder term R_n in (3) and showing that it is negligible in a mode consistent with the type of asymptotic result one wants to study for the U-statistics. As for the central limit theorem for U-statistic, it suffices to have $n^{1/2}R_n \to 0$, in probability, as $n \to \infty$, and hence, $E(R_n^2) = 0(n^{-1})$ is enough. However, in this context, a basic assumption is that $\theta(F)$ is *stationary of order zero* (at F), so that $\zeta_1 = \zeta_1(F) > 0$. On the other hand, if $\zeta_1 = 0$, then $g_1(x) = \theta(F)$ a.a.x., so that by (3), $n^{1/2}[U - \theta(F)]$ has a degenerate asymptotic distribution, although a nondegenerate limiting distribution may still hold for $n^d[U_n - \theta(F)]$ for some $d > 1/2$. Intrigued by this feature, Hoeffding (1948) formulated the concept of stationarity for $\theta(F)$ (at F) of various orders according to the positivity of the $\zeta_c: 0 < c \le m$. In a follow-up paper (1948), he actually showed that in the context of nonparametric tests for (bivariate) independence, under the null hypothesis, a class of U_n has such stationarity of order one (i.e., $\zeta_1 = 0 < \zeta_2$) and that $n[U_n - \theta(F)]$ has a nondegenerate asymptotic distribution that can be expressed as the distribution of $\sum_{j \ge 0}\lambda_j(Z_j^2 - 1)$, where the Z_j are independent standard normal variables and the λ_j are the eigenvalues corresponding to (complete) orthonormal functions relating to the second-order kernel $g_2(X_1, X_2) = E_F[g(X_1, \ldots, X_m)|X_1, X_2]$.

An elegant follow-up to this type of asymptotic distribution theory of "degenerate" U-statistics is due to Gregory (1977). Actually, this basic paper of Hoeffding's (under commentary) contains all the ingredients for a far deeper result that incorporates the orthogonal partition of U-statistics and provides the penultimate step for the asymptotic behavior of U-statistics depending on the order of stationarity. Hoeffding (1961) showed that for a symmetric kernel $g(X_1, \ldots, X_m)$ of degree m, one has for every $n \geq m$,

$$U_n = \theta(F) + \binom{m}{1} U_{n,1} + \cdots + \binom{m}{m} U_{n,m}, \qquad (4)$$

where the $U_{n,h}$ are themselves U-statistics (of order h), such that they are pairwise uncorrelated, $E[U_{n,h}] = 0$ and $E[U_{n,h}^2] = 0(n^{-h})$, for $h = 1, \ldots, m$. Actually, $U_{n,1}$ is the projection $\{$i.e., $n^{-1}\sum_{i=1}^n [g_1(X_i) - \theta(F)]\}$ referred to in (3). It is known (by now) that for different h, the normalized forms $n^{h/2}U_{n,h}$ have asymptotically nondegenerate and independent distributions. Thus, the stationarity of $\theta(F)$ can be linked directly to (4) to identify the appropriate nondegenerate distribution for U-statistics. In general, ζ_1 is positive, so that asymptotic normality holds. The representation in (4) is referred to as the *Hoeffding (or H−) decomposition of U-statistics*. The H-projection in (3) and, more generally, the H-decomposition in (4) have played a very significant role in the asymptotic distribution theory of (nonlinear) statistics, especially in the context of nonparametric statistics. By construction, U-statistics inherit a nonparametric flavor (where the d.f. F is allowed to be a member of a general class of distributions), and it is not surprising to see that in the field of nonparametrics in the 1950s and 1960s, the literature was flooded with estimators and test statistics that were expressible as U-statistics or their natural generalizations. We may refer to Chap.3 of Puri and Sen (1971) for a detailed account of these developments.

Although prior to Hoeffding (1948), these developments were spotty and piecemeal, soon after, the entire asymptotic theory of such nonparametric tests and estimates grew from Hoeffding's basic work. The most notable extension relates to a multisample situation. In a two-sample setup, with r.v.'s $X_1, \ldots, X_{n_1}$ from a distribution F and independent r.v.'s $Y_1, \ldots, Y_{n_2}$ from a distribution G, one may consider an estimable parameter $\theta(F, G) = E[g(X_1, \ldots, X_{m_1}; Y_1, \ldots, Y_{m_2})]$, where the kernel $g(.;.)$ is of degree (m_1, m_2). Without any loss of generality, we may assume that it is a symmetric function of its first m_1 (and last m_2) arguments, although the roles of these two sets may be different. Lehmann (1951) was the first to quote Hoeffding (1948) in this two-sample context and obtain the desired asymptotic theory as a direct corollary. Some discussion of the H-decomposition of such generalized U-statistics in a broader sense occurs in Chap. 3 of Sen (1981).

In a general multiparameter problem, typically arising in multisample and/

or multivariate nonparametric models, one needs to work with a vector of estimable parameters or regular functionals. Keeping this in mind, Hoeffding (1948) considered the general case of vector-valued U-statistics, discussed their estimability criteria, obtained suitable formulations of their covariance matrices (in similar series forms), and established their asymptotic (multi)normality in a very elegant manner. He actually went a step further in considering a "smooth" function of such a vector of U-statistics and deriving parallel results.

The methodology (i.e., the H-projection and H-decomposition) initiated by Hoeffding (1948) is by no means restricted to i.i.d. random vectors. In Sect. 7 of Hoeffding (1948), there is a detailed treatment of the general independent but not necessarily identically distributed random vectors' case, and the H-decomposition in (4) literally goes over to this case, with $U_{n,h}$ modified accordingly. These results proved to be very useful for the study of the asymptotic properties of various one- and multisample nonparametric test statistics that are either expressible or at least closely approximatable as a function of (generalized) U-statistics.

In this context, we refer to the last section of Hoeffding (1948) for a handful of important examples in the one-sample case, whereas multisample problems involving (generalized) U-statistics have been treated in Chap. 3 of Puri and Sen (1971). In nonparametric formulations, the parameters are functionals of the underlying distributions, and hence, such U-statistics are natural contenders to the classical parametric statistics. Dwass (1955) made a very nice attempt to approximate a two-sample rank-order test statistic by a (generalized) U-statistic and then studied the asymptotic properties of the former via those of U-statistics. For this, Dwass ended up with some restrictions on the scores (on which a rank-order test statistic is based), and later on, Hájek (1968) considered a far more general setup and showed that the H-projection referred to earlier works out neatly for general rank-order statistics under very mild regularity conditions. For a general statistic $S_n = S(X_1, \ldots, X_n)$ involving independent r.v.'s, Hájek (1968) considered the projection

$$S_n - ES_n = \sum_{i=1}^{n} [E(S_n | X_i) - ES_n] + R_n, \tag{5}$$

where R_n is the remainder term and the $E(S_n | X_i)$ are independent r.v.'s. In particular, if we set $S_n = U_n$, it is easy to verify that $E(S_n | X_i) - ES_n = mn^{-1}[g_1(X_i) - \theta(F)]$, so that (5) is a natural extension of the H-projection to a broader class of statistics. As such, I have the impression that Hájek's projection ideas might have germinated from the basic H-projection result in (3). To stress further the impact of the H-decomposition (and projection) in asymptotic theory, let us cite another important work by van Zwet (1984). He considered a symmetric statistic $T_n = \tau(X_1, \ldots, X_n)$ so normalized that $ET_n = 0$ and $ET_n^2 = 1$. He showed that parallel to (4) there exists an orthogonal decomposition $T_n = T_{n,0} + (T_{n,1} - T_{n,0}) + \cdots + (T_{n,n} - T_{n,n-1})$, where these n components are pairwise uncorrelated and they share properties similar to $U_{n,h}$ in (4). In the case of U-statistics, the degree m is fixed, so that in terms of a general T_n, there $T_{n,j} - T_{n,j-1} = 0$, with probability 1, for every $j \geq m + 1$. This clearly shows that the H-projection and H-decomposition may as well be applied to functionals of F for which the kernel may not have a finite degree. An important example of this type is an L-estimator (or a linear combination of a function of sample-order statistics) typically arising in robust estimation problems. An L-estimator with a smooth score function can be written in the form

$$L_n = \int_0^1 \psi(t) J_n[G_n(t)] \, dG_n(t), \tag{6}$$

where $\psi(t)$, $0 < t < 1$ is a smooth function, $J_n(\cdot)$ converges to a smooth $J(\cdot)$, and $G_n(\cdot)$ is the reduced empirical distribution function [corresponding to the uniform distribution on $(0, 1)$]. L_n estimates $\mu = \int_0^1 \psi(t) J(t) \, dt$, and this can be identified for suitable (location or scale) parameters of the d.f. F. For the asymptotic normality of L_n, an H-projection suffices, and this typically involves a form comparable to (3) where, in addition, R_n can be expressed as a U-statistic of degree 2 plus a higher-order term [see Chap. 7 of Sen (1981)]. The H-decomposition extended by van Zwet (1984) enables one to extend this to a form comparable to (4), so that deeper asymptotic results can be obtained. For general statistical functionals, the H-projection along with the second-order term in the H-decomposition have been incorporated in Sen (1988) in the study of the asymptotic bias and mean square error of jackknifed estimators. This H-decomposition lies in the roots of a simplified treatment of the speed of convergence to normality of the normalized version of suitable statistics (including U-statistics); this has been studied in increasing generality by a number of researchers and an up-to-date account of such Berry–Esseen bounds is contained in the van Zwet (1984) paper.

As has been mentioned earlier, the H-decomposition in (4) has its roots in the Hoeffding (1948) paper. In course of this study, he observed that for each $h(= 1, \ldots, m)$, $\left\{ \binom{n}{h} U_{n,h}, n \geq h \right\}$ forms a forward martingale sequence. Moreover, such a martingale structure can be extended readily to the case where the X_i are not necessarily identically distributed. If the X_i are i.i.d.r.v.'s, then one can also conclude that for each $h(= 1, \ldots, m)$, $\{U_{n,h}; n \geq h\}$ (and hence, $\{U_n; n \geq m\}$) form a reversed martingale sequence. This was explicitly addressed by Berk (1956) and Kingman (1969), although I have the feeling that Hoeffding was aware of it a while earlier. Back in the 1940s, martingales and reversed martingales used to reside in the remote corners of stochastic processes, and no wonder that Hoeffding did not incorporate them in his findings. However, nearly a quarter-century later on, they proved to be very useful tools for the study of some deeper asymptotic results for U-statistics. The reverse martingale property of U-statistics led to the almost sure convergence of U_n to $\theta(F)$ under the minimal condition (that EU_n exists), while Miller and Sen (1972) extended the asymptotic normality result to some weak invariance principles for U_n and V_n. The same H-projection and H-decomposition provided access not only for such weak convergence results, but also for certain stronger invariance principles that were studied later on [and reported in Chap. 3 of Sen (1981)]. Furthermore, the reverse martingale property of U-statistics provides an easily verifiable means for certain uniform integrability results arising in the context of the sequential estimation of such functionals. A detailed account of various invariance principles for U-statistics and their role in sequential analysis is given in Chap. 3, 9, and 10 of Sen (1981), and unequivocally, I must admit that the inspiration for this work germinated from the basic work of Hoeffding (1948). Serfling (1980) also contains a good deal of Hoeffding's work.

As has been mentioned earlier, Hoeffding (1948) did not restrict himself to independent and identically distributed random vectors. Rather, he showed that in a very natural way, his projection result works out equally well for not necessarily identically distributed random variables or vectors. In Sect. 8, he carefully lays out the necessary details toward this projection (leading to the desired asymptotic multinormality result), and in the preceding section, he showed how to handle a smooth function of several U-statistics. Under the usual first-order Taylor series expansion of such a function, Hoeffding was able to use his basic projection result to establish the asymptotic normality result. Nevertheless, in all these works, he assumed that the random vectors are independently distributed. Even this condition is not that crucial. The basic H-projection result whenever applicable yields a linear statistic for the projected term, and this statistic can be conveniently studied under suitable dependence conditions. For example, in sampling without replacement from a finite population, the sample observations are exchangeable (but not independent) random variables, and the projection technique works out well. The projection becomes a linear statistic in exchangeable random variables, and hence, known results can be incorporated in the study of the asymptotic (as well as exact) properties of U-statistics. Such results have been reported in Chap. 3 of Sen (1981). The theory of U-statistics works out well for m-dependent sequences. In various time-series models where autoregressive or ARMA models are usually adopted, Keenan (1983) formulated the asymptotic theory by incorporating the theory of U-statistics based on a spectral distributional setup. Again, the H-projection underlies the development of the asymptotic theory. In the concluding section of Hoeffding (1948), there is a wealth of illustrations, with special emphasis on nonparametric measures of correlation and association as well as tests for trends. The latter case brought-out the relevance of nonidentically distributed random variables. This is the application section of Hoeffding (1948) that opened up a vast area of research in nonparametrics, and the past 40 years have witnessed the phenomenal growth of literature on U-statistics.

To conclude, I suggest that Hoeffding (1948) has indeed opened up several novel avenues in the small as well as asymptotic distribution theory of symmetric statistics arising both in parametric and nonparametric models. It is no wonder that this article carries a banner of over 400 citations (during

1948–88), and an excellent source for all these references is a recent monograph [*Teoria U-Statistics*, in Russian] by Koroluk and Borovskih (1989), where a detailed account of U-statistics and their role in diverse problems of mathematical statistics is given. [Although in terms of unification, this monograph does not come close to the nice and lucid treatment in Hoeffding (1948).] The impact of Hoeffding (1948) is overwhelming at the present time, and it is very likely to continue in the years to come.

References

Berk, R.H. (1956). Limiting behavior of posterior distributions when the model is incorrect, *Ann. Math. Statist.*, **37**, 51–58.

Dwass, M. (1955). On the asymptotic normality of some statistics used in nonparametric tests, *Ann. Math. Statist.*, **26**, 334–339.

Gani, J. (ed.) (1982). *The Making of Statisticians*. Springer-Verlag, New York.

Gregory, G.G. (1977). Large sample theory for U-statistics and tests of fit, *Ann. Statist.*, **5**, 110–123.

Hájek, J. (1968). Asymptotic normality of simple linear rank statistics under alternatives, *Ann. Math. Statist.*, **39**, 325–346.

Halmos, P.R. (1946). The theory of unbiased estimation, *Ann. Math. Statist.*, **17**, 34–43.

Hoeffding, W. (1948). A nonparametric test for independence, *Ann. Math. Statist.*, **19**, 546–557.

Hoeffding, W. (1961). The strong law of large numbers for U-statistics. Institute of Statistics University of North Carolina Mimeo Series No. 302.

Keenan, D.M. (1983). Limiting behavior of functionals of the sample spectral distribution, *Ann. Statist.*, **11**, 1206–1217.

Kingman, J.F.C. (1969). An Ergodic theorem, *Bull. Lon. Math. Soc.*, **1**, 339–340.

Koroljuk, Yu.S., and Borovskih, Yu.V. (1989). *Teorija U-Statistik* (in Russian). Naukova Dumka, Kiev.

Lehmann, E.L. (1951). Consistency and unbiasedness of certain nonparametric tests, *Ann. Math. Statist.*, **22**, 165–179.

Miller, R.G., Jr., and Sen, P.K. (1972). Weak convergence of U-statistics and von Mises' differentiable statistical functions, *Ann. Math. Statist.*, **43**, 31–41.

Puri, M.L. and Sen, P.K. (1971). *Nonparametric Methods in Multivariate Analysis*. Wiley, New York.

Sen, P.K. (1981). *Sequential Nonparametrics: Invariance Principles and Statistical Inference*. Wiley, New York.

Sen, P.K. (1988). Functional jackknifing: Rationality and general asymptotics, *Ann. Statist.*, **16**, 450–469.

Serfling, R.J. (1980). *Approximation Theorems of Mathematical Statistics*. Wiley, New York.

Van Zwet, W.R. (1984). A Berry-Esseen bound for symmetric statistics, *Zeit. Wahrsch. verw. Gebiete*, **66**, 425–440.

von Mises, R.V. (1947). On the asymptotic distribution of differentiable statistical functions, *Ann. Math. Statist.*, **18**, 309–348.

A Class of Statistics with Asymptotically Normal Distribution[1]

Wassily Hoeffding
Institute of Statistics
University of North Carolina

1. Summary

Let $X_1, \ldots, X_n$ be n independent random vectors, $X_\nu = (X_\nu^{(1)}, \ldots, X_\nu^{(r)})$, and $\Phi(x_1, \ldots, x_m)$ a function of $m(\leq n)$ vectors $x_\nu = (x_\nu^{(1)}, \ldots, x_\nu^{(r)})$. A statistic of the form $U = \Sigma'' \Phi(X_{\alpha_1}, \ldots, X_{\alpha_m})/n(n-1) \ldots (n-m+1)$, where the sum Σ'' is extended over all permutations $(\alpha_1, \ldots, \alpha_m)$ of m different integers, $1 \leq \alpha_i \leq n$, is called a U-statistic. If $X_1, \ldots, X_n$ have the same (cumulative) distribution function (d.f.) $F(x)$, U is an unbiased estimate of the population characteristic $\theta(F) = \int \cdots \int \Phi(x_1, \ldots, x_m) \, dF(x_1) \ldots dF(x_m)$. $\theta(F)$ is called a regular functional of the d.f. $F(x)$. Certain optimal properties of U-statistics as unbiased estimates of regular functionals have been established by Halmos [9] (cf. Section 4).

The variance of a U-statistic as a function of the sample size n and of certain population characteristics is studied in Section 5.

It is shown that if $X_1, \ldots, X_n$ have the same distribution and $\Phi(x_1, \ldots, x_m)$ is independent of n, the d.f. of $\sqrt{n}(U - \theta)$ tends to a normal d.f. as $n \to \infty$ under the sole condition of the existence of $E\Phi^2(X_1, \ldots, X_m)$. Similar results hold for the joint distribution of several U-statistics (Theorem 7.1 and 7.2), for statistics U' which, in a certain sense, are asymptotically equivalent to U (Theorems 7.3 and 7.4), for certain functions of statistics U or U' (Theorem 7.5) and, under certain additional assumptions, for the case of the X_ν's having different distributions (Theorems 8.1 and 8.2). Results of a similar character,

---

[1] Research under a contract with the Office of Naval Research for development of multivariate statistical theory.

though under different assumptions, are contained in a recent paper by von Mises [18] (cf. Section 7).

Examples of statistics of the form U or U' are the moments, Fisher's k-statistics, Gini's mean difference, and several rank correlation statistics such as Spearman's rank correlation and the difference sign correlation (cf. Section 9). Asymptotic power functions for the non-parametric tests of independence based on these rank statistics are obtained. They show that these tests are not unbiased in the limit (Section 9f). The asymptotic distribution of the coefficient of partial difference sign correlation which has been suggested by Kendall also is obtained (Section 9h).

2. Functionals of Distribution Functions

Let $F(x) = F(x^{(1)}, \ldots, x^{(r)})$ be an r-variate d.f. If to any F belonging to a subset $\mathscr{D}$ of the set of all d.f.'s in the r-dimensional Euclidean space is assigned a quantity $\theta(F)$, then $\theta(F)$ is called a functional of F, defined on $\mathscr{D}$. In this paper the word functional will always mean functional of a d.f.

An infinite population may be considered as completely determined by its d.f., and any numerical characteristic of an infinite population with d.f. F that is used in statistics is a functional of F. A finite population, or sample, of size n is determined by its d.f., $S(x)$ say, and its size n. n itself is not a functional of S since two samples of different size may have the same d.f.

If $S(x^{(1)}, \ldots, x^{(r)})$ is the d.f. of a finite population, or a sample, consisting of n elements

$$x_\alpha = (x_\alpha^{(1)}, \ldots, x_\alpha^{(r)}), \qquad (\alpha = 1, \ldots, n), \tag{2.1}$$

then $nS(x^{(1)}, \ldots, x^{(r)})$ is the number of elements x_α such that

$$x_\alpha^{(1)} \leq x^{(1)}, \ldots, x_\alpha^{(r)} \leq x^{(r)}.$$

Since $S(x^{(1)}, \ldots, x^{(r)})$ is symmetric in $x_1, \ldots, x_n$, and retains its value for a sample formed from the sample (2.1) by adding one or more identical samples, the same two properties hold true for a sample functional $\theta(S)$. Most statistics in current use are functions of n and of functionals of the sample d.f.

A random sample $\{X_1, \ldots, X_n\}$ is a set of n independent random vectors

$$X_\alpha = (X_\alpha^{(1)}, \ldots, X_\alpha^{(r)}), \qquad (\alpha = 1, \ldots, n). \tag{2.2}$$

For any fixed values $x^{(1)}, \ldots, x^{(r)}$, the d.f. $S(x^{(1)}, \ldots, x^{(r)})$ of a random sample is a random variable. The functional $\theta(S)$, where S is the d.f. of the random sample, is itself a random variable, and may be called a random functional.

A remarkable application of the theory of functionals to functionals of d.f.'s has been made by von Mises [18] who considers the asymptotic distributions of certain functionals of sample d.f.'s. (Cf. also Section 7.)

3. Unbiased Estimation and Regular Functionals

Consider a functional $\theta = \theta(F)$ of the r-variate d.f. $F(x) = F(x^{(1)}, \ldots, x^{(r)})$, and suppose that for some sample size n, θ admits an unbiased estimate for any d.f. F in $\mathscr{D}$. That is, if $X_1, \ldots, X_n$ are n independent random vectors with the same d.f. F, there exists a function $\varphi(x_1, \ldots, x_n)$ of n vector arguments (2.1) such that the expected value of $\varphi(X_1, \ldots, X_n)$ is equal to $\theta(F)$, or

$$\int \cdots \int \varphi(x_1, \ldots, x_n) \, dF(x_1) \ldots dF(x_n) = \theta(F) \tag{3.1}$$

for every F in $\mathscr{D}$. Here and in the sequel, when no integration limits are indicated, the integral is extended over the entire space of $x_1, \ldots, x_n$. The integral is understood in the sense of Stieltjes-Lebesgue.

The estimate $\varphi(x_1, \ldots, x_n)$ of $\theta(F)$ is called unbiased over $\mathscr{D}$.

A functional $\theta(F)$ of the form (3.1) will be referred to as *regular over* $\mathscr{D}$.[2] Thus, the functionals regular over $\mathscr{D}$ are those admitting an unbiased estimate over $\mathscr{D}$.

If $\theta(F)$ is regular over $\mathscr{D}$, let $m(\leq n)$ be the smallest sample size for which there exists an unbiased estimate $\Phi(x_1, \ldots, x_m)$ of θ over $\mathscr{D}$:

$$\theta(F) = \int \cdots \int \Phi(x_1, \ldots, x_m) \, dF(x_1) \ldots dF(x_m) \tag{3.2}$$

for any F in $\mathscr{D}$. Then m will be called *the degree over* $\mathscr{D}$ of the regular functional $\theta(F)$.

If the expected value of $\varphi(X_1, \ldots, X_n)$ is equal to $\theta(F)$ whenever it exists, $\varphi(x_1, \ldots, x_n)$ will be called a *distribution-free unbiased estimate* (d-f.u.e.) of $\theta(F)$. The degree of $\theta(F)$ over the set $\mathscr{D}_0$ of d.f.'s F for which the right hand side of (3.1) exists will be simply termed the *degree* of $\theta(F)$.

A regular functional of degree 1 over $\mathscr{D}$ is called a linear regular functional over $\mathscr{D}$. If $\theta(F)$ has the same value for all F in $\mathscr{D}$, $\theta(F)$ may be termed a regular functional of degree zero over $\mathscr{D}$.

Any function $\Phi(x_1, \ldots, x_m)$ satisfying (3.2) will be referred to as a *kernel* of the regular functional $\theta(F)$.

For any regular functional $\theta(F)$ there exists a kernel $\Phi_0(x_1, \ldots, x_m)$ symmetric in $x_1, \ldots, x_m$. For if $\Phi(x_1, \ldots, x_m)$ is a kernel of $\theta(F)$,

$$\Phi_0(x_1, \ldots, x_m) = \frac{1}{m!} \Sigma \Phi(x_{\alpha_1}, \ldots, x_{\alpha_m}), \tag{3.3}$$

where the sum is taken over all permutations $(\alpha_1, \ldots, \alpha_m)$ of $(1, \ldots, m)$, is a symmetric kernel of $\theta(F)$.

If $\theta_1(F)$ and $\theta_2(F)$ are two regular functionals of degrees m_1 and m_2 over $\mathscr{D}$, then the sum $\theta_1(F) + \theta_2(F)$ and the product $\theta_1(F)\theta_2(F)$ are regular functionals of degrees $\leq m = \text{Max}(m_1, m_2)$ and $\leq m_1 + m_2$, respectively, over $\mathscr{D}$. For if $\Phi_i(x_1, \ldots, x_{m_i})$ is a kernel of $\theta_i(F)$, $(i = 1, 2)$, then

[2] This is an adaptation to functionals of d.f.'s of the term "regular functional" used by Volterra [21].

$$\theta_1(F) + \theta_2(F) = \int \cdots \int \{\Phi_1(x_1, \ldots, x_{m_1}) + \Phi_2(x_1, \ldots, x_{m_2})\} \, dF(x_1) \ldots dF(x_m)$$

and

$$\theta_1(F)\theta_2(F)$$
$$= \int \cdots \int \Phi_1(x_1, \ldots, x_{m_1})\Phi_2(x_{m_1+1}, \ldots, x_{m_1+m_2}) \, dF(x_1) \ldots dF(x_{m_1+m_2}).$$

More generally, *a polynomial in regular functionals is itself a regular functional*. Examples of linear regular functions are the moments about the origin,

$$\mu'_{v_1, \ldots, v_r} = \int \cdots \int (x^{(1)})^{v_1} \ldots (x^{(r)})^{v_r} \, dF(x^{(1)}, \ldots, x^{(r)}).$$

A moment about the mean is a polynomial in moments μ' about 0, and hence a regular functional over the set $\mathscr{D}_0$ of d.f.'s for which it exists (cf. Halmos [9]). For instance, the variance of $X^{(1)}$,

$$\sigma^2 = \int \int ((x_1^{(1)})^2 - x_1^{(1)}x_2^{(1)}) \, dF(x_1^{(1)}) \, dF(x_2^{(1)})$$

is a regular functional of degree 2. A symmetrical kernel of σ^2 is $(x_1^{(1)} - x_2^{(1)})^2/2$. If $\mathscr{D}$ is the set of univariate d.f.'s with mean μ and existing second moment, σ^2 is a linear regular functional of F over $\mathscr{D}$, since then we have

$$\sigma^2 = \int (x_1^{(1)} - \mu)^2 \, dF(x_1^{(1)}).$$

The function

$$v = \frac{1}{n(n-1)} \sum_{\alpha \neq \beta} \frac{1}{2}(x_\alpha^{(1)} - x_\beta^{(1)})^2 = \frac{1}{n-1} \sum_\alpha \left(x_\alpha^{(1)} - \frac{1}{n} \sum_\beta x_\beta^{(1)} \right)^2$$

is a distribution-free unbiased estimate of σ^2. The function

$$\Gamma\left(\frac{n-1}{2}\right)\sqrt{\frac{n-1}{2}}\sqrt{v}/\Gamma\left(\frac{n}{2}\right)$$

is known to be an unbiased estimate of σ over the set of univariate normal d.f.'s, but it is not a d.-f. u.e.

4. *U*-Statistics

Let $x_1, \ldots, x_n$ be a sample of n vectors (2.1) and $\Phi(x_1, \ldots, x_m)$ a function of $m(\leq n)$ vector arguments. Consider the function of the sample,

$$U = U(x_1, \ldots, x_n) = \frac{1}{n(n-1)\ldots(n-m+1)} \Sigma'' \Phi(x_{\alpha_1}, \ldots, x_{\alpha_m}), \quad (4.1)$$

where Σ'' stands for summation over all permutations $(\alpha_1, \ldots, \alpha_m)$ of m integers such that

$$1 \leq \alpha_i \leq n, \quad \alpha_i \neq \alpha_j \text{ if } i \neq j, \quad (i, j = 1, \ldots, m). \tag{4.2}$$

U is the average of the values of Φ in the set of ordered subsets of m members of the sample (2.1). U is symmetric in $x_1, \ldots, x_n$.

Any statistic of the form (4.1) will be called a U-statistic. Any function $\Phi(x_1, \ldots, x_m)$ satisfying (4.1) will be referred to as a *kernel* of the statistic U.

If $\Phi(x_1, \ldots, x_m)$ is a kernel of a regular functional $\theta(F)$ defined on a set $\mathscr{D}$, then U is an unbiased estimate of $\theta(F)$ over $\mathscr{D}$:

$$\theta(F) = \int \cdots \int U(x_1, \ldots, x_n) \, dF(x_1) \ldots dF(x_n) \tag{4.3}$$

for every F in $\mathscr{D}$.

For $n = m$, U reduces to the symmetric kernel (3.3) of $\theta(F)$.

From a recent paper by Halmos [9] it follows for the case of univariate d.f.'s $(r = 1)$:

If $\theta(F)$ is a regular functional of degree m over a set $\mathscr{D}$ containing all purely discontinuous d.f.'s, U is the only unbiased estimate over $\mathscr{D}$ which is symmetric in $x_1, \ldots, x_n$, and U has the least variance among all unbiased estimates over $\mathscr{D}$.

These results and the proofs given by Halmos can easily be extended to the multivariate case $(r > 1)$.

Combining (3.3) and (4.1) we may write a U-statistic in the form

$$U(x_1, \ldots, x_n) = \binom{n}{m}^{-1} \Sigma' \Phi_0(x_{\alpha_1}, \ldots, x_{\alpha_m}), \tag{4.4}$$

where the kernel Φ_0 is symmetric in its m vector arguments and the sum Σ' is extended over all subscripts α such that

$$1 \leq \alpha_1 < \alpha_2 < \cdots < \alpha_m \leq n.$$

Another statistic frequently used for estimating $\theta(F)$ is $\theta(S)$, where $S = S(x)$ is the d.f. of the sample (2.1). If S is substituted for F in (3.2), we have

$$\theta(S) = \frac{1}{n^m} \sum_{\alpha_1=1}^{n} \cdots \sum_{\alpha_m=1}^{n} \Phi(x_{\alpha_1}, \ldots, x_{\alpha_m}). \tag{4.5}$$

In particular, the sample moments have this form; their kernel Φ is obtained by the method described in section 3.

If $m = 1$, $\theta(S) = U$. If $m = 2$,

$$\theta(S) = \frac{n-1}{n} U + \frac{1}{n} \left\{ \frac{1}{n} \sum_{\alpha=1}^{n} \Phi(x_\alpha, x_\alpha) \right\},$$

and $\theta(S)$ is a linear function of U-statistics with coefficients depending on n. This is easily seen to be true for any m. In general $\theta(S)$ is not an unbiased estimate of $\theta(F)$. If, however, the expected value of $\theta(S)$ exists for every F in $\mathscr{D}$, we have

$$E\{\theta(S)\} = \theta(F) + O(n^{-1}),$$

and the estimate $\theta(S)$ of $\theta(F)$ may be termed unbiased in the limit over $\mathscr{D}$.

Numerous statistics in current use have the form of, or can be expressed in terms of U-statistics. From what was said above about moments as regular functionals, it is easy to obtain U-statistics which are d.-f. u.e.'s of the moments about the mean of any order (cf. Halmos [9]). Fisher's k-statistics are U-statistics, as follows from their definition as unbiased estimates of the cumulants, symmetric in the sample values. Another example is Gini's mean difference

$$\frac{1}{n(n-1)} \sum_{\alpha \neq \beta} |x_\alpha^{(1)} - x_\beta^{(1)}|.$$

More examples, in particular of rank correlation statistics, will be given in section 9.

5. The Variance of a U-Statistic

Let $X_1, \ldots, X_n$ be n independent random vectors with the same d.f. $F(x) = F(x^{(1)}, \ldots, x^{(r)})$, and let

$$U = U(X_1, \ldots, X_n) = \binom{n}{m}^{-1} \Sigma' \Phi(x_{\alpha_1}, \ldots, x_{\alpha_m}), \tag{5.1}$$

where $\Phi(x_1, \ldots, x_m)$ is symmetric in $x_1, \ldots, x_m$ and Σ' has the same meaning as in (4.4). Suppose that the function Φ does not involve n.

If $\theta = \theta(F)$ is defined by (3.2), we have

$$E\{U\} = E\{\Phi(X_1, \ldots, X_m)\} = \theta.$$

Let

$$\Phi_c(x_1, \ldots, x_c) = E\{\Phi(x_1, \ldots, x_c, X_{c+1}, \ldots, X_m)\}, \qquad (c = 1, \ldots, m), \tag{5.2}$$

where $x_1, \ldots, x_c$ are arbitrary fixed vectors and the expected value is taken with respect to the random vectors $X_{c+1}, \ldots, X_m$. Then

$$\Phi_{c-1}(x_1, \ldots, x_{c-1}) = E\{\Phi_c(x_1, \ldots, x_{c-1}, X_c)\}, \tag{5.3}$$

and

$$E\{\Phi_c(X_1, \ldots, X_c)\} = \theta, \qquad (c = 1, \ldots, m). \tag{5.4}$$

Define

$$\Psi(x_1, \ldots, x_m) = \Phi(x_1, \ldots, x_m) - \theta, \tag{5.5}$$

$$\Psi_c(x_1, \ldots, x_c) = \Phi_c(x_1, \ldots, x_c) - \theta, \qquad (c = 1, \ldots, m). \tag{5.6}$$

We have

$$\Psi_{c-1}(x_1, \ldots, x_{c-1}) = E\{\Psi_c(x_1, \ldots, x_{c-1}, X_c)\}, \tag{5.7}$$

$$E\{\Psi_c(X_1, \ldots, X_c)\} = E\{\Psi(X_1, \ldots, X_m)\} = 0, \qquad (c = 1, \ldots, m). \tag{5.8}$$

Suppose that the variance of $\Psi_c(X_1, \ldots, X_c)$ exists, and let

$$\zeta_0 = 0, \quad \zeta_c = E\{\Psi_c^2(X_1, \ldots, X_c)\}, \qquad (c = 1, \ldots, m). \tag{5.9}$$

We have

$$\zeta_c = E\{\Phi_c^2(X_1, \ldots, X_c)\} - \theta^2. \tag{5.10}$$

$\zeta_c = \zeta_c(F)$ is a polynomial in regular functionals of F, and hence itself a regular functional of F (of degree $\leq 2m$).

If, for some parent distribution $F = F_0$ and some integer d, we have $\zeta_d(F_0) = 0$, this means that $\Psi_d(X_1, \ldots, X_d) = 0$ with probability 1. By (5.7) and (5.9), $\zeta_d = 0$ implies $\zeta_1 = \cdots = \zeta_{d-1} = 0$.

If $\zeta_1(F_0) = 0$, we shall say that the regular functional $\theta(F)$ is *stationary*[3] for $F = F_0$. If

$$\zeta_1(F_0) = \cdots = \zeta_d(F_0) = 0, \quad \zeta_{d+1}(F_0) > 0, \qquad (1 \leq d \leq m), \tag{5.11}$$

$\theta(F)$ will be called *stationary of order d* for $F = F_0$.

If $(\alpha_1, \ldots, \alpha_m)$ and $(\beta_1, \ldots, \beta_m)$ are two sets of m different integers, $1 \leq \alpha_i$, $\beta_i \leq n$, and c is the number of integers common to the two sets, we have, by the symmetry of Ψ,

$$E\{\Psi(X_{\alpha_1}, \ldots, X_{\alpha_m})\Psi(X_{\beta_1}, \ldots, X_{\beta_m})\} = \zeta_c. \tag{5.12}$$

If the variance of U exists, it is equal to

$$\sigma^2(U) = \binom{n}{m}^{-2} E\{\Sigma'\Psi(X_{\alpha_1}, \ldots, X_{\alpha_m})\}^2$$

$$= \binom{n}{m}^{-2} \sum_{c=0}^{m} \Sigma^{(c)} E\{\Psi(X_{\alpha_1}, \ldots, X_{\alpha_m})\Psi(X_{\beta_1}, \ldots, X_{\beta_m})\},$$

where $\Sigma^{(c)}$ stands for summation over all subscripts such that

$$1 \leq \alpha_1 \leq \alpha_2 < \cdots < \alpha_m \leq n, \quad 1 \leq \beta_1 < \beta_2 < \cdots < \beta_m \leq n,$$

and exactly c equations

$$\alpha_i = \beta_j$$

are satisfied. By (5.12), each term in $\Sigma^{(c)}$ is equal to ζ_c. The number of terms in

---

[3] According to the definition of the derivative of a functional (cf. Volterra [21]; for functionals of d.f.'s cf. von Mises [18]), the function $m(m-1)\ldots(m-d+1)\Psi_d(x_1, \ldots, x_d)$, which is a functional of F, is a d-th derivative of $\theta(F)$ with respect to F at the "point" F of the space of d.f.'s.

$\Sigma^{(c)}$ is easily seen to be

$$\frac{n(n-1)\dots(n-2m+c+1)}{c!(m-c)!(m-c)!} = \binom{m}{c}\binom{n-m}{m-c}\binom{n}{m},$$

and hence, since $\zeta_0 = 0$,

$$\sigma^2(U) = \binom{n}{m}^{-1}\sum_{c=1}^{m}\binom{m}{c}\binom{n-m}{m-c}\zeta_c. \qquad (5.13)$$

When the distributions of $X_1, \dots, X_n$ are different, $F_\nu(x)$ being the d.f. of X_ν, let

$$\theta_{\alpha_1,\dots,\alpha_m} = E\{\Phi(X_{\alpha_1},\dots,X_{\alpha_m})\}, \qquad (5.14)$$

$$\Psi_{c(\alpha_1,\dots,\alpha_c)\beta_1,\dots,\beta_{m-c}}(x_1,\dots,x_c)$$
$$= E\{\Phi(x_1,\dots,x_c,X_{\beta_1},\dots,X_{\beta_{m-c}})\} - \theta_{\alpha_1,\dots,\alpha_c,\beta_1,\dots,\beta_{m-c}}, \quad (c=1,\dots,m), \qquad (5.15)$$

$$\zeta_{c(\alpha_1,\dots,\alpha_c)\beta_1,\dots,\beta_{m-c};\gamma_1,\dots,\gamma_{m-c}}$$
$$= E\{\Psi_{c(\alpha_1,\dots,\alpha_c)\beta_1,\dots,\beta_{m-c}}(X_{\alpha_1},\dots,X_{\alpha_c})\Psi_{c(\alpha_1,\dots,\alpha_c)\gamma_1,\dots,\gamma_{m-c}}(X_{\alpha_1},\dots,X_{\alpha_c})\} \qquad (5.16)$$

$$\zeta_{c,n} = \frac{c!(m-c)!(m-c)!}{n(n-1)\dots(n-2m+c+1)}\Sigma\zeta_{c(\alpha_1,\dots,\alpha_c)\beta_1,\dots,\beta_{m-c};\gamma_1,\dots,\gamma_{m-c}} \qquad (5.17)$$

where the sum is extended over all subscripts α, β, γ such that

$$1 \le \alpha_1 < \dots < \alpha_c \le n, \quad 1 \le \beta_1 < \dots < \beta_{m-c} \le n, \quad 1 \le \gamma_1 < \dots \gamma_{m-c} \le n,$$

$$\alpha_i \ne \beta_j, \quad \alpha_i \ne \gamma_j, \quad \beta_i \ne \gamma_j.$$

Then the variance of U is equal to

$$\sigma^2(U) = \binom{n}{m}^{-1}\sum_{c=1}^{m}\binom{m}{c}\binom{n-m}{m-c}\zeta_{c,n}. \qquad (5.18)$$

Returning to the case of identically distributed X's, we shall now prove some inequalities satisfied by $\zeta_1, \dots, \zeta_m$ and $\sigma^2(U)$ which are contained in the following theorems:

Theorem 5.1. *The quantities* $\zeta_1, \dots, \zeta_m$ *as defined by (5.9) satisfy the inequalities*

$$0 \le \frac{\zeta_c}{c} \le \frac{\zeta_d}{d} \quad \text{if } 1 \le c < d \le m. \qquad (5.19)$$

Theorem 5.2. *The variance* $\sigma^2(U_n)$ *of a U-statistic* $U_n = U(X_1,\dots,X_n)$, *where* $X_1,\dots,X_n$ *are independent and identically distributed, satisfies the inequalities*

$$\frac{m^2}{n}\zeta_1 \le \sigma^2(U_n) \le \frac{m}{n}\zeta_m. \qquad (5.20)$$

$n\sigma^2(U_n)$ is a decreasing function of n,

$$(n + 1)\sigma^2(U_{n+1}) \leq n\sigma^2(U_n), \tag{5.21}$$

which takes on its upper bound $m\zeta_m$ for $n = m$ and tends to its lower bound $m^2\zeta_1$ as n increases:

$$\sigma^2(U_m) = \zeta_m, \tag{5.22}$$

$$\lim_{n \to \infty} n\sigma^2(U_n) = m^2\zeta_1. \tag{5.23}$$

If $E\{U_n\} = \theta(F)$ is stationary of order $\geq d - 1$ for the d.f. of X_α, (5.20) may be replaced by

$$\frac{m}{d} K_n(m, d)\zeta_d \leq \sigma^2(U_n) \leq K_n(m, d)\zeta_m, \tag{5.24}$$

where

$$K_n(m, d) = \binom{n}{m}^{-1} \sum_{c=d}^{m} \binom{m-1}{c-1}\binom{n-m}{m-c}. \tag{5.25}$$

We postpone the proofs of Theorems 5.1 and 5.2.

(5.13) and (5.19) imply that a necessary and sufficient condition for the existence of $\sigma^2(U)$ is the existence of

$$\zeta_m = E\{\Phi^2(X_1, \ldots, X_m)\} - \theta^2 \tag{5.26}$$

or that of $E\{\Phi^2(X_1, \ldots, X_m)\}$.

If $\zeta_1 > 0$, $\sigma^2(U)$ is of order n^{-1}.

If $\theta(F)$ is stationary of order d for $F = F_0$, that is, if (5.11) is satisfied, $\sigma^2(U)$ is of order n^{-d-1}. Only if, for some $F = F_0$, $\theta(F)$ is stationary of order m, where m is the degree of $\theta(F)$, we have $\sigma^2(U) = 0$, and U is equal to a constant with probability 1.

For instance, if $\theta(F_0) = 0$, the functional $\theta^2(F)$ is stationary for $F = F_0$. Other examples of stationary "points" of a functional will be found in section 9d.

For proving Theorem 5.1 we shall require the following:

Lemma 5.1. If

$$\delta_d = \zeta_d - \binom{d}{1}\zeta_{d-1} + \binom{d}{2}\zeta_{d-2} \cdots + (-1)^{d-1}\binom{d}{d-1}\zeta_1, \tag{5.27}$$

we have

$$\delta_d \geq 0, \quad (d = 1, \ldots, m), \tag{5.28}$$

and

$$\zeta_d = \delta_d + \binom{d}{1}\delta_{d-1} + \cdots + \binom{d}{d-1}\delta_1. \tag{5.29}$$

PROOF. (5.29) follows from (5.27) by induction.

For proving (5.28) let

$$\eta_0 = \theta^2, \quad \eta_c = E\{\Phi_c^2(X_1, \ldots, X_c)\}, \qquad (c = 1, \ldots, m).$$

Then, by (5.10),

$$\zeta_c = \eta_c - \eta_0,$$

and on substituting this in (5.27) we have

$$\delta_d = \sum_{c=0}^{d} (-1)^{d-c} \binom{d}{c} \eta_c.$$

From (5.9) it is seen that (5.28) is true for $d = 1$. Suppose that (5.28) holds for $1, \ldots, d - 1$. Then (5.28) will be shown to hold for d.

Let

$$\bar{\Phi}_0(x_1) = \Phi_1(x_1) - \theta, \, \bar{\Phi}_c(x_1, x_2, \ldots, x_{c+1})$$
$$= \Phi_{c+1}(x_1, \ldots, x_{c+1}) - \Phi_c(x_2, \ldots, x_{c+1}), \qquad (c = 1, \ldots, d - 1).$$

For an arbitrary fixed x_1, let

$$\bar{\eta}_c(x_1) = E\{\bar{\Phi}_c^2(x_1, X_2, \ldots, X_{c+1})\}, \qquad (c = 0, \ldots, d - 1).$$

Then, by induction hypothesis,

$$\bar{\delta}_{d-1}(x_1) = \sum_{c=0}^{d-1} (-1)^{d-1-c} \binom{d-1}{c} \bar{\eta}_c(x_1) \geq 0$$

for any fixed x_1.

Now,

$$E\{\bar{\eta}_c(X_1)\} = \eta_{c+1} - \eta_c,$$

and hence

$$E\{\bar{\delta}_{d-1}(X_1)\} = \sum_{c=0}^{d-1} (-1)^{d-1-c} \binom{d-1}{c}(\eta_{c+1} - \eta_c) = \sum_{c=0}^{d} (-1)^{d-c} \binom{d}{c} \eta_c = \delta_d.$$

The proof of Lemma 5.1 is complete. □

PROOF OF THEOREM 5.1. By (5.29) we have for $c < d$

$$c\zeta_d - d\zeta_c = c \sum_{a=1}^{d} \binom{d}{a} \delta_a - d \sum_{a=1}^{c} \binom{c}{a} \delta_a$$

$$= \sum_{a=1}^{c} \left[c \binom{d}{a} - d \binom{c}{a} \right] \delta_a + c \sum_{a=c+1}^{d} \binom{d}{a} \delta_a. \qquad (5.30)$$

From (5.28), and since $c \binom{d}{a} - d \binom{c}{a} \geq 0$ if $1 \leq a \leq c \leq d$, it follows that each term in the two sums of (5.30) is not negative. This, in connection with (5.9) proves Theorem 5.1. □

PROOF OF THEOREM 5.2. From (5.19) we have

$$c\zeta_1 \leq \zeta_c \leq \frac{c}{m}\zeta_m, \qquad (c = 1, \ldots, m).$$

Applying these inequalities to each term in (5.13) and using the identity

$$\binom{n}{m}^{-1} \sum_{c=1}^{m} c\binom{m}{c}\binom{n-m}{m-c} = \frac{m^2}{n}, \tag{5.31}$$

we obtain (5.20).

(5.22) and (5.23) follow immediately from (5.13).

For (5.21) we may write

$$D_n \geq 0, \tag{5.32}$$

where

$$D_n = n\sigma^2(U_n) - (n+1)\sigma^2(U_{n+1}).$$

Let

$$D_n = \sum_{c=1}^{m} d_{n,c}\zeta_c.$$

Then we have from (5.13)

$$d_{n,c} = n\binom{m}{c}\binom{n-m}{m-c}\binom{n}{m}^{-1} - (n+1)\binom{m}{c}\binom{n+1-m}{m-c}\binom{n+1}{m}^{-1}, \tag{5.33}$$

or

$$d_{n,c} = \binom{m}{c}\binom{n-m+1}{m-c}(n-m+1)^{-1}\binom{n}{m}^{-1}\{(c-1)n - (m-1)^2\},$$

$$(1 \leq c \leq m \leq n).$$

Putting

$$c_0 = 1 + \left[\frac{(m-1)^2}{n}\right],$$

where $[u]$ denotes the largest integer $\leq u$, we have

$$d_{n,c} \leq 0 \qquad \text{if } c \leq c_0,$$

$$d_{n,c} > 0 \qquad \text{if } c > c_0.$$

Hence, by (5.19),

$$d_{n,c}\zeta_c \geq \frac{1}{c_0}\zeta_{c_0}cd_{n,c}, \qquad (c = 1, \ldots, m),$$

and

$$D_n \geq \frac{1}{c_0}\zeta_{c_0}\sum_{c=1}^{m} cd_{n,c}.$$

By (5.33) and (5.31), the latter sum vanishes. This proves (5.32).

For the stationary case $\zeta_1 = \cdots = \zeta_{d-1} = 0$, (5.24) is a direct consequence of (5.13) and (5.19). The proof of Theorem 5.2 is complete.   □

6. The Covariance of Two U-Statistics

Consider a set of g U-statistics,

$$U^{(\gamma)} = \binom{n}{m(\gamma)}^{-1} \Sigma' \Phi^{(\gamma)}(X_{\alpha_1}, \ldots, X_{\alpha_{m(\gamma)}}), \qquad (\gamma = 1, \ldots, g),$$

each $U^{(\gamma)}$ being a function of the same n independent, identically distributed random vectors $X_1, \ldots, X_n$. The function $\Phi^{(\gamma)}$ is assumed to be symmetric in its $m(\gamma)$ arguments $(\gamma = 1, \ldots, g)$.

Let

$$E\{U^{(\gamma)}\} = E\{\Phi^{(\gamma)}(X_1, \ldots, X_{m(\gamma)})\} = \theta^{(\gamma)}, \qquad (\gamma = 1, \ldots, g);$$

$$\Psi^{(\gamma)}(x_1, \ldots, x_{m(\gamma)}) = \Phi^{(\gamma)}(x_1, \ldots, x_{m(\gamma)}) - \theta^{(\gamma)}, \qquad (\gamma = 1, \ldots, g); \quad (6.1)$$

$$\Psi_c^{(\gamma)}(x_1, \ldots, x_c)$$
$$= E\{\Psi^{(\gamma)}(x_1, \ldots, x_c, X_{c+1}, \ldots, X_{m(\gamma)})\}, \quad (c = 1, \ldots, m(\gamma); \gamma = 1, \ldots, g); \quad (6.2)$$

$$\zeta_c^{(\gamma, \delta)} = E\{\Psi_c^{(\gamma)}(X_1, \ldots, X_c)\Psi_c^{(\delta)}(X_1, \ldots, X_c)\}, \quad (\gamma, \delta = 1, \ldots, g). \quad (6.3)$$

If, in particular, $\gamma = \delta$, we shall write

$$\zeta_c^{(\gamma)} = \zeta_c^{(\gamma, \gamma)} = E\{\Psi_c^{(\gamma)}(X_1, \ldots, X_c)\}^2. \quad (6.4)$$

Let

$$\sigma(U^{(\gamma)}, U^{(\delta)}) = E\{(U^{(\gamma)} - \theta^{(\gamma)})(U^{(\delta)} - \theta^{(\delta)})\}$$

be the covariance of $U^{(\gamma)}$ and $U^{(\delta)}$.

In a similar way as for the variance, we find, if $m(\gamma) \le m(\delta)$,

$$\sigma(U^{(\gamma)}, U^{(\delta)}) = \binom{n}{m(\gamma)}^{-1} \sum_{c=1}^{m(\gamma)} \binom{m(\delta)}{c}\binom{n - m(\delta)}{m(\gamma) - c} \zeta_c^{(\gamma, \delta)}. \quad (6.5)$$

The right hand side is easily seen to be symmetric in γ, δ.

For $\gamma = \delta$, (6.5) is the variance of $U^{(\gamma)}$ (cf. (5.13)).

We have from (5.23) and (6.5)

$$\lim_{n \to \infty} n\sigma^2(U^{(\gamma)}) = m^2(\gamma)\zeta_1^{(\gamma)},$$

$$\lim_{n \to \infty} n\sigma(U^{(\gamma)}, U^{(\delta)}) = m(\gamma)m(\delta)\zeta_1^{(\gamma, \delta)}.$$

Hence, if $\zeta_1^{(\gamma)} \ne 0$ and $\zeta_1^{(\delta)} \ne 0$, the product moment correlation $\rho(U^{(\gamma)}, U^{(\delta)})$ between $U^{(\gamma)}$ and $U^{(\delta)}$ tends to the limit

$$\lim_{n \to \infty} \rho(U^{(\gamma)}, U^{(\delta)}) = \frac{\zeta_1^{(\gamma, \delta)}}{\sqrt{\zeta_1^{(\gamma)}\zeta_1^{(\delta)}}}.$$

7. Limit Theorems for the Case of Identically Distributed X_α's

We shall now study the asymptotic distribution of U-statistics and certain related functions. In this section the vectors X_α will be assumed to be identically distributed. An extension to the case of different parent distributions will be given in section 8.

Following Cramér [2, p. 83] we shall say that a sequence of d.f.'s $F_1(x)$, $F_2(x)$, ... converges to a d.f. $F(x)$ if $\lim F_n(x) = F(x)$ in every point at which the one-dimensional marginal limiting d.f.'s are continuous.

Let us recall (cf. Cramér [2, p. 312]) that a g-variate normal distribution is called non-singular if the rank r of its covariance matrix is equal to g, and singular if $r < g$.

The following lemma will be used in the proofs.

Lemma 7.1. *Let V_1, V_2, ... be an infinite sequence of random vectors $V_n = (V_n^{(1)}, \ldots, V_n^{(g)})$, and suppose that the d.f. $F_n(v)$ of V_n tends to a d.f. $F(v)$ as $n \to \infty$. Let $V_n^{(\gamma)'} = V_n^{(\gamma)} + d_n^{(\gamma)}$, where*

$$\lim_{n \to \infty} E\{d_n^{(\gamma)}\}^2 = 0, \qquad (\gamma = 1, \ldots, g). \tag{7.1}$$

Then the d.f. of $V_n' = (V_n^{(1)'}, \ldots, V_n^{(g)'})$ tends to $F(v)$.

This is an immediate consequence of the well-known fact that the d.f. of V_n' tends to $F(v)$ if $d_n^{(\gamma)}$ converges in probability to 0 (cf. Cramér [2, p. 299]), since the fulfillment of (7.1) is sufficient for the latter condition.

Theorem 7.1. *Let $X_1, \ldots, X_n$ be n independent, identically distributed random vectors,*

$$X_\alpha = (X_\alpha^{(1)}, \ldots, X_\alpha^{(r)}), \qquad (\alpha = 1, \ldots, n).$$

Let

$$\Phi^{(\gamma)}(x_1, \ldots, x_{m(\gamma)}), \qquad (\gamma = 1, \ldots, g),$$

be g real-valued functions not involving n, $\Phi^{(\gamma)}$ being symmetric in its $m(\gamma)$ ($\leq n$) vector arguments $x_\alpha = (x_\alpha^{(1)}, \ldots, x_\alpha^{(r)})$, $(\alpha = 1, \ldots, m(\gamma); \gamma = 1, \ldots, g)$. Define

$$U^{(\gamma)} = \binom{n}{m(\gamma)}^{-1} \sum{}' \Phi^{(\gamma)}(X_{\alpha_1}, \ldots, X_{\alpha_{m(\gamma)}}), \qquad (\gamma = 1, \ldots, g), \tag{7.2}$$

where the summation is over all subscripts such that $1 \leq \alpha_1 < \cdots < \alpha_{m(\gamma)} \leq n$. Then, if the expected values

$$\theta^{(\gamma)} = E\{\Phi^{(\gamma)}(X_1, \ldots, X_{m(\gamma)})\}, \qquad (\gamma = 1, \ldots, g), \tag{7.3}$$

and

$$E\{\Phi^{(\gamma)}(X_1, \ldots, X_{m(\gamma)})\}^2, \qquad (\gamma = 1, \ldots, g), \tag{7.4}$$

exist, the joint d.f. of

$$\sqrt{n}(U^{(1)} - \theta^{(1)}), \ldots, \sqrt{n}(U^{(g)} - \theta^{(g)})$$

tends, as $n \to \infty$, to the g-variate normal d.f. with zero means and covariance matrix $(m(\gamma)m(\delta)\zeta_1^{(\gamma,\delta)})$, where $\zeta_1^{(\gamma,\delta)}$ is defined by (6.3). The limiting distribution is non-singular if the determinant $|\zeta_1^{(\gamma,\delta)}|$ is positive.

Before proving Theorem 7.1, a few words may be said about its meaning and its relation to well-known results.

For $g = 1$, Theorem 7.1 states that the distribution of a U-statistic tends, under certain conditions, to the normal form. For $m = 1$, U is the sum of n independent random variables, and in this case Theorem 7.1 reduces to the Central Limit Theorem for such sums. For $m > 1$, U is a sum of random variables which, in general, are not independent. Under certain assumptions about the function $\Phi(x_1, \ldots, x_m)$ the asymptotic normality of U can be inferred from the Central Limit Theorem by well-known methods. If, for instance, Φ is a polynomial (as in the case of the k-statistics or the unbiased estimates of moments), U can be expressed as a polynomial in moments about the origin which are sums of independent random variables, and for this case the tendency to normality of U can easily be shown (cf. Cramér [2, p. 365]).

Theorem 7.1 generalizes these results, stating that in the case of independent and identically distributed X_α's the existence of $E\{\Phi^2(X_1, \ldots, X_m)\}$ is sufficient for the asymptotic normality of U. No regularity conditions are imposed on the function Φ. This point is important for some applications (cf. section 9).

Theorem 7.1 and the following theorems of sections 7 and 8 are closely related to recent results of von Mises [18] which were published after this paper was essentially completed. It will be seen below (Theorem 7.4) that the limiting distribution of $\sqrt{n}[U - \theta(F)]$ is the same as that of $\sqrt{n}[\theta(S) - \theta(F)]$ (cf. (4.5)) if the variance of $\theta(S)$ exists. $\theta(S)$ is a differentiable statistical function in the sense of von Mises, and by Theorem I of [18], $\sqrt{n}[\theta(S) - \theta(F)]$ is asymptotically normal if certain conditions are satisfied. It will be found that in certain cases, for instance if the kernel Φ of θ is a polynomial, the conditions of the theorems of sections 7 and 8 are somewhat weaker than those of von Mises' theorem. Though von Mises' paper is concerned with functionals of univariate d.f.'s only, its results can easily be extended to the multivariate case.

For the particular case of a discrete population (where F is a step function), U and $\theta(S)$ are polynomials in the sample frequencies, and their asymptotic distribution may be inferred from the fact that the joint distribution of the frequencies tends to the normal form (cf. also von Mises [18]).

In Theorem 7.1 the functions $\Phi^{(\gamma)}(x_1, \ldots, x_{m(\gamma)})$ are supposed to be symmetric. Since, as has been seen in section 4, any U-statistic with non-symmetric kernel can be written in the form (4.4) with a symmetric kernel, this restriction is not essential and has been made only for the sake of convenience. Moreover, in the condition of the existence of $E\{\Phi^2(X_1, \ldots, X_m)\}$, the symmetric kernel may be replaced by a non-symmetric one. For, if Φ is non-symmetric, and Φ_0 is the symmetric kernel defined by (3.3), $E\{\Phi_0^2(X_1, \ldots, X_m)\}$ is a linear combination of terms of the form $E\{\Phi(X_{\alpha_1}, \ldots, X_{\alpha_m})\Phi(X_{\beta_1}, \ldots, X_{\beta_m})\}$, whose existence follows from that of $E\{\Phi^2(X_1, \ldots, X_m)\}$ by Schwarz's inequality.

If the regular functional $\theta(F)$ is stationary for $F = F_0$, that is, if $\zeta_1 = \zeta_1(F_0) = 0$ (cf. section 5), the limiting normal distribution of $\sqrt{n}(U - \theta)$ is, according to Theorem 7.1, singular, that is, its variance is zero. As has been seen in section 5, $\sigma^2(U)$ need not be zero in this case, but may be of some order n^{-c}, $(c = 2, 3, \ldots, m)$, and the distribution of $n^{c/2}(U - \theta)$ may tend to a limiting form which is not normal. According to von Mises [18], it is a limiting distribution of type c, $(c = 2, 3, \ldots)$.

According to Theorem 5.2, $\sigma^2(U)$ exceeds its asymptotic value $m^2\zeta_1/n$ for any finite n. Hence, if we apply Theorem 7.1 for approximating the distribution of U when n is large but finite, we underestimate the variance of U. For many applications this is undesirable, and for such cases the following theorem, which is an immediate consequence of Theorem 7.1, will be more useful.

Theorem 7.2 *Under the conditions of Theorem 7.1, and if*

$$\zeta_1^{(\gamma)} > 0, \qquad (\gamma = 1, \ldots, g),$$

the joint d.f. of

$$(U^{(1)} - \theta^{(1)})/\sigma(U^{(1)}), \ldots, (U^{(g)} - \theta^{(g)})/\sigma(U^{(g)})$$

tends, as $n \to \infty$, to the g-variate normal d.f. with zero means and covariance matrix $(\rho^{(\gamma, \delta)})$, where

$$\rho^{(\gamma, \delta)} = \lim_{n \to \infty} \frac{\sigma(U^{(\gamma)}, U^{(\delta)})}{\sigma(U^{(\gamma)})\sigma(U^{(\delta)})} = \frac{\zeta_1^{(\gamma, \delta)}}{\sqrt{\zeta_1^{(\gamma)}\zeta_1^{(\delta)}}}, \qquad (\gamma, \delta = 1, \ldots, g).$$

PROOF OF THEOREM 7.1. The existence of (7.4) entails that of

$$\zeta_m^{(\gamma)} = E\{\Phi^{(\gamma)}(X_1, \ldots, X_{m(\gamma)})\}^2 - (\theta^{(\gamma)})^2$$

which, by (5.19), (5.20) and (6.6), is sufficient for the existence of

$$\zeta_1^{(\gamma)}, \ldots, \zeta_{m-1}^{(\gamma)}, \quad \text{of} \quad \sigma^2(U^{(\gamma)}), \quad \text{and of} \quad \zeta_1^{(\gamma, \delta)} \le \sqrt{\zeta_1^{(\gamma)}\zeta_1^{(\delta)}}.$$

Now, consider the g quantities

$$Y^{(\gamma)} = \frac{m(\gamma)}{\sqrt{n}} \sum_{\alpha=1}^{n} \Psi_1^{(\gamma)}(X_\alpha), \qquad (\gamma = 1, \ldots, g)$$

where $\Psi_1^{(\gamma)}(x)$ is defined by (6.2). $Y^{(1)}, \ldots, Y^{(g)}$ are sums of n independent, random variables with zero means, whose covariance matrix, by virtue of (6.3), is

$$\{\sigma(Y^{(\gamma)}, Y^{(\delta)})\} = \{m(\gamma)m(\delta)\zeta_1^{(\gamma,\delta)}\}. \tag{7.5}$$

By the Central Limit Theorem for vectors (cf. Cramér [1, p. 112]), the joint d.f. of $(Y^{(1)}, \ldots, Y^{(g)})$ tends to the normal g-variate d.f. with the same means and covariances.

Theorem 7.1 will be proved by showing that the g random variables

$$Z^{(\gamma)} = \sqrt{n}(U^{(\gamma)} - \theta^{(\gamma)}), \qquad (\gamma = 1, \ldots, g), \tag{7.6}$$

have the same joint limiting distribution as $Y^{(1)}, \ldots, Y^{(g)}$.

According to Lemma 7.1 it is sufficient to show that

$$\lim_{n \to \infty} E(Z^{(\gamma)} - Y^{(\gamma)})^2 = 0, \qquad (\gamma = 1, \ldots, n). \tag{7.7}$$

For proving (7.7), write

$$E\{Z^{(\gamma)} - Y^{(\gamma)}\}^2 = E\{Z^{(\gamma)}\}^2 + E\{Y^{(\gamma)}\}^2 - 2E\{Z^{(\gamma)}Y^{(\gamma)}\}. \tag{7.8}$$

By (5.13) we have

$$E\{Z^{(\gamma)}\}^2 = n\sigma^2(U^{(\gamma)}) = m^2(\gamma)\zeta_1^{(\gamma)} + O(n^{-1}), \tag{7.9}$$

and from (7.5),

$$E\{Y^{(\gamma)}\}^2 = m^2(\gamma)\zeta_1^{(\gamma)}. \tag{7.10}$$

By (7.2) and (6.1) we may write for (7.6)

$$Z^{(\gamma)} = \sqrt{n}\left(\frac{n}{m(\gamma)}\right)^{-1} \Sigma'\Psi^{(\gamma)}(X_{\alpha_1}, \ldots, X_{\alpha_{m(\gamma)}})),$$

and hence

$$E\{Z^{(\gamma)}Y^{(\gamma)}\} = m(\gamma)\left(\frac{n}{m(\gamma)}\right)^{-1} \sum_{\alpha=1}^{n} \Sigma' E\{\Psi_1^{(\gamma)}(X_\alpha)\Psi^{(\gamma)}(X_{\alpha_1}, \ldots, X_{\alpha_{m(\gamma)}})\}.$$

The term

$$E\{\Psi_1^{(\gamma)}(X_\alpha)\Psi^{(\gamma)}(X_{\alpha_1}, \ldots, X_{\alpha_{m(\gamma)}})\}$$

is $= \zeta_1^{(\gamma)}$ if

$$\alpha_1 = \alpha \quad \text{or} \quad \alpha_2 = \alpha \cdots \quad \text{or} \quad \alpha_{m(\gamma)} = \alpha \tag{7.11}$$

and 0 otherwise. For a fixed α, the number of sets $\{\alpha_1, \ldots, \alpha_{m(\gamma)}\}$ such that $1 \leq \alpha_1 < \cdots < \alpha_{m(\gamma)} \leq n$ and (7.11) is satisfied, is $\binom{n-1}{m(\gamma)-1}$. Thus,

$$E\{Z^{(\gamma)}Y^{(\gamma)}\} = m(\gamma)\left(\frac{n}{m(\gamma)}\right)^{-1} n\binom{n-1}{m(\gamma)-1}\zeta_1^{(\gamma)} = m^2(\gamma)\zeta_1^{(\gamma)}. \tag{7.12}$$

On inserting (7.9), (7.10), and (7.12) in (7.8), we see that (7.7) is true.

The concluding remark in Theorem 7.1 is a direct consequence of the definition of a non-singular distribution. The proof of Theorem 7.1 is complete.

$\square$

Theorems 7.1 and 7.2 deal with the asymptotic distribution of $U^{(1)}, \ldots,$ $U^{(g)}$, which are unbiased estimates $\theta^{(1)}, \ldots, \theta^{(g)}$. The unbiasedness of a statistic is, of course, irrelevant for its asymptotic behavior, and the application of Lemma 7.1 leads immediately to the following extension of Theorem 7.1 to a larger class of statistics.

Theorem 7.3. *Let*

$$U^{(g)'} = U^{(g)} + \frac{b_n^{(\gamma)}}{\sqrt{n}}, \qquad (\gamma = 1, \ldots, g), \tag{7.13}$$

where $U^{(\gamma)}$ is defined by (7.2) *and $b_n^{(\gamma)}$ is a random variable. If the conditions of Theorem 7.1 are satisfied, and $\lim E\{b_n^{(\gamma)}\}^2 = 0$, $(\gamma = 1, \ldots, g)$, then the joint distribution of*

$$\sqrt{n}(U^{(1)'} - \theta^{(1)}), \ldots, \sqrt{n}(U^{(g)'} - \theta^{(g)})$$

tends to the normal distribution with zero means and covariance matrix

$$\{m(\gamma)m(\delta)\zeta_1^{(\gamma, \delta)}\}.$$

This theorem applies, in particular, to the regular functionals $\theta(S)$ of the sample d.f.,

$$\theta(S) = \frac{1}{n^m} \sum_{\alpha_1=1}^{n} \cdots \sum_{\alpha_m=1}^{n} \Phi(X_{\alpha_1}, \ldots, X_{\alpha_m}),$$

in the case that the variance of $\theta(S)$ exists. For we may write

$$n^m \theta(S) = \binom{n}{m} U + \Sigma^* \Phi(X_{\alpha_1}, \ldots, X_{\alpha_m}),$$

where the sum Σ^* is extended over all m-tuplets $(\alpha_1, \ldots, \alpha_m)$ in which at least one equality $\alpha i = \alpha_j (i \neq j)$ is satisfied. The number of terms in Σ^* is of order n^{m-1}. Hence

$$\theta(S) - U = \frac{1}{n} D,$$

where the expected value $E\{D^2\}$, whose existence follows from that of $\sigma^2\{\theta(S)\}$, is bounded for $n \to \infty$. Thus, if we put $U^{(\gamma)'} = \theta^{(\gamma)}(S)$, the conditions of Theorem 7.3 are fulfilled. We may summarize this result as follows:

Theorem 7.4. *Let $X_1, \ldots, X_n$ be a random sample from an r-variate population with d.f. $F(x) = F(x^{(1)}, \ldots, x^{(r)})$, and let*

$$\theta^{(\gamma)}(F) = \int \cdots \int \Phi^{(\gamma)}(x_1, \ldots, x_{m(\gamma)}) \, dF(x_1) \ldots dF(x_{m(\gamma)}^{(\gamma)}), \qquad (\gamma = 1, \ldots, g),$$

be g regular functionals of F, where $\Phi^{(\gamma)}(x_1, \ldots, x_{m(\gamma)})$ *is symmetric in the vectors* $x_1, \ldots, x_{m(\gamma)}$ *and does not involve n. If S(x) is the d.f. of the random sample, and if the variance of*

$$\theta^{(\gamma)}(S) = \frac{1}{n^m} \sum_{\alpha_1=1}^{n} \cdots \sum_{\alpha_{m(\gamma)}=1}^{n} \Phi^{(\gamma)}(X_{\alpha_1}, \ldots, X_{\alpha_{m(\gamma)}})$$

exists, the joint d.f. of

$$\sqrt{n}\{\theta^{(1)}(S) - \theta^{(1)}(F)\}, \ldots, \sqrt{n}\{\theta^{(g)}(S) - \theta^{(g)}(F)\}$$

tends to the g-variate normal d.f. with zero means and covariance matrix

$$\{m(\gamma)m(\delta)\zeta_1^{(\gamma,\delta)}\}.$$

The following theorem is concerned with the asymptotic distribution of a function of statistics of the form U or U'.

Theorem 7.5. *Let* $(U') = (U^{(1)'}, \ldots, U^{(g)'})$ *be a random vector, where* $U^{(\gamma)'}$ *is defined by (7.13), and suppose that the conditions of Theorem 7.3 are satisfied. If the function* $h(y) = h(y^{(1)}, \ldots, y^{(g)})$ *does not involve n and is continuous together with its second order partial derivatives in some neighborhood of the point* $(y) = (\theta) = (\theta^{(1)}, \ldots, \theta^{(g)})$, *then the distribution of the random variable* $\sqrt{n}\{h(U') - h(\theta)\}$ *tends to the normal distribution with mean zero and variance*

$$\sum_{\gamma=1}^{g} \sum_{\delta=1}^{g} m(\gamma)m(\delta) \left(\frac{\partial h(y)}{\partial y^{(\gamma)}}\right)_{y=\theta} \left(\frac{\partial h(y)}{\partial y^{(\delta)}}\right)_{y=\theta} \zeta_1^{(\gamma,\delta)}.$$

Theorem 7.5 follows from Theorem 7.3 in exactly the same way as the theorem on the asymptotic distribution of a function of moments follows from the fact of their asymptotic normality; cf. Cramér [2, p. 366]. We shall therefore omit the proof of Theorem 7.5. Since any moment whose variance exists has the form $U' = \theta(S)$ (cf. section 4 and Theorem 7.4), Theorem 7.5 is a generalization of the theorem on a function of moments.

8. Limit Theorems for $U(X_1, \ldots, X_n)$ When the X_α's Have Different Distributions

The limit theorems of the preceding section can be extended to the case when the X_α's have different distributions. We shall only prove an extension to this case of Theorem 7.1 (or 7.2), confining ourselves, for the sake of simplicity, to the distribution of a single U-statistic.

The extension of Theorems 7.3 and 7.5 with $g = 1$ to this case is immediate. One has only to replace the reference to Theorem 7.1 by that to the following Theorem 8.1, and θ and ζ_1 by $E\{U\}$ and $\zeta_{1,n}$.

Theorem 8.1. *Let* $X_1, \ldots, X_n$ *be n independent random vectors of r components,* X_a *having the d.f.* $F_a(x) = F_a(x^{(1)}, \ldots, x^{(r)})$. *Let* $\Phi(x_1, \ldots, x_m)$ *be a function symmetric in its m vector arguments* $x_\beta = (x_\beta^{(1)}, \ldots, x_\beta^{(r)})$ *which does not involve n, and let*

$$\bar{\Psi}_{1(v)}(x) = \binom{n-1}{m-1}^{-1} \sum_{(\neq v)}' \Psi_{1(v)\alpha_1, \ldots, \alpha_{m-1}}(x), \qquad (v = 1, \ldots, n), \qquad (8.1)$$

where Ψ *is defined by* (5.15), *and the summation is extended over all subscripts* α *such that*

$$1 \le \alpha_1 < \alpha_2 < \cdots < \alpha_{m-1} \le n, \quad \alpha_i \neq v, \qquad (i = 1, \ldots, m).$$

Suppose that there is a number A such that for every $n = 1, 2, \ldots$

$$\int \cdots \int \Phi^2(x_1, \ldots, x_m) \, dF_{\alpha_1}(x_1) \ldots dF_{\alpha_m}(x_m) < A,$$

$$(1 \le \alpha_1 \le \alpha_2 \le \cdots \le \alpha_m \le n), \qquad (8.2)$$

that

$$E|\bar{\Psi}_{1(v)}^3(X_v)| < \infty, \qquad (v = 1, 2, \ldots, n), \qquad (8.3)$$

and

$$\lim_{n \to \infty} \sum_{v=1}^{n} E|\bar{\Psi}_{1(v)}^3(X_v)| \Big/ \left\{ \sum_{v=1}^{n} E\{\bar{\Psi}_{1(v)}^2(X_v)\} \right\}^{3/2} = 0. \qquad (8.4)$$

Then, as $n \to \infty$, *the d.f. of* $(U - E\{U\})/\sigma(U)$ *tends to the normal d.f. with mean 0 and variance 1.*

The proof is similar to that of Theorem 7.1.
Let

$$W = \frac{m}{n} \sum_{v=1}^{n} \bar{\Psi}_{1(v)}(X_v).$$

It will be shown that
(a) the d.f. of

$$V = \frac{W - E\{W\}}{\sigma(W)}$$

tends to the normal d.f. with mean 0 and variance 1, and that
(b) the d.f. of

$$V' = \frac{U - E\{U\}}{\sigma(U)}$$

tends to the same limit as the d.f. of V.

Part (a) follows immediately from (8.3) and (8.4) by Liapounoff's form of the Central Limit Theorem.

According to Lemma 7.1, (b) will be proved when it is shown that

$$\lim_{n \to \infty} E\{V' - V\}^2 = \lim \left\{ 2 - 2 \frac{\sigma(U, W)}{\sigma(U)\sigma(W)} \right\} = 0$$

or

$$\lim_{n \to \infty} \frac{\sigma(U, W)}{\sigma(U)\sigma(W)} = 1. \tag{8.5}$$

Let c be an integer, $1 \leq c \leq m$, and write

$$x = (x_1, \ldots, x_c), \qquad y = (y_1, \ldots, y_{m-c}), \qquad z = (z_1, \ldots, z_{m-c})$$

$$F_{(\alpha)}(x) = F_{\alpha_1}(x_1) \ldots F_{\alpha_c}(x_c), \qquad F_{(\beta)}(y) = F_{\beta_1}(y_1) \ldots F_{\beta_{m-c}}(y_{m-c}),$$

$$F_{(\gamma)}(z) = F_{\gamma_1}(z_1) \ldots F_{\gamma_{m-c}}(z_{m-c}).$$

Then, by Schwarz's inequality,

$$\int \cdots \int \Phi(x, y)\Phi(x, z) \, dF_{(\alpha)}(x) \, dF_{(\beta)}(y) \, dF_{(\gamma)}(z)$$

$$\leq \left\{ \int \cdots \int \Phi^2(x, y) \, dF_{(\alpha)}(x) \, dF_{(\beta)}(y) \int \cdots \int \Phi^2(x, z) \, dF_{(\alpha)}(x) \, dF_{(\gamma)}(z) \right\}^{1/2},$$

which, by (8.2), is $< A$ for any set of subscripts.

By the inequality for moments, $\theta_{\alpha_1, \ldots, \alpha_m}$, as defined by (5.14), is also uniformly bounded, and applying these inequalities to (5.16), it follows that there exists a number B such that

$$|\zeta_{c(\alpha_1, \ldots, \alpha_c)\beta_1, \ldots, \beta_{m-c}; \gamma_1, \ldots, \gamma_{m-c}}| < B, \qquad (c = 1, \ldots, m), \tag{8.6}$$

for every set of subscripts satisfying the inequalities

$$\alpha_g \neq \alpha_h, \quad \beta_g \neq \beta_h, \quad \gamma_g \neq \gamma_h \quad \text{if} \quad g \neq h, \quad \alpha_i \neq \beta_j, \quad \alpha_i \neq \gamma_j,$$

$$(i = 1, \ldots, c; j = 1, \ldots, m - c).$$

Now, we have

$$E\{W\} = 0$$

and

$$\sigma^2(W) = \frac{m^2}{n^2} \sum_{v=1}^{n} E\{\bar{\Psi}^2_{1(v)}(X_v)\} \tag{8.7}$$

or, inserting (8.1) and recalling (5.16),

$$\sigma^2(W) = \frac{m^2}{n^2} \binom{n-1}{m-1}^{-2} \sum_{v=1}^{n} \sum_{(\neq v)}' \sum_{(\neq v)}' \zeta_{1(v)\alpha_1, \ldots, \alpha_{m-1}; \beta_1, \ldots, \beta_{m-1}}, \tag{8.8}$$

the two sums Σ' being over $\alpha_1 < \cdots < \alpha_{m-1}$, $(\alpha_i \neq v)$, and $\beta_1 < \cdots < \beta_{m-1}$, $(\beta_i \neq v)$, respectively. By (5.17), the sum of the terms whose subscripts $v, \alpha_1, \ldots,$

$\alpha_{m-1}, \beta_1, \ldots, \beta_{m-1}$ are all different is equal to

$$\frac{n(n-1)\ldots(n-2m+2)}{(m-1)!(m-1)!}\zeta_{1,n} = n\binom{n-1}{m-1}\binom{n-m}{m-1}\zeta_{1,n}.$$

The number of the remaining terms is of order n^{2m-2}. Since, by (8.6), they are uniformly bounded, we have

$$\sigma^2(W) = \frac{m^2}{n}\zeta_{1,n} + O(n^{-2}). \tag{8.9}$$

Similarly, we have from (5.18)

$$\sigma^2(U) = \frac{m^2}{n}\zeta_{1,n} + O(n^{-2}),$$

and hence

$$\sigma(U) = \sigma(W) + O(n^{-1}). \tag{8.10}$$

The covariance of U and W is

$$\sigma(U, W) = \binom{n}{m}^{-1}\frac{m}{n}\sum_{v=1}^{n}\sum{}' E\{\bar{\Psi}_{1(v)}(X_v)\Psi_{m(\alpha_1,\ldots,\alpha_m)}(X_{\alpha_1}, \ldots, X_{\alpha_m})\}.$$

All terms except those in which one of the α's $= v$, vanish, and for the remaining ones we have, for fixed $\alpha_1, \ldots, \alpha_m$,

$$E\{\bar{\Psi}_{1(v)}(X_v)\Psi_{m(\alpha_1,\ldots,\alpha_m)}(X_{\alpha_1}, \ldots, X_{\alpha_m})\}$$

$$= \binom{n-1}{m-1}^{-1}\sum_{(\neq v)}{}' E\{\Psi_{1(v)\beta_1,\ldots,\beta_{m-1}}(X_v)\Psi_{1(v)\gamma_1,\ldots,\gamma_{m-1}}(X_v)\}$$

$$= \binom{n-1}{m-1}^{-1}\sum_{(\neq v)}{}' \zeta_{1(v)\beta_1,\ldots,\beta_{m-1};\gamma_1,\ldots,\gamma_{m-1}}$$

where the summation sign refers to the β's, and $\gamma_1, \ldots, \gamma_{m-1}$ are the α's that arc $\neq v$. Inserting this in (8 11) and comparing the result with (8.8), we see that

$$\sigma(U, W) = \sigma^2(W). \tag{8.12}$$

From (8.12) and (8.10) we have

$$\frac{\sigma(U, W)}{\sigma(U)\sigma(W)} = \frac{\sigma(W)}{\sigma(U)} = \frac{n\sigma(W)}{n\sigma(W) + O(1)}.$$

Comparing condition (8.4) with (8.7), we see that we must have $n\sigma(W) \to \infty$ as $n \to \infty$. This shows the truth of (8.5). The proof of Theorem 8.1 is complete.

For some purposes the following corollary of Theorem 8.1 will be useful, where the conditions (8.2), (8.3), and (8,4) are replaced by other conditions which are more restrictive, but easier to apply.

Theorem 8.2. *Theorem 8.1 holds if the conditions* (8.2), (8.3), *and* (8.4) *are replaced by the following:*

There exist two positive numbers C, D such that

$$\int \cdots \int |\Phi^3(x_1, \ldots, x_m)| \, dF_{\alpha_1}(x_1) \ldots dF_{\alpha_m}(x_m) < C \qquad (8.13)$$

for $\alpha_i = 1, 2, \ldots, (i = 1, \ldots, m)$, and

$$\zeta_{1(v)\alpha_1, \ldots, \alpha_{m-1}; \beta_1, \ldots, \beta_{m-1}} > D \qquad (8.14)$$

for any subscripts satisfying

$$1 \le \alpha_1 < \alpha_2 < \cdots < \alpha_{m-1}, \quad 1 \le \beta_1 < \beta_2 < \cdots < \beta_{m-1}, \quad 1 \le v \ne \alpha_i, \beta_i.$$

We have to show that (8.2), (8.3), and (8.4) follow from (8.13) and (8.14).

(8.13) implies (8.2) by the inequality for moments. By a reasoning analogous to that used in the previous proof, applying Hölder's inequality instead of Schwarz's inequality, it follows from (8.13) that

$$E|\bar{\Psi}^3_{1(v)}(X_v)| < C'. \qquad (8.15)$$

On the other hand, by (8.7), (8.8), and (8.14),

$$\sum_{v=1}^{n} E\{\bar{\Psi}^2_{1(v)}(X_v)\} > nD. \qquad (8.16)$$

(8.15) and (8.16) are sufficient for the fulfillment of (8.4).

9. Applications to Particular Statistics

(a) Moments and Functions of Moments

It has been seen in section 4 that the k-statistics and the unbiased estimates of moments are U-statistics, while the sample moments are regular functionals of the sample d.f. By Theorems 7.1, 8.1, and 7.4 these statistics are asymptotically normally distributed, and by Theorem 7.5 the same is true for a function of moments, if the respective conditions are satisfied. These results are not new (cf., for example, Cramér [2]).

(b) Mean Difference and Coefficient of Concentration

If $Y_1, \ldots, Y_n$ are n independent real-valued random variables, Gini's mean difference (without repetition) is defined by

$$d = \frac{1}{n(n-1)} \sum_{\alpha \ne \beta} |Y_\alpha - Y_\beta|.$$

If the Y_α's have the same distribution F, the mean of d is

$$\delta = \int\int |y_1 - y_2|\, dF(y_1)\, dF(y_2),$$

and the variance, by (5.13) is

$$\sigma^2(d) = \frac{2}{n(n-1)}\{2\zeta_1(\delta)(n-2) + \zeta_2(\delta)\},$$

where

$$\zeta_1(\delta) = \int\left\{\int |y_1 - y_2|\, dF(y_2)\right\}^2 dF(y_1) - \delta^2, \tag{9.1}$$

$$\zeta_2(\delta) = \int\int (y_1 - y_2)^2\, dF(y_1)\, dF(y_2) - \delta^2 = 2\sigma^2(Y) - \delta^2. \tag{9.2}$$

The notation $\zeta_1(\delta)$, $\zeta_2(\delta)$ serves to indicate the relation of these functionals of F to the functional $\delta(F)$; δ is here merely the symbol of the functional, not a particular value of it. In a similar way we shall write $\Phi(y_1, y_2|\delta) = |y_1 - y_2|$, etc. When there is danger of confusing $\zeta_1(\delta)$ with $\zeta_1(F)$, we may write $\zeta_1(F|\delta)$.

U.S. Nair [19] has evaluated $\sigma^2(d)$ for several particular distributions.

By Theorem 7.1, $\sqrt{n}(d - \delta)$ is asymptotically normal if $\zeta_2(\delta)$ exists.

If $Y_1, \ldots, Y_n$ do not assume negative values, the coefficient of concentration (cf. Gini [8]) is defined by

$$G = \frac{d}{2\bar{Y}},$$

where $\bar{Y} = \Sigma Y_\alpha/n$. G is a function of two U-statistics. If the Y_α's are identically distributed, if $E\{Y^2\}$ exists, and if $\mu = E\{Y\} > 0$, then, by Theorem 7.5, $\sqrt{n}(G - \delta/2\mu)$ tends to be normally distributed with mean 0 and variance

$$\frac{\delta^2}{4\mu^4}\zeta_1(\mu) - \frac{\delta}{\mu^3}\zeta_1(\mu, \delta) + \frac{1}{\mu^2}\zeta_1(\delta),$$

where

$$\zeta_1(\mu) = \int y^2\, dF(y) - \mu^2 = \sigma^2(Y),$$

$$\zeta_1(\mu, \delta) = \int\int y_1|y_1 - y_2|\, dF(y_1)\, dF(y_2) - \mu\delta,$$

abd $\zeta_1(\delta)$ is given by (9.1).

(c) Functions of Ranks and of the Signs of Variate Differences

Let $s(u)$ be the signum function,

$$s(u) = \begin{cases} -1 & \text{if } u < 0; \\ 0 & \text{if } u = 0; \\ 1 & \text{if } u > 0, \end{cases} \tag{9.3}$$

and let

$$c(u) = \tfrac{1}{2}\{1 + s(u)\} = \begin{array}{l} 0 \text{ if } u < 0; \\ \tfrac{1}{2} \text{ if } u = 0; \\ 1 \text{ if } u > 0. \end{array} \qquad (9.4)$$

If

$$x_\alpha = (x_\alpha^{(1)}, \ldots, x_\alpha^{(r)}), \qquad (\alpha = 1, \ldots, n)$$

is a sample of n vectors of r components, we may define the *rank* $R_\alpha^{(i)}$ of $x_\alpha^{(i)}$ by

$$R_\alpha^{(i)} = \tfrac{1}{2} + \sum_{\beta=1}^{n} c(x_\alpha^{(i)} - x_\beta^{(i)})$$

$$= \frac{n+1}{2} + \frac{1}{2} \sum_{\beta=1}^{n} s(x_\alpha^{(i)} - x_\beta^{(i)}), \qquad (i = 1, \ldots, r). \qquad (9.5)$$

If the numbers $x_1^{(i)}, x_2^{(i)}, \ldots, x_n^{(i)}$ are all different, the smallest of them has rank 1, the next smallest rank 2, etc. If some of them are equal, the rank as defined by (9.5) is known as the mid-rank.

Any function of the ranks is a function of expressions $c(x_\alpha^{(i)} - x_\beta^{(i)})$ or $s(x_\alpha^{(i)} - x_\beta^{(i)})$.

Conversely, since

$$s(x_\alpha^{(i)} - x_\beta^{(i)}) = s(R_\alpha^{(i)} - R_\beta^{(i)}),$$

any function of expressions $s(x_\alpha^{(i)} - x_\beta^{(i)})$ or $c(x_\alpha^{(i)} - x_\beta^{(i)})$ is a function of the ranks.

Consider a regular functional $\theta(F)$ whose kernel $\Phi(x_1, \ldots, x_m)$ depends only on the signs of the variate differences,

$$s(x_\alpha^{(i)} - x_\beta^{(i)}), \qquad (\alpha, \beta = 1, \ldots, m; i = 1, \ldots, r). \qquad (9.6)$$

The corresponding U-statistic is a function of the ranks of the sample variates.

The function Φ can take only a finite number of values, $c_1, \ldots, c_N$, say. If $\pi_i = P\{\Phi = c_i\}, (i = 1, \ldots, N)$, we have

$$\theta = c_1 \pi_1 + \cdots + c_N \pi_N, \qquad \sum_{i=1}^{N} \pi_i = 1.$$

π_i is a regular functional whose kernel $\Phi_i(x_1, \ldots, x_m)$ is equal to 1 or 0 according to whether $\Phi = c_i$ or $\neq c_i$. We have

$$\Phi = c_1 \Phi_1 + \cdots + c_N \Phi_N.$$

In order that $\theta(F)$ exist, the c_i must be finite, and hence Φ is bounded. Therefore, $E\{\Phi^2\}$ exists, and if $X_1, X_2, \ldots$ are identically distributed, the d.f. of $\sqrt{n}(U - \theta)$ tends, by Theorem 7.1, to a normal d.f. which is non-singular if $\zeta_1 > 0$.

In the following we shall consider several examples of such functionals.

(d) Difference Sign Correlation

Consider the bivariate sample

$$(x_1^{(1)}, x_1^{(2)}), (x_2^{(1)}, x_2^{(2)}), \ldots, (x_n^{(1)}, x_n^{(2)}). \tag{9.7}$$

To each two members of this sample corresponds a pair of signs of the differences of the respective variables,

$$s(x_\alpha^{(1)} - x_\beta^{(1)}), s(x_\alpha^{(2)} - x_\beta^{(2)}), \qquad (\alpha \neq \beta; \alpha, \beta = 1, \ldots, n). \tag{9.8}$$

(9.8) is a population of $n(n-1)$ pairs of difference signs. Since

$$\sum_{\alpha \neq \beta} s(x_\alpha^{(i)} - x_\beta^{(i)}) = 0, \qquad (i = 1, 2),$$

the covariance t of the difference signs (9.8) is

$$t = \frac{1}{n(n-1)} \sum_{\alpha \neq \beta} s(x_\alpha^{(1)} - x_\beta^{(1)}) s(x_\alpha^{(2)} - x_\beta^{(2)}). \tag{9.9}$$

t will be briefly referred to as the *difference sign covariance* of the sample (9.7).

If all $x^{(1)}$'s and all $x^{(2)}$'s are different, we have

$$\sum_{\alpha \neq \beta} s^2(x_\alpha^{(i)} - x_\beta^{(i)}) = n(n-1), \qquad (i = 1, 2),$$

and then t is the product moment correlation of the difference signs.

It is easily seen that t is a linear function of the number of inversions in the permutation of the ranks of $x^{(1)}$ and $x^{(2)}$.

The statistic t has been considered by Esscher [6], Lindeberg [15], [16], Kendall [12], and others.

t is a U-statistic. As a function of a random sample from a bivariate population, t is an unbiased estimate of the regular functional of degree 2,

$$\tau = \iiiint s(x_1^{(1)} - x_2^{(1)}) s(x_1^{(2)} - x_2^{(2)}) \, dF(x_1) \, dF(x_2). \tag{9.10}$$

τ is the covariance of the signs of differences of the corresponding components of $X_1 = (X_1^{(1)}, X_1^{(2)})$ and $X_2 = (X_2^{(1)}, X_2^{(2)})$ in the population of pairs of independent vectors X_1, X_2 with identical d.f. $F(x) = F(x^{(1)}, x^{(2)})$. If $F(x^{(1)}, x^{(2)})$ is continuous, τ is the product moment correlation of the difference signs.

Two points (or vectors), $(x_1^{(1)}, x_1^{(2)})$ and $(x_2^{(1)}, x_2^{(2)})$ are called concordant or discordant according to whether

$$(x_1^{(1)} - x_2^{(1)})(x_1^{(2)} - x_2^{(2)})$$

is positive or negative. If $\pi^{(c)}$ and $\pi^{(d)}$ are the probabilities that a pair of vectors drawn at random from the population is concordant or discordant, respectively, we have from (9.10)

$$\tau = \pi^{(c)} - \pi^{(d)}.$$

If $F(x^{(1)}, x^{(2)})$ is continuous, we have $\pi^{(c)} + \pi^{(d)} = 1$, and hence

$$\tau = 2\pi^{(c)} - 1 = 1 - 2\pi^{(d)}. \tag{9.11}$$

If we put

$$\bar{F}(x^{(1)}, x^{(2)}) = \tfrac{1}{4}\{F(x^{(1)} - 0, x^{(2)} - 0) + F(x^{(1)} - 0, x^{(2)} + 0)$$
$$+ F(x^{(1)} + 0, x^{(2)} - 0) + F(x^{(1)} + 0, x^{(2)} + 0)\} \quad (9.12)$$

we have

$$\Phi_1(x|\tau) = 1 - 2\bar{F}(x^{(1)}, \infty) - 2F(\infty, x^{(2)}) + 4\bar{F}(x^{(1)}, x^{(2)}), \quad (9.13)$$

and we may write

$$\tau = E\{\Phi_1(X_1|\tau)\}. \quad (9.14)$$

The variance of t is, by (5.13),

$$\sigma^2(t) = \frac{2}{n(n-1)}\{2\zeta_1(\tau)(n-2) + \zeta_2(\tau)\}, \quad (9.15)$$

where

$$\zeta_1(\tau) = E\{\Phi_1^2(X_1|\tau)\} - \tau^2, \quad (9.16)$$

$$\zeta_2(\tau) = E\{s^2(X_1^{(1)} - X_2^{(1)})s^2(X_1^{(2)} - X_2^{(2)})\} - \tau^2. \quad (9.17)$$

If $F(x^{(1)}, x^{(2)})$ is continuous, we have $\zeta_2(\tau) = 1 - \tau^2$, and $\bar{F}(x^{(1)}, x^{(2)})$ in (9.13) may be replaced by $F(x^{(1)}, x^{(2)})$.

The variance of a linear function of t has been given for the continuous case by Lindeberg [15], [16].

If $X^{(1)}$ and $X^{(2)}$ are independent and have a continuous d.f., we find $\zeta_1(\tau) = \tfrac{1}{9}, \zeta_2(\tau) = 1$, and hence

$$\sigma^2(t) = \frac{2(2n+5)}{9n(n-1)}. \quad (9.18)$$

In this case the distribution of t is independent of the univariate distributions of $X^{(1)}$ and $X^{(2)}$. This is, however, no longer true if the independent variables are discontinuous. Then it appears that $\sigma^2(t)$ depends on $P\{X_1^{(i)} = X_2^{(i)}\}$ and $P\{X_1^{(i)} = X_2^{(i)} = X_3^{(i)}\}, (i = 1, 2)$.

By Theorem 7.1 the d.f. of $\sqrt{n}(t - \tau)$ tends to the normal form. This result has first been obtained for the particular case that all permutations of the ranks of $X^{(1)}$ and $X^{(2)}$ are equally probable, which corresponds to the independence of the continuous random variables $X^{(1)}$, $X^{(2)}$ (Kendall [12]). In this case t can be represented as a sum of independent random variables (cf. Dantzig [5] and Feller [7]). In the general case the asymptotic normality of t has been shown by Daniels and Kendall [4] and the author [10].

The functional $\tau(F)$ is stationary (and hence the normal limiting distribution of $\sqrt{n}(t - \tau)$ singular) if $\zeta_1 = 0$, which, in the case of a continuous F, means that the equation $\Phi_1(X|\tau) = \tau$ or

$$4F(X^{(1)}, X^{(2)}) = 2F(X^{(1)}, \infty) + 2F(\infty, X^{(2)}) - 1 + \tau \quad (9.19)$$

is satisfied with probability 1. This is the case if $X^{(2)}$ is an increasing function of $X^{(1)}$. Then $t = \tau = 1$ with probability 1, and $\sigma^2(t) = 0$. A case where (9.19) is fulfilled and $\sigma^2(t) > 0$ is the following: $X^{(1)}$ is uniformly distributed in the interval $(0, 1)$, and

$$X^{(2)} = X^{(1)} + \tfrac{1}{2} \quad \text{if} \quad 0 \le X^{(1)} < \tfrac{1}{2}, \quad X^{(2)} = X^{(1)} - \tfrac{1}{2} \quad \text{if} \quad \tfrac{1}{2} \le X^{(1)} \le 1.$$
(9.20)

In this case $\tau = 0$, $\zeta_2 = 1$, $\sigma^2(t) = 2/n(n-1)$.

(*Editors' note*: Sections e–h have been deleted.)

References

[1] H. Cramér, *Random Variables and Probability Distributions*, Cambridge Tracts in Math., Cambridge, 1937.

[2] H. Cramér, *Mathematical Methods of Statistics*, Princeton University Press, 1946.

[3] H.E. Daniels, "The relation between measures of correlation in the universe of sample permutations," *Biometrika*, Vol. 33 (1944), pp. 129–135.

[4] H.E. Daniels and M.G. Kendall, "The significance of rank correlations where parental correlation exists," *Biometrika*, Vol. 34 (1947), pp. 197–208.

[5] G.B. Dantzig, "On a class of distributions that approach the normal distribution function," *Annals of Math. Stat.*, Vol. 10 (1939) pp. 247–253.

[6] F. Esscher, "On a method of determining correlation from the ranks of the variates," *Skandinavisk Aktuar. tidş.*, Vol. 7 (1924), pp. 201–219.

[7] W. Feller, "The fundamental limit theorems in probability," *Am. Math. Soc. Bull.*, Vol. 51 (1945), pp. 800–832.

[8] C. Gini, "Sulla misura della concentrazione e della variabilità dei caratteri," *Atti del R. Istituto Veneto di S.L.A.*, Vol. 73 (1913–14), Part 2.

[9] P.R. Halmos, "The theory of unbiased estimation," *Annals of Math. Stat.*, Vol. 17 (1946), pp. 34–43.

[10] W. Höffding, "On the distribution of the rank correlation coefficient τ, when the variates are not independent," *Biometrika*, Vol. 34 (1947), pp. 183–196.

[11] H. Hotelling and M.R. Pabst, "Rank correlation and tests of significance involving no assumptions of normality," *Annals of Math. Stat.*, Vol. 7 (1936), pp. 20–43.

[12] M.G. Kendall, "A new measure of rank correlation," *Biometrika*, Vol. 30 (1938), pp. 81–93.

[13] M.G. Kendall, "Partial rank correlation," *Biometrika*, Vol. 32 (1942), pp. 277–283.

[14] M.G. Kendall, S.F.H. Kendall, and B. Babington Smith, "The distribution of Spearman's coefficient of rank correlation in a universe in which all rankings occur an equal number of times," *Biometrika*, Vol. 30 (1939), pp. 251–273.

[15] J.W. Lindeberg, "Über die Korrelation," VI *Skand. Matematikerkongres i København*, 1925, pp. 437–446.

[16] J.W. Lindeberg, "Some remarks on the mean error of the percentage of correlation," *Nordic Statistical Journal*, Vol. 1 (1929), pp. 137–141.

[17] H.B. Mann, "Nonparametric tests against trend," *Econometrica*, Vol. 13 (1945), pp. 245–259.

[18] R. v. Mises, "On the asymptotic distribution of differentiable statistical functions," *Annals of Math. Stat.*, Vol. 18 (1947), pp. 309–348.

[19] U.S. Nair, "The standard error of Gini's mean difference," *Biometrika*, Vol. 28 (1936), 428–436.

[20] K. Pearson, "On further methods of determining correlation," *Drapers' Company Research Memoirs*, Biometric Series, IV, London, 1907.

[21] V. Volterra, *Theory of Functionals*, Blackie, (authorized translation by Miss M. Long), London and Glasgow, 1931.

[22] G.U. Yule and M.G. Kendall, *An Introduction to the Theory of Statistics*, Griffin, 11th Edition, London, 1937.

Introduction to
Wald (1949) Statistical Decision Functions

L. Weiss
Cornell University

Abraham Wald was born on October 31, 1902 in Cluj, one of the main cities of Transylvania, which at the time belonged to Hungary. The official language was Hungarian, but the population was mixed, containing substantial numbers of Romanian, German, and Jewish inhabitants, as well as Hungarians. As a result, much of the population spoke more than one language, and the Jewish families used Yiddish as well as Hungarian. Wald's family would not allow their children to attend school on Saturday, the Jewish sabbath, and as a result, Wald was educated at home until he attended the local university, where the language of instruction was Hungarian. After graduating from the local university, he entered the University of Vienna in 1927 and received his Ph.D. in mathematics in 1931. His first research interest was in geometry, and he published 21 papers in that area between 1931 and 1937. Austria was in turmoil during much of this period as a result of Nazi agitation, and it was impossible for Wald, as a Jewish noncitizen, to obtain any academic appointment. He supported himself by tutoring a prominent Viennese banker and economist, Karl Schlesinger, in mathematics. As a result of this, Wald became interested in economics and econometrics, and published several papers and a monograph on these subjects. His first exposure to statistical theory was a natural result of his work on econometrics. Because of his publications in econometrics, Wald was invited to become a Fellow of the Cowles Commission and arrived in the United States in the summer of 1938. This invitation from the Cowles Commission saved Wald's life, for almost the whole of his family in Europe perished during the Holocaust. In the fall of 1938, Wald became a fellow of the Carnegie Corporation and started to study statistics at Columbia University with Harold Hotelling. Wald stayed at Columbia as a Fellow of the Carnegie Corporation until 1941, lecturing during the academic year 1939–40. In 1941, he joined the Columbia faculty and remained a member of that faculty for the rest of his life. During the war years, he was also a member of the Statistics Research Group at Columbia, doing research related to the war effort. In late 1950, Wald was giving a series of lectures in India at the invitation of the Indian government. On December 13, 1950, he and his wife were killed when their airplane crashed.

A more complete biography of Abraham Wald may be found in the *Encyclopedia of Statistical Sciences*.

Wald's 1949 paper, "Statistical Decision Functions," is notable for unifying practically all existing statistical theory by treating statistical problems as special cases of zero-sum two-person games. In 1950, a monograph of the same title was published. The 1949 paper is a condensation of that 1950 monograph.

The mathematical theory of games was described in the landmark book *Theory of Games and Economic Behavior* by Von Neumann and Morgenstern (1944). A "game" is a set of rules describing the alternatives available to a player at each move the player has to make, which player is to make each move when the past moves are given, when the play ends, and what happens when the play ends. It is assumed that at the end of the play, the outcome is measured by a payment (positive, negative, or zero) to each player. In a zero-sum two-person game, there are two players and the sum of payments made to them at the end of each play must be zero. Thus, the players must be antagonistic, since what one wins the other loses. As we can well imagine, this causes very conservative methods of play.

Zero-sum two-person games are further classified as being in "extensive form" or "normalized form." In the extensive form, one or both players have a sequence of moves to make, as in chess. In the normalized form, each player makes a move on only one occasion. In one of the key ideas of their theory, Von Neumann and Morgenstern show that any game in extensive form can be reduced mathematically to an equivalent game in normalized form, by using the notion of a "strategy." A strategy for a player in a game in extensive form is a complete set of instructions describing exactly what the player would do in any conceivable situation the player might encounter during the course of play. The game in extensive form is then reduced to normalized form by having each player choose a strategy (any one desired) at the beginning: This choice is the only move for the player, and so the resulting game is now in normalized form. Of course, as a practical matter, no strategy could be written down for a game like chess, but as a mathematical concept, the use of strategies does reduce games in extensive form to mathematically equivalent games in normalized form.

Now we are ready to describe the fundamental theorem of zero-sum two-person games given by Von Neumann and Morgenstern. Suppose player 1 has m possible strategies, and player 2 n possible strategies. m and n are assumed to be finite: Even in a game as complicated as chess, the standard stopping rules guarantee this. Let $a(i, j)$ denote the payment to player 1 if player 1 uses his strategy number i and player 2 uses her strategy number j. [Of course, then the payment to player 2 is $-a(i, j)$.] If there are chance moves during the play of the original game in extensive form, as in card games like poker, then $a(i, j)$ represents an expected payment. The fundamental theorem states there is a set of probabilities $(p_1^*, \ldots, p_m^*)$ and a set of probabilities $(q_1^*, \ldots, q_n^*)$ such that

$$\sum_{j=1}^{n} \sum_{i=1}^{m} p_i q_j^* a(i, j) \le \sum_{j=1}^{n} \sum_{i=1}^{m} p_i^* q_j^* a(i, j) \le \sum_{j=1}^{n} \sum_{i=1}^{m} p_i^* q_j a(i, j)$$

for *all* sets of probabilities $(p_1, \ldots, p_m)$ and $(q_1, \ldots, q_n)$.

We note that if player 1 chooses his strategy number i with probability p_i^*, he guarantees himself an expected payment of at least $\sum_{j=1}^{n} \sum_{i=1}^{m} p_i^* q_j^* a(i, j) \equiv V$, say, no matter what player 2 does. Similarly, if player 2 chooses her strategy number j with probability q_j^*, she guarantees that the expected payment to player 1 is no more than V, no matter what player 1 does. Then the recommended way to play the game is for player 1 to choose his strategy number i with probability p_i^* for $i = 1, \ldots, m$, and for player 2 to choose her strategy j with probability q_j^* for $j = 1, \ldots, n$. If one of the players departs from the recommendation, that player cannot gain anything and may be hurt. This is a satisfactory theoretical solution to the problem of how to play a zero-sum two-person game, but is of practical value only for the very simplest games in extensive form, such as tic-tac-toe and a simplified form of poker described by Von Neumann and Morgenstern.

Wald used the Von Neumann–Morgenstern theory by considering a statistical problem as a zero-sum two-person game, the first player being "Nature" and the second player the "Statistician." Nature chooses the probability distribution that will be followed by the random variables to be observed by the Statistician, keeping the choice hidden from the Statistician. The Statistician then observes the random variables and chooses a decision from a set of possible decisions. The payment to Nature depends on the combination of Nature's choice of the probability distribution and the Statistician's choice of a decision. This payment to Nature is, of course, the loss to the Statistician. Even with this brief description, we can see that most of statistical theory can be considered a special case of such a game between Nature and the Statistician. Let's look at testing hypotheses. Each possible distribution that could be chosen by Nature is in either H_0 (the null hypothesis) or H_1 (the alternative hypothesis). The Statistician has only two possible decisions: Decide the distribution is in H_0 or decide the distribution is in H_1. The payment to Nature is 0 if the Statistician decided correctly and 1 if the Statistician decided incorrectly. Thus, the expected loss to the Statistician is the probability of making the incorrect decision. Now let's look at point estimation. Here the possible distributions are indexed by a parameter θ. Nature chooses a value of θ, keeping the choice hidden from the Statistician. The Statistician then guesses the value of θ (after observing the random variables). The farther the Statistician's guess is from the value chosen by Nature, the greater the payment that must be made to Nature. A common method of deciding what the payment should be is "squared error loss": If the Statistician's guess is D, the payment to Nature is equal to $(D - \theta)^2$.

Before Wald could apply the Von Neumann–Morgenstern theory to statistical problems, he had to solve a difficult technical problem. The fundamental theorem of Von Neumann-Morgenstern, described above, assumes that each player has only a finite number of strategies. In most statistical problems, each player has an infinite number of strategies. For example, suppose the possible distributions for the random variables to be observed are normal distributions with a variance of 1 and a mean equal to any real number. Then Nature has an infinite number of possible choices for the mean. Wald had to extend the fundamental theorem to cases in which the players have an infinite number of possible strategies. Wald started work along these lines in a 1945 paper (1945a). Wald's 1949 paper starts out by extending the fundamental theorem even beyond his 1945 extension. This further extension was essential for handling the wide variety of statistical problems covered by Wald. Next, Wald sets up statistical decision problems as zero-sum two-person games between Nature (player 1) and the Statistician (player 2) in a way general enough to allow for sequential sampling. That is, the Statistician can take one observation at a time, and after each observation decide whether or not to take another observation, or cease sampling and choose a "terminal" decision. Of course, the Statistician is charged for the observations. Allowing sequential sampling meant that Wald was considering a game in extensive form. Remarkably, Wald had constructed an optimal way for the Statistician to play such a game with his sequential probability ratio test, without any reference to the theory of games (1945b).

The remainder of Wald's 1949 paper is devoted to studying the existence and construction of what might be considered "good" decision rules. A decision rule is any method used by the Statistician to play the game and may utilize randomization (randomization being the use of a random device to help in choosing a decision). If Nature has chosen the probability distribution F for the random variables to be observed by the Statistician, and the Statistician uses the decision rule D, let $r(F, D)$ denote the expected loss to the Statistician (that is, the expected payment to nature). The expectation must be used, since the random variables observed introduce an element of chance. A decision rule D_1 is defined to be uniformly better than a decision rule D_2 if $r(F, D_1) \leq r(F, D_2)$ for all possible distributions F, with strict inequality for at least one F. A class of decision rules is said to be complete if for any decision rule D not in the class there exists a decision rule D^* in the class that is uniformly better than D. If we can find a complete class of decision rules, we can limit our search for a good decision rule to the decision rules in that class, so Wald devotes a considerable portion of his paper to the construction of complete classes of decision rules by the methods we now describe.

Suppose that G is a probability distribution for the possible distributions F (G is called "an a priori distribution for nature"). Let $r^*(G, D)$ denote the expected value of $r(F, D)$ when F is considered a random variable with probability distribution G. If e is a given positive value, a decision rule D_0 is said to be an e-Bayes decision rule with respect to G if $r^*(G, D_0) \leq \text{Inf}_D\, r^*(G, D) + e$. If this last inequality holds for $e = 0$, then D_0 is said to be a Bayes decision rule with respect to G. Under certain conditions, Wald shows that for any $e > 0$, the class of all e-Bayes decision rules corresponding to all possible a priori distributions is a complete class. Also, it is shown that for any given a priori distribution G, there is a decision rule that is a Bayes decision rule with respect to G. A decision rule D_0 is said to be a Bayes decision rule in the wide sense if there exists a sequence $[G_i;\ i = 1, 2, \ldots]$ of a priori distributions such that $\lim_{i \to \infty} [r^*(G_i, D_0) - \text{Inf}_D\, r^*(G_i, D)] = 0$. A decision rule D is called a Bayes decision rule in the strict sense if there exists an a priori distribution G such that D is a Bayes decision rule with respect to G. It is shown that the class of all Bayes decision rules in the wide sense is a complete class. All the results just described were shown to hold under a set of highly technical conditions.

Wald also devoted a lot of attention to minimax decision rules. A decision rule D^* is called a minimax decision rule if $\text{Sup}_F\, r(F, D^*) \leq \text{Sup}_F r(F, D)$ for all decision rules D. Wald shows that under certain conditions there is a "least favorable a priori distribution" G_0 defined as satisfying $\text{Inf}_D r^*(G_0, D) = \text{Sup}_G \text{Inf}_D r^*(G, D)$, that there is a minimax decision rule D^*, and that D^* is a Bayes decision rule with respect to the least favorable a priori distribution G_0. The distribution G_0 is the analog of the set of probabilities $(p_1^*, \ldots, p_m^*)$ for player 1 in the description above of the Von Neumann–Morgenstern fundamental theorem, and a minimax decision rule for the Statistician is the analog of using the set of probabilities $(q_1^*, \ldots, q_n^*)$ for player 2. Thus, it is not surprising that a minimax decision rule can be extremely conservative. In his 1950 monograph, Wald states the following:

> Much attention is given to the theory of minimax solutions for two reasons: (1) a minimax solution seems, in general, to be a reasonable solution of the decision problem when an a priori distribution does not exist or is unknown to the experimenter; (2) the theory of minimax solutions plays an important role in deriving the basic results concerning complete classes of decision functions.

Having sketched the contents of Wald's 1949 paper, let's try to put its contributions into perspective by looking at what came before it and what came after it. (For this purpose, we treat the 1950 monograph as the 1949 paper itself, since the paper is a condensation of the book.) Almost all the developments leading to the 1949 paper were due to Wald himself. He combined the problems of testing hypotheses and estimation by making them both special cases of a more general problem in a 1939 paper. In this paper, Wald made no use of the theory of games and may not have been aware of the existence of the theory at the time he wrote the paper. But the idea of generalizing both estimation and testing hypotheses certainly predisposed him to utilizing the theory of games. Wald's first application of the theory of games to statistical theory was in another 1945 paper (1945c) and he wrote other papers using the theory of games before his 1949 paper. But in the introduction to the 1949 paper, Wald indicated his dissatisfaction with some of the conditions he had to impose in his earlier papers on the subject, so we may take the 1949 paper as the culmination of Wald's contributions to statistical decision theory.

The influence of Wald's work on statistical theory was enormous. Wald's death in 1950 meant that others had to generalize and apply the theory he developed. It was mentioned above that the conditions under which Wald proved his results are highly technical. Many others have undertaken the task of weakening Wald's conditions. Of course, the new and weaker conditions are also quite technical and difficult to describe in a compact manner. Johnstone (1988) refers to generalizations of Wald's results by several authors.

Many areas of statistics owe their existence and vitality, at least to some extent, to Wald's work on statistical decision theory. One such area is ranking and selection. In problems of ranking and selection, the statistician has to rank several populations according to the unknown values of their parameters or select populations with high or low values for the parameters. Such problems are neither problems of estimation nor of testing hypotheses, but fit right into Wald's formulation of the general statistical decision problem. There is an enormous literature on this subject, including the well-known monograph by Bechhofer, Kiefer, and Sobel (1968).

Another area that received a strong impetus from Wald's formulation is dynamic programming. In a dynamic programming problem, a sequence of decisions must be made over time, each decision affecting future possibilities and each decision affecting the overall loss. Wald's sequential probability ratio test is really a solution to a dynamic programming problem, one of the first.

The Bayesian approach to statistical problems is at least partly a reaction to minimax decision rules. As noted above, minimax decision rules are often very conservative. This was inevitable, considering the fact that they come from the theory of zero-sum two-person games, in which players are forced to be antagonistic. If a statistician is playing against nature, the question is whether nature is that malevolent. A Bayesian feels that he knows what a priori distribution is being used by nature. If so, the quotation from Wald's 1950 book given previously shows that Wald would have no objection to using this information.

The choice of which observations should be used to select a terminal decision was built into the Wald formulation. This certainly gave a strong impetus to the study of optimal experimental design.

More generally, the whole idea of using a general loss function and comparing decision rules to see if one is uniformly better than another permeates all of statistical theory. The whole subject of James–Stein estimators depends on this. James-Stein estimators were shown to be uniformly better than the classical maximum likelihood estimators when estimating several different normal means and using squared error loss. This led to a very extensive literature which studied similar phenomena when estimating other kinds of parameters and using other types of loss functions. At present, this is one of the most active areas of research in all of statistical theory.

Some well-known publications illustrating how thoroughly Wald's ideas have permeated all of statistical theory are Berger (1980), LeCam (1955), Lehmann (1981) and Lehmann (1986).

References*

Bechhofer, R.E., Kiefer, J.C., and Sobel, M. (1968). *Sequential Identification and Ranking Procedures*. University of Chicago Press.

Berger, J.O., (1980). *Statistical Decision Theory: Methods and Concepts*. Springer-Verlag, New York.

Johnstone, I. (1988). Wald's decision theory, in *Encyclopedia of Statistical Sciences*, Vol. 9, 518–522. Wiley, New York. Eds. S. Kotz, N.L. Johnson.

LeCam, L. (1955). An extension of Wald's theory of statistical decision functions, *Ann. Math. Statist.*, **26**, 69–81.

Lehmann, E.L. (1981). *Theory of Point Estimation*. Wiley, New York.

Lehmann, E. L. (1986). *Testing Statistical Hypotheses*, 2nd ed. Wiley, New York.

J. von Neumann and O. Morgenstern (1944). *Theory of Games and Economic Behavior*. Princeton University Press.

Wald, A. (1939). Contributions to the theory of statistical estimation and testing hypotheses, *Ann. Math. Statist.* **10**, pp. 299–326.

Wald, A. (1945a). Generalization of a theorem by von Neumann concerning zero-sum two-person games, *Ann. Math.* **46**, 281–286.

Wald, A. (1945b). Sequential tests of statistical hypotheses, *Ann. Math. Statist.*, **116**, pp. 117–186.

Wald, A. (1945c). Statistical decision functions which minimize the maximum risk, *Ann. Math.* **46**, pp. 265–280.

Wald, A. (1950). *Statistical Decision Functions* (monograph). Wiley, New York.

* As mentioned above, Wald's ideas have permeated all of statistics. This bibliography lists some publications illustrating this permeation.

Statistical Decision Functions

Abraham Wald
Columbia University[1]

Introduction and Summary

The foundations of a general theory of statistical decision functions, including
the classical non-sequential case as well as the sequential case, was discussed
by the author in a previous publication [3]. Several assumptions made in [3]
appear, however, to be unnecessarily restrictive (see conditions 1–7, pp. 297
in [3]). These assumptions, moreover, are not always fulfilled for statistical
problems in their conventional form. In this paper the main results of [3], as
well as several new results, are obtained from a considerably weaker set of
conditions which are fulfilled for most of the statistical problems treated in
the literature. It seemed necessary to abandon most of the methods of proofs
used in [3] (particularly those in section 4 of [3]) and to develop the theory
from the beginning. To make the present paper self-contained, the basic definitions already given in [3] are briefly restated in section 2.1.

In [3] it is postulated (see Condition 3, p. 207) that the space Ω of all
admissible distribution functions F is compact. In problems where the distribution function F is known except for the values of a finite number of parameters, i.e., where Ω is a parametric class of distribution functions, the compactness condition will usually not be fulfilled if no restrictions are imposed on
the possible values of the parameters. For example, if Ω is the class of all
univariate normal distributions with unit variance, Ω is not compact. It is true
that by restricting the parameter space to a bounded and closed subset of the
unrestricted space, compactness of Ω will usually be attained. Since such a
restriction of the parameter space can frequently be made in applied prob-

[1] Work done under the sponsorship of the Office of Naval Research.

lems, the condition of compactness may not be too restrictive from the point of view of practical applications. Nevertheless, it seems highly desirable from the theoretical point of view to eliminate or to weaken the condition of compactness of Ω. This is done in the present paper. The compactness condition is completely omitted in the discrete case (Theorems 2.1–2.5), and replaced by the condition of separability of Ω in the continuous case (Theorems 3.1–3.4). The latter condition is fulfilled in most of the conventional statistical problems.

Another restriction postulated in [3] (Condition 4, p. 297) is the continuity of the weight function $W(F, d)$ in F. As explained in section 2.1 of the present paper, the value of $W(F, d)$ is interpreted as the loss suffered when F happens to be the true distribution of the chance variables under consideration and the decision d is made by the statistician. While the assumption of continuity of $W(F, d)$ in F may seem reasonable from the point of view of practical application, it is rather undesirable from the theoretical point of view for the following reasons. It is of considerable theoretical interest to consider simplified weight functions $W(F, d)$ which can take only the values 0 and 1 (the value 0 corresponds to a correct decision, and the value 1 to a wrong decision). Frequently, such weight functions are necessarily discontinuous. Consider, for example, the problem of testing the hypothesis H that the mean θ of a normally distributed chance variable X with unit variance is equal to zero. Let d_1 denote the decision to accept H, and d_2 the decision to reject H. Assigning the value zero to the weight W whenever a correct decision is made, and the value 1 whenever a wrong decision is made, we have:

$$W(\theta, d_1) = 0 \quad \text{for} \quad \theta = 0, \quad \text{and} = 1 \quad \text{for} \quad \theta \neq 0;$$

$$W(\theta, d_2) = 0 \quad \text{for} \quad \theta \neq 0, \quad \text{and} = 1 \quad \text{for} \quad \theta = 0.$$

This weight function is obviously discontinuous. In the present paper the main results (Theorems 2.1–2.5 and Theorems 3.1–3.4) are obtained without making any continuity assumption regarding $W(F, d)$.

The restrictions imposed in the present paper on the cost function of experimentation are considerably weaker than those formulated in [3]. Condition 5 [3, p. 297] concerning the class Ω of admissible distribution functions, and condition 7 [3, p. 298] concerning the class of decision functions at the disposal of the statistician are omitted here altogether.

One of the new results obtained here is the establishment of the existence of so called minimax solutions under rather weak conditions (Theorems 2.3 and 3.2). This result is a simple consequence of two lemmas (Lemmas 2.4 and 3.3) which seem to be of interest in themselves.

The present paper consists of three sections. In the first section several theorems are given concerning zero sum two person games which go somewhat beyond previously published results. The results in section 1 are then applied to statistical decision functions in sections 2 and 3. Section 2 treats the case of discrete chance variables, while section 3 deals with the continuous case. The two cases have been treated separately, since the author was not able to find any simple and convenient way of combining them into a single more general theory.

1. Conditions for Strict Determinateness of a Zero Sum Two Person Game

The normalized form of a zero sum two person game may be defined as follows (see [1, section 14.1]): there are two players and there is a bounded and real valued function $K(a, b)$ of two variables a and b given where a may be any point of a space A and b may be any point of a space B. Player 1 chooses a point a in A and player 2 chooses a point b in B, each choice being made in complete ignorance of the other. Player 1 then gets the amount $K(a, b)$ and player 2 the amount $-K(a, b)$. Clearly, player 1 wishes to maximize $K(a, b)$ and player 2 wishes to minimize $K(a, b)$.

Any element a of A will be called a pure strategy of player 1, and any element b of B a pure strategy of player 2. A mixed strategy of player 1 is defined as follows: instead of choosing a particular element a of A, player 1 chooses a probability measure ξ defined over an additive class $\mathfrak{A}$ of subsets of A and the point a is then selected by a chance mechanism constructed so that for any element α of $\mathfrak{A}$ the probability that the selected element a will be contained in α is equal to $\xi(\alpha)$. Similarly, a mixed strategy of player 2 is given by a probability measure η defined over an additive class $\mathfrak{B}$ of subsets of B and the element b is selected by a chance mechanism so that for any element β of $\mathfrak{B}$ the probability that the selected element b will be contained in β is equal to $\eta(\beta)$. The expected value of the outcome $K(a, b)$ is then given by

$$K^*(\xi, \eta) = \int_B \int_A K(a, b) \, d\xi \, d\eta. \tag{1.1}$$

We can now reinterpret the value of $K(a, b)$ as the value of $K^*(\xi_a, \eta_b)$ where ξ_a and η_b are probability measures which assign probability 1 to a and b, respectively. In what follows, we shall write $K(\xi, \eta)$ for $K^*(\xi, \eta)$, $K(a, b)$ will be used synonymously with $K(\xi_a, \xi_b)$, $K(a, \eta)$ synonymously with $K(\xi_a, \eta)$ and $K(\xi, b)$ synonymously with $K(\xi, \eta_b)$. This can be done without any danger of confusion.

A game is said to be strictly determined if

$$\operatorname*{Sup}_{\xi} \operatorname*{Inf}_{\eta} K(\xi, \eta) = \operatorname*{Inf}_{\eta} \operatorname*{Sup}_{\xi} K(\xi, \eta). \tag{1.2}$$

The basic theorem proved by von Neumann [1] states that if A and B are finite the game is always strictly determined, i.e., (1.2) holds. In some previous publications (see [2] and [3]) the author has shown that (1.2) always holds if one of the spaces A and B is finite or compact in the sense of some intrinsic metric, but does not necessarily hold otherwise. A necessary and sufficient condition for the validity of (1.2) was given in [2] for spaces A and B with countably many elements. In this section we shall give sufficient conditions as well as necessary and sufficient conditions for the validity of (1.2) for arbitrary spaces A and B. These results will then be used in later sections....

2. Statistical Decision Functions: The Case of Discrete Chance Variable

2.1. The Problem of Statistical Decisions and Its Interpretation as a Zero Sum Two Person Game

In some previous publications (see, for example, [3]) the author has formulated the problem of statistical decisions as follows: Let $X = \{X^i\}$ $(i = 1, 2, \ldots,$ ad inf.) be an infinite sequence of chance variables. Any particular observation x on X is given by a sequence $x = \{x^i\}$ of real values where x^i denotes the observed value of X^i. Suppose that the probability distribution $F(x)$ of X is not known. It is, however, known that F is an element of a given class Ω of distribution functions. There is, furthermore, a space D given whose elements d represent the possible decisions that can be made in the problem under consideration. Usually each element d of D will be associated with a certain subset ω of Ω and making the decision d can be interpreted as accepting the hypothesis that the true distribution is included in the subset ω. The fundamental problem in statistics is to give a rule for making a decision, that is, a rule for selecting a particular element d of D on the basis of the observed sample point x. In other words, the problem is to construct a function $d(x)$, called decision function, which associates with each sample point x an element $d(x)$ of D so that the decision $d(x)$ is made when the sample point x is observed.

This formulation of the problem includes the sequential as well as the classical non-sequential case. For any sample point x, let $n(x)$ be the number of components of x that must be known to be able to determine the value of $d(x)$. In other words, $n(x)$ is the smallest positive integer such that $d(y) = d(x)$ for any y whose first n coordinates are equal to the first n coordinates of x. If no finite n exists with the above property, we put $n = \infty$. Clearly, $n(x)$ is the number of observations needed to reach a decision. To put in evidence the dependence of $n(x)$ on the decision rule used, we shall occasionally write $n(x; \mathfrak{D})$ instead of $n(x)$ where $\mathfrak{D}$ denotes the decision function $d(x)$ used. If $n(x)$ is constant over the whole sample space, we have the classical case, that is the case where a decision is to be made on the basis of a predetermined number of observations. If $n(x)$ is not constant over the sample space, we have the sequential case. A basic question in statistics is this: What decision function should be chosen by the statistician in any given problem? To set up principles for a proper choice of a decision function, it is necessary to express in some way the degree of importance of the various wrong decisions that can be made in the problem under consideration. This may be expressed by a non-negative function $W(F, d)$, called weight functions, which is defined for all elements F of Ω and all elements d of D. For any pair (F, d), the value $W(F, d)$ expresses the loss caused by making the decision d when F is the true distribution of X. For any positive integer n, let $c(n)$ denote the cost of making n observations. If the decision function $\mathfrak{D} = d(x)$ is used the expected loss plus the expected cost of experimentation is given by

$$r[F, \mathfrak{D}] = \int_M W[F, d(x)] \, dF(x) + \int_M c(n(x)) \, dF(x) \tag{2.1}$$

where M denotes the sample space, i.e. the totality of all sample points x. We shall use the symbol $\mathfrak{D}$ for $d(x)$ when we want to indicate that we mean the whole decision function and not merely a value of $d(x)$ corresponding to some x.

The above expression (2.1) is called the risk. Thus, the risk is a real valued non-negative function of two variables F and $\mathfrak{D}$ where F may be any element of Ω and $\mathfrak{D}$ any decision rule that may be adopted by the statistician.

Of course, the statistician would like to make the risk r as small as possible. The difficulty he faces in this connection is that r depends on two arguments F and $\mathfrak{D}$, and he can merely choose $\mathfrak{D}$ but not F. The true distribution F is chosen, we may say, by Nature and Nature's choice is usually entirely unknown to the statistician. Thus, the situation that arises here is very similar to that of a zero sum two person game. As a matter of fact, the statistical problem may be interpreted as a zero sum two person game by setting up the following correspondence:

Two Person Game	Statistical Decision Problem
Player 1	Nature
Player 2	Statistician
Pure strategy a of player 1	Choice of true distribution F by Nature
Pure strategy b of player 2	Choice of decision rule $\mathfrak{D} = d(x)$
Space A	Space Ω
Space B	Space Q of decision rules $\mathfrak{D}$ that can be used by the statistician.
Outcome $K(a, b)$	Risk $r(F, \mathfrak{D})$
Mixed strategy ξ of player 1	Probability measure ξ defined over an additive class of subsets of Ω (a priori probability distribution in the space Ω)
Mixed strategy η of player 2	Probability measure η defined over an additive class of subsets of the space Q. We shall refer to η as randomized decision function.
Outcome $K(\xi, \eta)$ when mixed strategies are used.	Risk $r(\xi, \eta) = \int_Q \int_\Omega r(F, \mathfrak{D}) \, d\xi \, d\eta$.

2.2. Formulation of Some Conditions Concerning the Spaces Ω, D, the Weight Function $W(F, d)$ and the Cost Function of Experimentation

A general theory of statistical decision functions was developed in [3] assuming the fulfillment of seven conditions listed on pp. 297–8.[4] The conditions listed there are unnecessarily restrictive and we shall replace them here by a considerably weaker set of conditions.

In this chapter we shall restrict ourselves to the study of the case where each of the chance variables X^1, X^2, ..., ad inf. is discrete. We shall say that a chance variable is discrete if it can take only countably many different values. Let a_{i1}, a_{i2}, ..., ad inf. denote the possible values of the chance variable X^i. Since it is immaterial how the values a_{ij} are labeled, there is no loss of generality in putting $a_{ij} = j(j = 1, 2, 3, \ldots,$ ad inf.$)$. Thus, we formulate the following condition.

Condition 2.1. *The chance variable X^i ($i = 1, 2, \ldots,$ ad inf.) can take only positive integral values.*

As in [3], also here we postulate the boundedness of the weight function, i.e., we formulate the following condition.

Condition 2.2 *The weight function $W(F, d)$ is a bounded function of F and d.*

To formulate condition 2.3, we shall introduce some definitions. Let ω be a given subset of Ω. The distance between two elements d_1 and d_2 of D relative to ω is defined by

$$\delta(d_1, d_2; \omega) = \operatorname*{Sup}_{F \in \omega} |W(F, d_1) - W(F, d_2)|. \tag{2.2}$$

We shall refer to $\delta(d_1, d_2; \Omega)$ as the absolute distance, or more briefly, the distance between d_1 and d_2. We shall say that a subset D^* of D is compact (conditionally compact) relative to ω, if it is compact (conditionally compact) in the sense of the metric $\delta(d_1, d_2; \omega)$. If D^* is compact relative to Ω, we shall say briefly that D^* is compact.

An element d of D is said to be uniformly better than the element d' of D relative to a subset ω of Ω if

$$W(F, d) \leqq W(F, d') \quad \text{for all } F \text{ in } \omega$$

and if

$$W(F, d) < W(F, d') \quad \text{for at least one } F \text{ in } \omega.$$

---

[4] In [3] only the continuous case is treated (existence of a density function is assumed), but all the results obtained there can be extended without difficulty to the discrete case.

A subset D^* of D is said to be complete relative to a subset ω of Ω if for any d outside D^* there exists an element d^* in D^* such that d^* is uniformly better than d relative to ω.

Condition 2.3. *For any positive integer i and for any positive ε there exists a subset $D^*_{i,\varepsilon}$ of D which is compact relative to Ω and complete relative to $\omega_{i,\varepsilon}$ where $\omega_{i,\varepsilon}$ is the class of all elements F of Ω for which $\mathrm{prob}\{X^1 \leqq i\} \geqq \varepsilon$.*

If D is compact, then it is compact with respect to any subset ω of Ω and Condition 2.3 is fulfilled. For any finite space D, Condition 2.3 is obviously fulfilled. Thus, Condition 2.3 is fulfilled, for example, for any problem of testing a statistical hypothesis H, since in that case the space D contains only two elements d_1 and d_2 where d_1 denotes the decision to reject H and d_2 the decision to accept H.

In [3] it was assumed that the cost of experimentation depends only on the number of observations made. This assumption is unnecessarily restrictive. The cost may depend also on the decision rule $\mathfrak{D}$ used. For example, let $\mathfrak{D}_1$ and $\mathfrak{D}_2$ be two decision rules such that $n(x; \mathfrak{D}_1)$ is equal to a constant n_0, while $\mathfrak{D}_2$ is such that at any stage of the experimentation where $\mathfrak{D}_2$ requires taking at least one additional observation the probability is positive that experimentation will be terminated by taking only one more observation. Let x^0 be a particular sample point for which $n(x^0; \mathfrak{D}_2) = n(x^0, \mathfrak{D}_1) = n_0$. There are undoubtedly cases where the cost of experimentation is appreciably increased by the necessity of having to look at the observations at each stage of the experiment before we can decide whether or not to continue taking additional observations. Thus in many cases the cost of experimentation when x^0 is observed may be greater for $\mathfrak{D}_2$ than for $\mathfrak{D}_1$. The cost may also depend on the actual values of the observations made. Thus, we shall assume that the cost c is a single valued function of the observations $x^1, \ldots, x^m$ and the decision rule $\mathfrak{D}$ used, i.e., $c = c(x^1, \ldots, x^m, \mathfrak{D})$.

Condition 2.4. *The cost $c(x^1, \ldots, x^m, \mathfrak{D})$ is non-negative and $\lim c(x^1, \ldots, x^m, \mathfrak{D}) = \infty$ uniformly in $x^1, \ldots, x^m, \mathfrak{D}$ as $m \to \infty$. For each positive integral value m, there exists a finite value c_m, depending only on m, such that $c(x^1, \ldots, x^m, \mathfrak{D}) \leqq c_m$ identically in $x^1, \ldots, x^m, \mathfrak{D}$. Furthermore, $c(x^1, \ldots, x^m, \mathfrak{D}_1) = c(x^1, \ldots, x^m, \mathfrak{D}_2)$ if $n(x; \mathfrak{D}_1) = n(x; \mathfrak{D}_2)$ for all x. Finally, for any sample point x we have $c(x^1, \ldots, x^{n(x, \mathfrak{D}_1)}, \mathfrak{D}_1) \leqq c(x^1, \ldots, x^{n(x, \mathfrak{D}_2)}, \mathfrak{D}_2)$ if there exists a positive integer m such that $n(x, \mathfrak{D}_1) = n(x, \mathfrak{D}_2)$ when $n(x, \mathfrak{D}_2) < m$ and $n(x, \mathfrak{D}_1) = m$ when $n(x, \mathfrak{D}_2) \geqq m$.*

2.3. Alternative Definition of a Randomized Decision Function, and a Further Condition on the Cost Function

In Section 2.1 we defined a randomized decision function as a probability measure η defined over some additive class of subsets of the space Q of all decision functions $d(x)$. Before formulating an alternative definition of a randomized decision function, we have to make precise the meaning of η by stating the additive class C_Q of subsets of Q over which η is defined. Let C_D be the smallest additive class of subsets of D which contains all subsets of D which are open in the sense of the metric $\delta(d_1, d_2; \Omega)$. For any finite set of positive integers $a_1, \ldots, a_k$ and for any element D^* of C_D, let $Q(a_1, \ldots, a_k, D^*)$ be the set of all decision functions $d(x)$ which satisfy the following two conditions: (1) If $x^1 = a_1$, $x^2 = a_2$, ..., $x^k = a_k$, then $n(x) = k$; (2) If $x^1 = a_1, \ldots, x^k = a_k$, then $d(x)$ is an element of D^*. Let C_Q^* be the class of all sets $Q(a_1, \ldots, a_k, D^*)$ corresponding to all possible values of k, $a_1, \ldots, a_k$ and all possible elements D^* of C_D. The additive class C_Q is defined as the smallest additive class containing C_Q^* as a subclass. Then with any η we can associate two sequences of functions

$$\{z_m(x^1, \ldots, x^m | \eta)\}$$

and

$$\{\delta_{x^1 \ldots x^m}(D^* | \eta)\} \, (m = 1, 2, \ldots, \text{ad inf.})$$

where $0 \le z_m(x^1, \ldots, x^m | \eta) \le 1$ and for any $x^1, \ldots, x^m$, $\delta_{x^1 \ldots x^m}$ is a probability measure in D defined over the additive class C_D. Here

$$z_m(x^1, \ldots, x^m | \eta)$$

denotes the conditional probability that $n(x) > m$ under the condition that the first m observations are equal to $x^1, \ldots, x^m$ and experimentation has not been terminated for $(x^1, \ldots, x^k)$ for $(k = 1, 2, \ldots, m-1)$, while

$$\delta_{x^1 \ldots x^m}(D^* | \eta)$$

is the conditional probability that the final decision d will be an element of D^* under the condition that the sample $(x^1, \ldots, x^m)$ is observed and $n(x) = m$. Thus

$$z_1(x^1 | \eta)z_2(x^1, x^2 | \eta)\ldots z_{m-1}(x^1, \ldots, x^{m-1} | \eta)[1 - z_m(x^1, \ldots, x^m | \eta)]$$

$$= \eta[Q(x^1, \ldots, x^m, D)] \tag{2.3}$$

and

$$\delta_{x^1 \ldots x^m}(D^* | \eta) = \frac{\eta[Q(x^1, \ldots, x^m, D^*)]}{\eta[Q(x^1, \ldots, x^m, D)]}. \tag{2.4}$$

We shall now consider two sequences of functions $\{z_m(x^1, \ldots, x^m)\}$ and $\{\delta_{x^1 \ldots x^m}(D^*)\}$, not necessarily generated by a given η. An alternative definition of a randomized decision function can be given in terms of these two sequences as follows: After the first observation x^1 has been drawn, the statistician determines whether or not experimentation be continued by a chance mechanism constructed so that the probability of continuing experimentation is equal to $z_1(x^1)$. If it is decided to terminate experimentation, the statistician uses a chance mechanism to select the final decision d constructed so that the probability distribution of the selected d is equal to $\delta_{x^1}(D^*)$. If it is decided to take a second observation and the value x^2 is obtained, again a chance mechanism is used to determine whether or not to stop experimentation such that the probability of taking a third observation is equal to $z_2(x^1, x^2)$. If it is decided to stop experimentation, a chance mechanism is used to select the final d so that the probability distribution of the selected d is equal to $\delta_{x^1 x^2}(D^*)$, and so on.

We shall denote by ζ a randomized decision function defined in terms of two sequences $\{z_m(x^1, \ldots, x^m)\}$ and $\{\delta_{x^1 \ldots x^m}(D^*)\}$, as described above. Clearly, any given η generates a particular ζ. Let $\zeta(\eta)$ denote the ζ generated by η. One can easily verify that two different η's may generate the same ζ, i.e., there exist two different η's, say η_1 and η_2 such that $\zeta(\eta_1) = \zeta(\eta_2)$.

We shall now show that for any ζ there exists a η such that $\zeta(\eta) = \zeta$. Let ζ be given by the two sequences $\{z_m(x^1, \ldots, x^m)\}$ and $\{\delta_{x^1 \ldots x^m}(D^*)\}$. Let b_j denote a sequence of r_j positive integers, i.e., $b_j = (b_{j1}, \ldots, b_{j, r_j}) (j = 1, 2, \ldots, k)$ subject to the restriction that no b_j is equal to an initial segment of $b_l (j \neq l)$. Let, furthermore, $D_1^*, \ldots, D_k^*$ be k elements of C_D. Finally, let $Q(b_1, \ldots, b_k, D_1^*, \ldots, D_k^*)$ denote the class of all decision functions $d(x)$ which satisfy the following condition: If $(x^1, \ldots, x^{r_j}) = b_j$ then $n(x) = r_j$ and $d(x)$ is an element of $D_j^* (j = 1, \ldots, k)$. Let η be a probability measure such that,

$$\eta[Q(b_1, \ldots, b_k, D_1^*, \ldots, D_k^*)]$$

$$= \delta_{b_1}(D_1^*) \ldots \delta_{b_k}(D_k^*) \prod_{m=1}^{\infty} \prod_{x^m=1}^{\infty} \prod_{x^{m-1}=1}^{\infty} \cdots \prod_{x^1=1}^{\infty}$$

$$\cdot \{z_m(x^1, \ldots, x^m)^{g_m(x^1, \ldots, x^m)} [1 - z_m(x^1, \ldots, x^m)]^{g_m^*(x^1, \ldots, x^m)}\} \quad (2.5)$$

holds for all values of k, b_1, ..., b_k, D_1^*, ..., D_k^*. Here $g_m(x^1, \ldots, x^m) = 1$ if $(x^1, \ldots, x^m)$ is equal to an initial segment of at least of one of the samples $b_1, \ldots, b_k$, but is not equal to any of the samples $b_1, \ldots, b_k$. In all other cases $g_m(x^1, \ldots, x^m) = 0$. The function $g_m^*(x^1, \ldots, x^m)$ is equal to 1 if $(x^1, \ldots, x^m)$ is equal to one of the samples $b_1, \ldots, b_k$, and zero otherwise. Clearly, for any η which satisfies (2.5) we have $\zeta(\eta) = \zeta$. The existence of such a η can be shown as follows. With any finite set of positive integers $i_1, \ldots, i_r$ we associate an elementary event, say $A_r(i_1, \ldots, i_r)$. Let $\bar{A}_r(i_1, \ldots, i_r)$ denote the negation of the event $A_r(i_1, \ldots, i_r)$. Thus, we have a denumerable system of elementary events by letting $r, i_1, \ldots, i_r$ take any positive integral values. We shall assume that the events $A_1(1), A_1(2), \ldots$, ad inf. are independent and the probability that $A_1(i)$ happens is equal $z_1(i)$. We shall now define the conditional probability of $A_2(i, j)$ knowing for any k whether $A_1(k)$ or $\bar{A}_1(k)$ happened. If $A_1(i)$ happened, the conditional probability of $A_2(i, j) = z_2(i, j)$ and 0 otherwise. The conditional probability of the joint event that $A_2(i_1, j_1), A_2(i_2, j_2) \ldots$, $A_2(i_r, j_r), \bar{A}_2(i_{r+1}, j_{r+1}), \ldots$, and $\bar{A}_2(i_{r+s}, j_{r+s})$ will happen is the product of the conditional probabilities of each of these events (knowing for each i whether $A_1(i)$ or $\bar{A}_1(i)$ happened). Similarly, the conditional probability (knowing for any i and for any (i, j), whether the corresponding event $A_2(i, j)$ happened or not) that $A_3(i_1, j_1, k_1)$ and $A_3(i_2, j_2, k_2)$ and $\ldots A_3(i_r, j_r, k_r)$ and $\bar{A}_3(i_{r+1}, j_{r+1},$

$k_{r+1})$ and ... and $\bar{A}_3(i_{r+s}, j_{r+s}, k_{r+s})$ will simultaneously happen is equal to the product of the conditional probabilities of each of them. The conditional probability of $A_3(i, j, k)$ is equal to $z_3(i, j, k)$ if $A_1(i)$ and $A_2(i, j)$ happened, and zero otherwise; and so on. Clearly, this system of probabilities is consistent.

If we interpret $A_r(i_1, \ldots, i_r)$ as the event that the decision function $\mathfrak{D} = d(x)$ selected by the statistician has the property that $n(x; \mathfrak{D}) > r$ when $x^1 = i_1, \ldots, x^r = i_r$, the above defined system of probabilities for the denumerable sequence $\{A_r(i_1, \ldots, i_r)\}$ of events implies the validity of (2.5) for $D_j^* = D(j = 1, \ldots, k)$. The consistency of the formula (2.5) for $D_j^* = D$ implies, as can easily be verified, the consistency of (2.5) also in the general case when $D_j^* \neq D$.

Let ζ_i be given by the sequences of $\{z_{mi}(x^1, \ldots, x^m)\}$ and $\{\delta_{x^1 \ldots x^m, i}\}$ $(m = 1, 2, \ldots,$ ad inf.). Let, furthermore, ζ be given by $\{z_m(x^1, \ldots, x^m)\}$ and $\{\delta_{x^1 \ldots x^m}\}$. We shall say that

$$\lim_{i=\infty} \zeta_i = \zeta \tag{2.6}$$

if for any $m, x^1, \ldots, x^m$ we have

$$\lim_{i=\infty} z_{mi}(x^1, \ldots, x^m) = z_m(x^1, \ldots, x^m) \tag{2.7}$$

and

$$\lim_{i=\infty} \delta_{x^1 \ldots x^m, i}(D^*) = \delta_{x^1 \ldots x^m}(D^*) \tag{2.8}$$

for any open subset D^* of D whose boundary has probability measure zero according to the limit probability measure $\delta_{x^1 \ldots x^m}$.

In addition to Condition 2.4, we shall impose the following continuity condition on the cost function.

Condition 2.5. *If*

$$\lim_{i=\infty} \zeta(\eta_i) = \zeta(\eta),$$

then

$$\lim_{i=\infty} \int_{Q(x^1, \ldots, x^m)} c(x^1, \ldots, x^m, \mathfrak{D}) \, d\eta_i = \int_{Q(x^1, \ldots, x^m)} c(x^1, \ldots, x^m, \mathfrak{D}) \, d\eta.$$

where $Q(x^1, \ldots, x^m)$ is the class of all decision functions $\mathfrak{D}$ for which $n(y, \mathfrak{D}) = m$ if $y^1 = x^1, \ldots, y^m = x^m$.

2.4. The Main Theorem

In this section we shall show that the statistical decision problem, viewed as a zero sum two person game, is strictly determined. It will be shown in subsequent sections that this basic theorem has many important consequences for the theory of statistical decision functions. A precise formulation of the theorem is as follows:

Theorem 2.1. *If Conditions 2.1–2.5 are fulfilled, the decision problem, viewed as a zero sum two person game, is strictly determined, i.e.,*

$$\text{Sup Inf } r(\xi, \eta) = \text{Inf Sup } r(\xi, \eta).\dots \tag{2.9}$$
$$\quad\; \xi \;\;\; \eta \qquad\qquad \eta \;\;\; \xi$$

2.5. Theorems on Complete Classes of Decision Functions and Minimax Solutions

For any positive ε we shall say that the randomized decision function η_0 is an ε-Bayes solution relative to the a priori distribution ξ if

$$r(\xi, \eta_0) \leq \text{Inf } r(\xi, \eta) + \varepsilon. \tag{2.37}$$
$$\qquad\qquad \eta$$

If η_0 satisfies (2.37) for $\varepsilon = 0$, we shall say that η_0 is a Bayes solution relative to ξ.

A randomized decision rule η_1 is said to be uniformly better than η_2 if

$$r(F, \eta_1) \leq r(F, \eta_2) \quad \text{for all } F \tag{2.38}$$

and if

$$r(F, \eta_1) < r(F, \eta_2) \quad \text{at least for one } F. \tag{2.39}$$

A class C of randomized decision functions η is said to be complete if for any η not in C we can find an element η^* in C such that η^* is uniformly better than η.

Theorem 2.2. *If Conditions 2.1–2.5 are fulfilled, then for any $\varepsilon > 0$ the class C_ε of all ε-Bayes solutions corresponding to all possible a priori distributions ξ is a complete class.…*

Theorem 2.3. *If D is compact, and if Conditions 2.1, 2.2, 2.4, 2.5 are fulfilled, then there exists a minimax solution, i.e., a decision rule η_0 for which*

$$\text{Sup } r(F, \eta_0) \leq \text{Sup } r(F, \eta) \quad \text{for all } \eta.\dots \tag{2.59}$$
$$\; F \qquad\qquad\quad F$$

Theorem 2.4. *If D is compact and if Conditions 2.1, 2.2, 2.4, 2.5 are fulfilled, then for any ξ there exists a Bayes solution relative to ξ.…*

We shall say that η_0 is a Bayes solution in the wide sense, if there exists a sequence $\{\xi_i\}$ ($i = 1, 2, \dots$, ad inf.) such that

$$\lim_{i=\infty} \left[r(\xi_i, \eta_0) - \text{Inf } r(\xi_i, \eta) \right] = 0. \tag{2.70}$$
$$\qquad\qquad\qquad\qquad \eta$$

We shall say that η_0 is a Bayes solution in the strict sense, if there exists a ξ such that η_0 is a Bayes solution relative to ξ.

Theorem 2.5. *If D is compact and Conditions 2.1–2.5 hold, then the class of all Bayes solutions in the wide sense is a complete class....*

We shall now formulate an additional condition which will permit the derivation of some stronger theorems. First, we shall give a convergence definition in the space Ω. We shall say that F_i converges to F in the ordinary sense if

$$\lim_{i=\infty} p_r(x^1, \ldots, x^r|F_i) = p_r(x^1, \ldots, x^r|F) \quad (r = 1, 2, \ldots, \text{ad inf.}). \quad (2.76)$$

Here $p_r(x^1, \ldots, x^r|F)$ denotes the probability, under F, that the first r observations will be equal to $x^1, \ldots, x^r$, respectively. We shall say that a subset ω of Ω is compact in the ordinary sense, if ω is compact in the sense of the convergence definition (2.76).

Condition 2.6. *The space Ω is compact in the ordinary sense. If F_i converges to F, as $i \to \infty$, in the ordinary sense, then*

$$\lim_{i=\infty} W(F_i, d) = W(F, d)$$

uniformly in d.

Theorem 2.6. *If D is compact and if Conditions 2.1, 2.2, 2.4, 2.5, 2.6 hold, then:*

(i) *there exists a least favorable a priori distribution, i.e., an a priori distribution ξ_0 for which*

$$\text{Inf}_{\eta} r(\xi_0, \eta) = \text{Sup}_{\xi} \text{Inf}_{\eta} r(\xi, \eta).$$

(ii) *A minimax solution exists and any minimax solution is a Bayes solution in the strict sense.*

(iii) *If η_0 is a decision rule which is not a Bayes solution in the strict sense and for which $r(F, \eta_0)$ is a bounded function of F, then there exists a decision rule η_1 which is a Bayes solution in the strict sense and is uniformly better than η_0....*

We shall now replace Condition 2.6 by the following weaker one.

Condition 2.6*. *There exists a sequence $\{\Omega_i\}$ $(i = 1, 2, \ldots, \text{ad inf.})$ of subsets of Ω such that Condition 2.6 is fulfilled when Ω is replaced by Ω_i, $\Omega_{i+1} \supset \Omega_i$ and $\lim_{i=\infty} \Omega_i = \Omega$.*

We shall say that η_i converges weakly to η as $i \to \infty$, if $\lim_{i=\infty} \zeta(\eta_i) = \zeta(\eta)$. We shall also say that η is a weak limit of η_i. This limit definition seems to be natural, since $r(\xi, \eta_1) = r(\xi, \eta_2)$ if $\zeta(\eta_2) = \zeta(\eta_1)$. We shall now prove the following theorem:

Theorem 2.7. *If D is compact and if Conditions 2.1, 2.2, 2.4, 2.5 and 2.6* are fulfilled, then:*

(i) *A minimax solution exists that is a weak limit of a sequence of Bayes solutions in the strict sense.*
(ii) *Let η_0 be a decision rule for which $r(F, \eta_0)$ is a bounded function of F. Then there exists a decision rule η_1 that is a weak limit of a sequence of Bayes solutions in the strict sense and such that $r(F, \eta_1) \leqq r(F, \eta_0)$ for all F in Ω. . . .*

3. Statistical Decision Functions: The Case of Continuous Chance Variables

3.1. Introductory Remarks

In this section we shall be concerned with the case where the probability distribution F of X is absolutely continuous, i.e., for any element F of Ω and for any positive integer r there exists a joint density function $p_r(x^1, \ldots, x^r | F)$ of the first r chance variables $X^1, \ldots, X^r$.

The continuous case can immediately be reduced to the discrete case discussed in section 2 if the observations are not given exactly but only up to a finite number of decimal places. More precisely, we mean this: For each i, let the real axis R be subdivided into a denumerable number of disjoint sets R_{i1}, $R_{i2}, \ldots$, ad inf. Suppose that the observed value x^i of X^i is not given exactly; it is merely known which element of the sequence $\{R_{ij}\}$ ($j = 1, 2, \ldots$, ad inf.) contains x^i. This is the situation, for example, if the value of x^i is given merely up to a finite number, say r, decimal places (r fixed, independent of i). This case can be reduced to the previously discussed discrete case, since we can regard the sets R_{ij} as our points, i.e., we can replace the chance variable X^i by Y^i where Y^i can take only the values $R_{i1}, R_{i2}, \ldots$, ad inf. (Y^i takes the value R_{ij} if X^i falls in R_{ij}). If $W(F_1, d) = W(F_2, d)$ whenever the distribution of Y under F_1 is identical with that under F_2, only the chance variables $Y^1, Y^2, \ldots$, etc. play a role in the decision problem and we have the discrete case. If the latter condition on the weight function is not fulfilled, i.e., if there exists a pair (F_1, F_2) such that $W(F_1, d) \neq W(F_2, d)$ for some d and the distribution of Y is the same under F_1 as under F_2, we can still reduce the problem to the discrete case, if in the discrete case we permit the weight W to depend also on a third extraneous variable G, i.e., if we put $W = W(F, G, d)$, where G is a variable about whose value the sample does not give any information. The results obtained in the discrete case can easily be generalized to include the situation where $W = W(F, G, d)$.

In practical applications the observed value x^i of X^i will usually be given only up to a certain number of decimal places and, thus, the problem can be reduced to the discrete case. Nevertheless, it seems desirable from the theoretical point of view to develop the theory of the continuous case, assuming that the observed value x^i of X^i is given precisely.

In section 2.3 an alternative definition of a randomized decision rule was given in terms of two sequences of functions $\{z_r(x^1, \ldots, x^r)\}$ and $\{\delta_{x^1 \ldots x^r}\}$ $(r = 1, 2, \ldots, \text{ad. inf.})$. We used the symbol ζ to denote a randomized decision rule given by two such sequences. It was shown in the discrete case that the use of a randomized decision function η generates a certain $\zeta = \zeta(\eta)$, and that for any given ζ there exists a η such that $\zeta = \zeta(\eta)$. Furthermore, because of Condition 2.5, in the discrete case we had $r(F, \eta_1) = r(F, \eta_2)$ if $\zeta(\eta_1) = \zeta(\eta_2)$. It would be possible to develop a similar theory as to the relation between ζ and η also in the continuous case. However, a somewhat different procedure will be followed for the sake of simplicity. Instead of the decision functions $d(x)$, we shall regard the ζ's as the pure strategies of the statistician, i.e., we replace the space Q of all decision functions $d(x)$ by the space Z of all randomized decisions rules ζ. It will then be necessary to consider probability measures η defined over an additive class of subsets of Z. It will be sufficient, as will be seen later, to consider only discrete probability measures η. A probability measure η is said to be discrete, if it assigns the probability 1 to some denumerable subset of Z. Any discrete η will clearly generate a certain $\zeta = \zeta(\eta)$. In the next section we shall formulate some conditions which will imply that $r(F, \eta_1) = r(F, \eta_2)$ if $\zeta(\eta_1) = \zeta(\eta_2)$. Thus, it will be possible to restrict ourselves to consideration of pure strategies ζ which will cause considerable simplifications.

The definitions of various notions given in the discrete case, such as minimax solution, Bayes solution, a priori distribution ξ in Ω, least favorable a priori distribution, complete class of decision functions, etc. can immediately be extended to the continuous case and will, therefore, not be restated here.

3.2 Conditions on Ω, D, $W(F, d)$ and the Cost Function

In this section we shall formulate conditions similar to those given in the discrete case.

Condition 3.1. *Each element F of Ω is absolutely continuous.*

Condition 3.2. *$W(F, d)$ is a bounded function of F and d.*

Condition 3.3. *The space D is compact in the sense of its intrinsic metric $\delta(d_1, d_2; \Omega)$ (see equation 2.2).*

This condition is somewhat stronger than the corresponding Condition 2.3. While it may be possible to weaken this condition, it would make the proofs of certain theorems considerably more involved.

Condition 3.4. *The cost of experimentation* $c(x^1, \ldots, x^m)$ *does not depend on* ζ. *It is non-negative and* $\lim\limits_{m=\infty} c(x^1, \ldots, x^m) = \infty$ *uniformly in* $x^1, \ldots, x^m$. *For each positive integral value* m, $c(x^1, \ldots, x^m)$ *is a bounded function of* $x^1, \ldots, x^m$.

This condition is stronger than Conditions 2.4 and 2.5 postulated in the discrete case. The reason for formulating a stronger condition here is that we wish the relation $r(F, \eta_1) = r(F, \eta_2)$ to be fulfilled whenever $\zeta(\eta_1) = \zeta(\eta_2)$ which will make it possible for us to eliminate the consideration of η's altogether. Since the ζ's are regarded here as the pure strategies of the statistician, it is not clear what kind of dependence of the cost on ζ would be consistent with the requirement that $r(F, \eta_1) = r(F, \eta_2)$ whenever $\zeta(\eta_1) = \zeta(\eta_2)$.

We shall say that $F_i \to F$ in the ordinary sense, if for any positive integral value m

$$\lim_{i=\infty} \int_{S_m} p_m(x^1, \ldots, x^m | F_i) \, dx^1 \ldots dx^m = \int_{S_m} p_m(x^1, \ldots, x^m | F) \, dx^1 \ldots dx^m$$

uniformly in S_m where S_m is a subset of the m-dimensional sample space.

Condition 3.5. *The space* Ω *is separable in the sense of the above convergence definition.*[6]

No such condition was formulated in the discrete case for the simple reason that in the discrete case Ω is always separable in the sense of the convergence definition given in (2.76).

(*Editors' note*: Section 3.3 has been omitted.)

3.4. Equality of Sup Inf r and Inf Sup r, and Other Theorems

In this section we shall prove the main theorems for the continuous case, using the lemmas derived in the preceding section.

Theorem 3.1. *If Conditions 3.1–3.5 are fulfilled, then*

$$\text{Sup}_{\xi} \text{Inf}_{\zeta} r(\xi, \zeta) = \text{Inf}_{\zeta} \text{Sup}_{\xi} r(\xi, \zeta). \ldots \tag{3.41}$$

(*Editors' note*: Proof has been omitted.)

Theorem 3.2. *If Conditions 3.1–3.5 are fulfilled, then there exists a minimax solution, i.e., a decision rule* ζ_0 *for which*

$$\text{Sup}_{F} r(F, \zeta_0) \leq \text{Sup}_{F} r(F, \zeta) \quad \text{for all } \zeta. \ldots \tag{3.64}$$

(*Editors' note*: Proof has been omitted.)

---

[6] For a definition of a separable space, see F. Hausdorff, *Mengenlehre* (3rd edition), p. 125.

Editors' note: See Addition at Proof Reading (Pg. 357 of this volume.)

Theorem 3.3. *If Conditions 3.1–3.5 are fulfilled, then for any ξ there exists a Bayes solution relative to ξ....*

Theorem 3.4. *If Conditions 3.1–3.5 are fulfilled, then the class of all Bayes solutions in the wide sense is a complete class....*

(*Editors' note*: Section 3.5 has been omitted.)

Addition at Proof Reading

After this paper was sent to the printer the author found that Ω is always separable (in the sense of the convergence definition in Condition 3.5) and, therefore, Condition 3.5 is unnecessary. A proof of the separability of Ω will appear in a forthcoming publication of the author....

Although not stated explicitly, several functions considered in this paper are assumed to be measurable with respect to certain additive classes of subsets. In the continuous case, for example, the precise measurability assumptions may be stated as follows: Let B be the class of all Borel subsets of the infinite dimensional sample space M. Let H be the smallest additive class of subsets of Ω which contains any subset of Ω which is open in the sense of at least one of the convergence definitions considered in this paper. Let T be the smallest additive class of subsets of D which contains all open subsets of D (in the sense of the metric $\delta(d_1, d_2, \Omega)$). By the symbolic product $H \times T$ we mean the smallest additive class of subsets of the Cartesian product $\Omega \times D$ which contains the Cartesian product of any member of H by any member of T. The symbolic product $H \times B$ is similarly defined. It is assumed that: (1) $W(F, d)$ is measurable $(H \times T)$; (2) $p_m(x^1, \ldots, x^m|F)$ is measurable $(B \times H)$; (3) $\delta_{x^1 \ldots x^r}(D^*)$ is measurable (B) for any member D^* of T; (4) $z_r(x^1, \ldots, x^r)$ and $c_r(x^1, \ldots, x^r)$ are measurable (B). These assumptions are sufficient to insure the measurability (H) of $r(F, \zeta)$ for any ζ.

References

[1] J. v. Neumann and Oskar Morganstern, *Theory of Games and Economic Behavior*, Princeton University Press, 1944.
[2] A. Wald, "Generalization of a theorem by v. Neumann concerning zero sum two-person games," *Annals of Mathematics*, Vol. 46 (April, 1945).
[3] A. Wald, "Foundations of a general theory of sequential decision functions," *Econometrica*, Vol. 15 (October, 1947).

Introduction to
Good (1952) Rational Decisions

D.V. Lindley

British statistics in 1951 was dominated by the ideas of Fisher and Neyman–Pearson. The decision-making ideas of Wald (1950), by then becoming popular in the United States, were just being noted by theoreticians but never became a significant force. Savage (1954) had yet to appear and even the name of de Finetti was virtually unknown. The work of Jeffreys (1939) lay largely unread. In September of that year, the Royal Statistical Society held a conference in Cambridge. An account of it is given in *J. Roy. Statist. Soc., Ser. A*, **115**, 568. At the time, I was a demonstrator, a title suggesting, but not actually implying, practicality, at the university, attended the meetings, and even gave a paper. Regrettably, I have no memory of them, only of the Masonic Hall in which they were held. Into this environment came I.J. Good with this extraordinary paper that is totally outside the Fisherian mold and a precursor of subsequent, Bayesian developments, The Cambridge audience could have had a foretaste in Good's (1950) book, but the paper is devoted to rational *action*, whereas the book was more concerned with rational *thinking*. I regrettably found the book's heterodoxy unimpressive and was presumably equally unappreciative of the talk. How wrong that attitude was! Good, extremely succinctly, describes an all-embracing philosophy of statistics and mentions several applications. Important ideas are scattered throughout and often only receive brief mention despite their importance. Many of them are original and even those that are not are treated in a highly original way that shows them in a new light.

Good was born in 1916 and from 1935–41 was at Cambridge University, reading mathematics as an undergraduate and then obtaining a doctorate under the leading, British, pure mathematician of the day, G.H. Hardy. There then followed four miraculous years at Bletchley Park working with other brilliant people, breaking the German codes, designing one of the first elec-

tronic computers, and engaging in philosophical speculations; see Good (1979).
This ability to embrace extremes of learning, from practice to theory, has
never left him. Although best known for his work on the theory of statistics,
papers like that on contingency tables (Crook and Good, 1980) show that
Good can firmly keep his feet on the ground. After a brief sojourn at Manchester University from 1945–48, he returned to confidential, government
work until his appointment at Virginia Polytechnic Institute and State University in 1967, where he is today, having recently officially retired. He has
been a prolific writer and the latest bibliography I have lists over 1800 publications. Good (1983) is a book containing some of his more important papers,
including this one. Lindley (1990) provides a short introduction to his work,
and the same issue of that journal includes papers given at a conference to
honor his retirement.

The paper is concerned with decision-making in the face of uncertainty and
how this can be done in a rational way. (Note the succinct title.) Rational
essentially means the avoidance of inconsistencies. It is argued that consistency is important in any scientific theory. (We would now replace "consistent"
by "coherent" to avoid confusion with the use of the former in terms like
"consistent estimator.") The basic idea behind consistency is the comparison
of procedures, or of the same procedure in different situations, to see if they
fit together, or cohere. A simple example is to compare performances at different sample sizes. It turns out that the usual interpretation of "significant at
5%" at one sample size is not consistent with that at another. Statisticians
have rarely been bothered by consistency, being content to look at situations
in isolation. Yet almost all currently-popular, statistical procedures are inconsistent. One recent exposition is by Basu (1988).

The main tool to achieve consistency is probability. To be consistent in
your uncertainty statements, those statements must combine according to the
rules of probability. This had first been demonstrated by Ramsey (1931) in a
paper unappreciated in 1951. Another account had been given by Jeffreys
(1939) and Good (1950) had produced an original and lucid presentation. The
axioms of subjective probability are set out in Good's book and there is only
brief mention of them in the paper (Sec. 5). Most statisticians deny a complete
role for probability in the description of uncertainty. It is always used for data
uncertainty, yet its use is typically denied for hypotheses or parameters. Thus,
the concept of a significance level or a confidence coefficient, although probability statements, are not about the hypotheses nor the parameters, but about
the data. It is this failure to use probability for hypotheses and parameters
that distinguishes the frequentist, sampling-theory approach to statistics from
the fully-coherent, Bayesian viewpoint. There have recently been attempts to
argue that the rules of combination of uncertainty statements should not necessarily be those of probability. A popular, alternative view is that of Shafer
(1976), with belief functions and Dempster's rule of combination. Another
system is based on fuzzy logic and is due to Zadeh with a journal, *Fuzzy Sets
and Systems*, devoted to it. But the role of probability is increasingly appreci-

ated, especially outside statistics, for example, in expert systems. The rules of Dempster and Zadeh are essentially arbitrary, whereas probability has an inevitability about it. [The best exposition is perhaps to be found in Chap. 6 of DeGroot (1970).]

The second tool needed to obtain consistency is utility. Curiously, this is not defined in the paper despite the fact that it is the introduction of the utility that distinguishes action, or decision-making, from thinking, or inference, the topic of Good's earlier book. A few words of explanation may not be out of place here [Chapter 7 of DeGroot (1970) is a good reference.] Utility is a measure of the worth of an outcome. If you take this action and that happens, then utility measures the value, or utility, to you of what happens. It is a measure of worth on a probability scale. If you take two outcomes as standards with utilities one and zero, an outcome has utility p if you are indifferent between that outcome for sure on the one hand, and a gamble with chance p of utility one (and $1 - p$ of utility zero) on the other. Similar gamble substitutes are available for outcomes outside the unit interval for utility. Good's paper is one of the first to include a discussion of utility in a statistical context, though the idea has a long history. It had recently come to prominence through the writings of von Neumann and Morgenstern (1947). It is related to Wald's concept of loss (see below). Notice that once probability has been introduced, utility is a simple consequence. An alternative approach constructs utility, from which probability easily follows. This was used by Ramsey, but Good was surely right to put probability first, and it is this order that has proved the more enduring.

The third feature needed to obtain a rational system of decision making is the combination of probability with utility in the evaluation of the expected utility of a decision, the expectation being with respect to the probability distribution. That decision is then selected that has maximum expected utility, MEU. I think Good's paper is the first occasion in which MEU is advocated as the solution to all statistical problems. Nowadays, its adoption is usually referred to as the Bayesian approach because of the central role played by Bayes' theorem. It is instructive to compare the paper with Savage's later book (1954). Savage dots all the i's and crosses every t in deriving MEU. Good slides over the surface with elegance and simplicity. Savage is the continental style of mathematician: Good follows the style of British, applied mathematics. Both have their merits. Detail is necessary for accuracy: simplicity is helpful for comprehension. There is another difference. Savage was initially sidetracked into attempting to justify the statistical practice of the 1950s by MEU, though he later came to admit his error. Good never fell into that trap, realizing immediately the very real distinction, yet seeing that some classical procedures could have a rational interpretation, for example, Good (1988). Another fine example of the breadth of Good's thinking is that he does not merely discuss the basic axioms (of probability and utility), but reinforces them with *rules* of application of the theory and *suggestions* referring to the techniques. He tries to relate the abstract model to the reality it is intended to examine. No theory, he argues, can be complete until it adds these two features.

There is a connection between MEU and the work of Wald (1950) that is discussed in the final section of the paper. Wald used the concept of loss, which is usually held to be negative utility, though Wald was vague on the point. His probabilities for the uncertain events are merely tools to obtain admissible solutions. Expected loss replaces expected utility and minimax replaces pure maximization. Good provides a masterly summary of Wald's complicated material in a few, lucid paragraphs. More important, Good has penetrating insight into the material that is totally lacking in Wald. For example, it is pointed out that minimax solutions are only acceptable if the least favorable distribution is in reasonable accord with your body of beliefs. This is surely correct and American statisticians would have been saved a lot of fruitless work had they recognized this earlier.

The bulk of the paper forms a commentary on MEU with descriptions of devices that make it easier to use. Two of the latter have proved to be important. These are scoring rules ("fair fees" in Sec. 8) and hierarchical probabilities (types 2 and 3 in the final paragraphs). A scoring rule is a device for getting people to state their probabilities in a reasonable manner. Suppose you assess your probability of an event to be p. Then the logarithmic scoring rule proposed by Good will score you $k \log(2p)$ if the event occurs and $k \log(2 - 2p)$ if not, where k is a suitable constant. Thus, if the event occurs and you had p near one, you will get a small score, but a p near zero will result in a large penalty. The later work of Savage (1971) on the topic was especially fruitful and de Finetti (1974, 1975) used a quadratic rule, with $(1 - p)^2$ and p^2, as one basis for his development. He showed that a Bayesian would always beat a non-Bayesian by getting a lower penalty score and that in finite time. Good shows his early appreciation of the practical value of scoring rules by suggesting their use in meteorology, a suggestion that has been adopted in the United States. I have long cherished the idea of scoring statisticians with their incoherent, tail-area probabilities. How many 1 in 20 errors actually occurred? It would be good to know.

Hierarchical probabilities have been even more successful. One of the practical difficulties in using the Bayesian paradigm is the assessment of probabilities. One often feels unsure of a probability distribution but can limit it to a class. Good's idea was to put a probability (of type 2) over the class. The notion can be repeated with probabilities of type 3, thereby building up a hierarchy. In its simplest form, probability for the data (type 1) can be described in terms of an unknown parameter, which can be given a distribution (type 2), or a class of distributions indexed by a hyperparameter, which is given a distribution of type 3. This idea occurs in Empirical Bayes methods [see Maritz and Lwin (1989)] and was exploited by Lindley and Smith (1972) in connection with the linear model to give a coherent approach to this popular model, replacing least squares that Stein (1956) had shown to be inadmissible.

The form of probability that Good espouses, a probability of belief, is subjective, dependent on the subject's beliefs. The pretense that science is objective is discussed in Sec. 3 (xi). Good interestingly distinguishes between probability and credibility. The latter is a logical consequence of the uncertain event A and the circumstances B under which it is being considered. Its value is shared by all people who are reasoning logically and is, to that extent, objective. The notion of credibility is basic to Jeffreys' (1939) approach to inference and, especially in later editions, he devoted much effort to the evaluation of the credibility of A when B contains little or no information, from which credibilities for general B can be found by Bayes' theorem. These, and other, attempts have not been entirely successful and Good's view that they do not exist has found increasing favor and is basic to de Finetti's (1974, 1975) approach. Rationality only applies to relationships between beliefs and not to their actual values. This allows for differences in beliefs between people, differences that will be reduced by experience.

There is one topic, or attitude of mind, that permeates the whole paper and, 40 years later, seems to me the most important aspect of the paper. It is encapsulated in the fourth sentence: "My main contention is that our methods of making rational decisions should *not depend on whether we are statisticians*" (emphasis added). What I think and hope Good meant by this is that the rational methods based on MEU are of general use and not confined to what we usually, and narrowly, think of as statistics. Here is an approach that is valid for *all* decision-making made by a single decision-maker, whether individual or corporate. Or, turning it around, if statisticians were to adopt MEU, then the role of the statistician in society would be greatly increased. We should not stop at our likelihood ratios or interval estimates but go on to consider utilities and decisions. As two illustrations of this, Good goes on to consider topics that in 1951 were unusual at statistical conferences: ethics and law. The latter is more common today as statistical ideas find increasing acceptance in legal situations, but ethics remains largely untouched.

The ethical point pondered is the relationship between public and private concepts of probability and utility. The specific case mentioned is that of a scientific adviser to a firm where, for example, the values to the firm and to the individual may differ. The difference could lead to disagreement over the best action. Good's resolution is for the adviser to estimate probabilities and utilities, for he has the knowledge and skills to do this, but that the firm should make the decision. "Leaders of industry should become more probability-conscious." There has been some recent work on how one person (the firm) can use another's (the adviser's) probabilities; for an example, see Genest and Zidek (1986), Good goes on to point out that statisticians would have to abandon their Neyman–Pearson significance tests and provide instead probabilities of hypotheses.

This is a brilliant paper, rich with novel ideas, that repays reading for its succinct statements, and because it does not elaborate on them, the linkages between the ideas become more apparent. It now reminds me of a duchess wearing her jewels; the many stones are brilliant but the wearer treats them as commonplace. Because the stones are brilliant, the reader will have to read slowly and ponder every sentence. But who minds if the rewards are so rich? Good (1983, p. xii) has provided his own commentary on the paper. Readers interested in subsequent developments can do no better than to read that book.

References

Basu, D. (1988). *Statistical Information and Likelihood*. Springer-Verlag, New York.

Crook, J.F., and Good, I.J. (1980). On the application of symmetric Dirichlet distributions and their mixtures to contingency tables, Part II, *Ann. Statist.*, **8**, 1198–1218 (also *ibid.*, **9**, 1133).

de Finetti, B. (1974–1975). *Theory of Probability*, 2 vols. Wiley, London.

DeGroot, M.H. (1970). *Optimal Statistical Decisions*. McGraw-Hill, New York.

Genest, C., and Zidek, J.V. (1986). Combining probability distributions: a critique and an annotated bibliography, *Statist. Sci.*, **1**, 114–148 (with discussion).

Good, I.J. (1950). *Probability and the Weighing of Evidence*. Griffin, London.

Good, I.J. (1979). Early work on computers at Bletchley, *Cryptologia*, **3**, 67–77.

Good, I.J. (1983). *Good Thinking*. University of Minnesota Press, Minneapolis.

Good, I.J. (1988). Statistical evidence, in *Encyclopedia of Statistical Sciences*, vol. **8** (S. Kotz, N.L. Johnson, and C.B. Read, eds.). Wiley, New York, pp. 651–656.

Jeffreys, H. (1939). *Theory of Probability*. Clarendon Press, Oxford.

Lindley, D.V. (1990). Good's work in probability, statistics and the philosophy of science, *J. Statist. Plann. Inf.*, **25**, 211–223.

Lindley, D.V., and Smith, A.F.M. (1972). Bayes estimates for the linear model, *J. Roy. Statist. Soc., Ser. B.*, **34**, 1–41 (with discussion).

Maritz, J.S., and Lwin, T. (1989). *Empirical Bayes Methods*. Chapman & Hall, London.

Von Neumann, J., and Morgenstern, O. (1947). *Theory of Games and Economic Behavior*. Princeton University Press.

Ramsey, F.P. (1931). *The Foundations of Mathematics and other Logical Essays*. Kegan, Paul, Trench, Trubner & Co., London, pp. 156–198.

Savage, L.J. (1954). *The Foundations of Statistics*. Wiley, New York.

Savage, L.J. (1971). Elicitation of personal probabilities and expectations, *J. Amer. Statist. Assoc.*, **66**, 783–801.

Shafer, G. (1976). *A Mathematical Theory of Evidence*. Princeton University Press.

Stein, C.M. (1956). Inadmissibility of the usual estimator for the mean of a multivariate normal distribution, in *Proceedings of 3rd Berkeley Symposium on Mathematical Statistics and Probability*, Vol. **1**. University of California Press, Berkeley, pp. 197–206.

Wald, A. (1950). *Statistical Decision Functions*. Wiley, New York.

Rational Decisions*

I.J. Good

Summary. This paper deals first with the relationship between the theory of probability and the theory of rational behaviour. A method is then suggested for encouraging people to make accurate probability estimates, a connection with the theory of information being mentioned. Finally Wald's theory of statistical decision functions is summarised and generalised and its relation to the theory of rational behaviour is discussed.

1. Introduction

I am going to discuss the following problem. Given various circumstances, to decide what to do. What universal rule or rules can be laid down for making rational decisions? My main contention is that our methods of making rational decisions should not depend on whether we are statisticians. This contention is a consequence of a belief that consistency is important. A few people think there is a danger that over-emphasis of consistency may retard the progress of science. Personally I do not think this danger is serious. The resolution of inconsistencies will always be an essential method in science and in cross-examinations. There may be occasions when it is best to behave irrationally, but whether there are should be decided rationally.

It is worth looking for unity in the methods of statistics, science and rational thought and behaviour; first in order to encourage a scientific approach

* This paper is based on a lecture delivered to the Royal Statistical Society on September 22nd, 1951, as a contribution to the week-end conference at Cambridge.

to non-scientific matters, second to suggest new statistical ideas by analogy with ordinary ideas, and third because the unity is aesthetically pleasing.

Clearly I am sticking my neck out in discussing this subject. In most subjects people usually try to understand what other people mean, but in philosophy and near-philosophy they do not usually try so hard.

2. Scientific Theories

In my opinion no scientific theory is really satisfactory until it has the following form:

(i) There should be a very precise set of axioms from which a purely abstract theory can be rigorously deduced. In this abstract theory some of the words or symbols may remain undefined. For example, in projective geometry it is not necessary to know what points, lines and planes are in order to check the correctness of the theorems in terms of the axioms.

(ii) There should be precise rules of application of the abstract theory which give meaning to the undefined words and symbols.

(iii) There should be suggestions for using the theory, these suggestions belonging to the technique rather than to the theory. The suggestions will not usually be as precisely formulated as the axioms and rules.

The adequacy of the abstract theory cannot be judged until the rules of application have been formulated. These rules contain indications of what the undefined words and symbols of the abstract theory are all about, but the indications will not be complete. It is the theory as a whole (i.e., the axioms and rules combined) which gives meaning to the undefined words and symbols. It is mainly for this reason that a beginner finds difficulty in understanding a scientific theory.

It follows from this account that a scientific theory represents a decision and a recommendation to use language and symbolism in a particular way (and possibly also to think and act in a particular way). Consider, for example, the principle of conservation of energy (or energy and matter). Apparent exceptions to the principle have been patched up by extending the idea of energy, to potential energy, for example. Nevertheless the principle is not entirely tautological.

Some theoreticians formulate theories without specifying the rules of application, so that the theories cannot be understood at all without a lot of experience. Such formulations are philosophically unsatisfactory.

Ordinary elementary logic can be regarded as a scientific theory. The recommendations of elementary logic are so widely accepted and familiar, and have had so much influence on the educated use of language, that logic is often regarded as self-evident and independent of experience. In the empirical sciences the selection of the theories depends much more on experience. The theory of probability occupies an intermediate position between logic and empirical sciences. Some people regard any typical theory of probability as self-evident, and others say it depends on experience. The fact is that, as in many philosophical disputes, it is a question of degree; the theory of probability does depend on experience, but does not require much more experience than does ordinary logic. There are a number of different methods of making the theory seem nearly tautological by more or less *a priori* arguments. The two main methods are those of "equally probable cases" and of limiting frequencies. Both methods depend on idealizations, but it would be extremely surprising if either method could be proved to lead to inconsistencies. When actually estimating probabilities, most of us use both methods. It may be possible in principle to trace back all probability estimates to individual experiences of frequencies, but this has not yet been done. Two examples in which beliefs do not depend in an obvious way on frequencies are (i) the estimation of the probability that a particular card will be drawn from a well-shuffled pack of 117 cards; (ii) the belief which newly-born piglings appear to have that it is a good thing to walk round the mother-pig's leg in order to arrive at the nipples. (This example is given for the benefit of those who interpret a belief as a tendency to act.)

3. Degrees of Belief

I shall now make twelve remarks about degrees of belief.

(i) I *define* the theory of probability as the logic of degrees of belief. Therefore degrees of belief, either subjective or objective, must be introduced. Degrees of belief are assumed (following Keynes) to be partially ordered only, i.e., some pairs of beliefs may not be comparable.

(ii) F.Y. Edgeworth, Bertrand Russell and others use the word "credibilities" to mean objective rational degrees of belief. A credibility has a definite but possibly unknown value. It may be regarded as existing independently of human beings.

(iii) A subjective theory of probability can be developed without assuming that there is necessarily a credibility of E given F for every E and F (where E and F are propositions). This subjective theory can be applied whether credibilities exist or not. It is therefore more general and economical not to assume the existence of credibilities as an axiom.

(iv) Suppose Jeffreys is right that there is a credibility of E given F, for every E and F. Then, either the theory will tell us what this credibility is, and we must adjust our degree of belief to be equal to the credibility. Or on the other hand the theory will not tell us what the credibility is, and then not much is gained, except perhaps a healthier frame of mind, by supposing that the credibility exists.

(v) A statistical hypothesis H is an idealized proposition such that for some E, $P(E|H)$ is a credibility with a specified value. Such credibilities may be called "tautological probabilities".

(vi) There is an argument for postulating the existence of credibilities other than tautological probabilities, namely that probability judgments by different people have some tendency to agree.

(vii) The only way to assess the cogency of this argument, if it can be assessed at all, is by the methods of experimental science whose justification is by means of a subjective theory.

(viii) My own view is that it is often quite convenient to accept the postulate that credibilities exist, but this should be regarded as a suggestion rather than an axiom of probability theory.

(ix) This postulate is useful in that it enables other people to do some of our thinking for us. We pay more attention to some people's judgment than to others'.

(x) If a man holds unique beliefs it is possible that everybody else is wrong. If we want him to abandon some of his beliefs we may use promises, threats, hypnotism and suggestion, or we may prefer the following more rational method: By asking questions we may obtain information about his beliefs. Some of the questions may be very complicated ones, of the form, "I put it to you that the following set of opinions is cogent:...". We may then show, by applying a subjective theory of probability, that the beliefs to which the man has paid lip-service are not self-consistent.

(xi) Some of you may be thinking of the slogan "science deals only with what is objective". If the slogan were true there would be no point for scientific purposes in introducing subjective judgments. But actually the slogan is false. For example, intuition (which is subjective) is the main instrument of original scientific research, according to Einstein. The obsession with objectivity arises largely from the desire to be convincing in published work. There are, however, several activities in which it is less important to convince other people than to find out the truth for oneself. There is another reason for wanting an objective theory, namely that there is a tendency to wishful thinking in subjective judgments. But objectivity is precisely what a subjective theory of probability is for: its function is to introduce extra rationality (and therefore objectivity) into your degrees of belief.

(xii) Once we have decided to objectify a rational degree of belief into a credibility it begins to make sense to talk about a degree of belief concerning the numerical value of a credibility. It is possible to use probability type-chains (to coin a phrase) with more than two links, such as a degree of belief equal to $\frac{1}{2}$ that the credibility of H is $\frac{1}{3}$ where H is a statistical hypothesis such that $P(E|H) = \frac{1}{4}$. It is tempting to talk about reasonable degrees of belief of higher and higher types, but it is convenient to think of all these degrees of belief as being of the same kind (usually as belonging to the same body of beliefs in the sense of section 4) by introducing *propositions* of different kinds. In the above example the proposition which asserts that the credibility of H is $\frac{1}{3}$ may itself be regarded as a statistical hypothesis "of type 2". Our type-chains can always be brought back ultimately to a subjective degree of belief. All links but the first will usually be credibilities, tautological or otherwise.

4. Utilities

The question whether utilities should be regarded as belonging to the theory of probability is very largely linguistic. It therefore seems appropriate to begin with a few rough definitions.

Theory of reasoning: A theory of logic plus a theory of probability.

Body of beliefs: A set of comparisons between degrees of belief of the form that one belief is held more firmly than another one, or if you like a set of judgments that one probability is greater than (or equal to) another one.

Reasonable body of beliefs: A body of beliefs which does not give rise to a contradiction when combined with a theory of reasoning.

A reasonable degree of belief is one which occurs in a reasonable body of beliefs. A *probability* is an expression of the form $P(E|F)$ where E and F are propositions. It is either a reasonable degree of belief "in E given F", or else it is something introduced for formal convenience. Degrees of belief may be called "probability estimates".

Principle of rational behaviour: The recommendation always to behave so as to maximize the expected utility per time unit.

Theory of rational behaviour: Theory of reasoning plus the principle of rational behaviour.

Body of decisions: A set of judgments that one decision is better than another. Hypothetical circumstances may be considered as well as real ones (just as for a body of beliefs).

Reasonable body of decisions: A body of decisions which does not give rise to a contradiction when combined with a theory of rational behaviour.

A reasonable decision is one which occurs in a reasonable body of decisions.

We see that a theory of reasoning is a recommendation to think in a particular way while a theory of rational behaviour is a recommendation to act in a particular way.

Utility judgments may also be called "value judgments". The notion of utility is not restricted to financial matters, and even in financial matters utility is not strictly proportional to financial gain. Utilities are supposed to include all human values such as, for example, scientific interest. Part of the definition of utility is provided by the theory of rational action itself.

It was shown by F.P. Ramsey* how one could build up the theory of probability by starting from the principle of maximizing expected utilities. L.J. Savage has recently adopted a similar approach in much more detail in some unpublished notes. The main argument for developing the subject in the Ramsey-Savage manner is that degrees of belief are only in the mind or expressed verbally, and are therefore not immediately significant operationally in the way that behaviour is. Actions speak louder than words. I shall answer this argument in four steps:

---

* *The Foundations of Mathematics* (London, 1931).

(i) It is convenient to classify knowledge into subjects which are given names and are discussed without very much reference to the rest of knowledge. It is possible, and quite usual, to discuss probability with little reference to utilities. If utilities are introduced from the start, the axioms are more complicated and it is debatable whether they are more "convincing". The plan which appeals to me is to develop the theory of probability without much reference to utilities, and then to adjoin the principle of rational behaviour in order to obtain a theory of rational behaviour. The above list of definitions indicates how easily the transition can be made from a theory of probability to a theory of rational behaviour.

(ii) People's value judgments are, I think, liable to disagree more than their probability judgments. Values can be judged with a fair amount of agreement when the commodity is money, but not when deciding between, say, universal education and universal rowing, or between your own life and the life of some other person.

(iii) The principle of maximizing the expected utility can be made to look fairly reasonable in terms of the law of large numbers, provided that none of the utilities are very large. It is therefore convenient to postpone the introduction of the principle until after the law of large numbers has been proved.

(iv) It is not quite clear that infinite utilities cannot occur in questions of salvation and damnation (as suggested, I think, by Pascal), and expressions like $\infty - \infty$ would then occur when deciding between two alternative religions. To have to argue about such matters as a necessasy preliminary to laying down any of the axioms of probability would weaken the foundations of that subject.

5. Axioms and Rules

The theory of probability which I accept and recommend is based on six axioms, of which typical ones are—

A1. $P(E|F)$ is a non-negative number (E and F being propositions).

A4. If E is logically equivalent to F then $P(E|G) = P(F|G)$, $P(G|E) = P(G|F)$.

There is also the possible modification—

A4'. If you have *proved* that E is logically equivalent to F then $P(E|G) = P(F|G)$, etc. (The adoption of $A4'$ amounts to a weakening of the emphasis on consistency and enables you to talk about the probability of purely mathematical propositions.)

The main rule of application is as follows: Let $P'(E|F) > P'(G|H)$ mean that you judge that your degree of belief in E given F (i.e., if F were assumed) would exceed that of G given H. Then in the abstract theory you may write $P(E|F) > P(G|H)$ (and conversely).

Axiom $A1$ may appear to contradict the assumption of section 3 that degrees of belief are only partially ordered. But when the axioms are combined with the above rule of application it becomes clear that we cannot necessarily effect the comparison between any pair of beliefs. The axioms are therefore stronger than they need be for the applications. Unfortunately if they are weakened they become much more complicated.†

6. Examples of Suggestions

(i) Numerical probabilities can be introduced by imagining perfect packs of cards perfectly shuffled, or infinite sequences of trials under essentially similar conditions. Both methods are idealizations, and there is very little to choose between them. It is a matter of taste: that is why there is so much argument about it.

(ii) Any theorem of probability theory and anybody's methods of statistical inference may be used in order to help you to make probability judgments.

(iii) If a body of beliefs is found to be unreasonable after applying the abstract theory, then a good method of patching it up is by being honest (using unemotional judgment). (This suggestion is more difficult to apply to utility judgments because it is more difficult to be unemotional about them.)

7. Rational Behaviour

I think that once the theory of probability is taken for granted, the principle of maximizing the expected utility per unit time is the only fundamental principle of rational behaviour. It teaches us, for example, that the older we become the more important it is to use what we already know rather than to learn more.

In the applications of the principle of rational behaviour some complications arise, such as—

(i) We must weigh up the expected time for doing the mathematical and statistical calculations against the expected utility of these calculations. Apparently less good methods may therefore sometimes be preferred. For example, in an emergency, a quick random decision is better than no decision. But of course theorizing has a value apart from any particular application.

(ii) We must allow for the necessity of convincing other people in some circumstances. So if other people use theoretically inferior methods we may be encouraged to follow suit. It was for this reason that Newton translated his calculus arguments into a geometrical form in the *Principia*. Fashions in modern statistics occur partly for the same reason.

† See B.O. Koopman, "The axioms and algebra of intuitive probability", *Annals of Math.*, **41** (1940), 269–92.

(iii) We may seem to defy the principle of rational action when we insure articles of fairly small value against postal loss. It is possible to justify such insurances on the grounds that we are buying peace of mind, knowing that we are liable to lapse into an irrational state of worry.

(iv) Similarly we may take on bets of negative expected financial utility because the act of gambling has a utility of its own.

(v) Because of a lack of precision in our judgment of probabilities, utilities, expected utilities and "weights of evidence" we may often find that there is nothing to choose between alternative courses of action, i.e., we may not be able to say which of them has the larger expected utility. Both courses of action may be reasonable and a decision may then be arrived at by the operation known as "making up one's mind". Decisions reached in this way are not usually reversed, owing to the negative utility of vacillation. People who attach too large a value to the negative utility of vacillation are known as "obstinate".

(vi) Public and private utilities do not always coincide. This leads to ethical problems.

EXAMPLE. An invention is submitted to a scientific adviser of a firm. The adviser makes the following judgments:

(1) The probability that the invention will work is p.
(2) The value to the firm if the invention is adopted and works is V.
(3) The loss to the firm if the invention is adopted and fails to work is L.
(4) The value to the adviser personally if he advises the adoption of the invention and it works is v.
(5) The loss to the adviser if he advises the adoption of the invention and it fails to work is l.
(6) The losses to the firm and to the adviser if he recommends the rejection of the invention are both negligible, because neither the firm nor the adviser have rivals.

Then the firm's expected gain if the invention is accepted is $pV - (1 - p)L$ and the adviser's expected gain in the same circumstances is $pv - (1 - p)l$. The firm has positive expected gain if $p/(1 - p) > L/V$, and the adviser has positive expected gain if $p/(1 - p) > l/v$. If $l/v > p/(1 - p) > L/V$, the adviser will be faced with an ethical problem, i.e., he will be tempted to act against the interests of the firm. Of course real life is more complicated than this, but the difficulty obviously arises. In an ideal society public and private expected utilities would always be of the same sign.

What can the firm do in order to prevent this sort of temptation from arising? In my opinion the firm should ask the adviser for his estimates of p, V and L, and should take the onus of the actual decision on its own shoulders. In other words, leaders of industry should become more probability-conscious.

If leaders of industry did become probability-conscious there would be quite a reaction on statisticians. For they would have to specify probabilities of hypotheses instead of merely giving advice. At present a statistician of the Neyman-Pearson school is not permitted to talk about the probability of a statistical hypothesis.

8. Fair Fees

The above example raises the question of how a firm can encourage its experts to give fair estimates of probabilities. In general this is a complicated problem, and I shall consider only a simple case and offer only a tentative solution. Suppose that the expert is asked to estimate the probability of an event E in circumstances where it will fairly soon be known whether E is true or false (e.g., in weather forecasts).

It is convenient at first to imagine that there are two experts A and B whose estimates of the probability of E are $p_1 = P_1(E)$ and $p_2 = P_2(E)$. The suffixes refer to the two bodies of belief, and the "given" propositions are taken for granted and omitted from the notation. We imagine also that there are objective probabilities, or credibilities, denoted by P. We introduce hypotheses H_1 and H_2 where H_1 (or H_2) is the hypothesis that A (or B) has objective judgment. Then

$$p_1 = P(E|H_1), \qquad p_2 = P(E|H_2).$$

Therefore, taking "H_1 or H_2" for granted, the factor in favour of H_1 (i.e., the ratio of its final to its initial odds) if E happens is p_1/p_2. Such factors are multiplicative if a series of independent experiments are performed. By taking logs we obtain an additive measure of the difference in the merits of A and B, namely $\log p_1 - \log p_2$ if E occurs or $\log(1 - p_1) - \log(1 - p_2)$ if E does not occur. By itself $\log p_1$ (or $\log(1 - p_1)$) is a measure of the merit of a probability estimate, when it is theoretically possible to make a correct prediction with certainty. It is never positive, and represents the amount of information lost through not knowing with certainty what will happen.

A reasonable fee to pay to an expert who has estimated a probability as p_1 is $k \log(2p_1)$ if the event occurs and $k \log(2 - 2p_1)$ if the event does not occur. If $p_1 > \frac{1}{2}$ the latter payment is really a fine. (k is independent of p_1 but may depend on the utilities. It is assumed to be positive.) This fee can easily be seen to have the desirable property that its expectation is maximized if $p_1 = p$, the true probability, so that it is in the expert's own interest to give an objective estimate. It is also in his interest to collect as much evidence as possible. Note that no fee is paid if $p_1 = \frac{1}{2}$. The justification of this is that if a larger fee were paid the expert would have a positive expected gain by saying that $p_1 = \frac{1}{2}$, without looking at the evidence at all. If the class of problems put to the expert has the property that the average value of p is x, then the factor 2 in the above formula for the fee should be replaced by $x^{-x}(1 - x)^{-(1-x)} = b$, say.* Another modification of the formula should be made in order to allow for the diminishing utility of money (as a function of the amount, rather than as a function of time). In fact if Daniel Bernouilli's logarithmic formula for the utility of money is assumed, the expression for the fee ceases to contain a logarithm and becomes $c\{(bp_1)^k - 1\}$ or $-c\{1 - (b - bp_1)^k\}$, where c is the initial capital of the expert.

The above would be a method of introducing piece-work into the Meteorological Office. The weather-forecaster would lose money whenever he made an incorrect forecast.

When making a probability estimate it may help to *imagine* that you are to be paid in accordance with the above scheme. (It is best to tabulate the amounts to be paid as a function of p_1.)

9. Legal and Statistical Procedures Compared

In legal proceedings there are two men A and B known as lawyers and there is a hypothesis H. A is paid to pretend that he regards the probability of H as 1 and B is paid to pretend that he regards the probability of H as 0. Experiments are performed which consist in asking witnesses questions. A sequential procedure is adopted in which previous answers influence what further questions are asked and what further witnesses are called. (Sequential procedures are very common indeed in ordinary life.) But the jury which has to decide whether to accept or to reject H (or to remain undecided) does not control the experiments. The two lawyers correspond to two rival scientists with vested interests and the jury corresponds to the general scientific public. The decision of the jury depends on their estimates of the final probability of H. It also depends on their judgments of the utilities. They may therefore demand a higher threshold for the probability required in a murder case than in a case of petty theft. The law has never bothered to specify the thresholds numerically. In America a jury may be satisfied with a lower threshold for condemning a black man for the rape of a white woman than vice versa (*News-Chronicle*, **32815** (1951), 5). Such behaviour is unreasonable when combined with democratic bodies of decisions.

The importance of the jury's coming to a definite decision (even a wrong one) was recognized in law at the time of Edward III (*c.* 1350). At that time it was regarded as disgraceful for a jury not to be unanimous, and according to some reports such juries could be placed in a cart and upset in a ditch (*Enc. Brit.*, 11th ed., **15**, 590). This can hardly be regarded as evidence that they

---

* For more than two alternatives the corresponding formula for b is $\log b = -\Sigma x_i \log x_i$, the initial "entropy".

believed in credibilities in those days. I say this because it was not officially recognized that juries could come to wrong decisions except through their stupidity or corruption.

10. Minimax Solutions

For completeness it would be desirable now to expound Wald's theory of statistical decision functions as far as his definition of Bayes solutions and minimax solutions. He gets as far as these definitions in the first 18 pages of *Statistical Decision Functions*, but not without introducing over 30 essential symbols and 20 verbal definitions. Fortunately it is possible to generalize and simplify the definitions of Bayes and minimax solutions with very little loss of rigour.

A number of mutually exclusive statistical hypotheses are specified (one of them being true). A number of possible decisions are also specified as allowable. An example of a decision is that a particular hypothesis (or perhaps a disjunction of hypotheses) is to be acted upon without further experiments. Such a decision is called a "terminal decision". Sequential decisions are also allowed. A sequential decision is a decision to perform further particular experiments. I do not think that it counts as an allowable decision to specify the final probabilities of the hypotheses, or their expected utilities. (My use of the word "allowable" here has nothing to do with Wald's use of the word "admissible".) The terminal and sequential decisions may be called "non-randomized decisions". A "randomized decision" is a decision to draw lots in a specified way in order to decide what non-randomized decision to make.

Notice how close all this is to being a classification of the decisions made in ordinary life, i.e., you often choose between (i) making up your mind, (ii) getting further evidence, (iii) deliberately making a mental or physical toss-up between alternatives. I cannot think of any other type of decision. But if you are giving advice you can specify the relative merits or expected utilities of taking various decisions, and you can make probability estimates.

A non-randomized decision function is a (single-valued) function of the observed results, the values of the function being allowable non-randomized decisions. A randomized decision function, δ, is a function of the observed results, the values of the function being randomized decisions.

Minus the expected utility for a given statistical hypothesis F and a given decision function δ is called the *risk* associated with F and δ, $r(F, \delta)$. (This is intended to allow for utilities including the cost of experimentation. Wald does not allow for the cost of theorizing.) If a distribution ξ of initial probabilities of the statistical hypotheses is assumed, the expected value of $r(F, \delta)$ is called $r^*(\xi, \delta)$, and a decision function δ which minimizes $r^*(\xi, \delta)$ is called a *Bayes solution* relative to ξ.

A decision function δ is said to be a *minimax solution* if it minimizes $\max_F r(F, \delta)$.

An initial distribution ξ is said to be *least favourable* if it maximizes $\min_\delta r^*(\xi, \delta)$. Wald shows under weak conditions that a minimax solution is a Bayes solution relative to a least favourable initial distribution. Minimax solutions seem to assume that we are living in the worst of all possible worlds. Mr. R.B. Braithwaite suggests calling them "prudent" rather than "rational".

Wald does not (in his book) explicitly recommend the adoption of minimax solutions, but he considered their theory worth developing because of its importance in the theory of statistical decision functions as a whole. In fact the book is more concerned with the theory than with recommendations as to how to apply the theory. There is, however, the apparently obvious negative recommendation that Bayes solutions cannot be applied when the initial distribution ξ is unknown. The word "unknown" is rather misleading here. In order to see this we consider the case of only two hypotheses H and H'. Then ξ can be replaced by the probability, p, of H. I assert that in most practical applications we regard p as bounded by inequalities something like $.1 < p < .8$. For if we did not think that $.1 < p$ we would not be prepared to accept H on a small amount of evidence. Is ξ unknown if $.1 < p < .8$? Is it unknown if $.4 < p < .5$; if $.499999 < p < .500001$? In each of these circumstances it would be reasonable to use the Bayes solution corresponding to a value of p selected arbitrarily within its allowable range.

In what circumstances is a minimax solution reasonable? I suggest that it is reasonable if and only if the least favourable initial distribution is reasonable according to your body of beliefs. In particular a minimax solution is always reasonable provided that only reasonable ξ's are entertained. But then the minimax solution is only one of a number of reasonable solutions namely all the Bayes solutions corresponding to the various ξ's.

It is possible to generalise Wald's theory by introducing a distribution ζ of the ξ's themselves. We would then be using a probability type-chain of type 3. (See section 3.) The expected value of $r^*(\xi, \delta)$ could be called $r^{**}(\zeta, \delta)$, and a decision function δ which minimizes $r^{**}(\zeta, \delta)$ could be called a "Bayes solution of type 3" relative to ζ. For consistency we could then call Wald's Bayes solutions "Bayes solutions of type 2". When there is only one available statistical hypothesis F we may define a "Bayes solution of type 1" (relative to F) as one which minimizes $r(F, \delta)$. The use of Bayes solutions of any type is an application of the principle of rational behaviour.

One purpose in introducing Bayes solutions of the third type is in order to overcome feelings of uneasiness in connection with examples like the one mentioned above, where $.1 < p < .8$. One feels that if $p = .09$ has been completely ruled out, then $p = .11$ should be *nearly* ruled out, and this can only be done by using probability type-chains of the third type. It may be objected that the higher the type the woollier the probabilities. It will be found, however, that the higher the type the less the woolliness matters, provided the calculations do not become too complicated. Naturally any lack of definition in ζ is reflected in ambiguity of the Bayes solution of type 3. This ambiguity can be avoided by introducing a *minimax solution of type* 2, i.e., a decision function which minimizes $\max_\xi r^*(\xi, \delta)$.

By the time we had gone as far as type-chains of type 3 I do not think we would be inclined to objectify the degrees of belief any further. It would therefore probably not be necessary to introduce Bayes solutions of type 4 and minimax solutions of type 3.

Minimax solutions (of type 1) were in effect originated by von Neumann in the theory of games, and it is in this theory that they are most justifiable. But even here in practice you would prefer to maximize your expected gain. You would probably use minimax solutions when you had a fair degree of belief that your opponent was a good player. Even when you use the minimax solution you may be maximizing your expected gain, since you may already have worked out the details of the minimax solution, and you would probably not have time to work out anything better once a game had started. To attempt to use a method other than the minimax method would then lead to too large a probability of a large loss, especially in a game like poker. (As a matter of fact I do not think the minimax solution has been worked out for poker.)

I am much indebted to the referee for his critical comments.

Introduction to
Robbins (1955) An Empirical Bayes Approach to Statistics

I.J. Good
Virginia Polytechnic Institute and
State University

Important ideas are liable to be overlooked unless they are named. The Empirical Bayes (EB) method, in its original form, was well named and well developed by Robbins and was given a further boost by the approval of Neyman (1962). The name was all the more appealing for seeming at first like an oxymoron. After 1962, there was a burgeoning of literature on the topic, helped along by Robbins, his graduate students, their graduate students, and so on, in a scholastic genealogical tree. There were 308 papers with "empirical Bayes" or "empirical Bayesian" in the title or as a keyword listed in *Current Index to Statistics* for the years 1975–1988, but as explained below, many of these papers do not deal with empirical Bayes as originally defined by Robbins. The literature is too vast to be comprehensively surveyed even in a short book, and I shall concentrate mainly on one example, the "Poisson case" defined below. This was the first case considered by Robbins and the main theme of his paper becomes clear when this case is well understood.

Robbins (1951) asked himself whether one could usefully apply Bayes's theorem even if the prior, say, $G(\Lambda)$, for a parameter Λ, is unknown but "exists." He meant that it exists as a physical rather than only as an epistemic distribution, but he didn't use this terminology because he didn't regard an epistemic probability as existing. (de Finetti took the opposite view!) Robbins considered a sequence of n occasions in which a random variable takes observed values $X_1, X_2, \ldots, X_n$, but with different values of the parameter λ, say, $\lambda_1, \lambda_2, \ldots, \lambda_n$. (It might have been better to have denoted the actual values of X by lowercase letters.) He assumed that $\lambda_1, \lambda_2, \ldots, \lambda_n$ are independently sampled from the prior $G(\Lambda)$ or G. Then, instead of regarding each of the n occasions as providing a separate self-contained problem of estimation of its corresponding value of Λ, he addressed the problem of estimating λ_n when all

n observations $X_1, X_2, \ldots, X_n$ are available. This is the EB problem attacked by Robbins (1955) where he assumed X to be a discrete variable for the sake of simplicity.

Note that observations on the n occasions must in some sense be interrelated in most applications because all the n realizations of Λ are obtained from a single prior.

Nearly all pioneering work is partially anticipated, and Robbins (1955) is no exception; indeed, that work itself cites two partial anticipations, Robbins (1951) and von Mises (1942) ("On the correct use of Bayes' formula"). Robbins (1983) cites Forcini (1982) who explains on his page 68 how Gini (1911, p. 138) had partially anticipated the EB method in a short passage. I shall discuss another partial anticipation, but let us first consider briefly what is in Robbins' paper.

For any assumed prior G, Robbins writes down, in his formulae (3) and (9), the marginal distribution of X, and the expected posterior distribution of Λ given $X = x$. In the important special case in which, for any given λ, X has a Poisson distribution, Robbins' formulae (3) and (9) reduce to (16) and (17), and formula (18) follows neatly. A natural statistic for estimating λ_n is then given by (19). Similar statistics are given for the cases where X has a geometric, a binomial, and a "Laplacian" distribution. Still other cases are discussed and classified by Rutherford and Krutchkoff (1969).

An especially interesting anticipation of the Poisson case was made by Turing (1941) but not published at that time. (Turing is regarded as a genius for other reasons.) He was concerned with what may be called the *species sampling* problem, but his application was to a kind of dictionary, namely, a code book. After the war, he didn't have time to write about statistics because he was too busy designing computers and computer languages, and speculating about artificial intelligence and the chemical basis of morphogenesis, so with his permission, I developed his idea in considerable detail in Good (1953) and later in a joint work [Good and Toulmin (1956)]. Turing's contribution was the vital formula (1) given below: Compare Robbins' formula (19). Neither Turing nor Robbins (1955) considered the smoothing requirement though it is clearly essential in many applications. Robbins was interested in whether his formula (22) is a theorem. He said he likes to prove theorems (Robbins, 1964a) and probably had his (1964b) in mind where he discusses in detail such questions as conditions under which EB methods are asymptotically optimal. This concentration on theorems is understandable coming from a joint author of that absolutely outstanding book, *What is Mathematics?* [Courant and Robbins (1941)]. Meanwhile Fisher (1943) had also written about the species sampling problem. I shall use this problem to exemplify a practical and interesting application of Robbins' Poisson case.

Suppose then that we have a random sample of N animals, and that n_r species are each represented exactly r times ($r = 1, 2, 3, \ldots$). I call the number n_r the *frequency of the frequency r*. In the application to dictionaries, species = type and animal = token. Turing realized that the frequencies of the frequencies are informative when the numbers n_r, for small r, are not themselves small. This usually requires both that the number N of animals in the sample and the number of categories (species) be large, at least in the hundreds or thousands, and the analysis is *qualitatively different* from most analyses of multinomial distributions. This point is sometimes overlooked in the literature when it is assumed that a multinomial is a multinomial is a multinomial.

We could reasonably define n_0 as the number of species not represented in the sample, but the sample does not give us the value of n_0. If r (the frequency of a specific species) is not large, it, meaning r, has approximately a Poisson distribution (when $r \geq 1$), and in any case, it has a binomial distribution for it is assumed that the animals, *not* the species, form a multinomial sample. (In practice, this too is only approximate.) The number of categories is equal to the number s of species in the *population*, but s is assumed to be large and unknown in nearly all the useful applications, and many of the species are not represented in the sample. In all real problems $s < 10!!!!!$, but in some models it can be regarded as infinite. We ask, what is a reasonable estimate $\hat{p}_r$ of the physical probability corresponding to a particular species that was represented r times in the sample ($r \geq 1$)? By symmetry, $\hat{p}_r$ will be the same for all the n_r species, each of which was represented r times. The maximum likelihood estimate would be r/N, but this is a poor estimate when r is small. For if we believed this estimate, we would infer that the next animal sampled would belong to a species that was already represented, whereas one of the most interesting questions is what is the probability π that the next animal sampled will belong to a *new* species, the total probability of all the events that have never occurred? We may call $1 - \pi$ the *coverage* of the sample, and it is of interest to intelligent ecologists, bacteriologists, and compilers of dictionaries, and even philosophers of probability. Turing's formula for $\hat{p}_r$ for $r \geq 1$ was

$$\frac{(r + 1)n_{r+1}}{n_r N} \tag{1}$$

while I recommended the slight, but often essential, modification,

$$\frac{(r + 1)n'_{r+1}}{n'_r N}, \tag{2}$$

where $n'_1, n'_2, n'_3, \ldots$ is a smoothing of the sequence $n_1, n_2, n_3, \ldots$, and a variety of smoothing methods were discussed in Good (1953). The smoothing is done to try to approximate

$$\frac{r + 1}{N} \cdot \frac{E(n_{r+1})}{E(n_r)} \tag{3}$$

(or with $N + 1$ replacing N though this is unimportant). A new proof of (3) was given by Good (1976, p. 134). The similarity of (1) to the fundamental relations (18), (19), (37), and (41) of Robbins (1955) is not a coincidence because the problems considered are almost identical, as is the philosophy that leads to these results. What Robbins calls "the number of terms $X_1, \ldots, X_n$ which are equal to x" is the *frequency of the frequency x*, and I shall call it v_x, just as n_r is the frequency of the frequency r.

It might be thought at first that species sampling is not EB because the animals are collected in a single experiment. But the theory is essentially the same as if each species in the *population* were sampled on a separate occasion, *N animals* being sampled each time, though this method would be fabulously expensive. The main difference is that in this "fabulously expensive" form of sampling, one would need to have a description of all s species before sampling, and n_0 would be known after sampling. The correspondence between the species sampling problem and Robbins' Poisson or binomial case is shown in the following table.

Species sampling problem	*Robbins' Poisson or binomial case*
A species.	The objects counted on a single occasion.
r, the number of times that some species is represented in the sample.	X_i ($i = 1, 2, \ldots, n$).
$\tilde{r}$, the corresponding r.v. that has (nearly enough) a Poisson distribution with mean λ, where λ varies from one species to another and so can be regarded as an r.v. with an unknown distribution G.	$\tilde{X}_i$, the r.v. occurring on a specific occasion, which has a Poisson distribution with mean λ, where λ varies from one occasion to another and so can be regarded as an r.v. with an unknown distribution G.
n_r or the frequency of the frequency r, that is, the number of species, each of which has frequency r in the sample.	v_x, the frequency of the frequency x, the number of occasions on which $X = x$.
$N = \sum\limits_{r=0}^{\infty} rn_r = \sum\limits_{r=1}^{\infty} rn_r$, the number of *animals* sampled.	$\sum\limits_{x=0}^{\infty} xv_x = \sum\limits_{x=1}^{\infty} xv_x = \sum\limits_{i=1}^{n} X_i$
s, the number of species in the *population*. It would be known in the "fabulously expensive" sampling procedure.	n, the number of occasions.
n_1/N, the simplest estimate of the total probability of unseen species [see formula (4) below].	$v_1 / \sum\limits_{x} xv_x$

n_1, the number of species that occurred exactly once, an estimate of the number of animals, in a new sample of size N, that will belong to a species not represented in the original sample.

v_1, the number of occasions where $X = 1$, an estimate of $\Sigma\lambda$, where the summation is over all those occasions where $X = 0$.

In a real species sampling problem, there is no *sequence* of occasions but rather a *set* of related data. Even in Robbins (1955), the fact that $X_1, X_2, \ldots, X_n$ forms a *sequence* is not taken seriously. It is not tested for trend or periodicity, but rather is regarded as permutable (exchangeable) so it too can be regarded as a set of occasions rather than a sequence.

An estimate of the coverage [Good (1953), formulae (9)] is

$$\sum_{r=1}^{\infty} n_r \cdot \frac{(r + 1)n_{r+1}}{n_r N} = \sum_{s=2}^{\infty} \frac{sn_s}{N} = 1 - \frac{n_1}{N}, \tag{4}$$

so the probability that the next animal sampled will belong to a new species is approximately n_1/N. This estimate, for a change, does *not* depend much on a smoothing of the frequencies of the frequencies. It is a reasonable estimate "ifif" n_1 is not small. For the "expensive" form of sampling, the formula n_1/N would also follow from Robbins' formula (19) with $x = 0$. For the variance of the estimates, see Good (1986). The theory can also be extended for estimating the coverage of a larger sample, if it is not too much larger [Good and Toulmin (1956)].

The species sampling problem was not expressed with the complete generality of Robbins' paper, but it covered an especially interesting example in great detail. Its basic philosophy might differ slightly from EB in that it makes no explicit reference to loss functions, but it sheds light on Robbins' paper. For example, it shows that Robbins was wise to question "whether formula (19) represents the best possible choice for minimizing in some sense the expected squared deviation." It is a very bad estimate unless the frequencies of the frequencies are replaced by smoothed values, and this requires that they can be adequately smoothed in the *vicinity* of x. This will usually require that the sample be very large. Robbins recognized this requirement, although he didn't discuss smoothing of the frequencies of the frequencies, and this recognition probably led him later to discuss the less ambitious linear EB method in which the estimates of λ are linear functions of the observations [Robbins 1983]. The special case in which X has a binomial distribution had previously been discussed by Griffin and Krutchkoff (1971). For multinomials, the linear assumption is equivalent to the assumption of a Dirichlet prior, see Good (1965, p. 25). The linear EB method is directly or indirectly constraining on the prior and is therefore closer to a Bayesian method than the general EB method.

Because EB is based on the assumption that a prior exists as a *physical* distribution, it is different from most interpretations of Bayesian statistics. It is, however, somewhat related to hierarchical Bayes [Good (1952, 1965, 1979, 1987) and Lindley and Smith (1972)], and to pseudo-Bayes or quasi-Bayes [for example, Good (1967); Griffin and Krutchkoff (1971), Bishop, Fienberg, and Holland (1975)]. The Bayes empirical Bayes method, as defined by Deely and Lindley (1981), is a special case of hierarchical Bayes. The self-explanatory notations E, B, EB, BB, EBB, BBB, EBBB, etc., depending on the number of levels in the hierarchy, would be helpful for classifying the literature. When the first letter is E, the notation is intended to indicate that the parameters, or hyperparameters, or hyperhyperparameters, etc. are estimated by a non-Bayesian method. For example, two-stage Bayes is BB, whereas Bayes empirical Bayes and pseudo-Bayes are EBB. When the hyperparameters are estimated by maximum likelihood, I call the method type-II ML [Good (1965) and several later publications, some joint with J.F. Crook]. This is a special case of EBB.

When considering the logic of EB methods, it is essential to distinguish between parametric (or rather, hyperparametric) and nonparametric (non-hyperparametric) EB. In parametric EB, the prior is assumed to depend on one or a few hyperparameters, whereas in nonparametric EB, the prior is much less constrained and can be regarded as having an (enumerable) infinity of parameters. The name *parametric EB*, though used here, is not very good, and misleading both historically and for indexing the literature. Other names used for essentially the same method are *semi-EB*, *semi-Bayesian* (half-full equals half-empty), and *pseudo-Bayesian.*

In *nonparametric* EB, it is more difficult to determine the prior for a given size of sample (just as nonparametric probability density estimation requires a larger sample than the parametric methods when they are applicable). The main kind of nonparametric EB enables one to make predictions without explicit estimation of the prior, and in that respect, it is more ingenious than parametric EB. But even the nonparametric method *implicitly* makes the assumption that the prior is reasonably smooth; otherwise, the frequencies of the frequencies could not be usefully smoothed and EB would be inapplicable. The implicit assumption in real applications, that the prior is not too rough, shows that EB is not free from fuzziness, apart from its asymptotic properties. (It would be interesting to know how EB methods would be affected by assuming that the prior is unimodal or by putting a bound on the roughness of the prior, defined, for example, by the integral of the square of its second derivative.)

For an early example of parametric EB, see Fisher's treatment of the species sampling problem [Fisher, Corbet, and Williams (1943)]. He assumed a prior of gamma form, though he avoided the expression "prior distribution" in case anyone thought he'd become a covert Bayesian. He chose the parameters of the prior to obtain a good fit to the frequencies of the frequencies of a sample of $N = 15,609$ moths trapped at Rothamsted, and containing 240 distinct species. The probability that the next moth captured would have been of a new species was about 1/400. The assumption of a parametric prior can be regarded as a device for smoothing the frequencies of the frequencies and was one of the methods used by Good (1953) for that purpose.

Another early example of parametric EB occurred in Good (1956), rejected in 1953. It dealt with the estimation of the probabilities in cells of large contingency tables, including the empty cells. This was also an early example of the EM algorithm and of a loglinear model. [See Gold (1989).]

Morris (1983, p. 49) tabulates six applications of parametric EB to real data, with references. The names of these applications were (1) revenue sharing, (2) quality assurance (where *quality* is a noun, not a commercialese adjective), (3) insurance rate-making, (4) law school admissions, (5) fire alarms, and (6) epidemiology. He also considers, on his p. 54, the advantages and dangers of the parametric EB method.

Let us very briefly discuss the law school admission problem. The aim was to predict, for each of 82 law schools, the first-year grade averages based on the Law School Aptitude Test scores and the undergraduate grade point averages. One could carry out a linear regression by least squares for each law school separately, or one could first lump together the data for all 82 schools and then carry out the regression calculation. Or one can improve on both methods by compromising between them. One such compromise is a parametric EB method described in detail by Rubin (1980) who avoids making extravagant claims for the method. He uses type-II ML and an iterative EM algorithm, so the philosophy is similar to the semi-Bayesian method used, for example, by Good (1956) for a different problem.

In the last page of Robbins' article, though that page is omitted from this printing because of pressure on space, he considers the problem of estimating the prior for $\tilde{\lambda}$, rather than just the expected value.

In Robbins (1983), he mentions that "Stein's (1956) inadmissibility result and the James–Stein (1961) estimator ... were put into an e.B. [EB] context by Efron & Morris, e.g. (1973)."

It is difficult to make an accurate estimate of the undoubtedly considerable influence of Robbins' paper because the historical waters have been muddied by the misleading expression "parametric empirical Bayes." I too have been misled by this expression. Applications of real forms of EB to important problems in reliability are mentioned in Chap. 13 of Martz and Waller (1982). They use the expression "empirical Bayes" in its correct historical sense in which a physical prior is assumed merely to *exist* (and implicitly that it is not absurdly rough). They emphasize especially a simple form of EB suggested by Krutchkoff (1972) who sat at Robbins' feet and at whose feet sat Martz. Krutchkoff's method ought to be used more than it has been. For reviews of much of the other literature of EB, see Maritz (1970), Maritz and Lwin (1989), Copas (1969), and Susarla (1982). But these writers do not always notice that parametric EB is essentially pseudo-Bayesian and often use type-II ML without noticing it.

Biography

Herbert Robbins was born in 1915. He entered Harvard in 1931 "with practically no high-school mathematics" (in his own words). In 1938, he worked as
a topologist at the Institute for Advanced Studies at Princeton under Marston
Morse. In 1939, he joined New York University and collaborated for two
years with Richard Courant on the book *What is Mathematics?*, which was
published in 1941. He joined the Navy in World War II, not as a mathematician. But his first article on probability arose because, while quietly reading
in a library and minding his own business, he overheard a conversation between two naval officers about inefficient bombing. (A typical example of
scientific information retrieval?) In 1953, he joined Columbia University in
New York, became Higgins Professor in 1974, and is now emeritus and a
special lecturer in the statistics department, as well as holding an appointment at Rutgers University.

For Robbins' opinion of his own work and for his philosophy of living,
see the interview, with a human touch, by Page (1984).

References*

Bishop, Y.M.M., Fienberg, S.E., and Holland, P.W. (1975). *Discrete Multivariate Anal-
ysis*. MIT Press, Cambridge, Mass.
Copas, J.B. (1969). Compound decisions and empirical Bayes, *J. Roy. Statist. Soc.,
Ser. B*, **31**, 397–425 (with discussion).
Courant, R., and Robbins, H. (1941). *What is Mathematics?* Oxford University Press,
London.
Deely, J.J., and Lindley, D.V. (1981). Bayes empirical Bayes, *J. Amer. Statist. Assoc.,*
76, 833–841.
Efron, B., and Morris, C. (1973). Stein's estimation rule and its competitors—an em-
pirical Bayes approach, *J. Amer. Statist. Assoc.,* **68**, 117–130.
Fisher, R.A., Corbet, A.S., and Williams, C.B. (1943). The relation between the number
of species and the number of individuals in a random sample of an animal popula-
tion, *J. Animal Ecol.,* **12**, 42–58.
Forcini, A. (1982). Gini's contributions to the theory of inference, *Int. Statist. Rev.,* **50**,
65–70.
Gini, C. (1911). Considerazioni sulla probabilità a posteriori e applicazioni al rapporto
dei sessi nelle nascite umane, reprinted in *Metron*, **15**, 133–172.
Gold, T. (1989). The inertia of scientific thought, *Speculations in Sci. and Technol.* **12**,
245–253.
†Good, I.J. (1952). Rational decisions, *J. R. Statist. Soc., Ser. B*, **14**, 107–114. (Based
on an invited paper at a conference of the society in Cambridge in 1951.)
Good, I.J. (1953). The population frequencies of species and the estimation of popula-
tion parameters, *Biometrika*, **40**, 237–264.
Good, I.J. (1956). On the estimation of small frequencies in contingency tables, *J. Roy.
Statist. Soc., Ser. B*, **18**, 113–124.

---

* Articles cited in the text are not listed here if they appear in the reference section of Robbins
(1955).
† This paper is reproduced in this Volume (pp. 365–377).

Good, I.J. (1965). *The Estimation of Probabilities: An Essay on Modern Bayesian Methods*. MIT Press, Cambridge, Mass.

Good, I.J. (1967). A Bayesian significance test for multinomial distributions, *J. Roy. Statist. Soc., Ser. B* **29**, 339–431 (with discussion); **36** (1974), 109.

Good, I.J. (1976). The Bayesian influence, or how to sweep subjectivism under the carpet, in *Foundations of Probability Theory, Statistical Inference, and Statistical Theories of Science* (proceedings of a conference in May 1973 at the University of Western Ontario; C.A. Hooker and W. Harper, eds.), *Vol. 2: Foundations and Philosophy of Statistical Inference*. Reidel, Dordrecht, Holland: pp. 125–174.

Good, I.J. (1979). Some history of the hierarchical Bayesian methodology, in *Bayesian Statistics: Proceedings of the 1st International Meeting Held in Valencia* (Spain), May 28–June 2, 1979 (J.M. Bernardo, M.H. DeGroot, D.V. Lindley, and A.F.M. Smith, eds.). University of Valencia 1981, pp. 489–510 and 512–519 (with discussion).

Good, I.J. (1986). The species sampling problem again, C261 in *J. Statist. Comput. Simul.*, **25**, 301–304.

Good, I.J. (1987). Hierarchical Bayesian and Empirical Bayesian methods, letter in *Amer. Statist.*, **41**, 92.

Good, I.J., and Toulmin, G.H. (1956). The number of new species, and the increase in population coverage, when a sample is increased, *Biometrika*, **43**, 45–63.

Griffin, B.S., and Krutchkoff, R.G. (1971). Optimal linear estimators: An empirical Bayes version with application to the binomial distribution, *Biometrika*, **58**, 195–201.

James, W., and Stein, C. (1961). Estimation with quadratic loss, in *Proceedings of 4th Berkeley Symposium on Mathematical and Statistical Probability* (J. Neyman and L. LeCam, eds.), Vol. 1, pp. 361–379.

Krutchkoff, R.G. (1972). Empirical Bayes estimation, *Amer. Statist.*, **26**, 14–16.

Lai, T.L., and Siegmund, D. (eds.) (1985). *Herbert Robbins: Selected Papers*. Springer-Verlag, New York.

Lindley, D.V., and Smith, A.F.M. (1972). Bayes estimates for the linear model, *J. Roy. Statist. Soc., Ser. B*, **34**, 1–18.

Maritz, J.S. (1970). *Empirical Bayes Methods*. Methuen, London.

Maritz, J.S., and Lwin, T. (1989). *Empirical Bayes Methods*. Chapman and Hall, London.

Martz, H.F., and Waller, R.A. (1982). *Bayesian Reliability Analysis*. Wiley, New York.

Morris, C.N. (1983). Parametric empirical Bayes inference: Theory and applications, *J. Amer. Statist. Assoc.*, **78**, 47–59.

Neyman, J. (1962). Two breakthroughs in the theory of statistical decision making, *Rev. Internat. Statist. Inst.*, **30**, 11–27.

Page, W. (1984). An interview with Herbert Robbins, *College Math. J.*, **15**, No. 1 2–24 [reprinted in Lai and Siegmund (1985)].

Robbins, H.E. (1964a). Oral communication, Feb. 5.

Robbins, H.E. (1964b). The empirical Bayes approach to statistical decision problems, *Ann. Math. Statist.*, **35**, 1–20.

Robbins, H.E. (1983). Some thoughts on empirical Bayes estimation (Jerzy Neyman Memorial Lecture), *Ann. Statist.*, **11**, 713–723.

Rubin, D.B. (1980). Using empirical Bayes techniques in the law school validity studies, *J. Amer. Statist. Assoc.*, **75**, 801–827 (with discussion).

Rutherford, J.R., and Krutchkoff, R.G. (1969). Some empirical Bayes techniques in point estimation, *Biometrika*, **56**, 133–137.

Stein, C. (1956). Inadmissibility of the usual estimator for the mean of a multivariate normal distribution, in *Proceedings of 3rd Berkeley Symposium on Mathematical and Statistical Probability* (J. Neyman and L. LeCam, eds.), Vol. 1, pp. 197–206.

Susarla, V. (1982). Empirical Bayes theory, in *Encyclopedia of Statistical Sciences*, Vol. 2 (S. Kotz, N.L. Johnson, and C.B. Read, eds.). Wiley, New York, pp. 490–503.

Turing, A.M. (1941). Private communication.

An Empirical Bayes Approach to Statistics[1]

Herbert E. Robbins
Columbia University

Let X be a random variable which for simplicity we shall assume to have discrete values x and which has a probability distribution depending in a known way on an unknown real parameter Λ,

$$p(x|\lambda) = Pr[X = x|\Lambda = \lambda], \tag{1}$$

Λ *itself being a random variable* with *a priori* distribution function

$$G(\lambda) = Pr[\Lambda \leq \lambda]. \tag{2}$$

The unconditional probability distribution of X is then given by

$$p_G(x) = Pr[X = x] = \int p(x|\lambda) \, dG(\lambda), \tag{3}$$

and the expected squared deviation of any estimator of Λ of the form $\varphi(X)$ is

$$E[\varphi(X) - \Lambda]^2 = E\{E[(\varphi(X) - \Lambda)^2|\Lambda = \lambda]\}$$

$$= \int \sum_x p(x|\lambda)[\varphi(x) - \lambda]^2 \, dG(\lambda)$$

$$= \sum_x \int p(x|\lambda)[\varphi(x) - \lambda]^2 \, dG(\lambda), \tag{4}$$

which is a minimum when $\varphi(x)$ is defined for each x as that value $y = y(x)$ for which

[1] Research supported by the United States Air Force through the Office of Scientific Research of the Air Research and Development Command, and by the Office of Ordnance Research, U.S. Army, under Contract DA-04-200-ORD-355.

$$I(x) = \int p(x|\lambda)(y - \lambda)^2 \, dG(\lambda) = \text{minimum.} \tag{5}$$

But for any fixed x the quantity

$$I(x) = y^2 \int p \, dG - 2y \int p\lambda \, dG + \int p\lambda^2 \, dG$$

$$= \int p \, dG \left(y - \frac{\int p\lambda \, dG}{\int p \, dG}\right)^2 + \left[\int p\lambda^2 \, dG - \frac{(\int p\lambda \, dG)^2}{\int p \, dG}\right] \tag{6}$$

is a minimum with respect to y when

$$y = \frac{\int p\lambda \, dG}{\int p \, dG}, \tag{7}$$

the minimum value of $I(x)$ being

$$I_G(x) = \int p(x|\lambda)\lambda^2 \, dG(\lambda) - \frac{[\int p(x|\lambda)\lambda \, dG(\lambda)]^2}{\int p(x|\lambda) \, dG(\lambda)}. \tag{8}$$

Hence the *Bayes estimator* of Λ corresponding to the *a priori* distribution function G of Λ [in the sense of minimizing the expression (4)] is the random variable $\varphi_G(X)$ defined by the function

$$\varphi_G(x) = \frac{\int p(x|\lambda)\lambda \, dG(\lambda)}{\int p(x|\lambda) \, dG(\lambda)}, \tag{9}$$

the corresponding minimum value of (4) being

$$E[\varphi_G(X) - \Lambda]^2 = \sum_x I_G(x). \tag{10}$$

The expression (9) is, of course, the expected value of the *a posteriori* distribution of Λ given $X = x$.

If the *a priori* distribution function G is known to the experimenter then φ_G defined by (9) is a computable function, but if G is unknown, as is usually the case, then φ_G is not computable. This trouble is not eliminated by the adoption of arbitrary rules prescribing forms for G (as is done, for example, by H. Jeffreys [1] in his theory of statistical inference). It is partly for this reason—that even when G may be assumed to exist it is generally unknown to the experimenter—that various other criteria for estimators (unbiasedness, minimax, etc.) have been proposed which have the advantage of not requiring a knowledge of G.

Suppose now that the problem of estimating Λ from an observed value of X is going to occur repeatedly with a fixed and known $p(x|\lambda)$ and a fixed but unknown $G(\lambda)$, and let

$$(\Lambda_1, X_1), (\Lambda_2, X_2), \dots, (\Lambda_n, X_n), \dots \tag{11}$$

denote the sequence so generated. [The Λ_n are independent random variables with common distribution function G, and the distribution of X_n depends only on Λ_n and for $\Lambda_n = \lambda$ is given by $p(x|\lambda)$.] If we want to estimate an unknown Λ_n from an observed X_n and if the previous values $\Lambda_1, \ldots, \Lambda_{n-1}$ are by now known, then we can form the empirical distribution function of the random variable Λ,

$$G_{n-1}(\lambda) = \frac{\text{number of terms } \Lambda_1, \ldots, \Lambda_{n-1} \text{ which are} \leq \lambda}{n-1}, \tag{12}$$

and take as our estimate of Λ_n the quantity $\psi_n(X_n)$, where by definition

$$\psi_n(x) = \frac{\int p(x|\lambda)\lambda \, dG_{n-1}(\lambda)}{\int p(x|\lambda) \, dG_{n-1}(\lambda)}, \tag{13}$$

which is obtained from (9) by replacing the unknown *a priori* $G(\lambda)$ by the empirical $G_{n-1}(\lambda)$. Since $G_{n-1}(\lambda) \to G(\lambda)$ with probability 1 as $n \to \infty$, the ratio (13) will, under suitable regularity conditions on the kernel $p(x|\lambda)$, tend for any fixed x to the Bayes function $\varphi_G(x)$ defined by (9) and hence, again under suitable conditions, the expected squared deviation of $\psi_n(X_n)$ from Λ_n will tend to the Bayes value (10).

In practice, of course, it will be unusual for the previous values $\Lambda_1, \ldots, \Lambda_{n-1}$ to be known, and hence the function (13) will be no more computable than the true Bayes function (9). *However, in many cases the previous values* $X_1, \ldots,$ X_{n-1} *will be available to the experimenter at the moment when* Λ_n *is to be estimated,* and the question then arises whether it is possible to infer from the set of values $X_1, \ldots, X_n$ the approximate form of the unknown G, or at least, in the present case of quadratic estimation, to approximate the value of the functional of G defined by (9). To this end we observe that for any fixed x the empirical frequency

$$p_n(x) = \frac{\text{number of terms } X_1, \ldots, X_n \text{ which equal } x}{n} \tag{14}$$

tends with probability 1 as $n \to \infty$ to the function $p_G(x)$ defined by (3), no matter what the *a priori* distribution function G. Hence there arises the following mathematical problem: from an approximate value (14) of the integral (3), where $p(x|\lambda)$ is a known kernel, to obtain an approximation to the unknown distribution function G, or at least, in the present case, to the value of the Bayes function (9) which depends on G. (This problem was posed in [4].) The possibility of doing this will depend on the nature of the kernel $p(x|\lambda)$ and on the class, say $\mathcal{G}$, to which the unknown G is assumed to belong. In order to fix the ideas we shall consider several special cases, the first being that of the *Poisson* kernel

$$p(x|\lambda) = e^{-\lambda}\frac{\lambda^x}{x!}; \qquad x = 0, 1, \ldots; \lambda > 0; \tag{15}$$

$\mathcal{G}$ being the class of all distribution functions on the positive real axis.

In this case we have

$$p_G(x) = \int p(x|\lambda) \, dG(\lambda) = \int_0^\infty e^{-\lambda} \lambda^x \, dG(\lambda)/x! \qquad (16)$$

and

$$\varphi_G(x) = \frac{\int_0^\infty e^{-\lambda} \lambda^{x+1} \, dG(\lambda)}{\int_0^\infty e^{-\lambda} \lambda^x \, dG(\lambda)}, \qquad (17)$$

and we can write the fundamental relation

$$\varphi_G(x) = (x+1) \cdot \frac{p_G(x+1)}{p_G(x)}. \qquad (18)$$

If we now define the function

$$\varphi_n(x) = (x+1) \frac{p_n(x+1)}{p_n(x)} = (x+1) \cdot \frac{\text{number of terms } X_1, \ldots, X_n \text{ which are equal to } x+1}{\text{number of terms } X_1, \ldots, X_n \text{ which are equal to } x} \qquad (19)$$

then no matter what the unknown G we shall have for any fixed x

$$\varphi_n(x) \to \varphi_G(x) \text{ with probability 1 as } n \to \infty. \qquad (20)$$

This suggests using as an estimate of the unknown Λ_n in the sequence (11) the computable quantity

$$\varphi_n(X_n), \qquad (21)$$

in the hope that as $n \to \infty$,

$$E[\varphi_n(X_n) - \Lambda_n]^2 \to E[\varphi_G(X) - \Lambda]^2. \qquad (22)$$

We shall not investigate here the question of whether (22) does actually hold for the particular function (19) or whether (19) represents the best possible choice for minimizing in some sense the expected squared deviation. (See [8].)

It is of interest to compute the value of (10) for various *a priori* distribution functions G in order to compare its value with the expected squared deviation of the usual (maximum likelihood, minimum variance unbiased) Poisson estimator, X itself, for which

$$E(X - \Lambda)^2 = E\Lambda = \int_0^\infty \lambda \, dG(\lambda). \qquad (23)$$

Suppose, for example, that G is a gamma type distribution function with density

$$G'(\lambda) = C\lambda^{b-1}e^{-h\lambda}; \qquad \lambda, b, h > 0; C = h^b/\Gamma(b). \qquad (24)$$

By elementary computation we find that

$$E\Lambda = \frac{b}{h}, \qquad \text{Var } \Lambda = \frac{b}{h^2} \tag{25}$$

and

$$\varphi_G(x) = \frac{x + b}{1 + h}, \qquad E[\varphi_G(X) - \Lambda]^2 = \frac{b}{h(1 + h)}; \tag{26}$$

hence

$$\frac{E[\varphi_G(X) - \Lambda]^2}{E(X - \Lambda)^2} = \frac{1}{1 + h}. \tag{27}$$

For example, if $b = 100$, $h = 10$ then

$$E\Lambda = 10, \quad \text{Var } \Lambda = 1, \quad \varphi_G(x) = \frac{x + 100}{11}, \quad \frac{E[\varphi_G(X) - \Lambda]^2}{E(X - \Lambda)^2} = \frac{1}{11}. \tag{28}$$

An even simpler case occurs when Λ has all its probability concentrated at a single value λ. In this case, of course, the Bayes function is

$$\varphi_G(x) = \lambda, \tag{29}$$

not involving x at all, and

$$E[\varphi_G(X) - \Lambda]^2 = 0, \tag{30}$$

while as before

$$E(X - \Lambda)^2 = E\Lambda = \lambda. \tag{31}$$

Here the sequence (11) consists of observations $X_1, \ldots, X_n, \ldots$ from the same Poisson population (although this fact may not be apparent to the experimenter at the beginning); the traditional estimator $\varphi(x) = x$ does not take advantage of this favorable circumstance and continues to have the expected squared deviation λ after any number n of trials.

As a second example we take the *geometric* kernel

$$p(x|\lambda) = (1 - \lambda)\lambda^x; \qquad x = 0, 1, \ldots; 0 < \lambda < 1; \tag{32}$$

for which

$$p_G(x) = \int_0^1 (1 - \lambda)\lambda^x \, dG(\lambda),$$

$$\varphi_G(x) = \frac{\int_0^1 (1 - \lambda)\lambda^{x+1} \, dG(\lambda)}{\int_0^1 (1 - \lambda)\lambda^x \, dG(\lambda)} = \frac{p_G(x + 1)}{p_G(x)}. \tag{33}$$

Here it is natural to estimate Λ_n by (21) with the definition

$$\varphi_n(x) = \frac{\text{number of terms } X_1, \ldots, X_n \text{ which are equal to } x + 1}{\text{number of terms } X_1, \ldots, X_n \text{ which are equal to } x}. \tag{34}$$

Our third example will be the *binomial* kernel

$$p_r(x|\lambda) = \binom{r}{x} \lambda^x (1 - \lambda)^{r-x}; \qquad x = 0, 1, \ldots, r; \qquad 0 \leq \lambda \leq 1. \quad (35)$$

Here r is a fixed positive integer representing the number of trials, X the number of successes, and Λ the unknown probability of success in each trial. G may be taken as the class of all distribution functions on the interval $(0, 1)$. In this case

$$\begin{cases} p_{G,r}(x) = \int p_r(x|\lambda)\, dG(\lambda) = \binom{r}{x} \int_0^1 \lambda^x (1 - \lambda)^{r-x}\, dG(\lambda), \\[2mm] \varphi_{G,r}(x) = \dfrac{\int_0^1 \lambda^{x+1}(1 - \lambda)^{r-x}\, dG(\lambda)}{\int_0^1 \lambda^x (1 - \lambda)^{r-x}\, dG(\lambda)}, \end{cases} \quad (36)$$

so that we can write the fundamental relation

$$\varphi_{G,r}(x) = \frac{x + 1}{r + 1} \cdot \frac{p_{G,r+1}(x + 1)}{p_{G,r}(x)}; \qquad x = 0, 1, \ldots, r. \quad (37)$$

Let

$$p_{n,r}(x) = \frac{\text{number of terms } X_1, \ldots, X_n \text{ which are equal to } x}{n}; \quad (38)$$

then $p_{n,r}(x) \to p_{G,r}(x)$ with probability 1 as $n \to \infty$. Now consider the sequence of random variables

$$X_1', X_2', \ldots, X_n', \ldots \quad (39)$$

where X_n' denotes the number of successes in, say, the first $r - 1$ out of the r trials which produced X_n successes, and let

$$p_{n,r-1}(x) = \frac{\text{number of terms } X_1', \ldots, X_n' \text{ which are equal to } x}{n}; \quad (40)$$

then $p_{n,r-1}(x) \to p_{G,r-1}(x)$ with probability 1 as $n \to \infty$. Thus if we set

$$\varphi_{n,r}(x) = \frac{x + 1}{r} \cdot \frac{p_{n,r}(x + 1)}{p_{n,r-1}(x)}, \quad (41)$$

then

$$\varphi_{n,r}(x) \to \frac{x + 1}{r} \cdot \frac{p_{G,r}(x + 1)}{p_{G,r-1}(x)} = \varphi_{G,r-1}(x) \quad (42)$$

with probability 1 as $n \to \infty$. If we take as our estimate of Λ_n the value

$$\varphi_{n,r}(X_n') \quad (43)$$

then for large n we will do about as well as if we knew the a priori G but confined ourselves to the first $r - 1$ out of each set of r trials. For large r this does not sacrifice much information, but it is by no means clear that

(43) is the "best" estimator of Λ_n that could be devised in the spirit of our discussion.

(*Editors' note*: The last section has been omitted.)

· · ·

I should like to express my appreciation to A. Dvoretzky, J. Neyman, and H. Raiffa for helpful discussions and suggestions.

References

[1] H. Jeffreys, *Theory of Probability*, 2d ed., Oxford, Clarendon Press, 1948.
[2] P.F. Lazarsfeld, "A conceptual introduction to latent structure analysis," *Mathematical Thinking in the Social Sciences*, Glencoe, Ill., Free Press, 1954, pp. 349–387.
[3] J. Neyman, *Lectures and Conferences on Mathematical Statistics and Probability*, 2d ed., Washington, U.S. Department of Agriculture Graduate School, 1952.
[4] H. Robbins, "Asymptotically subminimax solutions of compound statistical decision problems," *Proceedings of the Second Berkeley Symposium on Mathematical Statistics and Probability*, Berkeley and Los Angeles, University of California Press, 1951, pp. 131–148.
[5] ———, "A generalization of the method of maximum likelihood: estimating a mixing distribution," (abstract), *Annals of Math. Stat.*, Vol. 21 (1950), pp. 314–315.
[6] M.C.K. Tweedie, "Functions of a statistical variate with given means, with special reference to Laplacian distributions," *Proc. Camb. Phil. Soc.*, Vol. 43 (1947), pp. 41–49.
[7] R. von Mises, "On the correct use of Bayes' formula," *Annals of Math. Stat.*, Vol. 13 (1942), pp. 156–165.
[8] M.V. Johns, Jr., "Contributions to the theory of empirical Bayes procedures in statistics," doctoral thesis at Columbia University (unpublished), 1956.

Introduction to
Kiefer (1959) Optimum Experimental Designs

H.P. Wynn
City University of London

1. A Historial Perspective

Kiefer's (1959) paper was selected partly because of its historical interest in terms of the "sociology" of the subject of experimental design in the 20th century. The fact that the paper was a presented paper enabled a clash of statistical cultures to take place in an open forum and one with a declared tradition of critical discussion.

The opening paragraphs of the paper make clear Kiefer's lineage in the Wald school and indeed Jacob Wolfowitz, his co-worker on the early papers in design, is another outstanding figure in that tradition. The comments of Tocher show allegiance to "the whole school of British experimental designs ... a very practical race of people." Yates, one of the founders of this school, stresses the multiple purposes of design and criticizes Kiefer's criteria. Kiefer's reply is masterly. The division between the Rothamsted–British school and the Wald–Kiefer–Wolfowitz school is now more subdued, and one reason for this has been the general success of the optimum design methodologies, particularly in computer implementation. The present commentator feels that, in a way, there is justice on both sides. Mathematics, as in all sciences, has an important role to play (R.A. Fisher himself is responsible for bringing mathematics to bear in the first place), but further work is still needed to understand the real nature of the activity of experimentation. This struggle for understanding has strong links into epistemology and the philosophy of science and has roots back through John Stuart Mill to Francis Bacon, the much neglected father of the subject. [See Urbach (1987) for a recent assessment.]

2. The Theoretical Core

The central mathematical result that is the equivalence of D-optimality and G-optimality only just made it into the paper, which, mostly introduces the general decision theoretic approach. The key to the subject was the replacement of the usual $\mathbf{X}^T\mathbf{X}$ matrix in regression by a moment matrix that generalized $(1/n)\mathbf{X}^T\mathbf{X}$ to

$$\mathbf{M}(\xi) = \int \mathbf{f}(\mathbf{x})\mathbf{f}(\mathbf{x})^T \xi \, d\mathbf{x},$$

where ξ is a measure on the design space and the model is

$$E[Y_\mathbf{x}] = \mathbf{f}(\mathbf{x})^T\boldsymbol{\theta}.$$

The space of $\mathbf{M}(\xi)$ is, under suitable conditions, closed and convex and amenable to optimization problems of the kind

$$\max \phi(\mathbf{M}(\xi)),$$

where ϕ is a concave functional.

The importance of the General Equivalence Theorem (GET) was in showing that a useful response-based criterion, G-optimality,

$$\min_{\xi} \max_{x} \mathbf{f}(\mathbf{x})^T\mathbf{M}(\xi)^{-1}\mathbf{f}(\mathbf{x}),$$

was equivalent to the famous D-optimality

$$\max_{\xi} \det(\mathbf{M}(\xi)).$$

The first published proof was in Kiefer and Wolfowitz (1980).

Kiefer and his co-workers had hit a rich vein and worked it elegantly in two main areas:

(i) extension to other criteria
(ii) solutions for particular problems.

The extent of their success can best be seen by consulting the collected works [see Kiefer (1985)], in which additional commentaries can be found.

Among the powerful techniques which Kiefer brings to the subject of optimum design is that of invariance. The technique is explained simply as follows. If the design space is invariant under a group of transformations, then given an initial measure ξ, one can generate all other measures under the group in the usual way. Mixing all these measures uniformly produces a ξ' at least as good as ξ. But ξ' is an invariant measure. Thus, by restricting to invariant measures from the outset, one can hope to narrow the search for designs (in a similar fashion to invariant decision rules). These invariance "tricks" are particularly useful in optimum combinatorial design where there are natural groups such as the permutation group on treatments.

3. The Present Day

In an introduction of this kind, it is probably of the most benefit to sketch the scene today insofar as it can be traced back to the 1959 paper and the others that came out over that very productive four-year period. For another recent review, see Dodge et al. (1988).

As mentioned, one of the most important benefits of this heavy emphasis on optimization was the production of computer algorithms with which a range of authors are associated, the present commentator, Federov, Pazman, Mitchell, Silvey, Titterington, Wu, and more names one fears to leave out. The basis of these algorithms is to add and subtract observation points one at a time on "excursions." Kiefer himself in a series of joint papers with Galil contributed to this body of work; a sample paper is Galil and Kiefer (1980). A single augmentation with one point will add one row to the X matrix so that with obvious notation

$$X_{n+1}^T X_{n+1} = X_n^T X_n + f(x_{n+1}) f(x_{n+1})^T.$$

Unfortunately, some of the beauty of the continuous theory of Kiefer is lost, but the gains are in simple iterative procedures that converge swiftly to good designs (if not optimum, necessarily, in the continuous sense over all ξ). The optimization problems are now seen to belong to a wider class of optimization problems that might loosely be called optimum subset problems, for which similar algorithms have been used. The use of simulated annealing algorithms in image processing is an example.

To those who knew Jack Kiefer, it is a happy thought that the subject he devoted many years to has finally come of age in the realization of its importance to product development and quality improvement in industry. Some of this had been presaged by the parallel work on response surface designs by George Box, Norman Draper, and co-workers. But now a really hi-tech possibility is available in linking the optimization of experimental designs to the optimization of engineering design in areas like computer-aided design (CAD). Thus, the phrase computer-aided experimental design is one with a technological appeal undreamt of by early practitioners. Algorithms are now being built into major software packages and are available in smaller PC-based products.

There are, of course, dangers in making experimental design methods routine: the fundamental nature of the experimental process may be lost. To show that far from this happening—in fact, there is a deeper awareness of the importance of efficient data collection, it is worth listing some of the areas in which the spirit of the Kiefer approach is alive.

(i) computer experiments [Sacks et al. (1989)].
(ii) spatial observation [Sacks and Schiller (1988)].
(iii) nonlinear models [Titterington (1987)].
(iv) Bayesian methods [Chaloner (1984)].

4. More Theory?

The original Kiefer–Wolfowitz theory in a sense needs updating in light of the new practical uses to which experimental design is being put. Some of these are addressed in (i–iv) above, but some of the issues have remained and indeed were the subject of debates into which Kiefer entered with zeal. It is quality improvement that has brought these issues to the front.

The one really profound problem that has arisen is that of designing experiments, taking more accurately into account the real operating environment. The phrases, "random effects," "noise variables," and "robustness" have been used for this issue. Here is the problem in almost existentialist form. How do we experiment to ascertain the current behavior of a physical process when we have little knowledge of the process? Several attempts have been made. For example, additional terms in the model could be assumed [Box and Draper (1959)]. Correlation terms may be added to "absorb" the ignorance concerning the model [Sacks and Ylvisaker (1969)]. Can we reassess the process of decision making concerning experimentation under more extreme types of ignorance? The question can be rephrased; If we do not know what we are looking for, how do we know where to look? It is the requirement for awareness in the face of possibly hostile environments (manufacturing, ecological, geophysical) that will change the nature of the theory. What then is the best extension of the Kiefer theory to come to grips with such problems?

One can read the collected works for hints at solutions. Certainly, Kiefer was concerned, particularly in his work with Galil, on efficiency across a range of criteria (ϕ-functions) rather than optimality for a single criterion. Also, he was interested in models with correlated errors as early as 1960 [Kiefer (1960)].

Kiefer though concerned himself largely with controlled experimentation and the decision making which that control implies. Seeking a solution to the decision-making problem for more random environments asks for a change in the view of the nature of experimentation. There are indications in the work on design and also his later work on conditional inference that Kiefer was aware of these issues. His more or less complete solution to the classical optimum design problem remains a prerequisite for any serious discussion of them, a discussion that had he lived longer, he would have led with his particular scholarly and erudite technical style. It is this style that shines through his papers and makes them still such delightful reading.

5. Short Biography

Carl Jack Kiefer was born in 1924 in Cincinatti. His student career started in MIT in 1942 and was interrupted by military service during World War II. His masters thesis at MIT contained the Fibonacci search algorithm, which made him famous in optimization circles although it was not published until 1953.

Kiefer's formative years were spent in Abraham Wald's department at Columbia where he was Jack Wolfowitz's research student. He went with Wolfowitz to Cornell in 1951. During that particular period, Cornell and Columbia housed a good percentage of the better known names in theoretical statistics and probability. Jack Kiefer's more mature years can be divided into the period at Cornell until 1979 and the subsequent years as Miller Research Professor at Berkeley before his tragic death in 1981 at the age of 57. Although his work in experimental design intensified throughout his life, he is at least as well known in decision theory and probability theory. This is a remarkable achievement that puts him in the front line of 20th century statisticians. Examples of his other work are the "Kiefer–Wolfowitz procedure" in sequential analysis, his work with Sacks on sequential testing, his bound for the sample cdf, the related work on the law of the iterated logarithm for sample quantiles, and his research on minimaxity for Hotelling's T^2. He published over 100 papers, each one beautifully crafted and dense with ideas.

References

Box, G.E.P., and Draper, N.R. (1959). A basis for the selection of a response surface design, *J. Amer. Statist. Assoc.*, **54**, 622–654.

Chaloner, K. (1984). Optimal Bayesian experimental design for linear models, *Ann. Statist.*, **12**, 283–350.

Dodge, Y., Federof, V.V., and Wynn, H.P. (1988). Optimal design of experiment; An overview, in *Optimal Design and Analysis of Experiments*. North-Holland, Amsterdam.

Galil, Z., and Kiefer, J. (1980). Time and space-saving computer methods, related to Mitchell's DETMAX, for finding D-optimum designs.

Kiefer, J. (1960). Optimum experimental designs V, with applications to systematic and rotatable designs, in *Proceedings of 4th Berkeley Symposium on Mathematical and Statistical Probability*, Vol. 1. University of California Press, Berkeley, pp. 381–405.

Kiefer, J. (1985). *Collected Works, Vol. III*. Springer-Verlag, New York.

Kiefer, J., and Wolfowitz, J. (1980). The equivalence of two extremum problems, *Canad. J. Math.*, **12**, 363–366.

Sacks, J., and Schiller, S. (1988). Spatial designs, in *Statistical Design Theory and Related Topics*, Vol. IV. Springer-Verlag, New York, pp. 385–399.

Sacks, J., Welch, W.J., Mitchell, T.J., and Wynn, H.P. (1989). Design and analysis of computer experiments, *Statist. Sci.*, **4**, 409–435 (with discussion).

Sacks, J., and Ylvisaker, D. (1964). Designs for Regression Problems with correlated errors III, *Ann. Math. Statist.*, **41**, 2057–2074.

Titterington, D.M. (1987). Optimal design for nonlinear problems, in *Conference on Model-Oriented Data Analysis, Lecture Notes in Economics and Mathematical Systems*, Vol. 297. Springer-Verlag, Berlin.

Urbach, P. (1987). *Francis Bacon's Philosophy of Science*, Open Court, Lasalle, Ill.

Optimum Experimental Designs

J.C. Kiefer*
Cornell and Oxford Universities

Summary

After some introductory remarks, we discuss certain basic considerations such as the nonoptimality of the classical symmetric (balanced) designs for hypothesis testing, the optimality of designs invariant under an appropriate group of transformations, etc. In section 3 we discuss complete classes of designs, while in section 4 we consider methods for verifying that designs satisfy certain specific optimality criteria, or for computing designs which satisfy such criteria. Some of the results are new, while part of the paper reviews pertinent results of the author and others.

1. Introductory Remarks

This paper will survey various recent developments in the theory of the determination of optimum experimental designs, in the course of which a few new results will be given. Our purpose will not be to state things in the most general possible abstract setting, nor to try to apply the discussion to the more complicated practical problems; it is not only the computational difficulties which dictate this—rather, it is that our primary aim will be to stress certain ideas and principles which are most easily set forth in simple situations. These ideas and principles have just as much content in the complicated settings which one often encounters in practice, but their transparency is often obscured there by the arithmetic.

We shall be concerned with methods for verifying whether or not given

* Research sponsored by the U.S. Office of Naval Research.

designs satisfy certain optimality criteria, and for computing designs which satisfy such criteria. Our approach is in the spirit of Wald's decision theory. Thus, problems of constructing designs which satisfy certain *algebraic* conditions (resolvability, revolvability, etc.), as distinguished from problems of constructing designs which satisfy certain optimality criteria, will not primarily concern us; such algebraic conditions will be of interest to us only in so far as they may be proved to entail some property of optimality. Perhaps the traditional development of the subject of experimental design has been too much concerned with the former problems and too little with the latter, although recent developments show that progress in finding useful and efficient designs cannot and should not rest on this traditional approach of being satisfied to find designs which merely satisfy some intuitively appealing algebraic property.

Our discussion will not include all of the recent papers which fit into the above framework, but some of these excluded topics will now be mentioned, since they are important for future research. For example, we shall not treat any problems in the very important area of sequential design. In a sense, the first result in this area was really the proof of the optimum character of Wald's sequential probability ratio test (Wald and Wolfowitz, 1948), and other sequential decision theory can be viewed in the same way. However, in the case in which we are really interested in the present paper, wherein several possible experiments are available at each stage, there are very few explicit results, although the general setup was formulated some time ago (Wald, 1950). The earliest result of this kind was obtained by Stein (1948), but explicit results are so difficult to obtain that recent papers have obtained them in only a few simple settings such as those considered by Bradt and Karlin (1957), Bradt, Johnson and Karlin (1957), and Sobel and Groll (1959). There are also a few asymptotic optimality results for sequential designs, such as those obtained by Chung (1954), Hodges and Lehmann (1955), and Sacks (1958) for the stochastic approximation (sequential search) designs of Robbins and Monro (1951), Kiefer and Wolfowitz (1952), Blum (1954), and Sacks (1958), and the result of Robbins (1952). The much more difficult nonasymptotic problem of finding optimum designs with stopping rules (or even nonoptimum but reasonably efficient stopping rules with prescribed confidence properties) in these cases remains unanswered, although corresponding search problems where errors are negligible have been treated (Kiefer, 1953, 1957b). Although practical people often employ techniques like those of Box and Wilson (1951) in problems of this kind (such methods were first suggested by Friedman and Savage (1947) and were further developed by the author (1948)), such methods cannot in their *present* state have any role in satisfactorily solving these problems, since (as pointed out in the discussion to the paper by Box and Wilson) they have no guaranteed probability (confidence) properties (not even asymptotic ones such as those enjoyed by the stochastic approximation methods cited above), and in fact are often not even well-defined rules of operation; the main value of such intuitive considerations to the approach of the present paper is that they might suggest well-defined procedures which can be analysed precisely (under suitable restrictions on the regression function, Hotelling (1941) has obtained an optimum nonsequential solution to such a problem). These search problems usually fall into the area of *nonparametric* design problems, an important field which deserves more attention but which we shall not treat here.

We now discuss the assumptions we shall make in this paper, on the class of possible distributions which occur; this class can be nonparametric, but only trivially so: in the theory of point estimation, it is well known (see, e.g., Hodges and Lehmann, 1950) that best linear estimators possess certain more meaningful optimality properties than merely being best among *linear* estimators (e.g., they are minimax for certain weight functions) if the class of possible distributions contains certain normal distributions (this can also be shown for randomized designs, defined below). This fact can be used with any of the estimation criteria of the section 4 to yield a minimax result. We shall usually assume normality, so our results do have this nonparametric extension in point estimation problems. (Although best linear estimators and certain related confidence intervals are minimax in the normal case (see Stein and Wald, 1947, Wolfowitz, 1950; Kiefer, 1957a), if we add the risk functions of several estimators (rather than to consider the vector risk), they may some times have to be modified slightly in order to be admissible (Stein, 1955)). This justification for choosing a design on the basis of the obvious comparison between the covariance matrices of best linear estimators does not apply to such problems as those of testing hypotheses, as we shall see in Section 2A. It should be obvious that the considerations of that section apply with appropriate modifications under various other parametric assumptions (for example, the results for the classical randomized block setting are also mentioned in Section 2A); the striking thing is that these phenomena exist in the common, unpathological, normal case. A similar remark applies to other portions of this paper. For example, optimality results like those of Sections 4A–B can also be proved for appropriate components of variance (or mixed) models.

Throughout this paper, then, except where explicitly stated to the contrary, Y_d will denote an N element column vector whose components Y_{di} are independent (often normal) random variables with common variance σ^2 (unknown unless stated to the contrary, although this matters little); θ is an unknown m-vector with components θ_i, Θ is the m-space of possible θ's, X_d is a known $N \times m$ matrix depending on an index d (the "design") and which is, within limits, subject to choice by the experimenter; and the expected value of Y_d when θ and σ^2 are the parameter values and the design d is used is

$$E_{\theta,\sigma;d} Y_d = X_d \theta. \tag{1.1}$$

Additional restrictions of the form $A\theta = B$ can be assumed already to have been absorbed into (1.1). The consideration of various designs with different numbers of observations is easily accomplished; for example, if we are only interested in point estimation, or if σ^2 is known, we need only append a suitable number of rows of zeros to various X_d's. See Section 2D for further discussion of our assumptions.

We denote by Δ the set of all choices of the index d which are available to the experimenter, and by Δ_R a class of probability measures on Δ (usually these measures will have finite support, and measurability considerations will be trivial otherwise), the class of *randomized designs* (the classical use of this term is a special case of the present usage, and the reason for the classical use of such designs, under different probability models, is entirely different from ours) available to the experimenter. A randomized design δ is used by choosing a d from Δ according to this measure and then using the selected d. We think of Δ_R as including Δ.

Write $A_d = X'_d X_d$ (primes on vectors and matrices denote transposes) for the "information matrix" of the design d. If X_d (hence A_d) is of rank b_d, then there is a $b_d \times m$ matrix L_d, of rank b_d, such that the distribution of Y_d depends on θ only through the vector $L_d\theta$, whose best linear estimators (b.l.e.'s) are the components t_{di} of $t_d = L_d t$ where t is any solution of the normal equations $A_d t = X'_d Y_d$. If $b_d = m$, we can take L_d to be the identity matrix, and will always do so. If $\bar{S}_d$ is the usual best unbiased estimator of σ^2 (so that $(N - b_d)$ $\bar{S}_d/\sigma^2$ has the χ^2 distribution with $N - b_d$ degrees of freedom), then $(t_d, \bar{S}_d)$ is a minimal sufficient statistic.

The covariance matrix of t_d will be denoted by $\sigma^2 V_d$; of course, $V_d = A_d^{-1}$ if $b_d = m$.

If a randomized design δ assigns probability one to a set of d's all of which have the same L_d, then the covariance matrix of b.l.e.'s under δ is the expectation with respect to δ of $\sigma^2 V_d$, although the distribution is not generally normal. The role of such designs for problems of point estimation is considered in Section 3. If δ gives positive probability to a d for which a linear parametric function $q'\theta$ is not estimable, then the expected squared error for any estimator of $q'\theta$ is infinite when δ is used, although δ may still be useful for interval estimation, hypothesis testing, ranking, etc., involving $q'\theta$. In fact, as we shall see in Section 2A, the appropriate distribution theory—e.g., computation of δ's and associated tests with good power functions—is quite complicated, and the classical approach of resting considerations on the F-test and hence a comparison of V_d's (or of δ's which give probability one to a set of d's all of which have the same L_d and V_d) is untenable. Thus, for example, in the case of testing a hypothesis of equality of "treatment effects" in the setup of block designs, if we are interested in alternatives which are close to the hypothesis of equality, we shall see that it is not the average according to δ of V_d's, but rather of V_d^{-1}'s, which is relevant. More precisely, we shall have occasion to consider the average (with respect to δ) A_δ of a set of A_d's with common value b_d (this last condition is irrelevant if σ^2 is known). Of course, this A_δ is meaningless as far as variance is concerned; e.g., if A_δ is nonsingular, A_δ^{-1} is not a covariance matrix of b.l.e.'s unless δ gives probability one to a set of d's with identical A_d's. However, such an average will be meaningful as the inverse of a covariance matrix in another setting: if a new experiment is formed by replicating each of a set of d's a number of time $\eta(d)$, then the integral $A(\eta)$ of A_d with respect to the measure η is meaningful in the same way that A_d is. This circumstance will most often arise in regression problems, where we are given a set of row vectors x_d, each with m elements, the expected value of an observation corresponding to experiment d is $x_d\theta$, and we must choose an experiment d' consisting of N observations; if $X_{d'}$ has $\eta(d)$ rows equal to x_d, then $A_{d'}$ is just $A(\eta)$, but it will be more useful to think of the experiment in the latter form. See also Sections 2C and 3 in this connection. It will also be convenient in such regression problems to think of being given an m-vector f of real functions $f_1, \ldots, f_m$ on a space $\mathcal{X}$, an observation at the point x' of $\mathcal{X}$ yielding the experiment $x_d = (f_1(x') \ldots f_m(x'))$. In such settings we shall always assume $\mathcal{X}$ compact in a topology for which the f_i are continuous. This assumption merely insures that suprema of certain functions are attained, so that optimum designs exist; it is easily weakened, and is completely unnecessary in such places as our invariance considerations (Sections 2E, 3, and 4A). We also assume the f_i to be linearly independent; this is an assumption of identifiability which is easily dispensed with, as is often done in settings like those of Section 4B.

The reader is cautioned to keep clear the distinction between the measures δ and η; the first symbol will always refer to a randomized design; the second, to a non-randomized design (η is replaced by a probability measure $\xi = \eta/N$ in the approximate development described in Section 2C).

Many of the topics covered in this paper were treated also in a paper by the author (1958) and in a paper by the author and Wolfowitz (1959), which we will hereafter refer to as I and II, respectively.

2. Basic Considerations

Before taking up our complete class and optimality results, it is convenient to discuss a few topics which are relevant to considerations of this kind.

A. Randomization and Degrees of Freedom

It was indicated in the previous section that new considerations arise when the problem at hand is not one of point estimation. There are two phenomena here:

(i) It can happen that a design which is inefficient for point estimation is preferred for problems of testing hypotheses, etc., when σ^2 is *unknown*, due to the fact that the loss of accuracy in the relevant b.l.e.'s is more than offset by an increase in the number of degrees of freedom associated with $\bar{S}_d$. A trivial example of this occurs in the degenerate case of the "one-way analysis of variance" where $m = N$ and Δ consists of all matrices for which each row has

$m - 1$ zeros and a single one; here the only design in Δ_R which estimates all components of θ with finite variance takes one observation from each population (X_d is a permutation matrix), but this design is useless for testing hypotheses since $N - b_d = 0$; on the other hand, if r is a positive integer less than m and δ chooses at random among those X_d which have r columns of zeros (i.e., the m observations are taken from $m - r$ populations), we obtain an example of a randomized design for which, for example, we can obtain on the contour $\sigma^{-2}\theta'\theta = c > 0$ a power function for testing the hypothesis $\theta = 0$, whose infimum is greater than the size of the test. A slightly less trivial type of example, which has nothing to do with randomized designs, is given by the regression problem where in $m = 2$, we are interested in inference about θ_2, and the three possible x_d's are $x_1 = (0, 1)$, $x_2 = (1, 1)$, $x_3 = (1, 3.2)$. It is easy to show by the method of Section 4C that, when N is even, the unique best design d^* for minimizing the variance of the b.l.e. of θ_2 is given by $\eta(1) = 0$, $\eta(2) = \eta(3) = N/2$, the variance of the b.l.e. being $1/(1.21N)$. The design d^{**} for which $\eta(1) = N$, $\eta(2) = \eta(3) = 0$, yields the larger variance $1/N$, but also gives one more degree of freedom to error. If, e.g., we are testing the hypothesis that $\theta_2 = 0$ by using the two-tailed Student's test of size a, then for fixed N and small a the power at the alternative $|\theta_2|/\sigma = c$ is approximately $\alpha[1 + 1.21N(N - 2)c^2/2]$ under d^* and $\alpha[1 + N(N - 1)c^2/2]$ under d^{**} when $|c|$ is small (see I). Thus, if we are mainly interested in alternatives which are close to the null hypothesis, and if α is not too large, then the design d^{**} is superior to d^* when $N = 2, 4$, or 6, although d^* is better for point estimation of θ_2. (Throughout our discussion, "α sufficiently small" will include commonly employed values like 0.01 and 0.05.)

(If the reader dismisses these and subsequent examples on the grounds that they are not practical problems, let him be reminded that our examples are being chosen to be simple, and that similar examples exist in more complicated practical settings. Moreover, it seems to the author that the "burden of proof" in such practical settings rests on the proponents of various designs, who should seriously take up the question of whether or not the "intuitively appealing" designs being suggested by them are actually reasonably efficient or not.)

(ii) Even when σ^2 is known, it is generally true in settings where incomplete block designs, orthogonal arrays, Latin squares, etc., are used, that the symmetrical designs (like those just named) which are generally used are *not* optimum for testing the usual hypotheses against alternatives which are near the null hypothesis. This phenomenon can be seen in the following simple example: Suppose $N = 6$, $m = 3$, and again we have the "one-way analysis of variance" setup wherein each row of X_d must have a single one and two zeros, and an experiment d can thus be specified by the numbers of observations n_1, n_2, n_3 taken from each of the three populations. σ^2 is known, so we can assume $\sigma^2 = 1$. We consider three problems: (a) point estimation of θ, (b) testing the hypothesis $\theta_1 = \theta_2 = \theta_3 = 0$, (c) testing the hypothesis $\theta_1 = \theta_2 = \theta_3$; if in (b) and (c) we are mainly interested in alternatives close to the null hypothesis, and the size α in (c) is <0.3, then we shall see that *each of these three problems dictates the use of different design*. For problem (a), with any of the relevant definitions of optimality of Section 4, the nonrandomized design d^* for which $n_1 = n_2 = n_3 = 2$ is obviously optimum. For problem (b), let us compare d^* with the randomized design δ_1 which assigns probability $1/3$ to each of the three possible designs where one n_i equals 6; i.e., δ_1 chooses a population at random and takes all 6 observations from the single chosen population. If d^* is used, we use the usual test based on the χ^2 distribution with three degrees of freedom, and the power function is of the form $\alpha + c_3\lambda + O(\lambda^2)$ when λ is small, where λ, the parameter on which the power function depends, is just the excess of the expected value of the χ^2-distributed statistic over what it is under the null hypothesis; thus, $\lambda = 2(\theta_1^2 + \theta_2^2 + \theta_3^2)$. When δ_1 is employed, if population i is chosen we use the two-tailed normal test of the hypothesis $\theta_i = 0$, or, in other words, the test based on the χ^2-distribution with one degree of freedom, and with a notation parallel to that above we have for the conditional power (given that population i is chosen) $\beta_i = \alpha + c_1\lambda_i + O(\lambda_i^2)$, where $\lambda_i = 6\theta_i^2$; thus, the *un*conditional power associated with the design δ_1 is $(\beta_1 + \beta_2 + \beta_3)/3 = \alpha + c_1\lambda + O(\lambda^2)$. Thus, we are led to compare c_1 and c_3, and it is easily shown that $c_1 > c_3$ for all α (and $c_1/c_3 \to 3$ as $\alpha \to 0$). In fact, a simple way to prove this without computation is to consider the problem of testing whether or not a normal random variable Z, with unit variance, has mean $\mu = 0$, on the basis of Z, W_1, and W_2, where the W_i are normal, independent of each other and of Z, and have unit variances and means known to be zero: the test based on large values of Z^2 is well known to be superior to that based on $Z^2 + W_1^2 + W_2^2$, and c_1 and c_3 are just the derivatives of the two power functions with respect to μ^2, at the origin. One can compare other designs with δ_1 similarly, and conclude that δ_1 is the unique best design for maximizing the minimum power on the contour $\Sigma_i\theta_1^2 = b^2$ for all b in a neighbourhood of the origin. We must still consider problem (c). Here the behavior of the power functions is slightly more delicate. Let δ_2 be the design which chooses each possible pair of populations with probability $1/3$ and then takes 3 observations from each of the two chosen populations. Writing $\bar{\theta} = \Sigma\theta_i/3$ and $\lambda' = 2\Sigma(\theta_i - \bar{\theta})^2$, an argument similar to that used in discussing problem (b) yields $\alpha + c_2\lambda' + O(\lambda'^2)$ for the power function associated with d^* (where c_2 has the corresponding meaning when there are two degrees of freedom) and

$$\alpha + (3/4)c_1\lambda' + O(\lambda'^2)$$

for that associated with δ_2 and which arises from the usual test of $\theta_i = \theta_j$ for whichever pair (i, j) of populations is chosen. One computes easily that $(3/4)$ $c_1 > c_2$ if $\alpha < 0.3$, and a similar comparison of δ_2 with other designs thus shows that δ_2 maximizes the minimum power on the contour $\Sigma(\theta_i - \bar{\theta})^2 = b^2$ for all b in a neighbourhood of the origin, whenever $\alpha < 0.3$.

When σ^2 is unknown, the phenomena of (i) and (ii) reinforce each other. As was remarked earlier, these phenomena are present in all of the standard block and array design situations (see I); the intuitively appealing symmetrical designs which are good for point estimation are relatively poor for testing hypotheses against alternatives near the null hypothesis, while randomized designs which choose as few different "treatments" as possible (one in problems like (ii)(b), two in problems like (ii)(c)) for actual observation, are optimum against such alternatives. For various other problems, e.g., ranking problems and certain problems of interval estimation, similar results hold.

Randomized procedures of the type illustrated by δ_1 and δ_2 above have a certain unappealing property, despite their "local" optimality: in problem (ii)(b), for example, the minimum power on the contour $\Sigma\theta_i^2 = b^2$ when δ_1 is used approaches $(1 + 2\alpha)/3$ as $b \to \infty$. This lack of a desirable consistency property (which requires that this minimum power should approach one as $b \to \infty$) for δ_1 makes one wonder if d^* has some desirable *global* optimality properties, after all. For example, we can define the property of *stringency* of a design, analogous to that of a most stringent test of a hypothesis, as follows: under the restriction to procedures of size α for testing a given hypothesis, let the "envelope power function" $e(\theta')$ be the maximum possible power at the *particular* alternative $\theta = \theta'$ (with obvious modifications if σ^2 is unknown); if β_γ denotes the power function of the procedure γ (this is a design δ together with a test for each possible choice of d if δ is randomized), we define a *most stringent design of size α* to be one for which $L(\gamma) = \sup_\theta[e(\theta) - \beta_\gamma(\theta)]$ is a minimum.

The calculation of a most stringent design is in general quite a formidable task, and is usually much more difficult than the calculation of a most stringent test for a fixed design. However, the calculation can be performed easily in a few simple cases. For example, let us consider a slightly simpler problem than that of (ii)(b), where now $N = m = 2$. Here we shall let d' be the design for which $n_1 = n_2 = 1$, d_i will be the design for which $n_i = 2$, and d' will be the design for which d_1 and d_2 are each chosen with probability $1/2$; d' and δ' are thus the analogues of d^* and δ_1 in problem (ii)(b). The hypothesis is again $\theta_1 = \theta_2 = 0$. Here $e(\theta)$ is easy to compute: for example, if $\theta' = (\theta_1', \theta_2')$ with $\theta_1' > \theta_2' > 0$, then the best procedure against the particular alternative $\theta = \theta'$ is obviously to use design d_1 and reject the null hypothesis if the sum of the observations is large. Thus, we obtain $e(\theta) = \Phi[-k_\alpha + \sqrt{2} \max(|\theta_1|, |\theta_2|)]$, where Φ is the standard normal distribution function and $\Phi(-k_\alpha) = \alpha$. For the power function associated with δ' we obtain

$$\Sigma_i[\Phi(-k_{\alpha/2} + \sqrt{2}\,\theta_i) + \Phi(-k_{\alpha/2} - \sqrt{2}\,\theta_i)]/2,$$

while for d' it is the appropriate non-central χ^2 probability. Noting that $e(\theta)$ depends only on $\max(|\theta_1|, |\theta_2|)$ and that both d' and δ' attain their minimum power on the contour $\max(|\theta_1|, |\theta_2|) = c$ when $\theta_1 = c$ and $\theta_2 = 0$, one can compute directly that $e(\theta) - \beta_\gamma(\theta)$ has a maximum slightly greater than 0.5 when d' is used, and slightly less than 0.5 when δ' is used, when $\alpha = 0.01$. A similar calculation for other designs shows that δ' *is indeed the most stringent design in this case.* (How general this phenomenon is, is not known.) One can certainly conceive of situations in which consistency (in the sense used above) is not the most important factor in choosing a design.

Designs such as δ' have also been criticized on the ground that "they do not provide estimates of all θ_i". If the problem confronting the experimenter is really one of hypothesis testing, this intuitive objection must be rejected as being irrelevant.

In the setting where classical "randomization" is employed, e.g., to eliminate intrablock bias in randomized block designs, an argument like that of I shows that the locally best design for hypothesis testing must again choose (at random) as few treatments as possible to be used, and should then choose their positions at random in the blocks. .

What are the consequences of the facts discussed thus far in this section? Evidently, that if one is really interested in problems such as ones of hypothesis testing, and in criteria such as the local behaviour of the power function (which is a natural extension of various local Neyman-Pearson criteria for a fixed design) or stringency, then one must alter the widely held view that the design to use is necessarily the same one that would be used for point estimation. Moreover, we shall see in Section 3 that, even in problems of point estimation, randomized designs may be needed. (These phenomena also occur in regression problems like those discussed in Section 4C, although we shall see there that there are certain problems in which these phenomena do not occur.) Thus, essentially all of the known "optimality" results in the literature which refer to such designs must be understood to be incorrect for testing hypotheses unless we add the restriction that we limit our consideration to nonrandomized designs, or somewhat alter our optimality criteria. For example, if we use criteria of local optimality or stringency of the type considered above, but add the proviso that the design must yield a power function which is consistent in the sense described above, then the results of Section 4B imply that the classical designs have certain optimum properties. But one must be careful with this kind of hedging; for, if examples like those we have considered may sometimes help us to reconsider our goals more carefully and to realize, for example, that we may not really have a problem of hypothesis testing, nevertheless the rational approach is still to state the problem and the optimality criterion and then to find the appropriate design, and not to alter the statements of the problem and criterion just to justify the use of a design to which we are wedded by our prejudiced intuition.

B. The Use and Misuse of the F-Test

It is generally assumed in the normal case that, whatever design is employed, the appropriate F-test (or the corresponding χ^2-test if σ^2 is known) should be used in testing a hypothesis. Although this assumption is warranted if the design is appropriately symmetrical and we are equally concerned with alternatives in all directions from the null hypothesis, it can be drastically wrong in other cases. For an arithmetically simple example with important practical implications, suppose that with $m = 2$, $N = 4$, $\sigma^2 = 1$, and the problem of testing the hypothesis $\theta_1 = \theta_2 = 0$ in the one-way analysis of variance setup, we use the nonrandomized design d with 2 observations from each population. We are equally interested in alternatives in all directions, and might specify this fact by saying that we want a test of size α which maximizes the minimum power on the contour $\theta_1^2 + \theta_2^2 = c^2$. (Such considerations are discussed in detail by Neyman and Pearson, 1938.) It is well known that, for every $c > 0$, the usual test based on the χ^2-distribution with two degrees of freedom does the job (this is originally an unpublished result of Hunt and Stein; for a proof, see Kiefer, 1957a). Now suppose that we start with the design d, but are plagued by a missing observation and thus end up with the design d' for which

$$ A'_d = \begin{pmatrix} 1 & 0 \\ 0 & 2 \end{pmatrix}. $$

Now, presumably our equal interest in alternatives in all directions should not be "conveniently" changed in the way appropriate to justify the use of the χ^2-test (based on $t_{d'_1^2} + 2t_{d'_2^2}$), by the same accident which reduced us from d to d'. But if we still want a test of size a which maximizes the minimum power on the contour $\theta_1^2 + \theta_2^2 = c^2$, then this test no longer satisfies the criterion, and hence should not be used! Of course, similar considerations apply in examples such as that of A(ii)(c) of this section. We note also that if we had instead specified our interest in various alternatives above by saying that we were equally interested in all θ for which $\max(|\theta_1|, |\theta_2|) = c$ (one can certainly envisage this possibility), and hence wanted to maximize the minimum power on such a contour, then, even for d, the χ^2-test is not appropriate.

In this age of high-speed machinery, it should be easy to determine and table approximately optimum test procedures in situations like the above; it seems to be mainly the inertia of tradition which delays such computations.

Computational difficulties are somewhat reduced if we limit our considerations to the local theory (i.e., to the behavior of the power function near the null hypothesis). Thus, for either of the contours mentioned just above, if we let c approach 0 and rephrase our criterion in terms of derivatives of the power function, the problem reduces to that of finding a regular Neyman-Pearson test of type C (Neyman and Pearson, 1938; the author regrets that this reference was omitted from I through an oversight). In the case of design d, the appropriate test is then the usual one based on the χ^2 distribution, while in the case of design d' the critical region is of the form $at_{d'_1^2} + bt_{d'_2^2} > 1$, where the positive constants a and b can be obtained numerically. This was shown by Neyman and Pearson; a simple demonstration can be obtained by the usual minimax technique of considering Bayes procedures with respect to the *a priori* distribution which assigns probability p_1 to each of $\theta = (0, \varepsilon)$ and $(0, -\varepsilon)$, probability p_2 to $(\varepsilon, 0)$ and $(-\varepsilon, 0)$, and probability $1 - 2p_1 - 2p_2$ to $(0, 0)$, and then letting ε approach 0. It is useful to note that, when α is small, the local behaviour of the regular type C region is well approximated by that of a test with rectangular acceptance region, and the latter is easy to compute.

Thus, if we consider only nonrandomized designs and want to find that design and associated test which approximately maximize the minimum power on small spheres (with a similar treatment for a given family of ellipsoids) about the null hypothesis (we shall call these "designs of type L" in Section 4), we can conclude that a regular type C test is to be used with whatever design is chosen. For randomized designs, the situation is slightly more complicated.

Although it seems reasonable that the statistician should often be able to specify the relative importance of alternatives in each direction, the type C test has often been criticized on the grounds that it *does* depend on such a specification (although perhaps the subconscious motivation for such criticism was often the fact that this criterion did not always lead to the use of the classical test!). Thus, if we make a one-to-one transformation on the parameter space which is twice differentiable and has nonvanishing Jacobian at the null hypothesis, which hypothesis the transformation leaves fixed, then the regular type C tests for the new and old parametrizations will differ unless the transformation is locally a homogeneous stretching times an orthogonal transformation. Motivated by this, Isaacson (1951) defined a type D test of a simple null hypothesis to be a locally strictly unbiased test which maximizes the Gaussian curvature of the power function at the null hypothesis, and noted that a type D test remains invariant under all reparametrizations like the above. Thus, type D tests will appeal to statisticians who want an optimality criterion which does not require any specification of the relative importance of various alternatives, although obviously one must be warned against the use of such a criterion as a panacea for ignorance or laziness. It is shown in I that, in the normal case, the usual F-test is a type D test (or, more precisely, what Isaacson calls a type E test, which is a type D test for each fixed value of the nuisance parameters), with the analogous result for the test based on the χ^2-distribution if σ^2 is known.

It should be noted that this last criterion is the *only* simple optimality criterion which is satisfied by these classical tests no matter what V_d may be. Criteria like those associated with the theorems of Hsu (1941) and Wald (1942) (see also Wolfowitz, 1949), and which look at the power function on just those contours where the F-test has constant power, must be excluded as being unnatural in themselves, except in some cases where these contours are spheres or where, because of the relative importance of various errors, the design was chosen precisely to achieve the particular elliptical contours at hand.

C. Discreteness

A computational nuisance in the determination of optimum designs is the fact that such designs are often easy to characterize for many values of N, but that slight irregularities occur for other values of N. For example, to minimize the generalized variance in the case of cubic regression on the interval $[-1, 1]$, one should take $1/4$ of the observations at each of the values ± 1, $\pm 5^{-1/2}$, if N is divisible by 4. If $N = 5$, the appropriate design takes one observation at each of the values 0, ± 0.511, ± 1, and there are similar irregularities for the other values of N which are not divisible by 4. While it is of interest, especially for small values of N, to have a table of (exactly) optimum designs, the first task would seem to be the computation of a design pattern like that on the four points above, which will yield approximately optimum designs for *all* large N. Thus, in the above example, if N is not divisible by 4 and we take all observations at these four points with as equal a division as possible (many other similar designs will clearly be suitable), then we obviously obtain a design for which the generalized variance is at most $[1 + O(1/N)]$ times the minimum possible. The departure from exact optimality is far less than that of the commonly employed "equal spacing" designs in this situation.

Not only are approximately optimum designs of the type just illustrated convenient from the point of view of tabling designs; in addition, many useful algorithms will be seen below to exist for obtaining such designs, whereas the *exact* optimum designs are in general very difficult to compute in even quite simple situations. This is a familiar obstacle in extremum problems involving discrete variables.

We shall also see, in the discussion following equation (3.3), that randomized designs are unnecessary in point estimation, to within such approximations.

Thus, although we shall sometimes give examples in Sections 3 and 4 which illustrate the differences between the discrete (exact) theory and the continuous (sometimes approximate) theory, we shall most often consider the latter in problems of regression. In the notation introduced at the end of Section 1, these approximate considerations are accomplished by not requiring η to be integer valued. We shall usually write $\eta = N\xi$ and

$$M(\xi) = A(\eta)/N$$

(the information matrix of ξ, per observation), and thus shall be considering probability measures ξ on a space $\mathcal{X}$. These measures will always have finite support, this being the only type which it is practically meaningful or necessary to consider, although other measures can be included without theoretical difficulty (the support of ξ is the smallest closet set of unit ξ measure). The reader is cautioned again not to confuse such a probability measure ξ, which represents a nonrandomized design (more precisely, $N\xi$ does, approximately) with the probability measures δ on Δ, which are randomized designs.

In the setups where balanced block designs, orthogonal arrays, Latin square designs, etc., are customarily employed, we shall see in Section 4B that these classical symmetrical designs possess various optimum properties. What if the restrictions on the problem, e.g., the block sizes and number of blocks, are such that no such design exists? It is tempting to conjecture that a design which is in some sense as "near as possible" to being symmetrical (balanced) is optimum, but this is difficult to make precise in general. Thus, in simple cases like the one-way analysis of variance setup, it is obvious that a design which splits the N observations as equally as possible among the m populations minimizes both the average variance and the generalized variance, but in several more complex situations a corresponding result is not so easy to obtain, and optimum designs are often difficult to characterize when appropriately symmetrical ones do not exist. We shall not be concerned with the tedious optimality calculations in these cases, although for small N a (machine) enumeration of appropriate designs would be of practical use.

When N is large in a regression experiment, or the block size is large in a block experiment, discrete (exact) approximations to the ζ's in an approximate complete class will yield an (exact) ε-complete class (Wolfowitz, 1951) with ε small, with an analogous result for specific optimality criteria, as illustrated in the first paragraph of this subsection.

D. Heteroscedasticity and Variable Cost

If our assumption of constant variance is replaced by the assumption that the covariance matrix of Y_d under design d is $\sigma^2 R_d$, which we assume for convenience to be positive definite, and if B_d is the positive definite symmetric square root of R_d, then replacing Y_d by $B_d^{-1} Y_d$ and X_d by $B_d^{-1} X_d$ puts us in the previous framework. If the experiment d costs c_d and the total cost rather than total number of observations is to be kept constant, then replacing B_d by $c_d^{1/2} B_d$ in the above again returns us to a problem in the previous framework, whose solution yields the desired minimum cost for the problem at hand. (An obvious modification handles the case where the cost of *analyzing* each d is known in advance; such considerations are more difficult for randomized designs.) In the regression setup, analogous remarks apply with c_d and the scalar R_d referring to the observation with expectation $x_d \theta$.

Some further remarks on this subject appear just above the example of Section 3.

E. Invariance (Symmetry)

In many regression problems the functions f_i will be appropriately symmetric with respect to a group of transformations on $\mathscr{X}$, to enable us to conclude that there is a symmetric ζ which is optimum in a given sense. For example, in the case of polynomial regression on the interval $[-1, 1]$ ($f_i(x) = x^{i-1}$), the generalized variance is minimized (i.e., det $M(\xi)$ is maximized) by a ξ which is symmetric about 0 in the sense that $\xi(x) = \xi(-x)$ for each x in $\mathscr{X}$.

To give an example of one such general symmetry result, suppose G is a group of transformations on $\mathscr{X}$ such that, for each g in G, there is a linear transformation $\bar{g}$ on Θ which can be represented as an $m \times m$ matrix of determinant one for which

$$\Sigma_i \theta_i f_i(x) = \Sigma_i (\bar{g}\theta)_i f_i(gx)$$

for all x and θ, where we have written $(\bar{g}\theta)_i$ for the ith component of $\bar{g}\theta$. Then, for fixed g, the problem in terms of the parameters $\bar{\theta} = \bar{g}\theta$ and variable $\bar{x} = gx$, coincides with the original problem. Hence, if ξ minimizes the generalized variance (maximizes det $M(\xi)$) for the original problem in terms of x and θ, it is also optimum in this sense for the problem in terms of $\bar{x}$ and $\bar{\theta}$. But then the measure ξ_g defined by

$$\xi_g(L) = \xi(g^{-1}L)$$

is optimum for the problem in terms of $x = g^{-1}\bar{x}$ and $\bar{\theta}$. Thus, since $\bar{g}$ has determinant one, the inverse $\sigma^{-2}M(\xi_g)$ of the covariance matrix of b.l.e.'s of $\theta = \bar{g}^{-1}\bar{\theta}$ when ξ_g is used has the same determinant as the corresponding matrix $\sigma^{-2}M'(\xi)$ which is relevant when ξ is used. Suppose for the moment that G contains a finite number, say p, of elements. Write

$$\bar{\xi} = \sum_{g \in G} \xi_g/p.$$

Then $\bar{\xi}$ is a probability measure (design) on $\mathscr{X}$ which is symmetric with respect to (invariant under) G in the sense that $\bar{\xi}(L) = \bar{\xi}(gL)$ for every subset L of $\mathscr{X}$ and every g in G. We want to show that $\bar{\xi}$ is optimum.

Now, the experiment $\bar{\xi}$ can be "broken up" into the experiments ξ_g/p. Thus, for estimating θ under $\bar{\xi}$, we clearly have

$$M(\bar{\xi}) = p^{-1}\Sigma_g M(\xi_g), \tag{2.1}$$

and since the $M(\xi_g)$ all have the same determinant, the fact that $\bar{\xi}$ minimizes the generalized variance (maximizes det $M(\xi)$) will follow from the proposition that if det $M(\xi)$ is a maximum for $\xi = \xi'$ and for $\xi = \xi''$, then it is also a maximum for $\xi = \lambda\xi' + (1 - \lambda)\xi''$, where $0 < \lambda < 1$. To prove this last, let C be a nonsingular $m \times m$ matrix such that $CM(\xi')C'$ is the identity and $CM(\xi'')C'$ is diagonal with diagonal elements g_i and (since det $M(\xi') =$ det $M(\xi'')$)$\prod_i g_i = 1$. Since

$$M(\lambda\xi' + (1 - \lambda)\xi'') = \lambda M(\xi') + (1 - \lambda)M(\xi''),$$

our proposition comes down to proving that

$$\Pi_i(1 - \lambda + \lambda g_i) \geqslant 1, \tag{2.2}$$

since strict inequality here is impossible by the maximality of det $M(\xi')$. Now,

$$-\Sigma_i \log(1 - \lambda + \lambda g_i)$$

is convex in λ, and hence is $\leqslant (1 - \lambda)[-\Sigma_i \log 1] + \lambda[-\Sigma_i \log g_i] = 0$. Equation (2.2) follows at once. In fact, this last argument is valid for any optimum ξ' and ξ''; thus, we have proved:

There is a D-optimum ξ which is invariant under G; the set of all D-optimum ξ is a convex set which is invariant under G.

(*D*-optimality, which is discussed in Section 4, refers to minimizing the generalized variance, and invariance of a set of ξ's means that ξ_g is in the set whenever ξ is.) Of course, it may still require considerable computation to determine which symmetric ξ's are optimum.

Similar results hold for other optimality criteria. For example, from the orthogonality of the transformations $\bar{g}$ and the invariance under orthogonal transformations of the trace of the covariance matrix of best linear estimators we obtain that, for estimating θ, the average variance is the same under all ξ_g. Now, assuming G is finite, an unbiased estimator of θ_i under $\bar{\xi}$ can be obtained from the average of the b.l.e.'s of that θ_i under the subexperiments ξ_g/p which go to make up $\bar{\xi}$, and the b.l.e. under $\bar{\xi}$ must do at least as well. We conclude that $\bar{\xi}$ also minimizes the average variance, and again the set of all such minimizers is a convex invariant set.

Similar considerations for the case where we are interested in a subset of the components of θ are given in II. An improvement on these results will be given in Theorem 4.3, where the best possible result of this nature is proved.

Analogous results hold when G is compact or satisfies conditions which yield the usual minimax invariance theorem in statistics and games; see, e.g., Kiefer (1957a).

Throughout this paper, we shall limit our invariance considerations to nonrandomized designs; the reader will find it easy to obtain analogues for randomized designs. (For examples, note the designs δ_1 and δ_2 in Section 2A(ii).)

Of course, the validity of our results on invariant designs depends crucially on the invariance of the optimality criterion under the group of transformations. Thus, for example, in the trivial case of linear regression on $[-1, 1]$, where $f_i(x) = x^{i-1}$ for $i = 1, 2$, we can conclude that there is a symmetrical design which is optimum for estimating θ_1, one which is optimum for estimating θ_2, or one that minimizes the average (or the generalized) variance for estimating θ_1, θ_2; but for estimating $\theta_1 + \theta_2$, the unique optimum design is $\xi(1) = 1$. Thus, our symmetry considerations do not generally yield anything useful for the complete class results of Section 3 in the way that they do for particular optimality criteria; in fact, the polynomial example of Section 3 shows that ξ can be admissible without the corresponding $\bar{\xi}$ being admissible. However, we shall see in Theorem 3.3 that a subset of the designs which are invariant under a group of transformations each of which leaves fixed the same s parameters, say $\theta_1, \ldots \theta_s$, is an essentially complete class for any problem concerned only with $\theta_1, \ldots, \theta_s$.

It is to be noted that the optimum properties of balanced block designs, Latin square designs, orthogonal arrays, etc. (see Section 4B), are really symmetry results in the same sense as those we have considered here, but in a much more complex setting wherein we are interested only in certain linear functions of the θ_i; the proof that an appropriately symmetric design (if it exists) is optimum in the exact theory, is therefore more difficult.

The reader should keep in mind that our use of the word "symmetry" has nothing to do with its usage in reference to balanced incomplete block designs where the parameters b and u are equal. This is also true in Section 4B.

3. Admissibility and Complete Classes

We begin with some perfectly general considerations, and shall use the example of polynomial regression on a real interval for illustration at the end of the section.

By a *statistical problem* π we mean a specification of the possible states of nature, decisions, and losses (including costs of various possible experiments and of their analyses), with perhaps some restriction on the class of procedures we consider. We shall say of two designs δ and δ' in Δ_R that δ' is *at least as good* (π) as δ if for each risk function for problem π which is attainable by some procedure under δ, there is a risk function attainable by some procedure under δ' which is nowhere larger. If δ' is at least as good as δ for each of a collection P of problems π, we shall say that δ' is at least as good (P) as δ. If δ' is at least as good (P) as δ but not vice versa, we shall say that δ' is *better* (P) than δ, or that δ' *dominates* (P) δ. If δ' is such that no δ'' dominates (P) δ', then we say that δ' is *admissible* (P). A class C of designs is *complete* (P) (resp., *essentially complete* (P)) if for any δ not in C there is a δ' in C which is better (P) than (resp., as good (P) as) δ. If no proper subset of C has this property, C is said to be *minimal complete* (P) (resp., minimal essentially complete (P)).

The reader should be aware that the complete class considerations of Ehrenfeld (1955b) and of the present section, in the nonparametric case, have more restricted meaning than is usual in decision theory: we consider only best linear (minimax) estimators, and a complete class of experiments supplies us with a good experiment (and estimators) for any such problem. Broader complete class results which also compare nonlinear estimators depend on the extent of the nonparametric class.

Let $\mathscr{F} = \{F\}$ be the space of possible states of nature. For d in ∇, let $\mathscr{Y}_d$ be the space of possible complexes of observed values under d (thus, if Δ consists of experiments with N real observations, we could take each $\mathscr{Y}_d$ to be Euclidean N-space). An appropriate Borel field is given on each $\mathscr{Y}_d$. Let $H_F(A, d)$ be the probability that the outcome of the experiment will be in the measurable subset A of $\mathscr{Y}_d$, when experiment d is performed and F is the true state of nature. An experiment d' is said to be *sufficient* for the experiment d'' if there is a function $q(A, y)$ which for each y is a probability measure on $\mathscr{Y}_{d''}$ and for each A is a measurable function on $\mathscr{Y}_{d'}$, and such that

$$H_F(A, d'') = \int_{\mathscr{Y}_{d'}} q(A, y) H_F(dy, d') \tag{3.1}$$

for every measurable subset A of $\mathscr{Y}_{d''}$ and every F in $\mathscr{F}$. This is really the same notion of sufficiency that one is used to for comparing two random vectors for a fixed experiment; it says that d' is sufficient for d'' if we can imitate d'' probabilistically from d' in the way indicated here: observe the outcome y (say) of d', use the measure $q(A, y)$ to obtain a random variable Y'' with values in $\mathcal{Y}_{d''}$, and on integrating over possible values of y note that Y'' has the same distribution according to each F as would the outcome of experiment d''. Of course, q depends on d' and d'', but not on F.

Let $\mathcal{X}$ be the space of all pairs (d, y_d) for y_d in $\mathcal{Y}_d$ and d in Δ. If all $\mathcal{Y}_d$ can be taken to be the same space $\mathcal{Y}$, then $\mathcal{X}$ is just the cartesian product of Δ and $\mathcal{Y}$. In any event, we assume a Borel field to be given on $\mathcal{X}$ which includes the measurable subsets of $\mathcal{Y}_d$ as sections at d and which also includes each set which is the union over d in Δ' of all $\mathcal{Y}_d$, whenever Δ' is any subset of Δ which is a member of the Borel field with respect to which the δ's in Δ_R are measures. Since we are usually to be concerned with δ's which give measure one to a finite set, these considerations will be trivial. We now think of δ as being a probability measure on Δ. When δ is used and F is the true state of nature, the probability that the chosen design r (say) and subsequent outcome y_r of the experiment will satisfy $(r, y_r) \in B$, where B is a measurable subset of $\mathcal{X}$, is just

$$G_F(B; \delta) = \int_\Delta H_F(B_r, r)\delta(dr),$$

where $B_r = \{y_r : (r, y_r) \in B\}$. For randomized experiments, then, we say that δ' is sufficient for δ'' if there is a function $Q(B, z)$, which is a measure on $\mathcal{L}$ for each z and a measurable function on $\mathcal{L}$ for each B, such that

$$G_F(B; \delta'') = \int_{\mathcal{X}} Q(B, z)G_F(dz; \delta') \tag{3.2}$$

for each measurable B and each F in $\mathcal{F}$. This is the appropriate generalization of the previous definition, it now being necessary to think of δ as yielding an "observed value" in $\mathcal{X}$.

In the case of nonrandomized designs, sufficiency and its equivalence to certain other criteria have been treated extensively by D. Blackwell, S. Sherman, C. Stein, and others (see, e.g., Blackwell, 1950). The most important result for our considerations is that, under standard measurability assumptions, if P^* is the class of all possible problems (usually a much smaller subclass will suffice), then d' is at least as good (P^*) as d'' if and only if d' is sufficient for d''. Of course, the "if" part of this is an immediate consequence of the definition of sufficiency.

From the definition of sufficiency for randomized designs (in particular, the correspondence of (3.2) to (3.1) when $\mathcal{X}$ is made to correspond with $\mathcal{Y}_{d'}$ and $\mathcal{Y}_{d''}$, which enables us to think of a randomized design as a nonrandomized design with sample space $\mathcal{X}$), we see at once that all of the equivalence results just mentioned have natural analogues in the case of randomized designs; in particular, δ' is at least as good (P^*) as δ'' if and only if δ' is sufficient for δ''.

We shall now consider the setup of Section 1, where we shall discuss some results of Ehrenfeld (1955b). (Ehrenfeld's definition of completeness and "at least as good" differ from ours in that θ is fixed throughout. This does not matter in the linear theory now to be discussed, but yields different results in the nonlinear asymptotic case to be discussed a little later, where it yields a smaller "essentially complete class" which perhaps conforms less to the usual decision-theoretic meaning of such classes.)

If A and B are $m \times m$ matrices, we shall write $A \geqslant B$ if, for every m-vector w, we have $w'Aw \geqslant w'Bw$; or, equivalently, if $A - B$ is nonnegative definite. If $A \geqslant B$ and $A \neq B$, we write $A > B$. Ehrenfeld (1955b, pp. 59–61) shows by an algebraic argument that, if $A_{d''} \geqslant A_{d'}$, then any linear parametric function $\Sigma_i c_i \theta_i$ which is estimable under d' is estimable under d'', and its b.l.e. under d'' has a variance no greater than the variance of the b.l.e. under d'. We now give a transparent proof of this result: Assume normality and that σ^2 is known to be 1, since the linear theory (consideration of b.l.e.'s and their variances) does not depend on this. If $A_{d''} - A_{d'} = B$, which is nonnegative definite, let d^* be a "fictitious experiment" such that $X'_{d^*} X_{d^*} = B$ (such a matrix X_{d^*} clearly exists), with Y_{d^*} independent of $Y_{d'}$. Let $T_d = X'_d Y_d$; then T_d is a sufficient statistic under d, and $T_{d'} + T_{d^*}$ is sufficient under the combination of experiments d' and d^*. But $T_{d''}$ has the same distribution as $T_{d^*} + T_{d'}$, from which it is immediately clear that d'' is sufficient for d' and hence that Ehrenfeld's result holds. In fact, denoting by P_N^* the class of *all* problems under the assumption of normality and known variance, and by P_L the corresponding class of all nonparametric problems of *point estimation* of any linear parametric functions, where we restrict ourselves to linear estimators (as mentioned earlier in this section, the consequent limited meaning of complete (P_L) class results should be kept in mind), we have the following slightly more precise version of Ehrenfeld's result:

Theorem 3.1. *The following are equivalent*:

(a) $A_{d''} \geqslant A_{d'}$,
(b) d'' is at least as good (P_N^*) as d',
(c) d'' is at least as good (P_L) as d'.

These equivalences hold with "$\geqslant$" replaced by "$>$" and "at least as good as" by "better than".

For randomized designs, (c) is equivalent to (a) with A_d replaced by the inverse of the average of V_d with respect to δ, with an obvious modification in the singular case.

It is clear that we could also replace P_L in (c) by problems of vector estimation of all vectors of linear parametric functions; for example, we could treat one vector estimator as being at least as good as another if the variances of components for the former are all no greater than the corresponding variances of the latter (vector risk), or if the covariance matrix of the latter minus that of the former is nonnegative definite, or if the average (or generalized) variance is no greater for the former than the latter; no generality is gained in any case. The essential fact is that, if B and C are positive definite symmetric matrices, then $C \geqslant B$ if and only if $B^{-1} \geqslant C^{-1}$.

If we restrict ourselves to using nonrandomized designs, we see at once that d is admissible (P_N^* or P_L) if and only if there is no d' for which $A_{d'} > A_d$. Thus, if the class of possible A_d for d in Δ is compact as a set of points in m^2-space (this can be weakened), and if we restrict ourselves to designs in Δ, we have an immediate characterization of the minimal complete class (P_N^* or P_L).

If σ^2 is unknown and if we do not restrict ourselves to the point estimation problems in P_N^*, we must also consider the varying number of degrees of freedom associated with different designs, as discussed in Section 2A. This should also be kept in mind in evaluating the meaning of some further complete class results which will be discussed below.

What happens to the above results if we consider Δ_R rather than Δ? As noted in Section 2A(ii), further considerations are necessary in even the simplest problems of normal hypothesis testing; if we consider only the *local* theory there and σ^2 is known, and if we compare δ's which always yield the same number of degrees of freedom and use the F-test (see, however, Section 2B), then the matrices A_δ take the place of A_d in all of the above, although such considerations do not suffice outside of the local theory. For problems of linear (or normal) estimation with quadratic loss, it is the average of V_d with respect to δ that is relevant; more precisely, if c_d is the variance of the b.l.e. of a given linear parametric function when d is used, then the variance under δ is the average with respect to δ of c_d, with a corresponding statement for joint estimation of several linear parametric functions. If δ gives positive probability to a set of d's for which a given linear parametric function is not estimable, then the corresponding variance (more precisely, risk) under δ is infinite.

We have seen in Section 2A that randomized designs cannot in general be ruled out of consideration in hypothesis testing, and that is also true in estimation theory. For example, if $m = 2$ and $N = 3$ in the setup of Section 2A(ii), the generalized variance is a minimum for that δ which assigns probability $\frac{1}{2}$ to each of the two designs for which one n_i is 1 and the other n_i is 2. However, if Δ is sufficiently rich, such uses of randomized designs in estimation problems will be unnecessary. For suppose μ is a probability measure on Δ. Then μ may make sense as a *nonrandomized* experiment ξ (see Sections 1 and 2C), with corresponding information matrix $M(\mu)$, but we can also think of a *randomized* experiment δ as being represented by μ. We assert that *the former is always at least as good (P_L) as the latter*. It will be enough to consider the case where μ gives measure one to a set of d's for which A_d is nonsingular, other cases being handled by an appropriate passage to the limit, or by considering appropriate linear transformations and submatrices. The assertion is then a consequence of the following trivial lemma, which in fact tells us that, if μ makes sense as a nonrandomized experiment, then as a randomized experiment it is either *inadmissible (P_L)* or else gives measure one to a set of d's *all of which have the same A_d*:

Lemma. *If A and B are symmetric, positive definite $m \times m$ matrices and $0 < \lambda < 1$, then*

$$\lambda A + (1 - \lambda)B \geqslant (\lambda A^{-1} + (1 - \lambda)B^{-1})^{-1}, \tag{3.3}$$

with equality if and only if $A = B$.

This follows at once on simultaneously diagonalizing A and B. A generalization will be proved in Lemma 3.2 below.

Thus, to within the approximation adopted in Section 2C, if σ^2 is known it is unnecessary to use randomized designs for P_L; thus, the *usefulness of randomized designs for P_L will be greatest in the small sample-size or small block-size case*, where the discrete effects discussed in Section 2C are the largest.

Of course $M(\mu) = A_\mu$ in the above discussion, but this does not mean that randomized designs are unnecessary in normal hypothesis testing, even if σ^2 is known; we have discussed this in Section 2A.

It can sometimes happen that we are interested only in a subclass P'_L of problems in P_L which are concerned with a fixed, proper subset of the m parameters (or of a set of less than m linear parametric functions), say $\theta_1, \ldots, \theta_s$. Let

$$A_d = \begin{pmatrix} B_d & C_d \\ C'_d & D_d \end{pmatrix},$$

where B_d is $s \times s$. Then, if D_d and D'_d are nonsingular (with obvious modifications if they are not), we have that d is at least as good (P'_L) as d' if

$$B_d - C_d D_d^{-1} C'_d \geqslant B_{d'} - C_{d'} D_{d'}^{-1} C'_{d'},$$

the latter being the inverse (if it exists) of the covariance matrix of b.l.e.'s of $\theta_1, \ldots, \theta_s$ under d'. One can now imitate here our previous discussion in the case $s = m$. Instead of doing all this in detail, we shall only give a needed generalization of the previous lemma, and shall then apply this to obtain a generalization to the present case of the invariance considerations of Section 2E.

Lemma 3.2. *For $j = 1, \ldots, r$, let C_j be $s \times (m - s)$, let D_j be nonsingular positive definite and symmetric, and suppose $\lambda_j > 0$, $\Sigma \lambda_j = 1$. Then*

$$[\Sigma \lambda_j C_j][\Sigma \lambda_j D_j]^{-1}[\Sigma \lambda_j C_j]' \leqslant \Sigma \lambda_j C_j D_j^{-1} C'_j, \tag{3.4}$$

with equality if and only if all $C_j D_j^{-1}$ are equal.

Once the result is proved for the case $r = 2$, it follows easily for general r by induction. In the case $r = 2$, let F be such that $FD_1 F'$ is the $(m - s) \times (m - s)$ identity I and $FD_2 F'$ is a diagonal matrix Q, and let $E_i = C_i F'$. Write $\lambda_1 = \lambda$. Then (3.4) can be written

$$[\lambda E_1 + (1 - \lambda)E_2][\lambda I + (1 - \lambda)Q]^{-1}[\lambda E_1 + (1 - \lambda)E_2]'$$
$$\leq \lambda E_1 E_1' + (1 - \lambda)E_2 Q^{-1} E_2',$$

which, on writing out inverses and collecting terms, becomes

$$-\lambda(1 - \lambda)[E_1 - E_2 Q^{-1}][\lambda Q^{-1} + (1 - \lambda)I]^{-1}[E_1 - E_2 Q^{-1}]' \leq 0;$$

but this inequality is always true, with equality if and only if $E_1 = E_2 Q^{-1}$. $\cdot$
This proves the lemma.

Now suppose that G is a group satisfying the conditions of Section 2E, and in addition that each $\bar{g}$ acts trivially on $\theta_1, \ldots, \theta_s$; i.e. that $(\bar{g}\theta)_i = \theta_i$ for $1 \leq i \leq s$, and hence $f_i(gx) = f_i(x)$ for $1 \leq i \leq s$. Suppose ξ is any design, with $M(\xi)$ partitioned as

$$\begin{pmatrix} B(\xi) & C(\xi) \\ C(\xi)' & D(\xi) \end{pmatrix},$$

as above. For any g we have

$$M(\xi_g) = \begin{pmatrix} B(\xi) & C(\xi)F_g \\ F_g' C(\xi)' & F_g' D(\xi)F_g \end{pmatrix},$$

where F_g is a nonsingular $(m - s) \times (m - s)$ matrix. Hence, if $D(\xi)$ is non-singular (an obvious modification again sufficing otherwise), we have

$$B(\xi) - C(\xi)D(\xi)^{-1}C(\xi)' = B(\xi_g) - C(\xi_g)D(\xi_g)^{-1}C(\xi_g)'.$$

Suppose now that G is finite (other groups are handled as in Kiefer (1957a), the determinant of $\bar{g}$ no longer being required to be 1 as it was in Section 2E). An application of Lemma 3.2 and the fact that $B(\xi_g) = B(\xi)$ yield

$$B(\bar{\xi}) - C(\bar{\xi})D(\bar{\xi})^{-1}C(\bar{\xi})' \geq B(\xi) - C(\xi)D(\xi)^{-1}C(\xi)'.$$

Thus, recalling the remark which follows the proof of (3.3), we have the following result in the approximate theory:

Theorem 3.3. *If G is as specified above and every $\bar{g}$ leaves $\theta_1, \ldots, \theta_s$ fixed, then the designs which are invariant under G form an essentially complete (P_L') class.*

As noted in Section 2E we could extend this result to Δ_R and thus obtain an essentially complete class relative to problems in P_N^* concerned with θ_1, $\ldots, \theta_s$.

In Section 4A we shall give a corollary of Theorem 3.3 which refers to specific optimality criteria.

We now turn to the question of characterizing admissibility and complete-ness (P_L) in regression experiments in terms of the x_d defined in Section 1. As noted by Ehrenfeld, if $x_{d'} = a x_d$ with $|a| > 1$ (Ehrenfeld requires $a > 1$, but that is unnecessary), then $x_{d'}' x_{d'} > x_d' x_d$; hence, defining x_d to be maximal if there is no $x_{d'} = a x_d$ with $|a| > 1$, if $\{x_d : d \varepsilon \Delta\}$ is a compact set in R^m we conclude at once that, in either the approximate or exact theory, the class of all experiments η which use only maximal x_d is complete (P_L).

If the set of all A_d is convex (in particular, if we consider the approximate theory) and compact, there is a revealing game-theoretic approach: consider the zero-sum, two-person game with payoff function $K(d, t) = t'A_d t$, where the space of player 1's pure strategies is Δ and that of player 2 is $\{t: t't = 1\}$. It is well known (Wald, 1950) that a complete class for player 1 consists of all A_d which are maximal with respect to some mixed strategy of player 2, and this is clearly the same as the class of all A_d which are maximal with respect to some pure strategy of player 2. Thus (note also the remark preceding (3.3)), we have

Theorem 3.4. *If $\{A_d: d\varepsilon\Delta\}$ is convex, closed, and bounded, then*

$$\bigcup \{d' : t'A_{d'}t = \max_d t'A_d t\}$$

is complete (P_L).

The characterization of a minimal complete class can be given in the manner of Wald and Wolfowitz (1950).

In the setting of Theorem 3.4, it is also evident that the set of all X_d for which $t'X_d'$ is maximal for some t (i.e., for which there is a t such that for no a with $|a| > 1$ is there a d' with $t'X_{d'} = at'X_d$) is complete. It is obvious how to improve this result.

In the exact theory of P_L, Ehrenfeld considers further the case where the convex closure of $\{x_d: d\varepsilon\Delta\}$, or of the set of maximal x_d, has r extreme points (from the discussion of maximal x_d above, it is evident that it will suffice to consider a set which is often somewhat smaller than the set of all x_d; for example, each x_d can be replaced by $\pm x_d$ where the sign is chosen to make the first nonzero coordinate of x_d positive). In this case in the exact theory any N-observation experiment can be replaced by a better $(N + r)$-observation experiment using only the r extreme points (of course, $N + r$ would be replaced by N in the approximate theory). These developments are really all direct consequences of the simple inequality,

$$[\alpha X_1 + (1 - \alpha)X_2][\alpha X_1(1 - \alpha)X_2]' \leqslant \alpha X_1 X_1' + (1 - \alpha)X_2 X_2'$$
for $0 \leqslant \alpha \leqslant 1$.

In such a simple problem as that of quadratic regression on a finite interval (see below), there are infinitely many extreme points and the above criterion is useless in the exact theory. A more useful device in such cases would be a result like that claimed by de la Garza (1954), who stated in the polynomial case ($\mathscr{X} = [-1, 1], f_i(x) = x^{i-1}$ for $1 \leqslant i \leqslant m$) that for any N-observation experiment there is another N-observation experiment with the same information matrix, and which takes observations at no more than m different values in $\mathscr{X}$; unfortunately, this result appears to be incorrect (e.g., when

$m = 3$, $N = 4$, and observations are taken at -1, $-\frac{1}{2}$, $\frac{1}{2}$, 1), and the correct results seem complicated. The corresponding result in the approximate theory is trivial: it is an elementary fact in considerations of the moment problem (see, e.g., Shohat and Tamarkin, 1943) that for any probability measure ξ on $[-1, 1]$ there is a probability measure ξ' with the same moments of orders, $1, 2, \ldots, 2m - 2$ (these are elements of $M(\xi)$), and such that ξ' has support on at most m points. More generally, if all f_i are continuous functions on a compact $\mathcal{X}$, and if there are H different functions of the form $f_i f_j (i \leq j)$, then the set of all matrices $M(\xi)$ can be viewed as a closed subset S of m^2-dimensional space which is convex (since $M(\xi)$ is linear in ξ) and of dimension $J \leq H$ (here J is equal to the number of nonconstant, linearly independent functions $f_i f_j$, $i \leq j$). Any extreme point of S is attainable as the $M(\xi)$ corresponding to a ξ with support at a single point. Thus (see the remark following the proof of (3.3)), *the set of all ξ with support on at most $J + 1$ points is essentially complete (P_L) in the approximate theory.*

We shall investigate questions of admissibility (minimality of such classes) in an example, a little later. Of course, all these results are generally meaningful only as long as we do not consider problems such as ones of hypothesis testing.

Ehrenfeld has also considered *asymptotic essential completeness* in the case where the expected value of an observation is a *nonlinear* function of θ, say $F_d(\theta)$ corresponding to x_d (for convenience, we still assume normality). Under appropriate regularity conditions (and a precise statement of the results which takes account of Hodges' superefficiency phenomenon), one can characterize an asymptotically essentially complete class by replacing x_d in our previous (linear) considerations by the vector $x_d(\theta) = \{\partial F_d(\theta)/\partial\theta_i, 1 \leq i \leq m\}$ (note that, under suitable assumptions, $x_d(\theta)x_d(\theta)'$ is a normalized limiting covariance matrix for the corresponding maximum likelihood estimator). However, the class of all experiments on the set $\{d: x_d(\theta)$ is maximal for some $\theta\}$, which is what Ehrenfeld considered, is not asymptotically essentially complete in the usual decision-theoretic sense, but only in the sense that for each *fixed* θ it yields a set of $M(\theta, \xi)$ (corresponding to our earlier $M(\xi)$) which dominate all other $M(\theta, \xi)$. Thus, for example, if $m = 1$, $\mathcal{X} = [0, 1]$, $0 \leq \theta \leq 1$ (a slightly smaller interval will achieve the same result while keeping $\partial F_x(\theta)/\partial\theta$ bounded), and $F_x(\theta) = \sqrt{(\theta_x)} + \sqrt{[(1 - \theta)(1 - x)]}$, we obtain easily that $\partial F_x(\theta)/\partial\theta$ is maximal only for $x = 0$ (if $\theta \geq \frac{1}{2}$) or $x = 1$ (if $\theta \leq \frac{1}{2}$); but no allocation of N observations at the two values $x = 0$, $x = 1$ can achieve an asymptotic variance function which is at least as good *for all* θ as the experiment which puts all observations at $x = \frac{1}{2}$. What is needed here, in the same role as the Bayes procedures which enter into complete class theorems in the usual decision theory, is the class of all experiments ξ for which $\int M(\theta, \xi)\mu(d\theta)$ cannot be dominated for at least one probability measure μ on the parameter space, or perhaps the closure of this class in a suitable sense (see, e.g. Wald, 1950 or LeCam, 1955; it will usually suffice, for example, to take the closure of the set of ξ's corresponding to μ's with finite support, both here and also in an exact or nonasymptotic development). Thus, as in Theorem 3.4, we can obtain an asymptotically essentially complete class as *the class of all ξ which, for some t and μ, maximize $\int [t'M(\theta, \xi)t]\mu(d\theta)$* (or as a closure of this class, as indicated above). In these asymptotic considerations, there is essentially no difference between the exact and approximate theories.

If the linear problems P_L are modified by allowing the matrix R_d of Section 2D to depend on θ, then considerations of the above type are relevant.

We now turn to an important example, to illustrate some of the concepts of this section.

The Polynomial Case

Suppose $\mathscr{X}$ is a compact interval, which we can take to be $[-1, 1]$ without affecting the results below, and that $f_i(x) = x^{i-1}$, $1 \leqslant i \leqslant m$; we hereafter write $m = k + 1$ for convenience. For any design ξ, the $(i, j)^{th}$ element of $M(\xi)$ is the moment of order $i + j - 2$ of ξ:

$$m_{ij}(\xi) = \mu_{i+j-2}(\xi) = \int_{-1}^{1} x^{i+j-2}\xi(dx), \qquad 1 \leqslant i, j \leqslant k + 1.$$

Suppose now that $M(\xi') \geqslant M(\xi'')$. Of course, $m_{11}(\xi') = m_{11}(\xi'') = 1$. If t is an m-vector with first component $t_1 = 1$, $(r + 1)st$ component $t_{r+1} = u$, and all other components $t_i = 0$, we have

$$0 \leqslant t'[M(\xi') - M(\xi'')]t = 2u[\mu_r(\xi') - \mu_r(\xi'')] + u^2[\mu_{2r}(\xi') - u_{2r}(\xi'')]$$

for all u. Hence, $\mu_r(\xi') = \mu_r(\xi'')$ for $0 \leqslant r \leqslant k$. Repeating this argument with $t_{q+1} = 1$, $t_{s+1} = u$ (with $s > q$), and all other $t_i = 0$, first for $q = k$ and then for successively larger q, we obtain $\mu_r(\xi') = \mu_r(\xi'')$ for $0 \leqslant r \leqslant 2k - 1$. Thus, finally,

Lemma 3.5. $M(\xi'') > M(\xi'')$ *if and only if* $\mu_r(\xi') = \mu_r(\xi'')$ *for* $0 \leqslant r \leqslant 2k - 1$ *and* $\mu_{2k}(\xi'') > \mu_{2k}(\xi'')$.

Of course, this criterion can be used in both the exact and approximate developments, but we shall now see that we obtain more elegant results in the latter. (If we compare two procedures with different values of N, the two values of m_{11} differ and Lemma 3.5 does not apply.)

Consider in $2k$-dimensional space the set Q of all points $\mu(\xi) = (\mu_1(\xi), \ldots, \mu_{2k}(\xi))$ for ξ a probability measure on $\mathscr{X}$. Write Q^* for the corresponding set of points

$$\mu^*(\xi) = \mu_1(\xi), \ldots, \mu_{2k-1}(\xi)).$$

From the work of Karlin and Shapley (1953) (especially Theorems 11.1 and 20.2 and the discussion on pp. 60 and 64, where the results given for the interval $[0, 1]$ are easily transformed to the present setting) we have that Q is a convex body whose extreme points correspond to ξ's giving measure one to a single point; that μ in Q corresponds to a unique ξ if and only if μ is in the boundary of Q; that for μ^* in the interior of Q^* there is a nondegenerate line of points $\mu(\xi)$ for which $\mu^*(\xi) = \mu^*$, and among these $\mu_{2k}(\xi)$ is maximized by a ξ whose support is $k + 1$ points including 1 and -1; and that for μ^* in the boundary of Q^* there is a unique ξ for which $\mu^*(\xi) = \mu^*$, any such ξ being a limit of ξ's for which $\mu^*(\xi)$ is in the interior of Q^* and $\mu_{2k}(\xi)$ a maximum for that μ^*. Using these facts the well known result (Shohat and Tamarkin (1943), p. 42) that there is at most one ξ with given moments $\mu_1, \ldots, \mu_{2k}$ whose support is a set of at most $k + 1$ points containing a given point, the remark following the proof of (3.3), and Lemma 3.5, we have the following result in the continuous development:

Theorem 3.6. *The minimal complete and minimal essentially complete class of admissible ξ in the polynomial case consists of all ξ whose support consists of at most $k + 1$ points, at most $k - 1$ of which are in the interior of $\mathscr{X}$.*

Before turning to the exact theory, let us look briefly at the changes which occur in the above results if we replace the space $[-1, 1]$ in this example by a finite set of points, say $r_1, \ldots, r_L$. We can no longer limit our consideration to ξ's supported by $k + 1$ points. Writing $\xi = (\xi_1, \ldots, \xi_L)$, to determine whether or not there is a ξ'' better than a given ξ' we write $\xi'' - \xi' = \gamma = (\gamma_1, \ldots, \gamma_L)$ and find, as in Lemma 3.5, that such a ξ'' exists if and only if, for some $c > 0$, the linear equations

$$\sum_{i=1}^{L} \gamma_i r_i^j = 0, \qquad 0 \leqslant j < 2k,$$

$$\sum_{i=1}^{L} \gamma_i r_i^{2k} = c$$

have a solution $\gamma(c)$ such that all components of $\xi' + \gamma(c)$ are nonnegative. Let $c_{\xi'}$ be the largest value of c for which such a solution exists. The set of all ξ of the form $\xi' + \gamma(c_{\xi'})$ (for all ξ') is then the minimal complete (P_L) class. Thus, the determination of admissible ξ in this case becomes a linear programming problem.

Now let us consider the exact theory with $\mathscr{X} = [-1, 1]$. To avoid too much arithmetic, we shall limit our considerations to procedures in Δ. In the linear case ($k = 1$) it is easy to characterize the admissible procedures. For, applying Lemma 3.5, and noting that for a given value of μ_1 we maximize μ_2 if and only if at most one observation is in the interior of $\mathscr{X}$, we conclude that the class of designs with this property is the minimal essentially complete and minimal complete (P_L) class for any given N. For $k > 1$, the situation is more complicated, as we shall now indicate by mentioning a phenomenon which occurs when $k = 2$. When $N = 3$, it is easy to see that a given set of moments μ_1, μ_2, μ_3 is achievable by at most one design, so all designs are admissible.

For larger N, however, a design supported at three points may or may not be admissible. For a given N, we can think of the set of possible $\mu(\xi)$ as the subset Q' of Q for which ξ takes on only multiples of $1/N$ as values. For a given μ^* in Q^*, there may be zero or some positive number of points $\mu(\xi)$ in Q' for which $\mu^*(\xi) \in Q^*$, and for each μ^* for which such a ξ exists, we must again select that ξ for which $\mu^*(\xi) = \mu^*$ and $\mu_{2k}(\xi)$ is a maximum. Roughly speaking, the larger N, the more points will there be in Q' corresponding to a given μ^*. For example, consider the symmetrical ξ's on at most three points, say $\xi(b) = \xi(-b) = J/N$ and $\xi(0) = 1 - 2J/N$ (where J is an integer), which we represent as the pair (b, J). For a given value $\mu_2 > 0$, the designs

$$([N_{\mu 2}/2J]^{1/2}, J), \qquad N_{\mu 2}/2 \leqslant J \leqslant N/2,$$

all have $\mu^* = (0, \mu_2, 0)$. Thus, a design $(b', J - 1)$ is admissible among the symmetric distributions on at most three points if and only if no design $(b', J - 1)$ of this class exists with the same μ^*; i.e., if and only if $b > [(J - 1)/J]^{1/2}$. We see here the way in which the continuous theory approximates the exact for large N. The general considerations are arithmetically rather messy.

4. Specific Optimality Criteria

We now turn from the characterization of complete classes to the determination of designs which satisfy particular optimality criteria. We shall limit our considerations to designs in Δ; see the previous sections and I for discussions of necessary modifications if Δ_R is considered.

A. Various Criteria and Their Relationship

Suppose we are interested in inference concerning s given linearly independent parametric functions $\psi_j = \Sigma_i c_{ji}\theta_i$, $1 \leqslant j \leqslant s$. Let Δ' be the class of designs in Δ for which all ψ_j are estimable, and let $\sigma^2 v_d$ be the covariance matrix of b.l.e.'s of the ψ_j for a d in Δ'. For testing the hypothesis that all $\psi_j = 0$ under the assumption of normality, when a given design d is used, let $\bar{\beta}_\phi(c)$ be the infimum of the power function of the test ϕ over all alternatives for which $\Sigma_j \psi_j^2/\sigma^2 = c$, and let $\bar{\beta}(d, c, \alpha)$ be the supremum of $\bar{\beta}^\phi(c)$ over all ϕ of size α. As in I, we now consider five optimality criteria.

For $c > 0$ and $0 < \alpha < 1$, a design d^* is said to be $M_{\alpha,c}$-optimum in Δ if

$$\bar{\beta}(d^*, c, \alpha) = \max_{d \in \Delta} \bar{\beta}(d, c, \alpha).$$

A design d^* is said to be L_α-optimum in Δ if

$$\lim_{c \to 0} [\bar{\beta}(d^*, c, \alpha) - \alpha] \bigg/ \left[\sup_{d \in \Delta} \bar{\beta}(d, c, \alpha) - \alpha\right] = 1.$$

A design d^* is said to be *D-optimum in* Δ if $d^* \in \Delta'$ and

$$\det v_{d^*} = \min_{d \in \Delta'} \det v_d.$$

A design d^* is said to be *E-optimum in* Δ if $d^* \in \Delta'$ and

$$\pi(v_{d^*}) = \min_{d \in \Delta'} \pi(v_d),$$

where $\pi(v_d)$ is the maximum eigenvalue of v_d,

A design d^* is said to be *A-optimum in* Δ if $d^* \in \Delta'$ and

$$\text{trace } v_{d^*} = \min_{d \in \Delta'} \text{trace } v_d.$$

These definitions are meaningful whether or not σ^2 is known, and the last three do not require the assumption of normality. (Actually, normality has nothing to do with any of these definitions, but only with some of the interpretations below.) By replacing A_d by $NM(\xi)$, we obtain corresponding definitions for the approximate theory.

These criteria are discussed extensively in I and II, and we shall merely summarize some of the important points here. *M*-optimality is generally extremely difficult to characterize, even in very simple situations. *L*-optimality, which is a local version of *M*-optimality, involves the use of type C regions, as discussed in Section 2B. *E*-optimality was first considered in hypothesis testing (Wald (1943), Ehrenfeld (1955a)) because, if σ^2 is known or all b_d are equal for d in Δ', it is the design for which the associated *F*-test of size α maximizes the minimum power on the contour $\Sigma_j \psi_j^2 = c$, for every α and c; this throws serious doubt on the acceptability of this criterion for hypothesis testing, since (see Section 2B) the *F*-test may not be the one which, for a given design, maximizes this minimum power. For point estimation, an *E*-optimum design minimizes the maximum over all $(a_1, \ldots a_s)$ with $\Sigma a_i^2 = 1$ of the variance of the b.l.e. of $\Sigma a_i \psi_i$. An A-optimum design minimizes the average variance of the b.l.e.'s of $\psi_i, \ldots, \psi_s$, and thus of any s linear parametric functions obtained from the ψ_i by an orthogonal transformation. A *D*-optimum design minimizes the generalized variance of the b.l.e.'s of the ψ_i, and thus, under normality with σ^2 known or else all $N - b_d$ the same for d in Δ', minimizes the volume (or expected volume, if σ^2 is unknown) of the smallest invariant confidence region on $\psi_1, \ldots, \psi_s$ of any given confidence coefficient. For testing hypotheses under these same conditions of normality, it follows from the result on regions of type D discussed in Section 2B that, for each σ (and each set of values of the parameters other than $\psi_1, \ldots, \psi_s$), a D-optimum design achieves a test whose power function has maximum Gaussian curvature at the null hypothesis, among all locally unbiased tests of a given size.

Other criteria can be considered similarly. For example, the contour considered in hypothesis testing can be altered from $\Sigma \psi_i^2 = c$, or one can consider maximizing trace v_{d-1} in place of A-optimality (some examples in Section 4B throw a bad light on the latter possibility). The often-considered criterion of restricting oneself to designs for which the b.l.e.'s of $\psi_1, \ldots, \psi_s$ all have equal variances, and of minimizing this variance, will in unsymmetrical settings often produce a design inferior to that which minimizes the maximum diagonal element of v_d without restriction.

D-optimality has an appealing invariant property which is not possessed by the other criteria we have mentioned. Let $\psi'_1, \ldots, \psi'_s$ be related to $\psi_1, \ldots, \psi_s$ by a nonsingular linear transformation. Then, if d^* is D-optimum for the functions $\psi_1, \ldots, \psi_s$, it is also D-optimum for $\psi'_1, \ldots, \psi'_s$. The analogue for other criteria is false in even the simplest settings. For hypothesis testing, D-optimality is also invariant under nonlinear transformations, as discussed in Section 2B.

The invariance of D-optimality is well illustrated by the problem of polynomial regression (see Section 3) with $\psi_i = \theta_i$. For polynomial regression on the interval $[a, b]$, a D-optimum design is obtained from that on $[-1, 1]$ by simply transforming the loci of observations according to the linear transformation which takes $[-1, 1]$ onto $[a, b]$. For any other of the above criteria, even a simple change in units (consideration of $[-h, h]$ instead of $[-1, 1]$) will change the optimum design if $k > 1$. This is both intuitively unappealing (having the choice of design depend on whether measurements are recorded in inches or in feet), and also has the disadvantage of requiring us to give a table of designs which depend on a and b.

A precise statement of losses will obviously entail the use of any of a large number of designs, not always the D-optimum design. However, the discussion of the previous paragraphs should give some workers a good justification for using D-optimum designs in many settings. A further appealing property of D-optimum designs will now be described.

In the regression setup, suppose we are interested in estimating the whole regression function, $\Sigma \theta_i f_i$. As indicated in II, various criteria of optimality can be suggested. A design which minimizes the expected maximum deviation over $\mathscr{X}$ between estimated and true regression function (or the square of this deviation) will be different under normality from what it is under other assumptions, and will generally be very difficult to calculate. Another criterion which has been suggested is to minimize the integral of $\text{var}[\Sigma_i(t_{di} - \theta_i) f_i(x)]$ with respect to some measure on $\mathscr{X}$; the arbitrariness present in choosing the measure, the lack of invariance of seemingly "natural" measures on $\mathscr{X}$ under certain transformations, and the fact that the variance may be very large at some points while the average is small, are some of the shortcomings here. A criterion which has been considered by several authors is the minimization of $\sup_x \text{var}[\Sigma_i(t_{di} - \theta_i) f_i(x)]$ (see II for a discussion of the optimality of using $\Sigma t_{di} f_i$ rather than some other estimator of the function $\Sigma \theta_i f_i$ in this problem). Let us call a design which satisfied this last global criterion G-optimum in Δ.

The discussion of the present paragraph will refer to the approximate theory. In a remarkable paper written in 1918, perhaps the first systematic computation of optimum regression designs, K. Smith (1918) determined the G-optimum designs for the polynomial case with $k \leqslant 6$. Guest (1958) characterized the G-optimum designs in this case for arbitrary k in terms of the zeros of the derivative of a Legendre polynomial. Hoel (1958) computed the D-optimum design in this case, and noted that his D-optimum design coincided with Guest's G-optimum design. It was proved in II that this phenomenon holds more generally, and finally the following result in the approximate theory, which was announced in II and will appear elsewhere, was proved:

Theorem 4.1. *If $\mathcal{X}$ is a compact space on which the f_i are continuous and linearly independent, then ξ is D-optimum for $\theta_1, \ldots, \theta_m$ if and only if it is G-optimum.*

It is not possible to prove an analogue of Theorem 4.1 in the discrete case. For example, in the case of quadratic regression on $[-1, 1]$ with $N = 4$, a D-optimum design takes observations at ± 1, $\pm 2^{-1/2}$, while a G-optimum design takes observations at ± 1, $\pm 3^{-1/2}$. This is another illustration of the usefulness of considering the continuous theory, where many results are valid which are false in the discrete theory, but which are of practical value for large N.

Theorem 4.1 will also be helpful in the computation of D-optimum designs in problems like those of Section 4C, since it will permit us to exploit the interplay between different criteria for D- and G-optimality.

As Hoel also noted, if $x_1, \ldots, x_m$ are any m points of $\mathcal{X}$, writing $F = \|f_i(x_j)\|$ we see that the generalized variance of the b.l.e.'s of the regression function at the m points $x_1, \ldots, x_m$, which is merely the determinant of $\sigma^2 F v_d F'$, is minimized by a D-optimum design. (No analogous property for s points, $1 < s < m$, is generally valid.)

Having stated the relationship between G- and D-optimality in Theorem 4.1, we now turn to the question of the relationships among the other criteria. In general, D-, E-, A-, and L-optimality are unrelated, in either the exact or approximate development. For example, if $m = r = 2$ and there are two possible x_d's, $(1, 0)$ and $(1, 1)$, to which the design ξ assigns measures ξ_1 and $\xi_2 = 1 - \xi_1$, it is easy to compute that $\xi_2 = \frac{1}{2}$ for D-optimality, $2^{1/2} - 1$ for A-optimality, and $2/5$ for E-optimality; an L_a-optimum design is not explicitly known. However, in certain situations such as those where balanced block designs, orthogonal arrays, Latin squares, etc., are customarily employed, it happens that these criteria are related, due to the symmetric way in which the ψ_i enter into the problem (see the last paragraph of Section 2E). This is expressed in the following simple lemma, which was employed in I (it is not useful in most regression problems of the type treated in II):

Lemma 4.2. *If d^* maximizes the trace of $v_{d^{-1}}$ and also $N - b_d$, over all d for which v_d is nonsingular, and if v_{d^*} is a multiple of the identity, then d^* is A, D-, E-, and L_a-optimum (for all α) in Δ.*

Of course, the maximization of $N - b_d$ is unnecessary if σ^2 is known or if we are only interested in point estimation.

Under the stated hypothesis on the form of v_{d^*}, other similar results can be stated (e.g., D-optimality implies A-, E-, and L-optimality); but other relationships generally need additional conditions for their validity. For example, it is easy to find situations where an E-optimum design d for which v_d is the identity is neither A-, D-, nor L_α-optimum (see I).

We shall make further remarks on these relationships in Section 4B.

We now turn to the question of invariance. We shall extend the result of Section 2E to the case of s parameters ($1 \leqslant s \leqslant m$). Suppose in the setting of Section 2E that G is the direct product of groups G_1 and G_2, where G_1 is as in Section 2E, G_2 is as in Theorem 3.3, $\bar{g}$ leaves $\theta_1, \ldots, \theta_s$ fixed if $g \in G_2$, and $\bar{g}$ leaves $\theta_{s+1}, \ldots, \theta_m$ fixed if $g \in G_1$. By Theorem 3.3, we can restrict our consideration to ζ's which are invariant under G_2, and we hereafter do so. We shall use the decomposition of $M(\xi)$ employed in proving Theorem 3.3. Suppose that G_1 has p elements (which leave $\theta_{s+1}, \ldots, \theta_m$ fixed), and let ξ be a G_2-invariant D-optimum design. Then, for g in G_1 and $D(\xi)$ nonsingular (the singular case being treated as before), we clearly have $B(\xi_g) = H_g B(\xi) H_g'$, $C(\xi_g) = H_g C(\xi)$, and $D(\xi_g) = D(\xi)$ where H_g has determinant one. Hence,

$$B(\bar{\xi}) - C(\bar{\xi})D(\bar{\xi})^{-1}C(\bar{\xi})' = p^{-1}\Sigma_g H_g[B(\xi) - C(\xi)D(\xi)^{-1}C(\xi)']H_g'. \quad (4.1)$$

Since all H_g have determinant one, we can argue from (4.1) exactly as we did from (2.1) in Section 2E. Thus, we have (extending the result to nonfinite G as before),

Theorem 4.3. *If $G = G_1 \times G_2$ is as specified above, with $\bar{g}$ leaving $\theta_1, \ldots, \theta_s$ fixed if $g \in G_2$ and $\bar{g}$ leaving $\theta_{s+1}, \ldots, \theta_m$ fixed if $g \in G_1$, then there is a G-invariant ξ which is D-optimum for $\theta_1, \ldots, \theta_s$.*

Similarly, from the orthogonality of $\bar{g}$ for g in G_1, we obtain the same result for A-optimality.

B. Block Designs, Arrays, etc.

The settings where balanced block designs, Latin squares, orthogonal arrays, etc., are customarily employed are characterized by the fact that X_d is a matrix of 1's and 0's (and -1's, in the case of certain weighing experiments) satisfying certain restrictions. One of the first optimality results for such designs was proved by Wald (1943), who showed that, in the setting of two-way heterogeneity where $k \times k$ Latin square and higher Latin square designs are usually employed, these designs are actually D-optimum in Δ for inference on any full set of $k - 1$ linearly independent contrasts of "treatment effects". Shortly afterwards, Hotelling (1944) began the careful study of weighing problems, some ideas on this subject originating in earlier work of Yates (1935), Kishen, and Banerjee. A comprehensive treatment of weighing problems was given by Mood (1946), who considered problems of N weighings on m objects ($N \geq m$) on both spring and chemical balances, proved in the latter case the D-optimality of 2-level orthogonal arrays when they exist, as well as the minimization by them of the variances of b.l.e.'s of weights among all d for which the diagonal elements of v_d are all the same, and obtained optimum designs in the case of spring balances and also in certain cases where no orthogonal arrays exist (for small N), where the two optimality criteria just mentioned were noted not always to agree. At the same time, more general orthogonal arrays were considered independently by Plackett and Burman (1946) (see also Plackett, 1946), who proved their optimality in the multifactorial setup, in the second sense mentioned above in connection with Mood's results. An extensive study by Tocher (1952) considered also the settings where incomplete block designs and Youden squares are customarily employed, and proved that these designs are optimum in the sense of minimizing the variances of b.l.e.'s of treatment differences $\theta_i - \theta_j$, among all designs for which these differences are all estimated with the same variance. These last three papers also considered various methods of construction, which are not the subject of the present paper.

It should be mentioned at this point that although criteria like those mentioned in connection with the last two references happen to lead to designs which are optimum in other senses in situations where sufficiently symmetrical designs exist, these criteria are not intuitively acceptable in themselves; for there are many problems where the restriction to "equal precisions" is attainable only by relatively poor designs, and where there exist better designs which give unequal but better precisions to all estimates.

Ehrenfeld (1955a) proved the E-optimality of the Latin square design (for an appropriate set of treatment contrasts) and of orthogonal arrays in weighing problems. Under the restriction to designs for which each variety appears at most once in each block (or in each row and column), Mote (1958) proved the E-optimality of the balanced incomplete block design and Kshirsagar (1958) proved the A- and D-optimality of this design and of the Youden square. At the same time, without this restriction, more general results were obtained in I, although the approach described below, which leads to trivial proofs in the standard cases (e.g., Youden squares), can entail somewhat more arithmetic in the general settings (e.g., generalized Youden squares, defined below). We shall summarize some of these results.

We begin by indicating a simple approach to optimality proofs in all such symmetrical situations. In problems such as the usual k-way analysis of variance setup, weighing problems, and multifactorial problems, where the "treatment effects" $\theta_1, \ldots, \theta_u$ (say) in which we are customarily interested can all be estimated, the results are easily obtained using Lemma 4.2, since (with the partition of A_d used in Section 3) it is not difficult to find a bound on the trace of $B_d - C_d D_{d-1} C_d'$, and to show that this bound is attained by the appropriate symmetrical design, for which this matrix becomes a multiple of the identity. In settings like those where we are only interested in (or can only estimate) *contrasts* of $\theta_1, \ldots, \theta_u$, such as those where balanced incomplete block designs and Youden squares are employed, the above $u \times u$ matrix is singular. If the $u - 1$ linearly independent contrasts $\theta_1, \ldots, \theta_{u-1}$ which can be estimated are not chosen in a suitable way, the computation of v_{d-1} may be quite messy. The most expeditious choice in many settings is to let $\psi_i = \Sigma_{j=1}^u o_{ij}\theta_j, 1 \leqslant i \leqslant u - 1$, where $\|o_{ij}\| = \bar{O}$ is a $u \times u$ orthogonal matrix with $o_{uj} = u^{-1/2}, 1 \leqslant j \leqslant u$. This implies that our optimality criteria which refer to hypothesis testing are concerned with the power function on the contours $\Sigma_1^u (\theta_i - \bar{\theta})^2 = c\sigma^2$, where $\bar{\theta} = \Sigma_1^u \theta_i/u$.) The development can be carried out through direct consideration of the possible A_d's, but somewhat less arithmetic is needed if we use Bose's $\mathscr{C}$-matrices corresponding to the incomplete block (or h-way heterogeneity) setting at hand. In this approach, letting the first u θ_i's represent the treatment effects (the other θ_i's representing block effects or row and column effects, etc.), for any design there is a matrix $\mathscr{C}_d$ of rank at most $u - 1$ and a u-vector Z_d of linear functions of the Y_{di} such that the b.l.e. of any contrast $\Sigma_1^u c_i \theta_i$ (where $\Sigma c_i = 0$) which is estimable under d is given by $\Sigma c_i t_{di}^*$ where t_d^* is any solution of the reduced normal equations $\mathscr{C}_d t_d^* = Z_d$. Also, $\mathscr{C}_d$ has row and column sums equal to zero, and Z_d has covariance matrix $\sigma^2 \mathscr{C}_d$. Now, it is often easy to give a bound on the trace of all $\mathscr{C}_d$ in terms of N, etc. Moreover, appropriately symmetrical designs such as balanced incomplete block designs, Youden squares, etc., will generally have the property that all diagonal elements of $\mathscr{C}_d$ are equal (and $\mathscr{C}_d$ is not zero, and hence has rank $u - 1$). Suppose $\mathscr{C}_{d^*}$ is of this form and has maximum possible trace. Then $\bar{O}\mathscr{C}_{d^*}\bar{O}'$ is easily verified to have the same positive constant for each of its first $u - 1$ diagonal elements, and is zero elsewhere. Since the upper left hand $(u - 1) \times (u - 1)$ submatrix of $\bar{O}\mathscr{C}_d\bar{O}'$ is just the v_{d-1} for the b.l.e.'s of the $\psi_1, \ldots, \psi_{u-1}$ defined just above, we have the following elementary lemma:

Lemma 4.4. *If $\mathscr{C}_{d^*}$ has maximum possible trace, all diagonal elements equal, and all off-diagonal elements equal, then $v_{d^*}^{-1}$ has maximum possible trace and is a multiple of the identity.*

Thus, although any choice of $u - 1$ linearly independent contrasts $\psi_i, \ldots, \psi_{u-1}$ will lead to the same D-optimum design(s), the choice given above makes the arithmetic by far the least cumbersome in most applications. For if $\mathscr{C}_{d^*}$ has the form hypothesized in Lemma 4.4, we can combine the conclusion of this lemma with Lemma 4.2 and thus obtain the desired optimum properties.

This method can be employed to prove the D-, E-, L-, and A-optimality of balanced block designs (a generalization of balanced incomplete block designs, to be defined below), regular generalized Youden squares (defined below), higher Latin squares, orthogonal arrays, and other appropriately symmetrical designs in all of the customarily encountered settings where we are interested in contrasts of treatment effects; see I. As remarked earlier, the corresponding results are even simpler in the case where we are interested in all u effects, all of which are estimable.

In many optimality proofs, authors have not really proved the appropriate results, since they have restricted considerations to a subclass of Δ (for example, to incomplete block designs where each treatment occurs at most once per block). Such restrictions are quite unnecessary.

To illustrate these ideas, suppose we have b blocks of size k and u varieties in the usual incomplete block design setting, except that we do *not* assume $k \leqslant u$. Such situations arise often. Generalizing the notion of a balanced incomplete block design, we define a design d in the above setting to be a *balanced block design* (BBD) if (a) the number of times n_{dij} that variety i appears in block j is k/u if this is an integer, and is one of the two closest integers otherwise; (b) the number $\Sigma_j n_{dij}$ of replications of variety i is the same for all i; and (c) for each pair i_1, i_2 with $i_1 \neq i_2$, $\Sigma_j n_{di_1j} n_{di_2j}$ is the same. Certain designs of this type with $k > u$ have been considered by Tocher (1952) and others. Appropriate modifications of some of the constructive methods which are used when $k < u$ will sometimes work here. In the setting described here, it is easy to verify that, for any design d (not necessarily a BBD), the trace of $\mathscr{C}_d$ is

$$bk - \Sigma_{i,j} n_{dij}^2/k. \tag{4.2}$$

Since $\Sigma_{i,j} n_{dij} = bk$, expression (4.2) is clearly maximized by a BBD (if one exists), and the $\mathscr{C}_d$ of such a design is of the form hypothesized in Lemma 4.2. Thus, we have proved that BBD's are A-, D-, E-, and L-optimum.

It is interesting to note that many designs maximize (4.2) (all that is required is that the n_{dij} be as nearly equal as possible); however, of these designs, only a BBD will have the form required by Lemma 4.4. However, expression (4.2) and Lemmas 4.2 and 4.4 again suggest the idea, mentioned in Section 2C, that if no appropriately symmetrical (balanced) design (a BBD in this case) exists, then a design which is as close as possible to such symmetry in some sense will be optimum. Our next example shows, however, that considerable delicacy will be needed to make these notions precise, since optimality can be difficult to prove even for a design of maximum symmetry (balance).

In the setting of two-way heterogeneity (expected value of an observation = row effect + column effect + variety effect) with k_1 rows, k_2 columns, and u varieties, we say that d is a *generalized Youden square* (GYS) if it is a BBD when rows are considered to be blocks and also when columns are considered to be blocks. A GYS is said to be *regular* if at least one of k_1/u and k_2/u is an integer. Using Lemmas 4.2 and 4.4, it was shown in I that a regular GYS is A-, D-, E-, and L-optimum; the argument for proving that the trace of $\mathscr{C}_d$ is a maximum when d is a regular GYS is somewhat more complicated here than in the case of a BBD. In fact, if the GYS is not regular, its $\mathscr{C}_d$ may not have maximum trace, as was illustrated by an example in I in the case $k_1 = k_2 = 6$, $u = 4$. It still seems likely that such nonregular GYS's are optimum, but a different argument from that based on Lemmas 4.2 and 4.4 needs to be developed in this case and in certain other settings which represent extensions of the classical situations.

We mention one constructional aspect of GYS's: the method of Hartley and Smith (1948) can be extended to prove the following:

Lemma 4.5. *If* $L = k_2/u$ *is on integer and there is a BBD with parameters u,* $b = k_2, k = k_1$, *then there is a (regular) GYS with parameter values* u, k_1, k_2.

In fact, thinking of the BBD as a k_1 (rows) $\times k_2$ (columns) array, we let $m_{ij} = \max (0, [-L + \text{number of times variety } i \text{ appears in row } j])$ and $\Sigma_{i,j}m_{ij} = M$. Following Hartley and Smith, we can give a method for reducing M by at least one if $M > 0$, and then use induction: one has only to go through the demonstration of these authors, replacing the occurrence of a variety 0, 1, or more than 1 time in a row by its occurrence less than L, L, or more than L times, respectively. This method of construction cannot be modified to work for nonregular GYS's merely by trying to make the n_{dij} in rows as nearly equal as possible; for example, in the case $k_1 = 6, k_2 = 6, u = 4$, one can construct a design whose columns are a BBD and whose rows have all n_{dij} as nearly equal as possible, but which is not a GYS.

Similar optimality results for symmetrical designs can also be obtained for appropriate components of variance models and mixed models.

(*Editors' note*: Section c has been omitted.)

For problems of testing hypotheses, the considerations of Section 2A are again relevant: in such a simple problem as that of linear regression on $[-1, 1]$, the locally best design for testing the hypothesis that both parameters equal zero (or any other specified values) is the randomized design which takes all observations at 1 or -1, with equal probabilities.

We have already mentioned, in Section 4A, the results of Smith, Guest, and Hoel in the polynomial case. We mention here that Guest presents an interesting comparison of his designs with certain "equal spacing" designs. It should also be mentioned that Hoel compares various designs in the polynomial case for various types of dependence (see Cox (1952) for a discussion of related work in other settings); but the determination of optimum designs in these cases is still an open problem.

References

Achiezer, M. I. (1956), *Theory of Approximation*. New York: Ungar.

Blackwell, D. (1950), "Comparison of experiments", *Proc. Second Berkeley Symposium*, 93–102, Univ. of California Press.

Blum, J. (1954), "Multidimensional stochastic approximation methods", *Ann. Math. Statist.*, **25**, 737–744.

Box, G. E. P. & Wilson, K. B. (1951), "On the experimental attainment of optimum conditions", *J. R. Statist. Soc. B*, **13**, 1–45.

Box, G. E. P. (1952), "Multi-factor designs of first order", *Biometrika*, **39**, 49–57.

Bradt, R. N., Johnson, S. M. & Karlin, S. (1957), "On sequential designs for maximizing the sum of *n* observations", *Ann. Math. Statist.*, **28**, 1,060–1,074.

—— & Karlin, S. (1957), "On the design and comparison of certain dichotomous experiments", *Ann. Math. Statist.*, **28**, 390–409.

Chernoff, H. (1953), "Locally optimum designs for estimating parameters", *Ann. Math. Statist.*, **24**, 586–602.

Cox, D. R. (1952), "Some recent work on systematic experimental designs", *J. R. Statist. Soc. B*, **14**, 211–219.

Chung, K. L. (1954), "On a stochastic approximation method", *Ann. Math. Statist.*, **25**, 463–483.

De la Garza, A. (1954), "Spacing of information in polynomial regression", *Ann. Math. Statist.*, **25**, 123–130.

Ehrenfeld, S. (1955a), "On the efficiency of experimental designs", *Ann. Math. Statist.*, **26**, 247–255.

—— (1955b), "Complete class theorems in experimental design", *Proc. Third Berkeley Symposium*, vol. 1, 57–67. Univ. of California Press.

Elfving, G. (1952), "Optimum allocation in linear regression theory", *Ann. Math. Statist.*, **23**, 255–262.

—— (1955a), "Geometric allocation theory", *Skand. Akt.*, **37**, 170–190.

—— (1955b), "Selection of nonrepeatable observations for estimation", *Proc. Third Berkeley Symposium*, vol. 1, 69–75. Univ. of California Press.

Friedman, M. & Savage, L. J. (1947), "Experimental determination of the maximum of a function", *Selected Techniques of Statistical Analysis*, 363–372. New York: McGraw-Hill,

Guest, P. G. (1958), "The spacing of observations in polynomial regression", *Ann. Math. Statist.*, **29**, 294–299.

Hartley, H. O. & Smith, C. A. B. (1948), "The construction of Youden squares", *J. R. Statist. Soc. B*, **10**, 262–263.

Hodges, J. L., Jr. & Lehmann, E. L. (1950), "Some problems in minimax point estimation", *Ann. Math. Statist.*, **21**, 182–197.

—— (1955), "Two approximations to the Robbins-Monro process", *Proc. Third Berkeley Symposium*, vol. 1, 95–104. Univ. of California Press.

Hoel, P. G. (1958), "Efficiency problems in polynomial estimation", *Ann. Math. Statist.*, **29**, 1,134–1,145.

Hotelling, H. (1941), "Experimental determination of the maximum of a function", *Ann. Math. Statist.*, **12**, 20–46.

—— (1944), "Some improvements in weighing and other experimental techniques", *Ann. Math. Statist.*, **15**, 297–306.

Hsu, P. L. (1941), "Analysis of variance from the power function standpoint", *Biometrika*, **32**, 62.

Isaacson, S. (1951), "On the theory of unbiased tests of simple statistical hypotheses specifying the values of two or more parameters", *Ann. Math. Statist.*, **22**, 217–234.

Karlin, S. & Shapley, L. S. (1953), *Geometry of Moment Spaces*, vol. 12 of Amer. Math. Soc. Memoirs.

Kiefer, J. (1948), "Sequential determination of the maximum of a function", M.I.T. (thesis).

—— (1953), "Sequential minimax search for a maximum", *Proc. Amer. Math. Soc.*, **4**, 502–506.

—— (1957a), "Invariance, sequential minimax estimation, and continuous time processes", *Ann. Math. Statist.*, **28**, 573–601.

—— (1957b), "Optimum sequential search and approximation methods under minimum regularity conditions", *J. Soc. Ind. Appl. Math.*, **5**, 105–136.

—— (1958), "On the nonrandomized optimality and randomized nonoptimality of symmetrical designs", *Ann Math. Statist.*, **29**, 675–699. (Referred to in the present paper as I.)

———— & Wolfowitz, J. (1952), "Stochastic estimation of the maximum of a regression function", *Ann. Math. Statist.*, **23**, 462–466.

———— (1959), "Optimum designs in regression problems", *Ann. Math. Statist.*, **30**. (Referred to in the present paper as II.)

Kshirsagar, A. M. (1958), "A note on incomplete block designs", *Ann. Math. Statist.*, **29**, 907–910.

Le Cam, L. (1955), "An extension of Wald's theory of statistical decision functions", *Ann. Math. Statist.*, **26**, 69–78.

Mood, A. (1946), "On Hotelling's weighing problem", *Ann. Math. Statist.*, **17**, 432–446.

Mote, V. L. (1958), "On a minimax property of a balanced incomplete block design", *Ann. Math. Statist.*, **29**, 910–913.

Neyman, J. & Pearson, E. S. (1938), "Contributions to the theory of testing statistical hypotheses, III", *Stat. Res. Memoirs*, **2**, 25–27.

Plackett, R. L. (1946), "Some generalizations in the multifactorial design", *Biometrika*, **33**, 328–332.

———— & Burman, J. P. (1946), "The design of optimum multifactorial experiments", *Biometrika*, **33**, 296–325.

Robbins, H. (1952), "Some aspects of the sequential design of experiments", *Bull. Amer. Math. Soc.*, **58**, 527–535.

———— & Monro, S. (1951), "A stochastic approximation method", *Ann. Math. Statist.*, **22**, 400–407.

Sacks, J. (1958), "Asymptotic distribution of stochastic approximation procedures", *Ann. Math. Statist.*, **29**, 373–405.

Scheffé H. (1958), "Experiments with mixtures", *J. R. Statist. Soc. B*, **20**, 344–360.

Shohat, J. A. & Tamarkin, J. D. (1943), *The Problem of Moments*, Amer. Math. Soc. Surveys, No. 1.

Smith, K. (1918), "On the standard deviations of adjusted and interpolated values of an observed polynomial function and its constants and the guidance they give towards a proper choice of the distribution of observations", *Biometrika*, **12**, 1–85.

Sobel, M. & Groll, P. A. (1959), "On group testing with a finite population", (To be published.)

Stein, C. (1948), "On sequences of experiments" (abstract), *Ann. Math. Stat.*, **19**, 117–118.

———— (1955), "Inadmissibility of the usual estimator for the mean of a multivariate normal distribution", *Proc. Third Berkeley Symposium*, vol. 1, 197–206. Univ. of California Press.

———— & Wald, A. (1947), "Sequential confidence intervals for the mean of a normal distribution with known variance", *Ann. Math. Statist.*, **18**, 427–433.

Tocher, K. D. (1952), "The design and analysis of block experiments", *J. R. Statist. Soc. B*, **14**, 45–100.

Wald, A. (1942), "On the power function of the analysis of variance test", *Ann. Math. Statist.*, **13**, 434–439.

———— (1943), "On the efficient design of statistical investigations", *Ann. Math. Statist.*, **14**, 134–140.

———— (1950), *Statistical Decision Functions*, New York: John Wiley.

———— & Wolfowitz, J. (1948), "Optimum character of the sequential probability ratio test", *Ann. Math. Statist.*, **19**, 326–339.

———— (1950), "Characterization of the minimal complete class of decision functions when the number of distributions and decisions is finite", *Proc. Second Berkeley Symposium*, 149–157. Univ. of California Press.

Wolfowitz, J. (1949), "The power of the classical tests associated with the normal distribution", Ann. *Math. Statist.*, **20**, 540–551.

—— (1950), "Minimax estimates of the mean of a normal distribution with known variance", *Ann. Math. Statist.*, **21**, 218–230.

—— (1951), "On ε-complete classes of decision functions", *Ann. Math. Statist.*, **22**, 461–465.

Yates, F. (1935), "Complex experiments", *J. R. Statist. Soc. Sup.*, **2**, 181–247.

Introduction to
James and Stein (1961) Estimation with Quadratic Loss

Bradley Efron
Stanford University

Section 2 of this paper presents the most striking theorem of post-war mathematical statistics: Suppose $\mathbf{X}$ is a k-dimensional random vector, having a multivariate normal distribution with mean vector $\boldsymbol{\mu}$ and covariance matrix the identity,

$$\mathbf{X} \sim N_k(\boldsymbol{\mu}, I). \tag{1}$$

Then if the dimension k is greater than or equal to 3, the expected squared error of the "James–Stein" estimator of $\boldsymbol{\mu}$,

$$\hat{\boldsymbol{\mu}}^{JS} = \left[1 - \frac{k-2}{\|X\|^2} \right] \mathbf{X},$$

is less than k for every choice of $\boldsymbol{\mu}$. Here, $\|\mathbf{X}\|^2$ equals $\sum_{i=1}^{k} X_i^2$, the squared Euclidean length. More precisely,

$$E\|\hat{\boldsymbol{\mu}}^{JS} - \boldsymbol{\mu}\|^2 = k - E\frac{(k-2)^2}{k-2+2K}, \tag{2}$$

where K is a Poisson random variable with mean $\|\boldsymbol{\mu}\|^2/2$.

This result was, and sometimes still is, considered paradoxical. Before Stein's 1956 counterexample, there had been very reason to believe that the usual estimator $\hat{\boldsymbol{\mu}}^\circ$, the one that estimates each component μ_i with X_i,

$$\hat{\boldsymbol{\mu}}^\circ = \mathbf{X},$$

was at least admissible. (This is, it could not be beaten everywhere in the parameter space.) After all,

For the one-dimensional problem of estimating μ_i on the basis of $X_i \sim N(\mu_i, 1)$, with squared-error loss, it was known that $\hat{\mu}_i^\circ = X_i$ was admissible. See, for example, Blyth (1951).

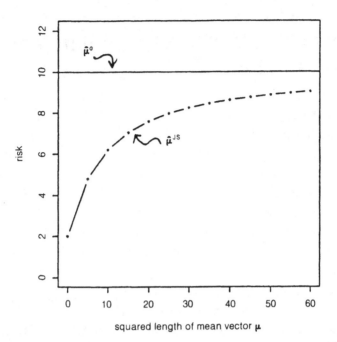

Figure 1. Risk of the James–Stein estimator μ^{JS}, dimension $k = 10$, compared to the risk of the usual estimator μ°.

Model (1) says that $X_i \sim N(\mu_i, 1)$ *independently* for $i = 1, 2, \ldots, k$.

The vector loss function $\|\hat{\mu} - \mu\|^2$ is just the sum of the one-dimensional squared-error losses $(\hat{\mu}_i - \mu_i)^2$.

The James–Stein theorem seems to say that there is some sort of latent information existing between independent decision problems that cannot be seen when they are considered separately, but comes into play when they are considered together. This is exactly what it does say, in the most specific terms, for the most natural of simultaneous decision problems.

The James–Stein effect is substantial. Figure 1, calculated from formula (2), shows $E\|\hat{\mu}^{JS} - \mu\|^2$ as a function of $\|\mu\|^2$, for $k = 10$. We see that it is always less than $E\|\hat{\mu}^\circ - \mu\|^2 = 10$, most emphatically so for small values of $\|\mu\|^2$. For $\|\mu\| = 0$, $E\|\hat{\mu}^{JS} - \mu\|^2 = 2$ in every dimension $k \geq 3$.

The James-Stein proof of (2) is long and intricate. Stein's (1981) paper gives a much easier argument, based on a simple integration-by-parts identity: If X_1 is a normally distributed random variable with variance 1, $X_1 \sim N(\mu_1, 1)$, and $g_1(X_1)$ is a differentiable function satisfying $E\{|g_1'(X_1)|\} < \infty$, then

$$E\{g_1'(X_1)\} = E\{(X_1 - \mu_1)g_1(X_1)\}. \tag{3}$$

Next suppose that $X \sim N_k(\mu, I)$ as in (1), and that we intend to estimate μ with the statistic $X - g(X)$, where $g(X)$ is a vector function of X satisfying

$E\{|\sum_{i=1}^{k}\partial g_i(X_i)/\partial X_i|\} < \infty$. It is easy to deduce from (3) that

$$E\|\mathbf{X} - \mathbf{g}(\mathbf{X}) - \boldsymbol{\mu}\|^2 = k - E\left\{2\sum_{i=1}^{k}\frac{\partial g_i(X_i)}{\partial X_i} + \|\mathbf{g}(\mathbf{X})\|^2\right\}. \qquad (4)$$

The choice

$$\mathbf{g}(\mathbf{X}) = \frac{k-2}{\|\mathbf{X}\|^2}\mathbf{X} \qquad (5)$$

makes $\mathbf{X} - \mathbf{g}(\mathbf{X}) = \hat{\boldsymbol{\mu}}^{JS}$, the James–Stein estimator. Applying (4) gives

$$E\|\hat{\boldsymbol{\mu}}^{JS} - \boldsymbol{\mu}\|^2 = k - E\left\{\frac{(k-2)^2}{\|\mathbf{X}\|^2}\right\}. \qquad (6)$$

The denominator $\|\mathbf{X}\|^2$ has a noncentral chi-squared distribution. $\|\mathbf{X}\|^2 \sim \chi_k^2(\|\boldsymbol{\mu}\|^2)$. The method of Pitman and Robbins says that this is also the distribution of a central chi-squared variate with a Poisson choice for the degrees of freedom,

$$\|\mathbf{X}\|^2 \sim \chi_{k+2K}^2 \qquad \text{where} \qquad K \sim \text{Poisson}(\|\boldsymbol{\mu}\|^2/2).$$

But $E\{1/\chi_j^2\} = 1/(j-2)$ for any choice of j, so (6) is really the same as (2).

Expression (4) does more than verify the James–Stein theorem; it allows us to investigate the squared-error risk of general biased estimates of the form $\mathbf{X} - \mathbf{g}(\mathbf{X})$. Later we will see how the James–Stein choice of $\mathbf{g}(\mathbf{X})$, [Eq. (5)] is naturally suggested by Bayesian considerations.

The course of statistical theory was profoundly affected by the James–Stein theorem. It called into question almost every existing principle of "goodness" for statistical estimators: invariance, the minimax principle, least squares, linear regression, maximum likelihood, and unbiasedness. In problem (1), for example, the usual estimator $\hat{\boldsymbol{\mu}}^\circ$ is uniformly minimum covariance unbiased, maximum likelihood, best invariant, and minimax, and yet its performance in terms of expected squared-error loss is demonstrably inferior to that of $\hat{\boldsymbol{\mu}}^{JS}$.

Or consider the familiar linear model

$$\mathbf{Y} = M\boldsymbol{\beta} + \mathbf{e}, \qquad (7)$$

where $\mathbf{Y}$ is an $n \times 1$ vector of observables, M an $n \times k$ matrix of known constants, $\boldsymbol{\beta}$ a $k \times 1$ vector of unknown parameters, and $\mathbf{e}$ a normal error vector, $\mathbf{e} \sim N_n(\mathbf{0}, \sigma^2 \mathbf{I})$. A change of coordinates reduces situation (7) to (1). The usual estimate $\hat{\boldsymbol{\mu}}^\circ$ now corresponds to the Gauss–Markov least-squares estimate for β, whereas $\hat{\boldsymbol{\mu}}^{JS}$ corresponds to an everywhere better estimator.

The James–Stein estimate for μ_i, the ith coordinate of $\boldsymbol{\mu}$, depends on X_i of course, but it also depends on X_j for $j \neq i$, even though $X_1, X_2, \ldots, X_k$ are statistically independent. This can seem paradoxical indeed when the coordinates refer to physically disparate quantities. Suppose μ_1 is the atomic weight of tin, μ_2 the height of Mount Ranier, and μ_3 the number of Chevrolets in

Chicago. Can X_2 and X_3 really help with the estimation of μ_1 (as the James–Stein theorem seems to say)? Efron and Morris (1977) give a nontechnical discussion of James–Stein "paradoxes."

The practical importance of James–Stein estimation has grown as the answers to this type of question have become more clear. Substantial practical gains cannot be expected from the James–Stein estimator when disparate estimation problems are combined. Big gains for $\hat{\mu}^{JS}$, and its more practical generalizations, are realized when the coordinates of μ are of similar magnitude; see Efron and Morris (1975).

This last statement has a Bayesian flavor. Robbins' (1956) empirical Bayes theory turns out to be closely related to the James–Stein theorem. Both theorems show that in certain simultaneous estimation problems it is possible to do considerably better than a classical analysis of the component estimation problems would suggest. In both cases, the statistician gets the advantage of using a Bayes estimation rule, without the trouble of choosing a Bayes prior distribution. The data effectively choose the correct prior. The empirical Bayes interpretation of $\hat{\mu}^{JS}$ is exposited in Efron and Morris (1973). Good (1953) presents a result, due originally to Turing, related to Robbins' empirical Bayes theorem. (Turing's result is a brilliant special example of the powerful general theory Robbins developed in the early 1950s.)

The James–Stein estimator is best motivated by a Bayesian argument. Suppose that the true means μ_i, $i = 1, 2, \ldots, k$, are independently selected according to a normal prior, after which each X_i is selected according to the normal distribution centered at μ_i,

$$\mu_i \sim N(0, A) \qquad \text{and} \qquad x_i | \mu_i \sim N(\mu_i, 1) \quad i = 1, 2, \ldots, k. \tag{8}$$

The Bayes estimate of the vector μ is well known to be

$$\hat{\mu}^{\text{Bayes}} = \left[1 - \frac{1}{A + 1} \right] \mathbf{X}. \tag{9}$$

The marginal distribution of the X_i, under model (8), is

$$X_i \sim N(0, A + 1) \qquad \text{independently for} \qquad i = 1, \ldots, k.$$

Then $\|\mathbf{X}\|^2$ is distributed as $(A + 1)\chi_k^2$, so that

$$E \frac{k - 2}{\|\mathbf{X}\|^2} = \frac{1}{A + 1}.$$

If we do not know the prior variance A, it is overwhelmingly tempting to use the "empirical Bayes" version of (9).

$$\hat{\mu}^{\text{Bayes}} = \left[1 - \frac{k - 2}{\|\mathbf{X}\|^2} \right] \mathbf{X},$$

which is exactly the James–Stein estimator.

Results like (2) have now been shown to hold for a wide variety of non-

normal estimation problems, see, in particular, Brown (1966). The key ingre-
dient is that the loss function, like $\|\hat{\boldsymbol{\mu}} - \boldsymbol{\mu}\|^2$, must add up the component
losses in a fairly even-handed way. This is the main limiting factor in the
practical use of James–Stein estimation: even if there are many related estimation problems to solve simultaneoulsy, the statistician may feel uncomfort-
able combining their losses. (This would be the case for the atomic weight of
tin problem posed above.) James–Stein estimation has become popular in
situations like the setting of insurance rates for different geographical areas,
where it is natural to combine the individual losses.

Section 4 of this paper (authored by Stein alone, only Sec. 2 was joint)
contains a much-quoted result on the *admissibility* of estimation rules. Under
quite general conditions, Stein shows that Pitman estimators are admissible
under squared-error loss, when the dimension k equals 1 or 2. It is worth
noting that Stein spent many years trying to verify the admissibility of $\hat{\boldsymbol{\mu}}^\circ$ in
situation (1), among other things giving the first proof of its admissibility in
dimension $k = 2$. Stein's 1956 paper thanks John Tukey and Herbet Robbins
"who helped break down my conviction that the usual procedure must be
admissible."

In 1955, Charles Stein was 35 years old and a professor of statistics of
Stanford University. A student of Abraham Wald, Stein had already done
fundamental research in mathematical statistics, including the famous
"Hunt–Stein" theorem relating group theory and statistical invariance.
Willard James was a mathematics graduate student from California State,
Long Beach spending the summer at Stanford. Stein, who had considered
estimations of the form $[1 + c/(a + \|\mathbf{X}\|^2)]\mathbf{X}$ in his 1955 paper, suspected
that the case $a = 0$ might lead to better numerical results. He assigned James
the task of computationally evaluating the risk of such estimations.

Instead, James reported back a theoretical result, which led quickly to (2).
(It is shown in the paper that $k - 2$ is the optimum choice of c.) This was to
be James' only publication in the statistics literature. He has continued on at
California State, teaching in its mathematics department. Stein, who retired
from the Stanford statistics department in 1989, is generally considered to be
the world's premier mathematical statistician.

References

Blyth, C. (1951). On minimax statistical decision procedures and their admissibility,
Ann. Math. Statist., **22**, 22–42.

Brown, L. (1966). On the admissibility of invariant estimators of one or more location
parameters, *Ann. Math. Statist.*, **37**, 1087–1136.

Efron, B., and Morris, C. (1973). Stein's estimation rule and its competitors—an
empirical Bayes approach, *JASA*, **68**, 117–130.

Efron, B., and Morris, C. (1975). Data analysis using Stein's estimator and its generalizations, *JASA*, **70**, 311–319.

Efron, B., and Morris, C. (1977). Stein's paradox in statistics, *Sci. Amer.*, **236**, 119–127.

Good, I. J. (1953). The population frequencies of species and the estimation of population parameters, *Biometrika*, **40**, 237–264.

Robbins, H. (1956). An empirical Bayes approach to statistics, *Proceedings of the 3rd Berkeley Symposium on Mathematical Statistics and Probability* (J. Neyman and L. LeCam, eds.), University of California Press, Berkeley and Los Angeles, Vol. 1, pp. 157–164.

Stein, C. (1956). Inadmissibility of the usual estimator for the mean of a multivariate normal distribution, in *Proceedings of the 3rd Berkeley Symposium on Mathematical Statistics and Probability* (J. Neyman and L. LeCam, eds.), University of California Press, Berkeley and Los Angeles, Vol. 1, pp. 197–206.

Stein, C. (1981). Estimating the mean of a multivariate normal distribution, *Ann. Stat.*, **9**, 1135–1151.

Estimation with Quadratic Loss[1]

W. James
Fresno State College

Charles Stein
Stanford University

1. Introduction

It has long been customary to measure the adequacy of an estimator by the smallness of its mean squared error. The least squares estimators were studied by Gauss and by other authors later in the nineteenth century. A proof that the best unbiased estimator of a linear function of the means of a set of observed random variables is the least squares estimator was given by Markov [12], a modified version of whose proof is given by David and Neyman [4]. A slightly more general theorem is given by Aitken [1]. Fisher [5] indicated that for large samples the maximum likelihood estimator approximately minimizes the mean squared error when compared with other reasonable estimators. This paper will be concerned with optimum properties or failure of optimum properties of the natural estimator in certain special problems with the risk usually measured by the mean squared error or, in the case of several parameters, by a quadratic function of the estimators. We shall first mention some recent papers on this subject and then give some results, mostly unpublished, in greater detail.

Pitman [13] in 1939 discussed the estimation of location and scale parameters and obtained the best estimator among those invariant under the affine transformations leaving the problem invariant. He considered various loss functions, in particular, mean squared error. Wald [18], also in 1939, in what may be considered the first paper on statistical decision theory, did the same for location parameters alone, and tried to show in his theorem 5 that the estimator obtained in this way is admissible, that is, that there is no estimator whose risk is no greater at any parameter point, and smaller at some point.

[1] This work was supported in part by an ONR contract at Stanford University.

However, his proof of theorem 5 is not convincing since he interchanges the order of integration in (30) without comment, and it is not clear that this integral is absolutely convergent. To our knowledge, no counterexample to this theorem is known, but in higher-dimensional cases, where the analogous argument seems at first glance only slightly less plausible, counterexamples are given in Blackwell [2] (which is discussed briefly at the end of section 3 of this paper) and in [14], which is repeated in section 2 of this paper. In the paper of Blackwell an analogue of Wald's theorem 5 is proved for the special case of a distribution concentrated on a finite arithmetic progression. Hodges and Lehmann [7] proved some results concerning special exponential families, including Wald's theorem 5 for the special problem of estimating the mean of a normal distribution. Some more general results for the estimation of the mean of a normal distribution including sequential estimation with somewhat arbitrary loss function were obtained by Blyth [3] whose method is a principal tool of this paper. Girshick and Savage [6] proved the minimax property (which is weaker than admissibility here) of the natural estimator in Wald's problem and also generalized the results of Hodges and Lehmann to an arbitrary exponential family. Karlin [8] proved Wald's theorem 5 for mean squared error and for certain other loss functions under fairly weak conditions and also generalized the results of Girshick and Savage for exponential families. The author in [16] proved the result for mean squared error under weaker and simpler conditions than Karlin. This is given without complete proof as theorem 1 in section 3 of the present paper.

In section 2 of this paper is given a new proof by the authors of the result of [14] that the usual estimator of the mean of a multivariate normal distribution with the identity as covariance matrix is inadmissible when the loss is the sum of squares of the errors in the different coordinates if the dimension is at least three. An explicit formula is given for an estimator, still inadmissible, whose risk is never more than that of the usual estimator and considerably less near the origin. Other distributions and other loss functions are considered later in section 2. In section 3 the general problem of admissibility of estimators for problems with quadratic loss is formulated and a sufficient condition for admissibility is given and its relation to the necessary and sufficient condition [15] is briefly discussed. In section 4 theorems are given which show that under weak conditions Pitman's estimator for one or two location parameters is admissible when the loss is taken to be equal to the sum of squares of the errors. Conjectures are discussed for the more difficult problem where unknown location parameters are also present as nuisance parameters, and Blackwell's example is given. In section 5 a problem in multivariate analysis is given where the natural estimator is not even minimax although it has constant risk. These are related to the examples of one of the authors quoted by Kiefer [9] and Lehmann [11]. In section 6 some unsolved problems are mentioned.

The results of section 2 were obtained by the two authors working together. The remainder of the paper is the work of C. Stein.

2. Inadmissibility of the Usual Estimator for Three or More Location Parameters

Let us first look at the spherically symmetric normal case where the inadmissibility of the usual estimator was first proved in [14]. Let X be a normally distributed p-dimensional coordinate vector with unknown mean $\xi = EX$ and covariance matrix equal to the identity matrix, that is, $E(X - \xi)(X - \xi)' = I$. We are interested in estimating ξ, say by $\hat{\xi}$ and define the loss to be

$$L(\xi, \hat{\xi}) = (\xi - \hat{\xi})'(\xi - \hat{\xi}) = \|\xi - \hat{\xi}\|^2, \tag{1}$$

using the notation

$$\|x\|^2 = x'x. \tag{2}$$

The usual estimator is φ_0, defined by

$$\varphi_0(x) = x, \tag{3}$$

and its risk is

$$\rho(\xi, \varphi_0) = EL[\xi, \varphi_0(X)] = E(X - \xi)'(X - \xi) = p. \tag{4}$$

It is well known that among all unbiased estimators, or among all translation-invariant estimators (those φ for which $\varphi(x + c) = \varphi(x) + c$ for all vectors x and c), this estimator φ_0 has minimum risk for all ξ. However, we shall see that for $p \geq 3$,

$$E\left\|\left(1 - \frac{p-2}{\|X\|^2}\right)X - \xi\right\|^2 = p - E\frac{(p-2)^2}{p-2+2K} < p, \tag{5}$$

where K has a Poisson distribution with mean $\|\xi\|^2/2$ Thus the estimator φ_1 defined by

$$\varphi_1(X) = 1 - \frac{p-2}{\|X\|^2}X \tag{6}$$

has smaller risk than φ_0 for all ξ. In fact, the risk of φ_1 is 2 at $\xi = 0$ and increases gradually with $\|\xi\|^2$ to the value p as $\|\xi\|^2 \to \infty$. Although φ_0 is not admissible it seems unlikely that there are spherically symmetrical estimators which are appreciably better than φ_1. An analogous result is given in formulas (19) and (21) for the case where $E(X - \xi)(X - \xi)' = \sigma^2 I$, where σ^2 is unknown but we observe S distributed independently of X as σ^2 times a χ^2 with n degrees of freedom. For $p \leq 2$ it is shown in [14] and also follows from the results of section 3 that the usual estimator is admissible.

We compute the risk of the estimator φ_2 defined by

$$\varphi_2(X) = \left(1 - \frac{b}{\|X\|^2}\right)X, \tag{7}$$

where b is a positive constant. We have

$$\rho(\xi, \varphi_2) = E \left\| \left(1 - \frac{b}{\|X\|^2}\right)X - \xi \right\|^2$$

$$= E\|X - \xi\|^2 - 2bE\frac{(X - \xi)'X}{\|X\|^2} + b^2 E\frac{1}{\|X\|^2}$$

$$= p - 2bE\frac{(X - \xi)'X}{\|X\|^2} + b^2 E\frac{1}{\|X\|^2}. \tag{8}$$

It is well known that $\|X\|^2$, a noncentral χ^2 with p degrees of freedom and noncentrality parameter $\|\xi\|^2$, is distributed the same as a random variable W obtained by taking a random variable K having a Poisson distribution with mean $1/2\|\xi\|^2$ and then taking the conditional distribution of W given K to be that of a central χ^2 with $p + 2K$ degrees of freedom. Thus

$$E\frac{1}{\|X\|^2} = E\frac{1}{\chi^2_{p+2K}} = E\left(E\frac{1}{\chi^2_{p+2K}}\bigg|K\right) = E\frac{1}{p - 2 + 2K}. \tag{9}$$

To compute the expected value of the middle term on the right side of (8) let

$$U = \frac{\xi'X}{\|\xi\|}, \qquad V = \left\|X - \frac{\xi'X}{\|\xi\|^2}\xi\right\|^2. \tag{10}$$

Then

$$W = \|X\|^2 = U^2 + V, \tag{11}$$

and U is normally distributed with mean $\|\xi\|$ and variance 1, and V is independent of U and has a χ^2 distribution with $p - 1$ d.f. The joint density of U and V is

$$\frac{1}{\sqrt{2\pi}}\exp\left\{-\frac{1}{2}(u - \|\xi\|)^2\right\}\frac{1}{2^{(p-1)/2}\Gamma[(p-1)/2]}v^{(p-3)/2}e^{-v/2} \tag{12}$$

if $v \geqq 0$ and 0 if $v < 0$. Thus the joint density of U and W is

$$\frac{1}{\sqrt{2\pi}\,2^{(p-1)/2}\Gamma[(p-1)/2]}(w - u^2)^{(p-3)/2}\exp\left\{-\frac{1}{2}\|\xi\|^2 + \|\xi\|u - \frac{1}{2}w\right\} \tag{13}$$

if $u^2 \leqq w$ and 0 elsewhere. It follows that

$$E\frac{U}{W} = \frac{\exp\left\{-\frac{1}{2}\|\xi\|^2\right\}}{\sqrt{2\pi}\,2^{(p-1)/2}\Gamma[(p-1)/2]}\int_0^\infty dw$$

$$\int_{-\sqrt{w}}^{\sqrt{w}}\frac{u}{w}(w - u^2)^{(p-3)/2}\exp\left\{\|\xi\|u - \frac{1}{2}w\right\}du. \tag{14}$$

Making the change of variable $t = u/\sqrt{w}$ we find

$$\int_0^\infty dw \int_{-\sqrt{w}}^{\sqrt{w}} \frac{u}{w}(w - u^2)^{(p+3)/2} \exp\left\{\|\xi\|u - \frac{1}{2}w\right\} du$$

$$= \int_0^\infty w^{(p-3)/2} \exp\left\{-\frac{1}{2}w\right\} dw \int_{-1}^1 t(1 - t^2)^{(p-3)/2} \exp\{\|\xi\|t\sqrt{w}\} dt$$

$$= \int_0^\infty w^{(p-3)/2} \exp\left\{-\frac{1}{2}w\right\} dw \sum_{i=0}^\infty \frac{(\|\xi\|\sqrt{w})^i}{i!} \int_{-1}^1 t^{i+1}(1 - t^2)^{(p-3)/2} dt$$

$$= \sum_{j=0}^\infty \frac{\|\xi\|^{2j+1}}{(2j+1)!} \int_0^\infty w^{\frac{p}{2}+j-1} \exp\left\{-\frac{1}{2}w\right\} dw \int_{-1}^1 t^{2(j+1)}(1 - t^2)^{(p-3)/2} dt$$

$$= \sum_{j=0}^\infty \frac{\|\xi\|^{2j+1}}{(2j+1)!} 2^{\frac{p}{2}+j} \Gamma\left(\frac{p}{2}+j\right) B\left(j + \frac{3}{2}, \frac{p-1}{2}\right)$$

$$= \sum_{j=0}^\infty \frac{\|\xi\|^{2j+1} 2^{\frac{p}{2}+j} \Gamma\left((j + \frac{3}{2})\right) \Gamma\left(\frac{p-1}{2}\right)}{(2j+1)! \left(\frac{p}{2}+j\right)}$$

$$= \Gamma\left(\frac{p-1}{2}\right) 2^{p/2} \sum_{j=0}^\infty \frac{2^j \Gamma\left(\frac{1}{2}\right) \|\xi\|^{2j+1}}{2^{2j+1} \Gamma(j+1)\left(\frac{p}{2}+j\right)}. \tag{15}$$

It follows from (10), (11), (14), an (15) that

$$E\frac{(X - \xi)'X}{\|X\|^2} = 1 - \|\xi\|E\frac{U}{W}$$

$$= 1 - \|\xi\| \frac{\exp\left\{-\frac{1}{2}\|\xi\|^2\right\}}{\sqrt{\pi}} \sum_{j=0}^\infty \frac{\Gamma\left(\frac{1}{2}\right) \|\xi\|^{2j+1}}{2^{j+1}\Gamma(j+1)\left(\frac{p}{2}+j\right)}$$

$$= \exp\left\{-\frac{1}{2}\|\xi\|^2\right\} \left\{\sum_{i=0}^\infty \frac{\left(\frac{1}{2}\|\xi\|^2\right)^i}{i!} - \|\xi\|^2 \sum_{j=0}^\infty \frac{\left(\frac{1}{2}\|\xi\|^2\right)^j}{j!(p+2j)}\right\}$$

$$= \exp\left\{-\frac{1}{2}\|\xi\|^2\right\} \sum_{i=0}^\infty \frac{\left(\frac{1}{2}\|\xi\|^2\right)^i}{i!} \frac{p-2}{p-2+2i}$$

$$= (p-2)E\frac{1}{p-2+2K}, \tag{16}$$

where K again has a Poisson distribution with mean $\|\xi\|^2/2$. Combining (8),

(9), and (16) we find

$$\rho(\xi, \varphi_2) = E \left\| \left(1 - \frac{b}{\|X\|^2} \right) X - \xi \right\|^2$$

$$= p - 2(p-2)bE \frac{1}{p-2+2K} + b^2 E \frac{1}{p-2+2K}. \quad (17)$$

This is minimized, for all ξ, by taking $b = p - 2$ and leads to the use of the estimator φ_1 defined by (6) and the formula (5) for its risk.

Now let us look at the case where X he mean ξ and covariance matrix given by

$$E(X - \xi)(X - \xi)' = \sigma^2 I \quad (18)$$

and we observe S independent of X distributed as σ^2 times a χ^2 with n degrees of freedom. Both ξ and σ^2 are unknown. We consider the estimator φ_3 defined by

$$\varphi_3(X, S) = \left(1 - \frac{aS}{\|X\|^2} \right) X, \quad (19)$$

where a is a nonnegative constant. We have

$$\rho(\xi, \varphi_3) = E \| \varphi_3(X, S) - \xi \|^2$$

$$= E\|X - \xi\|^2 - 2aE \frac{S(X-\xi)'X}{\|X\|^2} + a^2 E \frac{S^2}{\|X\|^2}$$

$$= \sigma^2 \left\{ p - 2aE \frac{S}{\sigma^2} E \frac{\left(\frac{X}{\sigma} - \frac{\xi}{\sigma} \right)' \frac{X}{\sigma}}{\left\| \frac{X}{\sigma} \right\|^2} + a^2 E \left(\frac{S}{\sigma^2} \right)^2 E \frac{1}{\left\| \frac{X}{\sigma} \right\|^2} \right\}$$

$$= \sigma^2 \left\{ p - 2an(p-2)E \frac{1}{p-2+2K} + a^2 n(n+2)E \frac{1}{p-2+2K} \right\} \quad (20)$$

by (9) and (16) with X and ξ replaced by X/σ and ξ/σ respectively. Here K has a Poisson distribution with mean $\|\xi\|^2/2\sigma^2$. The choice $a = (p-2)/(n+2)$ minimizes (20) for all ξ, giving it the value

$$\rho(\xi, \varphi_3) = \sigma^2 \left\{ p - \frac{n}{n+2}(p-2)^2 E \frac{1}{p-2+2K} \right\}. \quad (21)$$

We can also treat the case where the covariance matrix is unknown but an estimate based on a Wishart matrix is variable. Let X and S be independently distributed, X having a p-dimensional normal distribution with mean ξ and covariance matrix Σ and S being distributed as a $p \times p$ Wishart matrix with n degrees of freedom and expectation $n\Sigma$, where both ξ and Σ are unknown and Σ is nonsingular. Suppose we want to estimate ξ by $\hat{\xi}$ with loss function

$$L[(\xi, \Sigma), \hat{\xi}] = (\xi - \hat{\xi})'\Sigma^{-1}(\xi - \hat{\xi}). \tag{22}$$

We consider estimators of the form

$$\varphi(X, S) = \left(1 - \frac{c}{X'S^{-1}X}\right)X. \tag{23}$$

The risk function of φ is given by

$$\rho[(\xi, \Sigma), \varphi] = E_{\xi, \Sigma}[\varphi(X, S) - \xi]'\Sigma^{-1}[\varphi(X, S) - \xi]$$

$$= E_{\xi, \Sigma}\left[\left(1 - \frac{c}{X'S^{-1}X}\right)X - \xi\right]'\Sigma^{-1}\left[\left(1 - \frac{c}{X'S^{-1}X}\right)X - \xi\right]$$

$$= E_{\xi^*, I}\left[\left(1 - \frac{c}{X'S^{-1}X}\right)X - \xi^*\right]'\left[\left(1 - \frac{c}{X'S^{-1}X}\right)X - \xi^*\right], \tag{24}$$

where $\xi^{*\prime} = [(\xi'\Sigma^{-1}\xi)^{1/2}, 0, \ldots, 0]$. But it is well known (see, for example Wijsman [19]) that the conditional distribution of $X'S^{-1}X$ given X is that of $X'X/S^*$, where S^* is distributed as χ^2_{n-p+1} independent of X. Comparing (24) and (20) we see that the optimum choice of c is $(p - 2)/(n - p + 3)$ and, for this choice, the risk function is given by

$$\rho[(\xi, \Sigma), \varphi] = p - \frac{n - p + 1}{n - p + 3}(p - 2)^2 E \frac{1}{p - 2 + 2K}, \tag{25}$$

where K has a Poisson distribution with mean $(1/2)\xi'\Sigma^{-1}\xi$.

The improvement achieved by these estimators over the usual estimator may be understood better if we break up the error into its component along X and its component orthogonal to X. For simplicity we consider the case where the covariance matrix is known to be the identity. If we consider any estimator which lies along X, the error orthogonal to X is $\xi - (\xi'X/\|X\|^2)X$ and its mean square is

$$E_\xi\left\|\xi - \frac{\xi'X}{\|X\|^2}X\right\|^2 = \|\xi\|^2 - E_\xi\frac{(\xi'X)^2}{\|X\|^2}$$

$$= \|\xi\|^2\left(1 - E_\xi\frac{1 + 2K}{p + 2K}\right)$$

$$= (p - 1)\|\xi\|^2 E_\xi\frac{1}{p + 2K}$$

$$= (p - 1)\left[1 - (p - 2)E_\xi\frac{1}{p - 2 + 2K}\right]. \tag{26}$$

Thus the mean square of the component along X of the error of $[1 - (p - 2)/\|X\|^2]X$ is

$$\left[p - (p-2)^2 E_\xi \frac{1}{p-2+2K} \right] - (p-1)\left[1 - (p-2)E_\xi \frac{1}{p-2+2K} \right]$$

$$= 1 + (p-2)E_\xi \frac{1}{p-2+2K} \leq 2. \tag{27}$$

On seeing the results given above several people have expressed fear that they were closely tied up with the use of an unbounded loss function, which many people consider unreasonable. We give an example to show that, at least qualitatively, this is not so. Again, for simplicity we suppose X a p-variate random vector normally distributed with mean ξ and covariance matrix equal to the identity matrix. Suppose we are interested in estimating ξ by $\hat{\xi}$ with loss function

$$L(\xi, \hat{\xi}) = F(\|\hat{\xi} - \xi\|^2), \tag{28}$$

where F has a bounded derivative and is continuously differentiable and concave (that is, $F''(t) \leq 0$ for all $t > 0$). We shall show that for sufficiently small b and large a (independent of ξ) the estimator φ defined by

$$\varphi(X) = \left(1 - \frac{b}{a + \|X\|^2} \right) X \tag{29}$$

has smaller risk than the usual estimator X. We have (with $Y = X - \xi$)

$$\rho(\xi, \varphi) = E_\xi F\left[\left\|\left(1 - \frac{b}{a + \|X\|^2} \right) X - \xi \right\|^2 \right]$$

$$= E_\xi F\left[\|X - \xi\|^2 - 2b\frac{X'(X - \xi)}{a + \|X\|^2} + b^2 \frac{\|X\|^2}{(a + \|X\|^2)^2} \right]$$

$$\leq E_\xi F(\|X - \xi\|^2) - 2bE_\xi \left[\frac{X'(X - \xi) - \dfrac{b}{2}}{a + \|X\|^2} \right] F'(\|X - \xi\|^2)$$

$$= EF(\|Y\|^2) - 2bE \left[\frac{(Y + \xi)'Y - \dfrac{b}{2}}{a + \|Y + \xi\|^2} \right] F'(\|Y\|^2). \tag{30}$$

Just as in [14] one can obtain by a Taylor expansion

$$E\left[\frac{(Y + \xi)'Y - \dfrac{b}{2}}{a + \|Y + \xi\|^2} \right] F'(\|Y\|^2)$$

$$= E\frac{(Y + \xi)'Y - \dfrac{b}{2}}{a + \|\xi\|^2 + \|Y\|^2} \left(1 - \frac{2\xi'Y}{a + \|\xi\|^2 + \|Y\|^2} \right) F'(\|Y\|^2)$$

$$+ o\left[\left(\frac{1}{a + \|\xi\|^2} \right) \right]$$

$$
= EE \left\{ \frac{(Y + \xi)'Y - \dfrac{b}{2}}{a + \|\xi\|^2 + \|Y\|^2} \left(1 - \frac{2\xi'Y}{a + \|\xi\|^2 + \|Y\|^2} \right) \middle| \|Y\|^2 \right\} F'(\|Y\|^2)
$$

$$
+ o\left[\left(\frac{1}{a + \|\xi\|^2} \right) \right]
$$

$$
= E \left\{ \frac{\|Y\|^2 - \dfrac{b}{2}}{a + \|\xi\|^2 + \|Y\|^2} - \frac{2\|\xi\|^2 \|Y\|^2/p}{(a + \|\xi\|^2 + \|Y\|^2)^2} \right\} F'(\|Y\|^2)
$$

$$
+ o\left(\frac{1}{a + \|\xi\|^2} \right)
$$

$$
\geqq E \frac{\left(1 - \dfrac{2}{p} \right) \|Y\|^2 - \dfrac{b}{2}}{a + \|\xi\|^2 + \|Y\|^2} F''(\|Y\|^2) + o\left(\frac{1}{a + \|\xi\|^2} \right). \tag{31}
$$

To see that this is everywhere positive if b is sufficiently small and a sufficiently large we look at

$$
AE \frac{\|Y\|^2 F'(\|Y\|^2)}{A + \|Y\|^2} = f(A) \tag{32}
$$

and

$$
AE \frac{F'(\|Y\|^2)}{A + \|Y\|^2} = g(A). \tag{33}
$$

It is clear that f and g are continuous strictly positive valued functions on $[1, \infty)$ with finite nonzero limits at ∞. It follows that

$$
c = \inf_{A \geqq 1} f(A) = 0, \tag{34}
$$

$$
d = \sup_{A \geqq 1} g(A) < \infty. \tag{35}
$$

Then, if b is chosen so that

$$
\left(1 - \frac{2}{p} \right) c - \frac{b}{2} d > 0 \tag{36}
$$

it follows from (30) and (31) that, for sufficiently large a

$$
\rho(\xi, \varphi) < E_\xi F(\|X - \xi\|^2) \tag{37}
$$

for all ξ.

The inadmissibility of the usual estimator for three or more location parameters does not require the assumption of normality. We shall give the following result without proof. Let X be a p-dimensional random coordinate vector with mean ξ, and finite fourth absolute moments:

$$
E_\xi (X_i - \xi_i)^4 \leqq C < \infty. \tag{38}
$$

For simplicity of notation we assume the X_i are uncorrelated and write

$$\sigma_i^2 = E_\xi(X_i - \xi_i)^2. \tag{39}$$

Then for $p \geqq 3$ and

$$b < 2(p - 2) \min \sigma_i^2 \tag{40}$$

and sufficiently large a depending only on C

$$E_\xi \sum \left[\left(1 - \frac{b}{\sigma_i^2 [a + \sum X_i^2/\sigma_i^2]} \right) X_i - \xi_i \right]^2 < E_\xi \sum (X_i - \xi_i)^2 = \sum \sigma_i^2. \tag{41}$$

It would be desirable to obtain explicit formulas for estimators one can seriously recommend in the last two cases considered above.

3. Formulation of the General Problem of Admissible Estimation with Quadratic Loss

Let $\mathcal{X}$ be a set (the sample space), $\mathcal{B}$ a σ-algebra of subsets of $\mathcal{X}$ and λ a σ-finite measure on $\mathcal{B}$. Let Θ be another set (the parameter space), $\mathcal{C}$ a σ-algebra of subsets of Θ, and $p(\cdot|\cdot)$ a nonnegative valued $\mathcal{BC}$-measurable function on $\mathcal{X} \times \Theta$ such that for each $\theta \in \Theta$, $p(\cdot|\theta)$ is a probability density with respect to λ, that is,

$$\int p(z|\theta) \, d\lambda(z) = 1. \tag{42}$$

Let A, the action space, be the k-dimensional real coordinate space, α a $\mathcal{C}$-measurable function on Θ to the set of positive semidefinite symmetric $k \times k$ matrices, and η a $\mathcal{C}$-measurable function on Θ to A. We observe Z distributed over $\mathcal{X}$ according to the probability density $p(\cdot|\theta)$ with respect to λ, where θ is unknown, then choose an action $a \in A$ and suffer the loss

$$L(\theta, a) = [a - \eta(\theta)]'\alpha(\theta)[a - \eta(\theta)]. \tag{43}$$

An estimator φ of $\eta(\theta)$ is a $\mathcal{B}$-measurable function on $\mathcal{X}$ to A, the interpretation being that after observing Z one takes action $\varphi(Z)$, or in other words, estimates $\eta(\theta)$ to be $\varphi(Z)$. The risk function $\rho(\cdot, \varphi)$ is the function on Θ defined by

$$\rho(\theta, \varphi) = E_\theta[\varphi(Z) - \eta(\theta)]'\alpha(\theta)[\varphi(Z) - \eta(\theta)]. \tag{44}$$

Roughly speaking, we want to choose φ so as to keep $\rho(\cdot, \varphi)$ small, but this is not a precise statement since, for any given φ it will usually be possible to modify φ so as to decrease $\rho(\theta, \varphi)$ at some θ but increase it at other θ. In many problems there is a commonly used estimator, for example, the maximum likelihood estimator or one suggested by invariance, linearity, unbiasedness, or some combination of these. Then it is natural to ask whether this estimator is admissible in the sense of Wald. The estimator φ_0 is said to be admissible if there is no estimator φ for which

$$\rho(\theta, \varphi) \leqq \rho(\theta, \varphi_0) \tag{45}$$

for all θ with strict inequality for some θ. If there does exist such a φ then φ is said to be better than φ_0 and φ_0 is said to be inadmissible. We shall also find it useful to define an estimator φ_0 to be almost admissible with respect to a measure Π on the σ-algebra $\mathscr{C}$ of subsets of Θ if there is no estimator φ for which (45) holds for all θ with strict inequality on a set having positive Π-measure.

Next we give a simple sufficient condition for almost-admissibility of certain estimators. Although we do not discuss the necessity of the condition, its similarity to the necessary and sufficient condition of [15] leads us to apply it with confidence. It should be remarked that in [15] the condition of boundedness of the risk of the estimator (or strategy, b_0 in the notation of that paper) was inadvertently omitted in theorems 3 and 4. It is needed in order to justify the reduction to (48). If Π is a σ-finite measure on $\mathscr{C}$ we shall define

$$\varphi_\Pi(x) = \left[\int \alpha(\theta) p(x|\theta) \, d\Pi(\theta) \right]^{-1} \int \alpha(\theta) \eta(\theta) p(x|\theta) \, d\Pi(\theta) \tag{46}$$

provided the integrals involved are finite almost everywhere. Observe that if Π is a probability measure, φ_Π is the Bayes' estimator of $\eta(\theta)$, that is, $\varphi = \varphi_\Pi$ minimizes

$$\int d\Pi(\theta) E_\theta [\varphi(X) - \eta(\theta)]' \alpha(\theta) [\varphi(X) - \eta(\theta)]. \tag{47}$$

If q is a probability density with respect to Π we shall write (q, Π) for the induced probability measure, that is,

$$(q, \Pi)S = \int_S q(\theta) \, d\Pi(\theta). \tag{48}$$

Theorem 3.1. *If φ is an estimator of $\eta(\theta)$ with bounded risk such that for each set C in a denumerable family $\mathscr{F}$ of sets whose union is Θ*

$$\inf_{q \in \mathscr{S}(C)} \frac{\int q(\theta) \, d\Pi(\theta) E_\theta [\varphi(X) - \varphi_{(q,\Pi)}(X)]' \alpha(\theta) [\varphi(X) - \varphi_{(q,\Pi)}(X)]}{\int_C q(\theta) \, d\Pi(\theta)} = 0, \tag{49}$$

where $\mathscr{S}(C)$ is the set of probability densities with respect to Π which are constant (but not 0) on C, then φ is almost admissible with respect to Π.

(Editors' note: The proof has been omitted.)

4. Admissibility of Pitman's Estimator for Location Parameters in Certain Low Dimensional Cases

The sample space $\mathscr{Z}$ of section 3 is now of the form $\mathscr{X} \times \mathscr{Y}$, where $\mathscr{X}$ is a finite-dimensional real coordinate space, and $\mathscr{Y}$ arbitrary and the σ-algebra $\mathscr{B}$ is a product σ-algebra $\mathscr{B} = \mathscr{B}_1 \mathscr{B}_2$ where $\mathscr{B}_1$ consists of the Borel sets in $\mathscr{X}$ and $\mathscr{B}_2$ is an arbitrary σ-algebra of subsets of $\mathscr{Y}$. Here λ is the product measure $\lambda = \mu_\nu$ where μ is a Lebesgue measure on $\mathscr{B}_1$ and ν is an arbitrary probability measure on $\mathscr{B}_2$. The parameter space Θ and the action space A coincide with $\mathscr{X}$. The loss function is

$$L(\theta, a) = (a - \theta)'(a - \theta). \tag{54}$$

We observe (X, Y) whose distribution, for given θ, is such that Y is distributed according to ν and the conditional density of $X - \theta$ given Y is $p(\cdot | Y)$, a known density. We assume

$$\int p(x, y)\, dx = 1 \tag{55}$$

and

$$\int xp(x, y)\, dx = 0 \tag{56}$$

for all y. Condition (56) is introduced only for the purpose of making the natural estimator X (see [6] or [16]). The condition (49) for the natural estimator X to be almost admissible becomes the existence of a sequence π_σ of densities with respect to Lebesgue measure in $\mathscr{X}$ such that

$$\lim_{\sigma \to \infty} \frac{1}{\inf_{\|x\| \leq 1} \pi_\sigma(x)} \int d\nu(y) \int dx \frac{\left\| \int \eta\pi_\sigma(x - \eta)p(\eta | y)\, d\eta \right\|^2}{\int \pi_\sigma(x - \eta)p(\eta | y)\, d\eta} = 0, \tag{57}$$

This is derived in formula (63) below.

In [16] one of the authors proved the following theorem.

Theorem 4.1. *When* $\dim X = 1$, *if in addition to the above conditions*

$$\int d\nu(y)\left[\int x^2 p(x | y)\, dx\right]^{3/2} < \infty, \tag{58}$$

then X *is an admissible estimator of* θ, *that is, there does not exist a function* φ *such that*

$$\int d\nu(y) \int [\varphi(x, y) - \theta]^2 p(x - \theta | y)\, dx \leq \int d\nu(y) \int x^2 p(x | y)\, dx \tag{59}$$

for all θ *with strict inequality for some* θ.

This is proved by first showing that (57) holds with

$$\pi_\sigma(x) = \frac{1}{\pi\sigma\left(1 + \dfrac{x^2}{\sigma^2}\right)},$$

(60)

so that X is almost admissible, and then proving that this implies that X is admissible. It is not clear that the condition (58) is necessary.

We shall sketch the proof of a similar but more difficult theorem in the two-dimensional case.

Theorem 2. *When* dim $\mathcal{X} = 2$, *if in addition to* (55) *and* (56)

$$\int dv(y)\left[\int \|x\|^2 \log^{1+\delta} \|x\|^2 p(x, y)\, dx\right]^2 < \infty,$$

(61)

then X is an admissible estimator of θ, that is, there does not exist a function φ on $\mathcal{X} \times \mathcal{Y}$ to $\mathcal{X}$ such that

$$\int dv(y) \int \|\varphi(x, y) - \theta\|^2 p(x - \theta|y)\, dx \leqq \int dv(y) \int \|x\|^2 p(x|y)\, dx$$

(62)

for all θ with strict inequality for some θ.

(*Editors' note*: The Sketch of Proof has been omitted.)

Theorems 1 and 2 together with the results of section 2 settle in a fairly complete manner the question of admissibility of Pitman's estimator of a number of location parameters with positive definite translation-invariant quadratic loss function. If enough moments exist, Pitman's estimator is admissible if the number of parameters to be estimated is one or two, but inadmissible if this number is at least three. In the case where there are also parameters which enter as nuisance parameters, the only known results are an example of Blackwell [2] and some trivial consequences of the results given above.

Blackwell formulates his example as a somewhat pathological case involving one unknown location parameter where Pitman's estimator is almost admissible in the class of Borel-measurable estimators but not admissible (and not even almost admissible if measurability is not required). However, it can be reformulated as a nonpathological example involving four unknown location parameters with a quadratic loss function of rank one.

Now consider the problem where we have p unknown location parameters and the positive semidefinite translation-invariant quadratic loss function has rank r. From the result given without proof at the end of section 2, it follows that if $r \geqq 3$ and all fourth absolute moments exist, then Pitman's estimator is inadmissible. This follows from the application of the result at the end of section 2 to the problem we obtain if we look only at Pitman's estimator of the r parameters which enter effectively into the loss function, ignoring the rest of the observation. If $p = 2$ and $r = 1$, then, subject to the conditions of theorem 2, Pitman's estimator is admissible. If it were not, we could obtain a better estimator than Pitman's for a problem with $r = 2$ (contradicting theorem 2) by using the better estimator for the parameter which occurs in the original loss function of rank 1 and Pitman's estimator for another linearly independent parameter, with the new loss function equal to the sum of squares of the errors in estimating the parameters defined in this way. If r is 1 or 2 but p arbitrary, Pitman's estimator is admissible (subject to the existence of appropriate moments) if, for some choice of the coordinate system defining the $r - p$ nuisance parameters, Pitman's estimator coincides with what would be Pitman's estimator if these $r - p$ nuisance parameters were known. This is always the case if $r \leq 2$ and the observed random vector X is normally distributed.

These are all the results known to me for the problem of estimating unknown location parameters with positive semidefinite translation-invariant quadratic loss function. In the following conjectures it is to be understood that in all cases sufficiently many moments are assumed to exist. I conjecture that if $p = 3$ and $r = 1$, in the case of a single orbit (that is, when $\mathcal{Y}$, analogous to that of theorems 1 and 2, reduces to a single point) Pitman's estimator is admissible, but this does not hold in general when $\mathcal{Y}$ does not reduce to a point. In the other cases not covered in the preceding paragraph, that is, if $p \geq 3$ and $r = 2$ or if $p \geq 4$ and $r = 1$, I conjecture that Pitman's estimator is, in general, inadmissible, but of course there are many exceptions, in particular those mentioned at the end of the last paragraph. Blackwell's example supports this conjecture for $p \geq 4$ and $r = 1$.

5. Some Problems Where the Natural Estimator Is not Minimax

Kiefer [9] and Kudo [10] have shown that under certain conditions, a statistical problem invariant under a group of transformations possesses a minimax solution which is also invariant under this group of transformations. However, these conditions do not hold for the group of all nonsingular linear transformations in a linear space of dimension at least two. I shall give here a problem in multivariate analysis for which I can derive a minimax solution and show that the natural estimator (invariant under the full linear group) is not minimax.

Consider the problem in which we observe $X_1, \ldots, X_n$ independently normally distributed p-dimensional random vectors with mean 0 and unknown covariance matrix Σ where $n \geq p$. Suppose we want to estimate Σ, say by $\hat{\Sigma}$ with loss function

$$L(\Sigma, \hat{\Sigma}) = \operatorname{tr} \Sigma^{-1} \hat{\Sigma} - \log \det \Sigma^{-1} \hat{\Sigma} - p. \tag{72}$$

The problem is invariant under the transformations $X_i \to aX_i$, $\Sigma \to a\Sigma a'$, $\hat{\Sigma} \to a\hat{\Sigma}a'$ where a is an arbitrary nonsingular $p \times p$ matrix. Also

$$S = \sum_{i=1}^{n} X_i X_i' \tag{73}$$

is a sufficient statistic and if we make the transformation $X_i \to aX_i$, then $S \to aSa'$. We may confine our attention to estimators which are functions of S alone. The condition of invariance of an estimator φ (a function on the set of positive definite $p \times p$ symmetric matrices to itself) under transformation by the matrix a is

$$\varphi(asa') = a\varphi(s)a' \quad \text{for all } s. \tag{74}$$

Let us look for the best estimator φ satisfying (74) for all lower triangular matrices a, that is, those satisfying $a_{ij} = 0$ for $j > i$. We shall find that this $\varphi(S)$ is not a scalar multiple of S. At the end of the section we shall sketch the proof that such an estimator is minimax. Similar results hold for the quadratic loss function

$$L^*(\Sigma, \hat{\Sigma}) = \text{tr}(\Sigma^{-1}\hat{\Sigma} - I)^2 \tag{75}$$

but I have not been able to get an explicit formula for a minimax estimator in this case.

Putting $s = I$ in (74) we find

$$\varphi(aa') = a\varphi(I)a'. \tag{76}$$

When we let a range over the set of diagonal matrices with ± 1 on the diagonal, this yields

$$\varphi(I) = a\varphi(I)a', \tag{77}$$

which implies that $\varphi(I)$ is a diagonal matrix, say Δ, with ith diagonal element Δ_i. This together with (74) determines φ since any positive definite symmetric matrix S can be factored as

$$S = KK' \tag{78}$$

with K lower triangular (with positive diagonal elements) and we then have

$$\varphi(S) = K\Delta K'. \tag{79}$$

Since the group of lower triangular matrices operates transitively on the parameter space, the risk of an invariant procedure φ is constant. Thus we compute the risk only for $\Sigma = I$. We then have

$$\rho(I, \varphi) = E[\text{tr } \varphi(S) - \log \det \varphi(S) - p]$$

$$= E(\text{tr } K\Delta K' - \log \det K\Delta K' - p)$$

$$= E \text{ tr } K\Delta K' - \log \det \Delta - E \log \det S - p. \tag{80}$$

But

$$E \operatorname{tr} K\Delta K' = \sum_{i,k} \Delta_i EK_{ki}^2$$

$$\sum \Delta_i E\chi_{n-i+1+p-i}^2 = \sum \Delta_i(n + p - 2i + 1) \qquad (81)$$

since the elements of K are independent of each other, the ith diagonal element being distributed as χ_{n-i+1} and the elements below the diagonal normal with mean 0 and variance 1. Also, for the same reason,

$$E \log \det S = \sum_{i=1}^{p} E \log \chi_{n-i+1}^2. \qquad (82)$$

It follows that

$$\rho(\Sigma, \varphi) = \rho(I, \varphi)$$

$$= \sum_{i=1}^{p} [(n + p - 2i + 1)\Delta_i - \log \Delta_i] - \sum_{i=1}^{p} E \log \chi_{n-i+1}^2 - p.$$
$$(83)$$

This attains its minimum value of

$$\rho(\Sigma, \varphi^*) = \sum_{i=1}^{p} \left[1 - \log \frac{1}{n + p - 2i + 1} - E \log \chi_{n-i+1}^2 \right] - p$$

$$= \sum [\log(n + p - 2i + 1) - E \log \chi_{n-i+1}^2] \qquad (84)$$

when

$$\Delta_i = \frac{1}{n + p - 2i + 1}. \qquad (85)$$

We have thus found the minimax estimator in a class of estimators which includes the natural estimators (multiples of S) to be different from the natural estimators. Since the group of lower triangular matrices is solvable it follows from the results of Kiefer [9] that the estimator given by (79) and (85) is minimax. However, it is not admissible. One can get a better estimator by averaging this estimator and one obtained by permuting the coordinates, applying the method given above and then undoing the permutation. It must be admitted that the problem is somewhat artificial.

6. Some More Unsolved Problems

In section 4 several conjectures and unsolved problems concerning estimation of location parameters have been mentioned. Some other problems are listed below. Of course, one can combine these in many ways to produce more difficult problems.

(i) What are the admissible estimators of location parameters? In particular, what are the admissible minimax estimators of location parameters?

(ii) What results can be obtained for more general loss functions invariant under translation?

(iii) For a problem invariant under a group other than a translation group, when is the best invariant estimator admissible? In particular, is Pitman's estimator admissible when both locations and scale parameters are unknown?

(iv) What can we say in the case of more complicated problems where there may be no natural estimator? For example, consider the problem in which we observe $S_1, \ldots, S_n$ independently distributed as $\sigma_n^2 \chi_k^2$, and want to estimate $\sigma_1^2, \ldots, \sigma_n^2$ by $\hat{\sigma}_1^2, \ldots, \hat{\sigma}_n^2$ with loss function

$$L(\sigma_1^2, \ldots, \sigma_n^2; \hat{\sigma}_1^2, \ldots, \hat{\sigma}_n^2) = \frac{\sum (\sigma_i^2 - \hat{\sigma}_i^2)^2}{\sum \sigma_i^4}. \tag{86}$$

It is clear that

$$\hat{\sigma}_i^2 = \frac{1}{k+2} S_i \tag{87}$$

is a minimax estimator since the risk for this estimator is constant and it is minimax when all except one of the σ_i^2 are 0 (see Hodges and Lehmann [7]). But this estimator is clearly very poor if k is small and n is large. This problem arises in the estimation of the covariances in a finite stationary circular Gaussian process.

References

[1] A.C. Aitken, "On least squares and linear combination of observations," *Proc. Roy. Soc. Edinburgh, Sect. A*, Vol. 55 (1935), pp. 42–48.

[2] D. Blackwell, "On the translation parameter problem for discrete variables," *Ann. Math. Statist.*, Vol. 22 (1951), pp. 393–399.

[3] C. Blyth, "On minimax statistical decision procedures and their admissibility," *Ann. Math. Statist.*, Vol. 22 (1951), pp. 22–42.

[4] F.N. David and J. Neyman, "Extension of the Markoff theorem of least squares," *Statist. Res. Mem.*, Vol. 1 (1938), pp. 105–116.

[5] R.A. Fisher, "On the mathematical foundations of theoretical statistics," *Philos. Trans. Roy. Soc. London, Ser. A*, Vol. 222 (1922), pp. 309–368.

[6] M.A. Girshick and L.J. Savage, "Bayes and minimax estimates for quadratic loss functions," *Proceedings of the Second Berkeley Symposium on Mathematical Statistics and Probability*, Berkeley and Los Angeles, University of California Press, 1951, pp. 53–73.

[7] J.L. Hodges, Jr., and E.L. Lehmann, "Some applications of the Cramér-Rao inequality," *Proceedings of the Second Berkeley Symposium on Mathematical Statistics and Probability*, Berkeley and Los Angeles, University of California Press, 1951, pp. 13–22.

[8] S. Karlin, "Admissibility for estimation with quadratic loss," *Ann. Math. Statist.*, Vol. 29 (1958), pp. 406–436.

[9] J. Kiefer, "Invariance, minimax sequential estimation, and continuous time processes," *Ann. Math. Statist.*, Vol. 28 (1957), pp. 573–601.

[10] H. Kudo, "On minimax invariant estimators of the transformation parameter,"
      Nat. Sci. Rep. Ochanomizu Univ., Vol. 6 (1955), pp. 31–73.

[11] E.L. Lehmann, *Testing Statistical Hypotheses*, New York, Wiley, 1989, pp. 231
      and 338.

[12] A. Markov, *Calculus of Probability*, St. Petersburg, 1908 (2nd ed.). (In Russian.)

[13] E.J.G. Pitman, "Location and scale parameters," *Biometrika*, Vol. 30 (1939),
      pp. 391–421.

[14] C. Stein, "Inadmissibility of the usual estimator for the mean of a multivariate
      normal distribution," *Proceedings of the Third Berkeley Symposium on Mathe-
      matical Statistics and Probability*, Berkeley and Los Angeles, University of Cali-
      fornia Press, 1956, Vol. 1, pp. 197–206.

[15] ———, "A necessary and sufficient condition for admissibility," *Ann. Math.
      Statist.*, Vol. 26 (1955), pp. 518–522.

[16] ———, "The admissibility of Pitman's estimator for a single location parame-
      ter," *Ann. Math. Statist.*, Vol. 30 (1959), pp. 970–979.

[17] ———, "Multiple regression," *Contributions to Probability and Statistics, Essays
      in Honor of Harold Hotelling*, Stanford, Stanford University Press, 1960, pp.
      424–443.

[18] A. Wald, "Contributions to the theory of statistical estimation and testing hy-
      potheses," *Ann. Math. Statist.*, Vol. 10 (1939), pp. 299–326.

[19] R.A. Wijsman, "Random orthogonal transformations and their use in some
      classical distribution problems in multivariate analysis," *Ann. Math. Statist.*, Vol.
      28 (1957), pp. 415–423.

Introduction to
Birnbaum (1962) On the Foundations of Statistical Inference

Jan F. Bjørnstad
The University of Trondheim

1. Summary

The paper was presented at a special discussion meeting of the ASA on December 27, 1961 in New York City. It studies the likelihood principle (LP) and how the likelihood function can be used to measure the evidence in the data about an unknown parameter. Essentially, the LP says that if two experiments about the same unknown parameter produce proportional likelihood functions, then the same inference should be made in the two cases.

The LP developed mainly from the ideas of Fisher and Barnard. It caught widespread attention in 1962 when Birnbaum showed that the LP was a consequence of the more accepted principles of sufficiency (that a sufficient statistic summarizes all evidence from an experiment) and conditionality (that experiments not actually performed are irrelevant to inference). Since then the LP has been debated extensively with regard to its place in the foundations of statistical theory. The radical consequences of the LP to statistical analysis have been discussed in many areas. Two of the most important implications are that at the *inference* stage, stopping rules and sampling designs in survey sampling are irrelevant.

By far, most of the work on the LP's implications and applications has appeared after the 62-paper. During the last 20 years or so, likelihood approaches have been proposed in areas like (1) estimation with nuisance parameters, (2) prediction, (3) survey sampling, (4) missing data problems, and (5) meta-analysis. The monograph by Berger and Wolpert (1984) (hereafter denoted BW) gives an extensive and incisive presentation of the LP, discussing validity, generalizations, and implementations. Otherwise, general discussions of the LP and its consequences can be found in Cox and Hinkley (1974), Basu (1975), Dawid (1977), Dawid (1981), and Barnett (1982).

2. Impact of the Paper

Birnbaum's main result, that the LP follows from (and implies) sufficiency and conditionality principles that most statisticians accept, must be regarded as one of the deepest theorems of theoretical statistics, yet the proof is unbelievably simple. The result had a decisive influence on how many statisticians came to view the likelihood function as a basic quantity in statistical analysis. Still, even though the impact of this result alone has made a major contribution to the theory of statistics as illustrated in Sec. 5, the paper's contribution is not limited to this fundamental achievement. It has also affected in a general way how we view the science of statistics. Birnbaum introduced principles or axioms of equivalence within and between experiments, showing various relationships between these principles. This made it possible to discuss the different concepts from alternative viewpoints, thereby discovering weaknesses and strengths of the concepts. Birnbaum's approach also meant that various statistical "philosophies" could be discussed on a firm theoretical basis. Hence, the paper changed our way of thinking about statistical theories, giving all of us a most important and lasting contribution whether we agree with the LP or not.

3. The Development of Likelihood Ideas Prior to 1962

At the time when Birnbaum's paper appeared in 1962, likelihood ideas and methods did not attract much attention in the statistical community. The Neyman–Pearson school and Wald's decision theory were the dominating approaches, also for statistical problems not of a decision theoretic nature.

The major proponents of likelihood-based inference before Birnbaum's paper were Fisher and Barnard. Fisher's theory of estimation (excluding fiducial interval estimation) was essentially a pure likelihood approach, developed in the papers of 1922, 1925, and 1934. Barnard (1947) gave the first version of the LP.

1. Fisher's Contributions

The term "likelihood" first appeared in Fisher (1921) where the different nature between likelihood and probability was emphasized. The likelihood function as a basis for estimation was introduced by Fisher (1922) when the concepts of information and sufficiency and the method of maximum likelihood were presented. Here, Fisher also used the likelihood to measure the relative support the data give to different values of the parameter. When this paper appeared, the Bayesian theory of Laplace was the main approach to statistical inference. Fisher's likelihood-based alternative, together with his sharp criticism of the use of prior distributions, especially priors to represent ignorance [see Fisher (1932, p. 258)], led to a lesser interest in Bayesian inference.

After 1922, Fisher's work on the foundations of statistical inference emphasized likelihood and conditionality, in particular, in Fisher (1925, 1934). The concept of likelihood-based information played a central role in Fisher's estimation theory and in the development of conditional inference.

Although he came close to asserting a principle of likelihood inference in the theory of estimation [see Fisher (1973, p.73) where he states that in the theory of estimation it has appeared that the whole of the information is comprised in the likelihood function], it seems Fisher never actually stated the likelihood principle in general and may not have been thinking of it as a separate statistical principle [see Fraser (1976)]. Fisher did, however, state a conditionality principle when in 1934, the theory of exact conditional inference based on ancillary statistics was developed for translation families. Fisher's conditionality principle was motivated by finding the right measure of precision for the maximum likelihood estimate (m.l.e.), and this was attained by conditioning on the maximal ancillary statistic. This conditioning also recovered the information lost by the m.l.e. It should be noted that an ancillary statistic, as used by Fisher, was by definition a part of the minimal sufficient statistic and an index of precision of the m.l.e., not just any statistic whose distribution is independent of the parameter. Since Fisher also supported the principle of sufficiency, his theory of estimation, in effect, agreed with the LP. In general, however, Fisher did not follow the LP. For example, in tests of significance, he advocated using the P-value [see Fisher (1973)], which violates the LP.

Fisher, in his 1956 book, also proposed a likelihood function for prediction [see Fisher (1973, p. 135)]. More than 20 years would pass before the idea of likelihood-based prediction was developed further by Hinkley (1979).

2. Other Contributors

The first formal statement of a version of the likelihood principle was by Barnard (1947, 1949) in the case where the parameter has only two possible values and the LP reduces to stating that two experiments with the same likelihood ratio should lead to the same decision. This was in strong disagreement with the Neyman–Pearson theory that at the time was the dominant approach to hypothesis testing.

Barnard (1949) argued for the LP from an abstract theory of likelihood concepts based on log (odds) instead of probability. From this theory, it was deduced that the relevant measure of evidence of one parameter value θ against another θ' is the likelihood ratio. It was also shown, under certain assumptions, from the frequentist point of view (with the usual probability model as a basis) that in choosing between θ and θ', the decision must be based on the likelihood ratio.

Likelihood concepts were also employed by several other statisticians. Some references are listed in Kendall (1946, pp. 45, 83). For example, Bartlett (1936) used conditional and marginal likelihood for estimating one parameter in the presence of nuisance parameters, and Bartlett (1953) considered approximate confidence intervals based on the derivative of the log likelihood.

A conditionality principle plays a major role in Birnbaum's paper. A weaker version of this principle appeared in Cox (1958). Cox challenged several of the usual frequentistic approaches to inference, emphasizing the importance of conditioning with several illuminating examples. Cox's view on conditioning seems to have been essentially the same as Fisher's.

4. Contents of the Paper

The main aim of the paper is to show and discuss the implication of the fact that the LP is a consequence of the concepts of conditional frames of reference and sufficiency. To this aim, principles of sufficiency, conditionality, and likelihood are defined in terms of the concept of the evidential meaning of an outcome of an experiment.

A second aim of the paper is to describe how and why these principles are appropriate ways to characterize statistical evidence in parametric models for inference purposes. The paper is concerned primarily with approaches to inference that do not depend on a Bayesian model.

After the introduction, the paper is divided into two parts. Part I deals with the mentioned principles of inference and how they relate to each other. Part II deals mainly with how the likelihood function can be interpreted as a measure of evidence about the unknown parameter and considers how commonly used concepts like significance levels and confidence levels can be reinterpreted to be compatible with the LP, leading to intrinsic methods and levels. A discussion section follows the paper. The discussion shows that the participants are very much aware of the vast implications for statistical inference of adopting the LP.

In summarizing the contents of the paper, the sections are named as in the paper.

1. Introduction, Summary, and General Conclusions

An experiment E is defined as $E = \{\Omega, S, f(x, \theta)\}$, where f is a density with respect to a σ-finite measure μ and θ is the unknown parameter. Ω is the parameter space and S the sample space of outcomes x of E. The likelihood function determined by an observed outcome is then $L_x(\theta) = f(x, \theta)$.

Birnbaum restricts attention to problems of informative statistical inference, where one is interested in summarization of evidence or information about θ as provided by E and x alone [denoted by $Ev(E, x)$], and distinguishes two main problems of informative inference: (1) principles that statistical evidence should follow (part I) and (2) interpretation of statistical evidence in accordance with accepted principles (part II). The principles all prescribe conditions under which we should require the same inference for (E, x) and (E', x').

The introduction summarizes the principles of sufficiency (S), conditionality (C), and likelihood (L) defined in Secs. 3–5. (S) and (C) are derived from Fisher's ideas on sufficiency and ancillarity. Birnbaum gives the following formal LP:

(L) Let E and E' be two experiments (with common parameter θ). Assume x, y are outcomes of E, E' respectively with proportional likelihood functions, i.e., $L_x(\theta) = cL_y(\theta)$ for all θ in Ω, for some constant (in θ) $c > 0$. Then: $Ev(E, x) = Ev(E', y)$.

Note that the case $E = E'$ is included. Birnbaum states, without proof, that (L) implies that $Ev(E, x)$ depends on E and x only through $L_x(\theta)$. A proof of this result in the discrete case can be found in BW (p. 28). The three principles are described informally in the following way: (S) asserts the irrelevance of observations independent of a sufficient statistic. (C) asserts the irrelevance of experiments not actually performed. (L) asserts the irrelevance of outcomes not actually observed.

This section concludes with stating the main result of the paper [(S) and (C) *together are equivalent to* (L)] and a discussion of the radical consequences of (L) to the theory and practice of statistical inference. One aspect of the main result is that the likelihood function is given new support, independent of the Bayesian point of view.

Part I

2. Statistical Evidence

This section introduces the concept of the evidential meaning of an observation from an experiment and discusses the term informative inference, where Cox (1958) is a main reference.

Birnbaum states that the central purpose of the paper is to clarify the essential structure and properties of statistical evidence, termed the evidential meaning of (E, x) and denoted by $Ev(E, x)$, in various instances. We can say that $Ev(E, x)$ is the evidence about θ supplied by x and E. Nothing is assumed about what $Ev(E, x)$ actually is. It can be a report of the experimental results, the inferences made, the methods used, or a collection of different measures of evidence. Birnbaum restricts attention to problems of informative statistical inference, but as remarked by BW (p. 25), since no assumptions are made about $Ev(E, x)$, the "evidence" may also depend on a loss function. Hence, the theory should also apply to decision problems.

To illustrate the concept of informative inference, methods and concepts from the different perspectives of Fisher, the Neyman–Pearson school, and the Bayesian approach are discussed.

3. The Principle of Sufficiency

Here a principle of sufficiency (S) is defined in terms of $Ev(E, x)$:

(S): Let $t(x)$ be a sufficient statistic for E, and let E' be the experiment of observing $t = t(x)$. Then: $Ev(E, x) = Ev(E', t(x))$.

The following result is shown by using a result from Bahadur (1954):

Lemma 1. Assume x, x' are two outcomes of E with proportional likelihood functions. Then : $(S) \Rightarrow Ev(E, x) = Ev(E, x')$.

We note that this does not mean that (S) implies (L) since x, x' are from the same experiment.

4. The Principle of Conditionality

This section considers conditional frames of reference and defines a principle of conditionality (C) in terms of $Ev(E, x)$, first stated in Birnbaum (1961) and related to a discussion by Cox (1958) who provided the crucial idea of a mixture experiment. According to Birnbaum [in a discussion of Barnard et al. (1962)], it was this paper by Cox that made him appreciate the significance of conditionality concepts in statistical inference. The principle is as follows:

(C): Let E be a mixture of experiments with components $\{E_h\}$ (with common unknown θ), where E_h is selected by a known random mechanism. I.e., E consists of first selecting a component experiment E_h and then observing the outcome x_h of E_h such that the outcome of E can be represented as (h, x_h). Then: $Ev(E, (h, x_h)) = Ev(E_h, x_h)$.

(C) asserts that the experiments not actually performed are irrelevant. It is stated, without proof, that $(C) \Rightarrow (S)$. This is not correct. Birnbaum (1972) considers the discrete case and shows that (S) is implied by (C) *and* the principle of mathematical equivalence (M) that states: If x and x' are two outcomes of the same experiment E with identical likelihood functions, then $Ev(E, x) = Ev(E, x')$.

The main part of this section serves to illustrate, through various examples, why (C) seems unavoidable in interpreting evidence. It is shown that (C) alone implies quite radical consequences for inference theory, and Birnbaum believes that (C) will be generally accepted. The only serious criticism of (C) in the literature seems to have been Durbin (1970). BW (p. 45) illustrates why

Durbin's objection is not convincing. (C) or some version of it seems today to be accepted by most statisticians of various schools of inference.

5. The Likelihood Principle

This section contains the main result of the paper.

Lemma 2. (S) and (C) ⇔ (L).

The proof is surprisingly simple. Because the proof itself of (⇒) has played an important part in the discussion of this result, we shall give the reader a brief outline. Let (E_1, x_1) and (E_2, x_2) have proportional likelihood functions and construct then the mixture experiment E^* that chooses E_1 with probability 1/2. Then from (C), it follows that it is enough to show that $Ev(E^*, (1, x_1)) = Ev(E^*, (2, x_2))$, which follows from (S) since $(1, x_1)$ and $(2, x_2)$ have proportional likelihood functions in E^*.

The implication (⇒) is the most important part of the equivalence, because this means that if you do not accept (L), you have to discard either (S) or (C), two widely accepted principles. The most important consequence of (L) seems to be that evidential measures based on a specific experimental frame of reference (like P-values and confidence levels) are somewhat unsatisfactory (in Birnbaum's own words). In other words, (L) eliminates the need to consider the sample space or any part of it once the data are observed. Lemma 2 truly was a "breakthrough" in the foundations of statistical inference and made (L) stand on its own ground, independent of a Bayesian argument. As Savage (1962) noted in his discussion of the paper,

> Without any intent to speak with exaggeration or rhetorically, it seems to me that this is really a historic occasion. This paper is a landmark in statistics because it seems improbable to me that many people will be able to read this paper or to have heard it tonight without coming away with considerable respect for the likelihood principle.

Part II

6. Evidential Interpretations of Likelihood Functions

This is a short section describing the purposes of Sec. 7–9 that are mainly concerned with the question: What are the qualitative and quantitative properties of statistical evidence represented by $L_x(\theta)$?

7. Binary Experiments

This section covers the case $\#(\Omega) = 2$ and is closely related to parts of Birnbaum (1961). Let $\Omega = (\theta_1, \theta_2)$. In this case, (L) means that all information lies in the likelihood ratio, $\lambda(x) = f(x, \theta_2)/f(x, \theta_1)$. The question is now what evidential meaning [in accordance with (L)] we can attach to the number $\lambda(x)$. To answer this, Birnbaum first considers a binary experiment in which the sample space has only two points, denoted $(+)$ and $(-)$, and such that $P(+|\theta_1) = P(-|\theta_2) = \alpha$ for an $\alpha \leq 1/2$. Such an experiment is called a symmetric simple binary experiment and is characterized by the "error" probability α. For such an experiment, $\lambda(+) = (1 - \alpha)/\alpha$ and $\alpha = 1/(1 + \lambda(+))$. The important point now is that according to (L), two experiments with the same value of λ have the same evidential meaning about the value of θ. Therefore, the evidential meaning of $\lambda(x) \geq 1$ from *any* binary experiment E is the same as the evidential meaning of the $(+)$ outcome from a symmetric simple binary experiment with $\alpha(x) = 1/(1 + \lambda(x))$. $\alpha(x)$ is called the *intrinsic* significance level and is a measure of evidence that satisfies (L), while usual observed significance levels (P-values) violate (L). $1 - \alpha(x)$ is similarily called the intrinsic power at x.

8. Finite Parameter Spaces

Section 8.1 illustrates that some likelihood functions on a given Ω can be ordered in a natural way by constructing equivalent experiments with sample spaces consisting only of two points.

Let $k = \#(\Omega)$. In Sec. 8.2, intrinsic confidence methods and intrinsic confidence levels for an outcome x are defined. This is done in a similar fashion as in Sec. 7 by constructing an experiment E' with $\#(S) = k$ based on $L_x(\theta)$ such that the likelihood function for one outcome in E' is equal to $L_x(\theta)$. Then intrinsic confidence methods and levels are defined as regular confidence methods and levels in E'.

Sections 7 and 8 show that for finite parameter spaces, significance levels, confidence sets, and confidence levels can be based on the observed $L_x(\theta)$ [hence, satisfying (L)], defined as regular such methods and concepts for a constructed experiment with a likelihood function identical to $L_x(\theta)$. Therefore, in the case of finite parameter spaces, a clear and logical evidential interpretation of the likelihood function can be given through intrinsic methods and concepts.

9. More General Parameter Spaces

This section deals mainly with the case where Ω is the real line. Given E, x, and $L_x(\theta)$, a location experiment E' consisting of a single observation of Y with density $g(y, \theta) \propto L_x(\theta - y)$ is then constructed. Then (E, x) has the same likelihood function as $(E', 0)$, and (L) implies that the same inference should be used in (E, x) as in $(E', 0)$. For example, if a regular $(1 - \alpha)$ confidence interval in E' is used, then this intervalestimate (for $y = 0$) should be the one used also for (E, x) and is called a $(1 - \alpha)$ intrinsic confidence interval for (E, x). There is, however, one major problem with this approach: A nonlinear transformation of θ will lead to a different $g(y, \theta)$ and hence different intrinsic statements. This problem does not arise in Sec. 7 and 8, where Ω is finite.

Birnbaum considers the case where $L_x(\theta)$ has the form of a normal density and defines as an index of the precision of the maximum likelihood estimate, $\hat{\theta}(x)$, the standard deviation in $g(y, \theta)$, calling it the intrinsic standard error of $\hat{\theta}(x)$. Of course, according to (L), the usual standard error is not a proper measure of precision.

As a general comment, Birnbaum emphasizes that intrinsic methods and concepts can, in light of (L), be nothing more than *methods* of expressing evidential meaning already implicit in $L_x(\theta)$ itself. In the rejoinder in the discussion, Birnbaum does not recommend intrinsic methods as statistical methods in practice. The value of these methods is conceptual, and the main use of intrinsic concepts is to show that likelihood functions as such are evidentially meaningful.

Sequential experiments are also considered, and it is noted that (L) implies that the stopping rule is irrelevant (the stopping rule principle).

10. Bayesian Methods: An Interpretation of the Principle of Insufficient Reason

Birnbaum views the Bayes approach as not directed to informative inference, but rather as a way to determine an appropriate final synthesis of available information based on prior available information and data. It is observed that in determining the posterior distribution, the contribution of the data and E is $L_x(\theta)$ only, so the Bayes approach implies (L). In this section, the case of uniform priors to represent the absence of prior information is discussed.

11. An Interpretation of Fisher's Fiducial Argument

The aim of Fisher's fiducial approach is the same as the Bayes approach: Statements of informative inference should be in the form of probabilities. However, no ignorance priors are involved. In fact, as mentioned earlier, Fisher argued strongly against the use of noninformative priors. Birnbaum suggests that the frames of reference in which fiducial probabilities are considered may coincide, in general, with the constructed experiments in which intrinsic methods are defined and hence that fiducial confidence methods may coincide with intrinsic confidence methods. He formulates a fiducial argument compatible with (L) and shows how, in the case of finite Ω, this modified fiducial approach corresponds to the intrinsic approach in Sec. 8. However, Birnbaum's fiducial argument seems hard to generalize to the usual type of parameter spaces.

12. Bayesian Methods in General

In this section, Birnbaum considers what may separate Bayesian methods (with proper informative priors) from methods based only on $L_x(\theta)$. It is observed that in binary experiments, the class of "likelihood methods" is identical to the class of Bayesian methods for the problem of deciding between the two parameter values θ_1, θ_2.

13. Design of Experiments for Informative Inference

One may think that (L) has nothing to say in design problems since no data and hence no likelihood function are available. However, as Birnbaum mentions in this section, according to (L), the various experimental designs are to be evaluated and compared only in terms of inference methods based solely on the likelihood functions the designs determine, along with the costs. Illustrations are given using intrinsic methods.

5. Developments After Birnbaum's Paper

1. Discussion of Lemma 2, (S) and $(C) \Leftrightarrow (L)$

Lemma 2 has been widely discussed in the literature and, as stated, is relevant only in the discrete case. Also, the proof of $(L) \Rightarrow (S)$ was not stated correctly. A correct proof in the discrete case is given by Birnbaum (1972). Birnbaum (1961) had shown that (S) and $(C) \Rightarrow (L)$ for binary experiments (i.e., Ω consists of two points only).

In nondiscrete cases, there are some problems. Joshi (1976) raised objections to the definition of (S) in the continuous case and showed that a trivial application of (S) would suggest that in a continuous model, $Ev(E, x)$ is identically the same for all x. BW (Sec. 3.4.3) considers the nondiscrete case generally and suggests modifications of (C), (S) and (L) that lead to the same implication of (C) and (S). However, as Basu (1975) notes, the sample space in any realizable experiment must be finite due to our inability to measure with infinite precision, and continuous models are to be considered as mere approximations. One may therefore consider arguments for (L) in the discrete case as all that is needed in order to argue for (L) as a *principle* for inference in all experiments [also, Barnard et al. (1962) and Birnbaum (1972) discuss this point].

Note that Lemma 2 applies only to experiments that can be represented completely and realistically by the form $E = \{\Omega, S, f(x, \theta)\}$. Savage, Barnard, Bross, Box, Levene, and Kempthorne all commented on this aspect in the discussion of the paper. Later, objections to this representation of an experiment have been raised in the theories of pivotal and structural inference; see, e.g., Barnard (1980) and Fraser (1972). Birnbaum in his rejoinder discusses the possibility of a likelihood approach to robustness problems by enlarging the parametric model to a class of models labeled by nuisance parameters.

Basu (1975) considered the discrete case and defined the following weaker versions of (S) and (C):

Weak sufficiency principle [named by Dawid (1977)]: (S'): Let $t(x)$ be sufficient and assume $t(x) = t(x')$. Then $Ev(E, x) = Ev(E, x')$.

Weak conditionality principle (named by Basu): (C'): Let E be a mixture of E_1 and E_2 with known mixture probabilities π and $1 - \pi$. Then $Ev(E, (h, x_h)) = Ev(E_h, x_h)$.

Basu recognized that the proof of Lemma 2 requires only (S') and (C') and that Birnbaum in fact showed (S') and $(C') \Leftrightarrow (L)$. Actually, Birnbaum's proof shows that this result is true for (C') with $\pi = 1/2$.

Statisticians using sampling-theory-based inference do not act in accordance with (L) and must reject or at least modify (S), (S') or (C), (C'). Durbin (1970) and Kalbfleisch (1975) attempt such modification. Durbin suggests that in (C), the ancillary statistic h must depend on the minimal sufficient statistic. It is shown that the proof of Lemma 2 fails when the domain of (C) is restricted in this way. Arguments against Durbin's suggestion have been made by Savage (1970), Birnbaum (1970), and BW. As mentioned earlier, an example given by BW (p. 45) illustrates that Durbin's restriction seems unreasonable. Kalbfleisch (1975) distinguishes between experimental and mathematical ancillaries (an experimental ancillary is determined by the experimental design, and a mathematical ancillary is determined by the model of the problem) and suggests that (S) [or (S')] should apply only to experiments with no experimental ancillaries. Then (L) does not follow from (C) and (S).

Kalbfleisch's suggestion was criticized in the discussion of his paper, especially by Birnbaum and MacLaren. The main problems with such a restriction of sufficiency are (as mentioned by BW, p. 46) the following (1) It seems artificial to restrict principles of inference to certain types of experiments. (2) It is difficult to distinguish between mixture and nonmixture experiments. (3) Mixture experiments can often be shown to be equivalent to nonmixture experiments [Birnbaum and MacLaren illustrate (2) and (3)]. (4) In almost any situation, behavior in violation of sufficiency can be shown to be inferior.

Joshi (1990) claims there is an error in the proof of Lemma 2 in the discrete case (as presented in BW, p. 27), but as made clear by Berger (1990) in his response, the proof is in fact correct. Joshi simply argues for the same type of restriction of sufficiency as Kalbfleisch (1975).

By restricting consideration to the discrete case, various alternative principles to (C') and (S') also imply (L). Birnbaum (1972) showed that (M) and $(C') \Leftrightarrow (L)$. (M) was scrutinized by Godambe (1979) who disagreed with Birnbaum's interpretation of it. Pratt (1962) advanced an alternative justification of (L) based on a censoring principle (Ce) and (S). (Ce) was formalized by Birnbaum (1964) and is given by

(Ce) For a given experiment E with sample space S, let E^* be a censored version of E with sample space S^* such that certain points in S cannot be observed. If x is an outcome in both S and S^*, then: $Ev(E, x) = Ev(E^*, x)$.

Birnbaum (1964) proved that (Ce) and (S) imply (L) [see also Dawid (1977)] and in his 1972 paper finds (Ce) simpler and at least as plausible as (C). Dawid (1977) and Berger (1984) show that also other principles lead to (L).

2. Other Developments

Another major paper on likelihood inference, Barnard et al. (1962), appeared right after Birnbaum's paper and was read before the Royal Statistical Society in March 1962. They stated the same likelihood principle as Birnbaum in terms of making an inference about θ, and tried to argue that (L) follows from (S'). However, the argument was fallacious, as shown by Armitage and Birnbaum in the discussion of the paper. Likelihood methods consisting essentially of plotting the whole likelihood function were proposed and applied to autoregressive series, Markov chains, and moving average processes. As Birnbaum did, they showed that the stopping rule principle (SRP) is a consequence of (L). Pratt (1965) also discusses the SRP. A general study of the SRP is given by BW (Sec. 4.2).

Various examples have been constructed with the aim of showing that the likelihood principle leads to unreasonable inferences. These include "the stopping rule paradox," first discussed it seems by Armitage (1961), and the examples of Stein (1962), Stone (1976), and Joshi (1989). BW and Hill (1984) discuss the first three of these examples (and also a version of Joshi's example), Basu (1975) considers Stein's example and "the stopping rule paradox," and Good (1990) examines the example by Joshi. They all argue essentially that none of these examples speak against (L) itself, but rather against certain implementations of (L).

During the last three decades, the implementation of (L) by considering methods based on the observed likelihood function only (non-Bayesian likelihood methods) has been considered by many statisticians. We have already mentioned, in addition to Birnbaum, Barnard et al. (1962). Most of the likelihood methods that have been proposed depend on the interpretation that $L_x(\theta_1)/L_x(\theta_2)$ measures the relative support of the data for θ_1 and θ_2. Development of this idea can be found in Hacking (1965) and Edwards (1972). In the case of nuisance parameters, likelihood approaches have been suggested by, among others, Kalbfleisch and Sprott (1970), Sprott (1975), Cox (1975), Dawid (1975), Barndorff–Nielsen (1978), Barndorff-Nielsen (1986), Cox and Reid (1987), Fraser and Reid (1988), and McCullagh and Tibshirani (1990). Barnard (1967) discusses the use of the likelihood function in inference with applications to particle physics and genetics. Rubin (1976) considers likelihood-based inference for missing data problems [see also Little and Rubin (1987)]. Goodman (1989) suggests a likelihood approach to meta-analysis (the science of combining evidence from different studies) based on log-likelihood ratios. Other references can be found in BW (Sec. 5.2)

In maximum likelihood estimation, the expected Fisher information as a precision index of the estimate is not appropriate according to (L). The suggestion by Fisher (1925, 1934) of using instead the observed Fisher information [named by Edwards (1972)] does satisfy (L) and is also supported from a frequentist point of view as shown by Hinkley (1978) and Efron and Hinkley (1978).

Several writers have discussed the fact that (L) leads to a rejection of significance tests as valid measures of inferential evidence. We refer to BW (Sec. 4.4) for references.

One of the areas where the LP has had a major impact is in survey sampling. Two of the most important implications of (L) to survey sampling are: (1) : $(L) \Rightarrow$ sampling design is irrelevant at the inference stage, and (2) : $(L) \Rightarrow$ modeling of the population. In the next two paragraphs, we outline the development of these two implications.

(1) It was first shown by Godambe (1966) and Basu (1969) that, with usual noninformative sampling design, the likelihood function is flat for all possible values of the parameter (the population vector). Hence, (L) implies that the inference should not depend on the sampling design. However, Godambe (1966, 1982) claims that (L) may not be appropriate here since there is a relationship between the parameter and data (which is a part of the parameter.) It should be noted that Lemma 2 in Birnbaum's paper is valid also in this case, of course. Basu does not find Godambe's argument convincing and concludes that the sampling design is irrelevant at the inference stage. This was in dramatic disagreement with the classical approach.

(2) The fact that the likelihood function is flat have by some been viewed as a failure of (L) [see, e.g., Rao (1971)], but can also be seen as clarifying some of the limitations of the conventional model as noted by Royall (1976). Royall (1976) seems also to have been the first to recognize that (L) makes it necessary in a sense to model the population (see also BW, p. 114). From the likelihood principle point of view, the data do not, in fact, contain any information about the unobserved part of the population. To make inference, it is therefore necessary to relate the data to the unobserved values somehow, and a natural way of doing this is to formulate a model. Also, as noted by Royall (1971), modeling the population is as objective as any modeling usually done in statistics. General discussions and applications of population modeling can be found in Smith (1976) and Thomsen and Tesfu (1988).

Prediction is another area where a likelihood approach has been attempted. Kalbfleisch (1971) and Edwards (1974) considered Fisher's suggestion of a likelihood function for prediction and Hinkley (1979) coined the term "predictive likelihood" suggesting several such likelihoods. Since then several papers have appeared on the subject. A list of references can be found in Bjørnstad (1990).

The problem of nonresponse in survey sampling represents a prediction case where predictive likelihood may give valuable contributions. Little (1982) considers some likelihood aspects of the nonresponse problem.

Finally, it should be mentioned that Birnbaum (1968, 1977) came to view

(*L*) rather critically, because of its conflict with the so-called confidence principle. (For a discussion of this principle see BW, Sec. 4.1.5).

6. Biography

Allan D. Birnbaum was born on May 27, 1923, in San Francisco of Russian Jewish parents. He died in London in July 1976. He studied as an undergraduate at the University of California in Berkeley and Los Angeles, completing a premedical program in 1942 and receiving a bachelor's degree in mathematics in 1945. For the next two years, he took graduate courses at UCLA in science, mathematics, and philosophy. In 1947, Birnbaum went to Columbia where he obtained his Ph.D. in mathematical statistics in 1954 under the guidance of Erich L. Lehmann.

By then, he had been a faculty member at Columbia for three years. He stayed at Columbia until 1959 while also visiting Imperial College, London, and Stanford during this time. In 1959, he moved to the Courant Institute of Mathematical Sciences at New York University, becoming a full professor of statistics in 1963. He remained at the institute until 1972, when he left for an extended visit to Britain. In 1975, he accepted the chair of statistics at the City University of London where he remained until his death.

Birnbaum had several other professional interests, including medicine and philosophy. Four memorial articles about Birnbaum have been published; see Norton (1977), Godambe (1977), Lindley (1978), and Barnard and Godambe (1982).

Acknowledgments

Berger and Wolpert (1984) have been an invaluable help in writing this introduction. I thank the editors for their help in locating some of the memorial articles and some of the recent literature.

References

Armitage, P. (1961). Comment on "Consistency in statistical inference and decision" by C.A.B. Smith, *J. Roy. Statist. Soc., Ser. B*, **23**, 1–37.

Bahadur, R.R. (1954). Sufficiency and statistical decision functions, *Ann. Math. Statist.*, **25**, 423–462.

Barnard, G.A. (1947). A review of "Sequential Analysis" by Abraham Wald, *J. Amer. Statist. Assoc.*, **42**, 658–669.

Barnard, G.A. (1949). Statistical inference (with discussion), *J. Roy. Statist. Soc., Ser. B*, **11**, 115–139.

Barnard, G.A. (1967). The use of the likelihood function in statistical practice, in *Proceedings of 5th Berkeley Symposium on Mathematical Statistics and Probability*. pp. 27–40. J. Neyman and L. LeCam, (eds.) Vol. 1, Univ. of California Press.

Barnard, G.A. (1980). Pivotal inference and the Bayesian controversy (with discussion), in *Bayesian Statistics*. J.M. Bernardo, M.H. DeGroot, D.V. Lindley, and A.F.M. Smith, (eds.) University Press, Valencia, Spain.

Barnard, G.A., and Godambe, V.P. (1982). Memorial Article. Allan Birnbaum 1923–1976, *Ann. Statist.*, **10**, 1033–1039.

Barnard, G.A., Jenkins, G.M., and Winsten, C.B. (1962). Likelihood inference and time series (with discussion), *J. Roy. Statist. Soc., Ser. A*, **125**, 321–372.

Barndorff-Nielsen, O. (1978). *Information and Exponential Families in Statistical Theory*. Wiley, New York.

Barndorff-Nielsen, O. (1986). Inference on full or partial parameters, based on the standardized log likelihood ratio, *Biometrika*, **73**, 307–322.

Barnett, V. (1982). *Comparative Statistical Inference*, 2nd ed. Wiley, Chichester.

Bartlett, M.S. (1936). Statistical information and properties of sufficiency, *Proc. Roy. Soc. Lon., Ser. A*, **154**, 124–137.

Bartlett, M.S. (1953). Approximate confidence intervals, I & II, *Biometrika*, **40**, 13–19, 306–317.

Basu, D. (1969). Role of the sufficiency and likelihood principles in sample survey theory, *Sankhyā*, **31**, 441–454.

Basu, D. (1975). Statistical information and likelihood (with discussion), *Sankhyā*, **37**, 1–71.

Berger, J.O. (1984). In defense of the likelihood principle: Axiomatics and coherency, in *Bayesian Statistics II*. J.M. Bernardo, M.H. DeGroot, D.V. Lindley, and A.F.M. Smith, (eds.) University Press, Valencia, Spain.

Berger, J.O. (1990). Birnbaum's theorem is correct: A reply to a claim by Joshi (F27), F28 in discussion forum in *J. Statist. Plann. Inference*, **26**, 112–113.

Berger, J.O. and Wolpert, R.L. (1984) [BW]. *The Likelihood Principle*. IMS, Hayward, Calif.

Birnbaum, A. (1961). On the foundations of statistical inference: Binary experiments. *Ann. Math. Statist.*, **32**, 414–435.

Birnbaum, A. (1964). The anomalous concept of statistical evidence. Courant Institute of Mathematical Science, Tech. Rep. IMM-NYU 332.

Birnbaum, A. (1968). Likelihood, *in International Encyclopedia of the Social Sciences*, Vol. 9, D.L. Sills, (ed.). MacMillan and Free Press, New York, pp. 299–301.

Birnbaum, A. (1970). On Durbin's modified principle of conditionality, *J. Amer. Statist. Assoc.*, **65**, 402–403.

Birnbaum, A. (1972). More concepts of statistical evidence, *J. Amer. Statist. Assoc.*, **67**, 858–861.

Birnbaum, A. (1977). The Neyman–Pearson theory as decision theory and as inference theory: With a criticism of the Lindley-Savage argument for Bayesian theory, *Synthese*, **36**, 19–49.

Bjørnstad, J.F. (1990). Predictive likelihood: A review (with discussion), *Statist. Sci.*, **5**, 242–265.

Cox, D.R. (1958). Some problems connected with statistical inference, *Ann. Math. Statist.*, **29**, 357–372.

Cox, D.R. (1975). Partial likelihood, *Biometrika*, **62**, 269–276.

Cox, D.R., and Hinkley, D. (1974). *Theoretical Statistics*. Chapman and Hall, London.

Cox, D.R., and Reid, N. (1987). Parameter orthogonality and approximate conditional inference (with discussion), *J. Roy. Statist. Soc., Ser. B*, **49**, 1–39.

Dawid, A.P. (1975). On the concepts of sufficiency and ancillarity in the presence of nuisance parameters, *J. Roy. Statist. Soc., Ser. B*, **37**, 248–258.

Dawid, A.P. (1977). Conformity of inference patterns, in: *Recent Developments in Statistics* (J.R. Barra, et al., eds.). North Holland, Amsterdam, pp. 245–256.

Dawid, A.P. (1981). Statistical inference, in: *Encyclopedia of Statistical Sciences*, (S. Kotz, N.L. Johnson, and C.B. Read, eds.). Wiley, New York, pp. 89–105.

Durbin, J. (1970). On Birnbaum's theorem on the relation between sufficiency, conditionality and likelihood, *J. Amer. Statist. Assoc.*, **65**, 395–398.

Edwards, A.W.F. (1972). *Likelihood.* Cambridge University Press.

Edwards, A.W.F. (1974). A problem in the doctrine of chances, in *Proceedings of the conference on foundational questions in statistical inference at Aarhus.* O. Barndorff-Nielsen, P. Blæsild, and G. Schou (eds.) Department of Theoretical Statistics, University of Aarhus, Denmark.

Efron, B., and Hinkley, D.V. (1978). Assessing the accuracy of the maximum likelihood estimator: Observed versus expected Fisher information (with discussion), *Biometrika*, **65**, 457–482.

Fisher, R.A. (1921). On the "probable error" of a coefficient of correlation deduced from a small sample, *Metron*, vol. 1, part 4, 3–32.

Fisher, R.A. (1922). On the mathematical foundations of theoretical statistics, *Philos. Trans. A*, **222**, 309–368.

Fisher, R.A. (1925). Theory of statistical estimation, *Proc. Cambridge Philos. Soc.*, **22**, 700–725.

Fisher, R.A. (1932). Inverse probability and the use of likelihood, *Proc. Cambridge Philos. Soc.*, **28**, 257–261.

Fisher, R.A. (1934). Two new properties of mathematical likelihood, *Proc. Roy. Soc. Lon., Ser. A*, **144**, 285–307.

Fisher, R.A. (1973). *Statistical Methods and Scientific Inference*, 3rd ed. Hafner Press, New York.

Fraser, D.A.S. (1972). Bayes, likelihood or structural, *Ann. Math. Statist.*, **43**, 777–790.

Fraser, D.A.S. (1976). Comment on "On rereading R.A. Fisher" by L.J. Savage, *Ann. Statist.*, **4**, 441–500.

Fraser, D.A.S., and Reid, N. (1988). On conditional inference for a real parameter: A differential approach on the sample space, *Biometrika*, **75**, 251–264.

Godambe, V.P. (1966). A new approach to sampling from finite populations, In. *J. Roy. Statist. Soc., Ser B*, **28**, 310–319.

Godambe, V.P. (1977). Allan Birnbaum, *Amer. Statist.*, **31**, 178–179.

Godambe, V.P. (1979). On Birnbaum's mathematically equivalent experiments, *J. Roy. Statist. Soc., Ser. B*, **41**, 107–110.

Godambe, V.P. (1982). Likelihood principle and randomization, in *Statistics and Probability: Essays in Honor of C.R. Rao.* (G. Kallianpur, P.R. Krishnaiah, and J.K. Ghosh, eds.) North-Holland, Amsterdam pp. 281–294.

Good, I.J. (1990). In defense of "the" likelihood principle, F38 in discussion forum in *J. Statist. Plann. Inference*, **26**, 122–123.

Goodman, S.N. (1989). Meta-analysis and evidence, *Controlled Clinical Trials*, **10**, 188–204.

Hacking, I. (1965). *Logic in Statistical Inference.* Cambridge University Press.

Hill, B.M. (1984). Discussion in *The Likelihood Principle* by J.O. Berger, and R.L. Wolpert, IMS, Hayward CA, pp. 161–174.

Hinkley, D.V. (1978). Likelihood inference about location and scale parameters, *Biometrika*, **65**, 253–262.

Hinkley, D.V. (1979). Predictive likelihood, *Ann. Statist.*, **7**, 718–728 (corrigendum, **8**, 694).

Joshi, V.M. (1976). A note on Birnbaum's theory of the likelihood principle, *J. Amer. Statist. Assoc.*, **71**, 345–346.

Joshi, V.M. (1989). A counter-example against the likelihood principle, *J. Roy. Statist. Soc., Ser. B*, **51**, 215–216.

Joshi, V.M. (1990). Fallacy in the proof of Birnbaum's theorem, F27 in discussion forum in *J. Statist. Plann. Inference*, **26**, 111–112.

Kalbfleisch, J.D. (1971). Likelihood methods of prediction (with discussion), in *Foundations of Statistical Inference* (V.P. Godambe, and D.A. Sprott, (eds.) Holt, Rinehart and Winston, New York, pp. 378–392.

Kalbfleisch, J.D. (1975). Sufficiency and conditionality (with discussion), *Biometrika*, **62**, 251–268.

Kalbfleisch, J.D., and Sprott, D.A. (1970). Application of likelihood methods to models involving large numbers of parameters (with discussion), *J. Roy. Statist. Soc., Ser. B*, **32**, 175–208.

Kendall, M.G. (1946). *The Advanced Theory of Statistics*. C. Griffin & Co., London.

Lindley, D.V. (1978). Birnbaum, Allan, in *International Encyclopedia of Statistics* (W.H. Kruskal, and J.M. Tanur, eds.). The Free Press, New York, pp. 22–24.

Little, R.J.A. (1982). Models for nonresponse in sample surveys., *J. Amer. Statist. Assoc.*, **77**, 237–250.

Little, R.J.A., and Rubin, D.B. (1987). *Statistical Analysis with Missing Data*. Wiley, New York.

McCullagh, P., and Tibshirani, R. (1990). A simple method for the adjustment of profile likelihoods, *J. Roy. Statist. Soc., Ser. B*, **52**, 325–344.

Norton, B. (1977). Obituary. Allan D. Birnbaum 1923–1976, *J. Roy. Statist. Soc., Ser. A*, **140**, 564–565.

Pratt, J.W. (1962). Comment on "On the foundations of statistical inference" by A. Birnbaum, *J. Amer. Statist. Assoc.*, **57**, 269–326.

Pratt, J.W. (1965). Bayesian interpretation of standard inference statements (with discussion). *J. Roy. Statist. Soc., Ser. B*, **27**, 169–203.

Rao, C.R. (1971). Some aspects of statistical inference in problems of sampling from finite populations (with discussion), in *Foundations of Statistical Inference*, (V.P. Godambe, and D.A. Sprott, eds.). Holt, Rinehart and Winston, New York, pp. 177–202.

Royall, R. (1971). Comment on "An essay on the logical foundations of survey sampling, part one" by D. Basu, in *Foundations of Statistical Inference* (V.P. Godambe, and D.A. Sprott, eds.). Holt, Rinehart and Winston, New York, pp. 203–242.

Royall, R. (1976). Likelihood functions in finite population sampling survey, *Biometrika*, **63**, 605–617.

Rubin, D.B. (1976). Inference and missing data (with discussion by R.J.A. Little), *Biometrika*, **63**, 605–614.

Savage, L.J. (1962). Comment on "On the foundations of statistical inference" by A. Birnbaum, *J. Amer. Statist. Assoc.*, **57**, 269–326.

Savage, L.J. (1970). Comments on a weakened principle of conditionality, *J. Amer. Statist. Assoc.*, **65**, 399–401.

Smith, T.M.F. (1976). The foundations of survey sampling: A review, *J. Roy. Statist. Soc., Ser. A*, **139**, 183–204.

Sprott, D.A. (1975). Marginal and conditional sufficiency, *Biometrika*, **62**, 599–605.

Stein, C. (1962). A remark on the likelihood principle, *J. Roy. Statist. Soc., Ser. A*, **125**, 565–568.

Stone, M. (1976). Strong inconsistency from uniform priors (with discussion), *J. Amer. Statist. Assoc.*, **71**, 114–125.

Thomsen, I., and Tesfu, D. (1988). On the use of models in sampling from finite populations, in *Handbook of Statistics*, vol. 6 (P.R. Krishnaiah, and C.R. Rao, eds.) North-Holland, Amsterdam. pp. 369–397.

On the Foundations of Statistical Inference[1]

Allan Birnbaum
New York University

Abstract

The concept of conditional experimental frames of reference has a significance for the general theory of statistical inference which has been emphasized by R.A. Fisher, D.R. Cox, J.W. Tukey, and others. This concept is formulated as a *principle of conditionality*, from which some general consequences are deduced mathematically. These include the *likelihood principle*, which has not hitherto been very widely accepted, in contrast with the conditionality concept which many statisticians are inclined to accept for purposes of "informative inference." The likelihood principle states that the "evidential meaning" of experimental results is characterized fully by the likelihood function, without other reference to the structure of an experiment, in contrast with standard methods in which significance and confidence levels are based on the complete experimental model. The principal writers supporting the likelihood principle have been Fisher and G.A. Barnard, in addition to Bayesian writers for whom it represents the "directly empirical" part of their standpoint. The likelihood principle suggests certain systematic reinterpretations and revisions of standard methods, including "intrinsic significance and confidence levels" and "intrinsic standard errors," which are developed and illustrated. The close relations between non-Bayesian likelihood methods and Bayesian methods are discussed.

[1] This paper was presented at a special discussion meeting of the American Statistical Association on Wednesday, December 27, 1961 in the Roosevelt Hotel, New York City. George E.P. Box presided. Preprints of the paper were available several weeks before the meeting. Research on which the paper is based was supported by the Office of Naval Research.

1. Introduction, Summary, and General Conclusions

This paper treats a traditional and basic problem-area of statistical theory, which we shall call *informative inference*, which has been a source of continuing interest and disagreement. The subject-matter of interest here may be called *experimental evidence*: when an experimental situation is represented by an adequate mathematical statistical model, denoted by E, and when any specified outcome x of E has been observed, then (E, x) is an instance of *statistical evidence*, that is, a mathematical model of an instance of experimental evidence. Part of the specification of E is a description of the range of unknown parameter values or of statistical hypotheses under consideration, that is, the description of a parameter space Ω of parameter points θ. The remaining part of E is given by description of the sample space of possible outcomes x of E, and of their respective probabilities or densities under respective hypotheses, typically by use of a specified probability density function $f(x, \theta)$ for each θ.

Methods such as significance tests and interval estimates are in wide standard use for the purposes of reporting and interpreting the essential features of statistical evidence. Various approaches to statistical theory have been concerned to an appreciable extent with this function. These include: Bayesian approaches, including those utilizing the principle of insufficient reason; some approaches using confidence methods of estimation and related tests of hypotheses; the fiducial approach of R.A. Fisher; and approaches centering on the direct inspection and interpretation of the likelihood function alone, as suggested by Fisher and G.A. Barnard. However the basic concepts underlying this function seem in need of further clarification.

We may distinguish two main general problems of informative inference: The problem of finding an appropriate *mathematical characterization* of statistical evidence as such; and the problem of *evidential interpretation*, that is, of determining concepts and terms appropriate to describe and interpret the essential properties of statistical evidence. It is useful sometimes to think of these problems, especially the first one, in connection with the specific function of reporting experimental results in journals of the empirical sciences.

The present analysis of the first problem begins with the introduction of the symbol Ev(E, x) to denote the *evidential meaning* of a specified instance (E, x) of statistical evidence; that is, Ev(E, x) stands for the essential properties (which remain to be clarified) of the statistical evidence, as such, provided by the observed outcome x of the specified experiment E. The next steps involve consideration of conditions under which we may recognize and assert that two instances of statistical evidence, (E, x) and (E', y), are equivalent in all relevant respects; such an assertion of *evidential equivalence* between (E, x) and (E', y) is written: Ev(E, x) = Ev(E', y).

A first condition for such equivalence, which is proposed as an *axiom*, is related to the concept of sufficient statistic which plays a basic technical role in each approach to statistical theory. This is:

The Principle of Sufficiency (S). If E is a specified experiment, with outcomes x; if $t = t(x)$ is any sufficient statistic; and if E' is the experiment, derived from E, in which any outcome x of E is represented only by the corresponding value $t = t(x)$ of the sufficient statistic; then for each x, $\mathrm{Ev}(E, x) = \mathrm{Ev}(E', t)$, where $t = t(x)$.

A familiar illustration of the concept formulated here is given by the problem of determining confidence limits for a binomial parameter: It is well known that exact confidence levels in this problem are achieved only with use of an auxiliary randomization variable, and that such confidence limits cannot be represented as functions of only the binomial sufficient statistic; the reluctance or refusal of many statisticians to use such confidence limits for typical purposes of informative inference is evidently an expression, within the context of this approach, of the principle formulated above. (S) may be described informally as asserting the "irrelevance of observations independent of a sufficient statistic."

A second condition for equivalence of evidential meaning is related to concepts of conditional experimental frames of reference; such concepts have been suggested as appropriate for purposes of informative inference by writers of several theoretical standpoints, including Fisher and D.R. Cox. This condition concerns any experiment E which is mathematically equivalent to a *mixture* of several other *component* experiments E_h, in the sense that observing an outcome x of E is mathematically equivalent to observing first the value h of random variable having a known distribution (not depending upon unknown parameter values), and then taking an observation x_h from the component experiment E_h labeled by h. Then (h, x_h) or (E_h, x_h) is an alternative representation of the outcome x of E. The second proposed axiom, which many statisticians are inclined to accept for purposes of informative inference, is:

The Principle of Conditionality (C). If E is any experiment having the form of a mixture of component experiments E_h, then for each outcome (E_h, x_h) of E we have $\mathrm{Ev}(E, (E_h, x_h)) = \mathrm{Ev}(E_h, x_h)$. That is, the evidential meaning of any outcome of any mixture experiment is the same as that of the corresponding outcome of the corresponding component experiment, ignoring the over-all structure of the mixture experiment.

(C) may be described informally as asserting the "irrelevance of (component) experiments not actually performed."

The next step in the present analysis concerns a third condition for equivalence of evidential meaning, which has been proposed and supported as self-evident principally by Fisher and by G.A. Barnard, but which has not hitherto been very generally accepted. This condition concerns the likelihood function, that is, the function of θ, $f(x, \theta)$, determined by an observed outcome x of a specified experiment E; two likelihood functions, $f(x, \theta)$ and $g(y, \theta)$, are called the same if they are proportional, that is if there exists a positive constant c such that $f(x, \theta) = cg(y, \theta)$ for all θ. This condition is:

The Likelihood Principle (L). If E and E' are any two experiments with the same parameter space, represented respectively by density functions $f(x, \theta)$ and $g(y, \theta)$; and if x and y are any respective outcomes determining the same likelihood function; then $\text{Ev}(E, x) = \text{Ev}(E', y)$. That is, the evidential meaning of any outcome x of any experiment E is characterized fully by giving the likelihood function $cf(x, \theta)$ (which need be described only up to an arbitrary positive constant factor), without other reference to the structure of E.

(L) may be described informally as asserting the "irrelevance of outcomes not actually observed."

The fact that relatively few statisticians have accepted (L) as appropriate for purposes of informative inference, while many are inclined to accept (S) and (C), lends interest and significance to the result, proved herein, that (S) *and* (C) *together are mathematically equivalent to* (L). When (S) and (C) are adopted, their consequence (L) constitutes a significant solution to the first problem of informative inference, namely that a mathematical characterization of statistical evidence as such is given by the likelihood function.

For those who find (S) and (C) compellingly appropriate (as does the present writer), their consequence (L) has immediate radical consequences for the every-day practice as well as the theory of informative inference. One basic consequence is that reports of experimental results in scientific journals should in principle be descriptions of likelihood functions, when adequate mathematical-statistical models can be assumed, rather than reports of significance levels or interval estimates. Part II of this paper, Sections 6–13, is concerned with the general problem of evidential interpretation, on the basis of the likelihood principle.

(L) implies that experimental frames of reference, whether actual, conditional, or hypothetical, have no necessary essential role to play in evidential interpretations. But most current statistical practice utilizes concepts and techniques of evidential interpretation (like significance level, confidence interval, and standard error) based on experimental frames of reference. Hence it seems of considerable practical and heuristic value, as well as of theoretical interest, to consider how far the commonly used concepts and techniques can be reinterpreted or revised to provide modes of describing and interpreting likelihood functions as such, utilizing experimental frames of reference in a systematic but clearly conventional manner compatible with (L). This approach leads to concepts and techniques of evidential interpretation called "intrinsic significance levels," "intrinsic confidence sets, with intrinsic confidence levels," and "intrinsic standard error of an estimate"; these are illustrated by examples. Perhaps the principal value of this approach will be to facilitate understanding and use of likelihood functions as such, in the light of the likelihood principle, by relating them to concepts and techniques more familiar to many statisticians.

Bayesian methods based on the principle of insufficient reason, and a version of Fisher's fiducial argument, are interpreted as alternative partly-conventional modes of description and evidential interpretation of likelihood functions. Many points of formal coincidence between these and intrinsic confidence methods are noted.

This analysis shows that when informative inference is recognized as a distinct problem-area of mathematical statistics, it is seen to have a scope including some of the problems, techniques, and applications customarily subsumed in the problem-areas of point or set estimation, testing hypotheses, and multi-decision procedures. In fact the course of development of the latter areas of statistics seems to have been shaped appreciably by the practice of formulating problems of informative inference as problems of one of these kinds, and developing techniques and concepts in these areas which will serve adequately for informative inference. At the same time each of these methods can serve purposes distinct from informative inference; the inclusion of problems of two distinct kinds, one of them traditional but not clearly enough delineated, seems to have forced a certain awkwardness of formulation and development on these areas. For example, problems of estimation of a real-valued parameter have traditionally been dealt with by techniques which supply a point estimate supplemented by an index of precision, or an interval estimate, and such techniques serve the purposes of informative inference fairly adequately, particularly in problems of simple structure. However, in modern generalizations and refinements of theories of estimation it becomes clear that no single formulation is appropriate in general to serve the distinct functions of informative inference on the one hand and either point or interval estimation on the other hand; and that the attempt to serve both functions by a single formal theory and set of techniques makes for awkwardness and indistinctness of purpose.

Recognition of informative inference as a distinct problem-area with its own basic concepts and appropriate techniques should help unburden the other problem-areas of statistics, particularly statistical decision theory, for freer developments more clearly and deeply focused on the problems in their natural mathematical and practical scope. Tukey [20, pp. 450, 468–74]; [21] has recently emphasized that the "elementary" problems of mathematical statistics are still with us as live problems. Among these must be included questions of specification of "what are the problems and the problem-areas of mathematical statistics, what is their formal mathematical and extramathematical content, and what are their scopes of application?" For example, what are the typical substantial functions of point and interval estimation, and of tests of hypotheses, apart from the function of informative inference?

The fact that the likelihood principle follows from the principles of sufficiency and conditionality, which many find more acceptable than Bayes' principle, seems to provide both some comfort and some challenge to Bayesian viewpoints: The "directly empirical" part of the Bayesian position concerning the role of the likelihood function is given new support independent of Bayes' principle itself. But this suggests the question: What are the specific contributions of the Bayesian concepts and techniques to the interpretation and use of statistical evidence, above and beyond what is possible by less formalized interpretations and applications based on direct consideration of the likelihood function in the light of other aspects of the inference situation, without formal use of prior probabilities and Bayes' formula? Specifically, what are the precise contributions of quantitative prior probabilities, and of the other formal parts of the Bayesian methods? Evidently in the present state of our understanding there can be interesting collaboration between Bayesian and non-Bayesian statisticians, in exploring the possibilities and limitations of both formal and informal modes of interpreting likelihood functions, and in developing the important problem-areas of experimental design and of robustness from the standpoint of such interpretations.

These considerations also present some challenge to non-Bayesian statisticians accustomed to use of standard techniques of testing and estimation, in which error-probabilities appear as basic terms of evidential interpretation in a way which is incompatible with the principle of conditionality. The writer has not found any apparent objections to the latter principle which do not seem to stem from notions of "conditional" distinct from that considered here, or else from purposes other than the modest but important one of informative inference.

Part I

2. Statistical Evidence

A traditional standard in empirical scientific work is accurate reporting of "what was observed, and under what experimental plan and conditions." Such reports are an essential part of the literature and the structure of the empirical sciences; they constitute the body of observational or *experimental evidence* available at any stage to support the practical applications and the general laws, theories, and hypotheses of the natural sciences. (Cf. Wilson [25], especially Ch. 13, and references therein.)

In some circumstances the "experimental plan and conditions" can be represented adequately by a mathematical-statistical model of the experimental situation. The adequacy of any such model is typically supported, more or less adequately, by a complex informal synthesis of previous experimental evidence of various kinds and theoretical considerations concerning both subject-matter and experimental techniques. (The essential place of working "conclusions" in the fabric and process of science has been discussed recently by Tukey [22].) We deliberately delimit and *idealize* the present discussion by considering only models whose adequacy is postulated and is not in question.

Let E denote a mathematical-statistical model of a given experimental situation: When questions of experimental design (including choice of sample size or possibly a sequential sampling rule) have been dealt with, the sample space of possible outcomes x of E is a specified set $S = \{x\}$. We assume that each of the possible distributions of X is labeled by a parameter point θ in a specified parameter space $\Omega = \{\theta\}$, and is represented by a specified elementary probability function $f(x, \theta)$. The probability that E yields an outcome x in A is

$$P(A|\theta) \equiv \text{Prob}(X \in A|\theta) = \int_A f(x, \theta)\, d\mu(x),$$

where μ is a specified (σ-finite) measure on S, and A is any (measurable) set. Thus any mathematical model of an experiment, E, is given by specifying its mathematical ingredients: (Ω, S, f, μ). (No methods of advanced probability theory are used in this paper. The reader familiar only with probabilities defined by

$$P(A|\theta) = \sum_{x \in A} f(x, \theta),$$

for discrete distributions, and by

$$P(A|\theta) = \int_A f(x, \theta)\, dx,$$

for continuous distributions (with dx possibly representing $dx_1 \ldots dx_n$), can regard the symbol $\int_A f(x, \theta) d\mu(x)$ as a generalization including those two important cases and some others.)

In an experimental situation represented by such a model E, the symbol (E, x) denotes an instance of statistical evidence. The latter term will be used here to denote any such mathematical model of an instance of experimental evidence: x represents "what was observed," and E represents "under what experimental plan and conditions."

The central purpose of this paper is to clarify the essential structure and properties of statistical evidence in various instances. We use the symbol $\text{Ev}(E, x)$, and the term *evidential meaning* (of a specified outcome x of a specified experiment E), to refer to these essential properties and structure, whose precise nature remains to be discussed.

The first general problem to be considered (throughout Part I) is whether a satisfactory *mathematical characterization* can be found for evidential meaning in various instances. The second general purpose (in the following sections, Part II) is to consider what concepts, terms, and techniques are appropriate for representing, interpreting, and expressing evidential meaning in various instances; in other words, to consider critically the function of *evidential interpretation* of experimental results. The broad but delimited part of mathematical statistics which is concerned with these two problems, the characterization and the interpretation of statistical evidence as such, will be termed here the problem-area of *informative* (statistical) *inference*. While such problems and methods have broad and varied relevance and use, it will be helpful sometimes to focus attention on the specific and relatively simple function referred to above: the formal reporting of experimental results, in empirical scientific journals, in terms which are appropriate to represent their character as evidence relevant to parameter values or statistical hypotheses of interest. We restrict present consideration to situations in which all questions of characterizing and interpreting statistical evidence will have been considered in full generality before an experiment is carried out: Our discussion concerns all possible outcomes x and possible interpretations thereof, as these can in principle be considered at the outset of a specified experiment; such discussion can subsequently be broadened to include questions of appraisal, comparison, and design of experiments for purposes of informative inference. Our discussion will not touch on tests or other modes of inference in cases where the set of possible alternative distributions is not specified initially [9, Ch. 3].

Since the problem-area of informative inference has not received a generally accepted delineation or terminology, it will be useful to note here some of the terms and concepts used by writers representing several different approaches:

a) R.A. Fisher [9, pp. 139–41] has employed the term "estimation" to refer to this problem-area, in contrast with the widely current usage of this term to refer to problems of interval (or set) or point estimation. Fisher's paper [10, pp. 175–6] includes in its introductory section ("On the nature of the problem") the following interpretation of Gossett's fundamental work on testing a normal mean:

> In putting forth his test of significance "Student" (1908) specified that the problem with which he is concerned is that of a *unique* sample. His clear intention in this is to exclude from his discussion all possible suppositions as to the "true" distribution of the variances of the populations which might have been sampled. If such a distribution were supposed known, "Student's" method would be open to criticism and to correction. In following his example it is not necessary to deny the existence of knowledge based on previous experience, which might modify his result. It is sufficient that we shall deliberately choose to examine the evidence of the sample on its own merits only.

The last two sentences may be taken to be descriptive of the problem-area of informative inference, even though the context refers to significance tests. It is clear that many of the principal modern statistical concepts and methods developed by Fisher and other non-Bayesian writers have been directed to problems of informative inference. This applies in particular to

Fisher's description of three modes of statistical inference, significance tests, estimation (in the broad sense indicated above), and the fiducial argument [9, Ch. 3, especially p. 73].

While such phrases as "specification of uncertainty" and "measure of the rational grounds for ... disbelief" have sometimes been used [9, pp. 43–4] to describe the purpose and nature of informative inference, it is possible and it seems desirable to discuss these problems without use of terms having specifically subjective or psychological reference. The latter course will be followed throughout the present paper; our discussion of the structure and properties of statistical evidence will not involve terms or concepts referring to "reactions to evidence" in any sense.

b) Many of the developments and applications of statistical methods of testing and estimation which stem from the work of Neyman and Pearson have been directed to informative inference. Such methods are widely considered to serve this purpose fairly adequately and soundly. The basic terms of such applications and interpretations are probabilities of the errors of various kinds which could be made in connection with a given experiment. (Measures of precision of estimators can be interpreted as referring to probabilities of various possible errors in estimation.) It is considered an essential feature of such interpretations that these basic error-probability terms are objective, in the mathematical sense (and in the related physical sense) that conceptually-possible repetitions of an experiment, under respective hypotheses, would generate corresponding relative frequencies of errors. In typical current practice, some reference to such error-probabilities accompanies inference statements ("assertions," or "conclusions") about parameter values or hypotheses. If an inference is thus accompanied by relevant error-probabilities which are fairly small, the inference is considered supported by fairly strong evidence; if such relevant error-probabilities are all very small, the evidence is considered very strong. These remarks simply describe the general nature of evidential interpretations of experimental results, which is traditionally and widely recognized in scientific work; here the concepts and techniques of testing and estimation serve as frameworks for such evidential interpretations of results. Such evidential interpretations do not seem to differ in kind from those associated with the less technical notion of circumstantial evidence when all relevant hypotheses are considered (cf. for example Cohen & Nagel [7], pp. 347–51); they differ sharply in degree, in that precisely specified frameworks for such interpretations are provided by the mathematical models of experiments and by the formal definitions and properties of the inference methods employed.

The usefulness for informative inference of tests and especially of confidence set estimates has been emphasized recently by several writers, including Cox [8], Tukey [22], and Wallace [23], [24]. At the same time these writers have been concerned also with technical and conceptual problems related to such use and interpretation of these methods. Cox [8, p. 359] has cited the term "summarization of evidence" to indicate the function of informative inference, and like some other writers has described it as concerned with "statistical inferences" or "conclusions," in contrast with statistical decision problems for which the basic mathematical structure and interpretations seem relatively clear. As Cox writes [8, p. 354],

> It might be argued that in making an inference we are "deciding" to make a statement of a certain type about the populations and that, therefore, provided the word decision is not interpreted too narrowly, the study of statistical decisions embraces that of inferences. The point here is that one of the main general problems of statistical inference consists in deciding what types of statement can usefully be made and exactly what they mean. In statistical decision theory, on the other hand, the possible decisions are considered as already specified.

c) Approaches to statistical inference problems based upon Bayes' principle of inverse probability (with any interpretation) obtain on that basis clear and simple answers to questions of informative inference, as will be reviewed below. Writing from his own Bayesian standpoint, Savage [18] has recently described as follows the difficulties and prospects of non-Bayesian approaches such as those discussed above:

> Rejecting both necessary and personalistic views of probability left statisticians no choice but to work as best they could with frequentist views.... The frequentist is required, therefore, to seek a concept of evidence, and of reaction to evidence, different from that of the primitive, or natural, concept that is tantamount to application of Bayes' theorem.
>
> Statistical theory has been dominated by the problem thus created, and its most profound and ingenious efforts have gone into the search for new meanings for the concepts of inductive inference and inductive behavior. Other parts of this lecture will at least suggest concretely how these efforts have failed, or come to a stalemate. For the moment, suffice it to say that a problem which after so many years still resists solution is suspect of being ill formulated, especially since this is a problem of conceptualization, not a technical mathematical problem like Fermat's last theorem or the four-color problem.

The present paper is concerned primarily with approaches to informative inference which do not depend upon the Bayesian principle of inverse probability.

3. The Principle of Sufficiency

As the first step of our formal analysis of the structure of evidential meaning, $Ev(E, x)$, we observe that certain cases of *equivalence of evidential meaning* can be recognized, even in advance of more explicit characterization of the nature of evidential meaning itself. We shall write $Ev(E, x) = Ev(E', y)$ to denote that two instances of statistical evidence, (E, x) and (E', y), have the same (or equivalent) evidential meaning.

For example, let (E, x) and (E', y) be any two instances of statistical evidence, with E and E' having possibly different mathematical structures but the same parameter space $\Omega = \{\theta\}$. Suppose that there exists a one-to-one transformation of the sample space of E onto the sample space of E': $y = y(x)$, $x = x(y)$, such that the probabilities of all corresponding (measurable) sets under all corresponding hypotheses are equal: $\text{Prob}(Y \in A'|\theta) = \text{Prob}(X \in A|\theta)$ if $A' = y(A)$. Then the models E and E' are *mathematically equivalent*, one being a relabeling of the other. If respective outcomes x of E and y of E' are related by $y = y(x)$, they also are mathematically equivalent, and the two instances of statistical evidence (E, x) and (E', y) may be said to have the same evidential meaning: $\text{Ev}(E, x) = \text{Ev}(E', y)$. A simple concrete example is that of models of experiments which differ only in the units in which measurements are expressed.

Again, consider (E, x) and (E', t), where $t(x)$ is any *sufficient* statistic for E, and where E' represents the possible distributions of $t(x)$ under the respective hypotheses of E. Then, for reasons which are recognized within each approach to statistical theory, we may say that $\text{Ev}(E, x) = \text{Ev}(E', t)$ if $t = t(x)$. An example which occurs within the approach to informative inference which utilizes confidence intervals (and related tests) involves the possible use of randomized confidence limits (or tests), for example for a binomial parameter. The view, held by many, that randomized forms of such techniques should not be used seems to stem from an appreciation that sufficiency concepts must play a certain guiding role in the development of methods appropriate for informative inference. (For a recent discussion and references, cf. [21].)

Such considerations may be formalized as follows to provide an *axiom* which we adopt to begin our mathematical characterization of evidential meaning:

Principle of Sufficiency (S). Let E be any experiment, with sample space $\{x\}$, and let $t(x)$ be any sufficient statistic (not necessarily real-valued). Let E' denote the derived experiment, having the same parameter space, such that when any outcome x of E is observed the corresponding outcome $t = t(x)$ of E' is observed. Then for each x, $\text{Ev}(E, x) = \text{Ev}(E', t)$, where $t = t(x)$.

It is convenient to note here for later use certain definitions and a mathematical consequence of (S): If x is any specified outcome of any specified experiment E, the *likelihood function determined by* x is the function of θ: $cf(x, \theta)$, where c is assigned arbitrarily any positive constant value. Let E and E' denote any two experiments with the same parameter space (E' could be identical with E), and let x and y be any specified outcomes of these respective experiments, determining respective likelihood functions $f(x, \theta)$ and $g(y, \theta)$; if for some positive constant c we have $f(x, \theta) = cg(y, \theta)$ for all θ, x and y are said to determine *the same likelihood function*. It has been shown in the general theory of sufficient statistics (cf. [1]) that if two outcomes x, x'

of *one* experiment E determine the same likelihood function (that is, if for some positive c we have $f(x, \theta) = cf(x', \theta)$ for all θ), then there exists a (minimal) sufficient statistic t such that $t(x) = t(x')$. (In the case of any discrete sample space, the proof is elementary.) This, together with (S), immediately implies

Lemma 1. *If two outcomes* x, x' *of any experiment* E *determine the same likelihood function, then they have the same evidential meaning:* $\mathrm{Ev}(E, x) = \mathrm{Ev}(E, x')$.

4. The Principle of Conditionality

4.1.

The next step in our analysis is the formulation of another condition for equivalence of evidential meaning, which concerns conditional experimental frames of reference. This will be stated in terms of the following definitions:

An experiment E is called a *mixture* (or a mixture experiment), with *components* $\{E_h\}$, if it is mathematically equivalent (under relabeling of sample points) to a two-stage experiment of the following form:

(a) An observation h is taken on a random variable H having a fixed and known distribution G. (G does not depend on unknown parameter values.)

(b) The corresponding component experiment E_h is carried out, yielding an outcome x_h.

Thus each outcome of E is (mathematically equivalent to) a pair (E_h, x_h). (Each component experiment E_h, and E, all have the same parameter space. Every experiment is a mixture in the trivial sense that all components may be identical; the non-trivial cases, with non-equivalent components, are of principal interest. Examples will be discussed below.)

As a second proposed *axiom* concerning evidential meaning, we take the

Principle of Conditionality (*C*). If an experiment E is (mathematically equivalent to) a mixture G of components $\{E_h\}$, with possible outcomes (E_h, x_h), then

$$\mathrm{Ev}(E, (E_h, x_h)) = \mathrm{Ev}(E_h, x_h).$$

That is, the evidential meaning of any outcome (E_h, x_h) of any experiment E having a mixture structure is the same as: the evidential meaning of the corresponding outcome x_h of the corresponding component experiment E_h, ignoring otherwise the over-all structure of the original experiment E.

4.2.

A number of writers have emphasized the significance of conditionality concepts for the analysis of problems of informative inference. Fisher recently wrote [9, pp. 157–8] "The most important step which has been taken so far to complete the structure of the theory of estimation is the recognition of Ancillary statistics." (Evidently a statistic like h above, whose distribution is known and independent of unknown parameters, is an example of an ancillary statistic. "Estimation" is used here by Fisher in the broad sense of informative inference, rather than point or interval estimation.) Other relevant discussions have been given by Cox [8, pp. 359–63], Wallace [23, especially p. 864 and references therein], and Lehmann [14, pp. 139–40].

The following sections will be largely devoted to the deduction of some mathematical consequences of (C) and (S), and to their interpretation. The remainder of the present section is devoted to discussion and illustration of the meaning of (C); and to illustration of the considerations which seem to many statisticians, including the writer, to give compelling support to adoption of (C) as an appropriate extra-mathematical assertion concerning the structure of evidential meaning.

It can be shown that (S) is implied mathematically by (C). (The method of proof is the device of interpreting the conditional distribution of x, given $t(x) = t$, as a distribution $G_t(h)$ defining a mixture experiment equivalent to the given experiment.) This relation will not be discussed further here, since there seems to be little question as to the appropriateness of (S) in any case.

4.3. Example

A simple concrete (but partly hypothetical) example is the following: Suppose that two instruments are available for use in an experiment whose primary purpose is informative inference, for example, to make observations on some material of general interest, and to report the experimental results in appropriate terms. Suppose that the experimental conditions are fixed, and that these entail that the selection of the instrument to be used depends upon chance factors not related to the subject-matter of the experiment, in such a way that the instruments have respective known probabilities $g_1 = .73$ and $g_2 = .27$ of being selected for use. The experimental conditions allow use of the selected instrument to make just one observation, and each instrument gives only dichotomous observations, $y = 1$ ("positive") or 0 ("negative").

(We recall that discussion of design of experiments for informative inference has been deferred; but we stress that any satisfactory general analysis of evidential meaning must deal adequately with artificial and hypothetical experiments as well as with those of commonly-encountered forms. Even the present example is not very artificial, since the alternative instruments are simple analogues of observable experimental conditions (like independent variables in some regression problems) which may be uncontrollable and which have known effects on experimental precision.) If the instruments are labeled by $h = 1$ or 2, respectively, then each outcome of this experiment E is represented by a symbol (h, y) or (h, y_h), where $h = 1$ or 2, and $y = y_h = 0$ or 1. We assume that the material under investigation is known to be in one of just two possible states, H_1 or H_2 (two simple hypotheses). Each instrument has equal probabilities of "false positives" and of "false negatives." For the first instrument these are

$$\alpha_1 = \text{Prob}(Y_1 = 1 | H_1) = \text{Prob}(Y_1 = 0 | H_2) = \frac{1}{730} \doteq .0014,$$

and for the second instrument

$$\alpha_2 = \text{Prob}(Y_2 = 1 | H_1) = \text{Prob}(Y_2 = 0 | H_2) = .10.$$

As an instance of the general proposition (C), consider the assertion: $\text{Ev}(E, (E_1, 1)) = \text{Ev}(E_1, 1)$. This assertion is apparently not necessary on mathematical grounds alone, but it seems to be supported compellingly by considerations like the following concerning the nature of evidential meaning: Granting the validity of the model E and accepting the experimental conditions which it represents, suppose that E leads to selection of the first instrument (that is, $H = h = 1$ is observed). Then by good fortune the experimenter finds himself in the same position as if he had been assured use of that superior instrument (for one observation) as an initial condition of his experiment. In the latter hypothetical situation, he would be prepared to report either $(E_1, 0)$ or $(E_1, 1)$ as a complete description of the statistical evidence obtained. In the former actual situation, the fact that the first instrument might not have been selected seems not only hypothetical but completely irrelevant: For purposes of informative inference, if $Y = 1$ is observed with the first instrument, then the report $(E_1, 1)$ seems to be an appropriate and complete description of the statistical evidence obtained; and the "more complete" report $(E, (E_1, 1))$ seems to differ from it only by the addition of recognizably redundant elements irrelevant to the evidential meaning and evidential interpretation of this outcome of E. The latter redundant elements are the descriptions of other component experiments (and their probabilities) which might have been carried out but in fact were not. Parallel comments apply to the other possible outcomes of E.

4.4.

As formulated above, (C) is not a recommendation (or directive or convention) to replace unconditional by conditional experimental frames of reference wherever (C) is seen to be applicable. However if (C) is adopted it tends to invite such application, if only for the advantage of parsimony, since a conditional frame of reference is typically simpler and seems more appropriately refined for purposes of informative inference. Writers who have seen value in such conditionality concepts have usually focused attention on their use in this way. However, even the range of such applications has not been fully investigated in experiments of various structures. And the implications of such conditionality concepts for problems of informative inference in general appear considerably more radical than has been generally anticipated, as will be indicated below. We shall be primarily concerned with the use of (C) as a tool in the formal analysis of the structure of evidential meaning; and in such use, (C) as formulated above also sanctions the replacement of a conditional experimental frame of reference by an appropriately corresponding unconditional one (by substitution of $\mathrm{Ev}(E, (E_h, x_h))$ for an equivalent $\mathrm{Ev}(E_h, x_h)$).

4.5.

Another aspect of such interpretations can be discussed conveniently in terms of the preceding example. The example concerned an experiment whose component experiments are based on one or another actual experimental instrument. Consider next an alternative experiment plan (of a more familiar type) which could be adopted for the same experimental purpose: Here just one instrument is available, the second one described above, which gives observations $Y = 1$ with probabilities .1 and .9 under the same respective hypotheses H_1, H_2, and otherwise gives $Y = 0$. The present experimental plan, denoted by E_B, calls for 3 independent observations by this instrument; thus the model E_B is represented by the simple binomial distributions of

$$X = \sum_{j=1}^{3} Y_j:$$

$$H_1: f_1(x) = \binom{3}{x}(.1)^x(.9)^{3-x},$$

$$H_2: f_2(x) = \binom{3}{x}(.9)^x(.1)^{3-x},$$

for $x = 0, 1, 2, 3$. E_B will provide one of the instances of statistical evidence (E_B, x), $x = 0, 1, 2,$ or 3. The *physical* experimental procedures represented respectively by E and E_B are manifestly different. But we verify as follows that the mathematical-statistical models E and E_B are *mathematically* equivalent: Each experiment leads to one of four possible outcomes, which can be set in the following one-to-one correspondence:

E yields: $(E_h, y_h) =$	$(E_1, 0)$	$(E_2, 0)$	$(E_2, 1)$	$(E_1, 1)$
E_B yields $\quad x =$	0	1	2	3

It is readily verified that under each hypothesis the two models specify identical probabilities for corresponding outcomes. For example,

$$\text{Prob}((E_1, 0)|H_1, E) = (.73)\frac{729}{730} = .729$$

$$= \binom{3}{0}(.1)^0(.9)^3 = f_1(0)$$

$$= \text{Prob}(X = 0|H_1, E_B).$$

Thus $(E_B, 0)$ and $(E, (E_1, 0))$ are *mathematically equivalent* instances of statistical evidence. We therefore write $\text{Ev}(E_B, 0) = \text{Ev}(E, (E_1, 0))$. Is the latter assertion of equivalence of evidential meanings tenable here, because of the *mathematical* equivalence of $(E_B, 0)$ and $(E, (E_1, 0))$ alone, and despite the gross difference of *physical* structures of the experiments represented by E_B and E? An affirmative answer seems necessary on the following *formal* grounds: Each of the models E and E_B was assumed to be an adequate mathematical-statistical model of a corresponding physical experimental situation; this very strong assumption implies that there are no physical aspects of either situation which are relevant to the experimental purposes except those represented in the respective models E and E_B. The latter models may be said to represent adequately and completely the assumed *physical* as well as mathematical structures of the experiments in all *relevant* respects; for example, the usual conceptual frequency interpretations of all probability terms appearing in each model may be taken to characterize fully the physical structure and meaning of each model. Hence the assumed adequacy and the mathematical equivalence of the two models imply that the two experimental situations have in effect been assumed to be physically equivalent in all *relevant* respects. This interpretative conclusion can be illustrated further by considering the rhetorical question: On what theoretical or practical grounds can an experimenter reasonably support any definite preference between the experiments represented by E and E_B, for *any* purpose of statistical inference or decision-making, assuming the adequacy of each model?

Combining this discussion with section 4.3 above, we find that (C) implies that $\text{Ev}(E_B, 0) = \text{Ev}(E_1, 0)$, although no mixture structure was apparent in the physical situation represented by E_B, nor in the binomial model E_B as usually interpreted.

4.6.

We note that (C) above differs in meaning and scope from the purely technical use which is sometimes made of conditional experimental frames of reference, as in the development of similar tests of composite hypotheses (as in Lehmann [14, p. 136]) or of best unbiased estimators.

4.7.

We note also that (C) above does not directly involve, or ascribe meaning to, any notion of evidential interpretations "conditional on an observed sample point x." Rather, (C) ascribes equivalence to certain instances of evidential meaning of respective outcomes, each referred to a specified mathematically complete experimental frame of reference. (The phrase in quotes can be given a precise mathematical meaning under postulation of the principle of inverse probability, in which case it refers to a posterior distribution, given x. However our discussion is not based on such postulation.)

4.8.

In considering whether (C) seems appropriate for all purposes of informative inference, it is necessary to avoid confusion with still another usage of "conditional" which differs from that in (C). A familiar simple example of this other usage occurs in connection with a one-way analysis of variance experiment under common normality assumptions. Results of such an experiment may be interpreted either "conditionally" (Model I) or "unconditionally" (Model II), and in some situations there are familiar purposes of informative inference (focusing on a component of variance) in which the "unconditional" interpretation is useful and necessary. However, the latter important point is not relevant to the question of the general appropriateness of (C) for informative inference, because the "conditional" frame of reference in this example cannot be interpreted as a component experiment within a mixture experiment as required for applicability of (C).

4.9.

It is the opinion of the writer (among others) that upon suitable consideration the principle of conditionality will be generally accepted as appropriate for purposes of informative inference, and that apparent reservations will be found to stem either from purposes which can usefully be distinguished from informative inference, or from interpretations of "conditionality" different from that formulated in (C), some of which have been described above. (Of course purposes of several kinds are frequently represented in one experimental situation, and these are often served best by applying different concepts and techniques side by side as appropriate for the various purposes.) In any case, the following sections are largely devoted to examination of the *mathematical* consequences of (C) and their interpretation.

5. The Likelihood Principle

5.1.

The next step in our analysis concerns a third condition for equivalence of evidential meaning:

The Likelihood Principle (L). If E and E' are any two experiments with a common parameter space, and if x and y are any respective outcomes which determine likelihood functions satisfying $f(x, \theta) = cg(y, \theta)$ for some positive constant $c = c(x, y)$ and all θ, then $\mathrm{Ev}(E, x) = \mathrm{Ev}(E', y)$. That is, the evidential meaning $\mathrm{Ev}(E, x)$ of any outcome x of any experiment E is characterized completely by the likelihood function $cf(x, \theta)$, and is otherwise independent of the structure of (E, x).

5.2.

(L) is an immediate consequence of Bayes' principle, when the latter (with any interpretation) is adopted. Our primary interest, as mentioned, is in approaches which are independent of this principle.

5.3.

Fisher [9, pp. 68–73, 128–31, and earlier writings] and Barnard [2, and earlier writings] have been the principal authors supporting the likelihood principle on grounds independent of Bayes' principle. (The principle of maximum likelihood, which is directed to the problem of point-estimation, is not to be identified with the likelihood principle. Some connections between the distinct problems of point-estimation and informative inference are discussed below.) Self-evidence seems to be essential ground on which these writers support (L).

5.4.

Other modes of support for (L), such as the basic technical role of the likelihood function in the theory of sufficient statistics and in the characterization of admissible statistical decision functions, seem heuristic and incomplete, since (as in the formulation of (S), and its consequence Lemma 1, in Section 3 above) they do not demonstrate that evidential meaning is independent of the structure of an experiment apart from the likelihood function.

5.5.

Far fewer writers seem to have found (L) as clearly appropriate, as an extra-mathematical statement about evidential meaning, as (C). It is this fact which seems to lend interest to the following:

Lemma 2. (L) *implies, and is implied by,* (S) *and* (C).

PROOF. That (L) implies (C) follows immediately from the fact that in all cases the likelihood functions determined respectively by $(E, (E_h, x_h))$ and (E_h, x_h) are proportional. That (L) implies (S) follows immediately from Lemma 1 of Section 3.

The relation of principal interest, that (S) and (C) imply (L), is proved as follows: Let E and E' denote any two (mathematical models of) experiments, having the common parameter space $\Omega = \{\theta\}$, and represented by probability density functions $f(x, \theta)$, $g(y, \theta)$ on their respective sample spaces $S = \{x\}$, $S' = \{y\}$. (S and S' are to be regarded as distinct, disjoint spaces.) Consider the (hypothetical) mixture experiment E^* whose components are just E and E', taken with equal probabilities. Let z denote the generic sample point of E^*, and let C denote any set of points z; then $C = A \cup B$, where $A \subset S$ and $B \subset S'$, and

$$\text{Prob}(Z \in C|\theta) = \frac{1}{2}\text{Prob}(A|\theta, E) + \frac{1}{2}\text{Prob}(B|\theta, E')$$

$$= \frac{1}{2}\int_A f(x, \theta)\, d\mu(x) + \frac{1}{2}\int_B g(y, \theta)\, dv(y)$$

(where A and B are measurable sets). Thus the probability density function representing E^* may be denoted by

$$h(z, \theta) = \begin{cases} \frac{1}{2}f(x, \theta), & \text{if } z = x \in S, \\ \frac{1}{2}g(y, \theta), & \text{if } z = y \in S'. \end{cases}$$

Each outcome z of E^* has a representation

$$z = \begin{cases} (E, x), & \text{if } z = x \in S, \\ (E', y), & \text{if } z = y \in S'. \end{cases}$$

From (C), it follows that

$$\text{Ev}(E^*, (E, x)) = \text{Ev}(E, x), \qquad \text{for each } x \in S, \text{ and}$$
$$\text{Ev}(E^*, (E', y)) = \text{Ev}(E', y), \qquad \text{for each } y \in S'. \tag{5.1}$$

Let x', y' be any two outcomes of E, E' respectively which determine the same likelihood function; that is, $f(x', \theta) = cg(y', \theta)$ for all θ, where c is some positive constant. Then we have $h(x', \theta) \equiv ch(y', \theta)$ for all θ; that is, the two outcomes (E, x'), (E', y') of E^* determine the same likelihood function. Then it follows from (S) and its consequence Lemma 1 in Section 3 that

$$\text{Ev}(E^*, (E, x')) = \text{Ev}(E^*, (E', y')). \tag{5.2}$$

From (5.1) and (5.2) it follows that

$$Ev(E, x') = Ev(E', y'). \tag{5.3}$$

But (5.3) states that any two outcomes x', y' of any two experiments E, E' (with the same parameter space) have the same evidential meaning if they determine the same likelihood function. This completes the proof of equivalence of (L) with (S) and (C). □

5.6.

For those who adopt (C) and (S), their consequence (L) gives an explicit solution to our first general problem, the mathematical characterization of statistical evidence as such. The question whether different likelihood functions (on the same parameter space) represent different evidential meanings is given an affirmative answer in the following sections, in terms of evidential interpretations of likelihood functions on parameter spaces of limited generality; and presumably this conclusion can be supported quite generally.

5.7.

The most important general consequence of (L) (and of (C)) for problems of *evidential interpretation* seems to be the following: Those modes of representing evidential meaning which include reference to any specific experimental frame of reference (including the actual one from which an outcome was obtained) are somewhat unsatisfactory; in particular, they tend to conceal equivalences between instances of evidential meaning which are recognizable under (L). Various modes of *interpretation* of evidential meaning will be discussed in the following sections, with particular attention to their relations to (L).

5.8.

The scope of the role of ancillary statistics in informative inference seems altered in the light of the result that (C) and (S) imply (L). As mentioned, the usual use of (C) has depended on recognition of an ancillary statistic (or mixture structure) in the model of an actual experiment under consideration; and has consisted primarily of the adoption of conditional frames of reference, when thus recognized, for evidential interpretations. But the range of existence of ancillary statistics in experiments of various structures has not been completely explored; indeed, in the simple case of binary experiments (those with two-point parameter spaces), the fact that they exist in all but the simplest cases has been seen only very recently in reference [3]. Thus the potential scope and implications, which even such usual applications of (C) might have for informative inference, have not been fully seen.

Moreover, the question of conditions for uniqueness of ancillary statistics, when they exist, has received little attention. But simple examples have been found, some of which are described in reference [3], in which one experiment admits several essentially different ancillary statistics; when (C) is applied in the usual way to each of these alternative ancillary statistics in turn, one can obtain quite different conditional experimental frames of reference for evidential interpretation of a single outcome. Even isolated examples of this kind seem to pose a basic problem for this approach: it would seem that, in the face of such examples, the usual use of (C) must be supplemented either by a convention restricting its scope, or by a convention for choice among alternative conditional frames of reference when they exist, or by some radical interpretation of the consequences of (C), in which the role of experimental frames of reference in general in evidential interpretations is reappraised. The adoption of a convention to avoid certain possible applications of (C) would seem artificial and unsatisfactory in principle; on the other hand, the need for a radical reappraisal of the role of experimental frames of reference, which is apparent in the light of such examples, is confirmed quite generally by the above result, that (C) and (S) imply (L). For according to (L), reference to any particular experimental frame of reference, even an actual or a conditional one, for evidential interpretations, has necessarily a partly-conventional character.

Earlier proofs that (C) and (S) imply (L), restricted to relatively simple classes of experiments, utilized recognition of mixture structures in experiments [3], [4]. But in the above proof that (C) and (S) imply (L) for all classes of experiments, no existence of mixture structures in the experiments E, E', under consideration was required; the ancillary used there was constructed with the hypothetical mixture E^*. The conclusion (L) takes us beyond the need to examine specific experiments for possible mixture structure, since it eliminates the need to regard any experimental frames of reference, including actual or conditional ones, as essential for evidential interpretations. The possible usefulness of experimental frames of reference in a partly conventional sense for evidential interpretations will be discussed in some of the following sections.

Part II

6. Evidential Interpretations of Likelihood Functions

We have seen above that on certain grounds, the likelihood principle (L) gives a solution of the first general problem of informative inference, that of mathematical characterization of evidential meaning. On this basis the second general problem of informative inference, that of evidential interpretations in general, can be described more precisely as the problem of evidential interpretations of likelihood functions. The remaining sections of this paper are devoted to the latter problem, that is, to consideration of questions like the following: When any instance (E, x) of statistical evidence is represented by just the corresponding likelihood function $L(\theta) = cf(x, \theta)$ (c an arbitrary positive constant), what are the qualitative and quantitative properties of the statistical evidence represented by $L(\theta)$? What concepts and terms are appropriate for describing and interpreting these evidential properties? How are such modes of evidential interpretation related to those in current general use?

The principal writers supporting the use of just the likelihood function for informative inference have not elaborated in very precise and systematic detail the nature of evidential interpretations of the likelihood function. Fisher has recently given a brief discussion and examples of such interpretations [9, especially pp. 68–73, 128–31]. He describes the relative likelihoods of alternative values of parameters as giving "a natural order of preference among the possibilities" (p. 38); and states that inspection of such relative likelihoods "shows clearly enough what values are implausible" (p. 71). Such interpretations were also recently discussed and illustrated by Barnard [2]. Both writers stress that point estimates, even when supplemented by measures of precision, have limited value for these purposes. For example when $\log L(\theta)$ has (at least approximately) a parabolic form, then a point estimate (maximum likelihood) and a measure of its precision (preferably the curvature of $\log L(\theta)$ at its maximum) constitute a convenient mode of description of the complete likelihood function (at least approximately); but more generally, with very different forms of $L(\theta)$, such descriptive indices have less descriptive value.

More detailed discussion of evidential interpretations of likelihood functions, and clarification of the meanings of terms appropriate for such discussion, seems desirable if possible, as has been remarked by Cox [8, p. 366]. These are the purposes of the following sections. Since any non-negative function $L(\theta)$, defined on an arbitrary parameter space, is a possible likelihood function, it is convenient to consider in turn parameter spaces of various forms, beginning for simplicity with the case of a two-point parameter space, followed by the case of any finite number of parameter points, and then more general and typical cases.

7. Binary Experiments

Parts of this section are closely related to reference [3, pp. 429–34].)

7.1.

The simplest experiments, mathematically, are *binary* experiments, that is, experiments with parameter spaces containing just two points, θ_1, θ_2, representing just two simple hypotheses, H_1, H_2. Any outcome x of any such experiment determines a likelihood function $L(\theta) = cf(x, \theta)$ which may be represented by the pair of numbers $(cf(x, \theta_1), cf(x, \theta_2))$, with c any positive constant. Hence $L(\theta)$ is more parsimoniously represented by $\lambda = \lambda(x) = f(x, \theta_2)/f(x, \theta_1)$. ($\lambda(x)$ is the likelihood ratio statistic, which appears with rather different interpretations in other approaches to statistical theory.) Each possible likelihood function arising from any binary experiment is represented in this way by a number λ, $0 \leq \lambda \leq \infty$. What sorts of evidential interpretations can be made of such a number λ which represents in this way an outcome of a binary experiment?

As a convenient interpretative step, consider for each number α, $0 \leq \alpha \leq \frac{1}{2}$, a binary experiment whose sample space contains only two points, denoted "positive" ($+$) and "negative" ($-$), such that $\text{Prob}(+|H_1) = \text{Prob}(-|H_2) = \alpha$. Any such *symmetric simple* binary experiment is characterized by the "error probability" α which is the common value of "false positives" and "false negatives." (α is the common value of error-probabilities of Types I and II of the test of H_1 against H_2 which rejects just on the outcome $+$.) The outcomes of such an experiment determine the likelihood functions $\lambda(+) = (1 - \alpha)/\alpha \geq 1$ and $\lambda(-) = \alpha/(1 - \alpha) = 1/\lambda(+) \leq 1$ respectively, with smaller error probabilities giving values farther above and below unity, respectively. According to the likelihood principle (L), when *any* binary experiment E gives *any* outcome x determining a likelihood function $\lambda(x) \geq 1$, the evidential meaning of $\lambda(x)$ is the same as that of the positive outcome of the symmetric simple binary experiment with error-probability α such that $\lambda(x) = (1 - \alpha)/\alpha$, that is, $\alpha = 1/(1 + \lambda(x))$. If the actual experiment E had the latter form, the outcome would customarily be described as "significant at the α level" (possibly with reference also to the Type II error-probability, which is again α). This currently standard usage can be modified in a way which is in accord with the likelihood principle by calling $\alpha = 1/(1 + \lambda(x))$ the *intrinsic significance level* associated with the outcome x, regardless of form of E. Here the probability α is defined in a specified symmetric simple binary experiment, and admits therein the usual conceptual frequency interpretations. The relations between such an experiment and the outcome $\lambda(x)$ being interpreted are conceptual, in a way which accords with the likelihood principle; the conventional element involved in adopting such an experimental frame of reference for evidential interpretations is clear, and is necessary in the light of the likelihood principle. (Alternative conventions of choice of experimental frames of reference are discussed in reference [3].) Outcomes giving $\lambda(x) \leq 1$ can be interpreted similarly: such outcomes support H_1 against H_2, with evidential strength corresponding to the intrinsic significance level $\alpha = \lambda(x)/(1 + \lambda(x))$.

In connection with the current use of significance levels in evidential interpretations, it has often been stressed that consideration of the power of tests is essential to reasonable interpretations. But no systematic way of considering power along with significance levels seems to have been proposed specifically for the purpose of informative inference. And current standard practice often fails to include such consideration in any form (cf. reference

[12]). The concept of intrinsic significance level incorporates automatic consideration of error-probabilities of both types, within its own experimental frame of reference, in a way which is also in accord with the likelihood principle.

7.2.

Tukey [22] has recently stressed the need for a critical reappraisal of the role of significance tests in the light of a history of the practice and theory of informative inference. The next paragraphs are a brief contribution in this direction from the present standpoint.

Because the function of informative inference is so basic to empirical scientific work, it is not surprising that its beginnings can be traced back to an early stage in the development of the mathematical theory of probability. As early as 1710, Dr. John Arbuthnot computed the probability of an event which had been observed, that in each of a certain 82 successive years more male than female births would be registered in London, on the hypothesis that the probability of such an event in a single year was $\frac{1}{2}$; and he interpreted the very small probability $(\frac{1}{2})^{82}$ as strong evidence against the hypothesis [19, pp. 196–8]. This was perhaps the earliest use of a formal probability calculation for a purpose of statistical inference, which in this case was informative inference. Other early writers considered problems involving mathematically similar simple statistical hypotheses, and alternative hypotheses of a statistically-degenerate kind under which a particular outcome was certain: a "permanent cause" or "certain cause," or non-statistical "law of nature," that is, a hypothesis "which always produces the event" [6, pp. 261, 358]. (It is not altogether clear that a simple non-statistical alternative would correspond to Arbuthnot's view of his problem.) Non-occurrence of such an outcome, even once in many trials, warrants rejection of such a hypothesis without qualification or resort to statistical considerations; but occurrence of such an outcome on each of n trials provides statistical evidence which requires interpretation as such. If the event in question has probability p of occurrence in one trial under the first hypothesis (and probability 1 under the second), then the probability of its occurrence in each of n independent trials is $P = p^n$ under the first hypothesis (and 1 under the second). (It is convenient to assume in our discussion that n was fixed; this may be inappropriate in some interpretations of these early examples.) In Arbuthnot's example, $P = (\frac{1}{2})^{82}$.

In such problems, the quantity on which evidential interpretations center is P, and small values of P are interpreted as strong evidence against the first hypothesis and for the second. What *general* concepts and basic terms are involved in these simple and "obviously sound" evidential interpretations? We can distinguish three mathematical concepts which do not coincide in general, but which assume the common form P in cases of the present extreme simplicity: Here P is not only the probability of "what was observed" under

H_1: (a) P is the probability of an outcome "at least as extreme as that observed" under H_1 (because here there are no outcomes which are "more extreme"); that is P is a *significance level* (or critical level); and (b) P is the ratio of the probabilities, under respective hypotheses, of "what was observed"; that is, P is a *likelihood ratio* λ. To determine whether (a) or (b) is the *appropriate* general concept of evidential interpretation which is represented here by the obviously-appropriate quantity P, we must turn to more general considerations, such as the analysis of the preceding sections. Since in more complex problems the two concepts no longer coincide, one may wonder whether early and current uses of the significance level concept have sometimes derived support by inappropriate generalization, to (a) as against (b), from such simple and perhaps deceptively "clear" examples.

7.3.

It is convenient to discuss here a reservation sometimes expressed concerning (L) itself, because this reservation involves significance levels. Experiments of different structures, for example experiments based on observations of the same kind but based on different sampling rules, may lead to respective outcomes which determine the same likelihood function but which are assigned different significance levels according to common practice. It is felt by many that such differences in significance levels reflect genuine differences between evidential meanings, corresponding to the different sampling rules; and therefore that (L) is unreasonable because it denies such differences of evidential meaning. The following discussion of a concrete example may throw further light on this point, while providing additional illustrations of (C) and (L) and their significance. Consider once more the binomial experiment E_B of Section 4.4 above, consisting of three independent observations on Y, which takes the values 0 or 1, with probabilities .9, .1, respectively under H_1, and with probabilities .1, .9, respectively under H_2. Consider also a sequential experiment E_S in which independent observations of the same kind Y are taken until for the first time $Y = 0$ is observed: Let Z denote the number of times $Y = 1$ is observed before termination of such an experiment. Then the distribution of Z is given by $f_1(z) = (.9)(.1)^z$, under H_1, and by $f_2(z) = (.1)(.9)^z$, under H_2, for $z = 0, 1, 2 \ldots$. The experiment E_S can be represented as a mixture of simple binary component experiments, among which is the component E_2 (described in Section 4.3) consisting of a single observation Y; this component is assigned probability .09 in the mixture experiment equivalent to E_S. We recall that E_B also admits a mixture representation, in which the component E_2 appears, assigned probability .27. We may imagine two experimenters, using E_B and E_S respectively for the same purpose of informative inference, and we may imagine a situation in which the mathematical component experiments are realized physically by alternative measuring instruments as in our discussion of E_B in Section 4.3. Then the first experimenter's design E_B includes the equivalent of a .27 chance of using the instrument represented by E_2 (for a single observation); and the second experimenter's sequential design E_S includes the equivalent of a .09 chance of using the same instrument (for one observation). If by chance each experimenter obtained this instrument and observed a positive outcome from it, then evidently the two results would have identical evidential meaning (as (C) asserts). However the customary assignment of significance levels would give such results the .028 significance level in the framework of E_B, and the .01 significance level in the framework of E_S. Both of these differ from the .10 error-probability which characterizes the common component experiment E_2. The latter value would be the intrinsic significance level assigned in the interpretation suggested above; this value would be indicated immediately, in any of the experimental frames of reference mentioned, by the common value 9 assumed by the likelihood ratio statistic λ on each of the outcomes mentioned.

8. Finite Parameter Spaces

If E is any experiment with a parameter space containing only a finite number k of points, these may conveniently be labeled $\theta = i = 1, 2, \ldots, k$. Any observed outcome x of E determines a likelihood function $L(i) = cf(x, i)$, $i = 1, \ldots, k$. We shall consider evidential interpretations of such likelihood functions in the light of the likelihood principle, in cases where

$$\sum_{i=1}^{k} f(x, i)$$

is positive and finite. (The remaining cases are special and artificial in a sense related to technicalities in the role of density functions in defining continuous distributions.) It is convenient here to choose c as the reciprocal of the latter sum, so that without loss of generality we can assume that

$$\sum_{i=1}^{k} L(i) = 1.$$

The present discussion formally includes the binary case, $k = 2$, discussed above.

Any experiment E with a finite sample space labeled $j = 1, \ldots, m$, and finite parameter space is represented conveniently by a stochastic matrix

$$E = (p_{ij}) = \begin{bmatrix} p_{11} & \cdots & p_{1m} \\ \vdots & & \vdots \\ p_{k1} & \cdots & p_{km} \end{bmatrix},$$

where

$$\sum_{j=1}^{m} p_{ij} = 1,$$

and $p_{ij} = \text{Prob}[j|i]$, for each i, j. Here the ith row is the discrete probability distribution p_{ij} given by parameter value i, and the jth column is proportional to the likelihood function $L(i) = L(i|j) = cp_{ij}$, $i = 1, \ldots, k$, determined by outcome j. (The condition that

$$\sum_{i=1}^{k} p_{ij}$$

be positive and finite always holds here, since each p_{ij} is finite, and since any j for which all $p_{ij} = 0$ can be deleted from the sample space without effectively altering the model E.)

8.1. Qualitative Evidential Interpretations

The simplest nontrivial sample space for any experiment is one with only two points, $j = 1, 2$. Any likelihood function $L(i)$ (with

$$\sum_{i=1}^{k} L(i) = 1,$$

which we assume hereafter) can represent an outcome of such an experiment, for we can define

$$\text{Prob}[j = 1|i] = L(i) \quad \text{and} \quad \text{Prob}[j = 2|i] = 1 - L(i),$$

for $i = 1, \ldots, k$.

For example, the likelihood function $L(i) \equiv \frac{1}{3}$, $i = 1, 2, 3$, represents the possible outcome $j = 1$ of the experiment

$$E = \begin{bmatrix} \frac{1}{3} & \frac{2}{3} \\ \frac{1}{3} & \frac{2}{3} \\ \frac{1}{3} & \frac{2}{3} \end{bmatrix}.$$

Since this experiment gives the same distribution on the two-point sample space under each hypothesis, it is completely uninformative, as is any outcome of this experiment. According to the likelihood principle, we can therefore conclude that the given likelihood function has a simple evidential interpretation, regardless of the structure of the experiment from which it arises, namely, that it represents a completely uninformative outcome. (The same interpretation applies to a constant likelihood function on a parameter space of any form, as an essentially similar argument shows.)

Consider next the likelihood function $(1, 0, 0)$. (That is, $L(1) = 1$, $L(2) = L(3) = 0$, on the 3-point parameter space $i = 1, 2, 3$.) This represents the possible outcome $j = 1$ of the experiment

$$E = \begin{bmatrix} 1 & 0 \\ 0 & 1 \\ 0 & 1 \end{bmatrix}.$$

The outcome $j = 1$ of E is impossible (has probability 0) under hypotheses

$i = 2$ and 3 (but is certain under $i = 1$). Hence, its occurrence supports without risk of error the conclusion that $i = 1$. According to the likelihood principle, the same certain conclusion is warranted when such a likelihood function is determined by an outcome of *any* experiment. (Similarly any likelihood function which is zero on a parameter space of any form, except at a single point, supports a conclusion of an essentially non-statistical, "deductive" kind.)

The likelihood function $(\frac{1}{2}, \frac{1}{2}, 0)$ could have been determined by outcome $j = 1$ of

$$E = \begin{bmatrix} \frac{1}{2} & \frac{1}{2} \\ \frac{1}{2} & \frac{1}{2} \\ 0 & 1 \end{bmatrix}.$$

This outcome of E is impossible under hypothesis $i = 3$, and hence supports without risk of error the conclusion that $i \neq 3$ (that is, that $i = 1$ or 2). Furthermore, E prescribes identical distributions under hypotheses $i = 1$ and 2, and hence the experiment E, and each of its possible outcomes, is completely uninformative as between $i = 1$ and 2. The likelihood principle supports the same evidential interpretations of this likelihood function regardless of the experiment from which it arose. (Parallel interpretations show that in the case of any parameter space, any bounded likelihood function assuming a common value on some set of parameter points is completely uninformative as between those points.)

In the preceding experiment, the distinct labels $i = 1$ and 2 would ordinarily be used to distinguish two hypotheses with distinct physical meanings, that is, two hypotheses about some natural phenomenon which could be distinguished at least in a statistical sense by a suitably designed experiment. The particular experiment E is, as mentioned, completely uninformative as between these hypotheses. Therefore if an experiment of the form E were conducted, it would be natural for some purposes to describe the actual experimental situation in terms of a two-point parameter space, labeled by $i' = 1$ or 2, and by the model

$$E' = (p'_{i'j}) = \begin{pmatrix} \frac{1}{2} & \frac{1}{2} \\ 0 & 1 \end{pmatrix}.$$

Here $i' = 2$ stands just for the same simple hypothesis previously denoted by $i = 3$ in E; $i' = 1$ represents a simple (one-point) hypothesis in this actual experimental situation, but also represents the composite hypothesis previously denoted by $i = 1$ or 2 in E. Such examples illustrate a sense in which even the specification of the number of points in the parameter space (of an adequate mathematical-statistical model of an experiment) sometimes involves an element of conventionality.

Consider the likelihood function $(.8, .1, .1)$ on the 3-point parameter space $i = 1, 2, 3$. The interpretation that this likelihood function (or the outcome it represents) has the qualitative evidential property of *supporting* the hypothe-

sis $i = 1$, against the alternatives $i = 2$ or 3, is supported by various considerations including the following: This likelihood function represents the outcome $j = 1$ of

$$E = \begin{bmatrix} .8 & .2 \\ .1 & .9 \\ .1 & .9 \end{bmatrix} = (p_{ij}).$$

With use of E, if one reports the outcome $j = 1$ as "supporting $i = 1$" (in a
qualitative, merely statistical sense), and if one reports the remaining outcome
differently, for example as "not supporting $i = 1$," then one makes inappropriate reports only with probability $.1$ when $i = 2$ or 3, and only with probability $.2$ if $i = 1$. (Without use of an informative experiment, such reports could
be arrived at only arbitrarily, with possible use of an auxiliary randomization
variable, and the respective probabilities of inappropriate reports would then
total unity.) This illustrates, in the familiar terms of error-probabilities of two
kinds defined in the framework of a given experiment, the appropriateness of
this qualitative evidential interpretation. According to the likelihood principle, the same qualitative interpretation is appropriate when this likelihood
function is obtained from any experiment. (It can be shown similarly that on
any parameter space, when any bounded likelihood function takes different
constant values on two respective "contours," each point of the contour with
greater likelihood is supported evidentially more strongly than each point
with smaller likelihood.)

Consider the respective likelihood functions $(.8, .1, .1)$ and $(.45, .275, .275)$;
the latter is "flatter" than the first, but qualitatively similar. The interpretation that the first is *more informative than* the second (and therefore that the
first supports $i = 1$ more strongly than the second) is supported as follows:
Consider

$$E = \begin{bmatrix} .8 & .2 \\ .1 & .9 \\ .1 & .9 \end{bmatrix} = (p_{ij}).$$

Consider also the experiment E' based on E as follows: When outcome $j = 2$
of E is observed, an auxiliary randomization device is used to report "$w = 1$"
with probability $\frac{1}{2}$, and to report "$w = 2$" with probability $\frac{1}{2}$; when outcome
$j = 1$ of E is observed, the report "$w = 1$" is given. Simple calculations verify
that E' has the form

$$E = \begin{bmatrix} .9 & .1 \\ .55 & .45 \\ .55 & .55 \end{bmatrix} = (p'_{iw}).$$

The outcome $w = 1$ of E' determines the likelihood function $(.45, .275, .275)$
given above (the latter is proportional to the first column of E'). The experiment E' is less informative than E, since it was constructed from E by "adding

pure noise" (randomizing to "dilute" the statistical value of reports of outcomes). In particular, the outcome $w = 2$ of E' is exactly as informative as the outcome $j = 2$ of E, since $w = 2$ is known to be reported only when $j = 2$ was observed. But the outcome $w = 1$ of E' is less informative than the outcome $j = 1$ of E, since $w = 1$ follows all outcomes $j = 1$ of E and some outcomes $j = 2$ of E.

The preceding example illustrates that some likelihood functions on a given parameter space can be compared and ordered in a natural way. It can be shown that some pairs of likelihood functions are not comparable in this way, so that in general only a partial ordering of likelihood functions is possible. (An example is the pair of likelihood functions $(\frac{1}{2}, \frac{1}{3}, \frac{1}{6})$ and $(\frac{1}{2}, \frac{1}{4}, \frac{1}{4})$.) The special binary case, $k = 2$, is simpler in that all possible likelihood functions admit the simple ordering corresponding to increasing values of λ.

8.2. Intrinsic Confidence Methods

(Parts of the remainder of this paper where finite parameter spaces are considered are closely related to reference [4].) Consider the likelihood function (.90, .09, .01) defined on the parameter space $i = 1, 2, 3$. This represents the possible outcome $j = 1$ of the experiment

$$
E = \begin{bmatrix} .90 & .01 & .09 \\ .09 & .90 & .01 \\ .01 & .09 & .90 \end{bmatrix} = (p_{ij}).
$$

In this experiment, a confidence set estimator of the parameter i is given by taking, for each possible outcome j, the two values of i having greatest likelihoods $L(i|j)$. Thus outcome $j = 1$ gives the confidence set $i = 1$ or 2; $j = 2$ gives $i = 2$ or 3; and $j = 3$ gives $i = 3$ or 1. It is readily verified that under each value of i, the probability is .99 that the confidence set determined in this way will include i; that is, confidence sets determined in this way have confidence coefficient .99. For those who find confidence methods a clear and useful mode of evidential interpretation, and who also accept the likelihood principle, it may be useful for some interpretive purposes to consider the given likelihood function, regardless of the actual experiment from which it arose, in the framework of the very simple hypothetical experiment E in which it is equivalent to the outcome $j = 1$, and where it determines the 99 per cent confidence set $i = 1$ or 2. According to the likelihood principle, considering this outcome in the hypothetical framework E does not alter its evidential meaning; moreover, any mode of evidential interpretation which disallows such consideration is incompatible with the likelihood principle. Of course the standard meanings of confidence sets and their confidence levels are determined with reference to actual experimental frames of reference (or sometimes actual-conditional ones) and not hypothetically-considered ones. Hence in the present mode of evidential interpretation, the hypothetical, conventional role of the experimental frame of reference E must be made clear. This can be done by use of the terms "intrinsic confidence set" and "intrinsic confidence coefficient (or level)" to refer to confidence statements based in this way on a specified conventionally-used experimental frame of reference such as E.

With the same experiment E, if for each j we take the single most likely parameter point, namely $i = j$, we obtain a one-point confidence set estimator with intrinsic confidence coefficient .90. Thus the given likelihood function, arising from an experiment of any form, determines the intrinsic confidence set $i = 1$, with intrinsic confidence coefficient .90; the latter terms, again, are fully defined only when the form of the conventionally-used experiment E is indicated.

The general form of such intrinsic confidence methods is easily described as follows, for any likelihood function $L(i)$ defined on a finite parameter space $i = 1, \ldots, k$, and such that

$$\sum_{i=1}^{k} L(i) = 1:$$

If there is a unique least likely value i_1 of i (that is, if $L(i_1) < L(i)$ for $i \neq i_1'$), let $c_1 = 1 - L(i_1)$. Then the remaining $(k - 1)$ parameter points will be called an intrinsic confidence set with intrinsic confidence coefficient c_1; if there is no unique least likely value of i, no such set will be defined (for reasons related to the earlier discussion of points with equal likelihoods). If there is a pair of values of i, say i_1, i_2, with likelihoods strictly smaller than those of the remaining $(k - 2)$ points, call the latter set of points an intrinsic confidence set, with intrinsic confidence level $c_2 = 1 - L(i_1) - L(i_2)$. And so on. The experiment in which such confidence methods are actual as well as intrinsic confidence methods will always be understood to be

$$E = \begin{bmatrix} L(1) & L(k) & \cdots & L(2) \\ L(2) & L(1) & & L(3) \\ L(3) & L(2) & & \\ \vdots & \vdots & & \vdots \\ L(k) & L(k-1) & & L(1) \end{bmatrix}.$$

E is determined uniquely from the given $L(i)$ by taking the latter to determine the respective first-column elements, and then by completing E so that it is a "cyclic-symmetric" matrix, as illustrated (satisfying $p_{ij} = p_{i-1,j-1}$ for all i, j, with a subscript i or $j = 0$ here replaced by the value k).

By using here the basic technical relations between (ordinary) confidence methods and significance tests, we can obtain certain interpretations of the hypothesis-testing form from intrinsic confidence methods. For example, if a simple hypothesis $i = 1$ is of interest, and if a likelihood function $L(i)$ from any experiment leads to an intrinsic .99 confidence set containing $i = 1$, the outcome can be interpreted as "not intrinsically significant at the .01 level."

If the same likelihood function determines an intrinsic .95 confidence set not containing $i = 1$, this can be interpreted as "intrinsically significant at the .05 level," or "supporting rejection of $i = 1$ at the .05 intrinsic significant level." Here, in contrast with the special binary case $k = 2$, a single interpretive phrase like the latter does not incorporate unambiguous reference to the power of a corresponding test defined in E; nor does a single intrinsic confidence set report automatically incorporate such reference. On the other hand, a report of the set of all intrinsic confidence sets, with their respective levels, as defined above, does incorporate such reference, for it is readily seen that such a report determines uniquely the form of the likelihood function which it interprets. (Systematic use of confidence methods rather than significance tests, when possible, and of sets of confidence sets at various levels has been recommended by a number of recent writers; cf. [22], [23], [24], [15], [5] and references therein.)

An important category of problems is that involving several real-valued parameters, in which suitable estimates or tests concerning one of the parameters are of interest, the remaining parameters being nuisance parameters. Many such problems can be considered in miniature in the case of a finite parameter space, for example by labeling the parameter points by $(u, v), u = 1, \ldots, k', v = 1, \ldots, k''$, giving $k = k'k''$ points in all. Then intrinsic confidence sets for the parameter u can be defined, despite presence of the nuisance parameter v, by a generalization of the preceding discussion which includes a more general scheme for defining convenient relatively simple conventional experimental frames of reference.

(*Editors' note*: Section 9 has been omitted.)

10. Bayesian Methods: An Interpretation of the Principle of Insufficient Reason

In the method of treating statistical inference problems which was initiated by Bayes and Laplace, it was postulated that some mathematical probability distribution defined on the parameter space, the "prior distribution," represents appropriately the background information, knowledge, or opinion, available at the outset of an experiment; and that this, combined with experimental results by use of Bayes' formula, determines the "posterior distribution" which appropriately represents the information finally available. This formulation is widely referred to as Bayes' principle (or postulate), and we shall denote it by (B). (In this general form it should perhaps be credited to Laplace.) The extra-mathematical content of this principle has been interpreted in several ways by various of its proponents as well as critics [11, pp. 6–12]. This approach in general is not directed to the problem of informative inference, but rather to the problem of using experimental results along with other available information to determine an appropriate final synthesis of available information. However it is interesting to note that within this formulation the contribution of experimental results to the determination of posterior probabilities is always characterized just by the likelihood function and is otherwise independent of the structure of an experiment; in this sense we may say that (B) implies (L).

10.1.

The principle of insufficient reason, which we shall denote by (P.I.R.), is the special case of (B) in which a "uniform prior distribution" is specified to represent absence of background information or specific prior opinion. Evidently the intention of some who have developed and used methods based on (P.I.R.) has been to treat, in suitably objective and meaningful terms, the problem of informative inference as it is encountered in empirical research situations. This case of (B) was of particular interest to early writers on Bayesian methods. Following Laplace, this approach was widely accepted during the nineteenth century. Analysis and criticism, notably by Boole [6] and Cournot, of the possible ambiguity of the notion of "uniformity" of prior probabilities, and of the unclear nature of "prior probabilities" in general, led later to a wide-spread rejection of such formulations. The principal contemporary advocate of this approach is Jeffreys [13].

It is at least a striking coincidence that when experiments have suitable symmetry (or analogous) properties, inference methods based upon (P.I.R.) coincide exactly in form (although they differ in interpretation) with various modern inference methods developed without use of prior probabilities. For example, if any experiment E with a finite parameter space happens to be cyclic-symmetric, then uniform prior probabilities ($1/k$ on each parameter point) determine posterior probability statements which coincide in form with ordinary confidence statements. As a more general example, if E' has a k-point parameter space but *any structure*, it is easily verified that such posterior probability statements coincide in form with the intrinsic confidence statements determined as in Section 8 above. It follows that, leaving aside questions of extra-mathematical interpretation of (P.I.R.) itself, (P.I.R.) can be taken as a formal algorithm for convenient calculation of intrinsic confidence statements in the many classes of problems where such agreement can be demonstrated.

When the parameter space is more general, the "uniform distribution" has usually been chosen as some measure which is mathematically natural, for example Lebesgue measure on a real-line parameter space, even when such a measure does not satisfy the probability axiom of unit measure for the whole (parameter) space. In such cases again the posterior probabilities determined by formal application of Bayes' formula agree in form with ordinary or conditional confidence statements when an experiment has suitable symmetry-like (translation-parameter) properties; and more generally, such posterior probability statements agree in form with the intrinsic confidence statements described in Section 9 above. Furthermore the questions of conventionality, concerning the specification of a "uniform" distribution in such cases, are exactly parallel in form to the features of conventionality of choice of experimental frame of reference discussed in Section 9.

10.2.

A posterior probability statement determined by use of (P.I.R.) can be interpreted formally as merely a partial description of the likelihood function itself; and a sufficient number of such statements, or specification of the full posterior distribution, determine the likelihood function completely (provided the definition of "uniform" prior distribution is indicated unequivocally). This interpretation of (P.I.R.) makes it formally acceptable (in accord with (L)) as a solution of the first problem of informative inference, the mathematical characterization of evidential meaning. But this interpretation does not ascribe to (P.I.R.) any contribution to the second problem of informative inference, evidential interpretation, and does not include any specific interpretation of prior and posterior probabilities as such. On the interpretation mentioned, a posterior probability distribution might as well be replaced by a report of just the likelihood function itself. (On the basis of (L), without adoption of (P.I.R.) or (B), the absence of prior information or opinion admits a natural formal representation by a likelihood function taking any finite positive constant value, for example $L(\theta) \equiv 1$. Such a likelihood function is determined formally, for example, by any outcome of a completely uninformative experiment. Since likelihood functions determined from independent experiments are combined by simple multiplication, such a "prior likelihood function" combines formally with one from an actual experiment, to give the latter again as a final over-all "posterior" one.)

10.3.

A more complete explication of (P.I.R.) is suggested by the close formal relations indicated above between intrinsic confidence statements and statements based on (P.I.R.). Writers who have recommended (P.I.R.) methods use the term "probability" in a broad sense, which includes both the sense of probabilities $\text{Prob}(A|\theta)$ defined within the mathematical model E of an experiment (which admit familiar conceptual frequency interpretations), and the sense in which any proposition which is supported by strong evidence is called "highly probable" (the latter sense, according to some writers, need not necessarily be given any frequency interpretation). It is in the latter sense that a high posterior probability seems to be interpreted by some writers who recommend (P.I.R.). Now the present analysis has led to the likelihood function as the mathematical characterization of statistical evidence, and to intrinsic confidence statements as a possible mode of evidential interpretation. In the latter, an intrinsic confidence coefficient plays the role of an index of strength of evidence; such a coefficient is determined in relation to probabilities defined in a mathematical model of an experiment (generally a hypothetical one), but such an index is not itself a probability *of* the confidence statement to which it is attached. However in the broad usage described above, such an index of strength of evidence can be called a probability. Such an index becomes *also* a (posterior) probability in the mathematical sense when a "uniform" prior distribution is specified; but we can alternatively regard the latter formalism as merely a convenient mathematical algorithm for calculating intrinsic confidence sets and their intrinsic confidence coefficients. Under the latter interpretation, the principle of insufficient reason does not constitute an extra-mathematical postulate, but stands just for a traditional mode of calculating and designating intrinsic confidence sets and their coefficients.

11. An Interpretation of Fisher's Fiducial Argument

Fisher's program of developing a theory of fiducial probability is evidently directed to the problem of informative inference. This approach agrees with the traditional one based on the principle of insufficient reason, that statements of informative inference should have the form of probability statements about parameter values; but disagrees concerning appropriateness of adopting the principle of insufficient reason for determination of such statements (Fisher [9]). Such probabilities are defined by a "fiducial argument" whose full scope and essential mathematical structure have not yet been fully formalized. Nevertheless some of the mathematical and extra-mathematical features of this approach seem clear enough for discussion in comparison with the approaches described above.

In experiments with suitable symmetry (or analogous) properties, it has been recognized that fiducial methods coincide in form (although they differ in interpretation) with ordinary or conditional confidence methods. In more complex experiments such a correspondence does not hold; and Fisher has stated that in general fiducial probabilities need not be defined in an actual or actual-conditional experimental frame of reference, but in general may be defined in different conceptually-constructed but appropriate frames of reference. This fact, and the fact that symmetry (or mathematical transformation-group) properties of experimental frameworks play a prominent part in the fiducial argument, suggest that the frames of reference in which fiducial probabilities are to be considered defined may coincide in general with those in which intrinsic confidence methods are defined as in Sections 8 and 9 above.

The claim that fiducial probabilities are probabilities of the same kind discussed by the early writers on probability can perhaps be understood in the same general sense that "posterior probabilities" calculated under the principle of insufficient reason were interpreted in Section 10, that is, a high fiducial probability for a parameter set may be interpreted as an index of strong evidential support for that set. And the claim that such probabilities can be defined and interpreted independently of any extra-mathematical postulate such as (P.I.R.) could be interpreted in the same general sense as in the explication of (P.I.R.) suggested above in which the latter principle does not constitute an extra-mathematical postulate. In the latter interpretation, the fiducial argument would appear to be another purely mathematical algorithm for calculating statements of evidential interpretation.

These interpretations suggest that fiducial probability methods may in general coincide in form as well as in general intention with intrinsic confidence methods (and hence also with those based on (P.I.R.) as interpreted above); and that these approaches may differ only in their verbal and mathematical modes of expression.

The fiducial argument has usually been formulated in a way which does not apply to experiments with discrete sample spaces, nor to experiments lacking suitable symmetry properties. However, it is possible to formulate a version of the fiducial argument compatible with (L) which is free of such restrictions: If $E = (p_{ij})$ is any cyclic-symmetric experiment with a k-point parameter space, consider for each i the sufficient statistic

$$t(j, i) = \begin{cases} j - i + 1, & \text{if the latter is positive,} \\ j - i + 1 + k, & \text{otherwise.} \end{cases}$$

When i is true, the corresponding statistic $t(j, i)$ has the distribution $\text{Prob}(t(J, i) = t|i) = p_{1t}, t = 1, \ldots, k$. The form of the latter distribution is the same for each value of i, and hence can be written $\text{Prob}(t(J, i) = t) = p_{1t}$. (A family of statistics $t(j, i)$ with the latter property is a "pivotal quantity" in the usual terminology of the fiducial argument.) For each possible outcome j of E we define a mathematical probability distribution on the parameter space, the "fiducial distribution" determined by the observed value j, by

$$\text{Prob}(i|j) = \text{Prob}(t(j, I) = t) \equiv p_{1t}, \quad \text{where} \quad t = t(j, i).$$

Using the definition of $t(j, i)$ and the cyclic symmetry of E, this simplifies to

$$\text{Prob}(i|j) = p_{ij}.$$

Thus the fiducial distribution coincides here with the posterior distribution determined from (P.I.R.) and also with the likelihood function itself. The fiducial probability statements here will thus agree in form with posterior probability statements based on (P.I.R.) and also with ordinary confidence statements.

Next, suppose that E' is any experiment with a k-point parameter space, and consider the problem of evidential interpretations of an outcome of E' which determines a likelihood function $L(i)$. Under (L), the evidential meaning of $L(i)$ is the same as if $L(i)$ were determined by an outcome of a simple cyclic-symmetric experiment; and in the latter case, the fiducial statements determined as above would be formally available. Thus it seems appropriate to the general intention of the fiducial approach, and in accord with (L), to define the fiducial distribution by

$$L(i)\bigg/ \sum_{i'=1}^{k} L(i')$$

where $L(i)$ is the likelihood function determined by any outcome of any experiment E' with a k-point parameter space, without restriction on the form of E'. Under this interpretation, the intrinsic confidence statements described in Section 8, and the posterior probability statements described in Section 10, would also correspond formally with fiducial probability statements. Perhaps similar correspondences can be traced in other classes of problems where the fiducial argument takes somewhat different forms.

12. Bayesian Methods in General

As was mentioned in Section 10, Bayesian methods in general entail adoption of (L) for the delimited purpose of characterizing experimental results as actually used in such methods. In particular, for communication of any instance (E, x) of statistical evidence to one who will use or interpret it by Bayesian methods, it is sufficient (and in general necessary) to communicate just the corresponding likelihood function.

Much discussion of the differences between Bayesian methods in general and non-Bayesian statistical methods has centered on the likelihood principle. Hence it is of interest to consider here those distinctions and issues which may separate Bayesian methods in general (apart from (P.I.R.)) from methods and interpretations based on (L) but not (B). Such differences are not related to problems of informative inference, but concern problems of interpretation and/or use of likelihood functions, along with appropriate consideration of other aspects of an experimental situation including background ("prior") information, for scientific and/or utilitarian purposes.

Consider any binary experiment E concerning the statistical hypotheses H_1, H_2, in any situation of inference or decision-making where a certain "conclusion" or decision d would be adopted if the experimental outcome provides evidence supporting H_2 with sufficient strength. Apart from the simplicity of the binary case, evidently many inference situations can be described appropriately in such terms. Then it follows, from (L) and from the discussion of the evidential properties of the statistic λ in the binary case, that there is some critical value λ' such that the decision d would be adopted if and only if the outcome λ of E satisfies $\lambda \geq \lambda'$. The latter formulation can be recognized as appropriate and adopted, with some choice of λ' which seems appropriate in the light of the various aspects and purposes of the inference situation, along with some appreciation of the nature of statistical evidence as such; evidently this can be done by experimenters who adopt the likelihood principle but do not adopt Bayes' principle.

Consider alternatively, in the same situation, another experimenter whose information, judgments, and purposes are generally the same but who adopts and applies Bayes' principle. He will formulate his judgments concerning prior information by specifying numerical prior probabilities p_1, p_2, for the respective hypotheses H_1, H_2. He might formulate his immediate experimental purpose, if it is of a general scientific sort, by specifying that he will adopt the working conclusion d provided the posterior probability q_2 of d is at least as large as a specified number q_2'. Or if his experimental purpose is of a more utilitarian sort, he might specify that he will adopt the decision d provided that $q_2 U_2 \geq q_1 U_1$, where q_1, q_2 are respective posterior probabilities and U_1, U_2 are numerical "utilities" ascribed respectively to non-adoption of d when H_1 is true and to adoption of d when H_2 is true. Each such formulation leads mathematically to a certain critical value λ'' of the statistic λ and to an inference or decision rule of the form: Adopt d provided E yields an outcome $\lambda \geq \lambda''$. Thus there is no difference between the "patterns of inference or decision-making behavior" of Bayesian statisticians and of non-Bayesian statisticians who follow the likelihood principle, at least in situations of relatively simple structure. And, at least for such simple problems, one might say that (L) implies (B) in the very broad and qualitative sense that *use* of statistical evidence as characterized by the likelihood function alone entails that inference- or decision-making behavior will be externally indistinguishable from (some case of) a Bayesian mode of inference.

Some writers have argued that the qualitative features of the Bayesian mode of inference seem plausible and appropriate, but that the specification of definite numerical prior probabilities and the interpretation of specific numerical posterior probabilities seem less clearly appropriate and useful. (This viewpoint has been presented interestingly, with some detailed examples, by Polya [16].) The present writer hopes to see more intensive discussion, with detailed illustration by concrete examples, of the specific contributions which qualitative-Bayesian and quantitative-Bayesian formulations may have to offer to those statisticians who adopt the likelihood principle and interpret likelihood functions directly, making informal judgments and syntheses of the various aspects of inference or decision-making situations.

13. Design of Experiments for Informative Inference

If an experiment is to be conducted primarily for purposes of informative inference, then according to (L) the various specific experimental designs E

which are available are to be appraised and compared just in terms of the likelihood functions they will determine, with respective probabilities, under respective hypotheses, along with consideration of experimental costs of respective designs.

In the case of binary experiments, what is relevant is just the distribution of the statistic λ, defined in any binary experiment, under the respective hypotheses. The simplest specification of a problem of experimental design is evidently that a binary experiment should, with certainty, provide outcomes λ with evidential strength satisfying: $|\lambda| \geq \lambda'$, where λ' is a specified constant; for example, $\lambda' = 99$ indicates that each possible outcome of the experiment is required to have evidential strength associated (as in Section 7) with error-probabilities not exceeding .01. In experimental situations allowing sequential observation, it was shown in reference [3] that such a specification is met efficiently, in terms of required numbers of observations, by a design based on the sampling rule of Wald's sequential probability ratio test (with nominal error-probabilities both .01). If this sequential design is not feasible, some modification of the specified design criterion is indicated. For example, if only non-sequential designs are allowed, and a sample-size is to be determined, then in general one can guarantee only more or less high probabilities, under each hypothesis, that an experimental outcome will have at least the specified evidential strength.

Similarly, to obtain an intrinsic .95 confidence interval for the mean of a normal distribution with unknown variance, of length not exceeding a given number D, an efficient fully-sequential sampling rule is one which terminates when for the first time the .95 confidence interval, computed from all observations as if sampling were non-sequential, has length not exceeding D.

In general, such considerations concerning the design of experiments for informative inference under (L) lead to mathematical questions whose answers will often be found within the mathematical structures of the statistical theories of Fisher, Neyman and Pearson, and Wald, although these theories are typically used and interpreted differently, even for purposes of informative inference. For example, the distributions of the statistic λ in any binary experiment (which under (L) are basic for experimental design but irrelevant to evidential interpretation) are represented mathematically by the a "α, β curve," which represents the binary experiment, and is the focus of attention in the Neyman-Pearson and Wald treatments of binary experiments. More generally, the power functions of various tests admit interpretations relevant to experimental design under (L). And Fisher's asymptotic distribution theory of maximum likelihood estimates can be interpreted, as Fisher has indicated, as describing the asymptotic distributions, under respective hypotheses, of the likelihood function itself (at least in an interval around its maximum).

Clearly the problems of experimental design under (L) are manifold and complex, and their fruitful formulation and solution will probably depend on increased interest in and use of likelihood functions as such. Some of these problems of experimental design coincide in form with design problems as formulated by Bayesian statisticians [17]. Thus there is scope for interesting collaboration here between statisticians with somewhat different over-all view-points.

References

[1] Bahadur, R.R., "Sufficiency and statistical decision functions," *Annals of Mathematical Statistics*, **25** (1954), 423–62.

[2] Barnard, G.A., discussion of C.R. Rao's paper, "Apparent anomalies and irregularities in maximum likelihood estimation," *Bulletin of the International Statistical Institute*, **38** (1961).

[3] Birnbaum, A., "On the foundations of statistical inference; binary experiments," *Annals of Mathematical Statistics*, **32** (1961), 414–35.

[4] Birnbaum, A., "Intrinsic confidence methods," *Bulletin of the International Statistical Institute*, Vol. 39 (to appear), Proceedings of the 33rd Session of the I.S.I., Paris, 1961.

[5] Birnbaum, A., "Confidence curves: an omnibus technique for estimation and testing statistical hypotheses," *Journal of the American Statistical Association*, **56** (1961), 246–9.

[6] Boole, G., *Studies in Logic and Probability*. La Salle, Illinois: Open Court Publishing Company, 1952.

[7] Cohen, M.R. and Nagel, E., *An Introduction to Logic and Scientific Method*. New York: Harcourt, Brace and Company, 1934.

[8] Cox, D.R., "Some problems connected with statistical inference," *Annals of Mathematical Statistics*, **29** (1958), 357–72.

[9] Fisher, R.A., *Statistical Methods and Scientific Inference*. Edinburgh: Oliver and Boyd, 1956.

[10] Fisher, R.A., "The comparison of samples with possibly unequal variances," *Annal of Eugenics*, **9** (1939), 174–80.

[11] Good, I.J., *Probability and the Weighing of Evidence*. New York: Hafner Publishing Company, 1950.

[12] Harrington, G.M., "Statistics' Logic," *Contemporary Psychology*, Vol. 6, No. 9 (September 1961), 304–5.

[13] Jeffreys, H., *Theory of Probability*, Second Edition. London: Oxford University Press, 1948.

[14] Lehmann, E., *Testing Statistical Hypotheses*. New York: John Wiley and Sons, Inc., 1959.

[15] Natrella, M.G., "The relation between confidence intervals and tests of significance," *The American Statistician*, **14** (1960), No. 1, 20–22 and 38.

[16] Polya, G., *Mathematics and Plausible Reasoning*, Volume Two. Princeton: Princeton University Press, 1954.

[17] Raiffa, H. and Schlaifer, R., *Applied Statistical Decision Theory*. Boston: Division of Research, Harvard Business School, 1961.

[18] Savage, L.J., "The foundations of statistics reconsidered," *Proceedings of the Fourth Berkeley Symposium on Mathematical Statistics and Probability*. Berkeley: University of California Press, 1961.

[19] Todhunter, I., *A History of the Mathematical Theory of Probability*. New York: Chelsea Publishing Company, 1949.

[20] Tukey, J., "A survey of sampling from a contaminated distribution," *Contributions to Probability and Statistics*, Ed. by I. Olkin, *et al*. Stanford: Stanford University Press, 1960, 448–85.

[21] Tukey, J., "The future of data analysis," *Annals of Mathematical Statistics*, **33** (1962), 1–67.
[22] Tukey, J., "Conclusions vs. decisions," *Technometrics*, **2** (1960), 423–33.
[23] Wallace, D.L., "Conditional confidence level properties," *Annals of Mathematical Statistics*, **30** (1959), 864–76.
[24] Wallace, D.L., "Intersection region confidence procedures with an application to the location of the maximum in quadratic regression," *Annals of Mathematical Statistics*, **29** (1958), 455–75.
[25] Wilson, E.B., *An Introduction to Scientific Research*. New York: McGraw-Hill Book Company, 1952.

Introduction to
Edwards, Lindman, and Savage (1963) Bayesian Statistical Inference for Psychological Research

William H. DuMouchel
BBN Software Products

Why This Is a Breakthrough Paper

The 1963 paper by Edwards, Lindman, and Savage introduces Bayesian inference to practically minded audiences. It provides a brief history and definition of Bayesian methods and explicates some of the key implications of Bayesian theory for statistical practice. In addition, the paper develops two key Bayesian topics further than had been previously done: the principle of "stable estimation" and the comparison of the Bayesian and classical approaches to "sharp null hypotheses."

The paper is a breakthrough in its practical attitude toward Bayesian methods. As the authors state in their introduction, "... Our real comparison is between such procedures as a Bayesian would employ in an article submitted to the *Journal of Experimental Psychology*, say, and those now typically found in that journal." (The subjunctive "would" was necessary here, since it is hardly to be imagined that articles employing Bayesian analyses could have been submitted to any scientific journal since the classical outlook achieved dominance in the 1920s and 1930s.) The paper is proposing a paradigm shift in thinking for researchers using statistical methods, where as most of the previous Bayesian literature had been oriented toward professional statisticians.

At the time the paper was written, Edwards was a professor and Lindman a graduate student in the psychology department, and Savage a professor in the department of mathematics at the University of Michigan. Edwards had begun research into how people make decisions in the face of uncertainty [see, for example, Edwards (1962)]. Savage was known as one of the principal founders of the newly revived Bayesian school of inference ever since the 1954 publication of *The Foundations of Statistics*. According to Edwards (personal communication), the project originated with a suggestion by Edwards that

Lindman write up a brief description of Bayesian statistics. After they took the resulting manuscript to Savage to get his advice, Savage suggested that they all collaborate on a revised version. A year of weekly meetings resulted in the paper to follow.

The paper falls short of being a handbook: It is not a Bayesian version of Fisher's 1925 classic, *Statistical Methods for Research Workers* (more's the pity!). But perhaps the authors were hoping to clear the way for such an ambitious goal, in that the philosophical and practical implications of Bayesianism are developed and defended.

Review of Contents

The paper itself consists of 10 unnumbered sections. The first two sections, an untitled Introduction and the other titled "Elements of Bayesian Statistics," respectively, provide a superb eight-page tutorial for the Bayesian novice. There is a bit of history, a bit of probability theory, and lots of discussion and interpretation of the personal definition of probability. The authors draw an interesting distinction between personally subjective probabilities and probabilities that, though personal, may also be relatively public, in the sense that a probabilistic model is accepted by the majority of scientists in the field of application. They even go so far as to suggest that Bayesian *statistics*, as opposed to other applications of Bayesian or subjective probability, is characterized by a relatively public definition of $P(D|H)$, where D is the observed data and H a hypothesized theoretical fact or state of nature. The subjective emphasis in Bayesian statistics enters through $P(H)$.

The next section called "Principle of Stable Estimation," is a tour de force. It plunges unhesitatingly into what is perhaps the thorniest objection to the use of Bayesian methods: If prior opinions can differ from one researcher to the next, what happens to scientific objectivity in data analysis? What follows is a unique blend of philosophy, homey example, practical science, axiomatics, and mathematical proofs complete with epsilons and deltas.

The theory is inspired by the fact, known as far back as Laplace and Edgeworth, that as the sample size goes to infinity, the information in the likelihood will eventually totally dominate virtually any prior distribution. But such an appeal to asymptotics would be wholly inappropriate for this audience of practical researchers, so the authors take a very different tack to explaining "when prior distributions can be regarded as essentially uniform." Their goal is to show you how you yourself can check whether your uncertainty about the exact form of your prior matters in a particular application. They use as an example what you might conclude upon taking your temperature with a household fever thermometer. Why are you satisfied with the temperature reading without agonizing over the exact form of your prior?

Suppose that $u(\lambda)$ is your prior density of your true temperature λ, and $u(\lambda|x)$ is the posterior density of λ after the thermometer reads x. Suppose that $w(\lambda|x)$ is the posterior density you would have had, had your prior distribution been uniform [$u(\lambda)$ constant over a wide range of λ]. Then the principle of stable estimation states that for $w(\lambda|x)$ to be a very good approximation to $u(\lambda|x)$, "it suffices that your actual prior density change gently in the region favored by the data and not itself too strongly favor some other region." The next two pages provide a rigorous probabilistic interpretation and proof of this assertion. The authors provide three assumptions, in the form of separate inequalities for each of $w(\lambda|x)$, $u(\lambda)$, and $u(\lambda|x)$, that, when true, imply that $w(\lambda|x)$ is practically equivalent to $u(\lambda|x)$.

Next the authors go back to the fever thermometer example and, after showing that the three assumptions are valid in that case, they interpret the implications of the mathematically derived assumptions.

The section finishes with an illuminating list of five situations in which the exact form of your prior distribution *will* be crucial. In such cases, you have no choice but to formulate your prior opinions as best as you can and proceed. They paraphrase de Finetti's characterization of non-Bayesians who are reluctant to do so: "We see that it is not secure to build on sand. Take away the sand, we shall build on the void."

The last two sentences of the section are worth quoting

> The method of stable estimation might casually be described as a procedure for ignoring prior opinions. Actually, far from ignoring prior opinion, stable estimation exploits certain well-defined features of prior opinion and is acceptable only insofar as those features are really present.

The principle of stable estimation has some features in common with the proposals of some Bayesian statisticians [Jeffreys (1939), Zellner (1971), Box and Tiao, (1973), Bernardo (1979, 1980)], who recommend various "reference priors" that can be used routinely in situations where there is not much prior information and that will help Bayesian methods gain acceptance among classical statisticians. But the last quotation should make it clear that stable estimation was not intended as a way to avoid elicitation of a prior distribution or to help the Bayesian come up with the results of frequentistic theory.

The next two sections, "A Smattering of Bayesian Distribution Theory" and "Point and Interval Estimation," contain a now-standard textbook explanation of conjugate families [see, for example, Lindley (1965) or Box and Tiao (1973)], with the only example presented that of a normally distributed measurement with the variance known. They remind the reader that the stable estimation calculations can be used to check whether a conjugate prior approximates your true prior well enough.

The section "Introduction to Hypothesis Testing" reviews the classical theory and various philosophies of hypothesis testing. After a decision theoretic overview, three hypothetical examples are used to bring out points associated with various concepts of testing. The examples (comparing two teaching machines, testing for the presence of ESP, and weighing a suitcase to see if it meets an airline's weight limit) are well chosen to illustrate three quite different objectives of hypothesis testing. In the first example, the null hypothesis has much less credibility than in the second example. The authors note that in the last example, stable estimation leads exactly to a one-tailed classical significance level, while no Bayesian procedure yet known looks like a two-tailed test. (However, see the discussion below.) Here also is contained the now-famous definition of the "interocular traumatic test, you know what the data mean when the conclusion hits you between the eyes," which the authors attribute to J. Berkson. However, the main purpose of this section seems to be to warm up the reader for the next section, "Bayesian Hypothesis Testing."

That section could stand alone as a major paper in its own right. Perhaps the authors got carried away with their subject; some readers of *Psychological Review* might have preferred to retitle it "More than You ever Wanted to Know About Bayesian Hypothesis Testing." It is easy to understand how they might have become carried away. Nothing grates on the typical Bayesian's nerves so much as the current dominance, in statistical practice, of classical hypothesis testing. Alas, in spite of the best efforts of Edwards, Lindman and Savage, this dominance only solidified over the quarter century following their paper. By now thoughtful statisticians of all stripes are appalled at how prevalent is the notion that statistical practice consists of testing and not much else. The Bayesian hypothesis tests discussed here were among many developed first by Jeffreys (1939) and later given theoretical elaboration by Raiffa and Schlaifer (1961).

The section starts out with definitions and use of the terms "odds" and "likelihood ratios," which are central to the Bayesian approach. Rolling a die to test for fairness is used as an example. The relation of Bayesian to classical decision theory is developed by deriving the likelihood ratio decision rule of classical theory. The authors claim that almost all Bayesian testing procedures could be developed from this classical point of view, but that unfortunately, "Classical statistics tends to divert attention from (the likelihood ratio) to the two conditional probabilities of making errors." Most of the rest of the section is devoted to documenting how the usual classical procedures based on an analysis of type I and type II errors often seriously conflict with the (to these authors obviously correct) analysis based on likelihood ratios.

They are being a bit disingenuous here. The paper defines the likelihood ratio as $L(A|D) = P(D|A)/P(D|\bar{A})$, where D is observed data and A a hypothesis. In the classical simple vs. simple testing situation, both A and $\bar{A}$ are a single point, and this is the only way that likelihood ratios occur in classical inference. If A consists of a set of parameter values, $P(D|A)$ must be defined as

$$P(D|A) = \int P(D|\lambda)p(\lambda|A)\,d\lambda,$$

where $p(\lambda|A)$ is the prior distribution of the parameter λ under the hypothesis

A. In situations where $L(A|D)$ is formed as the ratio of two averaged predictive densities, averaged with respect to two prior distributions $p(\lambda|A)$ and $p(\lambda|\bar{A})$, the later literature prefers the term *Bayes factor* to "likelihood ratio."

The focus is on the so-called "sharp null hypothesis," in which the null hypothesis is that a parameter is concentrated near a special value, whereas the alternative theory predicts no particular value for the parameter, just a relatively broad distribution of alternatives. In such situations, as Lindley (1957) proved, for any classical significance level for rejecting the null hypothesis (no matter how small) and for any likelihood ratio (Bayes factor) in favor of the null hypothesis (no matter how large), there exists a datum significant at that level and with that likelihood ratio. This fact is usually referred to as Lindley's paradox or Jeffrey's (1939, p. 194) paradox.

The authors give lots of examples to help the reader's intuition grasp this shocking state of affairs, and they sum up the argument as

hinging on assumptions about the prior distribution under the alternative hypothesis. The classical statistician usually neglects that distribution—in fact, denies its existence. He considers how unlikely a t as far from 0 as 1.96 is if the null hypothesis is true, but he does not consider that a t as close to 0 as 1.96 may be even less likely if the null hypothesis is false.

Thus, classical procedures are quite typically, from a Bayesian point of view, far too ready to reject the null hypothesis.

There follow two extended examples, one on testing a binomial probability and one on testing a normal mean. Each uses a plausible example of a research problem in psychology, one based on testing whether a certain motor skill activity prompts the subject to move to the right or left, the other based on testing whether Weber's law on tone discrimination applies equally in a lighted or dark room. A careful discussion of what a reasonable prior distribution might be in both cases leads to the conclusions that Lindley's paradox would indeed crop up in these examples. The discussions are backed up by tables calculating just how severe the discrepancy would be between the Bayesian and classical conclusions for various sample sizes and outcomes. Of course, the extent of the discrepancy depends somewhat on the exact form of the prior distribution assumed. If stable estimation applies, it can be used to justify the comparisons. But even if it does not, the authors derive lower bounds L_{min} on the likelihood ratio for a barely (classically) significant result and show that even in this worst-case prior the posterior odds carry a far different message than the usual interpretation of a significant deviation from the null hypothesis.

A broader discussion follows. The interesting point is made that a Bayesian hypothesis test can add extensive support to the null hypothesis whenever the likelihood ratio is large. The classical test can only reject hypotheses, and it is not clear just what sort of evidence classical statistics would regard as a strong confirmation of a null hypothesis.

A discussion of testing a multidimensional parameter points out what is similar and what is different about the case of many parameters. The biggest difference is that the conditions for allowing stable estimation are much less likely to apply. The geometry of many dimensions makes the uniform distribution much less attractive as a realistic prior. They give the example of an ANOVA situation involving three factors each at four levels. There are 27 degrees of freedom for two-factor interactions. If you didn't think that the model was additive, might you approximate your prior by a uniform distribution over 27 dimensions? To do so would imply that even if you found out any 26 of the parameters, you would not feel competent to guess the last one to within several standard errors. In the years since, the emergence and frequent use of systems of hierarchical priors, as in Lindley and Smith (1972) or DuMouchel and Harris (1983), has greatly improved our understanding and ability to handle these types of problems. Diaconis and Freedman (1986) show that stable estimation cannot be relied on in infinite-dimensional spaces.

The section closes with an excellent summary called "Some Morals About Testing Sharp Null Hypotheses." It also includes an attempt to answer this question: "If classical significance tests so frequently reject true null hypotheses without real evidence, why have they survived so long and so dominated certain empirical sciences?" Four possible answers are suggested. In brief, they are the following. (1) Often test results are so significant that even a Bayes procedure would arrive at the same conclusion. (2) Investigators often require more than .05 or .01 significance if the problem is important. (3) Replication of experiments is rare, so the unfairly rejected hypothesis is not given a chance to make a comeback. (4) Often the null hypothesis is so unbelievable that the more likely it is to be rejected the better.

With the benefit of hindsight, further discussion is in order. First, the sociological context of most hypothesis testing virtually dictates that procedures that encourage the rejection of null hypotheses will be favored. Researchers are usually looking hard for an effect, and statistical significance is often viewed merely as a hurdle to be overcome. It may be hard to publish without it: The phenomenon of publication bias and the "file drawer problem" was noted by Rosenthal (1979) and modeled by Iyengar and Greenhouse (1988). Procedures with impeccable classical statistics credentials, like Fisher's exact test for two by two tables and simultaneous confidence intervals, have met with resistance among practitioners on the grounds that they are too conservative. Given this situation, a Bayesian crusade for making it harder to reject null hypotheses may amount to tilting at windmills.

Second, the later Bayesian literature is divided on the acceptability of Lindley's paradox when testing sharp null hypotheses. Smith and Spiegelhalter (1980), among many others representing the mainstream Bayesian view, agree with Edwards, Lindman, and Savage, but to varying degrees, Dempster (1973), Akaike (1978), Atkinson (1978), Bernardo (1980), and Jaynes (1980) have either rejected this approach to sharp null hypotheses or found Bayesian rationales for test procedures that behave like classical two-tailed tests. Some of these authors report that classical statisticians often require a lower p-value if the sample size is very large, a variant of Edwards. Lindman and Savage's explanation (2) above, since the Lindley paradox has the greatest effect for large samples, when the precision of the data is high and the value of α in Eq. (18) is small. See Shafer (1982) and Berger and Delampady (1987) for more recent treatments and reviews of the controversy.

The next section, titled "Likelihood Principle," comes on the heels of Birnbaum's famous 1962 paper on the same subject. This principle is a cornerstone of the Bayesian argument. The irrelevance of stopping rules in sequential experimentation is another key difference between the Bayesian and classical philosophy and the authors develop the ideas clearly and succinctly.

Finally, only a one-page section, "In Retrospect," delivering a final pep talk in favor of the Bayesian approach, and the references remain. The last two sentences deliver the authors' two punch lines

> Estimation is best when it is stable. Rejection of a null hypothesis is best when it is interocular.

Impact of the Paper

The Science and Social Science Citation Indices for the last few years show the size and breadth of the audience this paper is still reaching. These indices show an average of 11 citations per year over a seven-year span. The 79 citations were distributed percentagewise by field approximately as follows:

Statistics	30%
Psychology	20%
Other social science	15%
Medicine	15%
Economics/Business	10%
Other technical fields	10%

Lindley (1980) writes,

> Although this paper is well-known, its influence has not, I think, been what the authors had hoped.... Designed as a paper to guide practicing psychologists on how to use personalistic ideas in lieu of significance tests and other frequentistic methods, [... the paper lacks the requisite] operational techniques.

The authors surely had few illusions about removing the significance test recipe from the menu without coming up with others to replace it. But they may have been overly optimistic as to how long it would take for Bayesian recipes to appear.

They could not have anticipated the present role of computer packages in the analysis of scientific data. The near-ubiquitous use of standard computer programs whenever data are analyzed and the increasing appearance of large data sets and complicated computer-intensive methods have created a thrust away from the personalized analysis that a Bayesian elicitation of prior distributions seems to require. On large research projects, the statistical analyses may be run by a team member who lacks deep knowledge of either the statistics or the science, and who would not be the natural person to supply a prior distribution, even if the computer program were capable of processing one.

Although the present role of computers in data analysis has arguably made the situation more difficult for Bayesian adherents, there remains the hope that computers will eventually help with a solution. As their interactions with users become more sophisticated, the potential for computer elicitation and use of prior information about the subject matter of the analysis arises. It remains to be seen whether a future generation of Bayesian computer packages will meet this challenge. See DuMouchel (1988) for one such proposal.

As it is, the impact of this paper has perhaps been greater on the psychology of statistical inference (and decision) than on statistical inference in psychology. Psychology researchers have been extensively interested in a theory of behavior that emphasizes the subjective bases of inference and decision. Research bearing on that theory, ongoing as the paper was written has exploded since then. The work of Kahneman and Tversky (1973 and many later papers) and of the numerous others who worked with Edwards at Michigan has clearly shown that such theories do not, in fact, describe typical human behavior, as it is. But those same theories, treated as prescriptive rather than descriptive, are the basis for the emerging field of *decision analysis*. Edwards continued to do research on how people do and should make decisions in the face of uncertainty, with a recent focus on the use of these ideas for prescriptive purposes. His 1986 book, *Decision Analysis and Behaviorial Research*, coauthored with von Winterfeldt, reviews and synthesizes both the descriptive and prescriptive uses of Bayesian decision theory.

Many biographical comments and sketches have appeared that give perspective on Savage's life and work. In particular, see Berry (1972), de Finetti (1972), Dubins (1976), Fienberg and Zellner (1975), Kruskal (1978), Lindley (1979, 1988), and von Winterfeldt and Edwards (1986, pp. 563–565). The Savage Memorial Volume [Ericson (1981)] also contains extensive biographical material.

Savage often acknowledged the influence of others, especially that of de Finetti, on his work. Lindley (1980, p. 44) quotes Savage

> I am reminded when I see expressions like post- and pre-Savage that you are turning too much limelight on me. A reader familiar with Ramsey, Jeffreys, de Finetti, and Good has not really so much to learn from Savage. I think, as you seem to, that my main contribution has been to emphasize that the theory of subjective probability finds its place in a natural theory of rational behavior.

On p. 532 of the present paper, references are made to Ramsey (1931), de Finetti (1930, 1937), Jeffreys (1931, 1939), and Good (1950, 1960) as being pioneering Bayesian papers. The introduction by Barlow in this volume explains why de Finetti (1937) was a breakthrough paper.

Besides Edwards, Lindman, and Savage (1963), Savage's other influential work included two papers coauthored with Milton Friedman (1948, 1952) on utility theory and a paper with Paul Halmos (1949) on the mathematical foundations of sufficiency. Two well-known articles developing his ideas on the foundations of inference are Savage (1951, 1961), and his definitive work in this area is his 1954 book, *The Foundations of Statistics*, which contains an axiomatic development of the personalistic approach to statistical inference. For this, Lindley (1980) says, "Savage was the Euclid of statistics." It is ironic that Savage spent the last half of that book trying to use his axioms to derive the conventional statistical procedures of the time. Of course, he failed, and it was only in his later writings, perhaps best exemplified by Edwards, Lindman, and Savage in their 1963 paper, that he makes plain just how contradictory these two statistical paradigms are. In a new preface to the 1972 edition of *The Foundations of Statistics*, he wrote, "Freud alone could explain how the rash and unfulfilled promise (made early in the first edition, to show how frequentistic ideas can be justified by means of personalistic probabilities) went unamended through so many revisions of the manuscript."

Savage's work in probability theory with Lester Dubins culminated in the 1965 book *How to Gamble If You Must: Inequalities for Stochastic Processes*, which showed how many probability problems can be reformulated (and more easily solved) as gambling problems. His 1970 Fisher memorial lecture "On rereading Fisher" (published posthumously in 1976) has been called a classic by statisticians of every philosophy. Finally, much of the biographical material referenced above emphasizes his powerful intellectual presence, his brilliance as a consultant, and his magnetic personality.

It is far beyond the scope of this introduction to review the advances made in Bayesian theory and applications during the 20 years since Savage's death. The Valencia series of conference proceedings edited by Bernardo et al. (1980, 1985, 1988), as well as Kanji (1983), are collections exemplifying the progress that has been made.

Ward Edwards is currently professor at the Social Science Research Institute at the University of Southern California. Harold Lindman is professor of psychology at Indiana University. Leonard "Jimmie" Savage was professor of statistics at Yale University when he died on November 1, 1971 at the age of 54.

Acknowledgment

I would like to thank W. Edwards, D. Lindley, R. Olshen, S. Stigler, the editors, and a referee for their comments on an earlier draft of this paper.

References

Akaike, H. (1978). A Bayesian analysis of the minimum AIC procedure, *Ann. Inst. Statist. Math., Tokyo*, **30**, 9–14.

Atkinson. A.C. (1978). Posterior probabilities for choosing a regression model, *Biometrika*, **65**, 39–48.

Berger, J.O., and Delampady, M. (1987). Testing precise hypotheses. *Statist. Sci.*, **2**, 317–352 (with discussion).

Berry, D.A. (1972). Letter to the editor. *Amer. Statist.* **26**, 47.

Bernardo, J.M. (1979). Reference posterior distributions Bayesian inference, *J. Roy. Statist. Soc., Ser. B*, **41**, 113–147 (with discussion).

Bernardo, J.M. (1980). A Bayesian analysis of classical hypothesis testing, *in Bayesian Statistics: Proceedings of First International Meeting* (J.M. Bernardo, M.H. DeGroot, D.V. Lindley, and A.F.M. Smith, eds.). University Press, Valencia, Spain, pp. 605–647 (with discussion).

Bernardo, J.M., DeGroot, M.H., Lindley, D.V., and Smith. A.F.M. (eds.) (1980). *Bayesian Statistics: Proceedings of First International Meeting*. University Press, Valencia, Spain.

Bernardo, J.M., DeGroot, M.H., Lindley, D.V., and Smith. A.F.M. (eds.) (1985). *Bayesian Statistics*, Vol. 2. North Holland, Amsterdam.

Bernardo, J.M., DeGroot, M.H., Lindley, D.V., and Smith. A.F.M. (eds.) (1988). *Bayesian Statistics*, Vol. 3. Oxford University Press.

Birnbaum, A. (1962). On the foundations of statistical inference, *J. Amer. Statist. Assoc.*, **57**, 269–306.

Box, G.E.P. and Tiao, G.C. (1973). *Bayesian Inference in Statistical Analysis*. Addison-Wesley, Reading, Mass.

de Finetti, B. (1930). Fondamenti logici del ragionamento probabilistico, *Boll. Un. mat. Ital, Ser. A*, **9**, 258–261.

de Finetti, B. (1937). La prévision: Ses lois logiques, ses sources subjectives, *Ann. Inst. Henri Poincaré*, **7**, 1–68.

de Finetti, B. (1972). *Probability, Induction and Statistics*. Wiley, New York, pp. v–vi.

Dempster, A.P. (1973). The direct use of likelihood for significance testing, *Proceedings of Conference on Foundational Questions in Statistical Inference* (Barndorff-Nielson, P. Blaesild, and G. Schou, eds.). University of Aarhus, pp. 335–352.

Diaconis, P. and Friedman, D. (1986). On the consistency of Bayes estimates, *Ann. Statist.* **14**, 1–67 (with discussion).

Dubins, L.E., and Savage, L.J. (1965). *How to Gamble If You Must: Inequalities for Stochastic Processes*. McGraw-Hill, New York.

Dubins, L.E., (1976). Preface to *Inequalities for Stochastic Processes: How to Gamble If You Must* by L.E. Dubins and L.J. Savage. Dover, New York.

DuMouchel, W. (1988). A Bayesian model and a graphical elicitation procedure for the problem of multiple comparisons, in *Bayesian Statistics*, vol. 3 (Bernardo et al. eds.). Oxford University Press.

DuMouchel, W., and Harris, J.E. (1983). Bayes methods for combining the results of cancer studies in humans and other species, *J. Amer. Statist. Assoc.*, **78**, 293–315 (with discussion).

Edwards, W. (1962). Subjective probabilities inferred from decisions, *Psychol. Rev.*, **69**, 109–135.

Ericson, W.A. (ed.) (1981) *The Writings of Leonard Jimmie Savage—A Memorial Selection*. American Statistical Association and Institute of Mathematical Statistics, Washington.

Fienberg, S., and Zellner, A. (eds.) (1975). *Studies in Bayesian Econometrics and Statistics in Honor of Leonard J. Savage.* North-Holland, Amsterdam, pp. 3–4.

Fisher, R.A. (1925). *Statistical Methods for Research Workers.* Oliver and Boyd, London.

Friedman, M., and Savage, L.J. (1948). The utility analysis of choices involving risk, *J. Political Economy*, **56**, 279–304.

Friedman, M., and Savage. L.J. (1952). The expected-utility hypothesis and the measurement of utility, *J. Political Economy*, **60**, 463–474.

Good, I.J. (1950). *Probability and the Weighing of Evidence.* Hafner, New York.

Good, I.J. (1960). Weight of evidence, corroboration, explanatory power, information and the utility of experiments, *J. Roy. Statist. Soc., Ser. B*, **22**, 319–331.

Halmos, P.R., and Savage, L.J. (1949). Application of the Radon–Nikodym theorem to the theory of sufficient statistics, *Ann. Math Statist.* **20**, 225–241.

Iyengar, S. and Greenhouse, J.B. (1988). Selection models and the file drawer problem, *Statist. Sci.*, **3**, 109–135 (with discussion).

Jaynes, E.T. (1980). Discussion of the paper by Bernardo, in *Bayesian Statistics: Proceedings of First International Meeting* (J.M. Bernardo, M.H. DeGroot, D.V. Lindley, and A.F.M. Smith. eds.). University Press, Valencia, Spain, pp. 618–629.

Jeffreys, H. (1931). *Scientific Inference.* Cambridge University Press, England.

Jeffreys, H. (1939). *Theory of Probability.* Clarendon Press, Oxford.

Kahneman, D., and Tversky, A. (1973). On the psychology of prediction, *Psychol. Rev.*, **80**, 237–251.

Kanji, G.K. (ed.) (1983). Proceedings of the 1982 I.O.S. Annual Conference on practical Bayesian statistics, *The Statistician*, **32**, 1–278.

Kruskal, W. (1978). Leonard Jimmie Savage, in *International Encyclopedia of Statistics* (W. Kruskal, and J. Tanur, eds.). The Free Press, New York, pp. 889–892.

Lindley, D.V. (1957). A statistical paradox, *Biometrika*, **44**, 187–192.

Lindley, D.V. (1965). *Introduction to Probability and Statistics from a Bayesian Viewpoint.* Cambridge University Press.

Lindley, D.V. (1979). Savage, Leonard Jimmie, in *International Encyclopedia of the Social Sciences*, **18**, Biographical Supplement (D.L. Sills. ed.). The Free Press, New York.

Lindley, D.V. (1980). L.J. Savage—His work in probability and statistics, *Ann. Statist.*, **8**, 1–24 [reprinted in *The Writings of Leonard Jimmie Savage—A Memorial Selection* (W.A. Ericson, ed.). American Statistical Association, Washington 1981.

Lindley, D.V. (1988). Savage, Leonard J. in *Encyclopedia of Statistical Sciences*, Vol. 8 (S. Kotz, N. Johnson, and C.B. Read eds.). Wiley-Interscience, New York.

Lindley, D.V. and Smith, A.F.M. (1972). Bayes estimates for the linear model, *J. Roy. Statist. Soc., Ser. B*, **34**, 1–41 (with discussion).

Raiffa, H., and Schlaifer, R. (1961). *Applied Statistical Decision Theory.* Harvard University Graduate School of Business Administration, Cambridge, Mass.

Ramsey, F.P. (1931). Truth and probability (1926) and Further considerations (1928), in *The Foundations of Mathematics and Other Essays.* Harcourt Brace, New York.

Rosenthal, R (1979). The "file drawer problem" and tolerance for null results, *Psychol. Bull.*, **86**, 638–641.

Savage, L.J. (1951). The theory of statistical decision, *J. Amer. Statist. Assoc.*, **46**, 55–67.

Savage, L.J. (1954). *The Foundations of Statistics.* Wiley, New York.

Savage, L.J. (1961). The foundations of statistics reconsidered, in *Proceedings of 4th Berkeley Symposium on Mathematical and Statistical Probability*, Vol. 1. University of California Press, Berkeley, pp. 575–586.

Savage, L.J. (1972). *The Foundations of Statistics*, 2nd ed. Dover, New York.

Savage, L.J. (1976). On rereading Fisher. J.W. Pratt ed., *Ann Statist.*, , 441–500.

Shafer, G. (1982). Lindley's paradox, *J. Amer. Statist. Assoc.*, **77**, 325–351 (with discussion).

Smith, A.F.M., and Spiegelhalter, D.J. (1980). Bayes factors and choice criteria for linear models, *J. Roy. Statist. Soc., Ser. B*, **42**, 213–220.

von Winterfeldt, D., and Edwards, W. (1986). *Decision Analysis and Behavioral Research*. Cambridge University Press.

Zellner, A. (1971). *An Introduction to Bayesian Inference in Econometrics*. Wiley, New York.

Bayesian Statistical Inference for Psychological Research[1]

Ward Edwards, Harold Lindman, and Leonard J. Savage
University of Michigan

Abstract

Bayesian statistics, a currently controversial viewpoint concerning statistical inference, is based on a definition of probability as a particular measure of the opinions of ideally consistent people. Statistical inference is modification of these opinions in the light of evidence, and Bayes' theorem specifies how such modifications should be made. The tools of Bayesian statistics include the theory of specific distributions and the principle of stable estimation, which specifies when actual prior opinions may be satisfactorily approximated by a uniform distribution. A common feature of many classical significance tests is that a sharp null hypothesis is compared with a diffuse alternative hypothesis. Often evidence which, for a Bayesian statistician, strikingly supports the null hypothesis leads to rejection of that hypothesis by standard classical procedures. The likelihood principle emphasized in Bayesian statistics implies, among other things, that the rules governing when data collection stops are irrelevant to data interpretation. It is entirely appropriate to collect data until a point has been proven or disproven, or until the data collector runs out of time, money, or patience.

The main purpose of this paper is to introduce psychologists to the Bayesian outlook in statistics, a new fabric with some very old threads. Although

[1] Work on this paper was supported in part by the United States Air Force under Contract AF 49 (638)-769 and Grant AF-AFOSR-62-182, monitored by the Air Force Office of Scientific Research of the Air Force Office of Aerospace Research (the paper carries Document No. AFOSR-2009); in part under Contract AF 19(604)-7393, monitored by the Operational Applications Laboratory, Deputy for Technology, Electronic Systems Division, Air Force Systems Command; and in part by the Office of Naval Research under Contract Nonr 1224(41). We thank H. C. A. Dale, H. V. Roberts, R. Schlaifer, and E. H. Shuford for their comments on earlier versions.

this purpose demands much repetition of ideas published elsewhere, even Bayesian specialists will find some remarks and derivations hitherto unpublished and perhaps quite new. The empirical scientist more interested in the ideas and implications of Bayesian statistics than in the mathematical details can safely skip almost all the equations; detours and parallel verbal explanations are provided. The textbook that would make all the Bayesian procedures mentioned in this paper readily available to experimenting psychologists does not yet exist, and perhaps it cannot exist soon; Bayesian statistics as a coherent body of thought is still too new and incomplete.

Bayes' theorem is a simple and fundamental fact about probability that seems to have been clear to Thomas Bayes when he wrote his famous article published in 1763 (recently reprinted), though he did not state it there explicitly. Bayesian statistics is so named for the rather inadequate reason that it has many more occasions to apply Bayes' theorem than classical statistics has. Thus, from a very broad point of view, Bayesian statistics dates back at least to 1763.

From a stricter point of view, Bayesian statistics might properly be said to have begun in 1959 with the publication of *Probability and Statistics for Business Decisions*, by Robert Schlaifer. This introductory text presented for the first time practical implementation of the key ideas of Bayesian statistics: that probability is orderly opinion, and that inference from data is nothing other than the revision of such opinion in the light of relevant new information. Schlaifer (1961) has since published another introductory text, less strongly slanted toward business applications than his first. And Raiffa and Schlaifer (1961) have published a relatively mathematical book. Some other works in current Bayesian statistics are by Anscombe (1961), de Finetti (1959), de Finetti and Savage (1962), Grayson (1960), Lindley (1961), Pratt (1961), and Savage et al. (1962).

The philosophical and mathematical basis of Bayesian statistics has, in addition to its ancient roots, a considerable modern history. Two lines of development important for it are the ideas of statistical decision theory, based on the game-theoretic work of Borel (1921), von Neumann (1928), and von Neumann and Morgenstern (1947), and the statistical work of Neyman (1937, 1938b, for example), Wald (1942, 1955, for example), and others; and the personalistic definition of probability, which Ramsey (1931) and de Finetti (1930, 1937) crystallized. Other pioneers of personal probability are Borel (1924), Good (1950, 1960), and Koopman (1940a, 1940b, 1941). Decision theory and personal probability fused in the work of Ramsey (1931), before either was very mature. By 1954, there was great progress in both lines for Savage's *The Foundations of Statistics* to draw on. Though this book failed in its announced object of satisfying popular non-Bayesian statistics in terms of personal probability and utility, it seems to have been of some service toward the development of Bayesian statistics. Jeffreys (1931, 1939) has pioneered extensively in applications of Bayes' theorem to statistical problems. He is one of the founders of Bayesian statistics, though he might reject identification with the viewpoint of this paper because of its espousal of personal probabilities. These two, inevitably inadequate, paragraphs are our main attempt in this paper to give credit where it is due. Important authors have not been listed, and for those that have been, we have given mainly one early and one late reference only. Much more information and extensive bibliographies will be found in Savage et al. (1962) and Savage (1954, 1962a).

We shall, where appropriate, compare the Bayesian approach with a loosely defined set of ideas here labeled the classical approach, or classical statistics. You cannot but be familiar with many of these ideas, for what you learned about statistical inference in your elementary statistics course was some blend of them. They have been directed largely toward the topics of testing hypotheses and interval estimation, and they fall roughly into two somewhat conflicting doctrines associated with the names of R.A. Fisher (1925, 1956) for one, and Jerzy Neyman (e.g. 1937, 1938b) and Egon Pearson for the other. We do not try to portray any particular version of the classical approach; our real comparison is between such procedures as a Bayesian would employ in an article submitted to the *Journal of Experimental Psychology*, say, and those now typically found in that journal. The fathers of the classical approach might not fully approve of either. Similarly, though we adopt for conciseness an idiom that purports to define *the* Bayesian position, there must be at least as many Bayesian positions as there are Bayesians. Still, as philosophies go, the unanimity among Bayesians reared apart is remarkable and an encouraging symptom of the cogency of their ideas.

In some respects Bayesian statistics is a reversion to the statistical spirit of the eighteenth and nineteenth centuries; in others, no less essential, it is an outgrowth of that modern movement here called classical. The latter, in coping with the consequences of its view about the foundations of probability which made useless, if not meaningless, the probability that a hypothesis is true, sought and found techniques for statistical inference which did not attach probabilities to hypotheses. These intended channels of escape have now, Bayesians believe, led to reinstatement of the probabilities of hypotheses and a return of statistical inference to its original line of development. In this return, mathematics, formulations, problems, and such vital tools as distribution theory and tables of functions are borrowed from extrastatistical probability theory and from classical statistics itself. All the elements of Bayesian statistics, except perhaps the personalistic view of probability, were invented and developed within, or before, the classical approach to statistics; only their combination into specific techniques for statistical inference is at all new.

The Bayesian approach is a common sense approach. It is simply a set of techniques for orderly expression and revision of your opinions with due regard for internal consistency among their various aspects and for the data. Naturally, then, much that Bayesians say about inference from data has been said before by experienced, intuitive, sophisticated empirical scientists and statisticians. In fact, when a Bayesian procedure violates your intuition, reflection is likely to show the procedure to have been incorrectly applied.

If classically trained intuitions do have some conflicts, these often prove transient.

Elements of Bayesian Statistics

Two basic ideas which come together in Bayesian statistics, as we have said, are the decision-theoretic formulation of statistical inference and the notion of personal probability.

Statistics and Decisions

Prior to a paper by Neyman (1938a), classical statistical inference was usually expressed in terms of justifying propositions on the basis of data. Typical propositions were: Point estimates; the best guess for the unknown number μ is m. Interval estimates; μ is between m_1 and m_2. Rejection of hypotheses; μ is not 0. Neyman's (1938a, 1957) slogan "inductive behavior" emphasized the importance of action, as opposed to assertion, in the face of uncertainty. The decision-theoretic, or economic, view of statistics was advanced with particular vigor by Wald (1942). To illustrate, in the decision-theoretic outlook a point estimate is a decision to act, in some specific context, as though μ were m, not to assert something about μ. Some classical statisticians, notably Fisher (1956, Ch. 4), have hotly rejected the decision-theoretic outlook.

While Bayesian statistics owes much to the decision-theoretic outlook, and while we personally are inclined to side with it, the issue is not crucial to a Bayesian. No one will deny that economic problems of behavior in the face of uncertainty concern statistics, even in its most "pure" contexts. For example, "Would it be wise, in the light of what has just been observed, to attempt such and such a year's investigation?" The controversial issue is only whether such economic problems are a good paradigm of all statistical problems. For Bayesians, all uncertainties are measured by probabilities, and these probabilities (along with the here less emphasized concept of utilities) are the key to all problems of economic uncertainty. Such a view deprives debate about whether all problems of uncertainty are economic of urgency. On the other hand, economic definitions of personal probability seem, at least to us, invaluable for communication and perhaps indispensable for operational definition of the concept.

A Bayesian can reflect on his current opinion (and how he should revise it on the basis of data) without any reference to the actual economic significance, if any, that his opinion may have. This paper ignores economic considerations, important though they are even for pure science, except for brief digressions. So doing may combat the misapprehension that Bayesian statistics is primarily for business, not science.

Personal Probability

With rare exceptions, statisticians who conceive of probabilities exclusively as limits of relative frequencies are agreed that uncertainty about matters of fact is ordinarily not measurable by probability. Some of them would brand as nonsense the probability that weightlessness decreases visual acuity; for others the probability of this hypothesis would be 1 or 0 according as it is in fact true or false. Classical statistics is characterized by efforts to reformulate inference about such hypotheses without reference to their probabilities, especially initial probabilities.

These efforts have been many and ingenious. It is disagreement about which of them to espouse, incidentally, that distinguishes the two main classical schools of statistics. The related ideas of significance levels, "errors of the first kind," and confidence levels, and the conflicting idea of fiducial probabilities are all intended to satisfy the urge to know how sure you are after looking at the data, while outlawing the question of how sure you were before. In our opinion, the quest for inference without initial probabilities has failed, inevitably.

You may be asking, "If a probability is not a relative frequency or a hypothetical limiting relative frequency, what is it? If, when I evaluate the probability of getting heads when flipping a certain coin as .5, I do not mean that if the coin were flipped very often the relative frequency of heads to total flips would be arbitrarily close to .5, then what do I mean?"

We think you mean something about yourself as well as about the coin. Would you not say, "Heads on the next flip has probability .5" if and only if you would as soon guess heads as not, even if there were some important reward for being right? If so, your sense of "probability" is ours; even if you would not, you begin to see from this example what we mean by "probability," or "personal probability." To see how far this notion is from relative frequencies, imagine being reliably informed that the coin has either two heads or two tails. You may still find that if you had to guess the outcome of the next flip for a large prize you would not lift a finger to shift your guess from heads to tails or vice versa.

Probabilities other than .5 are defined in a similar spirit by one of several mutually harmonious devices (Savage, 1954, Ch. 1–4). One that is particularly vivid and practical, if not quite rigorous as stated here, is this. For you, now, the probability $P(A)$ of an event A is the price you would just be willing to pay in exchange for a dollar to be paid to you in case A is true. Thus, rain tomorrow has probability 1/3 for you if you would pay just $.33 now in exchange for $1.00 payable to you in the event of rain tomorrow.

A system of personal probabilities, or prices for contingent benefits, is inconsistent if a person who acts in accordance with it can be trapped into accepting a combination of bets that assures him of a loss no matter what happens. Necessary and sufficient conditions for consistency are the following, which are familiar as a basis for the whole mathematical theory of probability:

$$0 \leq P(A) \leq P(S) = 1, \qquad P(A \cup B) = P(A) + P(B),$$

where S is the tautological, or universal, event; A and B are any two incompatible, or nonintersecting, events; and $A \cup B$ is the event that either A or B is true, or the union of A and B. Real people often make choices that reflect violations of these rules, especially the second, which is why personalists emphasize that personal probability is orderly, or consistent, opinion, rather than just any opinion. One of us has presented elsewhere a model for probabilities inferred from real choices that does not include the second consistency requirement listed above (Edwards, 1962b). It is important to keep clear the distinction between the somewhat idealized consistent personal probabilities that are the subject of this paper and the usually inconsistent subjective probabilities that can be inferred from real human choices among bets, and the words "personal" and "subjective" here help do so.

Your opinions about a coin can of course differ from your neighbor's. For one thing, you and he may have different bodies of relevant information. We doubt that this is the only legitimate source of difference of opinion. Hence the personal in personal probability. Any probability should in principle be indexed with the name of the person, or people, whose opinion it describes. We usually leave the indexing unexpressed but underline it from time to time with phrases like "the probability for you that H is true."

Although your initial opinion about future behavior of a coin may differ radically from your neighbor's, your opinion and his will ordinarily be so transformed by application of Bayes' theorem to the results of a long sequence of experimental flips as to become nearly indistinguishable. This approximate merging of initially divergent opinions is, we think, one reason why empirical research is called "objective." Personal probability is sometimes dismissed with the assertion that scientific knowledge cannot be mere opinion. Yet, obviously, no sharp lines separate the conjecture that many human cancers may be caused by viruses, the opinion that many are caused by smoking, and the "knowledge" that many have been caused by radiation.

Conditional Probabilities and Bayes' Theorem

In the spirit of the rough definition of the probability $P(A)$ of an event A given above, the conditional probability $P(D|H)$ of an event D given another H is the amount you would be willing to pay in exchange for a dollar to be paid to you in case D is true, with the further provision that all transactions are canceled unless H is true. As is not hard to see, $P(D \cap H)$ is $P(D|H)P(H)$ where $D \cap H$ is the event that D and H are both true, or the intersection of D and H. Therefore,

$$P(D|H) = \frac{P(D \cap H)}{P(H)}, \qquad (1)$$

unless $P(H) = 0$.

Conditional probabilities are the probabilistic expression of learning from experience. It can be argued that the probability of D for you—the consistent you—after learning that H is in fact true is $P(D|H)$. Thus, after you learn that H is true, the new system of numbers $P(D|H)$ for a specific H comes to play the role that was played by the old system $P(D)$ before.

Although the events D and H are arbitrary, the initial letters of Data and Hypothesis are suggestive names for them. Of the three probabilities in Equation 1, $P(H)$ might be illustrated by the sentence: "The probability for you, now, that Russia will use a booster rocket bigger than our planned Saturn booster within the next year is .8." The probability $P(D \cap H)$ is the probability of the joint occurrence of two events regarded as one event, for instance: "The probability for you, now, that the next manned space capsule to enter space will contain three men and also that Russia will use a booster rocket bigger than our planned Saturn booster within the next year is .2." According to Equation 1, the probability for you, now, that the next manned space capsule to enter space will contain three men, given that Russia will use a booster rocket bigger than our planned Saturn booster within the next year is .2/.8 = .25.

A little algebra now leads to a basic form of Bayes' theorem:

$$P(H|D) = \frac{P(D|H)P(H)}{P(D)}, \qquad (2)$$

provided $P(D)$ and $P(H)$ are not 0. In fact, if the roles of D and H in Equation 1 are interchanged, the old form of Equation 1 and the new form can be expressed symmetrically, thus:

$$\frac{P(D|H)}{P(D)} = \frac{P(D \cap H)}{P(D)P(H)}$$

$$= \frac{P(H|D)}{P(H)}, \qquad (3)$$

which obviously implies Equation 2. A suggestive interpretation of Equation 3 is that the relevance of H to D equals the relevance of D to H.

Reformulations of Bayes' theorem apply to continuous parameters or data. In particular, if a parameter (or set of parameters) λ has a prior probability density function $u(\lambda)$, and if x is a random variable (or a set of random variables such as a set of measurements) for which $v(x|\lambda)$ is the density of x given λ and $v(x)$ is the density of x, then the posterior probability density of λ given x is

$$u(\lambda|x) = \frac{v(x|\lambda)u(\lambda)}{v(x)}. \qquad (4)$$

There are of course still other possibilities such as forms of Bayes' theorem in which λ but not x, or x but not λ, is continuous. A complete and compact generalization is available and technically necessary but need not be presented here.

In Equation 2, D may be a particular observation or a set of data regarded as a datum and H some hypothesis, or putative fact. Then Equation 2 prescribes the consistent revision of your opinions about the probability of H in the light of the datum D—similarly for Equation 4.

In typical applications of Bayes' theorem, each of the four probabilities in Equation 2 performs a different function, as will soon be explained. Yet they are very symmetrically related to each other, as Equation 3 brings out, and are all the same kind of animal. In particular, all probabilities are really conditional. Thus, $P(H)$ is the probability of the hypothesis H for you conditional on all you know, or knew, about H prior to learning D; and $P(H|D)$ is the probability of H conditional on that same background knowledge together with D.

Again, the four probabilities in Equation 2 are personal probabilities. This does not of course exclude any of them from also being frequencies, ratios of favorable to total possibilities, or numbers arrived at by any other calculation that helps you form your personal opinions. But some are, so to speak, more personal than others. In many applications, practically all concerned find themselves in substantial agreement with respect to $P(D|H)$; or $P(D|H)$ is public, as we say. This happens when $P(D|H)$ flows from some simple model that the scientists, or others, concerned accept as an approximate description of their opinion about the situation in which the datum was obtained. A traditional example of such a statistical model is that of drawing a ball from an urn known to contain some balls, each either black or white. If a series of balls is drawn from the urn, and after each draw the ball is replaced and the urn thoroughly shaken, most men will agree at least tentatively that the probability of drawing a particular sequence D (such as black, white, black, black) given the hypothesis that there are B black and W white balls in the urn is

$$\left(\frac{B}{B+W}\right)^b \left(\frac{W}{B+W}\right)^w,$$

where b is the number of black, and w the number of white, balls in the sequence D.

Even the best models have an element of approximation. For example, the probability of drawing any sequence D of black and white balls from an urn of composition H depends, in this model, only on the number of black balls and white ones in D, not on the order in which they appeared. This may express your opinion in a specific situation very well, but not well enough to be retained if D should happen to consist of 50 black balls followed by 50 white ones. Idiomatically, such a datum convinces you that this particular model is a wrong description of the world. Philosophically, however, the model was not a description of the world but of your opinions, and to know that it was not quite correct, you had at most to reflect on this datum, not necessarily to observe it. In many scientific contexts, the public model behind $P(D|H)$ may include the notions of random sampling from a well-defined population, as in this example. But precise definition of the population may be difficult or impossible, and a sample whose randomness would thoroughly satisfy you, let alone your neighbor in science, can be hard to draw.

In some cases $P(D|H)$ does not command general agreement at all. What is the probability of the actual seasonal color changes on Mars if there is life there? What is this probability if there is no life there? Much discussion of life on Mars has not removed these questions from debate.

Public models, then, are never perfect and often are not available. Nevertheless, those applications of inductive inference, or probabilistic reasoning, that are called statistical seem to be characterized by tentative public agreement on some model and provisional work within it. Rough characterization of statistics by the relative publicness of its models is not necessarily in conflict with attempts to characterize it as the study of numerous repetitions (Bartlett, in Savage et al., 1962, pp. 36–38). This characterization is intended to distinguish statistical applications of Bayes' theorem from many other applications to scientific, economic, military, and other contexts. In some of these nonstatistical contexts, it is appropriate to substitute the judgment of experts for a public model as the source of $P(D|H)$ (see for example Edwards, 1962a, 1963).

The other probabilities in Equation 2 are often not at all public. Reasonable men may differ about them, even if they share a statistical model that specifies $P(D|H)$. People do, however, often differ much more about $P(H)$ and $P(D)$ than about $P(H|D)$, for evidence can bring initially divergent opinions into near agreement.

The probability $P(D)$ is usually of little direct interest, and intuition is often silent about it. It is typically calculated, or eliminated, as follows. When there is a statistical model, H is usually regarded as one of a list, or partition, of mutually exclusive and exhaustive hypotheses H_i such that the $P(D|H_i)$ are all equally public, or part of the statistical model. Since $\Sigma_i P(H_i|D)$ must be 1, Equation 2 implies that

$$P(D) = \Sigma_i P(D|H_i)P(H_i).$$

The choice of the partition H_i is of practical importance but largely arbitrary. For example, tomorrow will be "fair" or "foul," but these two hypotheses can themselves be subdivided and resubdivided. Equation 2 is of course true for all partitions but is more useful for some than for others. As a science advances, partitions originally not even dreamt of become the important ones (Sinclair, 1960). In principle, room should always be left for "some other" explanation. Since $P(D|H)$ can hardly be public when H is "some other explanation," the catchall hypothesis is usually handled in part by studying the situation conditionally on denial of the catchall and in part by informal appraisal of whether any of the explicit hypotheses fit the facts well enough to maintain this denial. Good illustrations are Urey (1962) and Bridgman (1960).

In statistical practice, the partition is ordinarily continuous, which means roughly that H_i is replaced by a parameter λ (which may have more than one dimension) with an initial probability density $u(\lambda)$. In this case,

$$P(D) = \int P(D|\lambda)u(\lambda) \, d\lambda.$$

Similarly, $P(D)$, $P(D|H_i)$, and $P(D|\lambda)$ are replaced by probability densities in D if D is (absolutely) continuously distributed.

$P(H|D)$ or $u(\lambda|D)$, the usual output of a Bayesian calculation, seems to be exactly the kind of information that we all want as a guide to thought and action in the light of an observational process. It is the probability for you that the hypothesis in question is true, on the basis of all your information, including, but not restricted to, the observation D.

Principle of Stable Estimation

Problem of Prior Probabilities

Since $P(D|H)$ is often reasonably public and $P(H|D)$ is usually just what the scientist wants, the reason classical statisticians do not base their procedures on Equations 2 and 4 must, and does, lie in $P(H)$, the prior probability of the hypothesis. We have already discussed the most frequent objection to attaching a probability to a hypothesis and have shown briefly how the definition of personal probability answers that objection. We must now examine the practical problem of determining $P(H)$. Without $P(H)$, Equations 2 and 4 cannot yield $P(H|D)$. But since $P(H)$ is a personal probability, is it not likely to be both vague and variable, and subjective to boot, and therefore useless for public scientific purposes?

Yes, prior probabilities often are quite vague and variable, but they are not necessarily useless on that account (Borel, 1924). The impact of actual vagueness and variability of prior probabilities differs greatly from one problem to another. They frequently have but negligible effect on the conclusions obtained from Bayes' theorem, although utterly unlimited vagueness and variability would have utterly unlimited effect. If observations are precise, in a certain sense, relative to the prior distribution on which they bear, then the form and properties of the prior distribution have negligible influence on the posterior distribution. From a practical point of view, then, the untrammeled subjectivity of opinion about a parameter ceases to apply as soon as much data become available. More generally, two people with widely divergent prior opinions but reasonably open minds will be forced into arbitrarily close agreement about future observations by a sufficient amount of data. An advanced mathematical expression of this phenomenon is in Blackwell and Dubins (1962).

When Prior Distributions Can Be Regarded as Essentially Uniform

Frequently, the data so completely control your posterior opinion that there is no practical need to attend to the details of your prior opinion. For example, consider taking your temperature.

Headachy and hot, you are convinced that you have a fever but are not sure how much. You do not hold the interval $100.5°–101°$ even 20 times more probable than the interval $101°–101.5°$ on the basis of your malaise alone. But now you take your temperature with a thermometer that you strongly believe to be accurate and find yourself willing to give much more than 20 to 1 odds in favor of the half-degree centered at the thermometer reading.

Your prior opinion is rather irrelevant to this useful conclusion but of course not utterly irrelevant. For readings of $85°$ or $110°$, you would revise your statistical model according to which the thermometer is accurate and correctly used, rather than proclaim a medical miracle. A reading of $104°$ would be puzzling—too inconsistent with your prior opinion to seem reasonable and yet not obviously absurd. You might try again, perhaps with another thermometer.

It has long been known that, under suitable circumstances, your actual posterior distribution will be approximately what it would have been had your prior distribution been uniform, that is, described by a constant density. As the fever example suggests, prior distributions need not be, and never really are, completely uniform. To ignore the departures from uniformity, it suffices that your actual prior density change gently in the region favored by the data and not itself too strongly favor some other region.

But what is meant by "gently," by "region favored by the data," by "region favored by the prior distribution," and by two distributions being approximately the same? Such questions do not have ultimate answers, but this section explores one useful set of possibilities. The mathematics and ideas have been current since Laplace, but we do not know any reference that would quite substitute for the following mathematical paragraphs; Jeffreys (1939, see Section 3.4 of the 1961 edition) and Lindley (1961) are pertinent. Those who would skip or skim the mathematics will find the trail again immediately following Implication 7, where the applications of stable estimation are informally summarized.

Under some circumstances, the posterior probability density

$$u(\lambda|x) = \frac{v(x|\lambda)u(\lambda)}{\displaystyle\int v(x|\lambda')u(\lambda')\,d\lambda'} \tag{5}$$

can be well approximated in some senses by the probability density

$$w(\lambda|x) = \frac{v(x|\lambda)}{\displaystyle\int v(x|\lambda')\, d\lambda'}, \tag{6}$$

where λ is a parameter or set of parameters, λ' is a corresponding variable of integration, x is an observation or set or observations, $v(x|\lambda)$ is the probability (or perhaps probability density) of x given λ, $u(\lambda)$ is the prior probability density of λ, and the integrals are over the entire range of meaningful values of λ. By their nature, u, v, and w are nonnegative, and unless the integral in Equation 6 is finite, there is no hope that the approximation will be valid, so these conditions are adopted for the following discussion.

Consider a region of values of λ, say B, which is so small that $u(\lambda)$ varies but little within B and yet so large that B promises to contain much of the posterior probability of λ given the value of x fixed throughout the present discussion. Let α, β, γ, and φ be positive numbers, of which the first three should in practice be small, and are formally taken to be less than 1. In these terms, three assumptions will be made that define one set of circumstances under which $w(\lambda|x)$ does approximate $u(\lambda|x)$ in certain senses, for the given x.

Assumption 1:

$$\int_{\bar{B}} w(\lambda|x)\, d\lambda \leqq \alpha \int_{B} w(\lambda|x)\, d\lambda,$$

where $\bar{B}$ means, as usual, the complement of B. (That is, B is highly favored by the data; α might be 10^{-4} or less in everyday applications.)

Assumption 2: For all $\lambda \in B$,

$$\varphi \leqq u(\lambda) \leqq (1 + \beta)\varphi.$$

(That is, the prior density changes very little within B; .01 or even .05 would be good everyday values for β. The value of φ is unimportant and is not likely to be accurately known.)

Assumption 3:

$$\int_{\bar{B}} u(\lambda|x)\, d\lambda \leqq \gamma \int_{B} u(\lambda|x)\, d\lambda.$$

(That is, B is also highly favored by the posterior distribution; in applications, γ should be small, yet a γ as large as 100α, or even $1,000\alpha$, may have to be tolerated.)

Assumption 3 looks, at first, hard to verify without much knowledge of $u(\lambda)$. Consider an alternative:

Assumption 3': $u(\lambda) \leqq \theta\varphi$ for all λ, where θ is a positive constant. (That is, u is nowhere astronomically big compared to its nearly constant values in B; a θ as large as 100 or 1,000 will often be tolerable.)

Assumption 3' in the presence of Assumptions 1 and 2 can imply 3, as is seen thus.

$$\int_{\bar{B}} u(\lambda|x)\, d\lambda \Big/ \int_B u(\lambda|x)\, d\lambda = \int_{\bar{B}} v(x|\lambda)u(\lambda)\, d\lambda \Big/ \int_B v(x|\lambda)u(\lambda)\, d\lambda$$

$$\leq \theta\varphi \int_{\bar{B}} v(x|\lambda)\, d\lambda \Big/ \varphi \int_B v(x|\lambda)\, d\lambda$$

$$\leq \theta\alpha.$$

So if $\gamma \geq \theta\alpha$, Assumption 3′ implies Assumption 3.

Seven implications of Assumptions 1, 2, and 3 are now derived. The first three may be viewed mainly as steps toward the later ones. The expressions in the large brackets serve only to help prove the numbered assertions.

Implication 1:

$$\int v(x|\lambda)u(\lambda)\, d\lambda \left[\geq \int_B v(x|\lambda)u(\lambda)\, d\lambda \geq \varphi \int_B v(x|\lambda)\, d\lambda \right]$$

$$\geq \frac{\varphi}{1+\alpha} \int v(x|\lambda)\, d\lambda.$$

Implication 2:

$$\int v(x|\lambda)u(\lambda)\, d\lambda$$

$$\left[= \int_B v(x|\lambda)u(\lambda)\, d\lambda + \int_{\bar{B}} v(x|\lambda)u(\lambda)\, d\lambda \leq (1+\gamma) \int_B v(x|\lambda)u(\lambda)\, d\lambda \right]$$

$$\leq (1+\gamma)(1+\beta)\varphi \int v(x|\lambda)\, d\lambda.$$

With two new positive constants δ and ε defined by the context, the next implication follows easily.

Implication 3:

$$(1-\delta) = \frac{1}{(1+\beta)(1+\gamma)} \leq \frac{u(\lambda|x)}{w(\lambda|x)} \leq (1+\beta)(1+\alpha) = (1+\varepsilon)$$

for all λ in B, except where numerator and denominator of $u(\lambda|x)/w(\lambda|x)$ both vanish. (Note that it α, β, and γ are small, so are δ and ε.)

Let $u(C|x)$ and $w(C|x)$ denote $\int_C u(\lambda|x)d\lambda$ and $\int_C w(\lambda|x)d\lambda$, that is, the probabilities of C under the densities $u(\lambda|x)$ and $w(\lambda|x)$.

Implication 4: $u(B|x) \geq 1 - \gamma$, and for every subset C or B,

$$1 - \delta \leq \frac{u(C|x)}{w(C|x)} \leq 1 + \varepsilon.$$

Implication 5: If t is a function or λ such that $|t(\lambda)| \leq T$ for all λ, then

$$\left| \int t(\lambda) u(\lambda|x) \, d\lambda - \int t(\lambda) w(\lambda|x) \, d\lambda \right|$$

$$\left[\leq \int_B |t(\lambda)| \, |u(\lambda|x) - w(\lambda|x)| \, d\lambda \right.$$

$$+ \int_{\bar{B}} |t(\lambda)| u(\lambda|x) \, d\lambda + \int_{\bar{B}} |t(\lambda)| w(\lambda|x) \, d\lambda$$

$$\leq T \int_B \left| \frac{u(\lambda|x)}{w(\lambda|x)} - 1 \right| w(\lambda|x) \, d\lambda + T(\gamma + \alpha) \Bigg]$$

$$\leq T[\max(\delta, \varepsilon) + \gamma + \alpha].$$

Implication 6:

$$|u(C|x) - w(C|x)| \leq \max(\delta, \varepsilon) + \gamma + \alpha$$

for all C.

It is sometimes important to evaluate $u(C|x)$ with fairly good percentage accuracy when $u(C|x)$ is small but not nearly so small as α or γ, thus Implication 7:

$$(1 - \delta)\left(1 - \frac{\alpha}{w(C|x)}\right)$$

$$\left[\leq (1 - \delta)\frac{w(C \cap B|x)}{w(C|x)} \leq \frac{u(C \cap B|x)}{w(C|x)} \right]$$

$$\leq \frac{u(C|x)}{w(C|x)} \left[\leq \frac{u(C \cap B|x) + \gamma}{w(C|x)} \leq (1 + \varepsilon)\frac{w(C \cap B|x)}{w(C|x)} + \frac{\gamma}{w(C|x)} \right]$$

$$\leq (1 + \varepsilon) + \frac{\gamma}{w(C|x)}.$$

What does all this epsilontics mean for practical statistical work? The overall goal is valid justification for proceeding as though your prior distribution were uniform. A set of three assumptions implying this justification was pointed out: First, some region B is highly favored by the data. Second, within B the prior density changes very little. Third, most of the posterior density is concentrated inside B. According to a more stringent but more easily verified substitute for the third assumption, the prior density nowhere enormously exceeds its general value in B.

Given the three assumptions, what follows? One way of looking at the implications is to observe that nowhere within B, which has high posterior probability, is the ratio of the approximate posterior density to the actual posterior density much different from 1 and that what happens outside B is not important for some purposes. Again, if the posterior expectation, or average, of some bounded function is of interest, then the difference between the expectation under the actual posterior distribution and under the approximating distribution will be small relative to the absolute bound of the function. Finally, the actual posterior probability and the approximate probability of any set of parameter values are nearly equal. In short, the approximation is a good one in several important respects—given the three assumptions. Still other respects must sometimes be invoked and these may require further assumptions. See, for example, Lindley (1961).

Even when Assumption 2 is not applicable, a transformation of the parameters of the prior distribution sometimes makes it so. If, for example, your prior distribution roughly obeys Weber's law, so that you tend to assign about as much probability to the region from λ to 2λ as to the region from 10λ to 20λ, a logarithmic transformation of λ may well make Assumption 2 applicable for a considerably smaller β than otherwise.

We must forestall a dangerous confusion. In the temperature example as in many others, the measurement x is being used to estimate the value of some parameter λ. In such cases, λ and x are measured in the same units (degrees Fahrenheit in the example) and interesting values of λ are often numerically close to observed values of x. It is therefore imperative to maintain the conceptual distinction between λ and x. When the principle of stable estimation applies, the normalized function $v(x|\lambda)$ as a function of λ, not of x, approximates your posterior distribution. The point is perhaps most obvious in an example such as estimating the area of a circle by measuring its radius. In this case, λ is in square inches, x is in inches, and there is no temptation to think that the form of the distribution of x's is the same as the form of the posterior distribution of λ's. But the same point applies in all cases. The function $v(x|\lambda)$ is a function of both x and λ; only by coincidence will the form of the parameters of $v(x|\lambda)$ considered as a function of λ be the same as its form of parameters considered as a function of x. One such coincidence occurs so often that it tends to mislead intuition. When your statistical model leads you to expect that a set of observations will be normally distributed, then the posterior distribution of the mean of the quantity being observed will, if stable estimation applies, be normal with the mean equal to the mean of the observations. (Of course it will have a smaller standard deviation than the standard deviation of the observations.)

How good should the approximation be before you can feel comfortable about using it? That depends entirely on your purpose. There are purposes for which an approximation of a small probability which is sure to be within fivefold of the actual probability is adequate. For others, an error of 1% would be painful. Fortunately, if the approximation is unsatisfactory it will often be possible to improve it as much as seems necessary at the price of collecting additional data, an expedient which often justifies its cost in other ways too. In practice, the accuracy of the stable-estimation approximation will seldom be so carefully checked as in the fever example. As individual and collective experience builds up, many applications will properly be judged fate at glance.

Far from always can your prior distribution be practically neglected. At least five situations in which detailed properties of the prior distribution are crucial occur to us:

1. If you assign exceedingly small prior probabilities to regions λ for which $v(x|\lambda)$ is relatively large, you in effect express reluctance to believe in values of λ strongly pointed to by the data and thus violate Assumption 3, perhaps irreparably. Rare events do occur, though rarely, and should not be permitted to confound us utterly. Also, apparatus and plans can break down and produce data that "prove" preposterous things. Morals conflict in the fable of the Providence man who on a cloudy summer day went to the post office to return his absurdly low-reading new barometer to Abercrombie and Fitch. His house was flattened by a hurricane in his absence.

2. If you have strong prior reason to believe that λ lies in a region for which $v(x|\lambda)$ is very small, you may be unwilling to be persuaded by the evidence to the contrary, and so again may violate Assumption 3. In this situation, the prior distribution might consist primarily of a very sharp spike, whereas $v(x|\lambda)$, though very low in the region of the prior spike, may be comparatively gentle everywhere. In the previous paragraph, it was $v(x|\lambda)$ which had the sharp spike, and the prior distribution which was near zero in the region of that spike. Quite often it would be inappropriate to discard a good theory on the basis of a single opposing experiment. Hypothesis testing situations discussed later in this paper illustrate this phenomenon.

3. If your prior opinion is relatively diffuse, but so are your data, then Assumption 1 is seriously violated. For when your data really do not mean much compared to what you already know, then the exact content of the initial opinion cannot be neglected.

4. If observations are expensive and you have a decision to make, it may not pay to collect enough information for the principle of stable estimation to apply. In such situations you should collect just so much information that the expected value of the best course of action available in the light of the information at hand is greater than the expected value of any program that involves collecting more observations. If you have strong prior opinions about the parameter, the amount of new information available when you stop collecting more may well be far too meager to satisfy the principle. Often, it will not pay you to collect any new information at all.

5. It is sometimes necessary to make decisions about sizable research commitments such as sample size or experimental design while your knowledge is still vague. In this case, an extreme instance of the former one, the role of prior opinion is particularly conspicuous. As Raiffa and Schlaifer (1961) show, this is one of the most fruitful applications or Bayesian ideas.

Whenever you cannot neglect the details of your prior distribution, you have, in effect, no choice but to determine the relevant aspects of it as best you can and use them. Almost always, you will find your prior opinions quite vague, and you may be distressed that your scientific inference or decision has such a labile basis. Perhaps this distress, more than anything else, discouraged statisticians from using Bayesian ideas all along (Pearson, 1962). To paraphrase de Finetti (1959, p. 19), people noticing difficulties in applying Bayes' theorem remarked "We see that it is not secure to build on sand. Take away the sand, we shall build on the void." If it were meaningful utterly to ignore prior opinion, it might presumably sometimes be wise to do so; but reflection shows that any policy that pretends to ignore prior opinion will be acceptable only insofar as it is actually justified by prior opinion. Some policies recommended under the motif of neutrality, or using only the facts, may flagrantly violate even very confused prior opinions, and so be unacceptable. The method of stable estimation might casually be described as a procedure for ignoring prior opinion, since its approximate results are acceptable for a wide range of prior opinions. Actually, far from ignoring prior opinion, stable estimation exploits certain well-defined features of prior opinion and is acceptable only insofar as those features are really present....

Introduction to Hypothesis Testing

No aspect of classical statistics has been so popular with psychologists and other scientists as hypothesis testing, though some classical statisticians agree with us that the topic has been overemphasized. A statistician of great experience told us, "I don't know much about tests, because I have never had occasion to use one." Our devotion of most of the rest of this paper to tests would be disproportionate, if we were not writing for an audience accustomed to think of statistics largely as testing.

So many ideas have accreted to the word "test" that one definition cannot even hint at them. We shall first mention some of the main ideas relatively briefly, then flesh them out a bit with informal discussion of hypothetical substantive examples, and finally discuss technically some typical formal examples from a Bayesian point of view. Some experience with classical ideas of testing is assumed throughout. The pinnacle of the abstract theory of testing from the Neyman-Pearson standpoint is Lehmann (1959). Laboratory thinking on testing may derive more from R. A. Fisher than from the Neyman-Pearson school, though very few are explicitly familiar with Fisher's ideas culminating in 1950 and 1956.

The most popular notion of a test is, roughly, a tentative decision between two hypotheses on the basis of data, and this is the notion that will dominate the present treatment of tests. Some qualification is needed if only because, in typical applications, one of the hypotheses—the null hypothesis—is known by all concerned to be false from the outset (Berkson, 1938; Hodges & Lehmann, 1954; Lehmann, 1959; I. R. Savage, 1957; L. J. Savage, 1954, p. 254); some ways of resolving the seeming absurdity will later be pointed out, and at least one of them will be important for us here.

The Neyman-Pearson school of theoreticians, with their emphasis on the decision-theoretic or behavioral approach, tend to define a test as a choice between two actions, such as whether or not to air condition the ivory tower so the rats housed therein will behave more consistently. This definition is intended to clarify operationally the meaning of decision between two hypotheses. For one thing, as Bayesians agree, such a decision resembles a potential dichotomous choice in some economic situation such as a bet. Again, wherever there is a dichotomous economic choice, the possible values of the unknown parameters divide themselves into those for which one action or the other is appropriate. (The neutral zone in which both actions are equally appropriate is seldom important and can be dealt with in various ways.) Thus a dichotomous choice corresponds to a partition into two hypotheses. Nonetheless, not every choice is like a simple bet, for economic differences within each hypothesis can be important.

Sometimes the decision-theoretic definition of testing is expressed as a decision to act as though one or the other of the two hypotheses were believed, and that has apparently led to some confusion (Neyman, 1957, p. 16). What action is wise of course depends in part on what is at stake. You would not take the plane if you believed it would crash, and would not buy flight insurance if you believed it would not. Seldom must you choose between exactly two acts, one appropriate to the null hypothesis and the other to its alternative. Many intermediate, or hedging, acts are ordinarily possible, flying after buying flight insurance, and choosing a reasonable amount of flight insurance, are examples.

From a Bayesian point of view, the special role of testing tends to evaporate, yet something does remain. Deciding between two hypotheses in the light of the datum suggests to a Bayesian only computing their posterior probabilities; that a pair of probabilities are singled out for special attention is without theoretical interest. Similarly, a choice between two actions reduces to choosing the larger of two expected utilities under a posterior distribution. The feature of importance for the Bayesian was practically lost in the recapitulation of general classical definitions. This happened, in part, because the feature would seem incidental in a general classical theory though recognized by all as important in specific cases and, in part, because expression of the feature is uncongenial to classical language, though implicitly recognized by classical statisticians.

In many problems, the prior density $u(\lambda)$ of the parameter(s) is often gentle enough relative to $x|\lambda$ to permit stable estimation (or some convenient variation of it). One important way in which $u(\lambda)$ can fail to be sufficiently gentle is by concentrating considerable probability close to some point (or line, or surface, or the like). Certain practical devices can render the treatment of such a concentration of probability relatively public. These devices are, or should be, only rather rarely needed, but they do seem to be of some importance and to constitute appropriate Bayesian treatment of some of the scientific situations in which the classical theory of hypothesis testing has been invoked. At least occasionally, a pair of hypotheses is associated with the concentration of probability. For example, if the squirrel has not touched it, that acorn is almost sure to be practically where it was placed yesterday. For vividness and to maintain some parallelism with classical expressions, we shall usually suppose concentration associated with a null hypothesis, as in this example; it is straightforward to extend the discussion to situations where there is not really such a pair of hypotheses. The theory of testing in the sense of dealing with concentrated probability as presented here draws heavily on Jeffreys (1939, see Ch. 5 and 6 of the 1961 edition) and Lindley (1961).

Examples

Discussion of a few examples may bring out some points associated with the various concepts of testing.

EXAMPLE 1. Two teaching-machine programs for sixth-grade arithmetic have been compared experimentally.

For some purposes each program might be characterized by a single number, perhaps the mean difference between pretest and posttest performance on some standardized test of proficiency in arithmetic. This number, an index of the effectiveness of the program, must of course be combined with economic and other information from outside the experiment itself if the experiment is to guide some practical decision.

If one of the two programs must be adopted, the problem is one of testing in the sense of the general decision-theoretic definition, yet it is likely to be such that practicing statisticians would not ordinarily call the appropriate procedure a test at all. Unless your prior opinion perceptibly favored one of the two programs, you should plainly adopt that one which seemed, however slightly, to do better in the experiment. The classical counterpart of this simple conclusion had to be discovered against the tendency to invoke "significance tests" in all testing situations (Bahadur & Robbins, 1950).

But suppose one program is much more expensive to implement than the other. If such information about costs is available, it can be combined with information provided by the experiment to indicate how much proficiency can be bought for how many dollars. It is then a matter of judgment whether to make the purchase. In principle the judgment is simply one of the dollar value of proficiency (or equivalently of the proficiency value of dollars); in practice, such judgments are often difficult and controversial.

If the experiment is indecisive, should any decision be risked? Of course it should be if it really must be. In many actual situations there are alternatives such as further experimentation. The choice is then really at least trichotomous but perhaps with dichotomous emphasis on continuing, as opposed to desisting from, experimentation. Such suggestions as to continue only if the difference is not significant at, say, the 5% level are sometimes heard. Many classical theorists are dissatisfied with this approach, and we believe Bayesian statistics can do better (see Raiffa & Schlaifer, 1961, for some progress in this direction).

Convention asks, "Do these two programs differ at all in effectiveness?" Of course they do. Could any real difference in the programs fail to induce at least some slight difference in their effectiveness? Yet the difference in effectiveness may be negligible compared to the sensitivity of the experiment. In this way, the conventional question can be given meaning, and we shall often ask it without further explanation or apology. A closely related question would be, "Is the superiority of Method A over Method B pointed to by the experiment real, taking due account of the possibility that the actual difference may be very small?" With several programs, the number of questions about relative superiority rapidly multiplies.

EXAMPLE 2. Can this subject guess the color of a card drawn from a hidden shuffled bridge deck more or less than 50% of the time?

This is an instance of the conventional question, "Is there any difference at all?" so philosophically the answer is presumably "yes," though in the last analysis the very meaningfulness of the question might be challenged. We would not expect any such ostensible effect to stand up from one experiment to another in magnitude or direction. We are strongly prejudiced that the inevitable small deviations from the null hypothesis will always turn out to be somehow artifactual—explicable, for instance, in terms of defects in the shuffling or concealing of the cards or the recording of the data and not due to Extra-Sensory Perception (ESP).

One who is so prejudiced has no need for a testing procedure, but there are examples in which the null hypothesis, very sharply interpreted, commands some but not utter credence. The present example is such a one for many, more open minded about ESP than we, and even we can imagine, though we do not expect, phenomena that would shake our disbelief.

EXAMPLE 3. Does this packed suitcase weigh less than 40 pounds? The reason you want to know is that the airlines by arbitrary convention charge overweight for more. The conventional weight, 40 pounds, plays little special role in the structure of your opinion which may well be diffuse relative to the bathroom scale. If the scale happens to register very close to 40 pounds (and you know its precision), the theory of stable estimation will yield a definite probability that the suitcase is overweight. If the reading is not close, you will have overwhelming conviction, one way or the other, but the odds will be very vaguely defined. For the conditions are ill suited to stable estimation if only because the statistical model of the scale is not sufficiently credible.

If the problem is whether to leave something behind or to put in another book, the odds are not a sufficient guide. Taking the problem seriously, you would have to reckon the cash cost of each amount of overweight and the cash equivalent to you of leaving various things behind in order to compute the posterior expected worth of various possible courses of action.

We shall discuss further the application of stable estimation to this example, for this is the one encounter we shall have with a Bayesian procedure at all harmonious with a classical tail-area significance test. Assume, then, that a normally distributed observation x has been made, with known standard deviation σ, and that your prior opinion about the weight of your suitcase is diffuse relative to the measurement. The principle of stable estimation applies, so, as an acceptable approximation,

$$P(\lambda \leq 40|x) = \Phi\left(\frac{x - 40}{\sigma}\right) = \Phi(t),$$

in case $|t|$ is not too great. In words, the probability that your suitcase weighs at most 40 pounds, in the light of the datum x, is the probability to the left of t under the standard normal distribution. Almost by accident, this is also the one-tailed significance level of the classical t test for the hypothesis that $\lambda \leq 40$. The fundamental interpretation of $\Phi(t)$ here is the probability for you that your suitcase weighs less than 40 pounds; just the sort of thing that classical statistics rightly warns us not to expect a significance level to be. Problems in which stable estimation leads exactly to a one-tailed classical significance level are of very special structure. No Bayesian procedure yet known looks like a two-tailed test (Schlaifer, 1961, p. 212).

Classical one-tailed tests are often recommended for a situation in which Bayesian treatment would call for nothing like them. Imagine, for instance, an experiment to determine whether schizophrenia impairs problem solving ability, supposing it all but inconceivable that schizophrenia enhances the ability. This is classically a place to use a one-tailed test; the Bayesian recommendations for this problem, which will not be explored here, would not be tail-area tests and would be rather similar to the Bayesian null hypothesis tests discussed later. One point recognized by almost all is that if schizophrenia can do no good it must then do some harm, though perhaps too little to perceive.

Before putting the suitcase on the bathroom scales you have little expectation of applying the formal arithmetic of the preceding paragraphs. At that time, your opinion about the weight of the suitcase is diffuse. Therefore, no interval as small as 6 or 8 σ can include much of your initial probability. On the other hand, if $|t|$ is greater than 3 or 4, which you very much expect, you will not rely on normal tail-area computations, because that would put the assumption of normality to unreasonable strain. Also Assumption 2 of the discussion of stable estimation will probably be drastically violated. You will usually be content in such a case to conclude that the weight of the suitcase is, beyond practical doubt, more (or less) than 40 pounds.

The preceding paragraph illustrates a procedure that statisticians of all schools find important but elusive. It has been called the interocular traumatic test;[2] you know what the data mean when the conclusion hits you between the eyes. The interocular traumatic test is simple, commands general agreement, and is often applicable; well-conducted experiments often come out that way. But the enthusiast's interocular trauma may be the skeptic's random error. A little arithmetic to verify the extent of the trauma can yield great peace of mind for little cost.

Bayesian Hypothesis Testing

Odds and Likelihood Ratios

Gamblers frequently measure probabilities in terms of odds. Your odds in favor of the event A are (aside from utility effects) the amount that you would just be willing to pay if A does not occur in compensation for a commitment from someone else to pay you one unit of money if A does occur. The odds $\Omega(A)$ in favor of A are thus related to the probability $P(A)$ of A and the probability $1 - P(A)$ of not A, or $\bar{A}$, by the condition,

$$\Omega(A)[1 - P(A)] = P(A).$$

Odds and probability are therefore translated into each other thus,

$$\Omega(A) = \frac{P(A)}{1 - P(A)} = \frac{P(A)}{P(\bar{A})}; \qquad P(A) = \frac{\Omega(A)}{1 + \Omega(A)}.$$

For example, odds of 1, an even-money bet, correspond to a probability of 1/2; a probability of 9/10 corresponds to odds of 9 (or 9 to 1), and a probability of 1/10 corresponds to odds or 1/9 (or 1 to 9). If $P(A)$ is 0, $\Omega(A)$ is plainly 0; and if $P(A)$ is 1, $\Omega(A)$ may be called ∞, if it need be defined at all.

From a Bayesian standpoint, part of what is suggested by "testing" is finding the posterior probability $P(A|D)$ of the hypothesis A in the light of the datum D, or equivalently, finding the posterior odds $\Omega(A|D)$.

According to Bayes' theorem

$$P(A|D) = \frac{P(D|A)P(A)}{P(D)}, \tag{7}$$

$$P(\bar{A}|D) = \frac{P(D|\bar{A})P(\bar{A})}{P(D)}. \tag{8}$$

Dividing each side of Equation 7 by the corresponding side of Equation 8, canceling the common denominators $P(D)$, and making evident abbreviations leads to a condensation of Equations 7 and 8 in terms of odds;

$$\Omega(A|D) = \frac{P(D|A)}{P(D|\bar{A})}\Omega(A)$$

$$= L(A; D)\Omega(A). \tag{9}$$

In words, the posterior odds in favor of A given the datum D are the prior odds multiplied by the ratio of the conditional probabilities of the datum given the hypothesis A and given its negation. The ratio of conditional probabilities $L(A; D)$ is called the likelihood ratio in favor of the hypothesis A on the basis of the datum D.

[2] J. Berkson, personal communication, July 14, 1958.

Plainly, and according to Equation 9, D increases the odds for A, if and only if D is more probable under A than under its negation $\bar{A}$ so that $L(A; D)$ is greater than 1.

If D is impossible under $\bar{A}$, Equation 9 requires an illegitimate division, but it can fairly be interpreted to say that A has acquired probability 1 unless $\Omega(A) = 0$, in which case the problem is ill specified. With that rather academic exception, whenever $\Omega(A)$ is 0 so is $\Omega(A|D)$; roughly, once something is regarded as impossible, no evidence can reinstate its credibility.

In actual practice, $L(A; D)$ and $\Omega(A)$ tend to differ from person to person. Nonetheless, statistics is particularly interested in examining how and when Equation 9 can lead to relatively public conclusions, a theme that will occupy several sections.

Simple Dichotomy

It is useful, at least for expositions to consider problems in which $L(A; D)$ is entirely public. For example, someone whose word you and we trust might tell us that the die he hands us produces 6's either (A) with frequency 1/6 or $(\bar{A})$ with frequency 1/5. Your initial opinion $\Omega(A)$ might differ radically from ours. But, for you and for us, the likelihood ratio in favor of A on the basis of a 6 is (1/6)/(1/5) or 5/6, and the likelihood ratio in favor of A on the basis of a non-6 is (5/6)/(4/5) or 25/24. Thus, if a 6 appears when the die is rolled, everyone's confidence in A will diminish slightly; specifically, odds in favor of A will be diminished by 5/6. Similarly, a non-6 will augment $\Omega(A)$ by the factor 25/24.

If such a die could be rolled only once, the resulting evidence $L(A; D)$ would be negligible for almost any purpose; if it can be rolled many times, the evidence is ultimately sure to become definitive. As is implicit in the concept of the not necessarily fair die, if $D_1, D_2, D_3, \ldots$ are the outcomes of successive rolls, then the same function $L(A; D)$ applies to each. Therefore Equation 9 can be applied repeatedly, thus:

$$\Omega(A|D_1) = L(A; D_1)\Omega(A)$$

$$\Omega(A|D_2, D_1) = L(A; D_2)\Omega(A|D_1)$$

$$= L(A; D_2)L(A; D_1)\Omega(A)$$

$$\ldots\ldots$$

$$\Omega(A|D_n, \ldots, D_1) = L(A; D_n)\Omega(A|D_{n-1}, D_{n-2}, \ldots, D_1)$$

$$= L(A; D_n)L(A; D_{n-1})\ldots L(A; D_1)\Omega(A)$$

$$= \prod_{j=1}^{n} L(A; D_j)\Omega(A).$$

This multiplicative composition of likelihood ratios exemplifies an important general principle about observations which are independent given the hypothesis.

For the specific example of the die, if x 6's and y non-6's occur (where of course $x + y = n$), then

$$\Omega(A|D_n, \ldots, D_1) = \left(\frac{5}{6}\right)^x \left(\frac{25}{24}\right)^y \Omega(A).$$

For large n, if A obtains, it is highly probable at the outset that x/n will fall close to 1/6. Similarly, if A does not obtain x/n will probably fall close to 1/5. Thus, if A obtains, the overall likelihood $(5/6)^x(25/24)^y$ will probably be very roughly

$$\left(\frac{5}{6}\right)^{n/6} \left(\frac{25}{24}\right)^{5n/6} = \left[\left(\frac{5}{6}\right)^{1/6} \left(\frac{25}{24}\right)^{5/6}\right]^n$$

$$= (1.00364)^n$$

$$= 10^{0.00158n}.$$

By the time n is 1,200 everyone's odds in favor of A will probably be augmented about a hundredfold, if A is in fact true. One who started very skeptical of A, say with $\Omega(A)$ about a thousandth, will still be rather skeptical. But he would have to start from a very skeptical position indeed not to become strongly convinced when n is 6,300 and the overall likelihood ratio in favor or A is about 10 billion.

The arithmetic for $\bar{A}$ is:

$$\left[\left(\frac{5}{6}\right)^{1/5} \left(\frac{25}{24}\right)^{4/5}\right]^n = (0.9962)^n = 10^{-0.00165n}.$$

So the rate at which evidence accumulates against A, and for $\bar{A}$, when $\bar{A}$ is true is in this case a trifle more than the rate at which it accumulates for A when A is true.

Simple dichotomy is instructive for statistical theory generally but must be taken with a grain of salt. For simple dichotomies—that is, applications of Equation 9 in which everyone concerned will agree and be clear about the values of $L(A; D)$—rarely; if ever, occur in scientific practice. Public models almost always involve parameters rather than finite partitions.

Some generalizations are apparent in what has already been said about simple dichotomy. Two more will be sketchily illustrated: Decision-theoretic statistics, and the relation of the dominant classical decision-theoretic position to the Bayesian position. (More details will be found in Savage, 1954, and Savage et al., 1962, indexed under simple dichotomy.)

At a given moment, let us suppose, you have to guess whether it is A or $\bar{A}$ that obtains and you will receive \$$I$ it you guess correctly that A obtains, \$$J$ if you guess correctly that $\bar{A}$ obtains, and nothing otherwise. (No real generality is lost in not assigning four arbitrarily chosen payoffs to the four possible combinations of guess and fact.) The expected cash value to you of guessing A is $\$IP(A)$ and that of guessing $\bar{A}$ is $\$JP(\bar{A})$. You will therefore prefer to guess A if and only if $\$IP(A)$ exceeds $\$JP(\bar{A})$; that is, just if $\Omega(A)$ exceeds J/I. (More rigorous treatment would replace dollars with utiles.)

Similarly, if you need not make your guess until after you have examined a datum D, you will prefer to guess A if, and only if, $\Omega(A|D)$ exceeds J/I. Putting this together with Equation 9, you will prefer to guess A if, and only if,

$$L(A; D) > \frac{J}{I\Omega(A)} = \Lambda,$$

where your critical likelihood ratio Λ is defined by the context.

This conclusion does not at all require that the dichotomy between A and $\bar{A}$ be simple, or public, but for comparison with the classical approach to the same problem continue to assume that it is. Classical statisticians were the first to conclude that there must be some Λ such that you will guess A if $L(A; D) > \Lambda$ and guess $\bar{A}$ if $L(A: D) < \Lambda$. (For this sketch, it is excusable to neglect the possibility that $\Lambda = L(A; D)$.) By and large, classical statisticians say that the choice of Λ is an entirely subjective one which no one but you can make (e.g., Lehmann, 1959, p. 62). Bayesians agree; for according to Equation 9, Λ is inversely proportional to your current odds for A, an aspect of your personal opinion.

The classical statisticians, however, have overlooked a great simplification, namely that your critical Λ will not depend on the size or structure of the experiment and will be proportional to J/I. Once the Bayesian position is accepted, Equation 9 is of course an argument for this simplification, but it can also be arrived at along a classical path, which in effect derives much, if not all, of Bayesian statistics as a natural completion of the classical decision-theoretic position. This relation between the two views, which in no way depends on the artificiality of simple dichotomy here used to illustrate it, can-not all, of Bayesian statistics as a natural completion of the classical decision-theoretic position. This relation between the two views, which in no way depends on the artificiality of simple dichotomy here used to illustrate it, can-not be overemphasized. (For a general demonstration, see Raiffa & Schlaifer, 1961, pp. 24–27.)

The simplification is brought out by the set of indifference curves among the various probabilities of the two kinds of errors (Lehmann, 1958). Of course, any reduction of the probability of one kind of error is desirable if it does not increase the probability of the other kind of error, and the implications of classical statistics leave the description of the indifference curves at that. But the considerations discussed easily imply that the indifference curves should be parallel straight lines with slope $-[J/I\Omega(A)]$. As Savage (1962b) puts it:

the subjectivist's position is more objective than the objectivist's, for the subjectivist finds the range of coherent or reasonable preference patterns much narrower than the objectivist thought it to be. How confusing and dangerous big words are [p. 67]!

Classical statistics tends to divert attention from Λ to the two conditional probabilities of making errors, by guessing A when $\bar{A}$ obtains and vice versa. The counterpart of the probabilities of these two kinds of errors in more general problems is called the operating characteristic, and classical statisticians suggest, in effect, that you should choose among the available operating characteristics as a method of choosing Λ, or more generally, your prior distribution. This is not mathematically wrong, but it distracts attention from your value judgments and opinions about the unknown facts upon which your preferred Λ should directly depend without regard to how the probabilities of errors vary with Λ in a specific experiment.

There are important advantages to recognizing that your Λ does not depend on the structure of the experiment. It will help you, for example, to choose between possible experimental plans. It leads immediately to the very important likelihood principle, which in this application says that the numerical value of the likelihood ratio of the datum conveys the entire import of the datum. (A later section is about the likelihood principle.)

Wolfowitz (1962) dissents.

Approaches to Null Hypothesis Testing

Next we examine situations in which a very sharp, or null, hypothesis is compared with a rather flat or diffuse alternative hypothesis. This short section indicates general strategies of such comparisons. None of the computations or conclusions depend on assumptions about the special initial credibility of the null hypothesis, but a Bayesian will find such computations uninteresting unless a nonnegligible amount of his prior probability is concentrated very near the null hypothesis value.

For the continuous cases to be considered in following sections, the hypothesis A is that some parameter λ is in a set that might as well also be called A. For one-dimensional cases in which the hypothesis A is that λ is almost surely negligibly far from some specified value λ_0, the odds in favor of A given the datum D, as in Equation 9, are

$$
\Omega(A|D) = \frac{P(A|D)}{P(\bar{A}|D)}
$$

$$
= \frac{v(D|\lambda_0)}{\int v(D|\lambda)u(\lambda|\bar{A})\,d\lambda}\Omega(A)
$$

$$
= L(A; D)\Omega(A).
$$

Natural generalizations apply to multidimensional cases. The numerator $v(D|\lambda_0)$ will in usual applications be public. But the denominator, the probability of D under the alternative hypothesis, depends on the usually far from public prior density under the alternative hypothesis. Nonetheless, there are some relatively public methods of appraising the denominator, and much of the following discussion of tests is, in effect, about such methods. Their spirit is opportunistic, bringing to bear whatever approximations and bounds offer themselves in particular cases. The main ideas of these methods are sketched in the following three paragraphs, which will later be much amplified by examples.

First, the principle of stable estimation may apply to the datum and to the density $u(\lambda|\bar{A})$ of λ given the alternative hypothesis $\bar{A}$. In this case, the likelihood ratio reflects no characteristics of $u(\lambda|\bar{A})$ other than its value in the neighborhood favored by the datum, a number that can be made relatively accessible to introspection.

Second, it is relatively easy, in any given case, to determine how small the likelihood ratio can possibly be made by utterly unrestricted and artificial choice of the function $u(\lambda|\bar{A})$. If this rigorous public lower bound on the likelihood ratio is not very small, then there exists no system of prior probabilities under which the datum greatly detracts from the credibility or the null hypothesis. Remarkably, this smallest possible bound is by no means always very small in those cases when the datum would lead to a high classical significance level such as .05 or .01. Less extreme (and therefore larger) lower bounds that do assume some restriction on $u(\lambda|\bar{A})$ are sometimes appropriate; analogous restrictions also lead to upper bounds. When these are small, the datum does rather publicly greatly lower the credibility of the null hypothesis. Analysis to support an interocular traumatic impression might often be of this sort. Inequalities stated more generally by Hildreth (1963) are behind most of these lower and upper bounds.

Finally, when $v(D|\lambda)$ admits of a conjugate family of distributions, it may be useful, as an approximation, to suppose $u(\lambda|\bar{A})$ restricted to the conjugate family. Such a restriction may help fix reasonably public bounds to the likelihood ratio.

We shall see that classical procedures are often ready severely to reject the null hypothesis on the basis of data that do not greatly detract from its credibility, which dramatically demonstrates the practical difference between Bayesian and classical statistics. This finding is not altogether new. In particular, Lindley (1957) has proved that for any classical significance level for rejecting the null hypothesis (no matter how small) and for any likelihood ratio in favor of the null hypothesis (no matter how large), there exists a datum significant at that level and with that likelihood ratio.

To prepare intuition for later technical discussion we now show informally, as much as possible from a classical point of view, how evidence that leads to classical rejection of a null hypothesis at the .05 level can favor that null hypothesis. The loose and intuitive argument can easily be made precise (and is, later in the paper). Consider a two-tailed t test with many degrees of freedom. If a true null hypothesis is being tested, t will exceed 1.96 with probability 2.5% and will exceed 2.58 with probability .5%. (Of course, 1.96 and 2.58 are the 5% and 1% two-tailed significance levels; the other 2.5% and .5% refer to the possibility that t may be smaller than -1.96 or -2.58.) So on 2% of all occasions when true null hypotheses are being tested, t will lie between 1.96 and 2.58. How often will t lie in that interval when the null hypothesis is false? That depends on what alternatives to the null hypothesis are to be considered. Frequently, given that the null hypothesis is false, all values or t between, say, -20 and $+20$ are about equally likely for you. Thus, when the null hypothesis is false, t may well fall in the range from 1.96 to 2.58 with at most the probability $(2.58 - 1.96)/[+20 - (-20)] = 1.55\%$. In such a case, since 1.55 is less than 2 the occurrence of t in that interval speaks mildly for, not vigorously against, the truth of the null hypothesis.

This argument, like almost all the following discussion or null hypothesis testing, hinges on assumptions about the prior distribution under the alternative hypothesis. The classical statistician usually neglects that distribution— in fact, denies its existence. He considers how unlikely a t as far from 0 as 1.96 is if the null hypothesis is true, but he does not consider that a t as close to 0 as 1.96 may be even less likely if the null hypothesis is false.

A Bernoullian Example

To begin a more detailed examination of Bayesian methods for evaluating null hypotheses, consider this example:

We are studying a motor skills task. Starting from a neutral rest position, a subject attempts to touch a stylus as near as possible to a long, straight line. We are interested in whether his responses favor the right or the left of the line. Perhaps from casual experience with such tasks, we give special credence to the possibility that his long-run frequency p of "rights" is practically $p_0 = 1/2$. The problem is here posed in the more familiar frequentistic terminology; its Bayesian translation, due to de Finetti, is sketched in Section 3.7 of Savage (1954). The following discussion applies to any fraction p_0 as well as to the specific value 1/2. Under the null hypothesis, your density of the parameter p is sharply concentrated near p_0, while your density of p under the alternative hypothesis is not concentrated and may be rather diffuse over much of the interval from 0 to 1.

If n trials are undertaken, the probability of obtaining r rights given that the true frequency is p is of course $C_r^n p^r (1 - p)^{n-r}$. The probability of obtaining r under the null hypothesis that p is literally p_0 is $C_r^n p_0^r (1 - p_0)^{n-r}$. Under the alternative hypothesis, it is

$$\int_0^1 C_r^n p^r (1 - p)^{n-r} u(p|H_1)\, dp,$$

that is, the probability of r given p averaged over p, with each value in the average weighted by its prior density under the alternative hypothesis. The likelihood ratio is therefore

$$L(p_0; r, n) = \frac{p_0^r(1 - p_0)^{n-r}}{\int_0^1 p^r(1 - p)^{n-r}u(p|H_1)\, dp}. \tag{10}$$

The disappearance of C_r^n from the likelihood ratio by cancellation is related to the likelihood principle, which will be discussed later. Had the experiment not been analyzed with a certain misplaced sophistication, C_r^n would never have appeared in the first place. We would simply have noted that the probability of any specific sequence or rights and lefts with r rights and $n - r$ lefts is, given p, exactly $p^r(1 - p)^{n-r}$. That the number or different sequences of this composition is C_r^n is simply irrelevant to Bayesian inference about p.

One possible way to reduce the denominator of Equation 10 to more tractable form is to apply the principle of stable estimation, or more accurately certain variants of it, to the denominator. To begin with, if $u(p|H_1)$ were a constant u', then the denominator would be

$$\int_0^1 p^r(1 - p)^{n-r}u(p|H_1)\, dp = u' \int_0^1 p^r(1 - p)^{n-r}\, dp$$

$$= \frac{u'}{(n + 1)C_r^n}. \tag{11}$$

The first equality is evident; the second is a known formula, enchantingly demonstrated by Bayes (1763). Of course u cannot really be a constant unless it is 1, but if r and $n - r$ are both fairly large $p^r(1 - p)^{n-r}$ is a sharply peaked function with its maximum at r/n. If $u(p|H_1)$ is gentle near r/n and not too wild elsewhere, Equation 11 may be a satisfactory approximation, with $u' = u(r/n|H_1)$. This condition is often met, and it can be considerably weakened without changing the conclusion, as will be explained next.

If the graph of $u(p|H_1)$ were a straight though not necessarily horizontal, line then the required integral would be

$$\int_0^1 p^r(1 - p)^{n-r}u(p|H_1)\, dp = \frac{u\left(\dfrac{r + 1}{n + 2}|H_1\right)}{(n + 1)C_r^n}. \tag{12}$$

This is basically a standard formula like the latter part of Equation 11, and is in fact rather easily inferred from that earlier formula itself. Consequently, for large r and $n - r$, Equation 12 can be justified as an approximation with $u' = u[(r + 1)/(n + 2)|H_1]$ whenever $u(p|H_1)$ is nearly linear in the neighborhood of $(r + 1)/(n + 2)$, which under the assumed conditions is virtually indistinguishable from r/n.

In summary, it is often suitable to approximate the likelihood ratio thus:

$$L(p_0; r, n) = \frac{n+1}{u'} C_r^n p_0^r (1 - p_0)^{n-r}$$

$$= \frac{(n+1)P(r|p_0, n)}{u'} \tag{13}$$

where $u' = u(r/n|H_1)$ or $u[(r+1)/(n+2)|H_1]$.

Does this approximation apply to you in a specific case? If so, what value of u' is appropriate? Such subjective questions can be answered only by self-interrogation along lines suggested by our discussion of stable estimation. In particular, u' is closely akin to the φ of our Condition 2 for stable estimation. In stable estimation, the value of φ cancels out of all calculations, but here, u' is essential. One way to arrive at u' is to ask yourself what probability you attach to a small, but not microscopic, interval of values of p near r/n under the alternative hypothesis. Your reply will typically be vague, perhaps just a rough order of magnitude, but that may be enough to settle whether the experiment has strikingly confirmed or strikingly discredited the null hypothesis.

In this particular example of a person aiming at a line with a stylus, structuring your opinion in terms of a sharp null hypothesis and a diffuse alternative is rather forced. More realistically, your prior opinion is simply expressed by a density with a rather sharp peak, or mode, at $p_0 = 1/2$, and your posterior distribution will tend to have two modes, one at p_0 and the other about at r/n. Nonetheless, an arbitrary structuring of the prior density as a weighted average, or probability mixture, of two densities, one practically concentrated at p_0 and the other somewhat diffuse, may be a useful approach.

Conversely, even if the division is not artificial, the unified approach is always permissible. This may help emphasize that determining the posterior odds is seldom the entire aim of the analysis. The posterior distribution of p under the alternative hypothesis is also important. This density $u(p|r, n, H_1)$ is determined by Bayes' theorem from the datum (r, n) and the alternative prior density $u(p|H_1)$; for this, what the hypothesis H_0 is, or how probable you consider it either before or after the experiment are all irrelevant. As in any other estimation problem, the principle of stable estimation may provide an adequate approximation for $u(p|r, n, H_1)$. If in addition, the null hypothesis is strongly discredited by the datum, then the entire posterior density $u(p|r, n)$ will be virtually unimodal and identifiable with $u(p|r, n, H_1)$ for many purposes. In fact, the outcome of the test in this case is to show that stable estimation (in particular our Assumption 3) is applicable without recourse to Assumption 3'.

The stable-estimation density for this Bernoullian problem is of course $p^r(1 - p)^{n-r}$ multiplied by the appropriate normalizing constant, which is implicit in the second equality of Equation 11. This is an instance of the beta density or indices a and b,

$$\frac{(a + b - 1)!}{(a - 1)!(b - 1)!} p^{a-1}(1 - p)^{b-1}.$$

In this case, $a = r + 1$ and $b = (n - r) + 1$.

In view of the rough rule of thumb that u' is of the order of magnitude of 1, the factor $(n + 1)P(r|p_0, n)$ is at least a crude approximation to $L(p_0; r, n)$ and is of interest in any case as the relatively public factor in $L(p_0; r, n)$ and hence in $\Omega(H_0|r, n)$. The first three rows of Table 1 show hypothetical data for four different experiments or this sort (two of them on a large scale) along with the corresponding likelihood ratios for the uniform alternative prior. The numbers in Table 1 are, for illustration, those that would, for the specified number of observations, barely lead to rejection of the null hypothesis, $p = .5$, by a classical two-tailed test at the .05 level.

How would a Bayesian feel about the numbers in Table 1? Remember that a likelihood ratio greater than 1 leaves one more confident of the null hypothesis than he was to start with, while a likelihood ratio less than 1 leaves him less confident of it than he was to start with. Thus Experiment 1, which argues against the null hypothesis more persuasively than the others, discredits it by little more than a factor of 1.27 to 1 (assuming $u' = 1$) instead of the 20 to 1 which a naive interpretation of the .05 level might (contrary to classical as well as Bayesian theory) lead one to expect. More important, Experiments 3 and 4, which would lead a classical statistician to reject the null hypothesis, leave the Bayesian who happens to have a roughly uniform prior, more confident of the null hypothesis than he was to start with. And Experiment 4 should reassure even a rather skeptical person about the truth of the null hypothesis. Here, then, is a blunt practical contradiction between conclusions produced by classical and Bayesian rules for statistical inference. Though the Bernoullian example is special, particularly in that it offers relatively general grounds for u' to be about 1, classical procedures quite typically are, from a Bayesian point of view, far too ready to reject null hypotheses.

Approximation in the spirit of stable estimation is by no means the last word on evaluating a likelihood ratio. Sometimes, as when r or $n - r$ are too

Table 1. Likelihood Ratios under the Uniform Alternative Prior and Minimum Likelihood Ratios for Various Values of n and for Values of r Just Significant at the .05 Level.

	Experiment number				
	1	2	3	4	∞
n	50	100	400	10,000	(very large)
r	32	60	220	5,098	$(n + 1.96\sqrt{n})/2$
$L(p_0; r, n)$	.8178	1.092	2.167	11.689	$.11689\sqrt{n}$
L_{min}	.1372	.1335	.1349	.1465	.1465

small, it is not applicable at all, and even when it might otherwise be applicable, subjective haze and interpersonal disagreement affecting u' may frustrate its application. The principal alternative devices known to us will be at least mentioned in connection with the present example, and most of them will be explored somewhat more in connection with later examples.

It is but an exercise in differential calculus to see that $p^r(1 - p)^{n-r}$ attains its maximum at $p = r/n$. Therefore, regardless of what $u(p)$ actually is, the likelihood ratio in favor of the null hypothesis is at least

$$L_{\min} = \frac{p_0^r(1 - p_0)^{n-r}}{\left(\dfrac{r}{n}\right)^r\left(1 - \dfrac{r}{n}\right)^{n-r}}.$$

If this number is not very small, then everyone (who does not altogether reject Bayesian ideas) must agree that the null hypothesis has not been greatly discredited. For example, since $L_{\min}$ in Table 1 exceeds .05, it is impossible for the experiments considered there that rejection at the 5% significance level should ever correspond to a nineteenfold diminution of the odds in favor of the null hypothesis. It is mathematically possible but realistically preposterous for $L_{\min}$ to be the actual likelihood ratio. That could occur only if your $u(p|H_1)$ were concentrated at r/n, and prior views are seldom so prescient.

It is often possible to name, whether for yourself alone or for "the public," a number $u*$ that is a generous upper bound for $u(p|H_1)$, that is, a $u*$ of which you are quite confident that $u(p|H_1) < u*$ for all p (in the interval from 0 to 1). A calculation much like Equations 11 and 13 shows that if $u*$ is substituted for u' in Equation 13, the resultant fraction is less than the actual likelihood ratio. If this method of finding a lower bound for L is not as secure as that of the preceding paragraph, it generally provides a better, that is, a bigger, one. The two methods can be blended into one which is always somewhat better than either, as will be illustrated in a later example.

Upper, as well as lower, bounds for L are important. One way to obtain one is to paraphrase the method of the preceding paragraph with a lower bound rather than an upper bound for $u(p)$. This method will seldom be applicable as stated, since $u(p)$ is likely to be very small for some values of p, especially values near 0 or 1. But refinements of the method, illustrated in later examples, may be applicable.

Another avenue, in case $u(p|H_1)$ is known with even moderate precision but is not gentle enough for the techniques or stable estimation, is to approximate $u(p|H_1)$ by the beta density for some suitable indices a and b. This may be possible since the two adjustable indices of the beta distribution provide considerable latitude and since what is required of the approximation is rather limited. It may be desirable, because beta densities are conjugate to Bernoullian experiments. In fact, if $u(p|H_1)$ is a beta distribution with indices a and b, then $u(p|r, n, H_1)$ is also a beta density, with indices $a + r$ and $b + (n - r)$. The likelihood ratio in this case is

$$\frac{(a - 1)!(b - 1)!(n + a + b - 1)!}{(a + r - 1)!(b + n - r - 1)!(a + b - 1)!} p_0^r (1 - p_0)^{n-r}.$$

These facts are easy consequences of the definite integral on which Equation 11 is based. More details will be found in Chapter 9 of Raiffa and Schlaifer (1961).

A One-Dimensional Normal Example

We examine next one situation in which classical statistics prescribes a two-tailed t test. As in our discussion of normal measurements in the section on distribution theory, we will consider one normally distributed observation with known variance; as before, this embraces by approximation the case of 25 or more observations of unknown variance and many other applications such as the Bernoullian experiments.

According to Weber's Law, the ratio or the just noticeable difference between two sensory magnitudes to the magnitude at which the just noticeable difference is measured is a constant, called the Weber fraction. The law is approximately true for frequency discrimination of fairly loud pure tones, say between 2,000 and 5,000 cps; the Weber fraction is about .0020 over this fairly wide range of frequencies. Psychophysicists disagree about the nature and extent of interaction between different sense modalities. You might, therefore, wonder whether there is any difference between the Weber fraction at 3,000 cps for subjects in a lighted room and in complete darkness. Since search for such interactions among modalities has failed more often than it has succeeded, you might give considerable initial credence to the null hypothesis that there will be no (appreciable) difference between the Weber fractions obtained in light and in darkness. However, such effects might possibly be substantial. If they are, light could facilitate or could hinder frequency discrimination. Some work on arousal might lead you to expect facilitation; the idea of visual stimuli competing with auditory stimuli for attention might lead you to expect hindrance. If the null hypothesis is false, you might consider any value between .0010 and .0030 of the Weber fraction obtained in darkness to be roughly as plausible as any other value in that range. Your instruments and procedure permit determination of the Weber fraction with a standard deviation of 3.33×10^{-5} (a standard deviation of .1 cps at 3,000 cps, which is not too implausible if your procedures permit repeated measurements and are in other ways extremely accurate). Thus the range of plausible values is 60 standard deviations wide—quite large compared with similar numbers in other parts or experimental psychology, though small compared with many analogous numbers in physics or chemistry. Such a small standard deviation relative to the range of plausible values is not indispensable to the example, but it is convenient and helps make the example congenial to both physical and social scientists. If the standard deviation were more than 10^{-4}, however, the eventual application of the principle of stable estimation to the example would be rather difficult to justify.

A full Bayesian analysis of this problem would take into account that each observation consists or two Weber fractions, rather than one difference between them. However, as classical statistics is even too ready to agree, little if any error will result from treating the difference between each Weber fraction determined in light and the corresponding Weber fraction determined in darkness as a single observation. In that formulation, the null hypothesis is that the true difference is 0, and the alternative hypothesis envisages the true difference as probably between $-.0010$ and $+.0010$. The standard deviation of the measurement of the difference, if the measurements in light and darkness are independent, is $1.414 \times 3.33 \times 10^{-5} = 4.71 \times 10^{-5}$. Since our real concern is exclusively with differences between Weber fractions and the standard deviation of these differences, it is convenient to measure every difference between Weber fractions in standard deviations, that is to multiply it by $21,200 (= 1/\sigma)$. In these new units, the plausible range of observations is about from -21 to $+21$, and the standard deviation of the differences is 1. The rest of the discussion of this example is based on these numbers alone.

The example specified by the last two paragraphs has a sharp null hypothesis and a rather diffuse symmetric alternative hypothesis with good reasons for associating substantial prior probability with each. Although realistically the null hypothesis cannot be infinitely sharp, calculating as though it were is an excellent approximation. Realism, and even mathematical consistency, demands far more sternly that the alternative hypothesis not be utterly diffuse (that is, uniform from $-\infty$ to $+\infty$); otherwise, no measurement of the kind contemplated could result in any opinion other than certainty that the null hypothesis is correct.

Having already assumed that the distribution of the true parameter or parameters under the null hypothesis is narrow enough to be treated as though it were concentrated at the single point 0, we also assume that the distribution of the datum given the parameter is normal with moderate variance. By moderate we mean large relative to the sharp null hypothesis but (in most cases) small relative to the distribution under the alternative hypothesis of the true parameter.

Paralleling our treatment of the Bernoullian example, we shall begin, after a neutral formulation, with an approximation akin to stable estimation, then explore bounds on the likelihood ratio L that depend on far less stringent assumptions, and finally explore normal prior distributions.

Without specifying the form of the prior distribution under the alternative hypothesis, the likelihood ratio in the Weber-fraction example under discussion is

$$L(\lambda_0; x) = \frac{\dfrac{1}{\sigma}\varphi\left(\dfrac{x - \lambda_0}{\sigma}\right)}{\int \dfrac{1}{\sigma}\varphi\left(\dfrac{x - \lambda}{\sigma}\right) u(\lambda|\mathrm{H}_1)\, d\lambda}. \tag{14}$$

The numerator is the density of the datum x under the null hypothesis; σ is the standard deviation of the measuring instrument. The denominator is the density of x under the alternative hypothesis. The values of λ are the possible values of the actual difference under the alternative hypothesis, and λ_0 is the null value, 0. $\varphi[(x - \lambda)/\sigma]$ is the ordinate of the standard normal density at the point $(x - \lambda)/\sigma$. Hereafter, we will use the familiar statistical abbreviation $t = (x - \lambda_0)/\sigma$ for the t of the classical t test. Finally, $u(\lambda|H_1)$ is the prior probability density of λ under the alternative hypothesis.

If $u(\lambda|H_1)$ is gentle in the neighborhood of x and not too violent elsewhere, a reasonable approximation to Equation 14, akin to the principle of stable estimation, is

$$L(\lambda_0; x) = \frac{\varphi(t)}{\sigma u(x)}. \tag{15}$$

According to a slight variation of the principle, already used in the Bernoullian example, near linearity may justify this approximation even better than near constancy does. Since σ is measured in the same units as x or λ, say, degrees centigrade or cycles per second, and $u(x)$ is probability per degree centigrade or per cycle per second, the product $\sigma u(x)$ (in the denominator of Equation 15) is dimensionless. Visualizing $\sigma u(x)$ as a rectangle of base σ, centered at x, and height $u(x)$, we see $\sigma u(x)$ to be approximately your prior probability for an interval of length σ in the region most favored by the data.

Lower Bounds on L

An alternative when $u(\lambda|H_1)$ is not diffuse enough to justify stable estimation is to seek bounds on L. Imagine all the density under the alternative hypothesis concentrated at x, the place most favored by the data. The likelihood ratio is then

$$L_{\min} = \frac{\varphi(t)}{\varphi(0)} = e^{-(1/2)t^2}.$$

This is of course the very smallest likelihood ratio that can be associated with t. Since the alternative hypothesis now has all its density on one side of the null hypothesis, it is perhaps appropriate to compare the outcome of this procedure with the outcome of a one-tailed rather than a two-tailed classical test. At the one-tailed classical .05, .01, and .001 points, $L_{\min}$ is .26, .066, and .0085, respectively. Even the utmost generosity to the alternative hypothesis cannot make the evidence in favor of it as strong as classical significance levels might suggest. Incidentally, the situation is little different for a two-tailed classical test and a prior distribution for the alternative hypothesis concentrated symmetrically at a pair of points straddling the null value. If the prior distribution under the alternative hypothesis is required to be not only symmetric around the null value but also unimodal, which seems very safe for many problems, then the results are too similar to those obtained later for the smallest possible likelihood ratio obtainable with a symmetrical normal prior density to merit separate presentation here.

Upper Bounds on L

In order to discredit a null hypothesis, it is useful to find a practical upper bound on the likelihood ratio L, which can result in the conclusion that L is very small. It is impossible that $u(\lambda|H_1)$ should exceed some positive number for all λ, but you may well know plainly that $u(\lambda|H_1) \geqq u^* > 0$ for all λ in some interval, say of length 4, centered at x. In this case,

$$
\begin{aligned}
L(\lambda_0; x) &\leqq \frac{\varphi(t)}{\displaystyle\int_{-2}^{+2} \varphi\left(\frac{x-\lambda}{\sigma}\right) u(\lambda|H_1)\, d\lambda} \\[2ex]
&\leqq \frac{\varphi(t)}{\sigma u_*[\Phi(2) - \Phi(-2)]} \\[2ex]
&\leqq \frac{1.05 e^{-(1/2)t^2}}{\sqrt{2\pi}\sigma u_*} \leqq \frac{0.42 e^{-(1/2)t^2}}{\sigma u_*} \\[2ex]
&= \frac{0.42 L_{\min}}{\sigma u_*}.
\end{aligned}
$$

If, for example, you attach as much probability as .01 to the intervals of length σ near x, your likelihood ratio is at most $42\, L_{\min}$.

For t's classically significant at the .05, .01, and .001 levels, your likelihood ratio is correspondingly at most 10.9, 2.8, and .36. This procedure can discredit null hypotheses quite strongly; t's of 4 and 5 lead to upper bounds on your likelihood ratio of .014 and .00016, insofar as the normal model can be taken seriously for such large t's.

Normal Alternative Priors

Since normal densities are conjugate to normal measurements, it is natural to study the assumption that $u(\lambda|H_1)$ is a normal density. This assumption may frequently be adequate as an approximation, and its relative mathematical simplicity paves the way to valuable insights that may later be substantiated with less arbitrary assumptions. In this paper we explore not all normal alternative priors but only those symmetrical about λ_0, which seem especially important.

Let $u(\lambda|H_1)$, then, be normal with mean λ_0 and with some standard deviation τ. Equation 14 now specializes to

$$
\begin{aligned}
L(\lambda_0; x) &= \frac{\dfrac{1}{\sigma}\varphi(t)}{\dfrac{1}{\sqrt{\sigma^2 + \tau^2}}\varphi\left(\dfrac{x - \lambda_0}{\sqrt{\sigma^2 + \tau^2}}\right)} \\[3ex]
&= \frac{\varphi(t)}{\alpha\varphi(\alpha t)}, \tag{18}
\end{aligned}
$$

Table 2. Values of $L(\alpha, t)$ for Selected Values of α and for Values of t Corresponding to Familiar Two-Tailed Significance Levels.

		t and Significance level				
		1.654	1.960	2.576	3.291	3.891
α	σ/τ	.10	.05	.01	.001	.0001
.0001	.0001	2,585	1,465	362	44.6	5.16
.001	.0010	259	147	36.2	4.46	.516
.01	.0100	25.9	14.7	3.63	.446	.0516
.025	.0250	10.4	5.87	1.45	.179	.0207
.05	.0501	5.19	2.94	.731	.0903	.0105
.075	.0752	3.47	1.97	.492	.0612	.00718
.1	.1005	2.62	1.49	.375	.0470	.00556
.15	.1517	1.78	1.02	.260	.0336	.00408
.2	.2041	1.36	.791	.207	.0277	.00349
.5	.5774	.725	.474	.166	.0345	.00685
.9	2.0647	.859	.771	.592	.397	.264
.99	7.0179	.983	.972	.946	.907	.869

where

$$\alpha = \frac{\sigma}{\sqrt{\sigma^2 + \tau^2}} = \frac{1}{\sqrt{1 + (\tau/\sigma)^2}}$$

Plainly, α is a function of σ/τ and vice versa; for small values of either, the difference between α and σ/τ is negligible. We emphasize α rather than the intuitively more appealing σ/τ because α leads to simpler equations. Of course, α is less than one, typically much less. Writing the normal density in explicit form,

$$L(\alpha, t) = \frac{1}{\alpha} \exp -\frac{1}{2}(1 - \alpha^2)t^2. \qquad [19]$$

Table 2 shows numerical values of $L(\alpha, t)$ for some instructive values of α and for values of t corresponding to familiar two-tailed classical significance levels. The values of α between .01 and .1 portray reasonably precise experiments; the others included in Table 2 are instructive as extreme possibilities. Table 2 again illustrates how classically significant values of t can, in realistic cases, be based on data that actually favor the null hypothesis.

For another comparison of Equation 18 with classical tests consider that (positive) value t_0 of t for which L is 1. If L is 1, then the posterior odds for the two hypotheses will equal the prior odds; the experiment will leave opinion about H_0 and H_1 unchanged, though it is bound to influence opinion about λ given H_1. Taking natural logarithms of Equation 19 for $t = t_0$,

$$\ln \frac{1}{\alpha} - \frac{1}{2}(1 - \alpha^2)t_0^2 = 0, \qquad t_0 = \left\{ \frac{-\ln \alpha^2}{1 - \alpha^2} \right\}^{1/2} \qquad [20]$$

Table 3. Values of t_0 and Their
Significance Levels for Normal
Alternative Prior Distributions
for Selected Values of α.

α	t_0	Significance level
.1	2.157	.031
.05	2.451	.014
.01	3.035	.0024
.001	3.718	.00020
.0001	4.292	.000018

If α is small, say less than .1, then $1 - \alpha^2$ is negligibly different from 1, and so $t_0 \simeq \sqrt{-\ln \alpha^2}$. The effect of using this approximation can never be very bad; for the likelihood ratio actually associated with the approximate value of t_0 cannot be less than 1 or greater than 1.202. Table 3 presents a few actual values of t_0 and their corresponding two-tailed significance levels. At values of t slightly smaller than the break-even values in Table 3 classical statistics more or less vigorously rejects the null hypothesis, though the Bayesian described by α becomes more confident of it than he was to start with.

If $t = 0$, that is, if the observation happens to point exactly to the null hypothesis, $L = \dfrac{1}{\alpha}$; thus support for the null hypothesis can be very strong, since α might well be about .01. In the example, you perhaps hope to confirm the null hypothesis to everyone's satisfaction, if it is in fact true. You will therefore try hard to make σ small enough so that your own α and those of your critics will be small. In the Weber-fraction example, $\alpha \simeq .077$ (calculated by assuming that 90% of the prior probability under the alternative hypothesis falls between -21 and $+21$; assuming normality, it follows that $r \simeq 12.9$). If $t = 0$, then L is 12.9—persuasive but not irresistible evidence in favor of the null hypothesis. For $\alpha = .977$, t_0 is 2.3—just about the .02 level of a classical two-tailed test. Conclusion: An experiment strong enough to lend strong support to the null hypothesis when $t = 0$ will mildly support the null hypothesis even when classical tests would strongly reject it.

If you are seriously interested in supporting the null hypothesis if it is true—and you may well be, valid aphorisms about the perishability of hypotheses notwithstanding—you should so design your experiment that even a t as large as 2 or 3 strongly confirms the null hypothesis. If α is .0001, L is more than 100 for any t between -3 and $+3$. Such small α's do not occur every day, but they are possible. Maxwell's prediction of the equality of the "two speeds of light" might be an example. A more practical way to prove a null hypothesis may be to investigate several, not just one of its numerical consequences. It is not clear just what sort of evidence classical statistics would

Table 4. Values of L_{normin} and of L_{min} for Values of t
Corresponding to Familiar Two-Tailed Significance
Levels.

t	Significance level	L_{normin}	L_{min}
1.960	.05	.473	.146
2.576	.01	.154	.0362
3.291	.001	.0241	.00445
3.891	.0001	.00331	.000516

regard as strong confirmation of a null hypothesis. (See however Berkson, 1942.)

What is the smallest likelihood ratio L_{normin} (the minimum L for a symmetrical normal prior) that can be attained for a given t by artificial choice of α? It follows from Equation 19 that L is minimized at $\alpha = |t|^{-1}$, provided $|t| \geq 1$, and at the unattainable value $\alpha = 1$, otherwise.

$$L_{normin} = e^{1/2}|t|e^{-(1/2)t^2} = 1.65|t|e^{-(1/2)t^2} \qquad \text{for } |t| \geq 1$$
$$= 1 \qquad \text{for } |t| \leq 1.$$

With any symmetric normal prior, any $|t| \leq 1$ speaks for the null hypothesis. So L_{normin} exceeds L_{min} in all cases and exceeds it by the substantial factor $1.65|t|$ if $|t| \geq 1$. Values of t corresponding to familiar two-tailed significance levels and the corresponding values of L_{normin} are shown in Table 4.

From this examination of one-dimensional normally distributed observations, we conclude that a t of 2 or 3 may not be evidence against the null hypothesis at all, and seldom if ever justifies much new confidence in the alternative hypothesis. This conclusion has a melancholy side. The justification for the assumption of normal measurements must in the last analysis be empirical. Few applications are likely to justify using numerical values of normal ordinates more than three standard deviations away from the mean. And yet without those numerical values, the methods of this section are not applicable. In short, in one-dimensional normal cases, evidence that does not justify rejection of the null hypothesis by the interocular traumatic test is unlikely to justify firm rejection at all.

Haunts of χ^2 and F

Classical tests of null hypotheses invoking the χ^2, and closely related F, distributions are so familiar that something must be said here about their Bayesian counterparts. Though often deceptively oversimplified, the branches of statistics that come together here are immense and still full of fundamental mysteries for Bayesians and classicists alike (Fisher, 1925, see Ch. 4 and 5 in the

1954 edition; Green & Tukey, 1960; Scheffé, 1959; Tukey, 1962). We must therefore confine ourselves to the barest suggestions.

Much of the subject can be reduced to testing whether several parameters λ_i measured independently with known variance σ^2 have a specified common value. This multidimensional extension of the one-dimensional normal problem treated in the last section is so important that we shall return to it shortly.

(*Editors' note*: The Section "Multidimensional Normal Measurements and a Null Hypothesis" has been omitted.)

Some Morals About Testing Sharp Null Hypotheses

At first glance, our general conclusion that classical procedures are so ready to discredit null hypotheses that they may well reject one on the basis of evidence which is in its favor, even strikingly so, may suggest the presence of a mathematical mistake somewhere. Not so; the contradiction is practical, not mathematical. A classical rejection of a true null hypothesis at the .05 level will occur only once in 20 times. The overwhelming majority of these false classical rejections will be based on test statistics close to the borderline value; it will often be easy to demonstrate that these borderline test statistics, unlikely under either hypothesis, are nevertheless more unlikely under the alternative than under the null hypothesis, and so speak for the null hypothesis rather than against it.

Bayesian procedures can strengthen a null hypothesis, not only weaken it, whereas classical theory is curiously asymmetric. If the null hypothesis is classically rejected, the alternative hypothesis is willingly embraced, but if the null hypothesis is not rejected, it remains in a kind of limbo of suspended disbelief. This asymmetry has led to considerable argument about the appropriateness of testing a theory by using its predictions as a null hypothesis (Grant, 1962; Guilford, 1942, see p. 186 in the 1956 edition; Rozeboom, 1960; Sterling, 1960). For Bayesians, the problem vanishes, though they must remember that the null hypothesis is really a hazily defined small region rather than a point.

The procedures which have been presented simply compute the likelihood ratio of the hypothesis that some parameter is very nearly a specified single value with respect to the hypothesis that it is not. They do not depend on the assumption of special initial credibility of the null hypothesis. And the general conclusion that classical procedures are unduly ready to reject null hypotheses is thus true whether or not the null hypothesis is especially plausible a priori. At least for Bayesian statisticians, however, no procedure for testing a sharp null hypothesis is likely to be appropriate unless the null hypothesis deserves special initial credence. It is uninteresting to learn that the odds in favor of the null hypothesis have increased or decreased a hundredfold if initially they were negligibly different from zero.

How often are Bayesian and classical procedures likely to lead to different conclusions in practice? First, Bayesians are unlikely to consider a sharp null hypothesis nearly so often as do the consumers of classical statistics. Such procedures make sense to a Bayesian only when his prior distribution has a sharp spike at some specific value; such prior distributions do occur, but not so often as do classical null hypothesis tests.

When Bayesians and classicists agree that null hypothesis testing is appropriate, the results of their procedures will usually agree also. If the null hypothesis is false, the interocular traumatic test will often suffice to reject it; calculation will serve only to verify clear intuition. If the null hypothesis is true, the interocular traumatic test is unlikely to be of much use in one-dimensional cases, but may be helpful in multidimensional ones. In at least 95% of cases when the null hypothesis is true, Bayesian procedures and the classical .05 level test agree. Only in borderline cases will the two lead to conflicting conclusions. The widespread custom of reporting the highest classical significance level from among the conventional ones actually attained would permit an estimate of the frequency of borderline cases in published work; any rejection at the .05 or .01 level is likely to be borderline. Such an estimate of the number of borderline cases may be low, since it is possible that many results not significant at even the .05 level remain unpublished.

The main practical consequences for null hypothesis testing of widespread adoption of Bayesian statistics will presumably be a substantial reduction in the resort to such tests and a decrease in the probability of rejecting true null hypotheses, without substantial increase in the probability of accepting false ones.

If classical significance tests have rather frequently rejected true null hypotheses without real evidence, why have they survived so long and so dominated certain empirical sciences? Four remarks seem to shed some light on this important and difficult question.

1. In principle, many of the rejections at the .05 level are based on values of the test statistic far beyond the borderline, and so correspond to almost unequivocal evidence. In practice, this argument loses much of its force. It has become customary to reject a null hypothesis at the highest significance level among the magic values, .05, .01, and .001, which the test statistic permits, rather than to choose a significance level in advance and reject all hypotheses whose test statistics fall beyond the criterion value specified by the chosen significance level. So a .05 level rejection today usually means that the test statistic was significant at the .05 level but not at the .01 level. Still, a test statistic which falls just short of the .01 level may correspond to much stronger evidence against a null hypothesis than one barely significant at the .05 level. The point applies more forcibly to the region between .01 and .001, and for the region beyond, the argument reverts to its original form.

2. Important rejections at the .05 or .01 levels based on test statistics which would not have been significant at higher levels are not common. Psychologists tend to run relatively large experiments, and to get very highly significant main effects. The place where .05 level rejections are most common is in testing interactions in analyses of variance—and few experimenters take

those tests very seriously, unless several lines of evidence point to the same conclusions.

3. Attempts to replicate a result are rather rare, so few null hypothesis rejections are subjected to an empirical check. When such a check is performed and fails, explanation of the anomaly almost always centers on experimental design, minor variations in technique, and so forth, rather than on the meaning of the statistical procedures used in the original study.

4. Classical procedures sometimes test null hypotheses that no one would believe for a moment, no matter what the data; our list of situations that might stimulate hypothesis tests earlicr in the section included several examples. Testing an unbelievable null hypothesis amounts, in practice, to assigning an unreasonably large prior probability to a very small region of possible values of the true parameter. In such cases, the more the procedure is biased against the null hypothesis, the better. The frequent reluctance of empirical scientists to accept null hypotheses which their data do not classically reject suggests their appropriate skepticism about the original plausibility of these null hypotheses.

Likelihood Principle

A natural question about Bayes' theorem leads to an important conclusion, the likelihood principle, which was first discovered by certain classical statisticians (Barnard, 1947; Fisher, 1956).

Two possible experimental outcomes D and D'—not necessarily of the same experiment—can have the same (potential) bearing on your opinion about a partition of events H_i, that is, $P(H_i|D)$ can equal $P(H_i|D')$ for each i. Just when are D and D' thus evidentially equivalent, or of the same import? Analytically, when is

$$[P(H_i|D) =]\frac{P(D|H_i)P(H_i)}{P(D)} = \frac{P(D'|H_i)P(H_i)}{P(D')}[= P(H_i|D')] \qquad [23]$$

for each i?

Aside from such academic possibilities as that some of the $P(H_i)$ are 0, Equation 23 plainly entails that, for some positive constant k and for all i,

$$P(D'|H_i) = kP(D|H_i). \qquad [24]$$

But Equation 24 implies Equation 23, from which it was derived, no matter what the initial probabilities $P(H_i)$ are, as is easily seen thus:

$$P(D') = \Sigma P(D'|H_i)P(H_i)$$

$$= k\Sigma P(D|H_i)P(H_i)$$

$$= kP(D).$$

This conclusion is the likelihood principle: Two (potential) data D and D' are of the same import if Equation 24 obtains.

Since for the purpose of drawing inference, the sequence of numbers $P(D|H_i)$ is, according to the likelihood principle, equivalent to any other sequence obtained from it by multiplication by a positive constant, a name for this class of equivalent sequences is useful and there is precedent for calling it the likelihood (of the sequence of hypotheses H_i given the datum D). (This is not quite the usage of Raiffa & Schlaifer, 1961.) The likelihood principle can now be expressed thus: D and D' have the same import if $P(D|H_i)$ and $P(D'|H_i)$ belong to the same likelihood—more idiomatically, if D and D' have the same likelihood.

If, for instance, the partition is two-fold, as it is when you are testing a null hypothesis against all alternative hypothesis, then the likelihood to which the pair $[P(D|H_0), P(D|H_1)]$ belongs is plainly the set of pairs of numbers $[a, b]$ such that the fraction a/b is the already familiar likelihood ratio $L(H_0; D) = P(D|H_0)/P(D|H_1)$. The simplification of the theory of testing by the use of likelihood ratios in place of the pairs of conditional probabilities, which we have seen, is thus an application of the likelihood principle.

Of course, the likelihood principle applies to a (possibly multidimensional) parameter λ as well as to a partition H_i. The likelihood of D, or the likelihood to which $P(D|\lambda)$ belongs, is the class of all those functions of λ that are positive constant multiples of (that is, proportional to) the function $P(D|\lambda)$. Also, conditional densities can replace conditional probabilities in the definition of likelihood ratios.

There is one implication of the likelihood principle that all statisticians seem to accept. It is not appropriate in this paper to pursue this implication, which might be called the principle of sufficient statistics, very far. One application of sufficient statistics so familiar as almost to escape notice will, however, help bring out the meaning of the likelihood principle. Suppose a sequence of 100 Bernoulli trials is undertaken and 20 successes and 80 failures are recorded. What is the datum, and what is its probability for a given value of the frequency p? We are all perhaps overtrained to reply, "The datum is 20 successes out of 100, and its probability, given p, is $C_{20}^{100}p^{20}(1-p)^{80}$." Yet it seems more correct to say, "The datum is this particular sequence of successes and failures, and its probability, given p, is $p^{20}(1-p)^{80}$." The conventional reply is often more convenient, because it would be costly to transmit the entire sequence of observations; it is permissible, because the two functions $C_{20}^{100}p^{20}(1-p)^{80}$ and $p^{20}(1-p)^{80}$ belong to the same likelihood; they differ only by the constant factor C_{20}^{100}. Many classical statisticians would demonstrate this permissibility by an argument that does not use the likelihood principle, at least not explicitly (Halmos & Savage, 1949, p. 235). That the two arguments are much the same, after all, is suggested by Birnbaum (1962). The legitimacy of condensing the datum is often expressed by saying that the number of successes in a given number of Bernoulli trials is a sufficient statistic for the sequence of trials. Insofar as the sequence of trials is not altogether accepted as Bernoullian—and it never is—the condensation is not legitimate. The practical experimenter always has some incentive to look over the sequence of his data with a view to discovering periodicities, trends, or other departures from Bernoullian expectation. Anyone to whom the sequence is not available, such as the reader of a condensed report or the experimentalist who depends on automatic counters, will reserve some doubt about the interpretation of the ostensibly sufficient statistic.

Moving forward to another application or the likelihood principle, imagine a different Bernoullian experiment in which you have undertaken to continue the trials until 20 successes were accumulated and the twentieth success happened to be the one hundredth trial. It would be conventional and justifiable to report only this fact, ignoring other details of the sequence of trials. The probability that the twentieth success will be the one hundredth trial is, given p, easily seen to be $C_{19}^{99} p^{20}(1 - p)^{80}$ This is exactly 1/5 of the probability of 20 successes in 100 trials, so according to the likelihood principle, the two data have the same import. This conclusion is even a trifle more immediate if the data are not condensed; for a specific sequence of 100 trials of which the last is the twentieth success has the probability $p^{20}(1 - p)^{80}$ in both experiments. Those who do not accept the likelihood principle believe that the probabilities of sequences that might have occurred, but did not, somehow affect the import of the sequence that did occur.

In general, suppose that you collect data of any kind whatsoever—not necessarily Bernoullian, nor identically distributed, nor independent of each other given the parameter λ—stopping only when the data thus far collected satisfy some criterion of a sort that is sure to be satisfied sooner or later, then the import of the sequence of n data actually observed will be exactly the same as it would be had you planned to take exactly n observations in the first place. It is not even necessary that you stop according to a plan. You may stop when tired, when interrupted by the telephone, when you run out of money, when you have the casual impression that you have enough data to prove your point, and so on. The one proviso is that the moment at which your observation is interrupted must not in itself be any clue to λ that adds anything to the information in the data already at hand. A man who wanted to know how frequently lions watered at a certain pool was chased away by lions before he actually saw any of them watering there; in trying to conclude how many lions do water there he should remember why his observation was interrupted when it was. We would not give a facetious example had we been able to think of a serious one. A more technical discussion of the irrelevance of stopping rules to statistical analysis is on pages 36–42 or Raiffa and Schlaifer (1961).

This irrelevance or stopping rules to statistical inference restores a simplicity and freedom to experimental design that had been lost by classical emphasis on significance levels (in the sense of Neyman and Pearson) and on other concepts that are affected by stopping rules. Many experimenters would like to feel free to collect data until they have either conclusively proved their point, conclusively disproved it, or run out of time, money, or patience. Classical statisticians (except possibly for the few classical defenders of the likelihood principle) have frowned on collecting data one by one or in batches, testing the total ensemble after each new item or batch is collected, and stopping the experiment only when a null hypothesis is rejected at some preset significance level. And indeed if an experimenter uses this procedure, then with probability 1 he will eventually reject any sharp null hypothesis, even though it be true. This is perhaps simply another illustration of the over-readiness of classical procedures to reject null hypotheses. In contrast, if you set out to collect data until your posterior probability for a hypothesis which unknown to you is true has been reduced to .01, then 99 times out of 100 you will never make it, no matter how many data you, or your children after you, may collect. (Rules which have nonzero probability of running forever ought not, and here will not, be called stopping rules at all.)

The irrelevance of stopping rules is one respect in which Bayesian procedures are more objective than classical ones. Classical procedures (with the possible exceptions implied above) insist that the intentions or the experimenter are crucial to the interpretation of data, that 20 successes in 100 observations means something quite different if the experimenter intended the 20 successes than if he intended the 100 observations. According to the likelihood principle, data analysis stands on its own feet. The intentions of the experimenter are irrelevant to the interpretation of the data once collected, though of course they are crucial to the design of experiments.

The likelihood principle also creates unity and simplicity in inference about Markov chains and other stochastic processes (Barnard, Jenkins, & Winsten, 1962), which are sometimes applied in psychology. It sheds light on many other problems of statistics, such as the role of unbiasedness and Fisher's concept of ancillary statistic. A principle so simple with consequences so pervasive is bound to be controversial. For dissents see Stein (1962), Wolfowitz (1962), and discussions published with Barnard, Jenkins, and Winsten (1962), Birnbaum (1962), and Savage et al. (1962) indexed under likelihood principle.

In Retrospect

Though the Bayesian view is a natural outgrowth of classical views, it must be clear by now that the distinction between them is important. Bayesian procedures are not merely another tool for the working scientist to add to his inventory along with traditional estimates of means, variances, and correlation coefficients, and the t test, F test, and so on. That classical and Bayesian statistics are sometimes incompatible was illustrated in the theory of testing. For, as we saw, evidence that leads to classical rejection of the null hypothesis will often leave a Bayesian more confident of that same null hypothesis than he was to start with. Incompatibility is also illustrated by the attention many classical statisticians give to stopping rules that Bayesians find irrelevant.

The Bayesian outlook is flexible, encouraging imagination and criticism in its everyday applications. Bayesian experimenters will emphasize suitably chosen descriptive statistics in their publications, enabling each reader to form his own conclusion Where an experimenter can easily foresee that his readers will want the results of certain calculations (as for example when the data seem sufficiently precise to justify for most readers application of the principle of stable estimation) he will publish them. Adoption of the Bayesian outlook should discourage parading statistical procedures, Bayesian or other, as symbols of respectability pretending to give the imprimatur of mathematical logic to the subjective process of empirical inference.

We close with a practical rule which stands rather apart from any conflicts between Bayesian and classical statistics. The rule was somewhat overstated by a physicist who said, "As long as it takes statistics to find out, I prefer to investigate something else." Of course, even in physics some important questions must be investigated before technology is sufficiently developed to do so definitively. Still, when the value of doing so is recognized, it is often possible so to design experiments that the data speak for themselves without the intervention of subtle theory or insecure personal judgments. Estimation is best when it is stable. Rejection of a null hypothesis is best when it is interocular.

References

Anscombe, F.J. Bayesian statistics. *Amer. Statist.*, 1961, **15**(1), 21–24.

Bahadur, R.R., & Robbins, H. The problem of the greater mean. *Ann. Math. Statist.*, 1950, **21**, 469–487.

Barnard, G.A. A review of "Sequential Analysis" by Abraham Wald. *J. Amer. Statist. Ass.*, 1947, **42**, 658–664.

Barnard, G.A., Jenkins, G.M., & Winsten, C.B. Likelihood, inferences, and time series. *J. Roy. Statist. Soc.*, 1962, **125**(Ser. A), 321–372.

Bayes, T. Essay towards solving a problem in the doctrine of chances. *Phil. Trans. Roy. Soc.*, 1763, **53**, 370–418. (Reprinted: *Biometrika*, 1958, **45**, 293–315.)

Berkson, J. Some difficulties of interpretation encountered in the application of the chi-square test. *J. Amer. Statist. Ass.*, 1938, **33**, 526–542.

Berkson, J. Tests of significance considered as evidence. *J. Amer. Statist. Ass.*, 1942, **37**, 325–335.

Birnbaum, A. On the foundations of statistical inference. *J. Amer. Statist. Ass.*, 1962, **57**, 269–306.

Blackwell, D., & Dubins, L. Merging of opinions with increasing information. *Ann. Math. Statist.*, 1962, **33**, 882–886.

Borel, E. La théorie du jeu et les équations intégrales à noyau symétrique. *CR Acad. Sci., Paris*, 1921, **173**, 1304–1308. (Trans. by L.J. Savage, *Econometrica*, 1953, **21**, 97–124)

Borel, E. A propos d'un traité de probabilités. *Rev. Phil.*, 1924, **98**, 321–336. (Reprinted: In: *Valeur pratique et philosophie des probabilités*. Paris: Gauthier-Villars, 1939. Pp. 134–146)

Bridgman, P.W. A critique of critical tables. *Proc. Nat. Acad. Sci.*, 1960, **46**, 1394–1401.

Cramér, H. *Mathematical methods of statistics*. Princeton: Princeton Univer. Press, 1946.

de Finetti, B. Fondamenti logici del ragionamento probabilistico. *Boll. Un. mat. Ital.*, 1930, **9**(Ser. A), 258–261.

de Finetti, B. La prévision: Ses lois logiques, ses sources subjectives. *Ann. Inst. Henri Poincaré*, 1937, **7**, 1–68.

de Finetti, B. La probabilità e la statistica nei rapporti con l'induzione, secondo i diversi punti da vista. In, *Induzione & statistica*. Rome, Italy: Istituto Matematico dell'Universita, 1959.

de Finetti, B., & Savage, L.J. Sul modo di scegliere le probabilità iniziali. In, *Biblioteca del "metron."* Ser. C, Vol. 1. *Sui fondamenti della statistica*. Rome: University of Rome, 1962. Pp. 81–154.

Edwards, W. Dynamic decision theory and probabilistic information processing. *Hum. Factors*, 1962, **4**, 59–73. (a)

Edwards, W. Subjective probabilities inferred from decisions. *Psychol. Rev.*, 1962, **69**, 109–135. (b)

Edwards, W. Probabilistic information processing in command and control systems. Report No. 3780-12-T, 1963. Institute of Science and Technology, University of Michigan.

Fisher, R.A. *Statistical methods for research workers.* (12th ed., 1954) Edinburgh: Oliver & Boyd, 1925.

Fisher, R.A. *Contributions to mathematical statistics.* New York: Wiley, 1950.

Fisher, R.A. *Statistical methods and scientific inference.* (2nd ed., 1959) Edinburgh: Oliver & Boyd, 1956.

Good, I.J. *Probability and the weighing of evidence.* New York: Hafner, 1950.

Good, I.J. Weight of evidence, corroboration, explanatory power, information and the utility of experiments. *J. Roy. Statist. Soc.*, 1960, **22**(Ser. B), 319–331.

Grant, D.A. Testing the null hypothesis and the strategy and tactics of investigating theoretical models. *Psychol. Rev.*, 1962, **69**, 54–61.

Grayson, C.J., Jr. *Decisions under uncertainty: Drilling decisions by oil and gas operators.* Boston: Harvard Univer. Press, 1960.

Green, B.J., Jr., & Tukey, J.W. Complex analysis of variance: General problems. *Psychometrika*, 1960, **25**, 127–152.

Guilford, J.P. *Fundamental statistics in psychology and education.* (3rd ed., 1956) New York: McGraw-Hill, 1942.

Halmos, P.R., & Savage, L.J. Application of the Radon-Nikodym theorem to the theory of sufficient statistics. *Ann. math. Statist.*, 1949, **20**, 225–241.

Hildreth, C. Bayesian statisticians and remote clients. *Econometrica*, 1963, **31**, in press.

Hodges, J.L., & Lehmann, E.L. Testing the approximate validity of statistical hypotheses. *J. Roy. Statist. Soc.*, 1954, **16**(Ser. B), 261–268.

Jeffreys, H. *Scientific inference.* (3rd ed., 1957) England: Cambridge Univer. Press, 1931.

Jeffreys, H. *Theory of probability.* (3rd ed., 1961) Oxford, England: Clarendon, 1939.

Koopman, B.O. The axioms and algebra of intuitive probability. *Ann. Math.*, 1940, **41**(Ser. 2), 269–292. (a)

Koopman, B.O. The bases of probability. *Bull. Amer. Math. Soc.*, 1940, **46**, 763–774. (b)

Koopman, B.O. Intuitive probabilities and sequences. *Ann. Math.*, 1941, **42**(Ser. 2), 169–187.

Lehmann, E.L. Significance level and power. *Ann. math. Statist.*, 1958, **29**, 1167–1176.

Lehmann, E.L. *Testing statistical hypotheses.* New York: Wiley, 1959.

Lindley, D.V. A statistical paradox. *Biometrika*, 1957, **44**, 187–192.

Lindley, D.V. The use of prior probability distributions in statistical inferences and decisions. In, *Proceedings of the fourth Berkeley symposium on mathematics and probability.* Vol. 1. Berkeley: Univer. California Press, 1961. Pp. 453–468.

Neyman, J. Outline of a theory of statistical estimation based on the classical theory of probability. *Phil. Trans. Roy. Soc.*, 1937, **236**(Ser. A), 333–380.

Neyman, J. L'estimation statistique, traitée comme un problème classique de probabilité. In, *Actualités scientifiques et industrielles*. Paris, France: Hermann & Cie, 1938. Pp. 25–57. (a)

Neyman, J. *Lectures and conferences on mathematical statistics and probability*. (2nd ed., 1952) Washington, D.C.: United States Department of Agriculture, 1938. (b)

Neyman, J. "Inductive behavior" as a basic concept of philosophy of science. *Rev. Math. Statist. Inst.*, 1957, **25**, 7–22.

Pearson, E.S. In L.J. Savage et al., *The foundations of statistical inference: A discussion*. New York: Wiley, 1962.

Pratt, J.W. Review of *Testing Statistical Hypotheses* by E.L. Lehmann. *J. Amer. Statist. Ass.*, 1961, **56**, 163–167.

Raiffa, H., & Schlaifer, R. *Applied statistical decision theory*. Boston: Harvard University, Graduate School of Business Administration, Division of Research, 1961.

Ramsey, F.P. "Truth and probability" (1926), and "Further considerations" (1928). In, *The foundation of mathematics and other essays*. New York: Harcourt, Brace, 1931.

Rozeboom, W.W. The fallacy of the null-hypothesis significance test. *Psychol. Bull.*, 1960, **57**, 416–428.

Savage, I.R. Nonparametric statistics. *J. Amer. Statist. Ass.*, 1957, **52**, 331–344.

Savage, I.R. *Bibliography of nonparametric statistics*. Cambridge: Harvard Univer. Press, 1962.

Savage, L.J. *The foundations of statistics*. New York: Wiley, 1954.

Savage, L.J. The foundations of statistics reconsidered. In, *Proceedings of the fourth Berkeley symposium on mathematics and probability*. Vol. 1. Berkeley: Univer. California Press, 1961. Pp. 575–586.

Savage, L.J. Bayesian statistics. In, *Decision and information processes*. New York: Macmillan, 1962. Pp. 161–194. (a)

Savage, L.J. Subjective probability and statistical practice. In L.J. Savage et al., *The foundations of statistical inference: A discussion*. New York: Wiley, 1962. (b)

Savage, L.J., et al. *The foundations of statistical inference: A discussion*. New York: Wiley, 1962.

Scheffé, H. *The analysis of variance*. New York: Wiley, 1959.

Schlaifer, R. *Probability and statistics for business decisions*. New York: McGraw-Hill, 1959.

Schlaifer, R. *Introduction to statistics for business decisions*. New York: McGraw-Hill, 1961.

Sinclair, H. Hiawatha's lipid. *Perspect. Biol. Med.*, 1960, **4**, 72–76.

Stein, C. A remark on the likelihood principle. *J. Roy. Statist. Soc.*, 1962, **125**(Ser. A), 565–568.

Sterling, T.D. What is so peculiar about accepting the null hypothesis? *Psychol. Rep.*, 1960, **7**, 363–364.

Tukey, J.W. The future of data analysis. *Ann. math. Statist.*, 1962, **33**, 1–67.

Urey, H.C. Origin of tektites. *Science*, 1962, **137**, 746.

von Neumann, J. Zur Theorie der Gesellschaftsspiele. *Math. Ann.*, 1928, **100**, 295–320.

von Neumann, J., & Morgenstern, O. *Theory of games and economic behavior*. (3rd ed., 1953) Princeton: Princeton Univer. Press, 1947.

Wald, A. On the principles of statistical inference. (Notre Dame Mathematical Lectures, No. 1) Ann Arbor, Mich.: Edwards, 1942. (Litho)

Wald, A. *Selected papers in statistics and probability*. New York: McGraw-Hill, 1955.

Walsh, J.E. *Handbook of nonparametric statistics*. Princeton, N.J.: Van Nostrand, 1962.

Wolfowitz, J. Bayesian inference and axioms of consistent decision. *Econometrica*, 1962, **30**, 470–479.

Introduction to
Fraser (1966) Structural Probability and a Generalization

Nancy Reid
University of Toronto

This paper defines structural probability and develops its interpretation through the introduction of an error variable. The formulation of the observed data as having been generated by a combination of an error variable and an unknown parameter has come to be called the structural model, and inference based on this model structural inference.

The structural probability distribution is a probability distribution for an unknown parameter, and is obtained from a structural model. The ingredients of a structural model are the observed data, an error variable with a specified probability distribution, and a transformation linking the error variable to the data by means of the unknown parameter. The idea of assigning a probability distribution to a parameter, outside the usual Bayesian framework, has caused some controversy. A similar controversy surrounded Fisher's development of fiducial probability, and as the abstract to the paper (hereafter to be referred to as [SP]) states, structural probability is "a reformulation of fiducial probability for transformation models." However, in cases where the parameter and data are very directly linked by a group of transformations, the resistance to constructing a frequency distribution for the parameter may be more a difficulty of familiarity than a foundational difficulty.

The introduction of an error variable as an essential component of the statistical model provides a rather clear interpretation of structural probability, but perhaps more important, it provides a substantial clarification of many aspects of inference. In particular, it provides a more direct justification for conditioning than does the conditionality or ancillarity principle.

To illustrate the interpretation of structural probability, we consider an artificially simple example [adapted from Fraser (1963a)] of a structural model with a Bernoulli error variable. A fair coin is to be tossed, with the values $e = +1, e = -1$ assigned to heads and tails, respectively. An unknown real value θ is added to e, and the sum is reported as x. The probability distribution for x has mass $1/2$ on the two values $\theta + 1$ and $\theta - 1$. This is inverted to obtain the structural probability distribution for θ: θ takes the values $x + 1$ and $x - 1$ with probability $1/2$. It is essential for the interpretation of the structural probability distribution that θ and x are directly related through a group of transformations, in this case the location group on the real line. The inversion of the probability mass function for x to that for θ is analogous to a very familiar conversion in vector spaces: When there is no special role attached to a fixed point such as the origin, it is immaterial whether we regard a point A as 10 units northwest of B, or B as 10 units southeast of A. This is explained in the context of an arbitrary distribution for e in Sec. 1 of [SP].

It is in the context of independent sampling, however, that the formulation of the model as generated from an error variable has important implications for inference. Although it seems a slight change from "$x_1, \ldots, x_n$ are independent and identically distributed as $f_0(\cdot - \theta)$" to "$x_i = \theta + e_i$; $e_1, \ldots, e_n$ are independent and identically distributed as $f_0(\cdot)$," this latter identification of the structure of the observations is very useful. In particular, it is easy to see that differences between observations are free of θ and are equal to the corresponding differences between the errors. Thus, the configuration of the errors represented by, for example, the vector $d = (e_2 - e_1, \ldots, e_n - e_1)$ is known, so the relevant probability distribution for the data is conditional on these known characteristics of the error. The conditional distribution of x, given d, is now a density on $\mathbb{R}$ in direct correspondence with the parameter, as in the Bernoulli case above. The structural probability distribution for θ can be obtained as before (this is given on page 589 of [SP]). The paper then derives the appropriate conditional distribution for any model involving an error variable and a group of transformations taking that variable to the observed data by means of an unknown parameter. An example of a more general setting is the location-scale family $x_i = \mu + \sigma e_i$, where e_i has a known distribution $f_0(e)$, discussed at the end of Sec. 1 of [SP].

The rather formidable looking equations on p. 590 of [SP] provide the general formulae for computing the structural probability distribution and are presented in group theoretic notation, since the group structure is essential for the existence of a structural distribution. There is a great deal of mathematical background absorbed in this one page, and the details are carefully outlined in Chap. 1 and 2 of Fraser (1968). One way to clarify the interpretation of the formulae is to consider the location-scale example in detail. The next paragraph is an expanded summary of the analysis on p. 591 of [SP].

As discussed above, the location-scale family can be presented in two equivalent ways. The error variable presentation consists of two parts: an error variable $(e_1, \ldots, e_n)$ with known density $f_0(e)$ and a transformation from error to data in the form

$$x_i = \mu + \sigma e_i. \tag{1}$$

Let $d_i = (e_i - \bar{e})/s_e$; $i = 1, \ldots, n$ with $\bar{e} = n^{-1} \sum e_i$ and $s_e^2 = (n-1)^{-1} \sum (e_i - \bar{e})^2$. Since $d_i = (x_i - \bar{x})/s_x$, the aspects of e given by $(d_1, \ldots, d_n)$ are known, once x is observed. The conditional density of $(\bar{e}, s_e)$ given d is obtained from that of e in the usual manner:

$$f(\bar{e}, s_e | d) = f(\bar{e}, s, d) \bigg/ \int f(\bar{e}, s_e, d) \, d\bar{e} \, ds_e$$

$$= k(d) f_0(\bar{e}) |J|, \tag{2}$$

where $|J|$ is the Jacobian of the transformation, and the inverse transformation $e_i = s_e d_i + \bar{e}$ specifies e as a function of $(\bar{e}, s_e, d)$. In the expression on p. 591 of [SP], d is denoted by $[e']^{-1} e'$, and $(\bar{e}, s_e)$ by $[e]$. The density functions in (2) are computed with respect to the invariant measure on the sample space. This avoids explicit determination of the Jacobian of the transformation from e to $(\bar{e}, s_e, d)$, which is not really needed in its entirety, but only in its dependence on $[e]$, or in this example, $(\bar{e}, s_e)$. From (1) we have that $\bar{e} = (\bar{x} - \mu)/\sigma$ and $s_e = s_x/\sigma$, in group notation $[e] = [\theta]^{-1}[x]$, and this provides a structural density for $[\theta] = (\mu, \sigma)$ by inversion, exactly as in the Bernoulli example above. The choice of $(\bar{e}, s_e)$ as measures of location and scale is quite arbitrary, and other location-scale choices are perfectly appropriate and do not affect the structural distribution. What is required is that $[e]$ is a group element that transforms an identified reference point to the error variable e, but the choice of reference point is immaterial.

There are two somewhat separate aspects of the derivation on p. 590–591. One is the derivation of the appropriate conditional distribution; as mentioned above, using the invariant measure avoids the technical difficulty of computing the Jacobian. The second is the inversion of the conditional distribution to a structural distribution. Underlying both of these, of course, is the group structure that carries the error variable (the randomness in the system) to the observed data. The formulation of statistical models by examination of the way the unknowns (parameters) and randomness (error) combine to produce the observed data is a very innovative and elegant approach to statistical modeling that really has little precursion in the literature. It is perhaps closest in spirit to the notion of "addition of errors" that appeared in early writing in statistics applied to astronomy.

Section 2 of [SP] establishes a consistency result for structural probability under conditioning. This is outlined first for location-scale models and then for the general transformation model. At that time, there was some discussion in the literature about the nature of fiducial and structural probability, so it was of interest to investigate some of their properties. Fraser (1962) had established some consistency properties of fiducial probability to address some concerns raised in Sprott (1960) and Lindley (1958).

A limitation of the definition of structural probability is that it requires the data and the parameter to be in direct, structural, relation to each other. In transformation models, this can be established by conditioning on the known configuration of the error variable, but this conditioning argument is external to the definition of structural probability. Section 3 of [SP] describes a method of approximating a general stochastically increasing model by a location model, at least locally near some fixed point on the sample space, thus enabling the structural distribution to be derived from the approximating model.

At the time that [SP] was published, there was interest in conditional methods of inference and fiducial probability, particularly in the British school of statistics. Cox (1958) had introduced and formalized what is now usually called the "weighing machine example," which draws attention to the need for conditioning, at least in certain types of problems. Several examples are discussed in Berger and Wolpert [(1984), Chap. 2.]. Birnbaum's first paper on the connection between conditioning, ancillarity, and likelihood appeared in 1962. There was also, of course, a considerable amount of discussion on the Bayesian and frequency-based theories of inference, and the appropriateness of describing parameters by means of probability distributions, for general scientific work.

However, there was increasing dissatisfaction with Fisher's fiducial probability, partly perhaps because it was not very well explained in Fisher's writing. In 1962, Fisher commented on having "observed during the last 25 years the angry resentment induced in some perhaps overconventional minds by the word 'fiducial.'" It seems that Fisher would not brook any criticism of fiducial probability, and perhaps as a result, he did not address concerns raised about the applicability of fiducial methods. An interesting overview of some aspects of the theory of inference, with emphasis on Bayesian, fiducial, and frequentist arguments, was presented in the first R.A. Fisher Memorial Lecture, given by Bartlett in 1964 and published in Bartlett (1965). The 1962 meeting of the International Statistical Institute included several papers on difficulties in the theory of fiducial inference. For many younger statisticians, Efron (1978) put the nail in the fiducial coffin when he wrote, in an introductory article in the *American Mathematical Monthly* "most, though not all, contemporary statisticians consider it [fiducial inference] either a form of objective Bayesianism, or just plain wrong."

The present paper represents a breakthrough in Fraser's efforts to show that fiducial inference cannot simply be dismissed as "wrong," and that there is an elegant and general setting in which it is possible to justify the construction of probability statements about unknown parameters outside the Bayesian context. His (1961) *Biometrika* paper had laid much of the groundwork for the development of structural distributions in transformation models; invariant measures were introduced there, and the notion of an error variable was informally introduced. Several further results were obtained in Fraser (1962, 1963a, 1963b, 1964a, 1964b), and in (1963b) he wrote, "these facts suggest strongly that the underlying substance of fiducial theory is to be found in the transformation parameter model."

During the 1960s, there was considerable emphasis, especially in the United States, on more rigorous mathematical approaches to statistical inference, with particular emphasis on optimality of inference procedures. Fisher's approach to statistics was quite different, and Fraser felt that Fisher was rather unjustly overlooked by many in the statistical profession. Since that time, Fisher's preeminence in the history of statistics has been established in many ways. One example is the book of collected papers edited by Fienberg and Hinkley (1980).

A possibly more important breakthrough, in light of subsequent developments in statistical theory, is the "automatic" conditioning that is provided by the structural model. This conditioning does not require any appeal to the ancillarity principle. It does, however, lead to the same conditional distribution that is obtained by appealing to the ancillarity principle, in the most familiar cases of location and location-scale models, so it is easy to incorrectly assume that structural model conditioning is not essentially different from ancillarity conditioning. A practical achievement of the conditioning is the reduction of dimension from that of the minimal sufficient statistic to the dimension of the parameters. This allows ready calculation of significance levels, confidence intervals, and so on, for the unknown parameters. Although it might be convenient to use the structural density for θ for this purpose, it is certainly not necessary.

The relation between structural and fiducial inference was recently discussed in Dawid and Stone (1982), with discussion by Barnard and by Fraser. It seems clear from that development that it is necessary to restrict attention to transformation models, in order to provide a consistent interpretation of fiducial probability.

Structural probability and structural inference, connected as they are to fiducial methods, have been somewhat ignored by statisticians, and the view of many contemporary statisticians is perhaps that it is either very technical and difficult, or very limited in applicability. Much of the development of structural inference is due to Fraser and his students. A very clear overview of structural probability and structural inference is provided in the entries by Fraser on structural inference and structural models in the *Encyclopedia of Statistical Sciences*.

In fact, the range of models that can be analyzed by structural methods does include a number of regression models, as well as several models arising in classical multivariate analysis. An introduction to structural inference and some of these applications are given in Fraser (1968) and the references therein. The development of the conditional approach to linear regression models is presented in Fraser (1979) and has been followed up in Lawless (1982) and DiCiccio (1988). This latter work emphasizes the technical aspect referred to above of computing conditional distributions geometrically, without the explicit computation of n-dimensional Jacobian matrices.

The structural model for regression is $x_i = z_i'\beta + \sigma e_i; i = 1, \ldots, n$, where β is a $p \times 1$ vector of regression parameters and z_i a vector of covariates.

The configuration of the error variable may be expressed as $d = (d_1, \ldots, d_n)$ with $d_i = (x_i - z_i'\hat{\beta})/s$, where $\hat{\beta}$ is the least-squares estimate of β and $s^2 = \sum (x_i - z_i'\hat{\beta})^2$ the residual sum of squares. The resulting conditional distribution can be converted to a joint structural distribution for $(\beta_1, \ldots, \beta_p, \sigma)$, and this distribution can be marginalized to obtain structural distributions for components of interest. The analysis can be extended to allow the error density $f_0(e)$ to depend on one or more parameters, although a structural distribution is not usually available for those parameters. This permits the generalization of normal theory linear models to a wide class of regression models, including the normal, gamma, Weibull, logistic, extreme-value, and Student t-distributions, and to some types of censored samples from these distributions. The class of models can be a useful complement to the family of generalized linear models, and the robustness of linear modeling using Student t errors is emphasized in Fraser (1979).

An important aspect of the development of structural probability for transformation models is the resulting emphasis on the use of the likelihood function to provide direct inference for the unknown parameter. Indeed, in the case of the transformation model, the structural distribution of the unknown parameter is obtained directly from the likelihood function. In multiparameter problems, marginal or conditional likelihood functions for component parameters can often be derived, and Fraser's (1968) development of structural inference emphasized this.

Interestingly, there has recently been renewed interest in the use of the likelihood function to provide probability distributions, through the discovery that the likelihood function can provide an approximate distribution for the maximum likelihood estimator. A series of papers in *Biometrika* [Cox (1980), Hinkley (1980), Durbin (1980a, b), Barndorff–Nielsen (1980)] and substantial further development by Barndorff-Nielsen (1988) and the references therein, and others has emphasized the role of likelihood functions and of marginal and conditional likelihood functions to provide highly accurate inference for unknown parameters.

Recently, techniques of asymptotic analysis, particularly Laplace's method, have been applied to enable accurate approximation of multidimensional integrals. An immediate application of this is to the computation of marginal distributions for component parameters in multiparameter problems. This was recommended as a method of inference for nonnormal regression in Fraser (1979), but at that time there was no convenient way to evaluate the integrals involved. This approach is developed and applied in DiCiccio, Field, and Fraser (1990); Fraser, Lee, and Reid (1990), and Fraser, Reid, and Wong (1991).

In the last paper mentioned, attention is focused on deriving accurate approximations to the cumulative distribution function of the maximum likelihood estimator in one-parameter exponential families, using only the likelihood function at the observed data point. The resulting approximate c.d.f., $F(\hat{\theta}; \theta)$, say, is plotted against θ in order to quickly identify the 5th and 95th percentiles as the endpoints of a 90% confidence interval. Although it is not necessary for the interpretation of the confidence interval, the function of θ being plotted is, in fact, an approximate structural c.d.f., and the essential motivation comes from the formulation outlined in this 1966 paper.

About the Author

D.A.S. Fraser studied mathematics and statistics at the University of Toronto and completed his graduate studies at Princeton University under the guidance of Tukey and Wilks. He joined the department of mathematics at the University of Toronto in 1949, and held a regular position there until 1986 (from 1978 on in the department of statistics). He presently holds adjunct positions at the University of Toronto and the University of Waterloo, and a regular position at York University. Throughout his career at Toronto, he has continued to support and influence a great number of students, many of whom are now well-known statisticians themselves. As of this writing, he has supervised 41 students (6 current), and has published five books and some 160 papers.* In 1985, he was awarded the first gold medal of the Statistical Society of Canada. In 1990, he was awarded the 37th Fisher Memorial Prize and delivered the Fisher Memorial Lecture to the American Statistical Association in Anaheim, California.

Acknowledgments

I would like to thank D.A.S. Fraser for helpful conversations about structural inference. This work was partially supported by the Natural Sciences and Engineering Research Council of Canada.

References

Barndorff–Nielsen, O.E. (1980). Conditionality resolutions, *Biometrika*, **67**, 293–311.
Barndorff–Nielsen, O.E. (1988). *Parametric Statistical Models and Likelihood*. Lecture Notes in Statistics, Vol. **50**, Springer-Verlag, New York.
Bartlett, M.S. (1965). R.A. Fisher and the last fifty years of statistical methodology, *J. Amer. Statist. Assoc.*, **60**, 395–409.
Berger, J.O., and Wolpert, R.L. (1984). *The Likelihood Principle*. I.M.S. Lecture Notes Series, Vol. 6. Institute of Mathematical Statistics, Hayward, California.
Birnbaum, A. (1962). On the foundations of statistical inference, *J. Amer. Statist. Assoc.*, **57**, 269–326.

---

* He is a Fellow of the Institute of Mathematical Statistics, the American Statistical Association, the Royal Statistical Society, the International Statistical Institute, the Royal Society of Canada, and the American Association for the Advancement of Science.

Cox, D.R. (1958). Some problems connected with statistical inference, *Ann. Math. Statist.*, **29**, 357–372.

Cox, D.R. (1980). Local ancillarity, *Biometrika*, **67**, 279–286.

Dawid, A.P., and Stone, M. (1982). The functional model basis of fiducial inference (with discussion), *Ann. Statist.*, **10**, 1054–1074.

DiCiccio, T.J. (1988). Likelihood inference for linear regression models, *Biometrika*, **75**, 29–34.

DiCiccio, T.J., Field, C.A., and Fraser, D.A.S. (1990). Approximations of marginal tail probabilities and inference for scalar parameters, *Biometrika*, **77**, 77–95.

Durbin, J. (1980a). Approximations for densities of sufficient estimators, *Biometrika*, **67**, 311–335.

Durbin, J. (1980b). The approximate distribution of partial serial correlation coefficients calculated from residuals from regression on Fourier series, *Biometrika*, **67**, 335–350.

Efron, B. (1978). Controversies in the foundations of statistics, *Amer. Math. Monthly*, **85**, 231–246.

Fienberg, S.E., and Hinkley, D.V. (1980). *R.A. Fisher: An Appreciation.* Springer-Verlag, New York.

Fisher, R.A. (1962). Bayes' method of determination of probabilities, *J. Roy. Statist. Soc., Ser. B*, **24**, 118–124.

*Fraser, D.A.S. (1961). The fiducial method and invariance, *Biometrika*, **48**, 261–280.

Fraser, D.A.S. (1962). On the consistency of the fiducial method, *J. Roy. Statist. Soc., Ser. B*, **24**, 425–434.

Fraser, D.A.S. (1963a). On the sufficiency and likelihood principles, *J. Amer. Statist. Assoc.*, **58**, 641–647.

Fraser, D.A.S. (1963b). On the definition of fiducial probability, *Bull. Internat. Statist. Inst.*, **40**, 842–856.

Fraser, D.A.S. (1964a). Local conditional sufficiency, *J. Roy. Statist. Soc., Ser. B*, **26**, 52–62.

Fraser, D.A.S. (1964b). On local inference and information, *J. Roy. Statist. Soc., Ser. B*, **26**, 253–260.

Fraser, D.A.S. (1968). *The Structure of Inference.* Wiley, New York.

Fraser, D.A.S. (1979). *Inference and Linear Models.* McGraw-Hill, New York.

Fraser, D.A.S. (1988). Structural inference, structural models, structural prediction, structural probability, in *Encyclopedia of Statistical Sciences* (S. Kutz, N.L. Johnson, and C.B. Read, eds.). Wiley, New York.

Fraser, D.A.S., Lee, H.-S., and Reid, N. (1990). Nonnormal linear regression; an example of significance levels in high dimensions, *Biometrika*, **77**, 333–341.

Fraser, D.A.S., Reid, N., and Wong, A. (1991). Exponential linear models: A two-pass procedure for saddlepoint approximation, *J. Roy. Statist. Soc., Ser. B*, **53**, 483–492.

Fisher, R.A. (1962). Bayes' method of determination of probabilities, *J. Roy. Statist. Soc., Ser. B*, **24**, 118–124.

Hinkley, D.V. (1980). Likelihood as approximate pivotal distribution, *Biometrika*, **67**, 287–292.

Lawless, J.F. (1982). *Statistical Models and Methods for Lifetime Data.* Wiley, New York.

Lindley, D.V. (1958). Fiducial distributions and Bayes' theorem, *J. Roy. Statist. Soc., Ser. B*, **20**, 102–107.

Sprott, D.A. (1960). Necessary restrictions for distributions *a posteriori*, *J. Roy. Statist. Soc., Ser. B*, **22**, 312–318.

* Note that Fraser, D.A.S. (1961) through (1964b) appear in the paper being described.

Structural Probability and a Generalization*

D.A.S. Fraser
University of Toronto

Summary

Structural probability, a reformulation of fiducial probability for transformation models, is discussed in terms of an error variable. A consistency condition is established concerning conditional distributions on the parameter space; this supplements the consistency under Bayesian manipulations found in Fraser (1961). An extension of structural probability for real-parameter models is developed; it provides an alternative to the local analysis in Fraser (1964b).

1. Introduction

Fiducial probability has been reformulated for location and transformation models (Fraser, 1961) and compared with the prescriptions in Fisher's papers (Fraser, 1963b). The transformation formulation leads to a frequency interpretation and to a variety of consistency conditions; the term *structural probability* will be used to distinguish it from Fisher's formulation.

Fiducial probability was introduced by Fisher in 1930 and developed along with other inference methods through many of his papers. Fisher's work in inference seems to be the main basis and stimulus for the present attention to aspects of inference, such as likelihood, conditionality, significance testing, and seems to have led thereby to the present substantial alternatives to deci-

* Prepared at the University of Copenhagen 1964; revised at the Mathematics Research Center, United States Army and at the Department of Statistics, University of Wisconsin.

sion theory. Dempster (1964) is somewhat alone in belittling this large contribution. His criticisms seem most to indicate disillusionment that Fisher's contributions are not organically whole and logically consistent in codified form.
This is certainly ignoring the magnitude of the actual contributions, but perhaps more dangerously it is ignoring possible developments from things but
lightly touched by Fisher.

In commenting on fiducial probability for transformation models Dempster remarks '... it simply transforms the problem of choosing pivoted variables into a slightly narrower problem of choosing a group, and the latter
problem seems to me unintuitive and far removed from the central issue'. The
basis for this opinion would seem to be against the *standard statistical model*:
a variable, a parameter, a probability density function, and nothing more. But
in many applications there is more and it should appear as additional structure in the statistical model. Some examples will be considered in this section.
Against the augmented model Dempster's remark is inappropriate.

The simple measurement model in ordinary form involves a variable x, a
parameter θ and a probability density function $f(x - \theta)$ with f specified. In
analyses based on this, there is emphasis on x as a *variable* and on θ as *fixed*.
The location group as used in Fraser (1961) provides a simple means of describing the fixed shape of the distribution with differing values for θ.

The simple measurement model can be presented alternatively in the form

$$x = \theta + e$$

where e, an error variable, has a fixed distribution given by $f(\cdot)$. In an analysis
involving a particular instance, emphasis can be placed on x as *known* and θ
as *unknown*. Particular values for x and θ would depend on the conventional
origin on the measurement scale and can be viewed separately from the error
variable with its known distribution. A value for e gives the position of x with
respect to θ and its negative gives the position of θ with respect to x. The
ordinary form for this model has gone too far in making distinctions—a
different variable for a different θ—and needs the group to partially recover
the essential variable by a symmetry argument.

Consider the measurement model in the alternative form

$$x = \theta + e.$$

In a particular instance there is a realized value e^r for the error variable.
Probability statements concerning the unobserved e^r can be made in exactly
the way a dealer at poker might calculate probabilities after dealing cards but
before observing any. The probability element

$$f(e) \, de$$

describes the distribution appropriate to e^r. With x observed and *known* and
θ *unknown*, the distribution of $-e$ which describes θ position with respect to
x provides the distribution of possible values for θ

$$f(x - \theta)\, d\theta,$$

the *structural distribution* for θ based on an observed x.

The referee of the original version of this paper commented

> I cannot see how, within the ordinary meaning of probability, a distribution for θ can be obtained. An extra principle is needed in order to transfer the distribution from e to θ. This principle is not contained within the ordinary probability calculus. My feeling is that the principle should be stated explicitly.

For someone committed to the ordinary statistical model, an extra principle is seemingly needed. But its introduction can only be to compensate for an inadequacy in the ordinary model. For the alternative form of the measurement model, an extra principle is not needed: *x is known, θ is unknown, the distribution of θ position with respect to x is known, and the origin of the scale of measurement is conventional.*

With multiple observations the measurement model takes the form

$$x_1 = \theta + e_1$$

$$\vdots$$

$$x_n = \theta + e_n,$$

where $(e_1, \ldots, e_n)$ is a vector sample of error variables from the distribution $f(e)$. In a particular instance there is a realized sample $(e_1^r, \ldots, e_n^r)$ from the error distribution. Some aspects of the realized errors *are* observable

$$e_2^r - e_1^r = x_2 - x_1$$

$$\vdots$$

$$e_n^r - e_1^r = x_n - x_1.$$

One aspect, however, is *not* observable; it describes the location of the realized errors and can be described by, say, e_1^r

$$x_1 = \theta + e_1^r.$$

Probability statements concerning the unobserved e_1^r can be made in exactly the way a dealer at poker might calculate probabilities after dealing the cards and after observing his own hand: the dealer would condition on the cards observed. Correspondingly, the statistician should condition on the observed components of the error sample; the probability element for e_1 given $e_2^r - e_1^r$, $\ldots, e_n^r - e_1^r$ is

$$g(e_1 | e_2^r - e_1^r, \ldots, e_n^r - e_1^r)\, de_1 = \frac{f(e_1)f(e_1 + e_2^r - e_1^r)\ldots f(e_1 + e_n^r - e_1^r)}{h(e_2^r - e_1^r, \ldots, e_n^r - e_1^r)}\, de_1,$$

where $h(y_2, \ldots, y_n)$ is the marginal density for the sample differences $e_2 - e_1$, $\ldots, e_n - e_1$. With x_1 known and θ unknown, the conditional distribution of $-e_1$, which describes θ position with respect to x provides the distribution of possible values for θ

$$g(x_1 - \theta | x_2 - x_1, \ldots, x_n - x_1) \, d\theta = c(x_1, \ldots, x_n) \prod_{i=1}^{n} f(x_i - \theta) \, d\theta$$

the *structural distribution* for θ based on an observed sample $(x_1, \ldots, x_n)$.

For the general case consider an error variable e on a space X. And suppose that an observable x is obtained by a transformation $[\theta]$ applied to the variable e

$$x = [\theta]e.$$

Suppose that the transformations $[\theta]$ are indexed by a parameter θ with values in a parameter space Ω and that the transformations $[\theta]$ are precisely the transformations of a group $G = \{g\}$ that is *unitary* in its application to the space X: if $gx = hx$ then $g = h$ (at most one transformation carrying any point into any other point).

The elements of G and Ω are in one-one correspondence:

$$[\theta] \leftrightarrow \theta.$$

Any element of G can produce a transformation on G by left multiplication; correspondingly there is a transformation on Ω. Let the same group element be used to designate this isomorphic transformation. This, then, permits the representation

$$\theta = [\theta]\theta_0,$$

where the identity element is in correspondence with $\theta_0 : e \leftrightarrow \theta_0$.

To avoid degeneracy suppose that $[\theta']\,e$ and $[\theta'']\,e$ have different distributions whenever $\theta' \neq \theta''$.

The measurement model with unknown scaling provides a simple example of this general model

$$x_1 = \mu + \sigma e_1$$
$$\vdots$$
$$x_n = \mu + \sigma e_n,$$

where $(e_1, \ldots, e_n)$ is a vector sample from a distribution $f(e)$. This can be represented in the form

$$\mathbf{x} = [\mu, \sigma]\mathbf{e},$$

where a transformation $[a, c]$ is a location-scale transformation

$$[a, c](x_1, \ldots, x_n) = (a + cx_1, \ldots, a + cx_n)$$

with $-\infty < a < \infty, 0 < c < \infty$. In this example a latent error distribution is relocated and rescaled; in a typical application this would reflect the conventional nature of the origin and unit of measurement.

For a model described by a density function it is natural to use a measure that has invariance under the group of transformations. Let M be an invariant measure on $X : M(A) = M(gA)$ for all g in G and all A contained in X; let μ be the left invariant measure on $G : \mu(H) = \mu(gH)$; and let Δ be the modular function satisfying

$$d\mu(g) = \Delta(g)\, d\mu(g^{-1}),$$

$$d\mu(gh) = \Delta(h)\, d\mu(g),$$

where g is the measure variable.

Suppose now that e has a probability density function f with respect to the invariant measure M. The general model then has the form

$$x = [\theta]e$$

where the error variable e has element $f(e)dM(e)$.

Under transformations in G applied to X a point x is carried into an orbit

$$S_x = \{gx \mid g \in G\}.$$

Suppose that a reference point is chosen on each orbit and let $[x]$ designate the unique transformation in G that carries the reference point on the orbit S_x into the point x. The reference point for the orbit through x can then be designated by

$$[x]^{-1}x.$$

And since there is precisely one reference point on each orbit, the expression $[x]^{-1}x$ can be used to label the orbit through x.

In a particular instance there is a realized value e^r from the error variable. Some aspects of the realized error are observable

$$[e^r]^{-1}e^r = [e^r]^{-1}[\theta]^{-1}[\theta]e^r$$

$$= [[\theta]e^r]^{-1}[\theta]e^r$$

$$= [x]^{-1}x.$$

One aspect, however, is not observable; it describes the location of e^r on the orbit $[e^r]^{-1}e^r$ and is given conveniently by $[e^r]$

$$[x] = [\theta][e^r].$$

Probability statements concerning $[e^r]$ are obtained from the distribution of $[e]$ conditional on the observed aspects $[e^r]^{-1}e^r$; this conditional distribution (Fraser, 1963b, §4.2) has element

$$k([e^r]^{-1}e^r)f([e] \cdot [e^r]^{-1}e^r)\, d\mu([e]),$$

where k is a normalizing constant on the orbit $[e^r]^{-1}e^r$. This conditional distribution can be transformed to obtain the distribution of $[e]^{-1}$, which describes θ position with respect to x

$$[\theta] = [x][e]^{-1}.$$

With x known and θ unknown the probability element can be manipulated:

$$k([e^r]^{-1}e^r)f([e][e^r]^{-1}e^r)\,d\mu([e])$$

$$= k([e^r]^{-1}e^r)f([e][e^r]^{-1}e^r)\Delta([e])\,d\mu([e]^{-1})$$

$$= k([x]^{-1}x)f([\theta]^{-1}[x][x]^{-1}x)\Delta([\theta]^{-1}[x])\,d\mu([x]^{-1}[\theta])$$

$$= k([x]^{-1}x)f([\theta]^{-1}x)\Delta([\theta]^{-1}[x])\,d\mu([\theta]);$$

this is the *structural distribution* for the unknown θ based on a known x.

For the measurement model with unknown scale parameter take as reference point on the orbit

$$S_x = \{(a + cx_1, \ldots, a + cx_n)\},$$

the point having $\bar{x} = 0$ and $s_x = 1$; then

$$[\mathbf{x}] = [\bar{x}, s_x].$$

The observable aspect of the error is

$$[\bar{x}, s_x]^{-1}\mathbf{x} = \left(\frac{x_1 - \bar{x}}{s_x}, \ldots, \frac{x_n - \bar{x}}{s_x}\right)$$

$$= \left(\frac{e_1 - \bar{e}}{s_e}, \ldots, \frac{e_n - \bar{e}}{s_e}\right)$$

and the unobservable aspect is $(\bar{e}, s_e]$. From Fraser (1963b) the conditional distribution of $[\bar{e}, s_e]$ is

$$k\left(\frac{e_i^r - \bar{e}^r}{s_e^r}\right)\prod f\left([\bar{e}, s_e]\frac{e_i^r - \bar{e}^r}{s_e^r}\right)s_e^n\frac{d\bar{e}\,ds_e}{s_e^2},$$

and the structural distribution for $[\mu, \sigma]$ is

$$k\left(\frac{x_i - \bar{x}}{s_x}\right)\prod f\left(\frac{x_i - \mu}{\sigma}\right)\left(\frac{s_x}{\sigma}\right)^n\left(\frac{\sigma}{s_x}\right)\frac{d\mu\,d\sigma}{\sigma^2}.$$

2. Consistency: Conditional Distributions

Some consistency properties of structural probability have been examined in Fraser (1961, 1962): that the structural distribution from one set of variables can be used as a prior distribution for a Bayesian analysis on another set of variables with a result independent of the choice of the first set (provided all variables generate the same transformation group on the parameter space); and that a structural distribution can be combined directly with a prior distribution and yield the same result as a Bayesian analysis. In this section a consistency property for conditioned structural distributions is considered.

Consider first an example. Let $(x_1, \ldots, x_n)$ be a sample from the model

$x = \mu + \sigma e$ where e is standard normal. The structural distribution for (μ, σ) can be represented by

$$\mu = \bar{x} - \frac{z}{(n-1)^{-1/2}\chi} n^{-1/2} s,$$

$$\sigma = \frac{s}{(n-1)^{-1/2}\chi},$$

where z and χ are independent and are respectively standard normal and chi on $n - 1$ degrees of freedom.

Suppose the information $\sigma = \sigma_0$ becomes available. The joint distribution for (μ, σ) is easily conditioned since $\sigma = \sigma_0$ implies that

$$\chi = (n-1)^{1/2}\frac{s}{\sigma}$$

is known in value; and since z is statistically independent of χ it follows that

$$\mu = \bar{x} - \frac{z\sigma_0}{s} n^{-1/2} s$$

$$= \bar{x} - zn^{-1/2}\sigma_0.$$

This conditioned distribution is exactly the structural distribution that is obtained from the model: $(x_1, \ldots, x_n)$ is a sample from $x = \mu + \sigma_0 e$ where e is standard normal.

Alternatively, suppose that the information $\mu = \mu_0$ becomes available. The joint probability element for (μ, σ)

$$\frac{A_{n-1}}{(2\pi)^{(1/2)n}} \exp\left\{-\frac{n(\bar{x}-\mu)^2 + (n-1)s^2}{2\sigma^2}\right\}\left((n-1)^{1/2}\frac{s}{\sigma}\right)^n \left(\frac{n}{n-1}\right)^{1/2}\frac{\sigma}{s}\frac{du\,d\sigma}{\sigma^2}$$

can be conditioned according to $\mu = \mu_0$ and yields

$$k \exp\left\{-\frac{nS^2}{2\sigma^2}\right\}\left(\frac{S}{\sigma}\right)^n \frac{d\sigma}{\sigma}$$

where $S^2 = n^{-1}\Sigma(x_i - \mu_0)^2$. This is the structural distribution as obtained from the model: $(x_1, \ldots, x_n)$ is a sample from $x = \mu_0 + \sigma e$ where e is standard normal.

For the general case consider the transformation model (X, G, Ω):

$$x = [\theta]e,$$

where e has the element

$$f(e)\,dM(e)$$

on X, where $[\theta]$ takes values in the transformation group G unitary on X, and where $\theta = [\theta]\theta_0$ takes values in Ω.

Theorem. *If (X, H, Ω_0) is a transformation model with H a subgroup of G and $\Omega_0 = H\theta_0$, then the structural distribution from the submodel is the same as the structural distribution from the full model as conditioned to Ω_0 with respect to partition sets $g\Omega_0$.*

PROOF. A structural distribution derives from a conditional distribution on an orbit. For the model (X, G, Ω) it suffices then to suppose that the sample space X consists of just one orbit. Let x_0 be a reference point in the sample space and let the error $e = gx_0$ be expressed in terms of a variable g on the group G. And let θ_0 be a reference point in Ω; for convenience take θ_0 in Ω_0.

The subgroup H generates orbits on the space X. Let $x_0(x)$ be a reference point on the orbit through the point x and let

$$x_0(x) = a_x x_0,$$

where a_x is an element of G; x can then be written

$$x = h_x a_x x_0,$$

where h_x is an element of H. The error variable e can correspondingly be expressed in terms of components

$$e = gx_0 = hax_0,$$

where a has the marginal distribution of the orbital variable and h has a conditional distribution given a. The full model can then be expressed in the form

$$x = [\theta]gx_0 = [\theta]hax_0.$$

The structural distribution for the full model has the form

$$\theta = [x]g^{-1}\theta_0$$
$$= [x]a^{-1}h^{-1}\theta_0.$$

This is a distribution $(h^{-1}\theta_0)$ on Ω_0 transformed by a random group element $f = [x]a^{-1}$; it is thus expressed in a form appropriate to the partition $f\Omega_0$. The conditional distribution on Ω_0 is then obtained from the condition $a = a_x$ and has the form

$$\theta = h_x h^{-1}\theta_0,$$

where h has the conditional distribution given $a = a_x$.

Consider now the submodel with group H. The orbits are given by a_x and position on an orbit by h_x. The structural distribution then has the form

$$\theta = h_x h^{-1}\theta_0,$$

where h has the conditional error distribution on the orbit given by a_x. The two structural distributions are the same. $\square$

3. An Extension of Structural Probability

For models involving a real variable and a real parameter, structural probability as discussed in the preceding sections is available only for the location model $f(x - \theta)$. One extension for stochastically increasing variables is considered in Fraser (1964b); it leads to local structural probability and a residual likelihood and is based on local conditional sufficiency (Fraser, 1964a). In this section an alternative extension will be considered; it leads to a global structural distribution but with possible non-uniqueness dependent on the choice of initiating variable.

Consider a real variable x with a stochastically increasing distribution:

$$F_\theta(x|\theta) = \frac{\partial}{\partial\theta}F(x|\theta) < 0.$$

Properties of the distribution for x near x_0 will be used to analyse the parameter space. First, consider the increment x_0, $x_0 + d$; the distribution function increases in value by the amount $F_x(x_0|\theta)d$. Next, consider the increment θ, $\theta + \delta$; the distribution function decreases by an amount $-F_\theta(x_0|\theta)\delta$. The distribution function value $F(x_0|\theta)$ will be approximately equal to $F(x_0 + h, \theta + \delta)$ if δ and h are in the ratio given by

$$F_x(x_0|\theta)h = -F_\theta(x_0|\theta)\delta.$$

A change in x at x_0 can thus be viewed as corresponding to a topological shift on the parameter space with rate at θ given by

$$\frac{1}{h(\theta, x_0)} = -\frac{F_x(x_0|\theta)}{F_\theta(x_0|\theta)};$$

the function h is, in a sense, the density of θ values with respect to such a shift. A new parameter $\tau(\theta)$ can be defined having a rate of shift equal to unity:

$$\frac{d\theta(\tau)}{d\tau} = -\frac{F_x(x_0|\theta)}{F_\theta(x_0|\theta)},$$

$$\tau = \int^\theta -\frac{F_\theta(x_0|\theta)}{F_x(x_0|\theta)}\,d\theta = \int^\theta h(\theta, x)\,d\theta.$$

Let $H(x, \tau) = F(x|\theta(\tau))$ be the distribution function in terms of the transformed parameter. The definition of τ then shows that in the neighbourhood of x_0 the distribution has location form with respect to τ.

Consider now structural inference based on the local location form of the distribution. For an observation x let τ be the transformed parameter corresponding to the neighbourhood x. The structural probability element is then

$$|H_\tau(x|\tau)|\,d\tau = |F_\theta(x|\theta)|\,d\theta;$$

the structural density function for θ has the form

$$-F_\theta(x|\theta) = F_x(x|\theta)\left(-\frac{F_\theta(x|\theta)}{F_x(x|\theta)}\right)$$

$$= F_x(x|\theta)h(\theta, x)$$

and is thus seen to be the likelihood function $F_x(x|\theta)$ modulated by the function $h(\theta, x)$.

Consider now a sample $(x_1, \ldots x_n)$ from the distribution $F(x|\theta)$. The structural distribution from x_1 has the density

$$F_x(x_1|\theta)h(\theta, x_1).$$

Using this in a Bayesian analysis on the remainder of the sample yields the following *relative* density function* for θ

$$\prod F_x(x_i|\theta)h(\theta, x_1)$$

with a normalizing constant, say $c_1(\mathbf{x})$, that can be determined by integration; it is the joint likelihood function modulated by the θ-density function from the first observation. This distribution has a frequency interpretation in terms of the local structure of the distribution at the first observation.

Alternatively, commencing from the observation x_j a structural distribution with relative density

$$\prod F_x(x_i|\theta)h(\theta, x_j)$$

is obtained.

The criterion of uniqueness has been applied more severely to fiducial probability than to other areas of inference, perhaps in part because of Fisher's claim of uniqueness. In structural probability the criterion is in large measure satisfied for transformation models. The criterion may, however, be too strong to invoke for more general models. An inference analysis for a stochastically increasing model $F(x|\theta)$ might examine each of the structural distributions

$$c_j(\mathbf{x})h(\theta, x_j)\prod F_x(x_i|\theta)$$

and in some contexts might go further and examine weighted combinations

$$\sum l_j c_j(\mathbf{x})h(\theta, x_j)\prod F_x(x_i|\theta),$$

in particular the symmetric combination

$$\sum n^{-1}c_j(\mathbf{x})h(\theta, x_j)\prod F_x(x_i|\theta).$$

In each of these the likelihood function is present and is modulated by a function based on the θ-densities $h(\theta, x_j)$. As noted by the referee these extended structural distributions will typically violate the likelihood principle;

---

* A distribution of this form has been proposed by Roy (1960).

for the author this is not viewed as being adverse to the extension (Fraser, 1963a).

References

Dempster, A.P. (1984). On the difficulties inherent in Fisher's fiducial argument. *J. Amer. Statist. Ass.* **59**, 56–66.

Fisher, R.A. (1930). Inverse probability. *Proc. Camb. Phil. Soc.* **26**, 528–38.

Fraser, D.A.S. (1981). The fiducial method and invariance. *Biometrika*, **48**, 261–80.

Fraser, D.A.S. (1962). On the consistency of the fiducial method. *J. R. Statist. Soc. B*, **24**, 425–34.

Fraser, D.A.S. (1963a). On the sufficiency and likelihood principles. *J. Amer. Statist. Ass.* **58**, 641–7.

Fraser, D.A.S. (1963b). On the definition of fiducial probability. *Bull. Int. Statist. Inst.* **40**, 842–56.

Fraser, D.A.S. (1964a). Local conditional sufficiency. *J. R. Statist. Soc. B*, **26**, 52–62.

Fraser, D.A.S. (1964b). On local inference and information. *J. R. Statist. Soc. B*, **26**, 253–60.

Roy, A.D. (1960). Some notes on pistimetric inference. *J. R. Statist. Soc. B*, **22**, 338–47.

Introduction to
Akaike (1973) Information Theory and an Extension of the Maximum Likelihood Principle

J. deLeeuw
University of California at Los Angeles

Introduction

The problem of estimating the dimensionality of a model occurs in various forms in applied statistics: estimating the number of factors in factor analysis, estimating the degree of a polynomial describing the data, selecting the variables to be introduced in a multiple regression equation, estimating the order of an AR or MA time series model, and so on.

In factor analysis, this problem was traditionally solved by eyeballing residual eigenvalues, or by applying some other kind of heuristic procedure. When maximum likelihood factor analysis became computationally feasible, the likelihoods for different dimensionalities could be compared. Most statisticians were aware of the fact that the comparison of successive chi squares was not optimal in any well-defined decision theoretic sense. With the advent of the electronic computer, the forward and backward stepwise selection procedures in multiple regression also became quite popular, but again there were plenty of examples around showing that the procedures were not optimal and could easily lead one astray. When even more computational power became available, one could solve the best subset selection problem for up to 20 or 30 variables, but choosing an appropriate criterion on the basis of which to compare the many models remains a problem.

But exactly because of these advances in computation, finding a solution of the problem became more and more urgent. In the linear regression situation, the C_p criterion of Mallows (1973), which had already been around much longer, and the PRESS criterion of Allen (1974) were suggested. Although they seemed to work quite well, they were too limited in scope. The structural covariance models of Joreskog and others, and the log linear models of Goodman and others, made search over a much more complicated set of models necessary, and the model choice problems in those contexts could not be attacked by inherently linear methods. Three major closely related developments occurred around 1974. Akaike (1973) introduced the information criterion for model selection, generalizing his earlier work on time series analysis and factor analysis. Stone (1974) reintroduced and systematized cross-validation procedures, and Geisser (1975) discussed predictive sample reuse methods. In a sense, Stone–Geisser cross-validation is the more general procedure, but the information criterion (which rapidly became Akaike's information criterion or AIC) caught on more quickly.

There are various reasons for this. Akaike's many students and colleagues applied AIC almost immediately to a large number of interesting examples (compare Sakamoto, Ishiguro, and Kitagawa, 1986). In a sense, the AIC was more original and more daring than cross-validation, which simply seemed to amount to a lot of additional dreary computation. AIC has a close connection to the maximum likelihood method, which to many statisticians is still the ultimate in terms of rigor and precision. Moreover, the complicated structural equations and loglinear analysis programs were based on maximum likelihood theory, and the AIC criterion could be applied to the results without any additional computation. The AIC could be used to equip computerized "instant science" packages such as LISREL with an automated model search and comparison procedure, leaving even fewer decisions for the user (de Leeuw, 1989). And finally, Akaike and his colleagues succeeded in connecting the AIC effectively to the always mysterious area of the foundations of statistics. They presented the method, or at least one version of it, in a Bayesian framework (Akaike, 1977, 1978). There are many statisticians who consider the possibility of such a Bayesian presentation an advantage of the method.

Akaike's 1973 Paper

Section 1. Introduction

We start our discussion of the paper with a quotation. In the very first sentence, Akaike defines his information criterion, and the statistical principle that it implies.

> Given a set of estimates $\hat{\theta}$'s of the vector of parameters θ of a probability distribution with density $f(x|\theta)$ we adopt as our final estimate the one which will give the maximum of the expected log-likelihood, which is by definition
>
> $$\mathbf{E}(\log f(X|\hat{\theta})) = \mathbf{E}\left(\int f(x|\theta) \log f(x|\hat{\theta}) \, dx\right),$$
>
> where X is a random variable following the distribution with the density function $f(x|\theta)$ and is independent of $\hat{\theta}$.

This is an impressive new principle, but its precise meaning is initially rather unclear. It is important to realize, for example, that in this definition the expected value on the left is with respect to the joint distribution of $\hat{\theta}$ and X, while the expected value on the right is with respect to the distribution of $\hat{\theta}$. It is also important that the expected log-likelihood depends both on the estimate $\hat{\theta}$ and the true value θ_0. We shall try to make this more clear by using the notation $\hat{\theta}(Z)$ for the estimate, where Z is the data, and Z is independent of X.

Akaike's principle now tells us to maximize over a class of estimates, but it does not tell us over which class, and it also does not tell us what to do about the problem when θ_0 is unknown. He points out this is certainly not the same as the principle of maximum likelihood, which adopts as the estimate the $\hat{\theta}(Z)$ that maximizes the log-likelihood $\log f(z|\theta)$ for a given realization of Z. For maximum likelihood, of course, we do not need to know θ_0.

What remains to be done is to further clarify the unclear points we mentioned above and to justify this particular choice of distance measure. This is what Akaike sets out to do in the rest of his paper.

Section 2. Information and Discrimination

In this section, Akaike justifies, or at least discusses, the choice of the information criterion. The model $f(\cdot|\theta)$ is a family of parametrized probability densities, with $\theta \in \Theta$. We shall simply refer to both θ and Θ as "models," understanding that the "model" Θ is a set of simple "models" θ. Suppose we want to compare a general model θ with the "true" model θ_0. From general decision theory, we know that comparisons can be based without loss of efficiency on the likelihood ratio $\tau(\cdot) = f(\cdot|\theta)/f(\cdot|\theta_0)$. This suggests that we define the *discrimination* between θ and θ_0 at x as $\Phi(\tau(x))$ for some function Φ, and to define the *mean discrimination* between θ and θ_0, if θ_0 is "true," as

$$\mathscr{D}(\theta, \theta_0, \Phi) = \int_{-\infty}^{+\infty} f(x|\theta_0)\Phi(\tau(x)) \, dx = \mathbf{E}_X[\Phi(\tau(X))],$$

where $\mathbf{E}_X$ is the expected value over X, which has density $f(\cdot|\theta_0)$.

Now how do we choose Φ? We study $\mathscr{D}(\theta, \theta_0, \Phi)$ for θ close to θ_0. Under suitable regularity conditions, we have

$$\mathscr{D}(\theta, \theta_0; \Phi) = \Phi(1) + \tfrac{1}{2}\ddot{\Phi}(1)(\theta - \theta_0)'\mathscr{I}(\theta_0)(\theta - \theta_0) + o(\|\theta - \theta_0\|^2),$$

where

$$\mathscr{I}(\theta_0) = \int_{-\infty}^{+\infty} \left[\left(\frac{\partial \log f(x|\theta)}{\partial \theta}\right)_{\theta=\theta_0} \left(\frac{\partial \log f(x|\theta)}{\partial \theta}\right)'_{\theta=\theta_0} \right] f(x|\theta_0) \, dx$$

is the *Fisher information* at θ_0. Thus, it makes sense to require that $\Phi(1) = 0$ and $\ddot{\Phi}(1) > 0$ in order to make $\mathscr{D}$ behave like a distance. Akaike concludes, correctly, that this derivation shows the major role played by $\log f(\cdot|\theta)$, and he also concludes, somewhat mysteriously, that consequently, the choice $\Phi(t) = -2 \log(t)$ makes good sense. Thus, he arrives at his entropy measure, known in other contexts as the *negentropy* or *Kullback–Leibler distance*.

$$\mathscr{D}(\theta, \theta_0) = 2 \int_{-\infty}^{+\infty} f(x|\theta_0) \log \frac{f(x|\theta_0)}{f(x|\theta)} \, dx$$

$$= 2\mathbf{E}_X[\log f(X|\theta_0)] - 2\mathbf{E}_X[\log f(X|\theta)].$$

It follows from the inequality $\ln t > 1 + t$ that the negentropy is always nonnegative, and it is equal to zero if and only if $f(\cdot|\theta) = f(\cdot|\theta_0)$ a.e. The negentropy can consequently be interpreted as a measure of *distance* between $f(\cdot|\theta)$ and the true distribution. The Kullback-Leibler distance was introduced in statistics as early as 1951, and its use in hypothesis testing and model evaluation was propagated strongly by Kullback (1959). Akaike points out that maximizing the expected log-likelihood amounts to the same thing as minimizing $\mathbf{E}_z[\mathscr{D}(\hat{\theta}(Z), \theta_0)]$, the expected value over the data of the Kullback-Leibler distance between the estimated density $f(\cdot|\hat{\theta}(Z))$ and the true density $f(\cdot|\theta_0)$. He calls $\mathscr{D}(\hat{\theta}(Z), \theta_0)$ the *probabilistic negentropy* and uses the symbol $\mathscr{R}(\theta_0)$ for its expected value.

The justification given by Akaike for using $\Phi(t) = -2 \log(t)$ may seem a bit weak, but the result is a natural distance measure between probability densities, which has strong connections with the Shannon–Wiener information criterion, Fisher information, and entropy measures used in thermodynamics. One particular reason why this measure is attractive is the situation in which we have n repeated independent trials according to $f(\cdot|\theta_0)$. This leads to densities $f_n(\cdot, \theta)$ and $f_n(\cdot, \theta_0)$ that are products of the densities of the individual observations. If $\mathscr{D}_n(\theta, \theta_0)$ is the Kullback-Leibler distance between these two product densities, then trivially $\mathscr{D}_n(\theta, \theta_0) = n \, \mathscr{D}(\theta, \theta_0)$. Obviously, the additivity of the negentropy in the case of repeated independent trials is an important point in its favour.

Section 3. Information and the Maximum Likelihood Principle

Now Akaike has to discuss what to do about the problem of the unknown θ_0. The solution he suggests is actually very similar to the approach of classical statistical large sample theory, but because of the context of the information principle, we see it in a new light.

Remember that the *entropy maximization principle* tells us to evaluate the success of our procedure, and the appropriateness of the model Θ, by computing the expectation $\mathscr{R}(\theta_0)$ of the probabilistic negentropy over the data. Also remember that

$$\mathscr{R}(\theta_0) = 2\mathbf{E}_X[\log f(X|\theta_0)] - 2\mathbf{E}_{X,z}[\log f(X|\hat{\theta}_0(Z))],$$

which means that minimizing the expected probabilistic negentropy does indeed amount to the same thing as maximizing the expected log-likelihood mentioned in Sec. 1. Akaike's program is to estimate $\mathcal{R}(\theta_0)$, and if several models are compared, to select the model with the smallest value.

Of course, it is still not exactly easy to carry out this program. Because θ_0 is unknown we cannot really minimize the negentropy, and we cannot compute the expectation of the minimum over Z either. There is an approximate solution to this problem, however, if we have a large number of independent replications (or, more generally, if the law of large numbers applies). Minus the *mean log-likelihood ratio*

$$\hat{\mathcal{D}}_n(\theta, \theta_0) = \frac{2}{n} \sum_{i=1}^{n} \log \frac{f(x_i|\theta_0)}{f(x_i|\theta)}$$

will converge in probability to the negentropy, and under suitable regularity conditions, this convergence will be uniform in θ. This makes it plausible that maximizing the mean log- likelihood ratio (i.e., computing the *maximum likelihood estimate*) will tend to maximize the entropy, and that in the limit, the maximum likelihood estimate is the maximum entropy estimate. We do not need to know θ_0 in order to be able to compute the maximum likelihood estimate. Thus, Akaike justifies the use of maximum likelihood by deriving it from his information criterion. From now on, we will substitute the maximum likelihood estimate $\hat{\theta}(Z)$ for the unknown θ_0.

Section 4. Extension of the Maximum Likelihood Principle

This is the main theoretical section of the paper. Akaike proposes to combine point estimation and the testing of model fit into the single new principle of comparing the values of the mean log-likelihood or negentropy. This is his "extension" of the maximum likelihood principle. We have seen in the previous section that negentropy is minimized, approximately, by using the maximum likelihood estimate for $\hat{\theta}(Z)$. What must still be done is to find convenient approximations for $\mathcal{R}(\theta_0)$ at the maximum likelihood estimate.

This section is not particular easy to read. It does not have the usual proof/theorem format, expansions are given without precise regularity conditions, exact and asymptotic identities are freely mixed, stochastic and deterministic expressions are not clearly distinguished, and there are some unfortunate notational and especially typesetting choices. This is an "ideas paper," promoting a new approach to statistics, not a mathematics paper concerned with the detailed properties of a particular technique. Although we follow the paper closely, we have tried to make the notation a bit more explicit, for instance by using matrices.

Akaike analyzes the situation in which we have a number of subspaces Θ_k of Θ, with $0 \leq k \leq m$, Θ_{k+1} a subspace of Θ_k, and $\Theta_0 = \Theta$. Let $d_k = \dim(\Theta_k)$. Actually, it is convenient to simplify this, by a change of coordinates, to the problem in which $d = m$, $d_k = k$, and Θ_k is the subspace of $\mathfrak{R}^m$, which has the last $m - k$ elements equal to zero. We assume $\theta_0 \in \Theta_0$, and we assume we have n independent replications in Z. Let $\hat{\theta}_k(Z)$ be the corresponding maximum likelihood estimates. Akaike suggests that we estimate the expectation of the probabilistic entropy $\mathscr{R}(\theta_0)$ by using $\hat{\mathscr{D}}_n(\hat{\theta}_k(Z), \hat{\theta}_0(Z))$. But $\hat{\mathscr{D}}_n(\hat{\theta}_k(Z), \hat{\theta}_0(Z))$ will be a biased estimator of $\mathscr{R}(\theta_0)$, because of the substitution of the maximum likelihood estimator for θ_0.

It is known that $n\hat{\mathscr{D}}_n(\hat{\theta}_k(Z), \hat{\theta}_0(Z))$ is asymptotically chi square with $m - k$ degrees of freedom if $\theta_0 \in \Theta_k$. In general, $\hat{\mathscr{D}}_n(\hat{\theta}_k(Z), \hat{\theta}_0(Z))$ will converge in probability to $\mathscr{D}(\Theta_k, \theta_0)$, i.e., the Kullback–Leibler distance between θ_0 and the model closest to θ_0 in Θ_k. Now if $n\,\mathscr{D}(\Theta_k, \theta_0)$ is much larger than m, then the mean likelihood ratio will be very much larger than expected from the chi square appoximation. If $n\,\mathscr{D}(\Theta_k, \theta_0)$ is much smaller than m, then we can do statistics on the basis of the chi square because the model is "true." But the intermediate case, in which the two quantities are of the same order, and the model Θ_k is "not too false," is the really interesting one. This is the case Akaike sets out to study. It is, of course, similar to studying the Pitman power of large-sample tests by using sequences of alternatives converging to the null value.

First, we offer some simplifications. Instead of studying $\mathscr{D}(\theta, \theta_0)$, Akaike uses the quadratic approximation $\mathscr{W}(\theta, \theta_0) = (\theta - \theta_0)' I(\theta_0)(\theta - \theta_0)$ discussed in Sec. 2. Asymptotically, this leads to the same conclusions to the order of approximation that is used. He uses the Fisher information matrix $I(\theta_0)$ to define an inner product $\langle \cdot, \cdot \rangle_0$ and a norm $\| \cdot \|_0$ on Θ, so that $\mathscr{W}(\theta, \theta_0) = \|\theta - \theta_0\|_0^2$. Define $\theta_{0|k}$ as the projection of θ_0 on Θ_k in the information metric. Then, by Pythagoras,

$$\mathscr{W}(\hat{\theta}_k(Z), \theta_0) = \|\theta_{0|k} - \theta_0\|^2 + \|\hat{\theta}_k(Z) - \theta_{0|k}\|^2. \tag{1}$$

The idea is to use $\mathbf{E}_Z[\mathscr{W}(\hat{\theta}_k(Z), \theta_0)]$ to estimate $\mathscr{R}(\theta_0)$.

The first step in the derivation is to expand the mean log-likelihood ratio in a Taylor series. This gives

$$n\hat{\mathscr{D}}_n(\hat{\theta}_0(Z), \theta_{0|k}) = n(\hat{\theta}_0(Z) - \theta_{0|k})' \mathscr{H}[\hat{\theta}_0(Z), \theta_{0|k}](\hat{\theta}_0(Z) - \theta_{0|k}),$$

$$n\hat{\mathscr{D}}_n(\hat{\theta}_k(Z), \theta_{0|k}) = n(\hat{\theta}_k(Z) - \theta_{0|k})' \mathscr{H}[\hat{\theta}_k(Z), \theta_{0|k}](\hat{\theta}_k(Z) - \theta_{0|k}),$$

where

$$\mathscr{H}[\theta, \zeta] = \frac{1}{n} \sum_{i=1}^{n} \frac{\partial^2 \log f(x_i|\theta + \rho(\zeta - \theta))}{\partial\theta\partial\theta'},$$

for some $0 \le \rho \le 1$. Subtracting the two expansions gives

$$n\hat{\mathscr{D}}_n(\hat{\theta}_k(Z), \hat{\theta}_0(Z)) = n(\hat{\theta}_0(Z) - \theta_{0|k})' \mathscr{H}[\hat{\theta}_0(Z), \theta_{0|k}](\hat{\theta}_0(Z) - \theta_{0|k})$$
$$- n(\hat{\theta}_k(Z) - \theta_{0|k})' \mathscr{H}[\hat{\theta}_k(Z), \theta_{0|k}](\hat{\theta}_k(Z) - \theta_{0|k}).$$

Let n and k tend to infinity in such a way that $n^{1/2}(\theta_{0|k} - \theta_0)$ stays bounded. Then, taking plims, we get

$$n\hat{\mathcal{D}}_n(\hat{\theta}_k(Z), \theta_0(Z)) \approx n\|\hat{\theta}_0(Z) - \theta_{0|k}\|_0^2 - n\|\hat{\theta}_k(Z) - \theta_{0|k}\|_0^2. \tag{2}$$

This can also be written as

$$n\hat{\mathcal{D}}_n(\hat{\theta}_k(Z), \hat{\theta}_0(Z)) \approx n\|\theta_{0|k} - \theta_0\|_0^2 + n\|\hat{\theta}_0(Z) - \theta_0\|_0^2 - n\|\hat{\theta}_k(Z) - \theta_{0|k}\|_0^2$$
$$- 2n\langle \hat{\theta}_0(Z) - \theta_0, \theta_{0|k} - \theta_0 \rangle \tag{3}$$

In the next step, Taylor expansions are used again. For this step, we use the special symbol $=_k$, where two vectors x and y satisfy $x =_k y$ if their first k elements are equal.

$$n^{-1/2} \sum_{i=1}^{n} \left[\frac{\partial \log f(x_i|\theta)}{\partial \theta} \right]_{\theta=\theta_{0|k}} =_k n^{1/2} \mathcal{H}[\hat{\theta}_k(Z), \theta_{0|k}](\theta_{0|k} - \hat{\theta}_k(Z))$$
$$=_k n^{1/2} \mathcal{H}[\hat{\theta}_0(Z), \theta_{0|k}](\theta_{0|k} - \hat{\theta}_0(Z))$$

Then let n and k tend to infinity again in such a way that $n^{1/2}(\theta_{0|k} - \theta_0)$ stays bounded and take plims. This gives

$$n^{1/2}I(\theta_0)(\hat{\theta}_k(Z) - \theta_{0|k}) \approx_k n^{1/2}I(\theta_0)(\hat{\theta}_0(Z) - \theta_{0|k}),$$

and because of the definition of $\theta_{0|k}$ also,

$$n^{1/2}I(\theta_0)(\hat{\theta}_k(Z) - \theta_{0|k}) \approx_k n^{1/2}I(\theta_0)(\hat{\theta}_0(Z) - \theta_0). \tag{4}$$

It follows that $(\hat{\theta}_k(Z) - \theta_{0|k})$ is approximately the projection of $(\hat{\theta}_0(Z) - \theta_0)$ on Θ_k.

This implies that $n\|\hat{\theta}_0(Z) - \theta_0\|_0^2 - n\|\hat{\theta}_k(Z) - \theta_{0|k}\|_0^2$ and $n\|\hat{\theta}_k(Z) - \theta_{0|k}\|_0^2$ are asymptotically independent chi squares, with degrees of freedom $m - k$ and k. Akaike then indicates that the last (linear) term on the right-hand side of (3) is small compared to the other (quadratic) terms. If we ignore its contribution, and then subtract (3) from (1), we find

$$n\mathcal{W}(\hat{\theta}_k(Z), \theta_0) - n\hat{\mathcal{D}}_n(\hat{\theta}_k(Z), \hat{\theta}_0(Z))$$
$$\approx n\|\hat{\theta}_k(Z) - \theta_{0|k}\|^2 - n\|\hat{\theta}_0(Z) - \theta_0\|_0^2 - n\|\hat{\theta}_k(Z) - \theta_{0|k}\|_0^2.$$

Replacing the chi squares by their expectations gives

$$n\mathbf{E}_Z[\mathcal{W}(\hat{\theta}_k(Z), \theta_0)] \approx n\hat{\mathcal{D}}_n(\hat{\theta}_k(Z), \hat{\theta}_0(Z)) + 2k - m. \tag{5}$$

This defines the AIC. Of course, in actual examples, m may not be known or may be infinite (think of order estimation or log-spline density estimation), but in comparing models, we do not actually need m anyway, because it is the same for all models. Thus, in practice we simply compute $-2 \sum_{i=1}^{n} \log f(x_i \hat{\theta}_k(Z)) + 2k$ for various values of k.

Section 5. Applications

In this section, Akaike discusses the possible applications of his principle to problems of model selection. As we pointed out in the introduction, the systematic approach to these problems and the simple answer provided by the AIC, at no additional cost, have certainly had an enormous impact. The theoretical contributions of the paper, discussed above, have been much less influential than the practical ones. The recipe has been accepted rather uncritically by many applied statisticians in the same way as the principles of least-squares or maximum likelihood or maximum posterior probability have been accepted in the past without much questioning.

Recipes for the application of the AIC to factor analysis, principal component analysis, analysis of variance, multiple regression, and autoregressive model fitting in time series analysis are discussed. It is interesting that Akaike already published applications of the general principle to time series analysis in 1969 and to factor analysis in 1971. He also points out the equivalence of the AIC to C_p proposed by Mallows in the linear model context.

Section 6. Numerical Examples

This section has two actual numerical examples, both estimating the order k of an autoregressive series. Reanalyzing data by Jenkins and Watts leads to the estimate $k = 2$, the same as that found by the orginal analysis using partial autocorrelation methods. A reanalysis of an example by Whittle leads to $k = 65$, while Whittle has decided on $k = 4$ using likelihood-ratio tests. Akaike argues that this last example illustrates dramatically that using successive log-likelihoods for testing can be quite misleading.

Section 7. Concluding Remarks

Here Akaike discusses briefly, again, the relations between maximum likelihood, the dominant paradigm in statistics, and the Shannon–Wiener entropy, the dominant paradigm in information and coding theory. As Sec. 3 shows, there are strong formal relationships, and using expected likelihood (or entropy) makes it possible to combine point-estimation and hypothesis testing in a single framework. It also gives "easy" answers to very important but very difficult multiple-decision problems.

Discussion

The reasoning behind using X, the independent replication, to estimate $\mathscr{R}(\theta_0)$, is the same as the reasoning behind *cross-validation*. We use $\hat{\theta}(Z)$ to predict X, using $f(X|\hat{\theta}(Z))$ as the criterion. If we use the maximum likelihood estimate, we systematically underestimate the distance between the data and the model, because the estimate is constructed by minimizing this distance. Thus, we need an independent replication to find out how good our fit is, and plugging in the independent replication leads to overestimation of the distance. The AIC corrects for both biases. The precise relationship between AIC and cross-validation has been discussed by Stone (1977). At a later stage, Akaike (1978) provided an asymptotic Bayesian justification of sorts. As we have indicated, AIC estimates the expected distance between the model and the true value. We could also formulate a related decision problem as estimating the dimensionality of the model, for instance by choosing from a nested sequence of models. It can be shown that the minimum AIC does not necessarily give a consistent estimate of the true dimensionality. Thus, we may want to construct better estimates, for instance choosing the model dimensionality with the highest posterior probability. This approach, however, has led to a proliferation of criteria, among them the BIC criteria of Schwartz (1978) and Akaike (1977), or the MDL principle of Rissanen (1978 and later papers). Other variations have been proposed by Shibata, Bozdogan, Hannan, and others. Compare Sclove (1987), or Hannan and Deistler (1988, Chap. 7), for a recent review. Recently, Wei (1990) proposed a new "F.I.C." criterion, in which the complexity of the selected model is penalized by its redundant Fisher informations, rather than by the dimensionality used in the conventional criteria. We do not discuss these alternative criteria here, because they would take us too far astray and entangle us in esoteric asymptotics and ad hoc inference principles. We think the justification based on cross-validation is by far the most natural one.

We have seen that the paper discussed here was an expository one, not a mathematical one. It seems safe to assume that many readers simply skipped Sec. 4 and rapidly went on to the examples. We have also seen that the arguments given by Akaike in this expository are somewhat heuristic, but in later work by him, and by his students such as Inagaki and Shibata, a rigorous version of his results has also been published. Although many people contributed to the area of model selection criteria and there are now many competing criteria, it is clear that Akaike's AIC is by far the most important contribution. This is due to the forceful presentation and great simplicity of the criterion, and it may be due partly to the important position of Akaike in Japanese and international statistics. But most of all, we like to think, the AIC caught on so quickly because of the enormous emphasis on interesting and very real practical applications that has always been an important component of Akaike's work.

Biographical Information

Hirotogu Akaike was born in 1927 in Fujinomiya-shi, Shizuoka-jen, in Japan. He completed the B.S. and D.S. degrees in mathematics at the University of Tokyo in 1952 and 1961. He started working at the Institute of Statistical

Mathematics in 1952, worked his way up through the ranks, and became its Director General in 1982. In 1976, he had already become editor of the *Annals of the Institute of Statistical Mathematics*, and he still holds both these functions, which are certainly the most important in statistics in Japan. Akaike has received many prizes and honors: He is a member of the I.S.I., Fellow of the I.M.S., Honorary Fellow of the R.S.S., and current (1990) president of the Japanese Statistical Society.

It is perhaps safe to say that Akaike's main contribution has been in the area of time series analysis. He developed in an early stage of his career the program package TIMSAC, for time series analysis and control, and he and his students have been updating TIMSAC, which is now in its fourth major revision and extension. TIMSAC has been used in many areas of science. In the course of developing TIMSAC, Akaike had to study the properties of optimization methods. He contributed the first theoretically complete study of the convergence properties of the optimum gradient (or steepest descent) method. He also analyzed and solved the identification problem for multivariate time series, using basically Kalman's state-space representation, but relating it effectively to canonical analysis. And in modeling autoregressive patterns, he came up with the FPE (or final prediction error) criterion, which later developed rapidly into the AIC.

References

Akaike, H. (1973). Information theory and the maximum likelihood principle in *2nd International Symposium on Information Theory* (B.N. Petrov and F. Csàki, eds.). Akademiai Kiàdo, Budapest.

Akaike, H. (1977). On the entropy maximization principle, in: *Applications of Statistics* (P.R. Krishnaiah, ed.). North- Holland, Amsterdam.

Akaike, H. (1978). A Bayesian analysis of the minimum A.I.C.. procedure, *Ann. Inst. Statist. Math., Tokyo*, **30**, 9–14.

Allen, D.M. (1974). The relationship between variable selection and data augmentation and a method of prediction, Technometrics, **22**, 325–331.

Bozdogan, H. (1987). Model selection and Akaike's information criterion (AIC): The general theory and its analytical extensions, *Psychometrika*, **52**, 345–370.

de Leeuw, J. (1989). Review of Sakamoto et al., *Psychometrika*, **54**, 539–541.

Geisser, S. (1975). The predictive sample reuse method with applications, *J. Amer. Statist. Assoc.*, **70**, 320–328.

Hannan, E.J., and Deistler, M. (1988). *The Statistical Theory of Linear System*. Wiley, New York.

Kullback, S. (1959). Information theory and statistics, New York, Wiley.

Mallows, C. (1973). Some comments on C_p. Technometrics, **15**, 661–675.

Rissanen, J. (1978). Modeling by shortest data description, *Automatica*, **14**, 465–471.

Sakamoto, Y., Ishiguro, M., and Kitagawa, G. (1986). *Akaike Information Criterion Statistics*. Reidel, Dordrecht, Holland.

Schwartz, G. (1978). Estimating the dimension of a model, *Ann. Statist.* **6**, 461–464.

Sclove, S.L. (1987). Application of model-selection criteria to some problems in multivariate analysis, *Psychometrika*, **52**, 333–344.

Stone, M. (1974). Cross-validatory choice and assessment of statistical predictions (with discussion), *J. Roy. Statist. Soc., Ser. B*, **36**, 111–147.

Stone, M. (1977). An asymptotic equivalence of choice of model by cross-validation and Akaike's criterion. *J. Roy. Statist. Soc., Ser, B*, **39**, 44–47.

Wei, C.Z. (1990). On predictive least squares. *Technical Report*, Department of Mathematics, University of Maryland, College Park, Md.

Information Theory and an Extension of the Maximum Likelihood Principle

Hirotogu Akaike
Institute of Statistical Mathematics

Abstract

In this paper it is shown that the classical maximum likelihood principle can be considered to be a method of asymptotic realization of an optimum estimate with respect to a very general information theoretic criterion. This observation shows an extension of the principle to provide answers to many practical problems of statistical model fitting.

1. Introduction

The extension of the maximum likelihood principle which we are proposing in this paper was first announced by the author in a recent paper [6] in the following form:

Given a set of estimates $\hat{\theta}$ of the vector of parameters θ of a probability distribution with density function $f(x|\theta)$ we adopt as our final estimate the one which will give the maximum of the expected log-likelihood, which is by definition

$$E \log f(X|\hat{\theta}) = E \int f(x|\theta) \log f(x|\hat{\theta}) \, dx, \qquad (1.1)$$

where X is a random variable following the distribution with the density function $f(x|\theta)$ and is independent of $\hat{\theta}$.

This seems to be a formal extension of the classical maximum likelihood principle but a simple reflection shows that this is equivalent to maximizing an information theoretic quantity which is given by the definition

$$E \log\left(\frac{f(X|\hat{\theta})}{f(X|\theta)}\right) = E \int f(x|\theta) \log\left(\frac{f(x|\hat{\theta})}{f(x|\theta)}\right) dx. \tag{1.2}$$

The integral in the right-hand side of the above equation gives the Kullback-Leibler's mean information for discrimination between $f(x|\hat{\theta})$ and $f(x|\theta)$ and is known to give a measure of separation or distance between the two distributions [15]. This observation makes it clear that what we are proposing here is the adoption of an information theoretic quantity of the discrepancy between the estimated and the true probability distributions to define the loss function of an estimate $\hat{\theta}$ of θ. It is well recognized that the statistical estimation theory should and can be organized within the framework of the theory of statistical decision functions [25]. The only difficulty in realizing this is the choice of a proper loss function, a point which is discussed in details in a paper by Le Cam [17].

In the following sections it will be shown that our present choice of the information theoretic loss function is a very natural and reasonable one to develop a unified asymptotic theory of estimation. We will first discuss the definition of the amount of information and make clear the relative merit, in relation to the asymptotic estimation theory, of the Kullback–Leibler type information within the infinitely many possible alternatives. The discussion will reveal that the log-likelihood is essentially a more natural quantity than the simple likelihood to be used for the definition of the maximum likelihood principle.

Our extended maximum likelihood principle can most effectively be applied for the decision of the final estimate of a finite parameter model when many alternative maximum likelihood estimates are obtained corresponding to the various restrictions of the model. The log-likelihood ratio statistics developed for the test of composite hypotheses can most conveniently be used for this purpose and it reveals the truly statistical nature of the information theoretic quantities which have often been considered to be probabilistic rather than statistical [21].

With the aid of this log-likelihood ratio statistics our extended maximum likelihood principle can provide solutions for various important practical problems which have hitherto been treated as problems of statistical hypothesis testing rather than of statistical decision or estimation. Among the possible applications there are the decisions of the number of factors in the factor analysis, of the significant factors in the analysis of variance, of the number of independent variables to be included into multiple regression and of the order of autoregressive and other finite parameter models of stationary time series.

Numerical examples are given to illustrate the difference of our present approach from the conventional procedure of successive applications of statistical tests for the determination of the order of autoregressive models. The results will convincingly suggest that our new approach will eventually be replacing many of the hitherto developed conventional statistical procedures.

2. Information and Discrimination

It can be shown [9] that for the purpose of discrimination between the two probability distributions with density functions $f_i(x)$ ($i = 0, 1$) all the necessary information are contained in the likelihood ratio $T(x) = f_1(x)/f_0(x)$ in the sense that any decision procedure with a prescribed loss of discriminating the two distributions based on a realization of a sample point x can, if it is realizable at all, equivalently be realized through the use of $T(x)$. If we consider that the information supplied by observing a realization of a (set of) random variable(s) is essentially summarized in its effect of leading us to the discrimination of various hypotheses, it will be reasonable to assume that the amount of information obtained by observing a realization x must be a function of $T(x) = f_1(x)/f_0(x)$.

Following the above observation, the natural definition of the mean amount of information for discrimination per observation when the actual distribution is $f_0(x)$ will be given by

$$I(f_1, f_0; \Phi) = \int \Phi\left(\frac{f_1(x)}{f_0(x)}\right) f_0(x) \, dx, \tag{2.1}$$

where $\Phi(r)$ is a properly chosen function of r and dx denotes the measure with respect to which $f_i(x)$ are defined. We shall hereafter be concerned with the parametric situation where the densities are specified by a set of parameters θ in the form

$$f(x) = f(x|\theta), \tag{2.2}$$

where it is assumed that θ is an L-dimensional vector, $\theta = (\theta_1, \theta_2, \ldots, \theta_L)'$, where $'$ denotes the transpose. We assume that the true distribution under observation is specified by $\theta = \theta = (\theta_1, \theta_2, \ldots, \theta_L)'$. We Will denote by $I(\theta, \theta; \Phi)$ the quantity defined by (2.1) with $f_1(x) = f(x|\theta)$ and $f_0(x) = f(x|\theta)$ and analyze the sensitivity of $I(\theta, \theta; \Phi)$ to the deviation of θ from θ. Assuming the regularity conditions of $f(x|\theta)$ and $\Phi(r)$ which assure the following analytical treatment we get

$$\frac{\partial}{\partial \theta_1} I(\theta, \theta; \Phi)|_{\theta=\theta} = \int \left(\frac{d}{dr}\Phi(r)\frac{\partial r}{\partial \theta_1}\right)_{\theta=\theta} f_\theta \, dx = \Phi(1) \int \left(\frac{\partial f_\theta}{\partial \theta_1}\right)_{\theta=\theta} dx \tag{2.3}$$

$$\frac{\partial^2}{\partial \theta_1 \partial \theta_m} I(\theta, \theta; \Phi)|_{\theta=\theta} = \int \left[\left(\frac{d^2}{dr^2}\Phi(r)\right)\left(\frac{\partial r}{\partial \theta_1}\right)\left(\frac{\partial r}{\partial \theta_m}\right)\right]_{\theta=\theta} f_\theta \, dx$$

$$+ \int \left[\left(\frac{d}{dr}\Phi(r)\right)\left(\frac{\partial^2 r}{\partial \theta_1 \partial \theta_m}\right)\right]_{\theta=\theta} f_\theta \, dx$$

$$= \ddot{\Phi}(1) \int \left[\left(\frac{\partial f_\theta}{\partial \theta_1}\frac{1}{f_\theta}\right)\left(\frac{\partial f_\theta}{\partial \theta_m}\frac{1}{f_\theta}\right)\right]_{\theta=\theta} f_\theta \, dx$$

$$+ \dot{\Phi}(1) \int \left(\frac{\partial^2 f_\theta}{\partial \theta_1 \partial \theta_m}\right)_{\theta=\theta} dx, \tag{2.4}$$

where r, $\dot{\Phi}(1)$, $\ddot{\Phi}(1)$ and f_θ denote $\dfrac{f(x|\theta)}{f(x|\theta)}$, $\dfrac{d\Phi(r)}{dr}\Big|_{r=1}$, $\dfrac{d^2\Phi(r)}{dr^2}\Big|_{r=1}$ and $f(x|\theta)$, respectively, and the meaning of the other quantities will be clear from the context. Taking into account that we are assuming the validity of differentiation under integral sign and that $\int f(x|\theta)\,dx = 1$, we have

$$\int\left(\frac{\partial f}{\partial\theta_l}\right)dx = \int\left(\frac{\partial^2 f}{\partial\theta_l\partial\theta_m}\right)dx = 0. \tag{2.5}$$

Thus we get

$$I(\theta, \theta; \Phi) = \Phi(1) \tag{2.6}$$

$$\frac{\partial}{\partial\theta_l}I(\theta, \theta; \Phi)|_{\theta=\theta} = 0 \tag{2.7}$$

$$\frac{\partial^2}{\partial\theta_l\partial\theta_m}I(\theta, \theta; \Phi)|_{\theta=\theta} = \ddot{\Phi}(1)\int\left[\left(\frac{\partial f_\theta}{\partial\theta_l}\frac{1}{f_\theta}\right)\left(\frac{\partial f_\theta}{\partial\theta_m}\frac{1}{f_\theta}\right)\right]_{\theta=\theta} f_\theta\,dx. \tag{2.8}$$

These relations show that $\ddot{\Phi}(1)$ must be different from zero if $I(\theta, \theta; \Phi)$ ought to be sensitive to the small variations of θ. Also it is clear that the relative sensitivity of $I(\theta, \theta; \Phi)$ is high when $\left|\dfrac{\ddot{\Phi}(1)}{\Phi(1)}\right|$ is large. This will be the case when $\Phi(1) = 0$. The integral on the right-hand side of (2.8) defines the (l, m)th element of Fisher's information matrix [16] and the above results show that this matrix is playing a central role in determining the behaviour of our mean information $I(\theta, \theta; \Phi)$ for small variations of θ around θ. The possible forms of $\Phi(r)$ are e.g. $\log r$, $(r - 1)^2$ and $r^{1/2}$ and we cannot decide uniquely at this stage.

To restrict further the form of $\Phi(r)$ we consider the effect of the increase of information by N independent observations of X. For this case we have to consider the quantity

$$I_N(\theta, \theta; \Phi) = \int\Phi\frac{\displaystyle\prod_{i=1}^{N}f(x_i|\theta)}{\displaystyle\prod_{i=1}^{N}f(x_i|\theta)}\prod_{i=1}^{N}f(x_i|\theta)\,dx_1\ldots dx_N. \tag{2.9}$$

Corresponding to (2.5), (2.6) and (2.7) we have

$$I_N(\theta, \theta; \Phi) = I(\theta, \theta; \Phi) \tag{2.10}$$

$$\frac{\partial}{\partial\theta_l}I_N(\theta, \theta; \Phi)|_{\theta=\theta} = 0 \tag{2.11}$$

$$\frac{\partial^2}{\partial\theta_l\partial\theta_m}I_N(\theta, \theta; \Phi)|_{\theta=\theta} = N\frac{\partial^2}{\partial\theta_l\partial\theta_m}I(\theta, \theta; \Phi)|_{\theta=\theta}. \tag{2.12}$$

These equations show that $I_N(\theta, \theta; \Phi)$ is not responsive to the increase of information and that $\dfrac{\partial^2}{\partial\theta_l\partial\theta_m} I_N(\theta, \theta; \Phi)|_{\theta=\theta}$ is in a linear relation with N. It can be seen that only the quantity defined by

$$\frac{\partial \prod\limits_{i=1}^{N} f(x_i|\theta)}{\partial\theta_l} \frac{1}{\prod\limits_{i=1}^{N} f(x_i|\theta)}\Bigg|_{\theta=\theta} = \sum_{i=1}^{N} \left(\frac{\partial f(x_i|\theta)}{\partial\theta_l} \frac{1}{f_\theta}\right)_{\theta=\theta} \qquad (2.13)$$

is concerned with the derivation of this last relation. This shows very clearly that taking into account the relation

$$\frac{\partial f(x|\theta)}{\partial\theta_l} \frac{1}{f_\theta} = \frac{\partial \log f(x|\theta)}{\partial\theta_l}, \qquad (2.14)$$

the functions $\dfrac{\partial}{\partial\theta_l} \log f(x|\theta)$ are playing the central role in the present defini-
tion of information. This observation suggests the adoption of $\Phi(r) = \log r$ for the definition of our amount of information and we are very naturally led to the use of Kullback-Leibler's definition of information for the purpose of our present study.

It should be noted here that at least asymptotically any other definition of $\Phi(r)$ will be useful if only $\Phi(1)$ is not vanishing. The main point of our present observation will rather be the recognition of the essential role being played by the functions $\dfrac{\partial}{\partial\theta_l} \log f(x|\theta)$ for the definition of the mean information for the discrimination of the distributions corresponding to the small deviations of θ from θ.

3. Information and the Maximum Likelihood Principle

Since the purpose of estimating the parameters of $f(x|\theta)$ is to base our decision on $f(x|\hat\theta)$, where $\hat\theta$ is an estimate of θ, the discussion in the preceding section suggests the adoption of the following loss and risk functions:

$$W(\theta, \hat\theta) = (-2) \int f(x|\theta) \log\left(\frac{f(x|\hat\theta)}{f(x|\theta)}\right) dx \qquad (3.1)$$

$$R(\theta, \hat\theta) = EW(\theta, \hat\theta), \qquad (3.2)$$

where the expectation in the right-hand side of (3.2) is taken with respect to the distribution of $\hat\theta$. As $W(\theta, \hat\theta)$ is equal to 2 times the Kullback-Leibler's information for discrimination in favour of $f(x|\theta)$ for $f(x|\hat\theta)$ it is known that $W(\theta, \hat\theta)$ is a non-negative quantity and is equal to zero if and only if $f(x|\theta) = f(x|\hat\theta)$ almost everywhere [16]. This property is forming a basis of the proof of consistency of the maximum likelihood estimate of θ [24] and indicates the close relationship between the maximum likelihood principle and the information theoretic observations.

When N independent realizations x_i $(i = 1, 2, \ldots, N)$ of X are available, (-2) times the sample mean of the log-likelihood ratio

$$\frac{1}{N} \sum_{i=1}^{N} \log\left(\frac{f(x_i|\hat{\theta})}{f(x_i|\theta)}\right) \tag{3.3}$$

will be a consistent estimate of $W(\theta, \hat{\theta})$. Thus it is quite natural to expect that, at least for large N, the value of $\hat{\theta}$ which will give the maximum of (3.3) will nearly minimize $W(\theta, \hat{\theta})$. Fortunately the maximization of (3.3) can be realized without knowing the true value of θ, giving the well-known maximum likelihood estimate $\hat{\theta}$. Though it has been said that the maximum likelihood principle is not based on any clearly defined optimum consideration [18; p. 15] our present observation has made it clear that it is essentially designed to keep minimum the estimated loss function which is very naturally defined as the mean information for discrimination between the estimated and the true distributions.

4. Extension of the Maximum Likelihood Principle

The maximum likelihood principle has mainly been utilized in two different branches of statistical theories. The first is the estimation theory where the method of maximum likelihood has been used extensively and the second is the test theory where the log-likelihood ratio statistic is playing a very important role. Our present definitions of $W(\theta, \hat{\theta})$ and $R(\theta, \hat{\theta})$ suggest that these two problems should be combined into a single problem of statistical decision. Thus instead of considering a single estimate of θ we consider estimates corresponding to various possible restrictions of the distribution and instead of treating the problem as a multiple decision or a test between hypotheses we treat it as a problem of general estimation procedure based on the decision theoretic consideration. This whole idea can be very simply realized by comparing $R(\theta, \hat{\theta})$, or $W(\theta, \hat{\theta})$ if possible, for various $\hat{\theta}$'s and taking the one with the minimum of $R(\theta, \hat{\theta})$ or $W(\theta, \hat{\theta})$ as our final choice. As it was discussed in the introduction this approach may be viewed as a natural extension of the classical maximum likelihood principle. The only problem in applying this extended principle in a practical situation is how to get the reliable estimates of $R(\theta, \hat{\theta})$ or $W(\theta, \hat{\theta})$. As it was noticed in [6] and will be seen shortly, this can be done for a very interesting and practically important situation of composite hypotheses through the use of the maximum likelihood estimates and the corresponding log-likelihood ratio statistics.

The problem of statistical model identification is often formulated as the problem of the selection of $f(x|_k\theta)$ $(k = 0, 1, 2, \ldots, L)$ based on the observations of X, where $_k\theta$ is restricted to the space with $_k\theta_{k+1} = {}_k\theta_{k+2} = \cdots = {}_k\theta_L =$

0. k, or some of its equivalents, is often called the order of the model. Its decision is usually the most difficult problem in practical statistical model identification. The problem has often been treated as a subject of composite hypothesis testing and the use of the log-likelihood ratio criterion is well established for this purpose [23]. We consider the situation where the results x_i ($i = 1, 2, \ldots, N$) of N independent observations of X have been obtained. We denote by $_k\hat{\theta}$ the maximum likelihood estimate in the space of $_k\theta$, i.e., $_k\hat{\theta}$ is the value of $_k\theta$ which gives the maximum of the likelihood function $\prod_{i=1}^{N} f(x_i|_k\theta)$. The observation at the end of the preceding section strongly suggests the use of

$$_k\omega_L = -\frac{2}{N} \sum_{i=1}^{N} \log\left(\frac{f(x_i|_k\hat{\theta})}{f(x_i|_L\hat{\theta})}\right) \tag{4.1}$$

as an estimate of $W(\theta, _k\hat{\theta})$. The statistics

$$_k\eta_L = N \times {_k\omega_L} \tag{4.2}$$

is the familiar log-likelihood ratio test statistics which will asymptotically be distributed as a chi-square variable with the degrees of freedom equal to $L - k$ when the true parameter θ is in the space of $_k\theta$. If we define

$$W(\theta, _k\theta) = \inf_{_k\theta} W(\theta, _k\theta), \tag{4.3}$$

then it is expected that

$$_k\omega_L \to W(\theta, _k\theta) \text{ w.p.1.}$$

Thus when $NW(\theta, _k\theta)$ is significantly larger than L the value of $_k\eta_L$ will be very much larger than would be expected from the chi-square approximation. The only situation where a precise analysis of the behaviour of $_k\eta_L$ is necessary would be the case where $NW(\theta, _k\theta)$ is of comparable order of magnitude with L. When N is very large compared with L this means that $W(\theta, _k\theta)$ is very nearly equal to $W(\theta, \theta) = 0$. We shall hereafter assume that $W(\theta, \theta)$ is sufficiently smooth at $\theta = \theta$ and

$$W(\theta, \theta) > 0 \quad \text{for} \quad \theta \neq \theta. \tag{4.4}$$

Also we assume that $W(\theta, _k\theta)$ has a unique minimum at $_k\theta = {_k\theta}$ and that $_L\theta = \theta$. Under these assumptions the maximum likelihood estimates $\hat{\theta}$ and $_k\hat{\theta}$ will be consistent estimates of θ and $_k\theta$, respectively, and since we are concerned with the situation where θ and $_k\theta$ are situated very near to each other, we limit our observation only up to the second-order variation of $W(\theta, _k\hat{\theta})$. Thus hereafter we adopt, in place of $W(\theta, _k\hat{\theta})$, the loss function

$$W_2(\theta, _k\hat{\theta}) = \sum_{l=1}^{L} \sum_{m=1}^{L} (_k\hat{\theta}_l - \theta_l)(_k\hat{\theta}_m - \theta_m)C(l, m)(\theta), \tag{4.5}$$

where $C(l, m)(\theta)$ is the (l, m)th element of Fisher's information matrix and is given by

$$C(l, m)(\theta) = \int \left(\frac{\partial f_\theta}{\partial \theta_l} \frac{1}{f_\theta}\right)\left(\frac{\partial f_\theta}{\partial \theta_m} \frac{1}{f_\theta}\right) f_\theta \, dx = -\int \left(\frac{\partial^2 \log f}{\partial \theta_l \partial \theta_m}\right) f_\theta \, dx. \quad (4.6)$$

We shall simply denote by $C(l, m)$ the value of $C(l, m)(\theta)$ at $\theta = \theta$. We denote by $\|\theta\|_c$ the norm in the space of θ defined by

$$\|\theta\|_c^2 = \sum_{i=1}^{L} \sum_{m=1}^{L} \theta_l \theta_m C(l, m). \quad (4.7)$$

We have

$$W_2(\theta, {}_k\hat{\theta}) = \|{}_k\hat{\theta} - \theta\|_c^2. \quad (4.8)$$

Also we redefine ${}_k\theta$ by the relation

$$\|{}_k\theta - \theta\|_c^2 = \underset{{}_k\theta}{\text{Min}} \, \|{}_k\theta - \theta\|_c^2. \quad (4.9)$$

Thus ${}_k\theta$ is the projection of θ in the space of ${}_k\theta$'s with respect to the metrics defined by $C(l, m)$ and is given by the relations

$$\sum_{m=1}^{k} C(l, m) {}_k\theta_m = \sum_{m=1}^{L} C(l, m)\theta_m \quad l = 1, 2, \ldots, k. \quad (4.10)$$

We get from (4.8) and (4.9)

$$W_2(\theta, {}_k\hat{\theta}) = \|{}_k\theta - \theta\|_c^2 + \|{}_k\hat{\theta} - {}_k\theta\|_c^2. \quad (4.11)$$

Since the definition of $W(\theta, \hat{\theta})$ strongly suggests, and is actually motivated by, the use of the log-likelihood ratio statistics we will study the possible use of this statistics for the estimation of $W_2(\theta, {}_k\hat{\theta})$. Taking into account the relations

$$\sum_i \frac{\partial \log f(x_i|\hat{\theta})}{\partial \theta_m} = 0, \quad m = 1, 2, \ldots, L,$$

$$\sum_i \frac{\partial \log f(x_i|{}_k\hat{\theta})}{\partial \theta_m} = 0, \quad m = 1, 2, \ldots, k, \quad (4.12)$$

we get the Taylor expansions

$$\sum_{i=1}^{N} \log f(x_i|{}_k\theta) = \sum_{i=1}^{N} \log f(x_i|\hat{\theta}) + \frac{1}{2} \sum_{m=1}^{L} \sum_{l=1}^{L} N({}_k\theta_m - \hat{\theta}_m)({}_k\theta_l - \hat{\theta}_l)$$

$$\times \frac{1}{N} \sum_{i=1}^{N} \frac{\partial^2 \log f(x_i|\hat{\theta} + \varrho({}_k\theta - \hat{\theta}))}{\partial \theta_m \partial \theta_l}$$

$$= \sum_{i=1}^{N} \log f(x_i|{}_k\hat{\theta}) + \frac{1}{2} \sum_{m=1}^{k} \sum_{l=1}^{k} N({}_k\theta_m - {}_k\hat{\theta}_m)({}_k\theta_l - {}_k\hat{\theta}_l)$$

$$\times \frac{1}{N} \sum_{i=1}^{N} \frac{\partial^2 \log f(x_1|{}_k\hat{\theta} + \varrho_k({}_k\theta - {}_k\hat{\theta}))}{\partial \theta_m \partial \theta_l},$$

where the parameter values within the functions under the differential sign denote the points where the derivatives are taken and $0 \le \varrho_k, \varrho \le 1$, a conven-

tion which we use in the rest of this paper. We consider that, in increasing
the value of N, N and k are chosen in such a way that $\sqrt{N}(_k\theta_m - \theta_m)$
$(m = 1, 2, \ldots, L)$ are bounded, or rather tending to a set of constants for the
ease of explanation. Under this circumstance, assuming the tendency towards
a Gaussian distribution of $\sqrt{N}(\hat{\theta} - \theta)$ and the consistency of $_k\hat{\theta}$ and $\hat{\theta}$ as the
estimates of $_k\theta$ and θ we get, from (4.6) and (4.13), an asymptotic equality in
distribution for the log-likelihood ratio statistic $_k\eta_L$ of (4.2)

$$_k\eta_L = N\|\hat{\theta} - _k\theta\|_c^2 - N\|_k\hat{\theta} - _k\theta\|_c^2. \tag{4.14}$$

By simple manipulation

$$_k\eta_L = N\|_k\theta - \theta\|_c^2 + N\|\hat{\theta} - \theta\|_c^2 - N\|_k\hat{\theta} - _k\theta\|_c^2 - 2N(\hat{\theta} - \theta, \theta - \theta)_c, \tag{4.15}$$

where $(,)_c$ denotes the inner product defined by $C(l, m)$. Assuming the validity
of the Taylor expansion up to the second order and taking into account the
relations (4.12) we get for $l = 1, 2, \ldots, k$

$$\frac{1}{\sqrt{N}} \sum_{i=1}^{N} \frac{\partial}{\partial \theta_l} \log f(x_i|_k\theta)$$

$$= \sum_{m=1}^{k} \sqrt{N}(_k\theta_m - _k\hat{\theta}_m) \frac{1}{N} \sum_{i=1}^{N} \frac{\partial^2 \log f(x_i|_k\hat{\theta} + \varrho_k(_k\theta - _k\hat{\theta}))}{\partial \theta_m \partial \theta_l} \tag{4.16}$$

$$= \sum_{m=1}^{L} \sqrt{N}(_k\theta_m - \hat{\theta}_m) \frac{1}{N} \sum_{i=1}^{N} \frac{\partial^2 \log f(x_i|\hat{\theta} + \varrho(_k\theta - \hat{\theta}))}{\partial \theta_m \partial \theta_l}.$$

Let C^{-1} be the inverse of Fisher's information matrix. Assuming the tendency
to the Gaussian distribution $N(0, C^{-1})$ of the distribution of $\sqrt{N}(\hat{\theta} - \theta)$
which can be derived by using the Taylor expansion of the type of (4.16)
at $\theta = \theta$, we can see that for N and k with bounded $\sqrt{N}(_k\theta_m - \theta_m)$
$(m = 1, 2, \ldots, L)$ (4.16) yields, under the smoothness assumption of $C(l, m)(\theta)$
at $\theta = \theta$, the approximate equations

$$\sum_{m=1}^{k} \sqrt{N}(_k\theta_m - _k\hat{\theta}_m)C(l, m) = \sum_{m=1}^{L} \sqrt{N}(_k\theta_m - \hat{\theta}_m)C(l, m) \quad l = 1, 2, \ldots, k. \tag{4.17}$$

Taking (4.10) into account we get from (4.17), for $l = 1, 2, \ldots, k$,

$$\sum_{m=1}^{k} \sqrt{N}(_k\theta_m - _k\hat{\theta}_m)C(l, m) = \sum_{m=1}^{L} \sqrt{N}(\theta_m - \hat{\theta}_m)C(l, m). \tag{4.18}$$

This shows that geometrically $_k\hat{\theta} - _k\theta$ is (approximately) the projection
of $\hat{\theta} - \theta$ into the space of $_k\theta$'s. From this result it can be shown that
$N\|\hat{\theta} - \theta\|_c^2 - N\|_k\hat{\theta} - _k\theta\|_c^2$ and $N\|_k\hat{\theta} - _k\theta\|_c^2$ are asymptotically independently distributed as chi-square variables with the degrees of freedom $L - k$
and k, respectively. It can also be shown that the standard deviation of the
asymptotic distribution of $N(\hat{\theta} - \theta, _k\theta - \theta)_c$ is equal to $\sqrt{N}\|_k\theta - \theta\|_c$. Thus

if $N\|_k\theta - \theta\|_c^2$ is of comparable magnitude with $L - k$ or k and these are large integers then the contribution of the last term in the right hand side of (4.15) remains relatively insignificant. If $N\|_k\theta - \theta\|_c^2$ is significantly larger than L the contribution of $N(\hat{\theta} - \theta, _k\theta - \theta)_c$ to $_k\eta_L$ will also relatively be insignificant. If $N\|_k\theta - \theta\|_c^2$ is significantly smaller than L and k again the contribution of $N(\hat{\theta} - \theta, _k\theta - \theta)_c$ will remain insignificant compared with those of other variables of chi-square type. These observations suggest that from (4.11), though $N^{-1}{}_k\eta_L$ may not be a good estimate of $W_2(\theta, _k\hat{\theta})$,

$$r(\hat{\theta}, _k\hat{\theta}) = N^{-1}(_k\eta_L + 2k - L) \tag{4.19}$$

will serve as a useful estimate of $EW_2(\theta, _k\hat{\theta})$, at least for the case where N is sufficiently large and L and k are relatively large integers.

It is interesting to note that in practical applications it may sometimes happen that L is a very large, or conceptually infinite, integer and may not be defined clearly. Even under such circumstances we can realize our selection procedure of $_k\hat{\theta}$'s for some limited number of k's, assuming L to be equal to the largest value of k. Since we are only concerned with finding out the $_k\hat{\theta}$ which will give the minimum of $r(\hat{\theta}, _k\hat{\theta})$ we have only to compute either

$$_k\nu_L = _k\eta_L + 2k \tag{4.20}$$

or

$$_k\lambda_L = -2 \sum_{i=1}^{N} \log f(x_i|_k\hat{\theta}) + 2k. \tag{4.21}$$

and adopt the $_k\hat{\theta}$ which gives the minimum of $_k\nu_L$ or $_k\lambda_L$ $(0 \le k \le L)$. The statistical behaviour of $_k\lambda_L$ is well understood by taking into consideration the successive decomposition of the chi-square variables into mutually independent components. In using $_k\lambda_L$ care should be taken not to lose significant digits during the computation.

5. Applications

Some of the possible applications will be mentioned here.

1. Factor Analysis

In the factor analysis we try to find the best estimate of the variance covariance matrix Σ from the sample variance covariance matrix using the model $\Sigma = AA' + D$, where Σ is a $p \times p$ dimensional matrix, A is a $p \times m$ dimensional $(m < p)$ matrix and D is a non-negative $p \times p$ diagonal matrix. The method of the maximum likelihood estimate under the assumption of normality has been extensively applied and the use of the log-likelihood ratio criterion is quite common. Thus our present procedure can readily be incorporated to help the decision of m. Some numerical examples are already given in [6] and the results are quite promising.

2. Principal Component Analysis

By assuming $D = \delta I (\delta \geq 0, I;$ unit matrix$)$ in the above model, we can get the necessary decision procedure for the principal component analysis.

3. Analysis of Variance

If in the analysis of variance model we can preassign the order in decomposing the total variance into chi-square components corresponding to some factors and interactions then we can easily apply our present procedure to decide where to stop the decomposition.

4. Multiple Regression

The situation is the same as in the case of the analysis of variance. We can make a decision where to stop including the independent variables when the order of variables for inclusion is predetermined. It can be shown that under the assumption of normality of the residual variable we have only to compare the values $s^2(k)\left(1 + \dfrac{2k}{N}\right)$, where $s^2(k)$ is the sample mean square of the residual after fitting the regression coefficients by the method of least squares where k is the number of fitted regression coefficients and N the sample size. k should be kept small compared with N. It is interesting to note that the use of a statistics proposed by Mallows [13] is essentially equivalent to our present approach.

5. Autoregressive Model Fitting in Time Series

Though the discussion in the present paper has been limited to the realizations of independent and identically distributed random variables, by following the approach of Billingsley [8], we can see that the same line of discussion can be extended to cover the case of finite parameter Markov processes. Thus in the case of the fitting of one-dimensional autoregressive model $X_n = \sum_{m=1}^{k} a_m X_{n-m} + \varepsilon_n$ we have, assuming the normality of the process X_n, only to adopt k which gives the minimum of $s^2(k)\left(1 + \dfrac{2k}{N}\right)$ or equivalently $s^2(k)\left(1 + \dfrac{k}{N}\right)\left(1 - \dfrac{k}{N}\right)^{-1}$, where $s^2(k)$ is the sample mean square of the residual after fitting the kth order model by the method of least squares or some of its equivalents. This last quantity for the decision has been first introduced by the present author and was considered to be an estimate of the quantity called the final prediction error (FPE) [1, 2]. The use of this approach for the estimation of power spectra has been discussed and recognized to be very useful [3]. For the case of the multi-dimensional process we have to replace $s^2(k)$ by the sample generalized variance or the determinant of the sample variance-covariance matrix of residuals. The procedure has been extensively used for the identification of a cement rotary kiln model [4, 5, 19].

These procedures have been originally derived under the assumption of linear process, which is slightly weaker than the assumption of normality, and with the intuitive criterion of the expected variance of the final one step prediction (FPE). Our present observation shows that these procedures are just in accordance with our extended maximum likelihood principle at least under the Gaussian assumption.

6. Numerical Examples

To illustrate the difference between the conventional test procedure and our present procedure, two numerical examples are given using published data.

The first example is taken from the book by Jenkins and Watts [14]. The original data are described as observations of yield from 70 consecutive batches of an industrial process [14, p. 142]. Our estimates of FPE are given in Table 1 in a relative scale. The results very simply suggest, without the help of statistical tables, the adoption of $k = 2$ for this case. The same conclusion has been reached by the authors of the book after a detailed analysis of significance of partial autocorrelation coefficients and by relying on a some-what subjective judgement [14, pp. 199–200]. The fitted model produced an estimate of the power spectrum which is very much like their final choice obtained by using Blackman-Tukey type window [14, p. 292].

The next example is taken from a paper by Whittle on the analysis of a seiche record (oscillation of water level in a rock channel) [26; 27, pp. 37–38]. For this example Whittle has used the log-likelihood ratio test statistics in successively deciding the significance of increasing the order by one and adopted $k = 4$. He reports that the fitting of the power spectrum is very poor. Our procedure applied to the reported sample autocorrelation coefficients obtained from data with $N = 660$ produced a result showing that $k = 65$ should be adopted within the k's in the range $0 \le k \le 66$. The estimates of

Table 1. Autoregressive Model Fitting.

k	0	1	2	3	4	5	6	7
FPE$_k^*$	1.029	0.899	0.895	0.921	0.946	0.097	0.983	1.012

$$* \ \mathrm{FPE}_k = s^2(k)\left(1 + \frac{k+1}{N}\right)\left(1 - \frac{k+1}{N}\right)^{-1} \Big/ s^2(0)$$

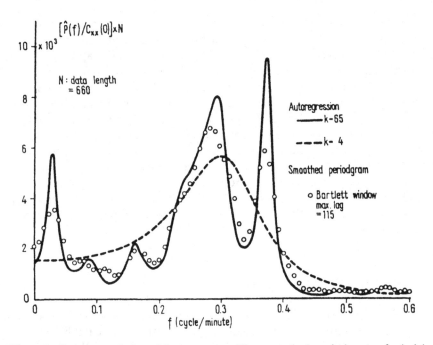

Figure 1. Estimates of the seiche spectrum. The smoothed periodogram of $x(n \Delta t)$ $(n = 1, 2, \ldots, N)$ is defined by

$$\Delta t \cdot \sum_{l}^{l} \left(1 - \frac{|s|}{l} \right) C_{xx}(s) \cos(2\pi f s \, \Delta t),$$

$$\text{where } l = \text{max. lag}, \quad C_{xx}(s) = \frac{1}{N} \sum_{n=1}^{N-|s|} \tilde{x}(|s| + n)\tilde{x}(n),$$

$$\text{where } \tilde{x}(n) = x(n \Delta t) - \bar{x} \quad \text{and} \quad \bar{x} = \frac{1}{N} \sum_{n=1}^{N} x(n \Delta t).$$

the power spectrum are illustrated in Fig. 1. Our procedure suggests that $L = 66$ is not large enough, yet it produced very sharp line-like spectra at various frequencies as was expected from the physical consideration, while the fourth order model did not give any indication of them. This example dramatically illustrates the impracticality of the conventional successive test procedure depending on a subjectively chosen set of levels of significance.

7. Concluding Remarks

In spite of the early statement by Wiener [28; p. 76] that entropy, the Shannon-Wiener type definition of the amount of information, could replace Fisher's definition [11] the use of the information theoretic concepts in the statistical circle has been quite limited [10, 12, 20]; The distinction between Shannon-Wiener's entropy and Fisher's information was discussed as early as in 1950 by Bartlett [7], where the use of the Kullback-Leibler type definition of information was implicit. Since then in the theory of statistics Kullback-Leibler's or Fisher's information could not enjoy the prominent status of Shannon's entropy in communication theory, which proved its essential meaning through the source coding theorem [22, p. 28].

The analysis in the present paper shows that the information theoretic consideration can provide a foundation of the classical maximum likelihood principle and extremely widen its practical applicability. This shows that the notion of informations, which is more closely related to the mutual information in communication theory than to the entropy, will play the most fundamental role in the future developments of statistical theories and techniques.

By our present principle, the extensions of applications 3) $\sim$ 5) of Section 5 to include the comparisons of every possible kth order models are straightforward. The analysis of the overall statistical characteristics of such extensions will be a subject of further study.

Acknowledgement

The author would like to express his thanks to Prof. T. Sugiyama of Kawasaki Medical University for helpful discussions of the possible applications

References

1. Akaike, H., Fitting autoregressive models for prediction. *Ann. Inst. Statist. Math.* **21** (1969) 243–217.
2. Akaike., H., Statistical predictor identification. *Ann. Inst. Statist. Math.* **22** (1970) 203–217.
3. Akaike, H., On a semi-automatic power spectrum estimation procedure. *Proc. 3rd Hawaii International Conference on System Sciences*, 1970, 974–977.
4. Akaike, H., On a decision procedure for system identification, Preprints, *IFAC Kyoto Symposium on System Engineering Approach to Computer Control.* 1970, 486–490.
5. Akaike, H., Autoregressive model fitting for control. *Ann. Inst. Statist. Math.* 23 (1971) 163–180.
6. Akaike, H., Determination of the number of factors by an extended maximum likelihood principle. Research Memo. 44, Inst. Statist. Math. March, 1971.
7. Bartlett, M. S., The statistical approach to the analysis of time-series. *Symposium on Information Theory* (mimeographed Proceedings), Ministry of Supply, London, 1950, 81–101.
8. Billingsley, P., *Statistical Inference for Markov Processes.* Univ. Chicago Press, Chicago 1961.
9. Blackwell, D., Equivalent comparisons of experiments. *Ann. Math. Statist.* 24 (1953) 265–272.
10. Campbell, L.L., Equivalence of Gauss's principle and minimum discrimination information estimation of probabilities. *Ann. Math. Statist.* 41 (1970) 1011–1015.

11. Fisher, R.A., Theory of statistical estimation. *Proc. Camb. Phil. Soc.* **22** (1925) 700–725, *Contributions to Mathematical Statistics.* John Wiley & Sons, New York, 1950, paper 11.
12. Good, I.J. Maximum entropy for hypothesis formulation, especially for multi-dimensional contingency tables. *Ann. Math. Statist.* **34** (1963) 911–934.
13. Gorman, J.W. and Toman, R.J., Selection of variables for fitting equations to data. *Technometrics* **8** (1966) 27–51.
14. Jenkins, G.M. and Watts, D.G., *Spectral Analysis and Its Applications.* Holden Day, San Francisco, 1968.
15. Kullback, S. and Leibler, R.A., On information and sufficiency. *Ann. Math Statist.* **22** (1951) 79–86.
16. Kullback, S., *Information Theory and Statistics.* John Wiley & Sons, New York 1959.
17. Le Cam, L., On some asymptotic properties of maximum likelihood estimates and related Bayes estimates. *Univ. Calif. Publ. in Stat.* 1 (1953) 277–330.
18. Lehmann, E.L., Testing Statistical Hypotheses. John Wiley & Sons, New York 1969.
19. Otomo, T., Nakagawa, T. and Akaike, H. Statistical approach to computer control of cement rotary kilns. 1971. *Automatica* **8** (1972) 35–48.
20. Rényi, A., Statistics and information theory. *Studia Sci. Math. Hung.* **2** (1967) 249–256.
21. Savage, L.J., The Foundations of Statistics. John Wiley & Sons, New York 1954.
22. Shannon, C.E. and Weaver, W., *The Mathematical Theory of Communication.* Univ. of Illinois Press, Urbana 1949.
23. Wald, A., Tests of statistical hypotheses concerning several parameters when the number of observations is large. *Trans. Am. Math. Soc.* **54** (1943) 426–482.
24. Wald, A., Note on the consistency of the maximum likelihood estimate. *Ann Math. Statist.* 20 (1949) 595–601.
25. Wald, A., Statistical Decision Functions. John Wiley & Sons, New York 1950.
26. Whittle, P., The statistical analysis of seiche record. *J. Marine Res.* **13** (1954) 76–100.
27. Whittle, P., *Prediction and Regulation.* English Univ. Press, London 1963.
28. Wiener, N., *Cybernetics.* John Wiley & Sons, New York, 1948.

Index

Springer Series in Statistics

(continued from p. ii)

Printed in the United States
by Baker & Taylor Publisher Services